Chemical Oceanography

VOLUME 1

Frontispiece

Frontispiece to Boohaave's Dutch translation (1786) of the French edition of 1725 of Count L. F. Marsilli's "Histoire Physique de la Mer". This was probably the first textbook of oceanography; it contained a section on the chemistry of the sea.

Chemical Oceanography

Edited by

J. P. RILEY

Department of Oceanography,
The University of Liverpool, England

and

G. SKIRROW

Department of Inorganic, Physical and Industrial Chemistry,
The University of Liverpool, England

VOLUME I

ACADEMIC PRESS
LONDON AND NEW YORK • 1965

ACADEMIC PRESS INC. (LONDON) LTD
BERKELEY SQUARE HOUSE
BERKELEY SQUARE
LONDON, W.1

U.S. Edition published by
ACADEMIC PRESS INC.
111 FIFTH AVENUE
NEW YORK, NEW YORK 10003

Second printing 1967
Third printing 1970

Library of Congress Catalog Card Number 65–15528

SBN 12-588701-9

REPRINTED PHOTOLITHO IN GREAT BRITAIN BY
J. W. ARROWSMITH LTD., WINTERSTOKE ROAD, BRISTOL 3

Contributors to Volume 1

F. A. J. Armstrong, *Marine Biological Association, Plymouth, Devon, England*

K. F. Bowden, *Department of Oceanography, The University of Liverpool, England*

R. A. Cox, *National Institute of Oceanography, Wormley, Surrey, England*

Frederick Culkin, *National Institute of Oceanography, Wormley, Surrey, England*

E. K. Duursma, *c/o I.A.E.A., Musée Océanographique, Principaute de Monaco*

Edward D. Goldberg, *Department of Earth Sciences, University of California, La Jolla, San Diego, California, U.S.A.*

Francis A. Richards, *Department of Oceanography, University of Washington, Seattle, U.S.A.*

J. P. Riley, *Department of Oceanography, The University of Liverpool, England*

Geoffrey Skirrow, *Department of Inorganic, Physical and Industrial Chemistry, Donnan Laboratories, The University of Liverpool, England*

J. D. H. Strickland, *Institute of Marine Resources, University of California, La Jolla, San Diego, California, U.S.A.*

Ralph F. Vaccaro, *Woods Hole Oceanographic Institute, Woods Hole, Massachusetts, U.S.A.*

Preface

During the last two decades, the rate of development of all branches of oceanography has been enormous. Although this general expansion is in fact in part a reflection of national prestige and self-interest (e.g. developments in submarine warfare) there is also a great interest in using the seas as a source of raw materials and particularly of food for the under-developed countries. Co-operation in any field is always desirable, and in this respect the recent oceanographic expeditions to the Indian Ocean and the Faroes-Iceland Channel are particularly welcome. These developments, of course, have led to a much more detailed knowledge of the events proceeding in the oceans than was available when the masterly work by Sverdrup, Johnson and Fleming entitled "The Oceans" was published in 1942. It is, therefore, surprising to note that there is an almost complete dearth of books dealing comprehensively with the chemistry of the sea and marine geochemistry. On the interaction between micro-nutrient elements and phytoplankton, Harvey's classic work—"The Chemistry and Fertility of Sea Water"—still stands out, and the work by Raymont is also of great value. Although several facets of chemical oceanography are excellently treated in the second volume of "The Sea" (edited by M. N. Hill), this work does not embrace the subject as a whole.

Bearing the above facts in mind, the Editors have invited a number of specialist authors to contribute chapters for a comprehensive textbook on chemical oceanography to cover the chemistry of the sea, the interaction between the components of sea water and marine life, and the geochemistry of marine sediments. The growth and complexity of the subject is shown by the fact that it has proved impossible to cover these subjects within a single volume.

Although this book is intended primarily for research workers and final year honours undergraduates in chemical oceanography and marine geochemistry, the presentation of each chapter is such that most of the material can be read and understood by readers at all levels, provided that they have a modest knowledge of chemistry. Thus, it is hoped that the book will be of value to workers in related fields, such as marine biology, and will give an authoritative and comprehensive coverage of the many facets of this rapidly expanding subject. Volume 1 deals with sea water, its chemistry and the processes controlling its composition, and with the interaction between marine organisms and the nutrient elements and compounds. Volume 2 contains topics of radio-chemical, analytical and geochemical interest.

No attempt has been made to deal specifically with physical oceanography except where and in so far as physical concepts are essential to a better understanding of the chemistry. The reader is referred to Defant's "Physical Oceanography" for a treatment of physical processes occurring in the sea. Since the geographical distribution of salinity variations is of more importance to the physical oceanographer than to the chemist and is well covered in Defant's treatise, it has not been considered in the present volume.

The work is not intended to be a practical handbook, but where practical details are required, reference should be made to the works indicated in the text.

We particularly wish to thank the authors for their ready co-operation which has greatly assisted us in the editing of the manuscripts. We are most grateful to the various publishers and authors for permission to use diagrams, figures and photographs which are their copyright. Without the critical assistance of Messrs. J. E. Portmann, J. Murphy and A. W. Morris, who acted as proof readers, many errors would have gone undetected and we would like to give them our sincere thanks; credit is also due to Mr. Morris for undertaking the arduous task of preparing the subject indexes. Finally, we would like to thank Academic Press for their understanding co-operation and for their willingness to add extra material, even at a late date, in order to ensure that the book is up to date.

Liverpool J. P. RILEY
February 1965 G. SKIRROW

CONTENTS

Contributors to Volume 1 v
Preface vii
Contents of Volume 2 xiv
Symbols Used in the Text xv

Chapter 1 *by* J. P. RILEY

Historical Introduction

I. Introduction 1
II. The Theories and Observations of Greek and Roman Philosophers 2
III. Arabian Contributions to the Chemistry of the Sea 3
IV. Developments in the Chemistry of the Sea During the Seventeenth and Eighteenth Centuries 5
V. Developments During the Nineteenth Century on the Abundances of the Major Components of Sea Water 9
VI. Dissolved Gases in Sea Water 19
VII. Trace Elements in the Sea 23
VIII. The Biologically Important Trace Nutrients 26
IX. Radioactivity 34
X. Raw Materials from the Sea 35
Bibliography 38
Acknowledgments 38
References 38

Chapter 2 *by* K. F. BOWDEN

Currents and Mixing in the Ocean

I. Introduction 43
II. Surface Currents of the Oceans 44
III. Salinity Distribution in the Surface Layer 49
IV. Deep Water Circulations 51
V. Currents in Coastal Seas 59
VI. Turbulent Mixing 61
VII. Quantitative Treatment of Mixing Problems 64
VIII. Overall Mixing Problems 70
References 71

Chapter 3 *by* R. A. COX

The Physical Properties of Sea Water

I. Introduction. 73
II. The Definition of Chlorinity and Salinity. 75
III. Physical Methods for the Determination of Salinity. 80
IV. The Density of Sea Water. 101
V. The Electrical Conductivity of Sea Water. 103
VI. The Refractive Index of Sea Water. 107
VII. The Velocity of Sound in Sea Water. 109
VIII. The Compressibility of Sea Water. 111
IX. The Transparency of Sea Water. 112
X. The Freezing Point and Osmotic Pressure of Sea Water. 115
XI. The Specific Heat of Sea Water. 116
XII. The Surface Tension of Sea Water. 117
References. 117

Chapter 4 *by* FREDERICK CULKIN

The Major Constituents of Sea Water

I. Introduction. 121
II. The Concept of Constant Composition. 123
III. Methods of Determination of Major Constituents. 126
IV. Principal Determinations of Major Constituents. 140
V. Isotopic Variations. 154
References. 158

Chapter 5 *by* EDWARD D. GOLDBERG

Minor Elements in Sea Water

I. Introduction. 163
II. Chemical Speciation. 166
III. Reactivities of Elements in the Oceans. 171
IV. Residence Times. 172
V. Degrees of Undersaturation. 175
VI. Activity Coefficients. 177
VII. Atmosphere – Hydrosphere Interactions. 178
VIII. Biosphere – Hydrosphere Interactions. 179
IX. Hydrosphere – Lithosphere Interactions. 187
References. 194

Chapter 6 *by* FRANCIS A. RICHARDS

Dissolved Gases Other than Carbon Dioxide

I. Introduction 197
II. Physical Factors Controlling the Gas Content of Sea Water ... 201
III. Changes of Gas Concentration caused by *in situ* Processes 214
IV. Advective and Diffusive Processes 218
V. Some Oceanographic Applications of Dissolved Gas Distributions 220
Acknowledgments 223
References 223

Chapter 7 *by* GEOFFREY SKIRROW

The Dissolved Gases—Carbon Dioxide

I. Introduction 227
II. Carbonate Equilibria in Sea Water 231
III. The Isotopes of Carbon 285
IV. Carbon Dioxide in the Atmosphere 288
V. The Radio-carbon Activity of Sea Water 300
VI. Circulation of Carbon Dioxide between the Atmosphere and the Sea and within the Sea 305
VII. Transfer of Carbon Dioxide across a Gas–Liquid Interface 312
References 317

Chapter 8 *by* F. A. J. ARMSTRONG

Phosphorus

I. Introduction 323
II. Occurrence 323
III. Abundance of Various Forms 326
IV. Seasonal Variations 329
V. Vertical and Horizontal Distribution of Phosphorus in the Sea 335
VI. Comparison of the Major Oceans 359
VII. Nitrogen : Phosphorus Ratio 360
References 362

Chapter 9 *by* RALPH F. VACCARO

Inorganic Nitrogen in Sea Water

I. Introduction 365
II. Geochemical Evolution of Inorganic Nitrogen in the Hydrosphere 366
III. Inorganic Nitrogen Exchanges across the Sea Surface 367
IV. Biological Factors Affecting the Distribution of Molecular and Combined Nitrogen 370

V. Combined Inorganic Nitrogen in the Sea.................. 383
VI. Qualitative and Quantitative Aspects.................. 401
References.................. 404

Chapter 10 *by* F. A. J. ARMSTRONG

Silicon

I. Introduction.................. 409
II. Occurrence.................. 410
III. Changes in Silicate Concentration.................. 415
IV. Distribution of Silicon in the Sea.................. 421
References.................. 430

Chapter 11 *by* E. K. DUURSMA

The Dissolved Organic Constituents of Sea Water

I. Introduction.................. 433
II. Historical.................. 434
III. Chemical Compounds possibly Formed by the Breakdown of Organic Matter.................. 437
IV. Identity of the dissolved Organic Compounds found in Sea Water 446
V. The Geographical Distribution and Seasonal Variation of Dissolved Organic Matter in the Sea.................. 462
VI. Some Additional Remarks.................. 470
References.................. 473

Chapter 12 *by* J. D. H. STRICKLAND

Production of Organic Matter in the Primary Stages of the Marine Food Chain

I. Introduction.................. 478
II. Definitions.................. 479
III. Phytoplankton Species.................. 483
IV. Chemical Composition of Phytoplankton and Methods for Measuring Plant Bio-mass.................. 487
V. Effect of Light, Temperature and Nutrition on Phytoplankton Growth and Composition.................. 518
VI. Field Measurements of Primary Production.................. 551
VII. The Growth and Distribution of Phytoplankton in Nature.... 561
VIII. Phytoplankton Utilization.................. 585
References.................. 595

Chapter 13 *by* FRANCIS A. RICHARDS

Anoxic Basins and Fjords

I. Introduction 611
II. Classification and Examples of Oxygen-deficient and Anoxic Marine Systems 613
III. Some Aspects of the Physical Oceanography of Anoxic Basins and Fjords 619
IV. Chemical and Biochemical Characteristics of Anoxic Basins and Fjords 623
V. Some Consequences of Anoxic Conditions 634
Acknowledgments 643
References 644

APPENDIX 647

(TABLE I. *Major ion composition of water of various salinities.* 647
TABLE II. *Preparation of artificial sea water.* .. 648
TABLE III. *First apparent dissociation constant of boric acid in sea water.* 648
TABLE IV. *Second apparent dissociation constant of boric acid in sea water.* 649
TABLE V. *Solubility of CO_2 in pure water and sea water.* 650
TABLE VI. *First apparent dissociation constant of H_2CO_3 in sea water.* 651
TABLE VII. *Second apparent dissociation constant of H_2CO_3 in sea water.* 651
TABLE VIII. *Molar fractions of H_2CO_3, HCO_3^- and CO_3^{2-} in fresh water.* 652
TABLE IX. *Molar fractions of H_2CO_3, HCO_3^- and CO_3^{2-} in sea water.* 660
TABLE X. *Collected conversion factors.* 671
TABLE XI. *Conversion of weights of N, P, and Si (μg./l.) to μg.-at./l.* 671
TABLE XII. *Solubility of oxygen in sea water.* .. 672
TABLE XIII. *Solubility of nitrogen in sea water.* .. 673)

Author Index 675
Subject Index 693

Contents of Volume 2

Chapter 14 *by* Ph. H. Kuenen
Geological Conditions of Sedimentation

Chapter 15 *by* R. Chester
Elemental Geochemistry of Marine Sediments

Chapter 16 *by* K. K. Turekian
Some Aspects of the Geochemistry of Marine Sediments

Chapter 17 *by* P. E. Cloud, Jr.
Carbonate Precipitation and Dissolution in the Marine Environment

Chapter 18 *by* H. Borchert
Formation of Marine Sedimentary Iron Ores

Chapter 19 *by* H. Borchert
Marine Evaporites—Principles of Oceanic Salt Deposition and Metamorphism

Chapter 20 *by* G. D. Nicholls
The Geochemical History of the Oceans

Chapter 21 *by* J. P. Riley
Analytical Chemistry of Sea Water

Chapter 22 *by* J. D. Burton
Radioactive Nuclides in Sea Water, Marine Sediments and Marine Organisms

Author Index

Subject Index

Symbols used in the text

A list of the more important symbols used in the text is given below. It is not exhaustive and contains some inevitable duplications of usage since some symbols have different accepted usages in two or more disciplines. The generally accepted symbols have been altered only when there was the possibility of ambiguity.

Concentration and activity. Several systems for the expression of concentration are in common use. The more important of these are the molarity scale (g.molecules/l. of solution) designated by c_i and the molality scale (g.molecules/1000g. of water) designated by m_i where the subscript i indicates the solute species. When i is an ion, charges are not included in the subscript except where confusion is possible. Other means of indicating concentration to be found in the text are: g. or mg/kg. of solution (for major components), μg. or mg./l. of solution (for minor elements or nutrients), and μg.-at./l. of solution (for nutrients). Factors for conversion of μg. to μg.-at. are to be found in the Appendix (Table X).

When an activity or activity coefficient is associated with a species, the symbol a_i and γ_i are used respectively regardless of the method of expressing concentration, where the subscript i has the significance indicated above. It is important to appreciate that the numerical values of the activity and activity coefficient depend on the method of expressing concentration. It should also be noted that since activity is a relative quantity it is dimensionless.

Units

Length

Å	= Ångstrom unit	$= 10^{-7}$ mm.
mμ	= millimicron	$= 10^{-6}$ mm.
μ	= micron	$= 10^{-3}$ mm.
mm.	= millimetre.	
cm.	= centimetre.	
m.	= metre.	
km.	= kilometre.	
fm.	= fathom (6 feet).	

Weight

$\mu\mu$g.	$= 10^{-12}$g.
mμg.	$= 10^{-9}$g.
μg.	$= 10^{-6}$g.

WEIGHT—*continued*

mg.	$= 10^{-3}$g.
g.	= gramme.
kg.	= kilogramme.
ton	$= 10^3$kg. $= 10^6$g.

VOLUME

ml.	= millilitre.
l.	= litre.

CONCENTRATION

p.p.m.	= part per million (μg./g. or mg./l.).
μg.-at./l.	= μg. atom/l. = μg./atomic weight/l.

VELOCITY

kt.	= 1 nautical mile (6080 feet)/hour.

RESISTANCE

Ω	= ohm.

LIGHT FLUX

ly	= langley (1 ly/min. = 1 g. cal./cm^2./min.)

GENERAL SYMBOLS

a_i	the activity of component i.
A	absorbance = optical density $\left(\log_{10} \frac{I_0}{I}\right)$.
Alk_{min}	the minimum alkalinity (meq./kg.).
Alk_{obs}	the observed alkalinity (meq./kg.).
Alk_X	the surplus alkalinity (meq./kg.).
A.O.U.	apparent oxygen utilization.
B.O.D.	biochemical oxygen demand.
c_i	the molar concentration of species i. Charges of the species i are omitted where these are obvious.
CA	the carbonate alkalinity (meq./l.).
Cl‰	chlorinity (g./kg.).
D_E	the depth of the euphotic layer.
D_g	the molecular diffusion coefficient in the gas phase.
D_l	the molecular diffusion coefficient in the liquid phase.
D_M	the depth of the layer in which plant cells are randomly transported by mixing processes.
d.p.m.	radioactive disintegrations per minute.
E	the E.M.F. of a cell. Also the extinction (optical density).
E^0	standard potential.

Eh	redox potential.
E_l	liquid junction potential.
$E^{1\%}_{1cm}$	specific extinction coefficient (extinction for a 1 cm. light path in 1% solution).
F	Faraday equivalent of electric charge (96,500 coulombs).
G	Gibb's free energy.
h	relative humidity.
H.P.P.U.	Harvey plant pigment unit (see p. 498).
k_a	the velocity constant/unit area for the transfer of carbon dioxide from the atmosphere to the sea.
k_s	the velocity constant/unit area for the transfer of carbon dioxide from the sea to the atmosphere.
k_w	concentration ionic product of water.
$k_{w(TD)}$	thermodynamic ionic product of water.
K_1	the first thermodynamic dissociation constant of carbonic acid (see Chapter 7, equation (20)).
K_2	the second thermodynamic dissociation constant of carbonic acid (see Chapter 7, equation (21)).
K'_1	the first apparent dissociation constant of carbonic acid after Buch (see Chapter 7, equation (22)).
K'_2	the second apparent dissociation constant of carbonic acid (see Chapter 7, equation (23)).
K'_{L1}	the first apparent dissociation constant of carbonic acid after Lyman (see Chapter 7, equation (24)).
K_f	first ionization constant of carbonic acid.
K_{H_2O}	thermodynamic dissociation constant of water.
K_{HA}	thermodynamic dissociation constant of the weak acid HA.
K_{in}	thermodynamic dissociation constant of an indicator.
K_0	hydration constant of dissolved carbon dioxide.
K'_B	the first apparent dissociation constant of boric acid in sea water.
K_x	coefficient of eddy diffusion in the x-direction (similarly for K_y and K_z).
I	attenuated light intensity.
I_λ	attenuated light intensity of wavelength λ.
I_0	initial light intensity.
$I_{0\lambda}$	initial light intensity of wavelength λ.
I_{comp}	compensation light intensity.
I_{max} or I_{sat}	saturation light intensity.
l	thickness of light attenuating medium.
m_i	molal concentration of the species i.

pH	$-\log a_H$ (see Chapter 7, Section II A(2)).
psH	pH on the Sørensen scale (see Chapter 7, Section II A(2)).
pH(s)	pH of a standard buffer.
pH(X)	pH of a test solution.
pH_X	proposed sea water pH scale (Smith and Hood, 1964) (see Chapter 7, Section II A(3)).
p_{CO_2}	partial pressure of carbon dioxide in the atmosphere.
p_0	saturated vapour pressure of pure water.
p_s	saturated vapour pressure of an aqueous solution.
P	photosynthetic rate.
$P_{euphotic}$	rate of production in the euphotic zone.
P_{max}	maximum rate of production at intensity I_{max}.
P_{100}, P_{10}, P_1	productivities at the surface and at depths where the light intensities are 10 and 1% of the surface values respectively (see Chapter 12, Section IV C(1d)).
P_t	total pressure.
P_{CO_2}	partial pressure of carbon dioxide in solution.
PQ	photosynthetic quotient.
P_z	hydrostatic pressure at depth z.
P_h	total hydrostatic pressure.
R	rate of production or loss of material.
R	the gas constant.
R	electrical resistance.
RQ	respiratory quotient.
$S‰$	salinity (g./kg.).
S.P.U.	specified pigments units (Chapter 12, Section IV A(3)).
t	temperature in °C.
t	time.
T	temperature in °K.
T	rate of transport/unit area.
u,v,w,	components of velocity of water in the x-, y- and z-directions respectively.
z	depth.

Greek Symbols

α	generalized solubility coefficient of a gas in a liquid (moles/l./atm).
α_0	solubility coefficient of a gaseous constituent in pure water (moles/l./atm).
α_s	solubility coefficient of a gaseous constituent in sea water (moles/l./atm).
α_f	isotopic fractionation factor (see Chapters 4, 7 and 22).

β	van Slyke buffer capacity.
γ_i	activity coefficient of species i.
$\delta^{13}C$	permillage enrichment of ^{13}C relative to a given standard.
$\delta^{14}C$	permillage enrichment of ^{14}C relative to a given standard.
$\Delta^{14}C$	^{14}C enrichment on the Lamont scale.
$\sum CO_2$	total carbon dioxide content of a solution $c_{H_2CO_3} + c_{CO_2} + c_{HCO_3} + c_{CO_3}$ (+ CO_2 involved in complexes).
ϵ	molar absorptivity.
κ	electrical conductivity.
λ	radio-isotope decay constant.
$\boldsymbol{\mu}_i$	chemical potential of species i.
μ	ionic strength.
ρ	density.
σ	density (σ_t at temperature t) relative to distilled water at 4°C.
τ	radio-isotope half-life.
τ	residence time.
χ	absorption coefficient (see Chapter 12, Section IV B(1a)).

Chapter 1

Historical Introduction

J. P. RILEY

Department of Oceanography
The University of Liverpool, England

Truly a rare and marvellous affair. A Philosopher not in his Study but on the High Seas, far away from all Scientists and among Sailors, not in Quiet and Solitude, but amidst savage shoutings, not in Peace but in Danger of Life! So much trouble it took to write this work to which he devoted his time from 1680 to 1725. Come on then, make use of the same and profit by all this which was compiled by such efforts, for which you would look in vain elsewhere and which is found here in such Abundance.

Translation of part of the preface by H. Boerhaave, to his Dutch edition of the works of Count L. F. Marsilli (1786).

I. Introduction 1
II. The Theories and Observations of Greek and Roman Philosophers 2
III. Arabian Contributions to the Chemistry of the Sea 3
IV. Developments in the Chemistry of the Sea During the Seventeenth and Eighteenth Centuries 5
V. Developments During the Nineteenth Century on the Abundances of the Major Components of Sea Water 9
VI. Dissolved Gases in Sea Water 19
A. Oxygen and Nitrogen 19
B. Carbon Dioxide and the Buffer Mechanism of Sea Water 20
C. Hydrogen Sulphide 23
VII. Trace Elements in the Sea 23
VIII. The Biologically Important Trace Nutrients 26
IX. Radioactivity 34
X. Raw Materials from the Sea 35
Bibliography 38
References 38

I. Introduction

Oceanography is the scientific study of the sea and of the processes occurring in it. In this science many disciplines such as physics, chemistry, geology and biology are united, and a knowledge of several of these sciences may be necessary in solving any particular problem. Thus, the distribution of certain chemical components in bodies of water may provide useful information for the physical oceanographer on water movements, or may give a guide to the factors controlling the fertility of the water mass for the growth of phytoplankton. Physicochemical studies may show how marine sediments have been produced and give a clue to their geochemistry, or may give information on such features as transfer of gases between the sea and the atmosphere.

The purpose of this book is to provide an account of modern work on the chemistry of the sea and of the influences of chemical factors on marine sediment formation and on the fertility of the sea. This chapter will be devoted to a review of the history of the subject. The development of scientific oceanography, as opposed to the purely speculative approach to the subject, has had to await the development of suitable experimental techniques, and in the chemical facets of the subject this has mainly depended on the production of adequate sampling apparatus and reliable analytical techniques. Thus, the provision of rapid and sensitive photometric methods of analysis was necessary before the role of nitrogen and phosphorus in the fertility of the sea could be investigated. The production of new physicochemical analytical methods is proceeding at an ever-increasing rate, and it is probable that in the future some of these techniques will find employment in chemical oceanography. Already some of them, such as neutron activation analysis and square wave polarography, have been used for the determination of trace inorganic components of sea water.

The future major developments in sea-water chemistry will probably centre around the dissolved organic components and their biochemistry. Here, highly sensitive methods of analysis, such as gas–liquid chromatography, in conjunction with newer methods of separation, such as gel-filtration, will play an important part.

II. The Theories and Observations of Greek and Roman Philosophers

Since prehistoric times, salt has been obtained from sea water by solar evaporation and even today, in certain parts of the world, much salt is obtained in this way. The cause of the saltiness of the sea puzzled the Greek philosophers who advanced many, to us, improbable explanations. According to Aristotle (384–322 B.C.), some classical authors thought that just as fresh water becomes saline when filtered through ashes, so the sea has become salty because of its content of similar material from the earth. Other writers (probably Empedocles and Democritus) considered that by analogy with sweat, which is induced by heat, the sea is a sort of perspiration exuded by the earth under the action of the heat of the sun; the logic of this argument was, to them, apparently confirmed by the fact that salt water, when exposed to the sun would disappear and leave crystals of salt. Pliny the Elder (died

A.D. 79), in his Natural History, also accepted that the sea derives its saline taste from the sun, but gave the rather different explanation that the heat of the sun removes from the sea the easily attracted sweet and light components and leaves behind the bitter constituents, and as a result of this, the deeper waters would be fresher than those at the surface.

The philosopher Seneca (3 B.C. to A.D. 65) summarized the largely empirical views of the ancients about natural science in his Quæstiones Naturales. He postulated that the world at its origin was in chaos, and that the elements dissolved in the primordial ocean separated out in the course of time. He classified the waters of the globe into (i) oceanic waters which have existed since the beginning of time; they constitute the bulk of the water on the earth and are the source from which all others are derived, (ii) subterranean waters which circulate in the earth and soil and which emerge at the surface as springs, (iii) waters which circulate or remain stagnant on top of the soil, (iv) water present in the atmosphere. Seneca also noted that the level and salinity of the ocean remained constant in spite of the continuous addition of river and rain water, and he considered that this constancy was caused by evaporation. He attributed the dissolving power that water possesses for even the hardest rocks to the presence of a gas; he thought that thermal springs had the most powerful solvent action and observed that the materials dissolved were frequently re-deposited, particularly by thermal waters. He explained the formation of calcareous tuffs as being brought about by the absorption or filtration of the saline substances from the water by earthy layers.

Seneca, who was a keen observer of nature, also noted the erosion of rocks by streams, currents and more particularly waves. He observed that the erosion products could be transported far from their source by the action of moving water, and that the fine material carried in suspension by rivers could be deposited at their mouths in the form of deltas.

III. Arabian Contributions to the Chemistry of the Sea

The extension of the Arabian Empire in the latter half of the first Millennium A.D. was followed by many voyages of discovery by Arabian Navigators which led to much interest in the sea and in geography. The contact of the Arabian scientists with the writings of the Greek philosophers at first had a hampering effect on the development of Arabian science. Later, the philosophical attitude to science was partly replaced by an experimental approach to the investigation of natural phenomena in which occurrences in nature were compared with those in the laboratory. Nevertheless, although Arabian science contained many accurate deductions and observations about natural phenomena, it was interspersed with mythological and fantastic explanations.

Mas'ūdī (lived *c.* A.D. 915), a prominent Arabian naturalist and geographer, summarized the ideas current in his time regarding the origin of the sea and its saltiness (see Sprenger, 1841):

> Some maintain that the sea is a remnant of the primitive humidity, the greater part of which has been dried up by fire and the surplus has been transformed under heat. Others allege that the whole of the primitive humidity having been submitted to the devouring action of the sun in its revolutions, all the pure parts have been removed and the present ocean is merely a saline and bitter residue. There are others who assert that the sea is simply the secretions which flow down from the earth scorched by the heat of the sun during its constant revolutions. Some believe that the sea is nothing less than the primitive humidity separated from all the coarse and terrestrial principles, just as fresh water mixed with ashes loses its sweetness and retains a saline taste even after it has been filtered. It is also supposed that in water the fresh and salt parts were mixed and that the sun volatilized the fresh portions because of their subtlety, be it that it absorbs these parts itself, be it that once arrived in high regions where cold condenses them and gives them, so to speak, a new form, they change a second time into water. . . . Several learned men have held that the mass of water which flows into the ocean either from the surface of the soil or from subterranean passages, having once arrived at that reservoir has absorbed everywhere the saline principles which the earth discharges into it. It is not to be wondered at, then, that the water of the sea preserves always the same weight and bulk, since the subtle parts, which the heat removes, change again into dew and water, whence arise mountain streams, which fall into rivulets and ponds and flow into the wet places of the land until they arrive at last in the great whirlpool of the ocean. It is thus that absolutely none of the water is lost and springs are like machines, which, drawing water from a river, return it to a rivulet that passes it again to the river. . . . It is evident, from experience, that all humid matters endowed with a certain relish, having passed through the retort and still preserve in their sublimate the same smell and taste like vinegar, date wine, rose saffron, gilly-flower water, except, however, saline substances, which change the taste and smell especially when they are submitted twice to the operation of the fire and the salt.

The Arabian scientists were also active in considering the geological action of water and of its action in the transport of sediment. Thus, Al-Bīrunī (*c.* A.D. 1000) was aware of the formation and destruction of coral islands, and had observed the sorting of materials carried down by streams (Ethe, 1869).

IV. Developments in the Chemistry of the Sea During the Seventeenth and Eighteenth Centuries

Arab science considerably influenced the course of science in Europe during the later middle ages, not only by its own observations, but by making the works of the Greek authors available to scholars. Although this period was a fruitful one for marine exploration and geography, little appears to have been done to investigate the chemistry of the sea, and it was not until the seventeenth century that the scientific study of the sea was commenced.

In about 1670, Robert Boyle, who may be considered the founder of scientific chemical oceanography, published a work entitled "Observations and Experiments on the Saltness of the Sea" (see Boulton, 1699). In this book he described a number of experiments which he had carried out on sea water, and related information which he had obtained from mariners and divers. Boyle introduced the silver nitrate test for salt water, and using it found that whereas sea water gave a dense white precipitate with this reagent, river and lake waters yielded only a turbidity. He concluded that all fresh waters contained a trace of salt which had been leached from the land and that the salt of the sea resulted from the accumulation of these leachings.

Boyle made the first attempts to determine the salinity of sea water by evaporating a known volume of sea water and igniting it in a weighed dish, which was reweighed after the water had evaporated. He found that if the experiment was repeated several times with the same sample of sea water the results seldom agreed, and he concluded that this was an unsatisfactory method for estimating the salt content of sea water. Measurement of the specific gravity of the sample, however, provided a more accurate measure of its salt content. For this purpose he either used a glass hydrometer, or weighed a glass tube full of the sea water and then weighed it again full of distilled water, or weighed a piece of sulphur both in distilled water and in the sea water.

By means of a valved box devised by his pupil Hooke (Hooke, 1667, see Fig. 1), he personally made a series of observations on the specific gravity of samples of English Channel water which had been collected at various depths. He also supplied a hydrometer to a traveller to Jamaica who observed that the specific gravity of the water:

FIG. 1. Hooke's sounding machine and water sampler (1667).

> grew heavier and heavier as he neared the line, till about 30° latitude, from whence to Jamaica he observed no alteration in the specific gravity in the least. And in confirmation of this I am likewise informed by one, who for his own satisfaction weighed the water, both under the Aequinoctial and at the Cape of Good Hope, and found that the weight of both was the same. To which may be added that it is commonly observed at Mosambique, one of the hottest places in the world, that the sea is so salt there that it bears up ships a considerable height out of the water; more than in other places; and that the water may be much salter in one place than another. . . .

In the early part of the eighteenth century, Marsilli (1711, 1725) (cf. frontispiece) using a hydrometer* found that the deep waters were denser than the surface water. He found by evaporation that surface and deep Mediterranean water, collected at a distance from rivers and other sources contained $\frac{1}{32}$ and $\frac{1}{29}$ of its own weight of salts respectively.

The composition of the surface waters of France engaged the attention of Antoine Lavoisier (1743–1794), and his investigations led him to the discovery that water is a compound of oxygen and hydrogen. He also had an interest in hydrometry which up to that time was the only quantitative method of assaying the salt content of waters. The results obtained by hydrometry gave no information about the nature and amounts of the dissolved substances in the water, and Lavoisier therefore devised techniques for the determination of these substances. In 1776 he published the first analysis of sea water (taken from the English Channel off Dieppe) (see Table I) and later, an analysis of the hypersaline waters of the Dead Sea (Macquer *et al.*, 1778). These analyses were made by slowly evaporating a known weight of sea water to dryness. During the earlier stages of the evaporation, calcium carbonate and sulphate precipitated and were removed, washed and weighed. When almost dry, the salts were extracted with alcohol to separate magnesium. The residue was heated with a 2:1 mixture of alcohol and water until it had all dissolved. On cooling, a mixture of sodium and magnesium salts crystallized and was removed and weighed.

*It is uncertain who invented the hydrometer, it may possibly have been Archimedes. The first clear description of the instrument is given in the fifteenth letter of Syrenius (A.D. 370–415) to his friend and mentor Hypatia (appointed Bishop of Ptolemāis, A.D. 410). "I am so unfortunate as to need a Hydroscope. Have a bronze one constructed and assembled for me. It consists of a cylindrical tube the size and shape of a flute. On this tube in a vertical line are the notches by which you tell the weight of the waters we are testing. A cone fits over the tube so evenly that cone and tube have a common base. . . . Now whenever you place the tube in the water it will remain erect." (Hercher, 1873). Robert Boyle (1675) designed a glass hydrometer very similar in design to the modern instrument.

TABLE I

The composition of sea water according to Lavoisier (1772)

	Pour chaque livre† d'eau de mer gros*	graines	Probable composition
1°. Terre calcaire soluble dans les acides et qui paroît ne pas différer de la terre calcaire commune; 2°. Selenite ou sel gypseux		$8\frac{1}{6}$	(calcium carbonate and sulphate)
Sel marine à base d'alkali fixé de la soude	1	$54\frac{4}{5}$	(sodium chloride)
Sel de Glauber et sel d'Epsom		$7\frac{17}{20}$	(sodium and magnesium sulphate)
Sel marine à base de sel d'Epsom		$14\frac{3}{4}$	(magnesium chloride)
Sel marine à base terreuse ordinaire, mête de sel marine à base de sel d'Epsom		$24\frac{13}{23}$	(calcium and magnesium chlorides)

*An ancient French measure of *c.* 3·83 g., a *grain* is $\frac{1}{72}$ part of a *gros*.
†1 livre = 453·6 g.

In his youth, Lavoisier assisted the mineralogist Jean Etienne Guettard with the preparations of a geological map of France. It is likely that this background induced Lavoisier to make his attempt to relate the observed abundances of salts in the rivers which he studied to the geology of the districts in which they flowed. Goldberg (1960) has quoted from Lavoisier's works to show the importance of the geochemical role of water in the sedimentary cycle:

> L'eau de mer est le résultat du lavage de toute la surface du globe; ce sont en quelque façon les rincures du grand laboratoire de la Nature, on doit donc s'attendre á trouver réunis dans cette eau, tout les sels qui peuvent se rencontrer dans le regne minéral et c'est ce qui arrive en effet: comme cette eau est la plus compliquée de toutes celles que j'ai occasion d'examiner, je l'ai choisie pour donner un exemple de l'application de l'esprit-de-vin à l'analyse des eaux minérales.

At about the same time that Lavoisier was carrying out his work on natural waters in France, Torbern Olaf Bergman (1735–1784) was devising a scheme for the qualitative analysis of such waters which he applied to the analysis of the surface waters of Europe. In 1779 he published (cf. also Bergman, 1788) an analysis of a water collected at a depth of 60 fm. using a glass bottle (Table II). The analyses were made by extracting magnesium chloride from a known weight of the dry salts with alcohol; the residual salts were weighed, and sodium chloride was extracted with cold water. The residue, which consisted of gypsum, was again weighed. He also observed that ocean waters were slightly alkaline since they turned Brasil-wood paper blue, but only slightly intensified the colour of turnsole paper. He erroneously associated this alkalinity with the presence of magnesium in the water.

TABLE II

Composition of sea water according to Bergman (1788)

Substance	*Weight per kanne**	
Common salt	2 oz.	433 grains
Salinited magnesia (magnesium chloride)		380 grains
Loose gypsum		45 grains

*An old Swedish measure of 2·617 litre.

A somewhat similar analysis was given by Bouillon-Lagrange and Vogel (1813) who found in 1 kg. of sea water: 25·1 g. of sodium chloride, 3·5 g. of magnesium chloride, 5·78 g. of magnesium sulphate, 0·2 g. of calcium and magnesium chloride and 0·15 g. of calcium sulphate.

V. DEVELOPMENTS DURING THE NINETEENTH CENTURY ON THE ABUNDANCES OF THE MAJOR COMPONENTS OF SEA WATER

The methods of analysis of sea water used in the eighteenth century left much to be desired since the salts produced by the evaporation of sea water vary somewhat according to the conditions used, and because the separations so obtained are never quantitative. At this period, long before the Ionic Theory of Arrhenius, there was considerable disagreement about the compounds that could be isolated from sea water and which were, therefore, thought to be present in it. Thus, no magnesium sulphate was found by Bergman although it had always been isolated by other chemists; similarly, Lavoisier reported the presence of calcium chloride and sodium sulphate.

The unsatisfactory nature of the results obtained by the crystallization process was realized by John Murray (1818) who analysed water from the Firth of Forth using Lavoisier's method. He examined each of the fractions so obtained, and demonstrated that they were far from pure. He therefore proposed the determination of the individual "acids and bases" in the water (anions and cations as we would now call them), since "this kind of analysis has the advantage that it can be executed with much more precision than the other: it is liable to fewer sources of error, and by finding the quantities, not of the compounds, but of the elements, any error that is introduced is discovered when the binary compounds are inferred." Gravimetric methods were used for the analysis; sulphate was precipitated by addition of barium chloride, and calcium was precipitated from the filtrate as oxalate and converted to sulphate for weighing. Magnesium was then precipitated from the filtrate as its ammonium phosphate and weighed as pyrophosphate, and sodium was determined by evaporation of the filtrate and weighed as sodium chloride; chloride was apparently determined by difference. It is interesting to note that most of these methods, with only slight modifications, have been used until recently for the analysis of sea water (see Chapter 4). Murray's analysis of sea water is shown in Table III.

Independently of Murray, Marcet (1819) also applied similar gravimetric methods to the determination of the same anions in sea water. However, chloride was determined by precipitation as silver chloride, and sodium was determined by the difference between the total equivalents of anions and cations. Using these techniques, he analysed 14 samples of sea water from a variety of seas and oceans, including the Atlantic and Arctic Oceans, and the Baltic, Black and Yellow Seas. (cf. Table III.) From the results it was concluded that:

> All the specimens of sea water which I have examined, however different their strength, contain the same ingredients all over the world, these bearing very nearly the same proportions to each other; so that they differ only as to the total amount of their saline contents.

Marcet also determined the density of more than 70 samples of sea water from a number of seas using a specific gravity bottle, which he modified by fitting a stopper containing a capillary tube. Many of these samples were collected from depths of up to 250 fm. by means of a water bottle triggered by a weight hanging below it making contact with the sea bed. In the same paper, he described a water bottle (Fig. 2) designed by a Mr. Newman which could be closed by means of a messenger dropped

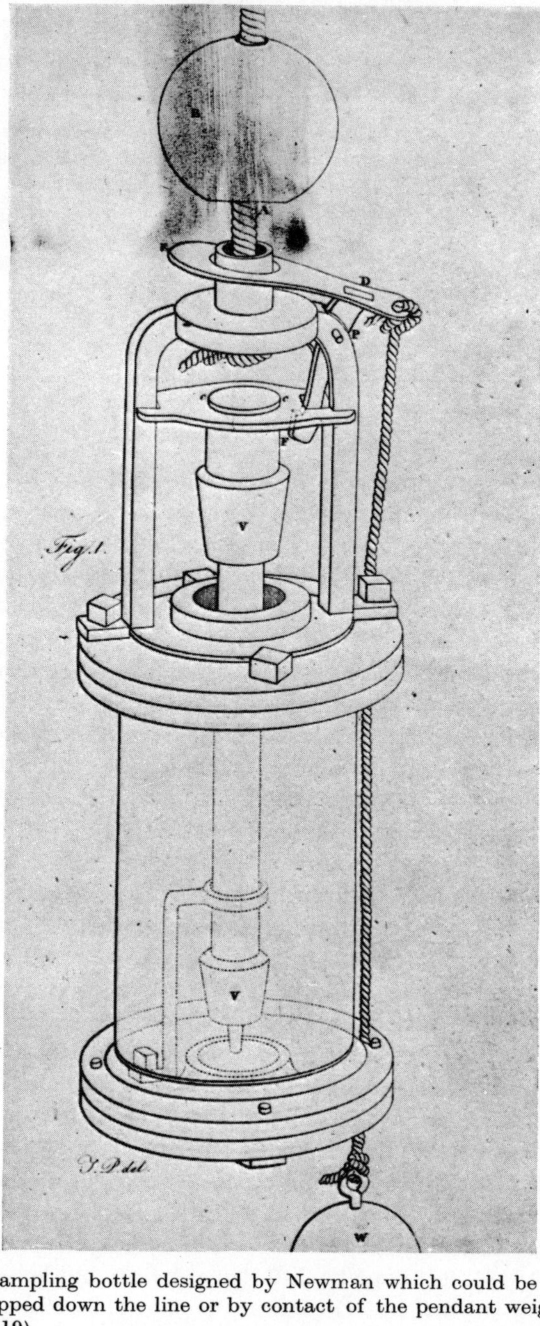

FIG. 2. Water sampling bottle designed by Newman which could be closed either by a messenger dropped down the line or by contact of the pendant weight with the sea floor (Marcet, 1819).

TABLE III

Composition of sea water according to Murray (1818) and Marcet (1819)

	Murray (1818) *grains per pint*	*Marcet (1819)** *grains in 500 grains of sea water*
Lime	2·9	0·314
Magnesia	14·8	1·08
Soda	96·3	8·11
Sulphuric acid	14·4	1·27
Muriatic acid	97·7	8·0

*Sample from North Atlantic.

down the suspension rope; this is thus the predecessor of the modern water bottle. Marcet attempted to show the existence of an underwater current of high density flowing from the Mediterranean through the Straits of Gibraltar into the Atlantic which had been postulated by earlier authors. He was, however, unable to reach any definite conclusions owing to the impossibility of obtaining samples from deeper than 250 fm. In the course of an investigation of the freezing of sea water, Marcet was also able to show that the density of sea water increases continuously with decrease in temperature until the water freezes, unlike distilled water in which maximum density is attained at 3·98°C.

In the early part of the nineteenth century, many investigations of the density of sea water were made, of which only two will be mentioned. J. Davy (1817) published a large number of densities determined on water samples taken at various latitudes. Gay-Lussac (1778–1850), well known for his Law of Combining Volumes and for his work with Humboldt on the composition of the atmosphere, examined the salinities and densities of waters collected between France and South America (Gay-Lussac, 1817). From the results of these experiments he was able to conclude that the large quantities of fresh water delivered by rivers do not mix rapidly with sea water; this extended the observation published in the same year by the Rev. F. Fleming (1818) that fresh water floats on top of sea water in estuaries. The amounts of evaporation and precipitation are probably out of balance over most of the sea and this is a principal cause of variations in salinity. However, ocean currents will tend to re-establish a salinity equilibrium. He also recognized (Gay-Lussac, 1818) that density differences would not lead to separation of the dissolved substances. On the subject of the influence of density upon the stability of water masses and temperature–density relationships he wrote:

> Pour que les eaux de la mer ne se mêlent pas entr'elles et restent en équilibre, il faut que les couches de niveau aient la même densité: or à cause de l'inegalité de température de l'équateur aux poles, qui seule ferait varier la densité de l'eau, on a conclu que l'eau de la mer devait contenir d'autant plus de matières salines que sa température est eleveé.

Until 1819 only sodium, calcium, magnesium, sulphate and chloride had been detected or determined in sea water. In this year, however, Wollaston (cited by Marcet, 1819) renowned for his work on platinum, detected and determined potassium in sea water. Of the other major components of sea water, bromine was discovered in bittern from the Mediterranean by Balard (1826), but the first determination was not attempted until some years later (Göbel, 1838). This analysis gave results which were low by an order of magnitude, and it was not until the precise work of Thorpe and Morton (1871) that the currently accepted value was recorded. The occurrence of boron in sea water appears to have been first reported by Strecker (1853) and its presence was confirmed by Forchhammer (1865), who also detected strontium and fluorine. Later determinations of these elements are discussed in Chapter 4.

The next advance in knowledge of the major components of sea water was made by Forchhammer* (1865), who over a period of twenty years from 1843 analysed several hundreds of samples of sea water for calcium, magnesium, chloride and sulphate, and a few for potassium. Sodium was determined by difference assuming that the cations were equivalent to the chloride plus sulphate, and ignoring the presence of bicarbonate and carbonate. Salinity was calculated as the sum of all these components. The results of these analyses (Table IV) are considered in Chapter 4; their accuracy is rather poor by modern standards. Forchhammer used differences in the salinities, and in some cases of the chloride contents of sea waters, for tracing the movements of water masses in the Atlantic Ocean and the Baltic Sea. He studied the influence of melting ice and river waters on the ocean. His original training as a geologist led him to consider the fate of inorganic material brought down to the sea by rivers. He noted that whereas calcium and silica were relatively abundant in river water, they were depleted in sea water; he attributed their removal

*Johann Georg Forchhammer (1794–1865) studied Chemistry and Pharmacy at Kiel University under Pfaff; Ph.D.1820. Attended lectures on Physics by Oersted whom he succeeded as Director of the Polytechnic Institute, Copenhagen. He became Secretary of the Danish Royal Academy of Science.

TABLE IV

Composition of sea salts according to Forchhammer (1865), Schmelck (1882) and Dittmar (1884) (g./100g. Cl + Br)

	Forchhammer	*Schmelck*	*Dittmar* *Ave.*	*Ave.**
Cl + Br	100·000	100·000	100·000	100·000†
SO_3	11·88	11·46	11·575	11·558
CO_2	—	—	0·274	0·275
CaO	2·93	2·99	3·027	3·025
MgO	11·03	11·40	11·212	11·269
K_2O	1·93	—	2·405	2·445
Na_2O	—	—	74·462	74·486

*Recalculated Lyman and Fleming's (1940) data. These data were themselves recalculated from the original figures using 1939 atomic weights.

†Br = 0·340.

from sea water to the action of organisms secreting them, and concluded that the quantity of the different elements in sea water is not proportional to the quantity of elements which river water pours into the sea,

FIG. 3. H.M.S. *Challenger* (Murray, 1895). By permission of H.M. Stationery Office, London.

but inversely proportional to the facility with which the elements in sea water are made insoluble by general chemical or organochemical actions in the sea.

The expedition of H.M.S. *Challenger* (Fig. 3) between 1873 and 1876 added greatly to the advancement of all branches of oceanography. The expedition's chemist, J. Y. Buchanan, was in charge of the ship's chemical laboratory (Fig. 4). His duties* included supervision of the collection at each station of samples from a variety of depths down to 800 fm. using a stopcock water bottle. The specific gravities of these samples and those of surface samples collected daily were measured by means of a hydrometer. Carbon dioxide was determined on board by titration. The dissolved gases were boiled out of many of these samples and analysed ashore. He also bottled many of the water samples and sent them home for analysis.

The analyses of 77 of these water samples and of the dissolved gases were conducted by Dittmar (1884)† (Fig. 5). Dittmar's analyses of the major component ions of the sea waters were carried out with great attention to detail using analytical methods which had been tested on synthetic samples. Innovations in the analysis included the titrimetric determination of halides by the Volhard method, the determination of bromine, the determination of potassium as platinichloride, and of sodium by difference from the weight of mixed sulphates. The titrimetric determination of alkalinity was also carried out in order to make allowance for the presence of carbonate and bicarbonate. Dittmar's results (Table IV) are a model of precision, and compare very favourably with modern figures. As discussed in Chapter 4, they confirmed Forchhammer's view that the ratios of the concentrations of the major dissolved constituents are constant, subject to very slight variations. The analyses were particularly important in indicating that the salinity of sea water could be characterized by the determination of any single major component. Dittmar himself suggested the use of chloride (+ bromide) for this purpose, and stated that the total solids could be determined by multiplying the chloride content (in g./kg.) by 1·8058.

Schmelk (1882) carried out analyses on 51 samples collected on the Norwegian Arctic expedition of 1876–1878; the results are shown in Table IV.

*Buchanan's autobiography (1919) gives a very interesting account of his work on the *Challenger*.

†William (Wilhelm?) Dittmar, F.R.S., F.R.S.E., was born near Darmstadt in 1833. He was educated there, and became assistant to Bunsen. Subsequently, after a period at Owens College, Manchester, under Roscoe, at Edinburgh and in Germany at Poppeldorf-bei-Bonn, he returned to Edinburgh as assistant to Crum Brown. He was appointed to the Chair of Chemistry in the Andersonian University, Glasgow, in 1874. He died in 1894.

FIG. 4. Chemical laboratory aboard H.M.S. *Challenger* (Murray, 1895). By permission of H.M. Stationery Office, London.

FIG. 5. Cartoon of Prof. W. Dittmar from "The Bathe" No. 322 for Dec. 18th, 1878.

At the suggestion of an International Commission, Forch, Knudsen and Sørensen thoroughly investigated the interrelationship of the chlorinity, salinity and density of sea water, and drew up definitions of chlorinity* and salinity (Forch *et al.*, 1902). It was found that chlorinity (Cl) was related to salinity (S) by the equation $S = 1{\cdot}8050\,Cl + 0{\cdot}03$. In 1899, Knudsen† (Fig. 6) published a description of an elegant routine titrimetric method for the precise determination of chlorinity and suggested at an International Congress in Stockholm the use of a Standard Sea Water for the standardization of the silver nitrate solution used in the determination. His method, or slight modifications of it, has remained in use in most oceanographical laboratories until recently, and it is only now being replaced by physical methods such as electrical conductivity, which are more rapid and precise (see Chapter 3).

Since Dittmar's work, few complete determinations of the major component ions in sea water have been made, and most recent workers have been content to determine particular ions, and to express their results as the ratio of concentration to chlorinity. The research School of Oceanography at the University of Washington, under T. G. Thompson‡, was particularly active both in this field and in that of trace element analysis between 1930 and 1958. For a summary of recent work on the major component see Chapter 4.

*According to this definition chlorinity was the total weight of chlorine, bromine and iodine in g./kg. of sea water, assuming bromide and iodide had been converted to their equivalent weight of chloride. It is therefore dependent upon the accepted atomic weights of silver, chloride and bromine. To overcome this difficulty chlorinity was re-defined in terms of the weight of silver necessary to precipitate the halogens in 0·3286707 kg. of sea water (Jacobsen & Knudsen, 1940).

†Martin Knudsen was born in Hasmark on the Isle of Funen, Denmark, in 1871. He entered the University of Copenhagen in 1890. After qualifying in physics in 1896, he was assistant at the Copenhagen Polytechnic High School; in 1899 he became an assistant at the University, at which he was later appointed lecturer (1901–1912) and Professor of Physics (1912–1914). He was *Rektor Magnificus* of the University from 1927–1928, and a member and secretary of The Royal Danish Academy of Science and Letters. He died in 1949. He was introduced to oceanography through the Danish Ingolf Expedition of 1895 to 1896, for which his chlorinity method was developed. From 1908 until his death he was in charge of the preparation and standardization of Standard Sea Water. He directed the work on the relationships between the salinity, chlorinity and density of sea water, the results of which were incorporated in his Hydrographic Tables (Knudsen, 1901). He also found time to take an active part in the founding of the International Council for the Exploration of the Sea and in its work.

‡Thomas Gordon Thompson (1888–1961) graduated as B.A. from Clark University, U.S.A. in 1914, and then proceeded to the University of Washington from which he graduated as Ph.D. in chemistry in 1918. He remained there until his retirement in 1959, first as Assistant Professor of Chemistry (1919), as Associate Professor (1923), and finally as Professor (1929).

FIG. 6. Prof. Martin Knudsen.

VI. Dissolved Gases in Sea Water

A. OXYGEN AND NITROGEN

The earliest analyses of the dissolved gases in sea water were made in the first part of the nineteenth century (Frémy, 1838; Morren, 1843) but the techniques used were rather crude. Jacobsen (1873) used the method developed by Bunsen for the removal of gases from sea water on board the *Pomerania* in an expedition to the North Sea. A modification of this process was used by Buchanan aboard the *Challenger*. The separated gases were analysed by Dittmar (1884) who devised an analysis train in which carbon dioxide was absorbed in potassium hydroxide, and oxygen was determined in the residual gas by the contraction resulting on explosion with a known volume of hydrogen; nitrogen was determined by difference. Dittmar also determined the absorption coefficients of oxygen and nitrogen in distilled water and sea water. He recognized that both oxygen and nitrogen are usually present in surface waters at about their saturation concentrations, and that solution was a slow process. In the deep waters, the concentrations of nitrogen varied within the same limits as those in surface waters, but the oxygen concentrations were often very much lower than would be calculated from the amount of nitrogen present. Although Dittmar was aware that the processes of oxidation could reduce the oxygen concentration, he attributed these low oxygen contents to incomplete removal of the dissolved gases from sea water or to malfunctioning of the sampling bottle. However, prior to this Carpenter (1872) had found in two deep stations in the Mediterranean that "nearly the whole available oxygen had been converted into carbonic acid – fully adequate to account for the paucity of animal life on the deep bottom."

It was not until 1909 that Legendre associated the diurnal variations of the oxygen contents of the surface waters with the photosynthesis and respiration of plants. Jacobsen (1912) confirmed these findings using the simple titrimetric method developed by Winkler (1888) for the determination of dissolved oxygen, which in modified form, is still the standard procedure. Jacobsen noted that water of the Mediterranean down to a depth of 25 to 75 m. is frequently supersaturated with oxygen in summer and attributed this fact to photosynthetic activity. He also confirmed the existence of the oxygen-minimum layer and related it to consumption of oxygen by the oxidation of organic matter. Similar distributions were found in the Atlantic Ocean by Brennecke (1909).

Subsequent work on the distribution of oxygen in the sea has been concerned not only with the biological aspects of the subject, but also with the use of oxygen analyses to supplement salinity determinations for the characterization of water masses. The determination of oxygen is normally a standard procedure on hydrographical cruises.

The comparatively few determinations of dissolved nitrogen which have been made since the first analyses by Dittmar (1884) have indicated that it is a conservative element, and that its concentration usually corresponds to the saturation value prevailing when it left the surface. Until recently, practically nothing was known of the occurrence of the inert gases in sea water, but the recent development of highly sensitive mass spectrometers has now made their determination feasible (König *et al.*, 1964).

B. CARBON DIOXIDE AND THE BUFFER MECHANISM OF SEA WATER

The presence of calcium carbonate in sea water was postulated by Lavoisier (1772), and its occurrence appeared to be confirmed by the alkalinity of sea water, which was first observed by Bergman (1788). The first determinations of dissolved carbon dioxide in the sea were made in the early part of the nineteenth century (e.g. Lichtenberg, 1811), but a systematic study was not carried out until 1873, when Jacobsen investigated the carbon dioxide content of the North Sea. He found that whereas the dissolved oxygen and nitrogen were readily removed from the water by boiling *in vacuo*, carbon dioxide was tenaciously retained by the water under these conditions, and was only removed by evaporating almost to dryness. He erroneously attributed this retention of carbon dioxide to the formation of a compound between it and the magnesium chloride thought to be present in the water. A similar view was accepted by Buchanan (Dittmar, 1884) who thought that the compound was formed with magnesium sulphate. However, Tørnoe (1880) was able to show, from the results of a long series of alkalinity and carbon dioxide determinations, that carbon dioxide was mainly present in sea water as bicarbonate.

Dittmar (1884) carried out alkalinity determinations on many of the samples from the *Challenger* Expedition and correlated them with the results of carbon dioxide determinations made on them by Buchanan. He concluded that "(i) free carbonic acid in sea waters is the exception. As a rule, carbonic acid is less than the proportion corresponding to bicarbonate, (ii) In surface waters the proportion of carbonic acid increases when the temperature falls and vice versa", and that ". . . the water of the ocean . . . even when it contains a minimum of carbonic acid is not yet saturated with carbonate of lime, but is ready to dissolve whatever of this compound the rivers send into it." In proof of this he cited the dissolution of Foraminiferal shells during their fall to the deepest parts of the oceans.

The first theoretical study of the carbon dioxide system in sea water was made by Hamberg (1885) who attempted to apply the law of mass action qualitatively to the problem. He also pointed out the relationship between the total carbon dioxide concentration, its partial pressure and the titration alkalinity. An unsuccessful attempt to examine this relationship quantitatively was made by Krogh (1904), who considered the possibility that other buffer systems such as that involving boric acid might be responsible for his failure. Krogh also devised a method for the determination of the partial pressure of carbon dioxide in sea water. Fox (1909) made determinations of (i) the total carbon dioxide and alkalinity at a constant partial pressure of carbon dioxide, and (ii) of partial pressure and concentration of carbon dioxide at constant alkalinity. He showed that the total bound carbon dioxide is proportional to the alkalinity, but owing to an erroneous assumption that alkalinity is a measure of the hydroxyl concentration of sea water his attempts to apply the law of mass action to the equilibrium led to the conclusion that sea water contains no carbonate and is slightly acid.

There was at this time no satisfactory method for the measurement of pH, and it was not until after the pioneer work of Sørensen and Palitzsch (1910) on the colorimetric estimation of pH that the relationship between pH and the partial pressure of carbon dioxide in sea water could be investigated (Henderson and Cohn, 1916; McClendon *et al.*, 1917).

In about 1915 Buch* commenced his monumental work on the carbonic acid equilibria in sea water by investigating the quantitative relationships between the total concentration and partial pressure of carbon dioxide, the alkalinity and the hydrogen ion concentration (Buch, 1917). He found that although the figures which he obtained from his results by applying the Law of Mass Action to the ionization of carbonic acid showed the desired proportionalities, there were certain deviations from what would have been expected from Arrhenius' theory. These deviations are now known to arise from changes in the activities of the ions present in sea water owing to its high ionic strength. Later, Buch (1930) reassessed the problem in light of modern theories of strong electrolytes and determined the first and second apparent dissociation constants of carbonic acid. Subsequently, at the suggestion of the International Council for the Exploration of the Sea, Buch, in collaboration with H. W. Harvey, H. Wattenberg and Miss Stina Gripenberg (Buch *et al.*, 1932), made very careful measurements of the first and second apparent dissociation constants of carbonic acid at various salinities and temperatures. These workers also derived expressions by means of which the concentration of the products of the carbonic acid ionization reactions could be calculated from these dissociation constants if the pH and alkalinity of the water were known. Further work by Buch (1933a, b) showed that boric acid makes a small but significant contribution to the buffer capacity of sea water. A reassessment of the dissociation constants of carbonic acid in sea water was given by Buch in 1951. While Buch and his co-workers were engaged in their work, a similar, independent investigation was being carried out at the Scripps Institution of Oceanography by Moberg *et al.* (1934), which in general confirmed the findings of the earlier workers.

In the last ten years interest in the marine chemistry of carbon dioxide has revived, and active research is again being carried out on the chemistry of the carbonic acid equilibria, on the transference processes

*Kurt Karl Vilhelm Buch (1881–) phil. lic. 1918. Hydrographer at the Institute Biologiska Havundersökningar (1910–1917). Head of the Chemical Department at the Institute of Marine Research, Helsinki (1919–1935). Professor of Chemistry at the Åbo Akademi (1935–1942) and at the University of Helsingfors (1942–1951). Member of the Finnish Academy of Science.

between the sea and atmosphere, and on mixing processes in the ocean using ^{14}C produced by nuclear explosions. These topics are discussed in Chapter 7 and Chapter 22 of Volume 2. Another important aspect of the carbon dioxide system in the sea is its influence on the complicated equilibria involved in the precipitation and dissolution of calcium carbonate; these facets of the subject are reviewed in Chapter 17 of Volume 2.

C. Hydrogen Sulphide

The existence of hydrogen sulphide in sea water which is rich in organic matter was observed by Forchhammer (1865). He found that the sulphate to chloride ratio in such waters was appreciably lower than that of ocean water and explained that "in this kind of decomposition where sulphuretted hydrogen is formed, the organic matter is changed into carbonic acid and water, while the oxygen which this change requires is taken from the sulphates . . .".

VII. Trace Elements in the Sea

Interest in the trace elements in the sea arose early in the nineteenth century and as early as 1822 Marcet wrote that in

> . . . the year 1819, I threw out the conjecture that the sea might contain minute quantities of every substance in nature, which is soluble in water. For the ocean having communication with every part of the earth through the rivers, all of which ultimately pour their waters into it; and soluble substances, even such as are theoretically incompatible with each other, being in almost every instance capable of coexisting in solution, provided the quantities be very minute. I could see no reason why the ocean should not be a general receptacle of all bodies which can be held in solution.

Up to this time, no trace elements had been definitely detected in sea water. In 1825 iodine was detected in Baltic sea water by Pfaff, and it was found in French coastal waters by Laurens in 1835. The results of the earlier quantitative determinations of iodine varied wildly, from 0·017 μg./l. found by Macadam (1852) to as much as 8000 μg./l. (Marchand, 1855). It was not until 1916 that analytical processes had improved sufficiently for Winkler (1916) not only to arrive at the currently accepted value of 38 μg./l., but also to show that both iodide and iodate were present.

Forchhammer (1865), in the course of his extensive studies on the composition of sea water, was able to detect ammonia, phosphate, silica, manganese, iron, aluminium and barium in addition to the major components strontium and fluorine, to which reference has already been made. He attempted without success to detect a number of elements, but he appreciated that "next to direct analyses of sea water, the analysis of seaweeds, and of animals living in the sea offers us precious means of determining those elements which occur in so small a quantity in sea water, that it hitherto has been impossible to ascertain their presence in the water by chemical tests." From the occurrence of silver, copper, lead and nickel in marine organisms he inferred that these elements were present in sea water although they could not be detected directly.

Other trace elements which were detected or determined in sea water during the nineteenth century include: silver (10 μg./l.) Malaguti and Durocher (1850, 1859); lithium (200 μg./l.) Marchand (1855); rubidium detected in sea salt, Schmidt (1878); zinc (1·6 to 2·2 μg./l.) Dieulafait (1880); arsenic (100 μg./l.) Daubrée (1851); gold (65 μg./l., Sonstadt (1872); 5 μg./l., Münster (1892); 10 to 65 μg./l., Liversidge (1895)). Although some of these figures agree well with the values accepted at present, many of the earlier analyses suffer from the inadequate sampling and the primitive analytical techniques employed. Advances in the study of trace elements in the sea have followed closely on the development of new and more sensitive methods of analysis.

The combination of co-precipitation concentration techniques with quantitative spectrographic analysis by Goldschmidt and his co-workers led to considerable advances in our knowledge of the concentration levels of many trace metals in the sea. Noddack and Noddack (1940) carried out an extensive investigation of the distribution of a range of heavy metals between sea water and marine organisms. Photometric methods of analysis have been used by the research schools of M. Ishibashi at Kyoto University, Japan and of T. G. Thompson at the University of Washington, U.S.A., and have added greatly to our knowledge of the abundances of trace elements in the sea. The recent development of neutron-activation analysis has enabled determinations of elements to be carried out at concentrations so low that normal physico-chemical methods of analysis would not be feasible. Thus, Goldberg *et al.* (1963) have been able to determine the whole sequence of the lanthanon elements in a 10-litre sample of sea water even though their concentrations average only about 2×10^{-3} μg./l.

Much of the work on trace metals in the sea in recent years has been directed towards the investigation of the processes whereby the products of rock weathering introduced to the sea by run-off from the land, are removed from it by a combination of chemical and biological mechanisms. The idea of a geochemical balance was postulated by Forchhammer (1865) who wrote:

> When we consider that the sea constantly loses a great quantity of pure water by evaporation, and that a large part of this water falls

on the land, dissolves a number of substances from it and carries them at last into the sea, where they constantly would increase in quantity if it were not for its organisms which deprive it again of them, we may suppose that these two effects, one of which acts to increase, and the other to diminish the quantity of mineral substances in sea water, are pretty equal, and leave the sea unchanged. I will, however, not dwell upon these mutual chemical decompositions and combinations, which partly depending upon organic life, partly upon inorganic mechanical and chemical forces, play such a great part in the changes of the earth. . . ."

Quantitative evaluations of the amounts of elements involved in the sedimentary cycle were first made by Joly (1899), but it was not until the work of Goldschmidt and his school at Göttingen in the period from 1930–1939 that a systematic investigation of the factors controlling the distribution of the elements in the sea was made (Goldschmidt, 1937). Goldschmidt prepared a balance sheet in which was set out the amount of the element available to the hydrosphere through the weathering of igneous rocks, the amount of the element present in the sea, and the amount contained in sediments. He classified the elements according to their ionic potentials (the ratio of ionic charge to ionic radius) and found that:

> Substances with low ionic potential, such as sodium, calcium, and magnesium remain in true ionic solution in the process of weathering and transportation; those with intermediate ionic potential are precipitated by hydrolysis, their ions being associated with hydroxyl radicals from aqueous solutions; and substances with still higher ionic potentials form anions containing oxygen and usually again soluble to give true ionic solution.

He showed that of the alkali elements (low ionic potential), only sodium remains to any extent in the sea; potassium, and to an even greater extent caesium and rubidium are adsorbed by hydrolysate sediments. A similar balance was found for the alkaline earth elements. Elements such as titanium, zirconium and beryllium are of intermediate ionic potential and tend to be strongly concentrated in the hydrolysate sediments. Elements of high ionic potential such as boron, silicon, sulphur and nitrogen form oxy-acids and generally remain in the aqueous phase. He was led to believe that the chlorine, boron and sulphur in the sea were not introduced by weathering, but were either primary constituents, or were perhaps derived from volcanic gases or hot springs.

Goldschmidt also stressed the importance of the adsorptive action of the particulate phases in the sea in preventing toxic concentrations of elements such as arsenic, selenium and lead building up in the hydrosphere. The residence times of the elements in the sea were evaluated by Goldberg and Arrhenius (1958), and were found to range from $2{\cdot}6 \times 10^8$ years for sodium, to relatively short periods of time for elements having intermediate ionic potentials (e.g. 100 years for aluminium, see Chapter 5).

Although most of the elements have now been detected in the sea, little is known about the variation of their concentrations with depth or position, or, in surface waters, with season.* Much further work will be necessary if we are to obtain a knowledge of the fundamental processes involved in their geochemistry and biogeochemistry. It will be essential to complement this with studies of their occurrence in sediments and marine organisms, and of the processes by which they are adsorbed by clay minerals, etc. and assimilated by marine organisms.

VIII. The Biologically Important Trace Nutrients

Even as early as the middle of the nineteenth century it was recognized that marine phytoplankton required nitrogen for their growth, thus, Forchhammer (1865) wrote:

> It can hardly be doubted that ammonia is partly formed by the living animals of the sea which exhale ammonia and partly by the putrefaction of their dead bodies. We might ask why we find so small quantities of ammonia, the cause for its formation being so general; but it is well known that plants will absorb it and that the circulation of nitrogen in the sea is between sea water, plants and animals as it is on dry land between soil, plants and animals.

Brandt (1899, 1902)† considered the cycles of nitrogen and phosphorus in the sea, and was able to show that marine plants assimilate ammonia, nitrite, nitrate and phosphate from the sea and build them into the organic compounds which they need for their metabolism. He found that dissolved inorganic nitrogen compounds were not formed in the

*Attempts to find a system in the variation of the concentrations of the elements in sea water have been made by Buljan (1949), and more successfully by Ishibashi and Shigematsu (1951). The latter workers found that if the log ratio of the percentage concentrations of the elements in the earth's crust and in the sea were plotted as a function of their ionic potentials, most of them fitted a peaked curve; the major component elements, however, did not fit this curve.

†Karl Brandt was born in Magdeburg in 1855 and studied science at Berlin and Halle. After short periods at the Berlin Physiological Institute, at the Zoological Station at Naples, and as Privatdozent at Königsberg (from 1878–1887) he went to the University of Kiel, where in 1888 he became Professor of Zoology, and from 1887–1913 also lectured at the Marine Academy. He was a member of the German Scientific Commission for International Sea Research and directed their Marine Biological Laboratory at Kiel. He retired from the University in 1922 but continued research until his death in 1931.

sea by bacterial nitrification of dissolved nitrogen, and concluded that ionic nitrogen was supplied to the sea by rivers or rain water. He (and also Feitel, 1903) showed that bacterial denitrification might occur in the sea. Brandt was led to the conclusion that the growth of plankton might considerably reduce the concentration of these micro-nutrients in the sea, and speculated on whether the concentration of phosphate was ever sufficiently small to limit the growth of plankton (1902, p. 67), but concluded that it was not. The seasonal variation of ionic nitrogen in the sea was first observed by Karsten (1907) who found that "ein Stickstoffminimum im nordischen Hochsommer käme also als Phytoplankton einschränkender Faktor in Betracht." A more systematic study of the occurrence of phosphate (in the North Sea and the Baltic) was made by Brandt (1916–1920) who was able to show that the phosphate contents of the surface waters of these seas reached a maximum in winter and fell to low values in the spring when phytoplankton activity was greatest; a secondary maximum occurred in October. Analogous seasonal variations were also found for nitrate, ammonia and silica; however, albuminoid nitrogen increased sharply in the spring and then gradually decreased. Matthews (1916) observed similar seasonal changes in the phosphate concentrations in the sea off Plymouth Sound, and found that the phosphate contents of the water samples were increased by treating them with a strong oxidizing agent. He correctly attributed this increase to the presence of a dissolved organic phosphorus compound which was present in similar amounts in both summer and winter.

Brandt (1916–1920, p. 329) discussed determinations of nitrate and phosphate made on samples of the deep waters of the North Atlantic collected on the German South Polar Expedition (Gazert, 1909) and on the Planet Expedition (Gräf, 1909). He found that the nitrate concentration increased steadily with depth down to *c.* 800 m.; below this depth it decreased slightly and then remained practically constant down to 3000 m. The phosphate content was observed to vary proportionally with the nitrate concentration.

The association of high productivity with areas of upwelling was probably first recognized by Puff (1890). Nathanson (1906) realized that the vertical currents increased the fertility by carrying up a nutrient, which he erroneously considered to be carbon dioxide; Brandt (*loc. cit*), however, correctly attributed the increased productivity to the upwelling of water rich in nitrate and phosphate.

The cycle of silica in the sea was demonstrated in the early years of the twentieth century. Karsten (1907) pointed out that the growth of diatoms in the sea would lead to an appreciable fall in the silicic acid content of the sea, since, as had been shown by Richter (1906), silicic acid is an essential growth element for diatoms. The return of silica to the water occurred through solution of thin walled types of diatoms in the neighbourhood of the thermocline (Karsten, 1907). There was, however, some loss of silica from the system, since thick walled diatoms sank to the bottom and were only slightly dissolved. Atkins and Harvey (1925) observed that compared with deep waters, surface waters were depleted in silica, but not sufficiently so as to limit the growth of diatoms.

Owing to the inadequacy of the analytical methods used, most of the earlier determinations of micro-nutrients in the sea were of only poor accuracy. The methods generally involved a considerable amount of manipulation and could be carried out only in a land-based laboratory; thus Brandt (1916–1920) concentrated phosphate from 10 litres of water by co-precipitation with ferric hydroxide and determined it gravimetrically. Furthermore, it was not generally appreciated that considerable changes in the concentrations of the micro-nutrients might occur in the long period which frequently elapses between sampling and analysis ashore.

The development of rapid colorimetric (or more correctly photometric) methods of analysis for phosphate (Atkins, 1923a), silicate (Atkins, 1923b), and nitrate (Harvey, 1926) initiated a new phase in the study of micro-nutrients in the sea. At the Laboratory of the Marine Biological Association of the United Kingdom at Plymouth, England, Harvey* and Atkins† carried out a long series of researches upon the effect of these nutrients on the fertility of the waters of the English Channel, and on their variation with season, locality, depth and with the prevailing hydrographic conditions. The results of these investigations were published in the Association's Journal commencing in 1923.

Laboratory experiments carried out at Plymouth on the micro-nutrient requirements of phytoplankton were supplemented by investigations both there and at Woods Hole, Massachusetts into the return of the nutrients to the water by bacterial decomposition of dead phyto- and zoo-plankton. Cooper (1933) showed that ammonia and

*H. W. Harvey, F.R.S. (1889–). Educated at Downing College, Cambridge, was Hydrographer at the Plymouth Laboratory from 1921 to 1958. Received the Alexander Agassiz Medal.

†W. R. G. Atkins, F.R.S. was born in 1884, he was educated at Trinity College, Dublin, and after graduating continued as a research assistant at first in the Chemistry Department and later in the Botany Department. In 1915 he became assistant chemist at Woolwich Arsenal and in 1916 was a volunteer assistant at the N.P.L. He served in the R.F.C. until 1919, when he returned to Dublin. After a short period in India as indigo research botanist, he was appointed (1921) as head of the Department of General Physiology at the Marine Biological Laboratory, Plymouth, where he remained until his retirement in 1955. Died 1959.

nitrite are formed transiently from the degradation of nitrogenous debris. Cooper's results were confirmed by von Brand and Rakestraw (1937–1942) who demonstrated that when diatoms were stored in aerated sea water in the dark, NH_3–N approximating in amount to the organic nitrogen present in the diatoms was set free. The ammonia was consecutively oxidized to nitrite ion and nitrate ion. In an analogous experiment, Cooper (1935a) found that after an initial time lag, phosphate was liberated, but even after three months, less than two-thirds of the added organic phosphorus had reappeared as phosphate. The decomposition of zoo-plankton was studied by Waksman *et al.* (1933, 1938) and by Cooper (1935a,b), who also gave detailed consideration to the nitrogen cycle in the sea (Cooper, 1937).

It is generally accepted that many marine organisms require small amounts of essential trace metals for their healthy growth. The sea contains concentrations in excess of 2 to 3 μg./l. of several of the trace elements known to be necessary to the growth of land plants, e.g. copper, zinc and boron, and it is likely, therefore, that these elements will not limit the growth of marine plants. There are, however, some grounds for suspecting that the growth of phytoplankton may be limited in some regions of the open oceans by shortage of iron and perhaps manganese, although this has never been proved. Harvey (1937) showed that diatoms can be produced in an iron deficient state, and that the addition of very small amounts of iron as its freshly precipitated hydroxide will lead to very rapid growth. Cooper (1935b, 1948) showed that most of the iron in the sea was present in the form of its hydroxide or phosphate, and was either colloidal or in the form of flocs. Recent work has suggested that an appreciable amount of dissolved chelated iron may also be present. Harvey (1945) has found that certain species of diatoms form auxospores and cease growing unless 1 to 2 μg. of manganese/l. is present.

The development of photometric methods of analysis for micronutrients meant that it was at last feasible to carry out routine determinations of these constituents on board ship. Among the first to make use of these procedures was Herman Wattenberg* who was appointed Chemist on the expedition of the *Meteor* (Fig. 7) to the Southern Atlantic

*Wattenberg was born in Berlin in 1901, in 1919 he graduated from the Gymnasium at Hanover. He obtained his Ph.D. from the University of Munich in 1923 after studying under Hönigschmid. After a short period in industry he became an assistant at the Technische Hochschule in Danzig. After the Meteor Expedition he took a Fellowship at the Chemical Institute of the University of Munich. In 1934 he was appointed as Head of the Laboratory of Marine Chemistry in the University of Kiel, and he later became Head of the Hydrographic and chemical division of the Institut für Meereskunde at Kiel. In 1944 he was appointed director of the Institute. He was killed in July 1944 when the Institute was destroyed by bombs. An interesting biography and assessment of Wattenberg's work has been given by Rakestraw (1956).

FIG. 7. The *Meteor*

(1925–1927). Among the objectives of the chemical programme were the study of the movements of the water masses, the investigation of the relationships between the distribution of phytoplankton and the concentration of phosphate, and between the processes of respiration and decomposition of organic matter and the production and removal of oxygen and carbon dioxide. In addition, the carbonic acid system was to be investigated. Wattenberg designed and set up a chemical laboratory (Fig. 8 (a) and (b)) on the *Meteor* in which analyses were carried out for salinity, dissolved oxygen, pH, phosphate, alkalinity and partial pressure of carbon dioxide. This expedition demonstrated that useful chemical work was possible at sea. It was highly successful and provided a vast amount of data useful to marine biologists and oceanographers which is still being evaluated.

Excellent laboratory facilities were also available aboard the *Discovery* on an Expedition from 1925 to 1927 to the South Atlantic and Southern Ocean. For the first time, some photometric determinations of nitrate and nitrite were carried out at sea, in addition to routine determinations of salinity, dissolved oxygen, pH and phosphate (Deacon, 1933). The results of these determinations enabled Deacon to deduce the movements of water masses in the South Atlantic, and to demonstrate the upwelling of the nutrient-rich water which is responsible for the high fertility of the Antarctic Seas.

The lead pioneered by the *Meteor* and *Discovery* Expeditions has been followed in many later oceanographic cruises. The results of routine water sampling and analysis have been used both to give a picture of the water masses present in the ocean and to show the interrelationship between micro-nutrients and productivity in the sea. The major deep-sea expeditions since 1873 have been reviewed by Wuest (1964).

The estimation of primary organic production has been greatly assisted by the development of the ^{14}C method for the determination of organic productivity (Steeman Nielsen, 1952) which was first used on the *Galathea* Expedition (Steeman Nielsen, 1953). Much useful additional information can be obtained by the photometric estimation of plant pigments (Richards and Thompson, 1952). Experiments at sea and in the laboratory have suggested that the standing crop of phytoplankton in a given area of the sea depends on the following factors:

(i) the incident light which, in conjunction with turbidity, determines the depth above which photosynthesis can take place;
(ii) the rate at which micro-nutrients are replenished in the photic layer;
(iii) the rate at which growing phytoplankton is carried below this level by turbulence or sinking.
(iv) the rate of grazing of phytoplankton by zooplankton.

FIG. 8(a). The Chemical Laboratory aboard the *Meteor*

FIG. 8(b). The Chemical Laboratory aboard the *Meteor*.

In many regions of the sea, grazing by zooplankton is one of the main factors preventing the very rapid multiplication of phytoplankton. Antia *et al.* (1963) carried out an ingenious series of experiments in which sea water, freed of zooplankton, was placed in a large plastic sphere and suspended in the sea. The standing crop of phytoplankton and the nutrient concentrations were compared at regular intervals with those in the sea surrounding the bag. In this way it was possible to make some assessment of the effect of grazing on the plankton crop and on the nutrients present.

It has been known for many years that the growth of plankton in the sea was very different from that in artificial culture medium. Synthetic sea water, or natural sea water, pasteurized at 70°C is incapable of supporting the growth of most species of phytoplankton, even if inorganic micro-nutrients are present in abundance, and it has been suggested that some organic growth-promoting substances were absent. The presence of differing amounts of such substances is probably responsible for the very different fertilities which may be shown by sea waters apparently identical with regard to their inorganic components. This difference was responsible for much of the confusion in the earlier literature on the growth of phytoplankton. Although the presence of dissolved organic matter in the sea has been recognized for at least a century (Forchhammer, 1865), very little is yet known of its nature (see Chapter 11). The nature of the growth-promoting substances has not yet been ascertained, but it is likely that certain vitamins, such as B_{12} and related compounds, may exert a powerful influence in deciding the species composition of the phytoplankton crop in the sea. It is probable, however, that shortage of nitrate is the main growth limiting factor. The development of highly sensitive methods of analysis, such as gas–liquid and thin-layer chromatography and bio-assay procedures, is certain to lead in the near future to great advances in our knowledge of the distribution of organic compounds in the sea and of their interaction with living matter.

IX. Radioactivity

Shortly after the discovery of radium several determinations were made of radium-226 in sea water (Joly, 1908a, 1909; Eve, 1909; Knocke, 1909; Satterly, 1911) and in the salts obtained by the evaporation of sea water (Strutt, 1906; Eve, 1907). The most interesting observation of the occurrence of this isotope in the marine environment was made by Joly (1908b) who found that it was concentrated in marine sediments, particularly red clay. Although Joly's results were erroneously high, his general conclusions were confirmed by two later series of analyses (Pettersson, 1930; Piggot, 1933).

Owing to the ease with which radium-226 could be determined at low levels by gas counting of radon, the emphasis on this isotope continued until the mid-twentieth century. The only radioactive element present in sea water in amounts sufficiently high to be determined directly by a chemical method is uranium. The first analysis for this element was performed by Hernegger and Karlick (1935a) using a fluorometric method. These workers (1935b) found the first evidence that the uranium decay series is in disequilibrium in sea water. Subsequently, Pettersson (1937) suggested that this disequilibrium, which is manifested by the unexpectedly low concentration of radium in sea water, was caused by the removal of its precursor, thorium-230, from the water, and its incorporation into the sediments.

With the development of low-level counting equipment and modern methods of concentration and separation, the determination of other members of the natural radioactive decay series has become possible. In this connection, the work of Goldberg's school on the occurrence of thorium isotopes in sea water and sediments is particularly important (Sackett *et al.*, 1958). Most of the recent work on these series has been directed towards the geochronology of marine sediments and on their use as geophysical tracers.

The increased interest in these nuclides in the post-war years has been paralleled by the intense development of work on cosmic ray induced activities and on artificially produced nuclides. In a famous paper Libby (1946) predicted that carbon-14 and tritium formed by cosmic ray activity would be found in nature. The discovery of carbon-14, and the development of methods for its routine determination at the extremely low levels at which it occurs, have provided a very valuable technique for the dating of water masses and sediments. Owing to its comparatively short half-life (12·7 years), tritium is of less value for dating purposes. Despite the numerous analytical difficulties in working with very large samples, the less abundant activities, such as beryllium-10 and silicon-32 have recently been receiving attention.

The development of the nuclear energy industry both in the industrial and military spheres has led to the introduction of considerable amounts of radioactive isotopes into the sea, both from fall-out and from the disposal of radioactive wastes. These additions of radioactive materials have proved of use in the study of mixing processes occurring in the sea.

X. Raw Materials from the Sea

The sea is potentially a source of great wealth both mineral and biological. The biological aspects of the subject, including fisheries, are too vast to be discussed in this work, and the reader is referred to specialized works on these subjects. Before passing on to consider the mineral resources of the sea, mention should be made of the growing use of chemicals derived from algae – alginic acid, laminarin and agar agar, etc., and of the now defunct industry based on the extraction of iodine from kelp.

Only major components of sea water are at present extracted commercially. Of these, common salt has been produced from sea water since the dawn of history in many countries with hot dry climates. Pliny, in his Natural History (30, 41), has mentioned solar evaporation of sea water at the time of the Ptolemies. G. Agricola, in his posthumous work De Re Metallica, published in 1556, has described the methods of extracting salt used in the late middle ages. Today, several million tons of salt are extracted from the sea every year by solar evaporation. This process can yield as by-products, magnesium and potassium, but it is unlikely that the latter can compete in price against that obtained from other sources such as the Stassfurt deposits or the Dead Sea.

The recovery of bromine from sea water was first achieved on a commercial scale in 1924 by the Ethyl Gasoline Corporation City, Maryland, U.S.A. The process used involved the treatment of the water with chlorine and aniline and collection of the precipitated tri-bromoaniline. Later, the company adapted a 4200 ton ship as a floating factory, but it was not a success and only made one voyage. The first successful bromine plant came into operation in 1934 at Kuri Beach, Freeport, Texas (Stewart, 1934). In this continuous process, sea water was adjusted to pH 3·5 with sulphuric acid and treated with chlorine. The liberated bromine was removed with a current of air and absorbed in sodium carbonate solution, from which it was recovered by acidification and stripping with steam. The crux of the first stage of the process is the adjustment of the pH to the optimum value, and much engineering skill was necessary to control the acidity of the enormous volumes of water treated in the plant. This process, and minor modifications of it, are now producing the bulk of the world's requirements of bromine, running at over 100,000 tons per annum.

Magnesia was first prepared from sea water at Aigues Mortes on the Mediterranean coast at the end of the last century by precipitation with milk of lime. Magnesium compounds were first obtained from sea water on a large scale in California as a by-product of the production of sodium chloride by solar evaporation (Seaton, 1931; Manning, 1938). As a result of the need for magnesium alloys for the construction of air-frames during the Second World War, several plants were set up for the production of magnesium from sea water. The first of these plants came into operation at Freetown, Texas, in 1941, and involved as a first stage the precipitation of magnesium hydroxide with milk of lime. After conversion to magnesium chloride and purification, magnesium metal was produced by electrolysis (Kirkpatrick, 1941). With the end of hostilities, the demand for magnesium fell rapidly, and most of the plants for its preparation from sea water were closed down on economic grounds.

In recent years, the preparation of potable water from sea water has become increasingly important, particularly in countries where the rainfall is slight. At present, stills heated by fossil fuels are mainly used. It is probable, however, that more economical stills using nuclear energy or solar heating will soon be developed. Currently, a number of other methods for desalting sea water are under investigation (e.g. solvent extraction).

It is probable that the sea contains traces at least of every naturally occurring element, and many of the artificially produced radio-isotopes. It also should be remembered that since the volume of the ocean is *c.* 1372×10^6 km.^{3}, the total weight of a trace element present in the sea is very large (e.g. if its concentration is only 0·1 μg./l., $1{\cdot}37 \times 10^8$ tons of it will be present). The concentrations of most elements are so low that their recovery from sea water is not an economic proposition at the present time because of the cost of treating the enormous volumes of water which it would be necessary to process for the recovery of worthwhile amounts. It is possible that in the future, the recovery of several valuable trace elements at once, e.g. uranium, silver and gold, might be feasible if porous barrages of ion exchange resin were used for their concentration, and if pumping was obviated by making use of tidal fall.

The idea of recovering gold from sea water seems to have tantalized many scientists and speculators. Caldwell (1938) in a survey of gold in sea water mentions that "nearly fifty patents on methods of securing gold from sea water are on record from several countries. Fraudulent promoters or overly hopeful persons have many times organized companies and sold stock therein, which companies, it was speculated, were to profitably extract gold from sea water." One of the most famous scientists to work on this problem was Fritz Haber who planned to repay the German War Debts after the First World War with gold extracted from the sea. Haber spent a considerable time developing an accurate method for the determination of gold in sea water and applied it to samples of sea water collected on the *Meteor* Expedition. He finally concluded (Haber, 1928) that sea water contained 0·004 mg. of gold/ton. Later work (Putnam, 1953; Robinson and Desbois, 1955) has suggested that the gold content depends upon the particular water mass and on biological activity; thus Putnam found amounts varying between 0·003 and 44 mg./m^3.

Many economically important elements, such as manganese, copper, cobalt, nickel, lead and zinc have been concentrated from sea water during the growth of the manganese nodules with which much of the deep Pacific floor is densely strewn (weights of up to 10 kg./m^3. have been reported). The concentrations of these elements which the nodules contain are sufficiently high to make the recovery of the metals an attractive possibility when the richer terrestrial ores become exhausted. Attention is already being given to the solution of the problem of dredging the nodules from depths of several km.

BIBLIOGRAPHY

Murray (1895) has provided a very interesting account of the earlier history of general oceanography, with emphasis on the geographical and descriptive aspects of the subject. The important paper by Goldberg on the history of chemical oceanography has proved of great assistance in the compilation of this chapter and is to be recommended for general reading. Other papers on the history of oceanography have been published by Carruthers (1963) and by Thompson (1956).

ACKNOWLEDGMENTS

The author wishes to thank Dr. A. Kent for providing biographical details of Professor Dittmar, and Dr. C. H. Mortimer for providing Fig. 5. He is also indebted to Dr. J. N. Carruthers for helpful advice, and to Dr. H. Postma for supplying the Frontispiece of this book and for bringing to his attention the passage from Boerhaave's introduction to Marsilli's Works with which this chapter is prefaced. Finally he is most grateful to Dr. K. Grasshoff for providing Figs. 8(a) and 8(b) and to Dr. S. Gripenberg for details of K. Buch's career.

REFERENCES

Antia, N. J., McAllister, C. D., Parsons, T. R., Stephens, K., and Strickland, J. D. H. (1963). *Limnol. Oceanogr.* **8**, 166.

Atkins, W. R. G. (1923a). *J. Mar. biol. Ass. U.K.* **14**, 72.

Atkins, W. R. G. (1923b). *J. Mar. biol. Ass. U.K.* **13**, 157.

Atkins, W. R. G., and Harvey, H. W. (1925). *Nature, Lond.* **116**, 784.

Balard, A. J. (1826). *Ann. Chim.* (*Phys.*) **32**, 337.

Bergman, T. O. (1779). "Opuscula physica et chemica pleraque antea seorsim edita jam ab auctore collecta revisa et aucta". Uppsala.

Bergman, T. O. (1788). "Physical and Chemical Essays", (translated from the original Latin by E. Cullen). Vol. III. J. Murray, London.

Bouillon-Lagrange, E. J., and Vogel, H. A. (1813). *Thompson's Ann.* **4**, 200.

Boulton, R. (1699). "The Works of the Honourable Robert Boyle Esq." (epitomized by R. Boulton). Vol. 1, p. 282. Phillips and Taylor, London.

Boyle, R. (1675). *Phil. Trans.* **9**, 31.

Brand, T. von, and Rakestraw, N. (1937–1942). *Biol. Bull. Woods Hole*, **72**, 165; **77**, 285; **70**, 231, **81**, 03; **83**, 273.

Brandt, K. (1899). *Wiss. Meeresuntersuch. Abt. Kiel*, N.F. **4**, 213.
Brandt, K. (1902). *Wiss. Meeresuntersuch. Abt. Kiel*, N.F. **6**, 23.
Brandt, K. (1916–1920). *Wiss. Meeresuntersuch. Abt. Kiel*, N.F. **18**, 185.
Brennecke, W. (1909). *Forschungsreise, S.M.S. Planet* (1906–7).
Buch, K. (1917). *Finn. hydrogr. biol. Untersuch. Nr. 14.*
Buch, K. (1930). *Rapp. Cons. Explor. Mer*, **67**, 51.
Buch, K. (1933a). *Rapp. Cons. Explor. Mer*, **85**, 71.
Buch, K. (1933b). *J. Cons. int. Explor. Mer*, **8**, 309.
Buch, K. (1951). *Merentutkimuslait. Julk.* N:O 151.
Buch, K., Harvey, H. W., Wattenberg, H., and Gripenberg, S. (1932). *Rapp. Cons. Explor. Mer*, **79**, 1.
Buchanan, J. Y. (1919). "Accounts Rendered of Work Done and Things Seen". Cambridge University Press, Cambridge, England.
Buljan, M. (1949). *Acta adriat.* **4**, 171.
Caldwell, W. E. (1938). *J. Chem. Educ.* **15**, 507.
Carpenter, W. B. (1872). *Proc. roy. Soc.* **20**, 535.
Carruthers, J. N. (1963). *J. Inst. Navig.* **16**, 180.
Cooper, L. H. N. (1933). *J. Mar. biol. Ass. U.K.* **18**, 677.
Cooper, L. H. N. (1935a). *J. Mar. biol. Ass. U.K.* **24**, 115.
Cooper, L. H. N. (1935b). *Proc. roy. Soc.* **B 118**, 419.
Cooper, L. H. N. (1937). *J. Mar. biol. Ass. U.K.* **22**, 183.
Cooper, L. H. N. (1948). *J. Mar. biol. Ass. U.K.* **27**, 314.
Daubrée, A. (1851). *Bull. Soc. géol. Fr.* **8**, 352.
Davy, J. (1817). *Phil. Trans.* **107**, 275.
Deacon, G. E. R. (1933). *'Discovery' Rep.* **7**, 173.
Dieulafait, L. (1880). *C.R. Acad. Sci., Paris*, **90**, 1573.
Dittmar, W. (1884). "Report on the Scientific Results of the Exploring Voyage of H.M.S. *Challenger*. Physics and Chemistry", Vol. 1. H.M. Stationery Office, London.
Ethe, H. (1869). "Kazwini's Kosmographie nach den Wüstfeldschen Textausgabe". Leipzig.
Eve, A. S. (1907). *Phil. Mag.* **13**, 248.
Eve, A. S. (1909). *Phil. Mag.* **18**, 102.
Feitel, R. (1903). *Wiss. Meeresunters. Abt. Kiel*, N.F. **7**, 91.
Fleming, J. (1818). *Trans. roy. Soc. Edinb.* **8**, 507.
Forch, C., Jacobsen, J. P., Knudsen, M. and Sørensen, S. P. L. (1902). *K. danske vidensk. Selsk.* **12**, 1.
Forchhammer, G. (1865). *Phil. Trans.* **155**, 203.
Fox, C. J. J. (1909). *Publ. Circ. Cons. Explor. Mer, No.* **44**, 1.
Frémy, (1838). *C.R. Acad. Sci., Paris*, **6**, 616.
Gay-Lussac, J. L. (1817). *Ann. Chim. Phys. 2nd Series*, **6**, 426.
Gay-Lussac, J. L. (1818). *Ann. Chim. Phys. 2nd Series*, **7**, 79.
Gazert, H. (1909). *Dtsch. Südpol. Exped.* **7**, 120.
Göbel, F. (1838). "Reise in die Steppen des südlichen Russlands internommen von Dr. Fr. Göbel", Vol. 2, p. 372. Dorpat.
Goldberg, E. D. (1960). *Chimiya*, **6**, 162.
Goldberg, E. D., and Arrhenius, G. O. S. (1958). *Geochim. et cosmoch. Acta*, **13**, 153.
Goldberg, E. D., Koide, M., Schmidt, R. A., and Smith, R. H. (1963). *J. geophys. Res.* **68**, 4209.
Goldschmidt, V. M. (1937). *J. Chem. Soc.* 655.

Gräf, O. (1909). "Forschungsreise, S.M.S. *Planet* 1906–7", Vol. 4, Biologie. K. Siegismund, Berlin.
Haber, F. (1928). *Z. Ges. Erdk. Berl.* **3**, 1.
Hamberg, A. (1885). *Bih. svensk. Vetensk. Akad. Handl.* **10**, 31.
Harvey, H. W. (1926). *J. Mar. biol. Ass. U.K.* **14**, 72.
Harvey, H. W. (1937). *J. Mar. biol. Ass. U.K.* **22**, 205.
Harvey, H. W. (1945). "Recent Advances in the Chemistry and Biology of Sea Water," p. 139. Cambridge University Press, Cambridge, England.
Henderson, L. J., and Cohn, E. J. (1916). *Proc. Nat. Acad. Sci., Wash.* **2**, 618.
Hercher, R. (1873). "Epistographia Graeci Recendsuit Recognovit Adnotatione Critica et Indicibus Instruxit," p. 649. R. Herchei, Paris.
Hernegger, F., and Karlick, B. (1935a). *S.B. Akad. Wiss. Wien*, **144**, 217.
Hernegger, F., and Karlick, B. (1935b). *Goteborgs. VetenskSamh. Handl.* **B 4**, 12.
Hooke, R. (1667). *Phil. Trans.* **2**, 439.
Ishibashi, M., and Shigematsu, T. (1951). *Bull. Inst. chem. Res. Kyoto Univ.* **27**, 42.
Jacobsen, J. P. (1912). *Rep. Danish Oceanogr. Exped. Medit.* (*1908–1910*), **1**, 207.
Jacobsen, J. P., and Knudsen, M. (1940). *Publ. Sci. Ass. Océanogr. phys. No.* **7**.
Jacobsen, O. (1873). *Liebigs Ann.* **167**, 1.
Joly, J. (1899). *Sci. Trans. R. Dublin Soc.* **7**, 23.
Joly, J. (1908a). *Phil. Mag.* **15**, 385.
Joly, J. (1908b). *Phil. Mag.* **16**, 190.
Joly, J. (1909). *Phil. Mag.* **18**, 396.
Karsten, G. (1907). *Wiss. Ergebn. 'Valdivia'*, **2**, 2.
Kirkpatrick, S. D. (1941). *Chem. metall. Engng.* **48**, 76.
Knocke, W. (1909). *Phys. Z.* **10**, 157.
Knudsen, M. (1901). "Hydrographic Tables". G.E.C. Gadd, Copenhagen.
König, H., Wanke, H., Bien, G. S., Rakestraw, N. W., and Suess, H. E. (1964). *Deep Sea Res.* **11**, 243.
Krogh, A. (1904). *Medd. Grönland*, **26**, 331.
Laurens, G. (1835). *J. Pharm. Sci. Access.* **21**, 90.
Lavoisier, A. L. (1772). *Hist. Acad. Roy. Sci. Pt.* **2**, 555, (published 1776).
Legendre, R. (1909). *Bull. Soc. Sci. Archachon*, 29 pp.
Libby, W. F. (1946). *Phys. Rev.* **69**, 671.
Lichtenberg, F. D. (1811). *J. für. Chemie u Physik.* **2**, 252.
Liversidge, A. (1895). *J. roy. Soc. N.S.W.* **29**, 335.
Lyman, J., and Fleming, R. H. (1940). *J. Mar. Res.* **3**, 135.
Macadam, S. (1852). *Chem. Gaz.* **10**, 281.
McClendon, J. F. (1917). *J. biol. Chem.* **30**, 265.
Macquer, P. J., Lavoisier, A. L., and Sage, B. G. (1778). *Hist. Acad. Roy. Sci.* **3**, 69, (published 1781).
Malaguti and Durocher (1850). *Ann. chim.* (*Phys.*) **87**, 184.
Malaguti and Durocher (1859). *C.R. Acad. Sci., Paris*, **49**, 463, 536.
Manning, P. D. V. (1938). *Chem. metall. Engng.* **45**, 478.
Marcet, A. M. (1819). *Phil. Trans.* **109**, 161.
Marcet, A. M. (1822). *Phil. Trans.* **112**, 448.
Marchand, E. (1855). *Mém. Acad. Méd. Paris*, **19**, 121.
Marsilli, L. F. (1711). "Brieve ristretto del saggio fisico intorno alla storia del Mare". Venice.
Marsilli, L. F. (1725). "Histoire Physique de la Mer", (translated by H. Boerhaave), p. 25. De'pens, Amsterdam.

Matthews, D. J. (1916). *J. Mar. biol. Ass. U.K.* **11**, 122, 251.
Moberg, E. G., Greenberg, D. M., Revelle, R., and Allen, E. C. (1934). *Bull. Scripps Instn. Oceanogr.* **3**, 321.
Morren, S. (1843). *Ann. chim. Phys. 3rd Series*, **12**, 5.
Münster, C. A. (1892). *Norsk. Tek. Tidsskr.* **10**, 13.
Murray, J. (1818). *Trans. roy. Soc. Edinb.* **8**, 259.
Murray, J. (1895). "Report on the Scientific Results of the Exploring Voyage of H.M.S. *Challenger*". Summary of scientific results, Part 1. H.M. Stationery Office, London.
Nathanson, A. (1906). *Kgl. Sächs. Ges. Wiss Leipzig, Math. Phys. Kl.* **29.**
Noddack, I., and Noddack, W. (1940). *Ark. Zool.* **32 A**, 1.
Petterson, H. (1930). *Imp. de Monaco, Fasc.* **81**, 1.
Petterson, H. (1937). *Anz. Akad. Wiss. Wien, Math-naturw. Kl.* **74**, 127.
Pfaff, C. (1825). *Jb. Chem. Phys.* **15**, 378.
Piggot, C. S. (1933). *Amer. J. Sci.* **25**, 229.
Puff, A. (1890). "Das kalte Auftreibswasser an der Ostseite des nordatlantischen und der Westseite des nordindischen Ozeans". Dissertation, Marburg.
Putnam, G. L. (1953). *J. Chem. Educ.* **30**, 576.
Rakestraw, N. W. (1956). *J. chem. Educ.* **33**, 217.
Richards, F. A., and Thompson, T. G. (1952). *J. Mar. Res.* **11**, 156.
Richter, O. (1906). "Zur Physiologie der Diatomen", p. 16. Berlin.
Robinson, C., and Desbois, C. (1955). *Trav. centre recherches et d'études oceanographiques*, **2**, No. 3, 4 pp.
Sackett, W. M., Portratz, H. A., and Goldberg, E. D. (1958). *Science*, **128**, 204.
Satterly, J. (1911). *Proc. Camb. phil. Soc.* **16**, 360.
Schmelck, L. (1882). *Norske Nordhavs. Exped. (1876–78) Chemi*, **3**, 12.
Schmidt, C. (1878). *Bull. Acad. St. Petersb.* **24**, 231.
Seaton, M. Y. (1931). *Chem. metall. Engng.* **38**, 638.
Sonstadt, E. (1872). *Chem. News*, **25**, 196, 231, 241.
Sørensen, S. P. L., and Palitzsch, S. (1910). *C. R. Lab. Carlsberg*, **9**, 8.
Sprenger, A. (1841). Mas'ūdī's "Meadows of Gold and Mines of Gems", (translated by A. Sprenger). London.
Steeman Nielsen, E. (1952). *J. Cons. int. Explor. Mer*, **18**, 117.
Steeman Nielsen, E. (1953). *J. Cons. int. Explor. Mer*, **19**, 309.
Stewart, L. C. (1934). *Industr. Engng. Chem.* **26**, 361.
Strecker, E. (1853). *Ann. Chem. u Pharm.* **87**, 184.
Strutt, R. J. (1906). *Proc. roy. Soc.* **A 78**, 150.
Thompson, T. G. (1956). *J. chem. Educ.* **33**, 108.
Thorpe, T. E., and Morton, E. H. (1871). *Mem. Manchr. lit. phil. Soc.* **4**, 287.
Tørnoe, H. (1880). "The Norwegian North Atlantic Expedition 1876–78. Chemistry", Vol. 24, Grøndahl. Christiania.
Waksman, S. A., Carey, C., and Reuszer, H. (1933). *Biol. Bull. Woods Hole*, **65**, 57.
Waksman, S. A., Reuszer, H., Carey, C., and Hardman, Y. (1938). *J. Bact.* **35**, 447.
Winkler, L. W. (1888). *Ber. dtsch. chem. Ges.* **21**, 2843.
Winkler, L. W. (1916). *Z. angew Chem.* **29**, 205.
Wuest, G. (1964). *In* "Progress in Oceanography" (M. Sears, ed.) Vol. II. Pergamon Press, Oxford. In the press.

Chapter 2

Currents and Mixing in the Ocean

K. F. BOWDEN

Department of Oceanography
The University of Liverpool, England

I. Introduction 43
II. Surface Currents of the Oceans 44
III. Salinity Distribution in the Surface Layer 49
IV. Deep Water Circulations 51
V. Currents in Coastal Seas 59
A. Tidal Currents 59
B. Local Wind-driven Currents 59
C. Density Currents 60
VI. Turbulent Mixing 61
A. Vertical Mixing 61
B. Horizontal Mixing 63
VII. Quantitative Treatment of Mixing Problems 64
A. Steady State 65
B. Variable (i.e. non-steady) State 68
VIII. Overall Mixing Problems 70
References 71

I. Introduction

There are two kinds of physical processes which influence the distribution of substances dissolved or suspended in sea water. In advective processes large scale movements of water occur, the dissolved or suspended matter being carried with the water, whereas in diffusive processes an exchange of properties takes place without any overall transport of water. The first class of processes includes ocean currents at all depths and also the vertical movements of upwelling or sinking of water masses. Diffusive effects are produced by turbulent mixing on a very wide range of scales, both in the vertical and horizontal directions. Molecular diffusion usually takes place so slowly compared with turbulent mixing that its direct influence may be neglected.

The oceans and seas of the world cover a total area of $360 \cdot 8 \times 10^6$ km.^{2}, which represents 70·8% of the earth's surface. Of this area, 7·6% is over the continental shelf, less than 200 m. deep; 8·5% has a depth between 200 m. and 2000 m., and the remaining 83·9% is more than 2000 m. deep. The greatest depth recorded to date is 11,034 m., in the Marianas Trench. The areas, volumes and depths of the three oceans and their adjacent seas are given in the following table (from Defant, 1961):

Ocean, including adjacent seas	*Area* ($10^6 km.^2$)	*Volume* ($10^6 km.^3$)	*Mean depth* (*m.*)	*Greatest depth* (*m.*)
Atlantic	106·2	353·5	3331	8526
Indian	74·9	291·9	3897	7450
Pacific	179·7	723·7	4028	11,034
All oceans	360·8	1369·1	3795	—

II. Surface Currents of the Oceans

The major oceanic circulations in the surface layers are driven by the prevailing wind systems, and are influenced by the form of the ocean boundaries. The general patterns in all the oceans are similar. In the North Atlantic for example (Fig. 1), there is a general clockwise circulation. The north-east trade winds drive the North Equatorial Current, which flows westwards from Africa towards the Windward and Leeward Islands, where it divides into two branches, one entering the Caribbean Sea and the Gulf of Mexico, from which it emerges through the Straits of Florida to form one source of the Gulf Stream. It is joined here by the other branch, which has turned northwards outside the Antilles arc of islands. The combined currents form the Gulf Stream, which flows parallel to the American coastline as far as Cape Hatteras and then turns eastwards across the ocean. In this part of its course, the current is being driven by the southwesterly winds of temperate latitudes, and on passing eastwards of the Grand Banks, off Newfoundland, it is known as the North Atlantic Current. The circulation is completed by the more diffuse, south-flowing Canaries Current, feeding the start of the North Equatorial Current. Another branch of the North Atlantic Current turns northwards past the British Isles and enters the Norwegian Sea.

In the South Atlantic there is an approximate mirror image of the above circulation, with the South Equatorial Current flowing westwards, driven by the south-east trades. The major part of this current turns southwards on approaching the coast of South America, and forms the Brazil Current, which leaves the neighbourhood of the coast between 30 and 40° S. to join the east-flowing South Atlantic Current or West Wind Drift (which is driven by the westerly winds of these latitudes). Nearing the coast of South Africa, part of the current turns northwards,

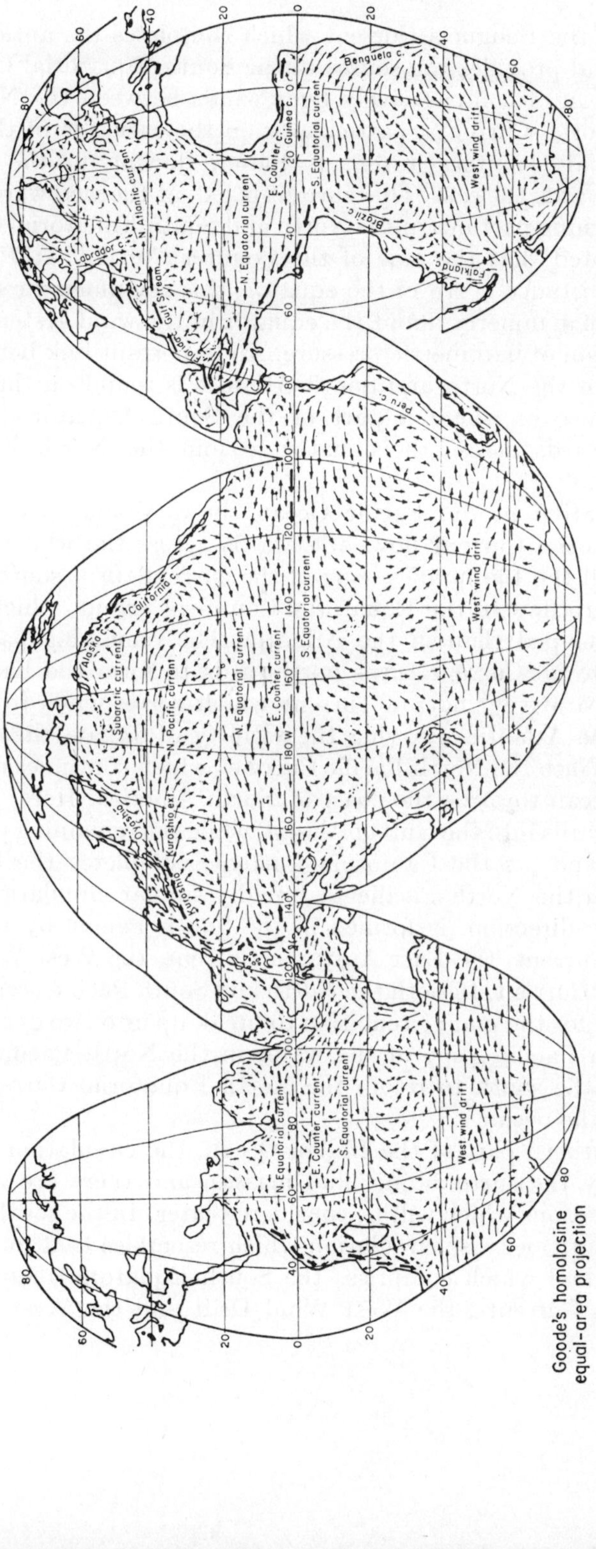

FIG. 1. Surface currents of the oceans in February–March (Sverdrup, 1945). By permission of George Allen & Unwin Ltd.

and becomes the Benguela Current which completes the anticlockwise circulation and provides the source of the South Equatorial Current.

In the region of comparatively light winds between the North and South Equatorial Currents, particularly on the east side of the ocean, the Equatorial Counter Current is found, flowing in the opposite direction, towards the east. This current is fed, at the westward end, by backward turning portions of the North and South Equatorial Currents. It will be noted that the axis of the Counter Current, at 7°N., is an appreciable distance north of the equator, this asymmetry reflecting a similar lack of symmetry about the equator in the wind systems and in the distribution of barometric pressure. An important link between the circulations in the North and South Atlantic is found on the western side of the ocean, where a part of the South Equatorial Current flows northwards across the equator to join the North Equatorial Current.

The circulation in the Pacific Ocean shows similar features. The North Equatorial Current flows westwards across the whole width of the ocean. Off the Philippines, part of the current turns southwards to feed the return flow in the Equatorial Counter Current, which is more strongly developed than in the Atlantic. The remaining part of the North Equatorial Current turns northwards to form the Kuroshio, a strong, comparatively narrow current which corresponds to the Gulf Stream in the Atlantic. Leaving the vicinity of Japan, the Kuroshio broadens out into the North Pacific Current, which continues eastwards across the ocean towards the coast of North America. Here, part of it turns northwards into the Gulf of Alaska, while the remaining part turns southwards and, as the California Current, completes the clockwise circulation in the North Pacific. A broadly similar circulation, in the anticlockwise direction, is formed in the South Pacific by the South Equatorial Current, the East Australia Current, the West Wind Drift and the Peru Current. In both the North and South Pacific oceans, there is a tendency for the main circulation to break up into two gyrals. Thus, in the North Pacific some of the water in the North Pacific Current turns off to the south to rejoin the North Equatorial Current in the mid-part of the ocean.

In the northern part of the Indian Ocean, the circulation is largely influenced by the surrounding land masses, and there are important seasonal variations which will be considered later. In the southern part, there is an anticlockwise circulation which resembles that in the other two oceans, and which comprises the South Equatorial, Mozambique and Agulhas Currents, the West Wind Drift and the West Australia Current.

All three oceans open out at their southern ends into the Antarctic, or Southern, Ocean, in which the West Wind Drift forms a circumpolar current. Its narrowest constriction is in the Drake Passage, between Cape Horn and the South Shetland Islands. The West Wind Drift in the Southern Ocean is, therefore, largely a closed circulation, although it has offshoots into the other three oceans and is joined by certain outflowing currents from them.

It is well known that the trade winds blow from the northeast and the southeast, rather than from directly north or south, as a consequence of the *geostrophic* or *Coriolis effect*, arising from the earth's rotation. For the same reason, wind-driven surface currents tend to flow in a direction 45° to the right of the wind in the northern hemisphere and 45° to the left in the southern hemisphere. A striking feature of the circulation in each of the oceans is the lack of symmetry between the currents on the western and eastern sides of the ocean. While the western boundary currents, such as the Gulf Stream and Kuroshio in the northern hemisphere and the Brazil Current and Agulhas Current in the southern hemisphere, are narrow, fast-flowing currents, the return flow on the eastern sides of the oceans is more diffuse. It was first shown by Stommel (1948) that this westward intensification of the circulation was a further consequence of the geostrophic effect. A comprehensive theoretical treatment of the wind-driven circulation was given by Munk (1950) and Munk and Carrier (1950).

The highest velocities are found in the western boundary currents. The Florida Current reaches speeds greater than 150 cm./second (3 kt) in flowing through the Straits of Florida, where the volume of water transported varies between 20 and 40×10^6 m.^{3}/second. The Gulf Stream itself attains its greatest volume off Chesapeake Bay, where it is transporting about 70×10^6 m.^{3}/second. On the basis of observations averaged over a considerable period, it appears to be about 150 km. wide and to have a maximum surface velocity of about 150 cm./second (3 kt). More detailed measurements made during research cruises in recent years, however, have shown that at a particular time its width may be as little as 50 km. and its velocity up to 300 cm./second (6 kt). After leaving Cape Hatteras, the Gulf Stream becomes much broader and correspondingly slower, so that in its continuation, the North Atlantic Current, its velocity is only about 15 cm./second, decreasing to 5 cm./second as it approaches Ireland. The structure of the Kuroshio in the North Pacific shows features similar to those of the Gulf Stream. In most of the other major ocean currents, the velocities seldom exceed 50 cm./second (1 kt) and are often considerably less, although because of their width and depth they transport very large volumes of water.

The North Equatorial Current in the Pacific, for example, transports 45×10^6 m.^{3}/second and the Countercurrent 25×10^6 m.^{3}/second. The greatest rate of transport is found in the Antarctic Circumpolar Current which, in the Drake Passage, is estimated to transport 110×10^6 m.^{3}/second although the surface velocity is usually less than 50 cm./second (1kt).

Seasonal variations. In certain regions, significant seasonal variations occur in the currents. In extreme cases, a current may disappear or reverse its direction at a particular season, while sometimes the variation is limited to a change in the velocity or in the volume of water transported, or to a shift in its position. The most striking seasonal variations occur in the Indian Ocean, where they are associated with the monsoons. During the winter months, when the north-east monsoon winds prevail over the northern part of the ocean, the North Equatorial Current is well developed, while to the south of it, the Equatorial Countercurrent flows eastwards, with its axis at 7°S. Along the African coast, south of the Gulf of Aden, the North Equatorial Current is deflected to flow southwards. In the summer months, when a low pressure area has become established over central Asia and the south-west monsoon wind blows across the ocean, the North Equatorial current is replaced by the Monsoon Current, which flows from west to east. Along the African coast, from 10°S. northwards, the flow is directed towards the north, and much water from the South Equatorial Current crosses the equator. In the Somali Current, flowing into the Arabian Gulf, the velocity may exceed 200 cm./second (4 kt). The Equatorial Countercurrent cannot be distinguished in this season. Certain changes also occur in the southern part of the Indian Ocean.

Another example of a seasonal current occurs off the west coast of North America, where the relatively cold California Current forms part of the general clockwise circulation in the North Pacific. During the months of November, December and January, a surface counter current, the Davidson Current, develops and flows northwards between the California Current and the coast. Upwelling occurs during the rest of the year, and is associated with a series of eddies in the surface layer, although a subsurface counter current flows northwards below about 200 m.

Irregular variations. Apart from seasonal changes, irregular variations of currents occur, sometimes with far-reaching results. An outstanding example is El Niño, the southward flowing current off the west coast of South America. This is a seasonal current which flows in January to March and which carries warm, low salinity water to the south, although its normal extent is limited to a few degrees south of the equator.

Occasionally this current penetrates beyond 12°S. with disastrous effects on the marine life, which in this region is accustomed to much lower temperatures.

A different kind of variation arises from the instability which appears to be inherent in certain current situations, as in the Gulf Stream east of Cape Hatteras. The current takes a meandering course, and the position of the axis of the current may shift by up to 60 km. in a week. From time to time the meanders may form a closed loop, which then develops into an eddy and becomes detached from the main current.

Most of the data from which charts of the surface currents have been constructed have been obtained from records of ship's drift collected over a number of years. In this method the vector distance travelled by a ship in a given time as determined by astronomical fixes, or, in recent years, by radio navigational techniques, is compared with that estimated from the ship's course and speed. The vector difference between these two quantities, corrected if necessary for the effect of wind on the ship, is attributed to the current. On research vessels other methods are available for measuring currents with greater accuracy and in more detail. Observations of this kind are necessary to determine comparatively rapid changes in the current, such as those occurring in the Gulf Stream. Descriptions of methods of measuring currents, and further details of the currents themselves will be found in the books by Sverdrup, Johnson and Fleming (1942), Defant (1961) or von Arx (1962).

III. Salinity Distribution in the Surface Layer

The distribution of salinity is determined by processes which take place at the sea surface and by currents and mixing. Although small but significant variations occur in some of the major constituents of sea water, these may be neglected when considering the general distribution of salinity. The processes taking place at the surface which need to be considered are, therefore, those which decrease or increase the salinity by the addition or removal of fresh water. The salinity is decreased by rainfall, or other forms of precipitation, the influx of fresh water from the land and by the melting of ice. Conversely it is increased by evaporation and by the formation of ice, since the ice has a lower salt content than the water from which it is formed, leaving a higher concentration of salt in the remaining water.

Figure 2 shows the surface salinity of the oceans in northern summer. Certain seasonal changes occur but, except in coastal waters, they are not large. In the open ocean the surface salinity ranges from 32–

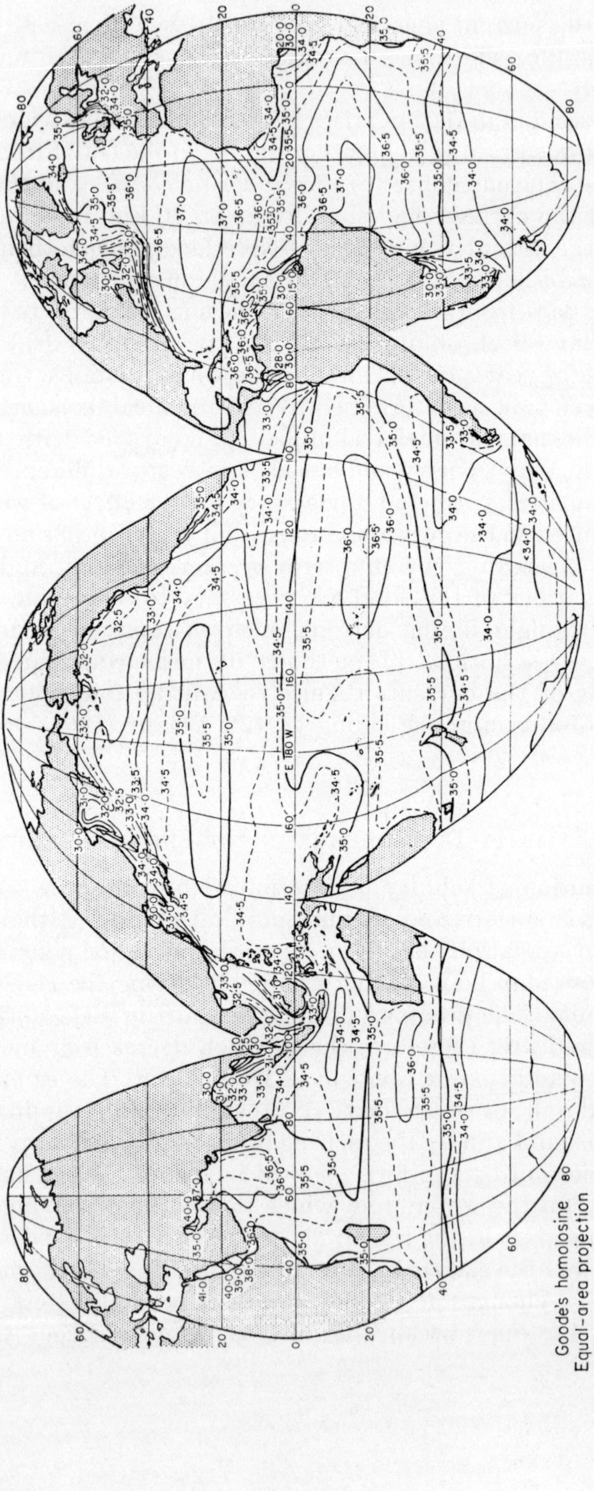

FIG. 2. Surface salinity of the oceans in the northern summer (Sverdrup, 1945). By permission of George Allen & Unwin Ltd.

37·5‰. The highest salinities are found in tropical latitudes, about 20–30° N. and 15–20° S., where the evaporation is high owing to the high temperatures and strong trade winds. In the equatorial zone, the salinity is lower because of the greater rainfall and lower wind speeds. Polewards of the tropical regions the salinity decreases because of the excess of rainfall over evaporation. Near the coasts of the continents the salinity is reduced owing to the influence of run-off from the land. In some areas, the effects of ocean currents are apparent: off the north-east coast of North America, for example, the salinity near the coast is low owing to the influence of the Labrador Current and there is a sharp increase in salinity across the Gulf Stream.

Salinities higher than those in the open ocean occur in partially enclosed seas in mid-latitudes, where evaporation greatly exceeds rainfall and run-off. Thus, the Mediterranean Sea has a salinity of 37–40‰ and that of the Red Sea is 40–41‰. In coastal seas in temperate latitudes the salinity is reduced by the influx of land water, and there may be a range of conditions, from almost completely fresh water, as in the inner parts of the Baltic Sea, to the salinity of the open ocean.

In the deeper layers of the ocean the variations in salinity are smaller than near the surface, but are nevertheless highly significant in their relation to the general circulation.

IV. Deep Water Circulations

The wind-driven circulation extends, on the average, only to a comparatively small fraction of the total depth of the ocean. In general, the depth is least in equatorial regions, where the surface currents extend through a layer varying from 20–200 m. in depth, and greatest in the western boundary currents of temperate latitudes. The Gulf Stream, for example, extends to a depth exceeding 1000 m. Deep currents are closely related to the distribution of density, which in turn depends on the distribution of temperature and salinity. Observations of these two variables at different depths can be made more easily than direct current measurements, so that charts or sections of temperature and salinity, or of density derived from them, are often used to indicate the positions and velocities of the currents. Other properties, such as the content of oxygen, silicate, or phosphate, have also been used as indicators of water movements.

In many parts of the oceans it is possible to distinguish five layers: surface, upper, intermediate, deep and bottom, each of which has a distinct water mass and movement of its own. The persistence of these

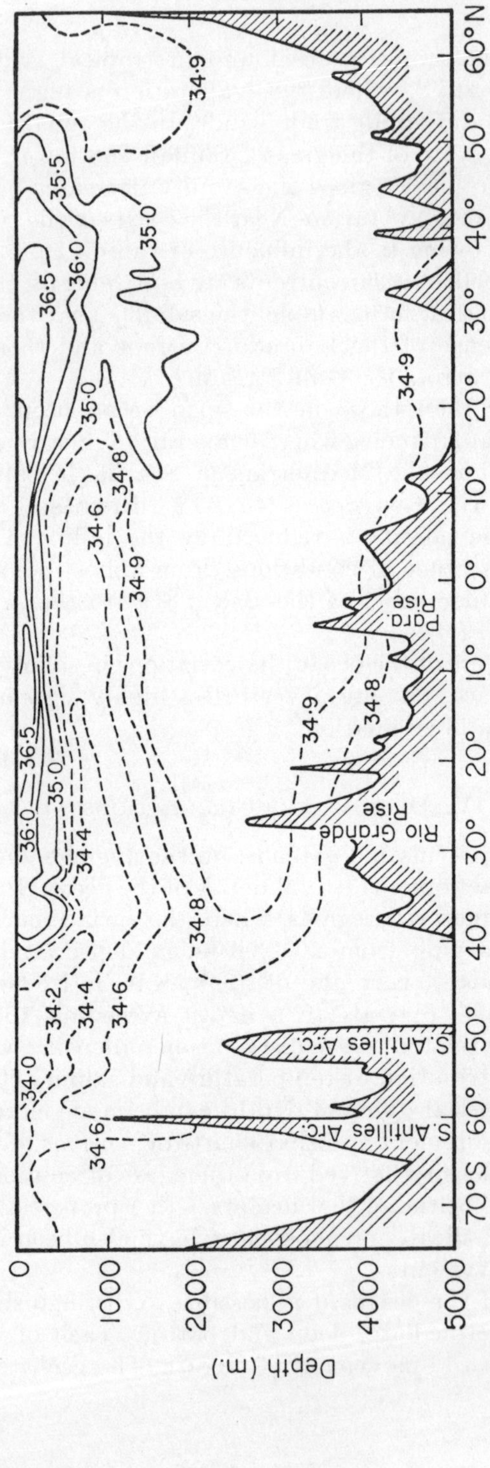

FIG. 3. Vertical section showing distribution of salinity in the Western Atlantic Ocean (von Arx, 1962). By permission of Addison-Wesley Publishing Co. Inc.

layers is due to the stable density gradient which greatly inhibits vertical mixing between water masses. Figure 3, which shows the salinity distribution in a north-south section of the Atlantic Ocean, provides evidence of these various layers. The surface layer, down to 100 or 200 m., is subject to considerable seasonal variations in temperature and salinity owing to direct interaction with the atmosphere. The upper layer extends from 100 or 200 m. down to about 800 m., and is not subject to these seasonal changes, but it does take part, with the surface layer, in the wind-driven surface circulation, its velocity usually decreasing with depth.

In the Atlantic, the main feature of the intermediate layer is the Antarctic Intermediate water, which sinks from the surface at the Antarctic Convergence at about 50°S. and then spreads northwards at about 1000 m. Its influence can be traced well north of the equator. Below it, at depths of 2000–4000 m., the North Atlantic Deep Water, formed by sinking in the Labrador and Irminger Seas, flows southwards. In the layer adjacent to the bottom, the Antarctic Bottom Water moves northwards. This water mass is formed by sinking at the edge of the Antarctic continental shelf, especially in the Weddell Sea. Its high density is a consequence of its very low temperature and also of its comparatively high salinity which results from the abstraction of water during the formation of ice.

It would be misleading to think that these deep water movements are spread uniformly over the whole width of the ocean. From the density distribution and certain reasonable assumptions about the depths at which the velocity was zero, Defant (1941) constructed charts of the currents at 800 m. and 2000 m. in the Atlantic. The former, reproduced in Fig. 4(a), shows the Antarctic Intermediate Water flowing northwards across the equator as a concentrated current with velocities of 6–12 cm./second on the west side of the ocean. Similarly, the chart for 2000 m. (Fig. 4(b)) shows the North Atlantic Deep Water travelling south in three distinct streams; one on the west side which follows the continental slope off North America, one on the east side of the Mid-Atlantic Ridge and another in the East Atlantic Trough. The velocities indicated are less than 2 cm./second in most areas but in a few localities exceed 10 cm./second. Stommel has shown that dynamical theory leads one to expect that deep currents, like surface ones, would be concentrated on the western side of the ocean basins. Swallow and Worthington (1957, 1960) showed by direct current measurements the existence of a counter current flowing southwards off South Carolina, under the Gulf Stream, at 2000 m., with speeds of up to 17 cm./second (0.3 kt).

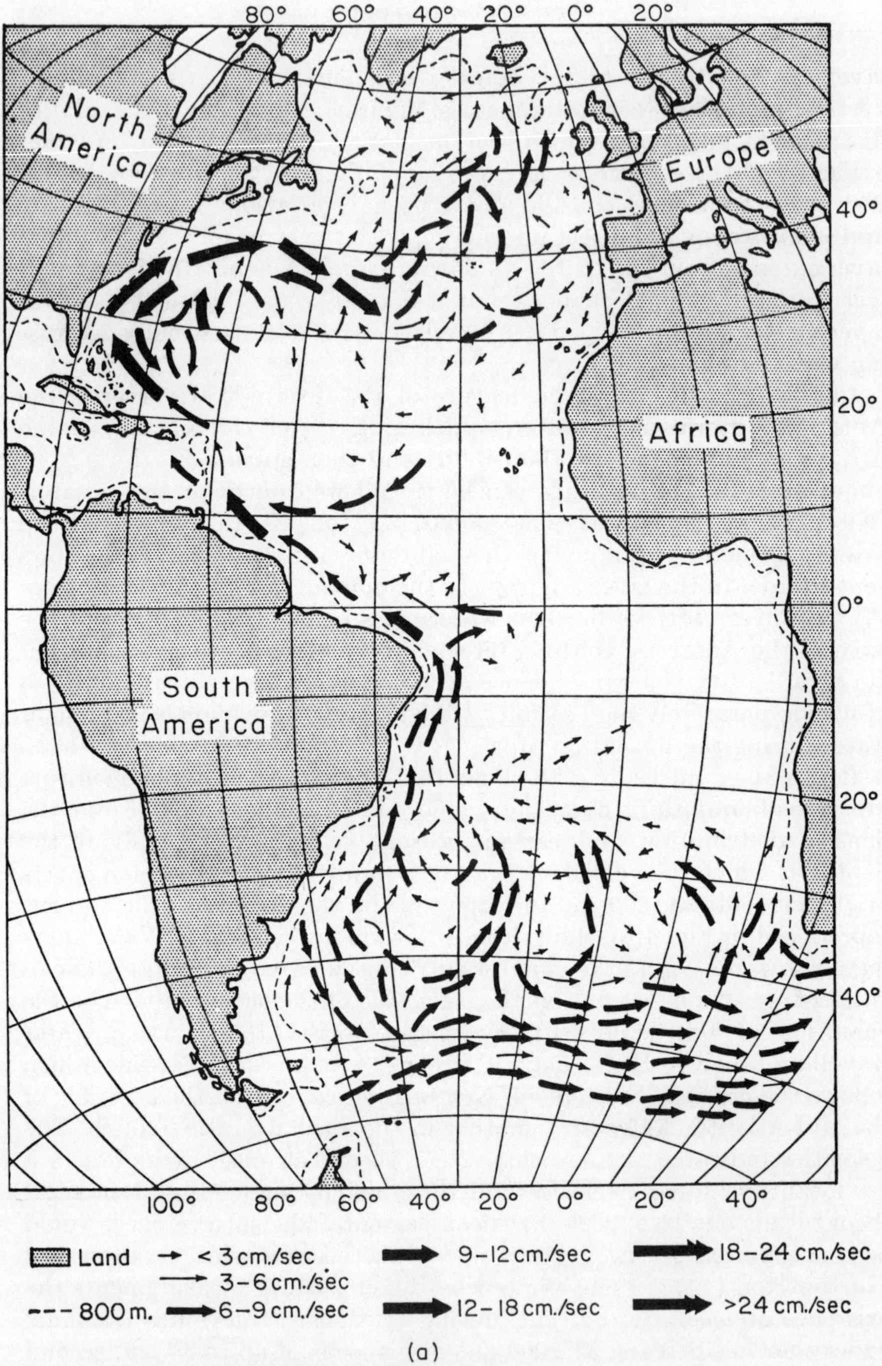

FIG. 4. (a) Current field at 800 m. in the Atlantic Ocean (Defant, 1961). By permission of Pergamon Press Ltd.

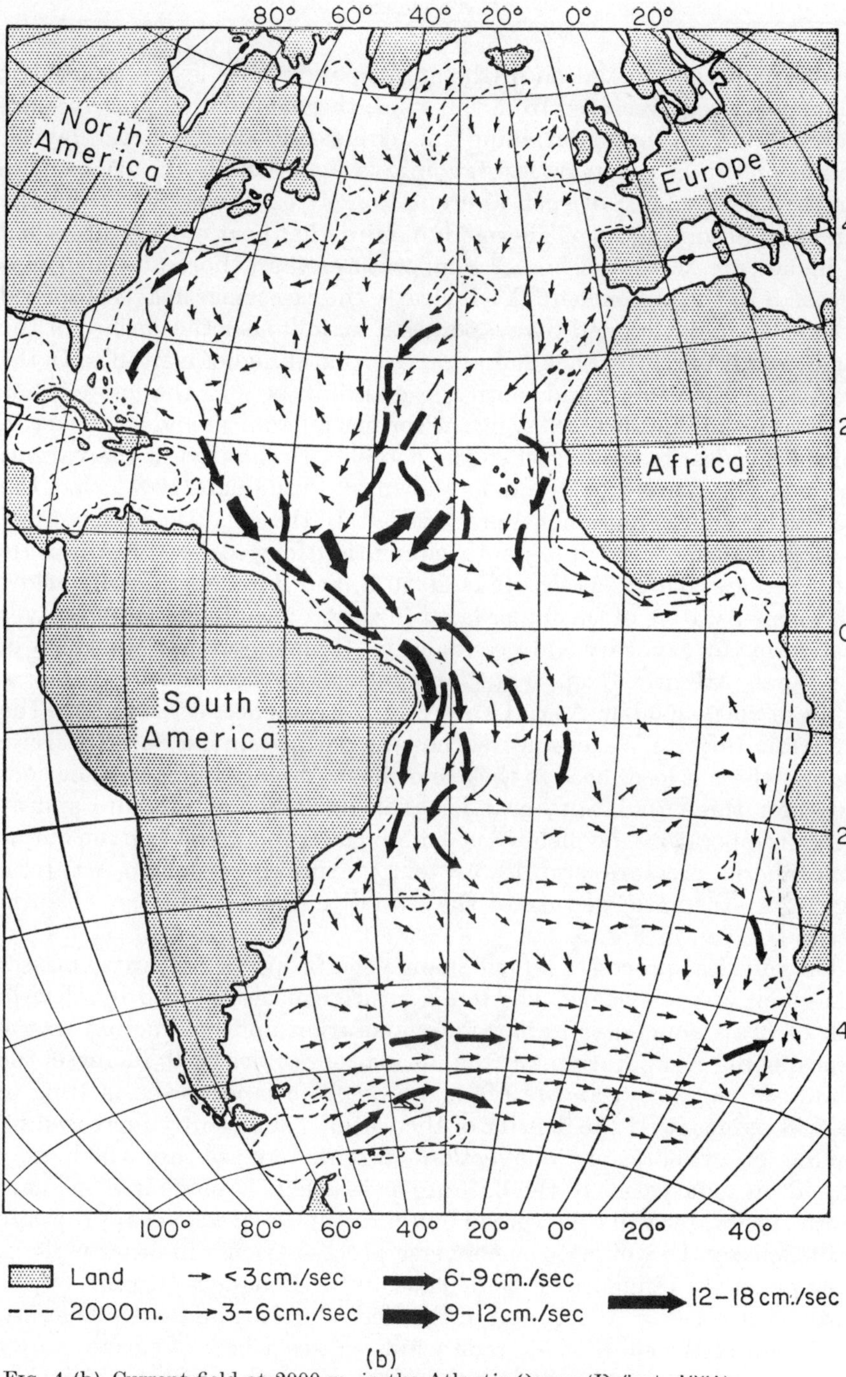

(b)

FIG. 4 (b) Current field at 2000 m. in the Atlantic Ocean (Defant, 1961). By permission of Pergamon Press Ltd.

Although the circulation of the surface and upper layers is brought about by energy received from the wind, thermodynamic processes are more important in maintaining the circulation of the lower layers. Conditions in the ocean are less favourable than those in the atmosphere for the maintenance of such a circulation. It can be shown, from the First and Second Laws of Thermodynamics, that heat must be absorbed at higher pressures and higher temperatures than those at which it is released. In the atmosphere this is clearly the case, since heat is absorbed from solar radiation and by evaporation mainly near the surface of the earth, and it is lost by back radiation to space at higher altitudes. In the oceans, on the other hand, both the main intake and the loss of heat occur in a shallow surface layer, the former in tropical and the latter in polar latitudes. However, when the complex circulation in the various layers is considered it is found, as shown by Proudman (1953), that the thermodynamic requirements are satisfied. In the Atlantic, for example, there is a surface and mid-depth circulation formed by the Antarctic Intermediate Water, which sinks at 50°S. and flows northwards below the warmer waters of the upper layer in the tropical region and receives heat from this layer by eddy conduction. It then becomes entrained in the North Atlantic Deep Water and returns to flow southwards at a depth of 2000–3000 m. At the Antarctic Convergence, it rises above the Antarctic Bottom Water and reaches the surface in the Sub-Antarctic region, where it loses heat to the atmosphere, mainly by back radiation. Thus, for this circulation, heat is taken in at higher pressure and at higher temperature, i.e. below 1000 m. in the tropics, and is given out at atmospheric pressure and lower temperature in the Sub-Antarctic region. A schematic picture of the overall circulation in the Atlantic Ocean is given in Fig. 5.

The physical processes which change the properties of water masses take place at the surface. The temperature can be changed by absorption of heat from solar radiation, evaporation and conduction to the atmosphere. Evaporation or the formation of ice will increase the salinity of the water, and rainfall, run-off from land or the melting of ice, will reduce it. If the density of the water is sufficiently increased by cooling or evaporation, convection currents are set up which may extend, in some cases, to the bottom, or in others to some intermediate depth. The water will then spread out horizontally at a level appropriate to its density. Upwelling, the converse process, occurs in other regions, often under the influence of wind. Thus, in the Benguela Current, off the south-west coast of Africa, and in the Peru Current, off South America, the action of the south-east trade winds causes a flow of surface water away from the coast. Its place is taken by sub-surface water which rises

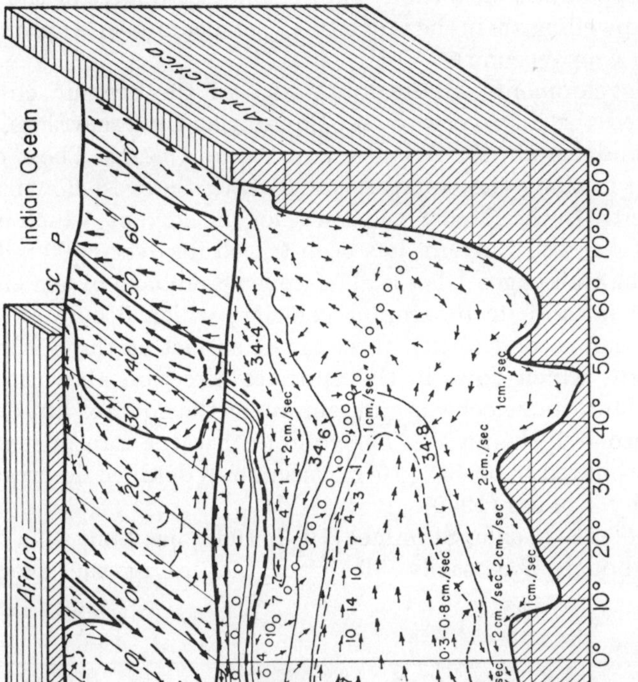

FIG. 5. Block diagram of the Atlantic Circulation (Defant, 1961). By permission of Pergamon Press Ltd.

from a depth of 200–300 m. This upwelling is very important from a biological point of view, since the upwelling water brings a fresh supply of nutrient substances into the euphotic zone. Estimates of the vertical velocity of upwelling are in the range 10–80 m. per month (approximately $0{\cdot}4-3 \times 10^{-3}$ cm./second).

Recent developments in the knowledge of the oceanic circulation have included the discovery of the Equatorial Undercurrents, first in the Pacific and then in the Atlantic and Indian Oceans. These currents flow eastwards along the equator, underneath the west-flowing Equatorial Currents. In the Pacific, the course of the current, which has been named the Cromwell Current, has been traced for over 3500 miles from 150°W. to the Galapagos Islands. The upper boundary of the current is at about 30 m. and its maximum velocity, which occurs at 100 m., reaches 150 cm./second (3 kt), three times as fast as the South Equatorial Current, which flows in the opposite direction above it. In the horizontal plane, its velocity is a maximum at the equator but decreases to almost zero at 1°N. and 1°S. The volume of water transported by the current is estimated to be 40×10^6 m.^{3}/second. A detailed description has been given by Knauss (1960).

On the theoretical side, Stommel (1957, 1958) has built up a comprehensive, although necessarily idealized, picture of the deep water

FIG. 6. The deep circulation of the world ocean, according to Stommel (1958). By permission of Pergamon Press Ltd.

circulation. He assumed only two sources of deep water, one in the Antarctic corresponding to the Antarctic Bottom Water and the other in the North Atlantic, corresponding to the North Atlantic Deep Water. The circulation deduced from dynamical theory is shown in Fig. 6. Its striking features are the strong western boundary currents and the broad diffuse flow over the greater part of the width of the oceans. Reference has been made above to the discovery, in 1957, of the deep counter current below the Gulf Stream which is in accordance with Stommel's scheme.

V. Currents in Coastal Seas

A. Tidal Currents

Tidal currents are present everywhere in the oceans and occur at all depths, but in the open ocean their velocities seldom exceed 10 cm./second and are often only 2 or 3 cm./second. In shallow waters they are stronger, frequently reaching 100–150 cm./second (2–3 kt), while in narrow straits they may reach 10 kt or more. Tidal currents are periodic, usually with a semidiurnal period of half a lunar day (12 hours 25 minutes), although an appreciable diurnal constituent is often superimposed and in some areas is predominant. In general, the current may be resolved into two components at right angles; these oscillate in different phases, so that the current vector traces out an ellipse during the tidal period. The tidal movements alone, whatever their amplitude, do not produce any net transport of water and so have an insignificant effect in transporting water masses by advection. The turbulence associated with their flow, however, has an important influence on both vertical and horizontal mixing.

B. Local Wind-Driven Currents

In many coastal areas, the permanent currents are weak or ill-defined and can be masked by currents produced locally by the wind. In the open sea the current tends to flow at an angle to the right of the wind (to the left in the southern hemisphere) and the wind factor, i.e. the ratio of surface current speed to wind speed, is about 0·02. In shallow water, the direction of the surface current deviates less from that of the wind, and the wind factor depends on the coastal boundaries and the degree of obstruction to the flow. It is largest where there is a free passage for the water, as, for example, with a southerly wind over the Irish Sea, or with a south-westerly wind over the English Channel. In the Straits of Dover the velocity of flow is increased by the throttling effect of the constriction of the channel, and with strong winds blowing along the channel a current of 100–150 cm./second (2–3 kts) may be produced. Even if the wind is blowing towards a land barrier, e.g. towards the head of a gulf, there will be a surface current in the direction of the wind but a compensating current will flow in the opposite direction below it.

C. DENSITY CURRENTS

Where fresh water flows into the sea, the reduction of density can give rise to density currents. In an estuary the effect is to produce an outflow of lower salinity water in the upper layer and an inflow of higher salinity water below it. The volume of water involved in this circulation may be about 10 or 20 times that of the river water discharged. The development of such a circulation in a particular estuary depends on the tidal currents, and on the vertical mixing which they produce, as well as on the river discharge. Typical velocities for these currents would be in the range 5–20 cm./second, but they may be superimposed on tidal currents whose amplitudes are an order of magnitude greater. Density currents may also develop off an open coast where there is an influx of fresh water. Near the coast there is again a surface flow away from the shore and a deep flow towards it. Owing to the earth's rotation there is a tendency for both of these currents to turn to the right in the northern hemisphere. A coastal current is set up, therefore, flowing with the fresh water inflow on its right, and with an undercurrent in the opposite direction. In the southern hemisphere these directions are reversed.

Currents which are similar in origin but larger in scale occur in the straits connecting a land-locked sea to the open ocean. The best known example is in the Straits of Gibraltar. Because of the intense evaporation in the Mediterranean Sea, the surface waters have salinities which are considerably higher than those in the Atlantic Ocean. There is, therefore, a surface inflow of lower salinity water from the Atlantic and a deep outflow of the higher salinity Mediterranean water. The average velocity of the surface inflow appears to be about 50 cm./second, with a similar velocity in the deep outflow, but there are large variations associated with changing wind conditions (Lacombe, 1961). Similar conditions occur in the Straits of Bab el Mandeb, connecting the Red Sea to the Gulf of Aden. The reverse type of flow is found in the Kattegat, where there is a surface outflow of low salinity water from the Baltic Sea (in which precipitation and run-off greatly exceed evaporation) and a deep inflow of higher salinity water from the North Sea.

VI. Turbulent Mixing

Because of the large scale of the motions involved, the flow of water in the sea is almost invariably turbulent. The turbulent movements, which may be spoken of rather loosely as "eddies", without implying that they have any regular structure, cover a very wide spectrum. There is usually, however, a difference of several orders of magnitude between the turbulent mixing effects in the vertical and the horizontal planes. This is due partly to the large ratio of width to depth of the ocean and also to the influence of a stable density gradient, which greatly inhibits vertical turbulence, without appreciably affecting turbulence in the horizontal plane. By reason of this difference it is convenient to treat vertical and horizontal mixing separately.

A. Vertical Mixing

The turbulence responsible for vertical mixing is generated in two main ways: (a) by the stress of the wind on the sea surface, acting through wave motion and the surface drift; (b) by vertical shear in currents arising, for example, from the action of bottom friction. Wind acts on the surface of the sea and generates waves, which cause oscillatory movements of the water particles as the wave form progresses across the sea; a further effect of the wind is to produce a steady drift of the surface water. In waves on deep water, the water particles move in circular orbits in a vertical plane, the water moving forwards at the crest of a wave and in the opposite direction in a trough. At the surface, the diameter of the orbit is equal to the height of the wave, but the movements decrease rapidly with depth, and fall to $e^{-\pi}$ (approximately 0·043) of their surface values at a depth equal to half a wavelength. The orbits are not completely closed, and the water particles near the surface have a small net movement in the direction of propagation of the waves. This "mass transport" velocity falls off even more rapidly with depth. A regular train of waves of comparatively small height would have very little mixing effect, the elements of water simply being distorted in shape as the waves passed. In a wind-driven sea, however, waves of many different periods will be superimposed, producing an ever-changing pattern of movements in the water and so enhancing the mixing effect. If the waves are also breaking, much of their energy is being converted into turbulence, although the scale is probably fairly small, and directly affects only a comparatively shallow layer of water.

The stress of the wind on the sea also produces a drift of the surface water which, in the simplified conditions assumed in Ekman's theory, is directed 45° to the right of the wind in the northern hemisphere and 45° to the left in the southern hemisphere. The water below the surface is brought into motion by shearing stresses communicated from the water above through the action of eddy viscosity, but its movement is slower and deviates further from the wind direction. At a certain depth, called the "depth of frictional influence" and denoted by D, the drift would be in the opposite direction to that at the surface but its velocity would be only $e^{-\pi}$ of the surface velocity. Below this depth, the movement is small, and for practical purposes one may consider that the mixing effect of the wind current is confined to a layer of depth D. The value of D depends, among other factors, on the strength of the wind, and estimates made for it in the open ocean are mostly between 50 and 200 m.

When the sea gains heat by the absorption of solar radiation, most of the heat is absorbed in the first few metres. Molecular conductivity alone could provide only a very slow transport of heat downwards and a steep temperature gradient would occur just below the surface. Turbulence gives rise to an "eddy conductivity" which is much larger, and the heat becomes distributed through a deeper layer. A certain temperature gradient will develop, however, and the downward decrease in temperature causes an increase in density. Under these conditions, vertical mixing involves an increase in the potential energy of a vertical column of water and this energy must be supplied by the turbulence. A stable temperature gradient has, therefore, an inhibiting effect on the turbulence. If the turbulence is being generated by the action of wind on the surface, it may happen that at a certain depth the turbulent energy is no longer great enough to overcome the stability effect. A steep temperature gradient will then develop at this depth, and thereby separate the warmer water above from the colder water below. In this transition layer, known as the "thermocline", vertical turbulence is almost entirely suppressed and mixing of water across it can take place only very slowly.

In many areas a thermocline is formed at a depth of 30–50 m. in summer by this process. In autumn, when the sea surface is cooled, convection currents are set up which break down the temperature gradient. In addition to seasonal thermoclines of this nature, a main or permanent thermocline occurs over widespread areas of the ocean. This is at a greater depth and is only affected a little by seasonal influences. The depth of the thermocline varies considerably from one region to another; its upper limit may be less than 100 m. and its lower limit may be deeper than 1000 m. A typical vertical profile of temperature, indicating the thermocline, is given in Fig. 7. The processes maintaining

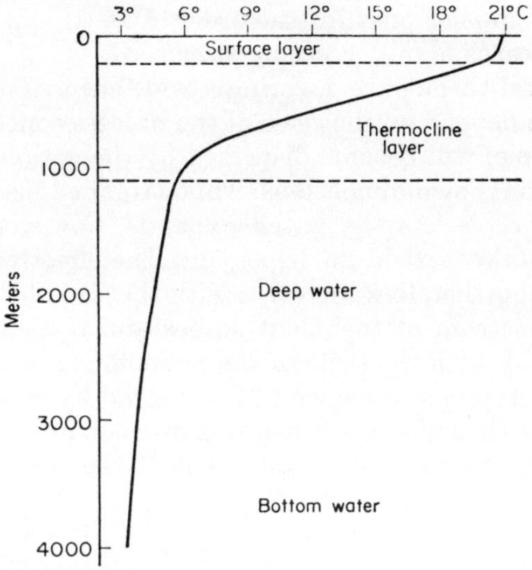

FIG. 7. The vertical subdivisions of the ocean (von Arx, 1962). By permission of Addison-Wesley Publishing Co. Ltd.

the permanent thermocline are closely related to the general circulation of the oceans.

In shallow water, in addition to wind-generated turbulence, which is at its greatest intensity near the surface, the effect of bottom friction on currents is to generate turbulence near the bottom, some of it diffusing upwards. There may then be a sufficient intensity of turbulence at all depths to maintain thorough vertical mixing and prevent the formation of a thermocline. This occurs, for example, in the English Channel and the southern part of the North Sea, where the tidal currents exceed about 100 cm./second (2 kt).

B. HORIZONTAL MIXING

A very wide spectrum of horizontal motions occurs in the ocean. On the largest scale are the main wind-driven circulations of ocean-wide dimensions. Developing from these are eddies of about 100 km. in diameter, such as those which occur in the Gulf Stream after it has left the continental slope and has turned eastwards across the Atlantic. Eddies of similar dimensions occur in latitudes 30–40° S., between the South Equatorial Current and the West Wind Drift. Water movements on this scale may also be set up by moving storm areas. Such eddies break down into smaller ones, of the order of 10 km. in diameter, and so on through a cascade of decreasing sizes.

When horizontal turbulence covering a wide spectrum is present, its mixing effect will depend on the scale of the process concerned. A patch of dye, for example, will become dispersed by the action of eddies of a scale smaller than its own dimensions, while larger eddies will transport the patch as a whole. As the patch expands, however, increasingly large eddies will take part in its dispersion. The effective coefficient of horizontal diffusion therefore increases with the size of the patch itself. The idea of a spectrum of turbulent eddies and a coefficient of eddy diffusion increasing with the scale of the phenomena is a feature of the theory of locally isotropic turbulence, introduced by Kolmogoroff. The application of this theory to the ocean was discussed by Stommel (1949) and Defant (1954). A review of turbulence in the sea has been published by Bowden (1962).

VII. Quantitative Treatment of Mixing Problems

Let rectangular axes OX, OY, OZ be taken, with OX and OY in a horizontal plane and OZ vertically downwards. Let u, v, w be the components of the velocity of the water parallel to these axes and consider the distribution of some property which has a concentration c per unit mass. By analogy with molecular diffusion, the turbulent transport of this property may be regarded as a process of eddy diffusion. Across a plane perpendicular to OZ, for example, the rate of transport per unit area may be represented by

$$T_z = -\rho K_z \frac{\partial c}{\partial z}$$

where ρ is the density of the water. K_z is defined as the coefficient of eddy diffusion in the vertical direction or, more simply, the vertical eddy diffusivity. Coefficients K_x and K_y may be defined similarly for turbulent transport across planes perpendicular to OX and OY respectively. The three coefficients K_x, K_y and K_z will, in general, differ from one another, vary with position, and depend on the type and scale of motion and on the stability. Their numerical values cover a very wide range, but are always large compared with the molecular diffusivity. In the sea, K_z is usually in the range $0{\cdot}1 - 10^3$ cm.^{2}/second, while K_x and K_y fall within the range $10^4 - 10^8$ cm.^{2}/second. For comparison, the coefficient of molecular diffusion in water is about 2×10^{-5} cm.^{2}/second.

The general equation governing the concentration, c, of a particular constituent at any point in the sea, may be written:

$$\frac{\partial c}{\partial t} + u\frac{\partial c}{\partial x} + v\frac{\partial c}{\partial y} + w\frac{\partial c}{\partial z} = \frac{\partial}{\partial x}\left(K_x\frac{\partial c}{\partial x}\right) + \frac{\partial}{\partial y}\left(K_y\frac{\partial c}{\partial y}\right) + \frac{\partial}{\partial z}\left(K_z\frac{\partial c}{\partial z}\right) + R \quad (1)$$

where R denotes the rate of production (or loss) of c within a small volume surrounding the point itself. For the oxygen content, for example, R will be positive if there is a net production of oxygen by photosynthesis and negative if there is a net consumption by oxidation processes. The decay of a radioactive substance would be represented by a negative R, i.e. equal to $-\lambda c$, where λ is the decay constant. For a conservative property, such as salinity, $R = 0$.

The above equation, which allows for both advection and turbulent mixing in three dimensions, is too complicated to be solved in its general form. In a particular problem, a preliminary examination will usually suggest that some of the terms are likely to be small compared with those representing the main processes and so can be neglected. A few particular solutions are given below.

A. STEADY STATE

If the distribution is not changing with time, $\partial c/\partial t = 0$.

(1) *Horizontal flow and vertical mixing*. It is assumed that the current is in the OX direction whereas turbulent mixing takes place in the OZ direction only. For a conservative property such as salinity S, $R = 0$ and equation (1) reduces to

$$u\frac{\partial S}{\partial x} = \frac{\partial}{\partial z}\left(K_z\frac{\partial S}{\partial z}\right). \quad (2)$$

Now

$$\frac{\partial}{\partial z}\left(K_z\frac{\partial S}{\partial z}\right) = K_z\frac{\partial^2 S}{\partial z^2} + \frac{\partial K_z}{\partial z}\cdot\frac{\partial S}{\partial z}. \quad (3)$$

If it is assumed that K_z does not vary with z, equation (2) becomes

$$u\frac{\partial S}{\partial x} = K_z\frac{\partial^2 S}{\partial z^2} \quad (4)$$

and an analytical solution can be found, corresponding to given boundary conditions. For example

$$S = S_0 + S_1\,\mathrm{e}^{-\alpha x}\cos\frac{\pi z}{2l} \quad (5)$$

where

$$\alpha = \frac{\pi^2 K_z}{4l^2 u}$$

represents the distribution of the property S in a current extending over a depth $2l$, flowing with velocity u, while mixing takes place with the water above and below it. The original value of S in the current is $S_0 + S_1$, while the water above and below has the value S_0. This solution is shown in Fig. 8.

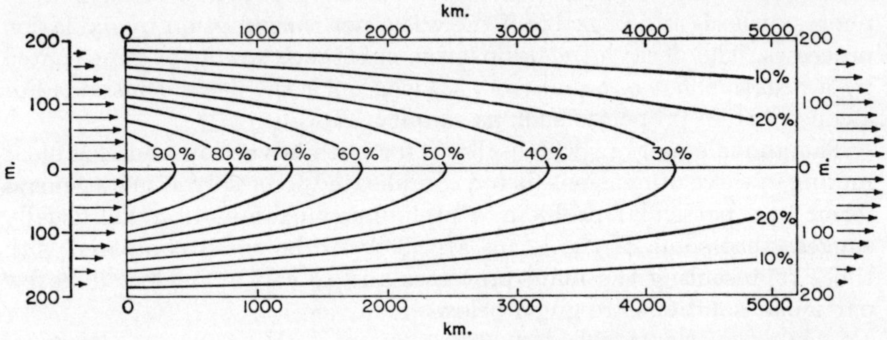

FIG. 8. Distribution of a property of sea water produced by horizontal flow and vertical mixing (Defant, 1961). By permission of Pergamon Press Ltd.

An example to which this solution applies approximately is the movement of the Antarctic Intermediate Water northwards, at a depth of about 1000 m., after sinking at the Antarctic Convergence at 50°S. This water is characterized by a salinity minimum (i.e. S_1 is negative), and its salinity increases slowly as it moves northwards, owing to vertical mixing with higher salinity water both above and below it.

In practice, K_z is usually unknown, and an observed distribution may be used to estimate its value. One cannot assume, in general, that $\partial K_z/\partial z = 0$, but it is seen from equation (3) that the second term on the right-hand side may also be neglected if $\partial S/\partial z = 0$. This implies that S is a maximum or a minimum at the depth considered, and equation (4) may then be applied. Then

$$\frac{K_z}{u} = \frac{\partial S}{\partial x} \bigg/ \frac{\partial^2 S}{\partial z^2} \tag{6}$$

and the differential coefficients may be evaluated if the distribution of S in the XOZ plane is known. Methods of applying this approach have been described by Proudman (1953). It should be noted that a knowledge of the distribution alone enables only the ratio K_z/u to be computed. In order to determine K_z itself, the current u must be determined independently. For the Antarctic Intermediate Water, mentioned above, K_z/u in the South Atlantic has been estimated as averaging 1 cm. Since u has been estimated by other means to be about 4 cm./second, $K_z = 4$ cm.^{2}/second.

If measurements of the current at various depths are available, as well as the salinity distribution, K_z may be determined as a function of z by integrating equation (2) directly. If $\partial S/\partial z = 0$ at some depth z_0, then K_z at depth z is given by

$$K_z = \frac{1}{\frac{\partial S}{\partial z}} \int_{z_0}^{z} u \frac{\partial S}{\partial x} \, dz. \tag{7}$$

This method was applied by Jacobsen in 1913 to the salinity distribution in the Kattegat, and yielded values of K_z lying between 0·04 and 0·1 cm.^{2}/second in these highly stable conditions. It has since been applied to many other areas.

(2) *Horizontal flow and lateral mixing.* If the current is flowing in the OX direction, but mixing is taking place laterally, in the OY direction, the equation becomes

$$u \frac{\partial S}{\partial x} = \frac{\partial}{\partial y}\left(K_y \frac{\partial S}{\partial y}\right). \tag{8}$$

This equation is applicable, for example, to a current such as the Equatorial Counter-current, flowing between two other water masses with which lateral mixing takes place. Mathematically this equation is identical with (2), with the y co-ordinate replacing z, and analogous solutions may be found. This method has given values of K_y of the order 3×10^7 cm.^{2}/second for the Equatorial Counter-current in the Atlantic.

(3) *Radioactive decay and vertical mixing.* If the concentration of a radioactive substance in the sea is c and its decay constant is λ, then in equation (1) $R = -\lambda c$. If vertical mixing is the only other process affecting the distribution and if a steady state is maintained, then

$$\frac{\partial}{\partial z}\left(K_z \frac{\partial c}{\partial z}\right) - \lambda c = 0. \tag{9}$$

If K_z is independent of depth

$$K_z \frac{\partial^2 c}{\partial z^2} - \lambda c = 0. \tag{10}$$

If the concentration is greatest at the bottom, where it has the value c_0, and decreases to very small values at a great distance from the bottom, the appropriate solution of (10) is

$$c = c_0 e^{-\alpha z'} \tag{11}$$

where $\alpha = (\lambda/K_z)^{\frac{1}{2}}$ and z' is the distance measured upwards from the bottom. The flux of the radioactive substance through the bottom would then be $(\lambda K_z)^{\frac{1}{2}} c_0$, in the appropriate units. Koczy (1956, 1958) has applied equation (9) directly to the observed distribution of radium at several oceanic stations and has derived estimates of K_z as a function of depth.

B. VARIABLE (i.e. NON-STEADY) STATE

(1) *Vertical mixing only*. Under these conditions, equation (1) reduces to

$$\frac{\partial c}{\partial t} = \frac{\partial}{\partial z}\left(K_z \frac{\partial c}{\partial z}\right). \tag{12}$$

If, again, K_z is independent of z

$$\frac{\partial c}{\partial t} = K_z \frac{\partial^2 c}{\partial z^2}. \tag{13}$$

This is essentially the heat conduction equation in one dimension and a number of solutions are possible, depending on the initial and boundary conditions.

In the particular case of a periodic oscillation of period T, the solution is

$$c = c_0 + c_1 e^{-\alpha z} \cos\left(\frac{2\pi t}{T} - \alpha z\right) \tag{14}$$

where $\alpha = (\pi K_z/T)^{\frac{1}{2}}$.

At $z = 0$,

$$c = c_0 + c_1 \cos \frac{2\pi t}{T}$$

and with increasing z, the oscillation decreases in amplitude and becomes later in phase. A practical example of this solution is the penetration downwards of a seasonal variation of temperature owing to an alternating gain and loss of heat through the sea surface, or of salinity arising from seasonal changes in precipitation and evaporation. The assumption that K_z is independent of depth is not very realistic but, if the seasonal variation at various depths is known for a particular area, equation (12) may be solved numerically to determine K_z.

(2) *Spreading of a patch by horizontal diffusion.* A particular case of interest is that of the spreading of a patch of some contaminant under the action of horizontal turbulence. It will be assumed that the turbulence is the same in all directions and that the patch has radial symmetry. If the distance of any point from the centre is denoted by r and $K_x = K_y = K_r$, the diffusion equation may be written

$$\frac{\partial c}{\partial t} = \frac{1}{r}\frac{\partial}{\partial r}\left(K_r \frac{\partial c}{\partial r}\right). \tag{15}$$

If K_r is constant, the solution of (15) is

$$c(r,t) = \frac{c_0}{4\pi K_r t}\, e^{-r^2/4K_r t} \tag{16}$$

where $c(r,t)$ is the concentration at a distance r from the centre at time t.

It was suggested by Sverdrup (1946, unpublished) that K_r should be taken as proportional to r, i.e. $K_r = Pr$ where P is a constant, which has the dimensions of velocity. Joseph and Sendner (1958) introduced postulates leading to the same form for K_r. If this is inserted in equation (15) the solution becomes

$$c(r,t) = \frac{c_0}{2\pi P^2 t^2}\, e^{-r/Pt}. \tag{17}$$

Joseph and Sendner gave a number of examples of horizontal mixing in the sea which appeared to confirm their theory over a range of r from 10–1500 km., with P having the common value of $1 \pm 0{\cdot}5$ cm./second. The theory has been used by a number of other investigators for treating problems such as the spread of effluents, including radioactive materials. Other relationships for K_r as a function of r have been suggested, based on various hypotheses. Thus, the solutions to the above problem differ in the rate of decrease of concentration with time at the centre of the patch and in the decrease of concentration with distance from the centre at a given time. A review of the various theories of horizontal diffusion and of their application to observed distributions has been given by Okubo (1962). At present there is no clear indication from observations of the optimum functional relationship to take for K_r. In fact, the characteristics of the turbulent processes may vary from one set of conditions to another so that no law of this kind is universally applicable.

VIII. Overall Mixing Problems

The thermocline has been described above as a transition layer in which the temperature gradient is large, the stability high and the coefficient of vertical eddy diffusion very small. Vertical mixing can take place more readily either below or above the thermocline than within it, so that it acts as a partial barrier between the water masses above and below. This fact led to the introduction of box-type models of the ocean when considering the overall distribution of a constituent, such as carbon dioxide. In the simplest model, the world ocean is represented by two boxes, or reservoirs, the first representing the surface layers to a depth of 100 or 200 m., and the second the deeper layers from this depth to the bottom, with the thermocline separating them. Complete mixing within each box is assumed to take place very rapidly, but exchange between the two boxes takes place comparatively slowly. Estimates of the rate of exchange, and hence the average residence time of a molecule in each box, can be made by considering the distribution of a natural radioactive tracer such as ^{14}C.

Atmospheric carbon dioxide, which contains a known concentration of ^{14}C, enters the surface layer through the air-sea interface and becomes evenly distributed throughout the layer. Passage downwards into the deep layer takes place more slowly and there is a corresponding passage upwards from the deep layer. The molecules escaping upwards, however, will have been, on the average, in the deep layer for a considerable number of years and their ^{14}C content will have been reduced appreciably by radioactive decay. The $^{14}C/^{12}C$ ratio is, therefore, lower in the deep water than in the atmosphere. In the surface layer, the $^{14}C/^{12}C$ ratio is intermediate between the two, because a fraction of the carbon atoms have come from the lower layer and the others directly from the atmosphere. These differences are measurable and from them the exchange rates and residence times may be calculated. Two-box models of this type were considered by Arnold and Anderson (1957) and Craig (1957) (see also Chapter 7).

The main discrepancy between the simple two-box model and the real ocean is that in the Arctic and Antarctic regions the thermocline is absent and the deep water extends upwards to the free surface. In these areas, where deep and bottom water is being formed, convection currents extend from the surface to the bottom. This situation has been allowed for in an outcrop model, in which the deep box extends to the surface over a certain proportion of its area, representing the polar regions. An alternative representation is by a three-box model, in which the polar regions are represented by a separate box. A more complex model includes separate boxes for the Arctic and Antarctic regions and for the Atlantic and Pacific + Indian Oceans. Each additional box involves at least one additional interface with an unknown exchange rate and so the more complex the model, the more detailed are the observational data required in order to make use of it. A survey of such models has been given by Broecker (1963). The present state of knowledge indicates that the average residence time of a molecule in the deep layer is at least 500 years in the Atlantic and 800 years in the Pacific. The estimated time in the surface layer is more sensitive to the assumptions made in the different models, but is likely to be in the range 10–100 years for dissolved solids. The time for carbon dioxide will be less by a factor of 2 or 3 because of the exchange with the atmosphere.

According to present ideas on the general circulation of the oceans, the renewal of the deep water by sinking from the surface layer takes place in a small number of fairly well-defined areas in polar regions. To preserve continuity, there is a slow rise of deep water over the rest of the ocean. Since the main thermocline remains at the same average depth from year to year, the uprising of the deep water must take place through it. The temperature structure in the thermocline itself is maintained by a balance between the downward flux of heat by eddy conduction and an upward advective flux by the rising water. Theories of the thermocline based on reasoning of this kind have been put forward by Robinson and Stommel (1959) and Welander (1959). The upward velocity required is very small (about $10^{-4}-10^{-5}$ cm./second), and there are no means of measuring it directly at the present time.

REFERENCES

Arnold, J. R., and Anderson, E. C. (1957). *Tellus*, **9**, 28.

Arx, W. S., von (1962). "Introduction to Physical Oceanography". Addison-Wesley, Reading (Mass.), London.

Bowden, K. F. (1962). *In* "The Sea" (M. N. Hill, ed.) Vol. 1, p. 802. Interscience Publishers, New York, London.

Broecker, W. (1963). *In* "The Sea" (M. N. Hill, ed.) Vol. 2, p. 88. Interscience Publishers, New York, London.

Craig, H. (1957). *Tellus*, **9**, 1.

Defant, A. (1941). *Meteor Rep.* **6**, (2), 5 Lief.

Defant, A. (1954). *Dtsch. hydrogr. Z.* **7**, 2.

Defant, A. (1961). "Physical Oceanography", Vol. 1. Pergamon Press, Oxford, London, New York, Paris.

Joseph, J., and Sendner, H. (1958). *Dtsch. hydrogr. Z.* **11**, 49.

Knauss, J. A. (1960). *Deep Sea Res.* **6**, 265.

Koczy, F. F. (1956). *Nature, Lond.* **178**, 585.

Koczy, F. F. (1958). *Proc. 2nd U.N. Internat. Conf. Peaceful Uses Atomic Energy* **18**, 336.

Lacombe, H. (1961). *Cah. Océanogr.* **13**, 73.
Munk, W. H. (1950). *J. Met.* **7**, 79.
Munk, W. H., and Carrier, G. F. (1950). *Tellus*, **2**, 158.
Okubo, A. (1962). *J. Oceanogr. Soc. Japan, 20th Anniv. Vol.* 286.
Proudman, J. (1953). "Dynamical Oceanography". Methuen, London; John Wiley, New York.
Robinson, A., and Stommel, H. (1959). *Tellus*, **11**, 295.
Stommel, H. (1948). *Trans. Amer. geophys. Un.* **29**, 202.
Stommel, H. (1949). *J. Mar. Res.* **8**, 199.
Stommel, H. (1957). *Deep Sea Res.* **4**, 149.
Stommel, H. (1958). *Deep Sea Res.* **5**, 80.
Sverdrup, H. U., Johnson, M. W., and Fleming, R. H. (1942). "The Oceans, their Physics, Chemistry and General Biology". Prentice Hall, New York.
Sverdrup, H. U. (1945). "Oceanography for Meteorologists". George Allen and Unwin, London.
Swallow, J. C., and Worthington, L. V. (1957). *Nature, Lond.* **179**, 1183.
Swallow, J. C., and Worthington, L. V. (1960). *Deep Sea Res.* **8**, 1.
Welander, P. (1959). *Tellus*, **11**, 309.

Chapter 3

The Physical Properties of Sea Water

R. A. COX

National Institute of Oceanography
Wormley, Surrey, England

I. Introduction 73
II. The Definition of Chlorinity and Salinity 75
A. Historical Summary 75
B. Introduction of Physical Methods for Salinity Determination 78
III. Physical Methods for the Determination of Salinity 80
A. General Introduction 80
B. Conductivity Methods 81
C. Direct Density Methods 97
D. Refractive Index Methods 99
E. Sound Velocity Methods 101
IV. The Density of Sea Water 101
V. The Electrical Conductivity of Sea Water 103
VI. The Refractive Index of Sea Water 107
VII. The Velocity of Sound in Sea Water 109
VIII. The Compressibility of Sea Water 111
IX. The Transparency of Sea Water 112
X. The Freezing Point and Osmotic Pressure of Sea Water 115
XI. The Specific Heat of Sea Water 116
XII. The Surface Tension of Sea Water 117
References 117

I. Introduction

All the physical properties of sea water depend on the concentration, and to a lesser extent on the relative proportions, of the ions present. Thus, although this is a book essentially devoted to chemical oceanography, some account of the physical properties of sea water is necessary, with particular reference to their variations with salinity.

The determination of salinity by way of the chlorinity titration has been, until recently, one of the most difficult problems confronting the marine chemist. However, the determination of salinity is of special importance, because of its use (in conjunction with temperature and pressure) as a definitive parameter for sea water. Consequently, no apology is made for devoting much of this chapter to the definition and significance of chlorinity and salinity, and to an account of the physical methods now in use for salinity determination. Other important physical properties are discussed more briefly, emphasis being given to the effects of salinity variations.

Tables showing the variation of some of the more important physical properties with salinity, temperature and occasionally pressure are included. Some of these tables were not previously available in a convenient form; they are based on the best observations available, and every effort has been made to ensure that they are accurate. Where values quoted in the tables are known to be extrapolations significantly beyond the range of the observations, this is indicated by parentheses. In all the tables, values are referred to salinity rather than chlorinity. This preference has been shown to be fundamentally sound (Cox *et al.*, 1962) for conductivity and density, and with the growing importance of physical determinations of salinity it is to be expected that chlorinity will gradually go out of use as a basis for computing physical properties.

In over 97% of the sea water in the world, the salinity lies between 33‰ and 37‰. The tables accordingly cover this range in greater detail, but when they are available, values have also been included for high and low salinities. It should be noted that for waters of low salinity, where sea water has been considerably diluted, the chemical composition will be affected by that of the diluting water. The principle of the constancy of relative proportions (probably first propounded by Marcet, 1819) is a reasonable approximation for ocean waters, but it is certainly less reliable at salinities below 30‰.

Present knowledge of the physical properties is not as complete as would be desirable. Some of the tables commonly given in reference books are, in fact, based on measurements having doubtful validity, and often it is difficult to locate the original data. The best determinations of the specific gravity of sea water were made around 1900, when the existence of isotopes was unknown, and the volume of 1 kg. of water was assumed to be exactly 1 dm^3. The figures usually designated "density" of sea water are, in fact, ratios of the true density to that of "pure water" at 4°C. No information is available on the source of the "pure water" which was used as reference in these earlier measurements; all that is known of its preparation is that it was "repeatedly redistilled". It is now known that repeated distillation can change the density of water by as much as 20 parts in 10^6, by causing fractionation of the isotopes of oxygen and hydrogen; nevertheless, oceanographers often report densities to 1 part in 10^6. As far as possible in this chapter, the uncertainties in the measurements have been assessed, and the probable errors in the data presented in the tables have been indicated.

II. The Definition of Chlorinity and Salinity

A. HISTORICAL SUMMARY

Salinity is the definitive property of sea water by which it is distinguished from most other natural waters. It is unfortunate that its direct determination is so difficult, as this has introduced into the oceanographic records a mass of data based on doubtful approximations and unchecked assumptions which it is not possible to disentangle completely. If sea water is evaporated to dryness, the residue is a complex mixture of inorganic salts, together with a small but variable proportion of organic material. If this residue is heated, water is progressively lost as the temperature is increased. The mixture of salts contains some magnesium chloride, which decomposes when heated, evolving hydrogen chloride and leaving a basic salt of uncertain composition. As the temperature increases, part of the organic matter evaporates, and the remainder is later oxidized. The weight of the residue will depend on the treatment, and is not easily made reproducible; several workers (see, for example, Marcet, 1819, 1822; Forchhammer, 1865; Dittmar, 1884) found difficulty in determining salinity. At an international conference held in Stockholm in 1899, a commission of experts was appointed to investigate and advise on the definition and determination of salinity. The practical work was done under the supervision of Professor Knudsen, and reported in 1902 (Forch, Knudsen and Sørensen, 1902). Sørensen devised a method for salinity determination which is slow and difficult, but gives reasonably reproducible results. The commission recommended a definition of salinity based on this method to a second international conference held in Kristiania (Knudsen, 1901):*

"Salinity is defined as the weight in grammes of the dissolved inorganic matter in 1 kg. of sea water, after all bromide and iodide have been replaced by the equivalent amount of chloride, and all carbonate converted to oxide."

* The definition was given in two slightly different forms. In the official report of the commission (Knudsen, 1901) the wording is:

"Als die Salzmenge ist die Gewichtsmenge der in 1 kg. Meerwasser enthaltenen anorganischen Salze zu verstehen, wenn alles Bromid und Jodid durch eine äquivalente Menge Chlorid und alles Carbonat durch eine äquivalente Menge Oxyd ersetz wird."

However, in the scientific report (Forch *et al.*, 1902), the wording is:

"Unter Salzmenge versteht mann dann im folgenden: Die Gewichtsmenge der in 1 kg. Meerwasser befindlichen, gelösten fester Stoffe, mit der Beschränkung, das man alles Brom durch eine äquivalente Menge von Chlor ersetz, alles Karbonat in Oxyd umgebildet und alle organischen Stoffe verbrannt dekt."

The slight differences between these definitions are not of practical significance.

This direct determination of salinity, in accordance with the definition, has remained of theoretical interest only, as it is too difficult and slow for routine use. The only serious attempt to improve the method was the ingenious suggestion of Guntz and Kocher (1952) that the determination could be simplified by adding an excess of sodium fluoride to the sample before evaporation. This precipitates the alkaline earth metals as insoluble and stable fluorides and allows the residue to be ignited without loss of hydrogen chloride. This method has recently been further developed by Morris and Riley (1964) whose findings are mentioned in Chapter 21 (Volume 2).

The international commission also investigated alternative means of determining salinity. In view of the recent developments in electrical conductivity measurements, it is of historical interest that the commission investigated this technique (Knudsen, 1900) but decided against it, probably because of the inadequacy of the equipment available at that time. However, the commission measured the density, salinity and chloride content of a number of samples of sea water, and found a reasonably constant relationship between chloride content and salinity. They defined a new parameter, chlorinity, as the halide concentration in parts per mille by weight, measured by reaction with silver nitrate, and computed on the assumption that all the halide is chloride. They quoted an arithmetical expression relating chlorinity to salinity:

$$S\text{‰} = 1{\cdot}805Cl\text{‰} + 0{\cdot}030.$$

This expression is based on determinations of chlorinity and salinity on nine samples of sea water (all from the surface). Two of these were low salinity samples from the Baltic Sea, four were intermediate, Baltic–North Sea mixed waters, one was a high salinity Red Sea sample, and only two were typical oceanic waters. These nine observations, when plotted, form a close approximation to a straight line; the standard deviation of the points from the best line is 0·01‰ in salinity, the largest individual deviation being 0·022‰. If Sørensen had included deep-water samples in his series, the deviations would probably have been greater; it is unfortunate that he did not make use of more of the samples which Knudsen was using for his density determinations. However, the commission seem to have concluded that the relative simplicity of the chlorinity determination represented so great an advantage that it would be best to standardize on this method, and in effect abandon the direct salinity determination as of theoretical interest only. The resolutions of the 2nd International Conference on the Exploration of the Sea (Reports, p. 4) make this quite clear.

> By salinity is to be understood the total weight in grammes of solid matter dissolved in 1000 grammes of water. The ratios between Salinity, Density and Chlorine given in Dr. Martin Knudsen's Hydrographic tables are to be adopted; and the salinity is to be calculated by the use of these tables from the determinations of chlorine or from the specific gravity.

The phrase "dissolved in 1000 g. of water" is a mistranslation, as the German version reads "in 1000 g. Seewasser gelöst". It is important to realize, of course, that the weight in 1 kg. of sea water is meant, rather than the weight in 1 kg. of pure water.

It is evident from this resolution that it was the decision of the International Conference that the Knudsen tables should be accepted as the practical definition of "salinity". Apparently, this resolution has never been revoked or superseded. It is unfortunate that the conference did not recommend that observations should be published as chlorinities. The wording of the resolution suggests that the chlorinity titration and direct measurements of density (by hydrometer) were to be considered as alternative methods, and the term "salinity" was retained as a common basis for reporting the results. In practice, the chlorinity method was generally adopted as standard throughout the world. In Europe and countries bordering on the Atlantic Ocean it has been usual to convert chlorinities to salinities using Knudsen's tables; in those countries bordering on the Pacific Ocean (including the West coast of the United States) it has been the accepted practice to report chlorinities as such.

This rather illogical situation has proved reasonably satisfactory, until recently. The Standard Sea-Water Service was established (initially under the control of the International Council for the Exploration of the Sea, and later of the International Association of Physical Oceanography), and has continued to supply the World's oceanographers with a chlorinity standard of exemplary reliability.

In 1937 a new basic standard sea water was under preparation, and it was then realized that changes in the accepted atomic weights of silver and chlorine since 1900 would make a significant difference to the chlorinity value. To overcome this difficulty, the definition of chlorinity was changed to make this concept independent of atomic weights (Jacobsen and Knudsen, 1940). The revised definition is:

> The chlorinity is the mass in grammes of pure silver necessary to precipitate the halogens in 328·5233 grammes of the sea water.

This definition preserves the Knudsen relationship between salinity and density, but it means that on the basis of current atomic weights, chlorinity does not now exactly represent the weight of "equivalent chloride" in the sample. On the basis of 1963 atomic weights, the chlorine equivalent is slightly greater than the chlorinity, in the ratio 1:1·00043. This ratio will be important if precise chlorinity determinations are ever made using sodium or potassium chloride as standard rather than standard sea water or silver.

Salinity and chlorinity are weight ratios, and thus are independent of temperature. However, determinations of ionic concentrations in sea water are usually reported in weight per unit volume. To facilitate the direct comparison of such results with the chloride concentration, some workers have preferred to report chloride determinations as *chlorosity*, defined as the weight of chloride and bromide in *one litre* of sea water at 20°C, when bromide is replaced by the equivalent amount of chloride. Chlorosity can be computed by multiplying the chlorinity by the specific gravity at 20°C. The term has seldom been used in recent years.

B. INTRODUCTION OF PHYSICAL METHODS FOR SALINITY DETERMINATION

During the last ten years, the "salinity" concept has been further complicated by the widespread adoption of physical methods for the determination of salinity. Of these, electrical conductivity is at present most favoured, but direct density, sound velocity and refractive index methods are all attracting attention. These developments have been encouraged by the demand for speed and convenience when large numbers of samples are taken. The chlorinity determination is an extremely precise measurement when made by a careful and skilful analyst, but such workers are scarce. Moreover, although it is quite possible to determine chlorinity at sea (this was done during all the *Meteor* surveys, and *Discovery* surveys up to 1955), the problems of ship motion undoubtedly make the operator's task more difficult.

Conductivity measurements were first used on a routine basis for the determination of salinity on the vessels of the International North Atlantic Ice Patrol, using an apparatus developed by Wenner *et al.* (1930). Although the original apparatus was slow and rather inconvenient to operate, it proved capable of yielding results as reproducible as those of a skilled analyst using the chlorinity titration, and it could be satisfactorily handled by relatively unskilled personnel. Since 1955, the conductivity method has been widely adopted, as will be described below; the introduction of this method however, presents several difficult problems.

Knowledge of the relationships between chlorinity and the physical properties of sea water is, at present, not entirely adequate. Results obtained by determining the halide alone (chlorinity) can be compared with those obtained by conductivity methods, only if certain assumptions are made. The first is that a constant relationship exists between chlorinity and conductivity or, in other words, all sea waters having a chlorinity of $z‰$ will have a conductivity of y mho/cm. This is likely to be true only if the relative proportions of the major ions are constant, for any given value of the salinity.

This principle of "constant relative proportions" is often stated in text books on oceanography as a proven fact. There is no doubt that it is a reasonable approximation, but it is no more than that. The most comprehensive analyses of sea water available until recently were those made on samples obtained on the *Challenger* Expedition and analysed under the direction of Professor Dittmar of Glasgow University (Dittmar, 1884). Dittmar's conclusions are worth quoting:

> When we compare the percentages of the several components with the respective means, we frequently meet with differences which lie decidedly beyond the probable limits of the analytical errors; hence, the variations must be owing partly to natural causes.

Although considerable variations were found, Dittmar could find no correlation between composition and the geographical distribution. However, he did find a strong correlation with depth, in that most deep samples contained relatively more calcium than did surface samples. This last observation has been confirmed by several later workers (Moberg and Revelle, 1937).

Some years ago the British National Institute of Oceanography started a major programme of investigation into the variations in chemical composition and physical properties of sea-water samples from all parts of the world. About 400 samples have been collected, and so far the analyses have, in general, confirmed Dittmar's conclusions. Variations exist in the relative proportions of some of the major ions in sea water, and these variations often exceed the analytical errors of the determinations (see Chapter 4). Probably because of this, it is not possible to convert a conductivity measurement into a chlorinity value within the precision of measurement. It seems that conductivity is very closely related to density, but there is much less certainty in the relationship between chlorinity and density. Taking a single sample of deep Mediterranean water, and computing its chlorinity from a conductivity measurement, the probable deviation between the result and the true chlorinity is 0·004‰. With a surface Pacific sample, it is about 0·008‰ (Cox *et al.*, 1962.)

The International Association of Physical Oceanography has recommended, on the advice of an international committee of scientists (UNESCO, 1962), that:

(1) standard sea water shall be certified in conductivity as well as chlorinity;

(2) salinity shall be redefined as a relationship to density at 0°, this relationship being so chosen that the Knudsen equation connecting chlorinity and salinity be preserved as closely as the natural scatter will permit;

(3) new oceanographic tables be made, based on the empirically determined relationships between salinity, so defined, and conductivity, and between salinity and refractive index.

An apparatus is under construction at the National Institute of Oceanography which, it is hoped, will be suitable for the calibration of the standard sea water in terms of conductivity. It is hoped that its use will place the conductivity method of salinity determination on a sound basis.

For a fuller account of the problems involved in salinity determination, and the definitions, the reader is referred to the papers by Carritt and Carpenter (1959), Lyman (1959) and Cox (1963).

III. Physical Methods for the Determination of Salinity

A. General Introduction

Physical methods of salinity determination offer the advantages that they do not require skilled operators and that they are convenient and rapid. Unless the analyst is exceptionally competent and conscientious, the standard deviation of chlorinity results obtained by the standard procedure will not be less than 0·015‰, and one analyst can make only about 50 determinations per day. The precision can be improved considerably by weighing the sea-water sample, by using a weight-burette and by other refinements, but the determination is still slow and expensive.

More than 100 samples per day may be collected on a modern research ship working closely spaced hydrographic stations. It is desirable that the salinity analyses should be completed within a few hours, so that the results are available to guide hydrologists when planning the next day's work. An *in situ* salinity recorder, which would dispense with most of the water sampling would be even more desirable.

At present (1964), this field is developing rapidly and details of experimental methods reported in this chapter will be outdated in a year or two. In consequence, most of this section is devoted to the theory behind the methods of measurement, and to the relationships between the various physical properties.

Provided that the physical relationships are adequately known, it is theoretically possible to compute salinity, density or any other desired parameter of sea water from three basic measurements, temperature, pressure and one other variable. The third variable could be chlorinity, density *in situ*, density at a defined temperature, conductivity *in situ*, etc. Thus, any apparatus which measures either the three essential parameters, or measures one or more while the others are maintained constant, may in a sense, be called a "salinometer". For this reason, some consideration is given in this section to a number of instruments, such as specific gravity meters and sound velocity meters, which come within this general classification, as well as those specifically designated as salinometers.

B. CONDUCTIVITY METHODS

(1) *General Discussion.* Conductivity is the parameter which has been most used in the physical determination of salinity. Although the technique has been known for at least 60 years (Knudsen, 1900), it has only come into favour with the development of reliable audio-frequency oscillators and power amplifiers.

Conductivity is roughly proportional to salinity, and therefore, to attain a precision of salinity measurement of 0·001‰, the conductivity must be measured to one part in 40,000. Conductivity is also dependent on temperature; with a sea water of salinity 35‰, a change of 1 in 40,000 in the conductivity is produced by a temperature change of 0·001°C. Thus, it is necessary either to make precise simultaneous measurements of the temperature, or to provide some form of temperature control.

It is theoretically possible to measure the conductivity of solutions with direct current, but it is difficult in practice. The solution becomes contaminated with products of electrolysis, and the electrolysis consumes energy, the amount of which is difficult to compute. Moreover, although direct currents and voltages are easy to measure, they are more difficult than alternating currents to amplify. Because of the capacity between the leads, it is not possible to lead a.c. signals down long cables without power loss; this can easily be done with d.c. Thus, although it has not as yet been applied, the d.c. method may have applications in deep water *in situ* measurements.

Alternating current measurements use either electrode or inductive methods, depending on how the current is generated in the solution. In the former system, which is more usual, two or more electrodes are immersed in the solution, and an a.c. current is passed between them. This system presents two serious practical problems—polarization and electrode fouling. In a variant of this system, the electrodes are isolated from the solution by a layer of glass or other insulating material which forms the dielectric of a capacitor. A very high frequency current, usually in the megacycle/second range, is necessary. By this means, electrode fouling is eliminated but polarization, while reduced in significance by the high frequency, is still present. In the second basic system, the current is induced by making the solution form a conducting loop in a transformer. This system eliminates both polarization and fouling, as there are no electrodes and there is no electrolysis. It is fundamentally sounder than the electrode system, and eventually will almost certainly replace it for all precision work.

For precise measurements of conductivity, "bridge" circuits are used. The solution is confined within a cell of fixed dimensions made of glass or other insulating material, and its resistance is compared with that of a reference or standard resistance. The bridge circuit is always arranged so that at the balance point the detector indicates either zero or minimum potential. Since no current passes through the detector at the balance point, a very sensitive indicator may be used which need not give a linear response. In simple bridges, headphones are generally used as the detector, but in more refined circuits, it is usual to use an amplifier linked to either a meter or a cathode-ray tube. The balance may be achieved by varying one or more resistors, as in the classical Wheatstone bridge, or by varying a voltage source, while keeping the resistance constant. Typical circuits will be discussed later in this chapter.

In precision conductivity measurements, one of the basic problems is temperature control. In "bench" measurements, it is usual to immerse the solution under test in a thermostat. Many salinometers take this principle one stage further, by immersing a second similar cell containing sea water in the same bath, and using this second sample as the reference resistance. Provided that the two samples have the same temperature coefficient of conductivity and that both respond at the same rate to the temperature fluctuations of the bath, then a relatively crude thermostat controlled to perhaps 0·2°C is adequate. Unfortunately, the temperature coefficient of conductivity of sea water varies appreciably with salinity (see Cox, 1962), and therefore if there is a considerable difference in salinity between the "reference" water and the "unknown", more precise temperature control is needed.

In a modification of this principle, a thermistor or a resistance of nickel or platinum wire is used instead of a second water sample. By suitably shunting either the water sample or the standard resistor with non-temperature-sensitive resistors, it is possible to make the temperature coefficients of both the same. However, complete compensation for temperature variations can be attained only for water of a particular salinity. If this system is used without a thermostat, so that not only salinity but also temperature may vary considerably, errors are introduced for which the simple arrangements described cannot compensate, and corrections may be necessary to the results.

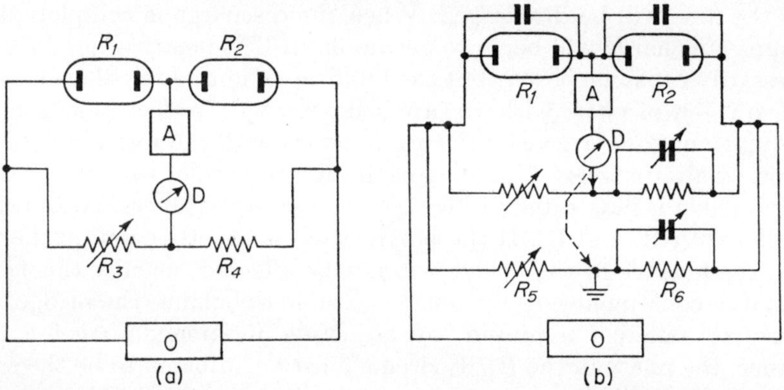

FIG. 1(a). Basic circuit of resistance bridge for conductivity measurement; (b) improved bridge circuit, incorporating Wagner earth.

Figure 1 illustrates the basic resistance bridge circuit, still used, with various modifications, for most measurements of conductivity. In the simple circuit shown in Fig. 1(a), R_1 and R_2 are similar glass cells fitted with platinum electrodes and immersed in the same thermostat. Normally, R_1 contains the standard, a reference sea water, and R_2 contains the "unknown" under test. Let us assume that cell R_1, full of sea water, has a resistance of exactly R_1 ohms (where resistance is indicated by italic type). R_4 is a fixed resistor of known value, and R_3 a variable resistor such as a resistance box. O is an oscillator (500 to 5000 cycles/second low voltage output), A is an amplifier (which may be omitted for approximate work) and D a detector, which may be an a.c. meter, a telephone receiver, a cathode-ray tube or any other device which will indicate potential. If R_3 is adjusted until the detector shows a zero (or minimum) reading, then $R_1/R_2 = R_3/R_4$. Since conductivity is the reciprocal of resistance, the conductivity ratio is given by $\kappa_2/\kappa_1 = R_3/R_4$. The term κ_2/κ_1, the ratio of the conductivity of the unknown to that of the standard, is termed the relative conductivity or conductivity ratio. Normally, for oceanographic work, the standard is chosen to have a salinity of exactly 35·000‰.

Unfortunately, although the circuit shown in Fig. 1(a) may suffice for simple measurements, it suffers from several sources of error which must now be considered. The most important of these is the polarization effect. During each half cycle, while the current is flowing, there is an accumulation of positive ions on one electrode and negative on the other. These ions act, in effect, as a primary cell, the E.M.F. of which opposes the applied potential. After the applied potential changes sign, a current caused by the back E.M.F. will flow in the reverse direction, and the ions will be discharged. When the discharge is complete, ions of opposite charge will begin to accumulate. This polarization does not necessarily consume power, but the build-up of ions always lags behind, and so is out of phase with, the applied potential. In electronic terms, it is equivalent to a large capacitance in series with the resistance of the water, or alternatively a small capacitance in parallel with it.

This effect is illustrated in Fig. 1(b), where capacitances are shown in parallel with R_1 and R_2. If the polarization is exactly equal in the two cells, the balance of the bridge will not be affected: usually this is not so, and in consequence it will not be possible to balance the bridge, but merely to attain a minimum by adjusting R_3 alone. To get a true balance, the phase in the R_3/R_4 circuit must be adjusted to be the same as that in R_1/R_2. This can be achieved by connecting a variable capacitor either across R_3 or R_4, as shown, or across either R_1 or R_2. Whichever system is used, suitable adjustment of R_3 and the capacitor will now give a zero reading at D.

Another appreciable source of error arises from leakage to earth. In effect, there is a small but finite capacitance from every part of the circuit to earth, and a small current will flow through these capacitances. For precision measurements, it is necessary to annul these capacitance leaks by the introduction of a third "arm" (Wagner earth circuit). This, as shown in Fig. 1(b), consists of two resistors, R_5 variable and R_6 fixed, with a variable capacitor which may be connected in parallel with either. The centre point of this arm is earthed. The bridge is first balanced by adjusting R_3 and the capacitor as already described; the detector is then disconnected from the R_3/R_4 arm and connected to earth, as shown by the dotted line in Fig. 1(b). R_5 and the associated capacitor are then adjusted until D shows a minimum signal; D is switched back to R_3/R_4 and the balancing repeated. This process equalizes the earth-leaks so that the bridge gives true readings. Although this adjustment is tedious, usually the earth leaks are small in comparison with the bridge currents, and a single adjustment of the Wagner earth is generally sufficient.

Many sources of error in resistance measurement may be kept within acceptable limits by attention to the following points.

1. The optimum oscillator frequency lies between 500 and 5000 cycles/second.

2. The signal must be pure. Harmonics make balance impossible.

3. The cell resistance should be at least 1,000 Ω.

4. Careful temperature control of both cells is essential. An oil-filled thermostat introduces a minimum capacity effect.

5. Although more difficult to make, thermostats working below room temperature eliminate the risk of bubble formation which may occur if the samples are warmed.

6. Standard resistors should have a low temperature coefficient; it may be necessary to put them in the thermostat.

7. Lead resistances should be minimized by using short, heavy wires.

8. The best detector is a cathode-ray tube, two plates of which should be fed from the oscillator, and the other two from the amplified bridge signal. The resulting display shows both phase and resistance balance clearly and independently (see Cox, 1958).

9. If the Wagner earth is used, the apparatus, including all connections, should have earthed screens. In a less precise circuit without a Wagner earth, screens may be harmful and are usually better avoided.

10. Self-heating of the water samples and resistors must be minimized; the resistances should be as high as is practicable, the volume of the conducting part of the cells should be large, and the voltage across the bridge as low as the sensitivity of the detector will permit.

Figure 2 shows the basic circuit of an alternative system (the transformer bridge) which is being increasingly adopted. T_1 is a transformer

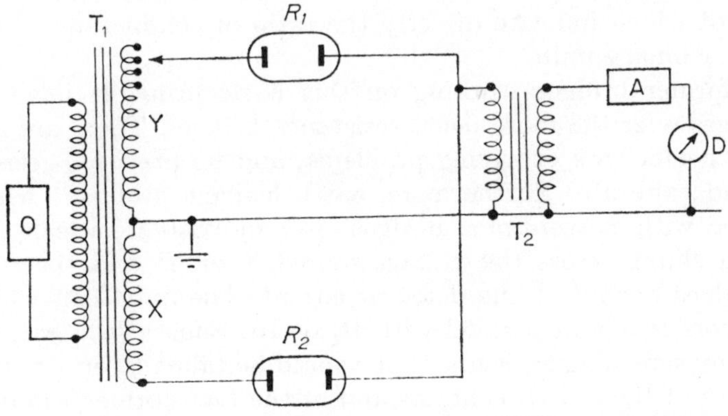

FIG. 2. Basic circuit of transformer bridge for conductivity measurement.

driven by an oscillator O; it has two secondary windings, X and Y, connected in series and earthed as shown. If the transformer is properly designed, these windings produce voltages which are in exactly the same ratio as the respective number of turns. R_1 and R_2 are cells containing the unknown and the standard waters respectively. If the number of turns in winding Y is now adjusted until $Y/X = R_1/R_2$, there will be no potential across the output transformer T_2, and the detector D will show zero signal. Since conductivity is the reciprocal of resistance, $Y/X = R_1/R_2 = \kappa_2/\kappa_1$.

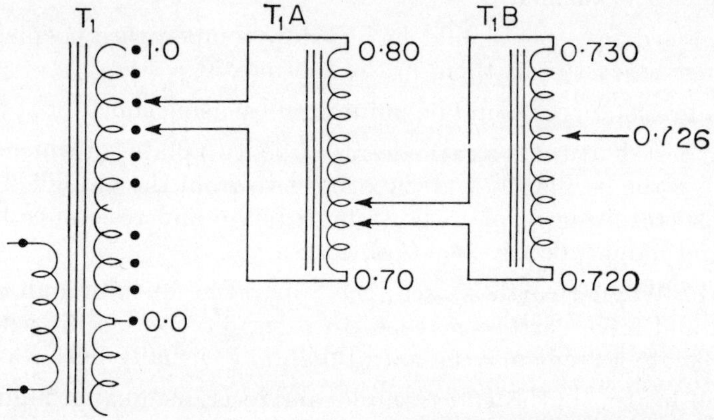

FIG. 3. Modified transformer bridge, showing decade transformers.

In practice, to attain the necessary discrimination without a very large number of tappings on T_1, the circuit is modified as shown in Fig. 3. A series of inter-tapped transformers is used in place of the single transformer T_1 in Fig. 2; these are connected to rotary switches, the settings of which indicate directly the ratio of conductances either in decimal or binary units.

Transformer bridges working on this basic principle have several advantages over the more usual resistance bridges. There are no temperature coefficients or ageing problems, and no precision resistors to make and calibrate. Furthermore, earth leakage problems which are associated with conventional bridges are non-existent; a leak to earth acts as a shunt across the voltage-source X or Y, and the resulting current does not affect the detector circuit. The only leak which can cause errors is one in parallel with R_1 or R_2. Such a leak can best be avoided by screening all leads. Care should be taken to ensure that the resistances of the switch contacts and of the transformer windings are low relative to R_1 and R_2.

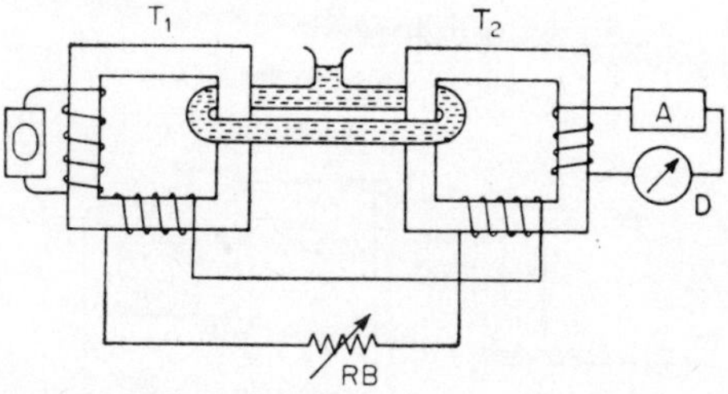

FIG. 4. Inductively-coupled conductivity cell (diagrammatic).

Both types of bridge can be used with inductive cells. Figure 4 shows the basic circuit of a simple inductive system, which uses standard resistors. T_1 and T_2 are two transformers, with cores of high-permeability metal. The oscillator O excites an alternating flux in T_1, which induces current in the loop of sea water confined within a glass tube. This current similarly excites T_2, which produces a signal amplified by A and indicated by the detector D. Another secondary winding on T_1, with x turns, produces a current which passes through the resistance box RB and a winding of y turns on T_2. The direction of the current through T_2 is chosen to oppose the flux generated by the sea-water loop. When RB is adjusted so that no signal is detected by D, the resistance of the sea-water loop equals RB/xy.

It would be quite feasible to use a similar system to make a direct comparison between two sea waters, as is customary with electrode cells; but perhaps the most promising field for the inductive system is in the construction of *in situ* salinometers. It is particularly simple to construct an inductive cell for such use; a typical arrangement is shown in Fig. 5, which shows a section through the cell. The two transformers consist of rings of metal with toroidal windings. The two rings are completely embedded in epoxy resin, one above the other, and are screened to prevent direct pick-up between them. A hole is left through the axis of the casting, or a short length of glass or silica tube may be cast in position; such an insert is easier to clean and more stable in dimensions than a cell made of resin alone. When the cell is immersed in the sea, the water forms a "loop", almost all the resistance being in the axial hole as indicated in Fig. 5.

(2) *Bench Salinometers*. At least twenty types of bench salinometer have been described in the last ten years, and about ten have been produced commercially. The salient points of some representative instruments will be discussed in some detail, giving particular emphasis to the principles of their construction and operation.

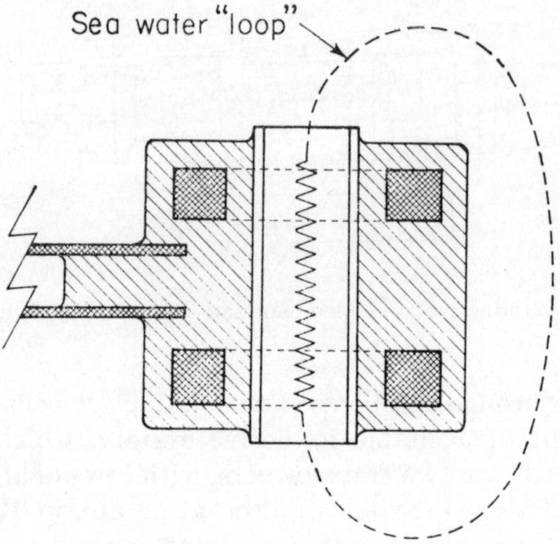

FIG. 5. Layout of inductive cell for *in situ* measurements.

A simple, but useful salinometer is the portable temperature–chlorinity bridge described by Hamon (1956) and later in a modified form by Cox and Moorey (1962). This instrument can be used both for bench and for *in situ* measurements. It has an immersion head (Fig. 6) of epoxy resin, in which is cast an open-ended tube about 12 cm. long and 0·5 cm. diameter bearing three cylindrical platinized-platinum electrodes. The outer two electrodes are connected together and earthed, and the bridge measures the resistance between them and the centre electrode. The reference resistor is a thermistor (T), also mounted in the epoxy resin head, the temperature-sensitive end being exposed to the sea water. By suitable selection of a fixed resistor in series (R_2), the temperature coefficient of the thermistor is made almost exactly equal to that of a chosen sea water. The basic circuit, together with a diagram of the measuring head, is shown in Fig. 6. The scale attached to the linear potentiometer P is computed from the known relationships between salinity or chlorinity and conductivity. The resistor R_1 is then adjusted until, with the measuring head immersed in a sea water of known concentration and the bridge balanced, the potentiometer reads correctly. If desired, P can be modified to cover several ranges of salinity by adjustment of the resistance train associated with P as shown in the detached "arm" in Fig. 6. The bridge is powered by an oscillator, and the detector is a two-stage amplifier feeding a centre-zero meter. A phase-sensitive arrangement is used so that the deflections of the meter indicate the direction in which P must be adjusted to give balance. The

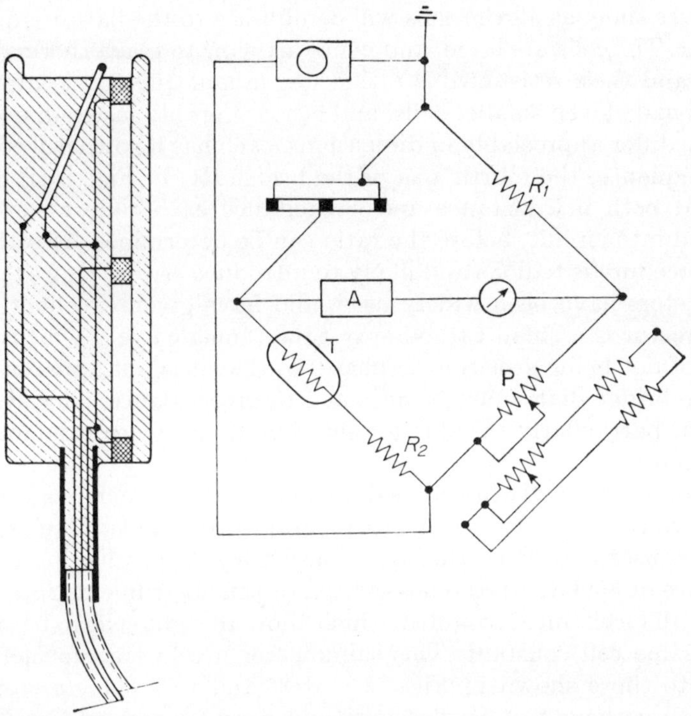

FIG. 6. Measuring head and basic circuit of the portable chlorinity-temperature bridge. (Hamon, 1955.)

latest form of this instrument, which is powered by a single small battery, uses transistor circuits and is very light and portable.

Under good conditions, this instrument has a precision similar to the routine Knudsen titration (about 0·02‰*S*), and each measurement takes about ten seconds. However, it requires fairly large samples (at least 200 ml., or more if the head is to be rinsed between measurements). Cable errors associated with this instrument are discussed in Section III, B(3).

The first precision salinometer was that designed for the Atlantic Ice Patrol (Wenner *et al.*, 1930). It is of great interest since it established some of the basic principles used in most later instruments, in particular the use of a thermostat holding the test sample and standard in similar cells. A greatly improved form of this instrument is that designed at the University of Washington (Paquette, 1959). This instrument has a 15°C thermostat housing six cells. One cell acts as the reference resistor and contains standard water, and the others which are filled with the test samples can be switched into circuit in turn. This multiplication of cells saves time, as all the cells will equilibrate to the bath temperature together. The cells are large, and hence are slow to reach thermal equilibrium, and their resistance is rather low (about 100 Ω). A later modification had eleven smaller cells, and is considerably faster in use. Since the cells differ appreciably in dimensions, each has its own cell constant; in consequence, the fourth arm of the bridge (R_4 in Fig. 1(a)) must be adjusted both in resistance and capacitance each time a new cell is switched into circuit, before the ratio can be determined by setting R_3. This procedure is tedious and likely to introduce errors. However, these salinometers have been widely used, and have given excellent results. The detector is a small cathode-ray tube ("magic eye") which has the failing of not being sensitive to phase, so that it is not apparent which way the bridge dials must be adjusted to give balance. This system is, however, particularly suitable for ship-board use, since it is unaffected by motion.

A variant of the "twin-cell" salinometer is that which has been built by the National Institute of Oceanography in England (Cox, 1958). This also uses a 15°C thermostat, housing eight cells having a typical resistance of 3500 Ω. In the latest form of this instrument, the cells are fitted with mechanical trimmers which allow all eight cells to be adjusted to the same cell constant. This salinometer uses a transformer bridge similar to those shown in Figs. 2 and 3, and two detectors are used —a small cathode-ray oscilloscope, and a centre-zero, phase-sensitive meter.

A more modern salinometer, which is tending to replace the earlier designs with their heavy and expensive thermostats, has been described by Brown and Hamon (1961). Versions of this are manufactured commercially in Australia and the U.S.A. One uses as reference resistor a thermistor, and the other a platinum resistance thermometer, immersed in the water sample. An inductive system similar to that illustrated in Fig. 5 is employed, and is mounted in a Perspex cell holding about 80 ml. of water. The sample is stirred vigorously during measurement to keep its temperature uniform and equal to that of the thermistor. The resistance of the water is compared with that of the thermistor by a transformer bridge resembling that used in the National Institute of Oceanography instrument. The lack of a thermostat and the use of a transistor oscillator and amplifier have made this salinometer smaller, lighter and less expensive than others of comparable precision. In good conditions, it is as precise as the National Institute of Oceanography model, but the precision may be limited by the variation of the temperature coefficient of conductivity with salinity. No simple compensating circuit can completely replace a thermostat if the standard water (with which the instrument is calibrated) and the unknown differ significantly in both temperature and salinity. The Australian salinometer is, however, provided with a correction table to allow for this relatively small error.

All these precision salinometers are capable of measuring conductivity, relative to that of a standard, to a few parts in 10^5. They should thus be precise to 0·001‰ in salinity, provided that the samples can be transferred to the conductivity cell without error. In routine work, it is generally found that the theoretical precision is not attained; a typical standard deviation would be about 0·002‰. This slight loss in precision arises mainly from transfer errors. For example, the sample may be contaminated by the previous sample unless great care is taken in rinsing the cell. Under conditions of low ambient humidity, the sample may evaporate during transfer. The care necessary in precise salinity determination may be indicated by the fact that if a sample of sea water is allowed to saturate its own volume of dry air with water vapour, the resulting increase in salinity is clearly detectable. Most samples of surface sea water contain an appreciable amount of oily matter which tends to accumulate in the cells in a rather irregular way and cause small shifts in cell constant. Changes in cell constant also occur with deep samples taken with Nansen reversing bottles, since the grease from the stopcocks tends to contaminate the samples.

(3) *Direct-reading or* in situ *salinometers.* This group includes the various instruments designed to record variations of salinity with time at a fixed location, or to measure horizontal or vertical variations of salinity in the sea. The developments in this important field have been particularly rapid during the last few years. As in the previous section on bench salinometers, this account will emphasize the special difficulties associated with *in situ* measurements of salinity, and show how various designers have attempted to overcome them.

All the problems inherent in the bench instrument are, of course, still present in the direct-reading salinometer, together with additional difficulties such as response time and the transmission of information. In the simplest form of the instrument, the sea water is pumped continuously through the measuring cell of a bench salinometer fitted with a temperature compensator. A basic problem is the time constant of the pumping system—the time before a change of salinity at the inlet is fully shown in the cell. Since this time is always considerable, such systems cannot record rapid fluctuations, and are best suited for showing slow processes such as tidal cycles in estuaries. In the simplest applications, it will be adequate to balance the salinometer bridge manually, exactly as for bench operation, and to note the salinity at suitable intervals of time or distance. For truly continuous operation, however, a recorder is needed.

The introduction of a recorder into a bridge circuit presents immediate problems. In Section III, B(1) it was noted that it is not necessary for the detector in a bridge circuit to give a proportional signal, but only to indicate zero. With a continuously recording instrument, this advantage can only be retained by automatically balancing the bridge by a servo-system; the servo-motor can then be arranged to operate a recorder. The servo-motor can be dispensed with in approximate measurements of conductivity by abandoning the bridge principle. A constant voltage is applied to the conductivity cell and the resulting current recorded. A more sophisticated solution is to use a circuit such as a Wien bridge in which the frequency of an oscillator is made to depend on a resistance (of the conductivity cell) and to record the frequency. If a simple bridge circuit is used and allowed to go out of balance as the salinity changes, the signal will be related to the salinity. However, it will not be a linear relationship, since as the variations become larger the signal is less in proportion to the variations. The apparently simple solution of using a bridge such as that shown in Fig. 1, but in which a recorder is used as a detector gives a record which is difficult to interpret quantitatively.

The other major problem facing the designer of an *in situ* salinometer can perhaps be called the transmission problem. It is necessary to consider the essential features of this complex subject, although an exhaustive discussion is outside the scope of this chapter. For most *in situ* applications, the sensing element or transducer must be in the sea, often a considerable distance from the ship or shore station where the record of salinity is to be made. The recorder can sometimes be located near the sensing element. Thus, in a deep-sea instrument, a tape or wire recorder, with its associated electronics, could be enclosed in a pressure-tight case and lowered on a cable from the ship. It is usually more convenient however, to locate the recorder in the ship. Then the information must be transmitted either by electrical signals along a cable, by sonic signals through the water, or occasionally (as with anchored buoys) by radio.

There is no difficulty in measuring a resistance at the other end of a long cable with direct current, but it is an entirely different problem with alternating current. Here, the capacity between the leads in the cable, and between the leads and the screen or the sea water outside, have to be considered. The longer the cable, and the higher the frequency, the more serious these effects become. These restrictions make it impossible in practice to make precision measurements of a.c. current or voltage over anything other than short distances. Frequency signals or a coded signal consisting of a series of pulses, can however be used for longer distances. Usable sound signals in the sea are also restricted in frequency, since the attenuation increases rapidly at higher frequencies. At ranges of 1 km. or more, the upper limit of frequency is about 50 kilocycles/second.

As with bench salinometers, several *in situ* instruments will now be considered in some detail, to illustrate how the various problems have been solved.

As a first example, it is appropriate to refer again to the chlorinity–temperature bridge as described by Hamon (1956) and later modified by Cox and Moorey (1962). Although this has been commonly used as a bench instrument (Section III, B(2)), it can be used for *in situ* measurements, despite the restriction of its depth range by errors which are introduced if too long a cable is used. In the original version, the cable was limited to about 50 m. Later versions with a modified circuit extended the range to 100 m. and occasionally, where the temperature range was small, to 200 m.

The errors introduced by the cable have two main causes. Firstly, the cable conductors are part of the bridge circuit, and their resistance is appreciable compared with that of the conductivity cell. Being made of copper, this resistance is temperature dependent. If the lead resistances are too large, this will cause serious errors if the water temperature changes. Secondly, the cable has an appreciable capacitance; it has an earthed screen which serves as the return lead for both cell and thermistor. The capacitance between screen and cores gives rise to a small capacity in parallel with the cell and the thermistor. As the measuring head is lowered into the sea, the pressure of the water squeezes the cable, and this capacity increases. It can be compensated at a chosen salinity and temperature by inserting a capacitance across R_1 (Fig. 6), but errors are still introduced at other settings.

To keep the cumulative errors within acceptable limits, the cable length must normally be restricted to 100 m. An exception was a salinometer designed for work in the Indian Ocean. For this investigation it was possible to specify narrow ranges of temperature and salinity, thereby reducing the cable errors, and this enabled 200 m. of cable to be used.

An earlier, but somewhat similar instrument, was that described by Wattenberg and Joseph (1943). This used a temperature compensator consisting of a cell similar to the measuring cell, filled with a reference water, sealed, and immersed in the sea. This system has been used by several other workers (Thuras, 1918; Hin Lew, 1945), and should give a very precise compensation, particularly if the reference water is similar in salinity to the sample. Unfortunately, the sealed tube is slow in reaching thermal equilibrium, and cannot follow accurately the rapid fluctuations of temperature which commonly occur in the sea. The thermistor compensator, although in theory less perfect, is in practice more satisfactory as it has a much more rapid response.

A "salinity–temperature–depth" recorder has been developed by Jacobsen (1948). This measured conductivity and temperature by means of an electrode cell and resistance thermometer respectively. Salinity was computed by a rather complex system of servo-driven non-linear potentiometers. Its depth limit was about 300 m., but cable errors and imperfections of the computing system limited its precision to 0·3‰S. The three parameters, salinity, temperature and depth, were recorded on the same chart by a three-pen recorder.

The first salinometer to use a thermistor temperature compensator was that described by Dorrestein (1956). This instrument was also the first to use a servo-operated self-balancing bridge to give direct recording on a chart. The servo-motor rotated a drum on which was wound a helix of resistance wire, which formed the variable resistor of the bridge. The out-of-balance signal was amplified to drive the motor. The motor simultaneously drove a lead-screw controlling the traverse of the recorder pen. The measuring head was attached by 50 m. of cable, and had a three-electrode cell, 70 cm. long and of 1 cm. bore, which used electrodes of gas-carbon. This relatively large cell had a resistance of about 450 Ω, higher than is usual for *in situ* meters. The bridge was supplied with a 50 cycles/second current. The high cell resistance helped to compensate for the polarization caused by this low frequency, and to reduce cable errors. This salinometer was designed primarily for towing, particularly for the study of estuarine conditions off the Dutch coast. Its accuracy was about 0·1‰ in salinity.

Several workers have described what might be called "shallow water" *in situ* salinometers which use inductive cells. Probably the first was the "Induction Conductivity Indicator" described by Esterson (1957). Esterson's report also includes an excellent study of the theory of polarization, and a complete bibliography of early work on conductivity. The instrument used an inductive head similar to that already described (see Fig. 5), and a metal resistance thermometer. It was capable of operating at depths down to about 60 m. Salinity was computed from tables. The inductive cell offers the advantage, previously discussed, of avoiding polarization. This leads to several secondary benefits in a direct-reading salinometer. Since there is no polarization, there is no advantage in keeping the water-path resistance high, and the cell may be relatively short and of wide bore. These proportions reduce self-heating and facilitate flushing and cleaning.

The design of a deep-water *in situ* salinometer is a particularly challenging problem. Transmission difficulties are extreme, since the partial solutions which are tolerable with shallow-water instruments are precluded by the greater range. Because of the much smaller variations of salinity found in the deep oceans, a considerably higher precision is necessary. Furthermore, the effect of pressure on conductivity, which is not significant at depths down to 100 m. becomes increasingly important at greater depths.

Although not primarily designed for deep-water use, the instrument described by German (1960) presents an ingenious solution to the transmission problem. An inductive head is used, and the bridge with its transistor oscillator and amplifier is in a pressure-tight housing adjacent to the head. The bridge is balanced by an electric motor which is controlled from the ship, and which drives a helical potentiometer. The electronics and motor are powered by d.c. current fed down the cable, and the out-of-balance a.c. signal is brought up the same cable to the amplifier and headphones on the ship. Since this signal is zero at balance, losses in the cable are immaterial. When the bridge is balanced, a relay disconnects the potentiometer from the bridge and connects it to a d.c. Wheatstone bridge on the ship, where its setting is easily determined. The cable resistances can easily be allowed for in a d.c. circuit or, if necessary, compensated by including an extra lead in the cable, and making this part of the opposite arm of the bridge.

The instrument also measures temperature, by means of a thermistor. No attempt has been made to provide automatic temperature compensation; readings are taken of temperature and conductivity, and salinity is determined by computation or from tables, allowance being made for pressure when necessary. The precision claimed for depths down to 180 m. is $0{\cdot}03‰ S$, and there seems to be no reason why this instrument could not be used without much modification at considerably greater depths, if a suitable loadbearing cable and slip-ring winch were available.

The first *in situ* salinometer specifically designed for deep water use was that of Hamon (1955). This was the first meter to use a frequency signal for telemetering the information to the surface. A pressure-tight sea-unit housed three transducers—a thermistor for temperature measurement, a three-electrode open cell for salinity (with a thermistor compensating circuit) and a Bourdon tube operating a potentiometer for pressure. A servo-operated rotary switch connected each transducer in turn to an oscillator. The oscillator frequency was modified by the change in resistance caused by temperature, conductivity or pressure variations, and the output was passed up the single-core screened cable to the ship. The same cable carried 50 cycles/second power to the electronic unit.

At the surface, a filter passed the high-frequency oscillator signal to the detector unit. A frequency meter converted the signal to a d.c. output proportional to the frequency, and this was fed to a single-pen recorder, which showed temperature, salinity and depth consecutively, with a cycle of about 5 seconds.

Precision was about 0·05‰ in salinity, 0·15°C in temperature, and 10 m. in depth, and the instrument could be used down to 1000 m. Thus, it was precise and versatile enough for most purposes in the upper 1000 m., but lacked the precision required for deep-sea measurements if dynamic computations of current were needed. Despite its imperfections, however, this instrument proved that the frequency telemeter was practicable, and that a sea-unit could be suspended on a relatively cheap and strong single-conductor cable.

In recent years, at least three deep-sea salinometers have been designed which work on similar principles to that described above. The only one which has been described in the literature is that built in Kiel, Germany (Siedler, 1963). The others, made in the U.S.A. are basically so similar that separate description is not necessary.

The German instrument, named the "Bathysonde", is a much more sophisticated instrument than that of Hamon. The principle advance is that three separate oscillators are used, with the three transducers for temperature, conductivity and pressure, each varying over a chosen range, but in one of three widely-spaced frequency bands. Thus, the three signals can be conveyed simultaneously up the single-conductor cable, to be separated at the surface by electronic filters. A second advance is a range-switching system for both salinity and temperature. Switches in the sea-unit which are actuated by a coded frequency signal from the ship select the appropriate resistors in the bridge circuits. A further improvement is the form of presentation of the results. The output is recorded on a flat-chart servo-operated x_1 x_2 y recorder which plots temperature and conductivity simultaneously against depth.

The three transducers comprise an inductive cell for conductivity measurement, a platinum wire mounted in a fine gold tube which acts as a resistance thermometer (Hinkelmann, 1956, 1957) and a Bourdon tube for pressure measurement. The precision of the first instrument was about 0·02°C, 0·04‰S and 1% of the depth range, with a limit of 2000 m. These values have been improved in later models.

This salinometer is a versatile and valuable tool for the oceanographer. It will dispense with much routine sampling and analysis, and allow salinity profiles to be recorded in greater detail than has hitherto been possible. The considerably greater detail available and the large saving in ship-time should more than off-set the high cost of the equipment.

One of the American instruments (Brown, personal communication), while generally similar to the "Bathysonde", has certain refinements. In particular, it has a computer which compensates both for temperature and pressure and thus provides a direct salinity reading which is independent of the other variables. This instrument, like that of Siedler, presents the data on an $x_1\ x_2\ y$ plotter.

It is difficult to forecast future developments in this field. A great research effort in space projects has been directed towards the development of radio-telemeter systems. In many ways, the problems are very similar to those facing the designer of *in situ* salinometers and include difficulties with power supplies, the need to reduce the electronics to the smallest possible bulk and weight, and the need to transmit accurately and reliably much information. It is likely that many of the technical advances accruing from space research will be applicable to oceanographic problems. If the necessary information could be coded into a series of pulses which, in turn, modulate a carrier wave, they could be passed to the ship by a sonic system. Vertical transmission of sonic signals needs very little power and it would, therefore, be possible to make the sea-unit self-contained and dispense with expensive electric cables and winches.

C. DIRECT DENSITY METHODS

No instrument has yet been designed for direct *in situ* density measurement. A possible system whereby this could be achieved has been described by Richardson (1959) in which a hollow rod, filled with the water, is induced to vibrate by an electromagnet driven by an oscillator. The frequency which induces resonance is related to the mass of the tube, and therefore to the density of the water. There are serious problems, the most difficult being the considerable effect of temperature on the dimensions and mechanical properties of the tube. Apparently the idea has not been developed.

Numerous methods have been devised for bench measurements of density. The water can be weighed in a pyknometer, as in the measurements on which Knudsen's tables are based (Forch *et al.*, 1902). This method is potentially accurate but slow. Ordinary stem hydrometers are not sufficiently accurate for anything but approximate work. Various types of immersion hydrometers have been described of which several use a plummet suspended from a balance (Cummings, 1932;

Rochford, 1954; Kalle, 1957). All these suffer from the effect of surface tension on the wire suspending the plummet. Even though this error can be reduced by using a very fine wire, the results are affected by grease on the surface of the water or on the wire. The apparatus described by Kalle, which is probably the most refined, uses a torsion balance from which is suspended a quartz plummet weighing about 6 g. Density can be determined to about 1 in 10^5, representing a precision in salinity of nearly 0·01‰.

Other workers have avoided the wire difficulty by using free-floating total immersion hydrometers. Nansen (1900) used a hydrometer ballasted with small weights. It was accurate, but too slow for routine use and needed a very large sample. Stott (1928) described an elegantly simple apparatus. A small glass tubular sinker was adjusted so as to be just lighter than water and was immersed in the sample, which was slowly heated until the density of the water and the float were identical, as indicated by neutral buoyancy of the float. A graph was prepared relating this "equilibrium temperature" to the salinity. Precision in the determination of salinity was about 0·02‰, but the method was slow and suffered from shortcomings arising from the evolution of air-bubbles which often adhered to, and buoyed up, the float. The most useful immersion hydrometer has proved to be that described by Cox (1954, 1957) which used a small sinker made either of glass or Monel alloy. This was made slightly less dense than water, but was slightly compressible; it was immersed in the sample and slight pressure was applied until the float was neutrally buoyant. This pressure, measured with a mercury manometer, is a measure of density. A simple mechanical device compensated for temperature variations, and gave a direct reading in salinity. This instrument was accurate to about 0·02‰, and was easy and rapid to use. It has been used extensively, particularly in the Soviet Union, but has been largly supplanted for routine use by conductivity salinometers. A refined form of this apparatus which employs a thermostat and an optical system to detect movement of the float, was used by Cox *et al.* (1962) to determine density at 0°C. It measured differences in density between samples to about 2 parts in 10^7.

The future use of direct density devices for routine measurements seems likely to be confined to those laboratories where only small numbers of measurements are required, and where use of the more expensive conductivity meters is not justified. With the possible exception of Stott's apparatus, none of the direct density meters will work satisfactorily on a moving ship.

D. REFRACTIVE INDEX METHODS

The refractive index of sea water is a function of salinity, temperature and wavelength. The variation with salinity is small, but with suitable apparatus the precision of measurement is high. Refractive index has certain advantages over conductivity as a measured property. The effect of temperature variations on refractive index is small, and the temperature coefficient does not vary much with the salinity.

These points have an important practical application. When comparing the refractive index of an unknown sea water with a standard, provided that both samples are at the *same* temperature, the actual temperature can vary by as much as $\pm 2°C$ without producing significant error. This makes possible the use of a precision comparison refractometer for most work, without the need for a thermostat. Usually a room maintained at temperature constant to 1°C is adequate.

Since an increase in salinity of 1·0‰ increases refractive index by approximately 0·0002, a refractometer precise to 0·00001 will determine salinity to about 0·05‰. This is usually taken as the limit of precision of the best Abbé and Pulfricht refractometers, which are thus not accurate enough for normal work with sea water. The maximum precision of these instruments can be attained only if care is taken to maintain the samples at a constant temperature. With the Abbé refractometer difficulties are also caused by water loss from the sample by evaporation.

To gain higher precision, a comparison instrument, as indicated above, must be used. Such instruments fall into two general classes—prismatic instruments and interferometers. Differential prismatic refractometers can be made with a precision of better than 3×10^{-6} (Brice and Halwer, 1951), and should therefore be capable of measuring salinity to 0·015‰. However, no such instrument seems to have been used for routine determination of salinity, despite its ease of operation compared with that of the interferometer. This may be a reflection of the general availability of commercial interferometers. It would seem that this is a useful field for further research.

Interferometers have been used extensively for salinity measurement, and Pape (1924) has described the use of such a meter. One was used extensively on the *Meteor* expedition (Bein *et al.*, 1935) and a commercial "water interferometer" was manufactured by the firm of Zeiss. More recently, the Russian-built ITR-2 interferometer has been used on the *Vityaz* and other Soviet research vessels (Vel' Mozhnaya, 1962). This instrument is claimed to distinguish between waters differing in salinity by only 0·001‰, although calibration difficulties limit its working accuracy to about 0·01‰.

All precision interferometers suffer from an optical defect termed *fringe jumping*. The difference in refractive index between the standard and the unknown is compensated by adjusting the angles of glass plates mounted in the two light beams. Since the dispersion of glass differs from that of water, the apparent centre of the fringe systems is displaced, and must be aligned before the measurement is made. This displacement may introduce an error into the reading corresponding to one or more fringe-widths. A typical calibration of an interferometer against salinity given in Fig. 7 shows the graph to be a series of discontinuous straight lines, separated by the fringe jumps.

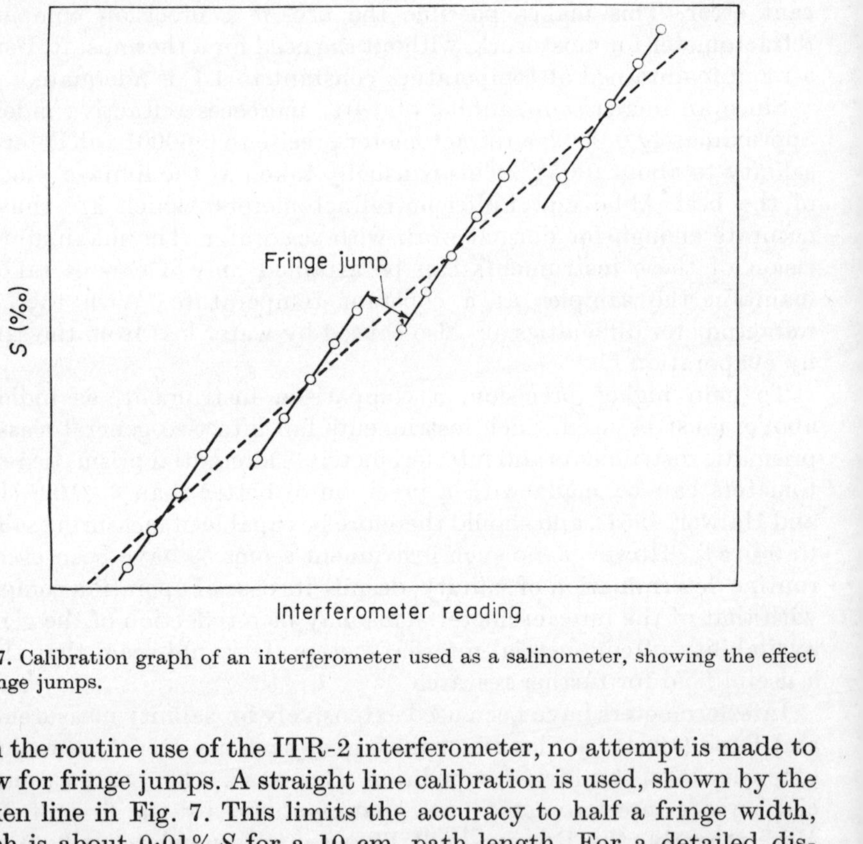

FIG. 7. Calibration graph of an interferometer used as a salinometer, showing the effect of fringe jumps.

In the routine use of the ITR-2 interferometer, no attempt is made to allow for fringe jumps. A straight line calibration is used, shown by the broken line in Fig. 7. This limits the accuracy to half a fringe width, which is about 0·01‰*S* for a 10 cm. path length. For a detailed discussion of the fringe jumping problem see Faust and Marrinan (1955).

The refractometer is potentially a most useful instrument for salinity determination, and it is difficult to see why it has not been more generally adopted. If certain minor problems in calibration could be overcome, it would compare favourably with the conductivity method in convenience, speed and economy, and is adequately precise for most purposes.

E. SOUND VELOCITY METHODS

The velocity of sound in sea water depends on salinity, temperature and pressure. However, the effect of salinity is relatively small compared with that of the other variables. The velocity in sea water is typically about 1500 m./sec., the change with temperature is 3 m./sec./°C, with pressure is 1·8 m./sec./100 decibars, and with salinity is only 1·3 m./sec./‰*S*. It is therefore unlikely that a sound velocity meter could be made to give precise salinity measurements.

In echo-ranging and echo-sounding problems, where the sound velocity is the parameter ultimately required, there are obvious advantages in measuring this property directly. Several direct velocity meters have been described (see, for example, Tschiegg and Hays, 1959; Shaffer, 1960), but in their present state of development, the limit in precision seems to be about ±0·2 m./sec. which corresponds to a precision of only ±0·2‰, in salinity.

In situ velocity meters generally use a transducer driven by a high-frequency oscillator. The sonic signal is received by a second transducer mounted a short distance away. The frequency of the oscillator is adjusted until the input and output signals are exactly in phase. If the distance between the transducers is known, the velocity can be computed from the frequency. The system is useful, since the frequency signal can easily be read at a considerable distance from the measuring head. The effects of pressure and temperature on the physical dimensions of the apparatus pose problems in the design and use of these instruments.

IV. THE DENSITY OF SEA WATER

The figures often reported as "densities" of sea water are not in fact densities in absolute units (g./cm.^{3}). No determination of the absolute density of sea water has been published in the literature. The various tables for computing density are based on measurements of the specific gravity of sea water by Forch *et al.* (1902). These measurements were based on the assumption that the density of "pure water" at 4°C was unity. The densities computed from these tables are thus dimensionless ratios. Attention was drawn to this fact by Pollack (1961), who also pointed out that the use of "specific volume" rather than density has introduced further confusion. When this term was introduced by

Bjerknes and Sandström (1910) they mistakenly applied the units of cm.^{3}/g. to the reciprocal of the specific gravity.

However, Pollack was mistaken in his suggestion that the difficulties could be solved by multiplying specific gravities (and dividing specific volumes) by 1·000028 (the internationally accepted figure for the volume of the litre in dm.^{3}). This correction would be valid only if both the specific gravity of sea water and the volume of the litre were based on "pure water" having the same absolute density. This assumption is open to considerable doubt, since both measurements were made before the discovery of isotopes, and it is now known that the density of pure water can be changed by as much as 20 parts in 10^6 by isotopic fractionation during repeated distillation. Thompson and Wirth (1931) in their determination of the specific gravity of sea water, found their samples to be heavier on average by 20 parts in 10^6 than the value computed from the chlorinity by Knudsen's tables. Some recent (unpublished) measurements at the National Institute of Oceanography agree more closely with those of Thompson and Wirth than with those of Forch *et al.* (1902).

This problem can be resolved only by a determination of the specific gravity of sea water relative to that of pure water of which the density is known in absolute units. From the results of such an experiment, the correction necessary to convert the specific gravities and specific volumes based on Knudsen's tables to absolute densities could readily be computed.

Part of the programme at the National Institute of Oceanography referred to above is a determination of specific gravity using a defined "pure water". It is hoped that the British National Physical Laboratory will measure the density of the reference water, which will make it possible to express the sea water densities in absolute units. It is strongly urged that for the present no units should be quoted when specific gravity values are reported.

Reference should be made to an excellent summary of the mathematics of the calculation of density from temperature, salinity and pressure, which is given by Sverdrup *et al.* (1942). The density is increased approximately linearly by increasing salinity. The effect of temperature is more complex. Thus, pure water has a maximum density at 4°C; addition of salt lowers the temperature of maximum density; above about 20‰S it lies below the normal freezing point. The rate of change of density with temperature increases rapidly as the temperature increases.

Typical values of the densities as a function of temperature and salinity are given in Table I. The reliability of these values when used to compute *differences* in density between two similar sea waters is probably ±1 part in 10^5. The above discussion suggests that the absolute value of the density is in doubt by two or three times this amount. The usual reservations apply for salinities below 30‰.

TABLE I

Specific gravity anomaly $\times 10^5$ *of sea water*

S‰	Temperature(°C)									
	0	*2*	*4*	*6*	*8*	*10*	*15*	*20*	*25*	*30*
0	−13	−3	0	−5	−16	−32	−87	−177	−293	−433
5	397	403	402	394	381	362	301	207	87	−57
10	801	804	799	788	772	750	685	586	462	315
15	1204	1204	1195	1181	1162	1138	1067	964	836	686
20	1607	1603	1589	1573	1551	1532	1450	1342	1210	1057
25	2008	2001	1988	1970	1947	1920	1832	1720	1585	1428
30	2410	2400	2384	2363	2340	2308	2215	2098	1960	1801
32	2571	2560	2543	2521	2494	2464	2364	2250	2110	1950
34	2732	2719	2701	2678	2651	2619	2522	2402	2261	2100
36	2893	2879	2860	2836	2808	2775	2676	2554	2412	2250
38	3055	3040	3019	2994	2965	2931	2830	2707	2563	2400
40	3216	3200	3179	3153	3122	3088	2985	2860	2714	2550
42	3377	3361	3337	3310	3279	3243	3138	3011	2864	2700

The values in Table I are expressed in terms of specific gravity anomaly, multiplied by 10^5. Thus, a figure of 1204 represents a specific gravity of 1·01204 or a σ_t of 12·04; a figure of −32 represents a specific gravity of (1 − 0·00032), or 0·99968.

The effect of pressure on the density of sea water is discussed in Section VIII.

V. THE ELECTRICAL CONDUCTIVITY OF SEA WATER

Electrical conductivity has become very important because of its increasing use for salinity determination. Unfortunately, the inadequacies of present knowledge of conductivity have led to much confusion in the calibration of conductivity salinometers. Conductivity is approximately proportional to salinity; the temperature coefficient varies with temperature. Thus, the conductivity increases by about 3%/°C increase in temperature at 0°C and only 2%/°C at 25°C. At 30°C the conductivity is about double the value at 0°C, for the same salinity.

The most comprehensive observations published for the specific conductivity are those of Thomas *et al.* (1934). These workers compared the resistances of natural and diluted sea waters with that of potassium chloride solution. The values published by Jones and Bradshaw (1933) for potassium chloride were used as reference. Pollack (1954) has queried certain details of the procedure, and has suggested minor corrections to the values, but it is doubtful if the measurements are precise enough to make the corrections worthwhile. A series of observations was made at 25°C, over the salinity range 3‰ to 38·7‰. Measurements were also made at 0°, 5°, 10°, 15° and 20°C at salinities up to 34·8‰, but unfortunately these series were not extended to the higher salinities. Mathematical expressions connecting chlorinity to conductivity at each temperature were computed. Agreement between the data and these expressions is poor, particularly at 15°C. Extrapolation to cover the salinity ranges up to 39‰ introduces errors of over 0·1‰.

A new mathematical treatment of the original data of Thomas *et al.* has been made by Weyl (1964). He has produced a single expression from which specific conductivity can be computed at any chlorinity and temperature. Comparison of values computed from this expression with those of Brown and Hamon (1961) suggests that they are probably correct to ±0·03‰ in the salinity range 30 – 40‰. Below this, there are uncertainties arising from the variable nature of the diluting water. Table II displays values of specific conductivity computed from Weyl's expression. The low salinity samples used by Thomas *et al.* were made by diluting ocean water with pure water. In consequence, in the computing of Table II, chlorinity has been converted to salinity by the expression $S‰ = 1{\cdot}8065\ Cl\ ‰$ instead of the usual Knudsen formula.

New determinations of the absolute conductivity of sea water are at present under way at the Chesapeake Bay Institute, Baltimore, and of the temperature effects by N. D. Brown of the Hytech Corporation, San Diego. More reliable data should, therefore, be available shortly to supplant Table II.

For the calibration of precision salinometers, which are standardized with standard sea water, the conductivities are most conveniently presented in terms of a conductivity ratio, or relative conductivity. This is the ratio of the conductivity to that of standard sea water having a chlorinity of 19·374‰ (salinity 35·000‰). The ratio depends on the temperature. Most thermostat salinometers are designed to run at 15°C; the thermistor-compensated meters have tables computed to be correct at 22·5°C. Table III gives conductivity ratios at the two temperatures. The 15°C figures are based on the National Institute of

TABLE II

*Specific conductivity of sea water**

S‰	Temperature (°C) 30	25	20	15	10	5	0
10	(19·127)	17·345	15·628	13·967	12·361	10·816	9·341
20	(35·458)	32·188	29·027	25·967	23·010	20·166	17·456
30	(50·856)	46·213	41·713	37·351	33·137	29·090	25·238
31	(52·360)	47·584	42·954	38·467	34·131	29·968	26·005
32	(53·859)	48·951	44·192	39·579	35·122	30·843	26·771
33	(55·352)	50·314	45·426	40·688	36·110	31·716	27·535
34	(56·840)	51·671	46·656	41·794	37·096	32·588	28·298
35	(58·323)	53·025	47·882	42·896	38·080	33·457	29·060
36	(59·801)	54·374	(49·105)	(43·996)	(39·061)	(34·325)	(29·820)
37	(61·274)	55·719	(50·325)	(45·093)	(40·039)	(35·190)	(30·579)
38	(62·743)	57·061	(51·541)	(46·187)	(41·016)	(36·055)	(31·337)
39	(64·207)	58·398	(52·754)	(47·278)	(41·990)	(36·917)	(32·094)
40	(65·667)	(59·732)	(53·963)	(48·367)	(42·962)	(37·778)	(32·851)

* Conductivity in milliohms/cm.

TABLE III

Relative conductivity of sea water

S ‰	15°C	22·5°C
31	0·89693	0·89725
32	0·92291	0·92312
33	0·94876	0·94887
34	0·97450	0·97450
35	1·00012	1·00000
36	1·02536	1·02537
37	1·05080	1·05062
38	1·07615	1·07575
39	1·10150	1·10080
40	1·12697	—

Oceanography world-wide survey, that is, measurements which were made on actual samples from the sea. The 22·5°C figures are based on measurements by Hamon (see Brown and Hamon 1961) on a single sample of high salinity water diluted by weight (see Cox and Folkard, 1964).

Table IV gives correction factors which enable ratios determined at temperatures between 10° and 25°C to be compared with the 15°C values given in Table III. Table IV is based on unpublished observations made at the National Institute of Oceanography. For the calibration of *in situ* salinometers, it is necessary to take into account the effect

TABLE IV

Effect of temperature on conductivity ratio

S‰	*Temperature(°C)*						
	10·0	*12·5*	*15·0*	*17·5*	*20·0*	*22·5*	*25·0*
30	42	20	0	−18	−35	−51	−70
31	34	16	0	−14	−29	−41	−58
32	26	12	0	−11	−22	−32	−44
33	17	8	0	−7	−14	−21	−28
34	9	4	0	−4	−7	−11	−14
35	0	0	0	0	0	0	0
36	−10	−5	0	5	8	11	16
37	−19	−9	0	9	16	23	30
38	−28	−13	0	14	25	35	45
39	−37	−17	0	19	34	48	61
40	−47	−22	0	24	43	60	77

of pressure on the conductivity. This effect has been measured by several workers (Hamon, 1958; Horne and Frysinger, 1962), but the most reliable values come from work recently completed at the Woods Hole Institution (Schleicher and Bradshaw, 1964). Table V shows the effect of pressure as a percentage increase in conductivity for pressures up to 10,000 db, approximately corresponding to a depth of 10,000 m. It will be seen that the effect of pressure is much greater at low temperatures. The effect of salinity on the compressibility is linear, but relatively slight.

TABLE V

*Effect of pressure on the conductivity of sea water**

	Pressure (db)	S ‰ 31	S ‰ 35	S ‰ 39		S ‰ 31	S ‰ 35	S ‰ 39
	1000	1·599	1·556	1·512		1·032	1·008	0·985
	2000	3·089	3·006	2·922		1·996	1·951	1·906
	3000	4·475	4·354	4·233		2·895	2·830	2·764
	4000	5·759	5·603	5·448		3·731	3·646	3·562
0°C	5000	6·944	6·757	6·569	15°C	4·506	4·403	4·301
	6000	8·034	7·817	7·599		5·221	5·102	4·984
	7000	9·031	8·787	8·543		5·879	5·745	5·612
	8000	9·939	9·670	9·401		6·481	6·334	6·187
	9000	10·761	10·469	10·178		7·031	6·871	6·711
	10000	11·499	11·188	10·877		7·529	7·358	7·187
	1000	1·368	1·333	1·298		0·907	0·888	0·868
	2000	2·646	2·578	2·510		1·755	1·718	1·680
	3000	3·835	3·737	3·639		2·546	2·492	2·438
	4000	4·939	4·813	4·686		3·282	3·212	3·142
5°C	5000	5·960	5·807	5·655	20°C	3·964	3·879	3·795
	6000	6·901	6·724	6·547		4·594	4·496	4·399
	7000	7·764	7·565	7·366		5·174	5·064	4·954
	8000	8·552	8·333	8·114		5·706	5·585	5·464
	9000	9·269	9·031	8·794		6·192	6·060	5·929
	10000	9·915	9·661	9·408		6·633	6·492	6·351
	1000	1·183	1·154	1·125		0·799	0·783	0·767
	2000	2·287	2·232	2·177		1·547	1·516	1·485
	3000	3·317	3·237	3·157		2·245	2·200	2·156
	4000	4·273	4·170	4·067		2·895	2·837	2·780
10°C	5000	5·159	5·034	4·910	25°C	3·498	3·429	3·359
	6000	5·976	5·832	5·688		4·056	3·976	3·896
	7000	6·728	6·565	6·402		4·571	4·481	4·390
	8000	7·415	7·236	7·057		5·045	4·945	4·845
	9000	8·041	7·847	7·652		5·478	5·369	5·261
	10000	8·608	8·400	8·192		5·872	5·756	5·640

* Percentage increase compared with the conductivity at one atmosphere.

VI. THE REFRACTIVE INDEX OF SEA WATER

The refractive index of sea water has been measured by Utterback *et al.* (1934), Bein *et al.* (1935) and Miyake (1939). The results reported by Utterback *et al.* are the most comprehensive. They cover determinations at temperature intervals of 5° from 0° to 25°C, over a wide range of salinities. The highest salinity was 38·6‰ in a sample from the

Mediterranean Sea. Samples with salinities below 30‰ were prepared by dilution with pure water. Thus, the refractive indices quoted for the lower salinities may not be reliable for such regions as the Baltic Sea, but they are the best available at present. In the important range of 32 – 38‰ salinity, the results at 25°C agree with those of Miyake within the claimed precision of measurement. The results of Bein *et al.* are presented in such a form that no direct comparison is possible.

TABLE VI

*Refractive index of sea water**

S‰	Temperature (°C) 0	5	10	15	20	25
0	1·3 3395	1·3 3385	1·3 3370	1·3 3340	1·3 3300	1·3 3250
5	3500	3485	3465	3435	3395	3345
10	3600	3585	3565	3530	3485	3435
15	3700	3685	3660	3625	3580	3525
20	3795	3780	3750	3715	3670	3620
25	3895	3875	3845	3805	3760	3710
30	3991	3966	3935	3898	3851	3798
31	4011	3985	3954	3916	3869	3816
32	4030	4004	3973	3934	3886	3834
33	4049	4023	3992	3953	3904	3851
34	4068	4042	4011	3971	3922	3868
35	4088	4061	4030	3990	3940	3886
36	4107	4080	4049	4008	3958	3904
37	4127	4099	4068	4026	3976	3922
38	4146	4118	4086	4044	3994	3940
39	4166	4139	4105	4062	4012	3958
40	(4185)	(4157)	(4124)	(4080)	(4031)	(3976)
41	(4204)	(4176)	(4143)	(4098)	(4049)	(3994)

* For sodium D light.

Table VI gives values for the refractive index at various salinities and temperatures. These are computed from the original observations (not the computed table) of Utterback *et al.* (1934), chlorinity being converted to salinity by the Knudsen equation. The scatter in the original data suggests the probable uncertainty of the results is ± 3 in the last figure between salinities 30‰ and 39‰, and rather more at the lower salinities. The values for salinities below 30‰ have been rounded off to ± 5 units.

All these results were obtained using the sodium D lines. Owing to the doublet character of this radiation, it is not suitable for use in precision work with interferometers, since it causes double fringes which reduce precision. For precision work, a more suitable source is the mercury green line, at 5461 Å. No measurements for sea water are available with green light, but the published figures for pure water suggest that an approximate figure could be obtained by multiplying the figure from Table VI by the factor 1·00113.

VII. The Velocity of Sound in Sea Water

Values for the velocity of sound in sea water are given in Table VII (a). It will be seen that the natural ranges of temperature and pressure produce much greater effects than those caused by normal salinity variations. This point has already been elaborated in Section III, E, with reference to the use of velocity meters as salinometers.

The velocity of sound can be computed on a thermodynamic basis; for example from the expression:

$$v^2 = \frac{\gamma}{\rho K}$$

where v is sound velocity, γ is the ratio of specific heats (C_p/C_v), ρ is the density, and K the compressibility. Numerous tables have been computed giving velocity as a function of temperature, salinity and pressure; probably the most useful are those of Matthews (1939). However, instruments have been designed recently which measure sound velocity directly (see Section III, E), and direct measurements of sound velocity have been made in sea water over a wide range of temperature, salinity and pressure by Wilson (1960a,b, 1962). These measurements are certainly more reliable than the computed values previously used. The differences at atmospheric pressure are not large, the measured values being about 3 m./sec. higher than the computed values, but the velocities at higher pressures differ rather more markedly, and have caused some reconsideration of our knowledge of the compressibility of sea water (see next section).

The tables given by Wilson (1960a) use an absolute scale in which kg./cm.^{2} is the unit of pressure, and consequently a small correction is necessary to obtain values at atmospheric pressure. Most oceanographers prefer to use tables in which atmospheric pressure is taken as the arbitrary zero, in order that pressure may be directly related to depth, and in which pressure is expressed in bars. It should be remembered that pressure in units of weight per unit area is dependent on the value of gravity, which may vary over a range of about 0·5%. Other tables

TABLE VII(a)*

Velocity of sound in sea water†

Pressure (db)	Temperature(°C) 0	5	10	15	20	25	30
0	1449·3	1471·0	1490·4	1507·4	1522·1	1534·8	1545·8
1000	1465·8	1487·4	1506·7	1523·7	1538·5	1551·3	1562·5
2000	1482·4	1504·0	1523·2	1540·2	1555·0	1567·9	1579·2
3000	1499·4	1520·7	1538·6	1555·6			
4000	1516·5	1537·7	1555·2	1572·2			
5000	1533·9	1554·8	1571·9	1588·9			
6000	1551·5	1572·1					
7000	1569·3						
8000	1587·3						
9000	1605·4						
10000	1623·5						

* Reproduced by permission of U.S. Navy Oceanographic Office.

† Velocities in m./sec.; pressures in decibars above atmospheric. Salinity 35‰. For other salinities see Table VII(b).

TABLE VII(b)*

Effect of salinity on sound velocity†

S‰	Temperature(°C) 0	5	10	15	20	25	30
30	−7·0	−6·7	−6·5	−6·2	−5·9	−5·6	−5·3
32	−4·2	−4·0	−3·9	−3·7	−3·5	−3·4	−3·2
33	−2·8	−2·7	−2·6	−2·5	−2·4	−2·2	−2·1
34	−1·4	−1·3	−1·3	−1·2	−1·2	−1·1	−1·1
35	0	0	0	0	0	0	0
36	1·4	1·3	1·3	1·2	1·2	1·1	1·1
37	2·8	2·7	2·6	2·5	2·4	2·3	2·1
38	4·2	4·1	3·9	3·7	3·6	3·4	3·2
40	7·0	6·8	6·5	6·2	6·0	5·7	5·3

* Reproduced by permission of U.S. Navy Oceanographic Office.

† Corrections to be applied to the values in Table VII(a) for salinities other than 35‰.

based on Wilson's measurements have been published (U.S. Navy, 1961). In these tables pressures are expressed in decibars, and the pressure at the sea surface, is taken as the arbitrary zero. Table VII is a brief extract from this source.

VIII. Compressibility of Sea Water

The numerical values of the compressibility of sea water have recently been disputed. The measurements which have been most used are those of Ekman (1908). These were relative measurements, the compressibility of sea water being compared with that of pure water, and the tabulated values are therefore dependent on the measurements made by Amagat (1893) on pure water. Eckart (1958), in a critical review of the properties of sea water, concluded that Ekman's data are significantly in error at high pressures. Recent measurements by Kennedy *et al.* (1958) appear to confirm Eckart's conclusions. However, Crease (1962) has recomputed the effect of pressure, using Wilson's (1960a) direct measurements of sound velocity, and more accurate values for the specific heat of sea water (Cox and Smith, 1959). His conclusions very closely confirm the observations of Ekman and of Amagat.

Most of the tables concerned with the effect of pressure on sea water tabulate either the specific gravity or its reciprocal, the specific volume, as a function of pressure, temperature and salinity. Since all these variables affect the final value of specific gravity, it is hardly practicable to provide full tables in close intervals of all the variables. Examples of tables requiring interpolation are those of Ekman (1910) for specific gravity, and of Matthews (1938) for specific volume.

In all deep ocean areas, the temperature of the water is close to 0°C and the salinity is about 35‰. Table VIII lists values of the specific

Table VIII

*Density and volume reduction of sea water under pressure**

Pressure (db)	*Specific gravity*	*% decrease in volume*
0	1·02812	0·000
100	1·02860	0·046
200	1·02908	0·093
500	1·03050	0·231
1000	1·03285	0·458
2000	1·03747	0·902
3000	1·04198	1·330
4000	1·04640	1·747
5000	1·05071	2·150
6000	1·05494	2·542
7000	1·05908	2·932
8000	1·06314	3·294
9000	1·06713	3·656
10000	1·07104	4·007

* Salinity, 35·0‰; temperature 0°C.

gravity and compression of sea water (salinity 35‰) at 0°C, at pressures up to 10,000 db. The compression is expressed as a percentage reduction in volume relative to the volume at atmospheric pressure. The work of Crease (1962) suggests that the tabulated values for specific gravity are precise to three digits in the last figure quoted. The compressibility varies significantly with temperature and salinity; these factors are usually important only in the upper 1000 m. of the ocean, and over this limited range, compressibility can be considered directly proportional to pressure. Table IX lists the volume reduction for a pressure of 1000 db at various temperatures and salinities. The values for lower pressures can be computed by simple proportion.

TABLE IX

Percentage reduction in volume of sea water under a pressure of 1000 db at various temperatures and salinities

S‰	Temperature (°C) 0	10	20	30
0	0·487	0·470	0·451	0·435
10	0·475	0·461	0·443	0·426
20	0·458	0·452	0·435	0·420
30	0·451	0·442	0·425	0·413
35	0·445	0·436	0·420	0·409
40	0·440	0·431	0·418	0·400

IX. THE TRANSPARENCY OF SEA WATER

Most studies of the transparency of sea water have been confined to wave-lengths of biological significance. Although of fundamental importance to productivity, this aspect of transparency is not of great interest to the chemist, and will not be discussed here. Almost all the absorption of visible light in the sea is caused by particulate, suspended material rather than dissolved substances.

The chemist is more concerned with the absorption of radiation by pure sea water, as this has an influence on photometric analysis in a sea water medium. The best determinations of the transparency of sea water are those of Clarke and James (1939). They made measurements of the percentage transmission by a 1 m. length of water at wavelengths from 3600 – 8000 Å, using both natural and artificial waters purified by various filtration processes. The results show that the absorption of pure sea water is never less than that of distilled water, but usually is significantly greater in two regions, as shown in Fig. 8. At wavelengths above 5600 Å there is no significant difference between sea water and pure water; at 5200 Å the extinction for 1 m. of water is 0·025 for pure water and 0·070 for artificial sea water; at 3600 Å it is 0·010 for pure water and 0·110 for sea water. This effect should be borne in mind when colorimetric measurements are made against blank solutions prepared in distilled water, particularly when low colour absorbances (which require a long path-length) are measured.

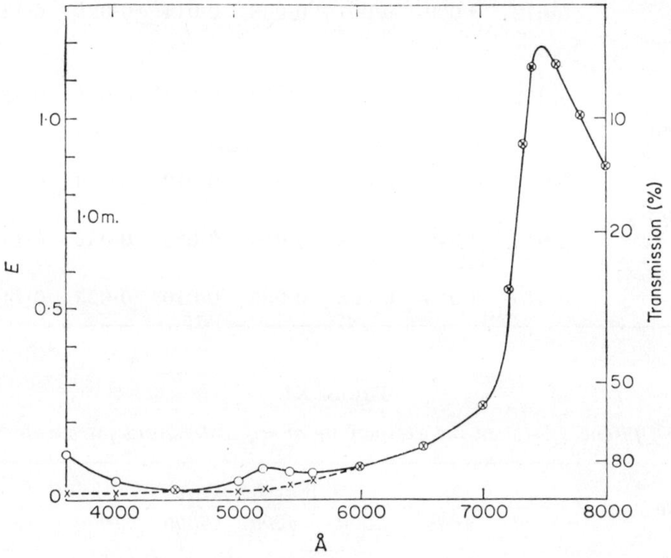

FIG. 8. Extinction (1 m. path) against wavelength. Solid line—sea water. Broken line—pure water.

The values indicated in Fig. 8 refer to sea water which is free from dissolved or suspended organic material. With natural unfiltered sea water, the absorption is always higher. Clarke and James (1939) give values for a number of typical oceanic and coastal samples, and these show that the absorption of oceanic water is reduced by passage through a micro-filter. With continental shelf water the effect is more pronounced. Filtration through paper scarcely affects the absorption; a micro-filter (see Chapter 21, Section IV, Volume 2), however, reduces the colour by removing fine particulate material. Even after filtration, coastal water still shows a considerable colour, since there is a strong absorption at all wavelengths below 6000 Å.

TABLE X

Light absorption of typical sea waters
Extinction for 10 cm. path length

Sample	Wavelength Å							
	3600	*4000*	*5000*	*5200*	*6000*	*7000*	*7500*	*8000*
Pure water	0·001	0·001	0·002	0·002	0·010	0·025	0·115	0·086
Artificial sea water	0·011	0·003	0·005	0·007	0·010	0·025	0·115	0·086
Ocean water, unfiltered	0·012	0·009	0·007	0·008	0·011	0·025	0·115	0·086
Continental slope waters, unfiltered	0·052	0·030	0·011	0·010	0·012	0·035	0·130	0·088
Continental slope waters, filtered	0·016	0·010	0·005	0·005	0·012	0·030	0·115	0·086
Inshore water unfiltered	0·055	0·042	0·028	0·026	0·035	0·052	0·140	0·100
Inshore water, filtered	0·015	0·010	0·005	0·005	0·010	0·025	0·110	0·086

TABLE XI

*Differences between the extinctions of sea waters and pure water**

Sample	Wavelength Å							
	3600	*4000*	*5000*	*5200*	*6000*	*7000*	*7500*	*8000*
Artificial sea water	0·010	0·002	0·003	0·005	nil	nil	nil	nil
Ocean water, unfiltered	0·011	0·008	0·005	0·006	0·001	nil	nil	nil
Continental slope water, unfiltered	0·051	0·029	0·009	0·008	0·002	0·010	0·015	0·002
Continental slope water, filtered	0·015	0·009	0·003	0·003	0·002	0·005	nil	nil
Inshore water, unfiltered	0·054	0·041	0·026	0·024	0·025	0·027	0·025	0·015
Inshore water, filtered	0·014	0·009	0·003	0·003	nil	nil	nil	nil

* $E_{10\text{ cm.}}$ (sea water) $-$ $E_{10\text{ cm.}}$ (pure water) as a function of wavelength.

The measurements of Clarke and James on coastal waters should be regarded as indicative only of the order of absorption to be expected. For precision photometric measurements with unfiltered water, a "blank" absorption should be determined on the sample before the colour is developed. Table X (after Clarke and James) gives the extinction of a 10 cm. light path of several samples, before and after filtration. Table XI gives the increase in extinction when 10 cm. of pure water is replaced by various samples.

X. The Freezing Point and Osmotic Pressure of Sea Water

The freezing point of sea water has been measured by Knudsen (1903) who gave an expression relating the freezing point, t (in °C), with the specific gravity at 0°C,

$$t = -0{\cdot}0086 - 0{\cdot}064633\ \sigma_0 - 0{\cdot}0001055\ \sigma_0^2.$$

Table XII is computed from this expression, and gives the freezing point at various salinities. Knudsen (1903) claimed these figures to be accurate to $\pm 0{\cdot}02$°C.

Table XII

The freezing point and osmotic pressure of sea water

Salinity ‰	*F.P. (°C)*	*O.P. at 0°C (atm.)*
5	−0·27	3·2
10	−0·53	6·4
15	−0·80	9·7
20	−1·07	13·0
25	−1·35	16·3
30	−1·63	19·7
35	−1·91	23·1
40	−2·20	26·6

No measurements have been published of the osmotic pressure of sea water. Stenius (1904) claimed that the osmotic pressure in atmospheres at 0°C can be computed from the expression

$$OP\ (0°) = -12{\cdot}08 t_0$$

where t_0 is the freezing point (in °C) at the particular salinity. The osmotic pressure at temperature t°C is given by the expression:

$$OP(t°) = OP(0°) \cdot \frac{(273 + t).}{273}$$

Values of $OP(0°)$ are given in Table XII.

XI. The Specific Heat of Sea Water

The specific heat of sea water has been measured at 17·5°C by Thoulet and Chevallier (1889) and at temperatures from 0 – 30°C by Cox and Smith (1959). Ekman (1914) recalculated the specific heat from Thoulet and Chevallier's measurements, and added an estimated correction for temperature variation. However, Cox and Smith have shown that Ekman's estimate of the temperature coefficient was wrong, being opposite in sign to the true correction. Tables based on his paper are misleading, and should not be used.

Cox and Smith have shown that the specific heat falls with increasing salinity from 4·21 Joules/g. for pure water to 3·96 Joules/g. at 40‰*S* (at 0°C). The temperature coefficient, which is negative for pure water, becomes zero at 20‰*S* and positive at higher salinities.

Table XIII

*Specific heat of sea water at constant pressure**

S‰	*Temperature (°C)*						
	0	*5*	*10*	*15*	*20*	*25*	*30*
0	4·217	4·202	4·192	4·186	4·182	4·179	4·178
5	4·179	4·168	4·161	4·157	4·154	4·153	4·152
10	4·142	4·135	4·130	4·128	4·126	4·126	4·126
15	4·107	4·103	4·100	4·099	4·098	4·099	4·100
20	4·074	4·072	4·071	4·071	4·071	4·072	4·074
25	4·043	4·042	4·042	4·043	4·045	4·046	4·048
30	4·013	4·014	4·015	4·016	4·018	4·020	4·023
32	4·002	4·003	4·004	4·006	4·008	4·010	4·013
34	3·990	3·992	3·993	3·995	3·998	4·000	4·003
35	3·985	3·986	3·988	3·990	3·993	3·995	3·999
36	3·979	3·981	3·983	3·985	3·988	3·991	3·994
38	3·968	3·970	3·972	3·975	3·978	3·981	3·985
40	3·957	3·959	3·962	3·965	3·968	3·972	3·976

* In absolute Joules/g.

Table XIII gives the specific heat of sea water at constant pressure, in absolute Joules/g., for temperatures from −2° to 30°C and salinities from 0 – 40‰. Cox and Smith claim a probable error for these values of 0·0015 Joules/g.

XII. The Surface Tension of Sea Water

The surface tension of sea water has been measured by Krümmel (1900). The expression given by him was revised by Fleming and Revelle (1939) to take account of more recent measurements on pure water. They gave the expression:

$$\text{Surface tension (dyne/cm.)} = 75{\cdot}64 - 0{\cdot}144t + 0{\cdot}0399\, Cl‰$$

where t is the temperature in °C and $Cl‰$ the chlorinity. Table XIV lists the surface tension, computed from this expression. The effect of salinity variations is small, but temperature changes produce a large effect.

Table XIV

Surface tension of clean sea water in dynes/cm.

$S‰$	*Temperature (°C)*			
	0	*10*	*20*	*30*
0	75·64	74·20	72·76	71·32
10	75·86	74·42	72·98	71·54
20	76·08	74·64	73·20	71·76
30	76·30	74·86	73·42	71·98
35	76·41	74·97	73·53	72·09
40	76·52	75·08	73·64	72·20

Krümmel's measurements were made on bubbles below the surface; they take no account of surface contamination. Lumby and Folkard (1956) have measured the effect of contamination on the surface of the sea, using the spreading-drop method of Adam (1937). They found that the surface tension is always lower than that of clean water because of contamination; the minimum depression they found was 2·5 dynes/cm. in clean, ruffled water. Where more oily matter was present, the depression was sometimes as great as 22 dynes/cm.

References

Adam, N. K. (1937). *Proc. roy. Soc.* **B122**, 134.

Amagat, E. H. (1893). *Ann. Chim.* (*Phys.*) **29**, 68, 505.

Bein, W., Hirsekorn, H., and Möller, L. (1935). *Veröff. Inst. Meeresk. Univ. Berl.* **28**, 240 pp.

Bjerknes, V., and Sandström, J. W. (1910). "Dynamic Meteorology and Hydrography". Part one, Statics. Pub. 88. Carnegie Institute, Washington, U.S.A.

Brice, B. A., and Halwer, H. (1951). *J. opt. Soc. Amer.* **41, 1033.**

Brown, N. L., and Hamon, B. V. (1961). *Deep Sea Res.* **8**, 65.

Carritt, D. E., and Carpenter, J. H. (1959). *Nat. Acad. Sci., Nat. Res. Coun. Pub.* **600**, 67.

Clarke, G. L., and James, H. R. (1939). *J. opt. Soc. Amer.* **29**, 43.

Cox, R. A. (1954). *J. Cons. int. Explor. Mer*, **20**, 9.

Cox, R. A. (1957). *J. Cons. int. Explor. Mer*, **22**, 38.

Cox, R. A. (1958). The thermostat salinity meter. N.I.O. Internal Rept. C2, Mar. 1958. (Unpublished manuscript).

Cox, R. A. (1962). *Deep Sea Res.* **9**, 504.

Cox, R. A. (1963). *In* "Progress in Oceanography" (M. Sears, ed.), Vol. 1, p. 243. Pergamon Press, Oxford, England.

Cox, R. A., and Folkard, A. R. (1964). *Bull. int. Comm. N.W. Atlan. Fisheries* (in the press).

Cox, R. A., and Moorey, J. (1962). The portable temperature salinity bridge. N.I.O. internal rept. C4, Oct. 1962. (Unpublished manuscript).

Cox, R. A., and Smith, N. D. (1959). *Proc. roy. Soc.* **A252**, 51.

Cox, R. A., Culkin, F., Greenhalgh, R., and Riley, J. P. (1962). *Nature, Lond.* **193**, 518.

Crease, J. (1962). *Deep Sea Res.* **9**, 209.

Cummings, N. W. (1932). *Bull. Scripps Instn. Oceanogr. tech.* **3**, 101.

Dittmar, W. (1884). Report on the scientific results of the exploring voyage of H.M.S. *Challenger* (Physics and Chemistry). Vol. 1. H.M. Stationery Office, London.

Dorrestein, R. (1956). *Int. hydr. Rev.* **33**, 1.

Eckart, C. (1958). *Amer. J. Sci.* **251**, 225.

Ekman, V. W. (1908). *Publ. Circ. Cons. Explor. Mer*, **43**, 47 pp.

Ekman, V. W. (1910). *Publ. Circ. Cons. Explor. Mer*, **49**, 48 pp.

Ekman, V. W. (1914). *Ann. Hydrogr., Berl.* **42**, 340.

Esterson, G. L. (1957). The Johns Hopkins University, Chesapeake Bay Institute technical report XIV–57–3.

Faust, R. C., and Marrinan, H. J. (1955). *J. appl. Phys.* **6**, 351.

Fleming, R. H., and Revelle, R. R. (1939). Recent Marine Sediments (Symposium) Amer. Assn. Petroleum Geol., Tulsa, Oklahoma, **48**.

Forch, C., Knudsen, M., and Sørensen, S. P. L. (1902). *K. danske Vidensk. Selsk.* **12**, 1.

Forchhammer, G. (1865). *Phil. Trans.* **155**, 203.

German, C. (1960). *J. Cons. int. Explor. Mer*, **26**, 21.

Guntz, A. A., and Kocher, J. (1952). *C. R. Acad. Sci., Paris*, **234**, 2300.

Hamon, B. V. (1955). *J. Cons. int. Explor. Mer*, **21**, 72.

Hamon, B. V. (1956). *J. sci. Instrum.* **33**, 329.

Hamon, B. V. (1958). *J. Mar. Res.* **16**, 83.

Hinkelmann, H. (1956). *Kieler Meeresforsch.* **12**, 200.

Hinkelmann, H. (1957). *Z. angew. Phys.* **9**, 505.

Hin Lew (1945). National defence council Canada, division of Physics and Electrical Engineering, report PS251, "The salinometer model XI". September, 1945.

Horne, R. A., and Frysinger, G. R. (1962). Arthur D. Little. Inc., Technical report 1270862. (U.S. Bureau of Ships contract NObsr-81564.)

Jacobsen, J. P., and Knudsen, M. (1940). *Publ. sci. Ass. Océanogr. phys.* **7**, 38 pp.

Jacobsen, A. W. (1948). *Trans. Amer. Inst. elect. Engrs*, **67**, 1.

Jones, G., and Bradshaw, B. C. (1933). *J. Amer. chem. Soc.* **55**, 1780.

Kennedy, G. C., Knight, L. W., and Holser, W. T. (1958). *Amer. J. Sci.* **256**, 590.
Kalle, K. (1957). *Dtsch. hydrogr. Z.* **10**, 99.
Knudsen, M. (1900). *Beretn. Kommission. fra Vidensk. Undersøgelser Danske Farvende, København,* **2**, 15 pp.
Knudsen, M. (1901). 2me. Conference International pour l'Exploration de la Mer, *Kristiania*. Report, Supplement 9.
Knudsen, M. (1903). *Pub. Circ. Cons. Explor. Mer,* **5**, 3 pp.
Krümmel, O. (1900). *Wiss. Meeresuntersuch.* **5**, 9.
Lumby, J. R., and Folkard, A. R. (1956). *Bull. Inst. oceanogr. Monaco,* **1080**, 1.
Lyman, J. (1959). *Nat. Acad. Sci., Nat. Res. Coun. Publ.* **600**, 87.
Marcet, A. (1819). *Phil. Trans.* **109**, 161.
Marcet, A. (1822). *Phil. Trans.* **112**, 448.
Matthews, D. J. (1938). Tables for the specific volume of sea water under pressure. *Cons. Perm. exp. Mer,* Copenhagen.
Matthews, D. J. (1939). Tables for the velocity of sound in pure water and sea water for use in echo-sounding and echo-ranging. H.D.282, 2nd ed., 1939. Hydrographic Dept., Admiralty, London.
Miyake, Y. (1939). *J. chem. Soc. Japan,* **14**, 239.
Moberg, E. G., and Revelle, R. R. (1937). *P. V. Ass. Océanogr. phys. Un. géol. int.* **2**, 153.
Morris, A. W., and Riley, J. P. (1964). *Deep Sea Res.* In the press.
Nansen, F. (1900). *Sci. Res. Norweg. N. Pol. Exped., 1893–96,* **3**, 88 pp.
Pape, C. (1924). *Ann. Hydrogr. Berl.* **52**, 249.
Paquette, R. G. (1959). University of Washington, Seattle, Dept. Oceanog. Tech. Rep. 59/2 (unpublished manuscript).
Pollack, M. J. (1954). *J. Mar. Res.* **13**, 228.
Pollack, M. J. (1961). *J. Mar. Res.* **19**, 28.
Richardson, W. S. (1959). *Nat. Acad. Sci., Nat. Res. Coun. Pub.* **600**, 113.
Rochford, D. (1954). *Nature, Lond.* **174**, 641.
Schleicher, K., and Bradshaw, A. (1964). *Deep Sea Res.* In the press.
Shaffer, J. D. (1960). *Rev. sci. Instrum.* **31**, 1318.
Siedler, G. (1963). *Deep Sea Res.* **10**, 269.
Stenius, S. (1904). *Öfvers. finska Vetensk Soc. Förh.* **46**, 1.
Stott, V. (1928). *J. Cons. int. Explor. Mer,* **3**, 328.
Sverdrup, H. U., Johnson, M. W., and Fleming, R. H. (1942). "The Oceans". Prentice Hall, New York.
Thomas, B. D., Thompson, T. G., and Utterback, C. L. (1934). *J. Cons. int. Explor. Mer,* **9**, 28.
Thompson, T. G., and Wirth, H. E. (1931). *J. Cons. int. Explor. Mer,* **6**, 232.
Thoulet, J., and Chevallier, A. (1889). *C. R. Acad. Sci., Paris,* **108**, 794.
Thuras, A. L. (1918). *J. Wash. Acad. Sci.* **8**, 676.
Tschiegg, C. E., and Hays, E. E. (1959). *J. acoust. Soc. Amer.* **31**, 1038.
UNESCO (1962). Report of the joint panel on the equation of state of sea water. NS/9/114B.
U.S. Navy (1961). Tables of the velocity of sound in sea water. Bureau of ships reference NObsr 81564 S–7001–0307. Washington, U.S.A.
Utterback, C. L., Thompson, T. G., and Thomas, B. D. (1934). *J. Cons. int. Explor. Mer,* **9**, 35.
Vel'Mozhnaya, Yu. A. (1960). *Trud. Morskogo Gidrofiz. Inst.* **22**, 26. U.S. Hydrographic office translation TRANS–158 (1962).

Wattenberg, H., and Joseph, J. (1943). *Ann. Hydrogr., Berl.* **71**, 240.
Wenner, F., Smith, E. H., and Soule, F. M. (1930). *Bur. Stand. J. Res., Wash.* **5**, 711.
Weyl, P. (1964). *Limnol. Oceanogr.* **9**, 75.
Wilson, W. D. (1960a). *J. acoust. Soc. Amer.* **32**, 641.
Wilson, W. D. (1960b). *J. acoust. Soc. Amer.* **32**, 1357.
Wilson, W. D. (1962). *J. acoust. Soc. Amer.* **34**, 866.

Chapter 4

The Major Constituents of Sea Water

FREDERICK CULKIN
National Institute of Oceanography
Wormley, Surrey, England

I. Introduction 121
II. The Concept of Constant Composition 123
III. Methods of Determination of Major Constituents 126
A. Calcium and Magnesium 126
B. Sodium and Potassium 132
C. Sulphate 134
D. Strontium 135
E. Boron 136
F. Bromine 138
G. Fluorine 139
IV. Principal Determinations of Major Constituents 140
A. Calcium 140
B. Magnesium 142
C. Sodium 144
D. Potassium 145
E. Sulphate 146
F. Strontium 148
G. Boron 149
H. Bromine 151
I. Fluorine 153
V. Isotopic Variations 154
A. Isotopes of Hydrogen and Oxygen 154
B. Isotopes of Sulphur 158
References 158

I. Introduction

The major constituents (sometimes called the conservative elements) of sea water are those which make a significant contribution to salinity. They are listed in Table I together with their concentrations in sea water of salinity 35‰.

Calculations by Garrels and Thompson (1962), based on the assumption that interactions between the major ions result only in the formation of ion pairs, showed that the major cations in sea water exist mainly as uncomplexed ions. About 9% of the calcium, 13% of the magnesium and 1% of the sodium and potassium are associated with anions. Of the anions, half of the sulphate, a third of the bicarbonate and 90% of the

TABLE I

Composition of sea water

Constituent	*gm./kg. of water of salinity 35‰*
Chloride	19·353
Sodium	10·76
Sulphate	2·712
Magnesium	1·294
Calcium	0·413
Potassium	0·387
Bicarbonate	0·142
Bromide	0·067
Strontium	0·008
Boron	0·004
Fluoride	0·001

carbonate are paired with cations, whereas the chloride is completely uncomplexed.

As a material for chemical analysis, sea water has the advantage of being a simple solution of eleven inorganic ions, and thus does not necessitate the tedious dissolution procedures encountered in the analysis of biological and geological samples. This advantage is partly outweighed by the difficulties of storage. Polythene containers of the high density type are satisfactory, but low density polythene tends to transmit water slowly. Most types of glass contain alkali and alkaline earth metals and consequently are often unsuitable for the construction of containers for the long-term storage of sea water because of the danger of ion exchange between the dissolved ions and the glass. The most satisfactory glass container for storage of sea water is the special ampoule used by the Standard Sea Water Service, Copenhagen. Water sealed in these ampoules does not change its composition over a period of several years.

From the point of view of the analytical chemist, the composition of sea water poses some difficult problems. The ratio of magnesium to calcium (three to one by weight) is higher than that in the lithosphere, and causes co-precipitation troubles in gravimetric methods. The chemical similarity of strontium and calcium results in the two elements being determined together in most gravimetric and volumetric methods, and necessitates correction of the calcium figure for strontium interference. Precipitation as barium sulphate is still the only satisfactory method for the determination of sulphate in sea water. As is well known, this procedure is susceptible to interference because of co-precipitation of foreign ions, and careful control of the conditions of precipitation is necessary if reproducible results are to be obtained in its application to sea water. Perhaps the least satisfactory determination at the present time is that of the major constituent, sodium, for which there is still no satisfactory procedure. The one constituent of sea water which can be satisfactorily determined is chloride (plus bromide), and the method used by the Laboratoire Hydrographique, Copenhagen, is one of the most accurate chemical determinations available today.

II. The Concept of Constant Composition

Since the earliest reported analysis of sea water by Bergman in 1779, numerous workers have determined the concentration of one or more of its constituents. Often the investigator restricted his analyses to water which was of particular interest to him, or to which he had easy access. Comparison of analyses by different workers for the same constituent is not always satisfactory since the analytical methods often involved subjective errors. Occasionally however, samples of water from different oceans were analysed by one worker, and it is instructive to examine the results of these investigations and the conclusions drawn from them.

In 1819, Marcet presented a paper to the Royal Society in which he made what was probably the first suggestion that the relative composition of sea salt is constant. Marcet had obtained samples of water from the Arctic and Atlantic Oceans and the Mediterranean, Black, Baltic, China and White Seas. He determined the dissolved matter in each of these samples by evaporation followed by drying the residue at 100°C. He also determined chloride as silver chloride, sulphate as barium sulphate, calcium as calcium oxalate and magnesium as magnesium ammonium phosphate, methods which have been used, with some modifications, ever since for the determination of these constituents. Marcet's results are not reliable since his analytical techniques were very crude and atomic weights were known only approximately at that time. However, his conclusion that "all the specimens of sea water contain the same ingredients all over the world, *these bearing very nearly the same proportions to each other*, so that they differ only as to the total amount of their saline contents" is worth quoting for historical reasons. This paper was also noteworthy in that it contained an announcement that Wollaston had discovered potassium in sea water.

Marcet's findings probably made little impact on the scientific world of his day. Analyses of sea water were carried out by other workers (e.g. Calamai, 1848; von Bibra, 1851) but these were mainly concerned with obtaining an average figure for the total salt content of sea water and the most probable state of combination of the constituent anions and cations. It is interesting to note however, that in his book, "The Physical Geography of the Sea", Maury (1855) stated: "As a general rule the sea is nearly of a uniform degree of saltiness and the constituents of sea water are as constant in their proportions as are the components of the atmosphere." Although no authority is cited for this somewhat ambiguous statement it was probably prompted by the work of Marcet, Calamai and von Bibra.

The first extensive investigation of the composition of sea water was carried out by Forchhammer (1865) who determined chloride, sulphate, magnesium, calcium and potassium directly, and sodium by difference in several hundred surface waters from all parts of the globe. He reported that the ratios of the major constituents were subject to only *very slight variations* for different parts of the ocean, if the results for the Mediterranean, Black, Caribbean and Baltic Seas and the German Ocean were omitted. Forchhammer's investigation has been justifiably criticized on the grounds that all his samples were surface waters and that his analytical methods were inaccurate, but his suggestion that variations in composition do occur cannot be completely ignored.

During the world-wide cruise of H.M.S. *Challenger* (1873–6), seventy-seven samples of sea water were collected at various depths from some of the major oceans and were analysed by Dittmar (1884). The major criticisms which can be made of these analyses are that the samples were stored for as long as two years in glass bottles before examination and that they did not include any samples of Arctic, Antarctic and Mediterranean waters. Dittmar's early approach to what are still very difficult problems of chemical analysis is, however, much less open to criticism. In each of the methods which he adopted for the seven major ions, chloride, calcium, magnesium, potassium, sulphate, bromide and sodium, he adhered rigidly to a *modus operandi*, so that any errors would be constant and might be eliminated by subsequent investigation. Each method was thoroughly tested on synthetic samples, thus giving an estimate of the reliability of his results.

Dittmar's results were self-consistent and agreed fairly well with those of Forchhammer. In general, Dittmar regarded his results as extending Forchhammer's proposition from surface waters to ocean waters from all depths. Some of the variations which Dittmar found in the ratios of magnesium, potassium, sulphate, bromide and sodium to chloride and total salts were greater than would have been expected from analytical errors alone. Although Dittmar failed to correlate them with depth or geographical position, it must be emphasized that variations were found. The variations found for calcium were appreciable and the calcium/chloride ratios for deep and medium depth waters were higher than for surface waters by *c.* 0·3%—five times more than the probable analytical errors. Of this, Dittmar wrote: "It really appears as if the deep sea waters did contain more lime than the shallow, on account of their coming from greater depth."

Since Dittmar's time, many workers have studied the ratio of single constituents to chlorinity and the principal results are discussed in the following sections. Assessment of these results is not easy, but some of the variations cannot be dismissed as experimental error.

In discussing the implications of variations in ionic ratios, Carpenter and Carritt (1959) have pointed out a tendency which has crept into the literature to regard the relative composition of sea water as constant, and to completely disregard the fact that variations have been reported. Admittedly the variations found are small and are often of the same magnitude as the experimental error of the analytical method used and for many purposes they may be regarded as negligible. A more accurate assessment of their magnitudes and, if possible, their causes is still required.

An investigation which may throw some light on the matter is now in progress at the National Institute of Oceanography and the University of Liverpool. Measurements of density, electrical conductivity and chlorinity on sea water samples from all over the world have shown that conductivity is a better measure of density than is chlorinity (Cox *et al.*, 1962). In particular, deep water samples have been found to have a slightly higher density, relative to chlorinity, than have surface waters. A probable explanation of this is that ionic ratios do vary, and the waters are now being analysed, using modern analytical techniques, in order to find out which of the constituents vary sufficiently to account for the anomalies. Preliminary results show that the ratios of both sulphate and bromide to chlorinity do not vary significantly with depth or position. The ratios of calcium to chlorinity for deep waters are slightly higher than those for surface waters. The differences are not as great as those found by Dittmar, but they are statistically significant. Magnesium shows no tendency to vary with depth, but a few regional variations have been noted. This work is not yet complete, but the results obtained so far tend to confirm that small variations in the relative composition of sea water do occur.

III. Methods of Determination of Major Constituents

A. Calcium and Magnesium

Calcium and magnesium in sea water can be determined by both gravimetric and volumetric methods. It is convenient to consider both elements together because the determination of one, by either method, depends to a certain extent on the determination of the other.

(1) *Gravimetric methods.* In the precipitation of one constituent from a complex solution such as sea water, numerous difficulties arise because of the danger of co-precipitation of the other components. Attempts to minimize co-precipitation by multiple precipitations introduce the additional hazards of loss caused by solubility and losses occurring during transference etc. These generalizations are amply illustrated by the precipitation of calcium and magnesium.

The classical gravimetric methods consisted basically of the precipitation of calcium as oxalate followed by the precipitation of magnesium as magnesium ammonium phosphate. The calcium oxalate precipitate was either ignited to oxide and weighed as such (Dittmar, 1884; Thompson and Wright, 1930; Matida and Yamauchi, 1950), or dissolved in acid and titrated with permanganate (Kirk and Moberg, 1933; Gripenberg, 1937). Ignition of the magnesium ammonium phosphate converted it to magnesium pyrophosphate, in which form it was weighed.

The finely divided precipitate which is obtained when calcium is precipitated as oxalate from cold, neutral or alkaline solutions can be filtered and washed only with difficulty. A much coarser precipitate can be obtained by adding ammonium oxalate to a hot, acid calcium solution and then neutralizing it slowly with ammonium hydroxide. Slight contamination of the precipitate with oxalic acid or acid oxalate does occur, but this does not affect the results if the precipitate is ignited to calcium oxide. Slightly high results are obtained, however, if the calcium is determined volumetrically with permanganate.

When calcium is to be precipitated from a solution containing other ions, the problems of co-precipitation become more serious. In general, co-precipitation of anions is greater in hot solutions than in cold, while for cations the converse is true. Sodium is more strongly co-precipitated than is potassium, and its presence usually necessitates a double precipitation. Co-precipitation of sulphate is caused by the low solubility of calcium sulphate and, although this is partially decomposed to the oxide when the oxalate precipitate is ignited, it is a source of error. Some contamination of calcium oxalate with magnesium might be expected since the solubility of magnesium oxalate is low. This danger is reduced by the tendency of the magnesium salt to form a soluble complex, but some postprecipitation may take place if the solution is allowed to stand too long before filtration owing to decomposition of the complex. The solubility of calcium oxalate increases in the presence of magnesium ions, presumably because of competition for oxalate ions. Errors caused by this effect can be avoided by addition of sufficient oxalate to complex both the calcium and magnesium present.

Elimination of magnesium contamination of the calcium oxalate precipitate is doubly important if determination of magnesium in the filtrate is to follow, and much emphasis has been placed on this point in the application of the method to the determination of calcium in sea water. Thompson and Wright (1930) recommended triple precipitation, Kirk and Moberg (1933) were satisfied with two precipitations, whereas Gripenberg (1937) claimed satisfactory results with only one precipitation at a pH of 4·0 to 7·4.

The effect of each of these procedures on the strontium content of the precipitate was ignored for some time. The view that strontium is completely co-precipitated with calcium as oxalate has been opposed by Banewicz and Kenner (1952), Chow and Thompson (1955b), and Carpenter (1957). If calcium figures are to be corrected for strontium interference, due regard must be paid to the precipitation procedure, and also to the method used in the final stage of the determination (i.e. gravimetric as calcium oxide or volumetric with permanganate).

In order to obtain reliable results for the determination of magnesium as pyrophosphate it is first necessary to obtain a precipitate of pure $MgNH_4PO_4$. The precipitation must be carried out at room temperature or lower, because of the fairly high solubility of the precipitate, and in the absence of all other metals except the alkalis. Co-precipitation of sodium is not serious, but potassium tends to form mixed crystals of $Mg(NH_4,K)PO_4$. However, the most serious contamination is caused by co-precipitation of ammonium phosphates and magnesium phosphates which, on ignition, are not converted to magnesium pyrophosphate. Ammonium salts are usually present in large amounts when calcium has been removed from the solution previously by oxalate precipitation, and ammonium interference can, therefore, be very significant in the initial precipitation. Excess of oxalate may also tend to retard precipitation of magnesium. It is usually possible to obtain quantitative precipitation of magnesium, but it is virtually impossible to obtain it in a pure state by a single precipitation. A second or third precipitation, when the amount of ammonium salt and excess of ammonium phosphate can be carefully controlled, is necessary if a precipitate of sufficient purity for conversion to magnesium pyrophosphate is to be obtained.

In some of the earlier determinations of the magnesium content of sea water, only one or two precipitations were used. However, Epperson (1928) found that three precipitations were necessary to remove potassium contamination from the precipitate, and Thompson and Wright (1930) confirmed this finding. The high results obtained by Bertrand (1922), using two precipitations on synthetic sea water also suggested that at least three precipitations are necessary. It has been pointed out by some workers (Schmelck, 1882; Ellis and Matthews, 1928) that multiple precipitations introduce losses, owing to the solubility of magnesium ammonium phosphate. Such losses are more probably caused by hydrolysis, and can be avoided by washing the precipitate with dilute ammonium hydroxide.

Magnesium can be quantitatively precipitated from alkaline solution with 8-hydroxyquinoline (oxine). After drying at 105°C, the precipitate has the formula $Mg(C_9H_6ON)_2 2H_2O$, but it becomes anhydrous if dried at 160°C, at which temperature co-precipitated oxine is volatilized. The method requires the absence of all other metals except the alkalis, and has been applied to the determination of magnesium in sea water only by Ellis and Matthews (1928). Thompson and Wright (1930) also tested this method, but rejected it on the grounds that it was far less accurate than the pyrophosphate method.

(2) *Volumetric methods.* The pioneer work of Kampitsch *et al.* (1945) on the use of aminopolycarboxylic acids (complexones) in complex-formation titrations, and the subsequent development of metal ion indicators which respond to metal ion activity in solution, has revolutionized the volumetric determination of metals.

The complexones have the characteristic grouping $-N(CH_2.COOH)_2$ and the most important one is ethylenediamine-N,N,N',N'-tetra-acetic acid (E.D.T.A.) which forms stable chelates with a wide range of polyvalent cations. At pH values above *c.* 8·5 calcium, magnesium and strontium form strong complexes (with log formation constants (K) of 10·70, 8·69 and 8·63 respectively at 20°C). If a solution of one of these elements, buffered to pH 9·5 to 10·0, is titrated with a solution of E.D.T.A. (di-sodium salt), and pM ($= -\log [M]$, where M = Ca, Mg or Sr) is determined potentiometrically, using a mercury electrode, then a plot of pM against the fraction of metal titrated yields a sigmoidal curve, similar to a pH titration curve.

The end-point in the titration can be detected with electrode systems, but it is more convenient to use metallochromic indicators. These are compounds which form coloured complexes with metal ions above certain characteristic pM values in the same way that acid-base indicators form hydrogen ion complexes within certain pH ranges. A detailed discussion of their use has been presented by Reilley and Schmid (1959). The general requirements of a metallochromic indicator are (i) that the free indicator must have a different colour from the metal-indicator complex, (ii) that the metal-indicator complex and the E.D.T.A.-metal complex must both be formed rapidly under the same conditions, (iii) that the metal-indicator complex must be less stable than the E.D.T.A.-metal complex.

The indicators used in the E.D.T.A. titration of calcium and magnesium are usually *o,o′*-dihydroxy-azo compounds, a typical example of which is Eriochrome Black T (E.B.T.) (I). In the pH range 8 to 11 this compound has a blue colour, and with calcium and magnesium it forms relatively weak, chelate ring complexes (II) which are red in colour.

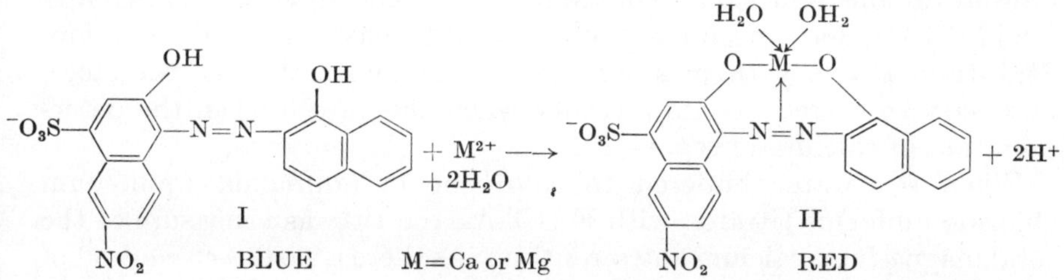

The colour of the magnesium complex is very intense, and a 10^{-6}M solution of magnesium will give a detectable colour with E.B.T. In the presence of an excess of metal ions, the solution has the red colour of the metal-E.B.T. complex. As the titration proceeds, the metal ions are complexed by the E.D.T.A., which forms much stronger complexes with the metal than does the indicator (ratio of formation constants of E.D.T.A. and indicator complexes $>10^4$) and this leads to decomposition of the red complex. At about the stoichiometric end-point, the red colour is replaced by the pure blue colour of the free indicator. Although the end-points of the titrations are visually reasonably sharp, considerably greater precision can be attained by the use of a photometric titrator (cf. Headridge, 1961).

The calcium-E.B.T. complex is less stable than that of magnesium, and E.B.T. is, therefore, of little use as an indicator for the titration of calcium. If a small quantity of the magnesium-E.D.T.A. complex is added to a calcium solution however, the calcium is preferentially complexed by the E.D.T.A. and a sharper end-point is obtained, since it is virtually that of the magnesium complex. Several other indicators have been recommended for use in the titration of calcium with E.D.T.A. Murexide (ammonium purpurate), which was introduced by Gysling and Schwarzenbach (1949), has been proposed by de Sousa (1954), Shtukovskaya *et al.* (1957) and Kobanov and Skopintsev (1958) for the determination of calcium in sea water. Pate and Robinson (1958) found that visual detection of the end-point was difficult, but Carpenter (1957) obtained satisfactory results using a photometric method, after separation of calcium from magnesium. Belcher *et al.* (1958), who have investigated the determination of calcium in the presence of different amounts of magnesium, have recommended the use of Calcon (Hildebrand and Reilley, 1957) and this indicator was used in a photometric method by Voipio (1959). The use of Calcon for the determination of calcium in sea water was rejected by Pate and Robinson (1958) in favour of Calred, which was found to have similar sensitivity but gave greater accuracy. For the determination of calcium in estuarine waters, Calcein (Molins and Rial, 1962) and Eriochrome Blue S.E. (Price and Priddy, 1961) have been employed, but they gave only low precision. Strontium reacts with most of these indicators and it is, therefore, necessary to correct for its presence when they are used in the determination of calcium in sea water.

When sea water buffered to a pH of 10 (ammonia–ammonium chloride buffer) is titrated with E.D.T.A., the titre is a measure of the total magnesium, calcium and strontium. As there is no direct method of titrating magnesium alone, it has to be determined by difference, after deducting the titre due to calcium and strontium. Titration of calcium plus strontium is usually carried out at a pH of 12 or more (sodium hydroxide or diethylamine buffer), under which conditions magnesium is precipitated as its hydroxide which does not react with E.D.T.A. When the ratio of magnesium to calcium is greater than 0·5:1, as it is in sea water, serious errors can arise owing to co-precipitation of calcium with the magnesium hydroxide. In much of the published work on the determination of calcium in sea water, no correction has been applied for this loss, which has been assessed by Voipio (1959) at 1% of the total calcium. Pate and Robinson (1958) have reduced this loss by adding 95% of the theoretical quantity of E.D.T.A. prior to precipitation of the magnesium.

Unfortunately, in most methods involving precipitation of magnesium hydroxide, the indicator colour tends to fade owing to adsorption by the precipitate. It is now possible, however, to titrate calcium selectively in the presence of magnesium with a complexone developed by Anderegg *et al.* (1957). This compound, 1, 2-bis-[2-di-(carboxymethyl)-aminoethoxyl]ethane (E.G.T.A.), forms a calcium complex of high stability ($\log K = 11{\cdot}0$) and a comparatively weak magnesium complex ($\log K = 5{\cdot}2$). Reilley and Schmid (1957) have used a mercury indicator electrode in the potentiometric titration of calcium in the presence of magnesium. The lack of a specific indicator for calcium was overcome by Pensar *et al.* (1958) who used an indirect indicator consisting of Zincon and Zn-E.G.T.A.

$$\underset{\text{ORANGE}}{Ca^{2+} + HIn^{3-}} + \text{Zn-E.G.T.A.}^{2-} \rightarrow \text{Ca-E.G.T.A.}^{2-} + \underset{\text{BLUE}}{ZnIn^{2-}} + H^+$$

If the Zn-E.G.T.A. complex is added to an ammoniacal calcium solution, an exchange reaction takes place and the solution turns blue. When the solution is titrated with E.G.T.A., all the calcium is complexed first, and finally the titrant reacts with the zinc–zincon complex, liberating the free indicator and changing the colour of the solution from blue to orange-yellow. This method has been successfully adapted to the direct titration of calcium in sea water, by F. Culkin (unpublished work), using photometric detection of the end-point.

Because of the above-mentioned difficulties in the direct titration of calcium and magnesium in the presence of each other, a number of workers have separated the cations by ion exchange techniques, and have then titrated them separately with E.D.T.A. Carpenter (1957) adsorbed the cations on a sulphonated polystyrenedivinylbenzene cation exchanger. Magnesium was eluted with 0·2 M ammonium acetyl acetonate, calcium with a 1 M solution of the same reagent, and finally strontium with 2 M ammonium chloride. Before calcium and magnesium can be titrated in the eluates, it is necessary to decompose the eluting agent. Carpenter carried out the decomposition by heating the solution, but it has been found by other workers that the yellow decomposition products interfere with visual (but not with the photometric) titration.

R. Greenhalgh, J. P. Riley and M. Tongudai (unpublished work) eluted sodium and potassium from a similar resin with 0·15 M ammonium chloride, magnesium with 0·35 M ammonium chloride, calcium with ammonium acetylacetonate and finally strontium with ammonium nitrate, but found it necessary to evaporate the calcium fraction to dryness with nitric acid to destroy the eluting agent prior to photometric titration. These workers have also found it possible to separate all the major cations in sea water using ammonium chloride, without recourse to acetylacetone, but the eluate volumes required are rather large.

Kitano and Tsubota (1960) used ammonium formate–formic acid buffer (pH 4·0) for the elution of first magnesium and then calcium. This eluting agent would probably not interfere with the E.D.T.A. titration of either element but is easily removed by heating. The separation of calcium and magnesium is not as clean as with acetylacetone, however.

(3) *Flame photometric methods.* Although flame photometry is widely used for the determination of calcium in natural waters, it is subject to serious interference from other constituents when applied to sea water. Chow and Thompson (1955b) made a study of these interferences and found that chloride, magnesium, potassium and sulphate ions showed a negative radiation interference, and that sodium gave a positive radiation interference in the calcium determination. They recommended an "internal standard" technique for the determination of calcium in sea water. In this method, known amounts of calcium were added to aliquots of sea water (diluted fifty times) and emission intensities at 422·7 mμ and background emissions at 418 mμ were measured. Net emission intensity was plotted against added calcium, and the calcium content of the sample was calculated from the slope and intercept of the graph. The method assumes that interfering substances affect the emission of standard and unknown calcium equally, and a coefficient of variation of *c.* 1% was claimed.

B. SODIUM AND POTASSIUM

For the determination of sodium and potassium in sea water, two general methods are available. One element (usually potassium) can be determined directly and the other found by difference from the sum of the two, or both can be determined directly.

The first procedure is less satisfactory. Determination of the sum of sodium and potassium as chlorides or sulphates requires a preliminary separation of the alkalis from the divalent metals. Usually, calcium is precipitated as oxalate and magnesium as phosphate, but in both of these methods there is a tendency for the alkalis to be co-precipitated. Dittmar (1884) determined the total calcium, magnesium, potassium and sodium as sulphates, the first three being obtained individually and sodium by difference. All four sulphates are stable to heat, and the major criticism which can be made of the method is that the sodium figure is subject to errors in the determination of the other three metals. A much more reliable determination of total alkalis has been developed by R. Greenhalgh *et al.* (unpublished work) who separated them from the divalent metals by ion exchange prior to weighing as sulphates. F. Culkin (unpublished work) has determined total cations by passing the sea water through an ion exchange resin (H^+ form) and titrating the liberated acid. The alkaline earths are determined by E.D.T.A. titration, and the alkalis are obtained by difference.

Direct determination of the alkali metals is probably one of the most difficult problems in analytical chemistry. Most alkali metal salts have an appreciable solubility in water. Determination of potassium in sea water is further complicated by the fact that about thirty times more sodium than potassium is present. Some of the less soluble salts can be used for gravimetric determination, however, if volumes of solutions are kept to a minimum and organic solvents are used to reduce solubility.

The only reagent which has been used for the direct determination of sodium in sea water is zinc uranyl acetate. With sodium, this reagent forms a precipitate with the formula $Na(C_2H_3O_2)Zn(C_2H_3O_2)_23UO_2(C_2H_3O_2)_26H_2O$, which has a solubility of 5·85 g. per 100 ml. of solution. The weight of the precipitate is thus sixty-seven times the weight of sodium from which it is derived and, therefore, only a small sample of sea water (usually 1 ml.) is required. It has been found that the water content of the precipitate is slightly higher than would correspond to the hexahydrate, but this is compensated by small solubility losses in the determination. The method has been used for the determination of sodium in sea water by Webb (1939) and by Knapman and Robinson (1941) who claimed a coefficient of variation of 0·1 – 0·2%.

Fortunately, there is a slightly wider choice of reagents for the determination of potassium. In all the determinations of potassium in sea water up to 1939 the chloroplatinate method was used. This method is based on the fact that potassium hexachloroplatinate is only sparingly soluble in 80% ethyl alcohol, whereas the sodium salt is very soluble. A solution of the alkali metals, in the form of their chlorides, is treated with sufficient chloroplatinic acid to convert both metals to their chloroplatinates. The solution is evaporated to a small volume and the sodium salt and excess of the reagent are extracted from the residue with 80% ethyl alcohol. Sulphate must be absent, but small amounts of calcium, magnesium, and strontium chlorides do not interfere. The dried precipitate does not have the exact composition K_2PtCl_6, and it is necessary to use an empirical factor in calculating the weight of potassium from the weight of the residue. Dittmar (1884), who appears to be the only worker who made careful trials of this method on synthetic sea water, found that the most satisfactory results could be obtained by reducing the precipitate with hydrogen, washing out soluble salts with water and weighing the residual platinum metal. This method has been used by other workers, but Webb (1939) has reported that it gives results which are *c.* 0·7% high.

A less accurate procedure is to precipitate the potassium as $K_2NaCo(NO_2)_6$ or $K_2AgCo(NO_2)_6$. Webb (1939) determined the precipitate volumetrically with ceric sulphate. Jentoft and Robinson (1956) dissolved the precipitate in acid and reprecipitated the potassium as meta-periodate, which they determined iodimetrically. Jentoft and Robinson claimed that, after correcting for a small solubility loss, the coefficient of variation of the K:Cl‰ ratios was *c.* 0·2%. The ratio of potassium to sodium in the precipitate varies with the conditions of precipitation, and the ratio of the two metals in the sample. Belcher and Robinson (1962), who have examined the precipitate formed under various conditions by reaction of sodium cobaltinitrite with potassium solutions using X-ray diffraction and electron microscopy, have concluded that it is an unsuitable weighing form for the determination of potassium.

Fukai and Shiokawa (1955) used the perchlorate method in which the alkali metals are converted to their perchlorates and then the more soluble sodium salt is extracted with an organic solvent. This method was found to have a standard deviation of ±5%.

The introduction of sodium tetraphenylboron as a reagent for the determination of potassium is one of the most important contributions to analytical chemistry in recent years. The sodium salt of this reagent is very stable in air and can be heated to 200°C without decomposition. It is readily soluble in water, forming almost neutral solutions which can be boiled without decomposition. In contrast, the potassium salt has a comparatively low solubility in water (5·4 mg./100 ml.), and this, together with its stoichiometric composition, high equivalent weight, and good thermal stability (minimum decomposition temperature 265°C) makes it an excellent weighing form for potassium.

The conditions to be used for the precipitation are governed mainly by the composition of the sample. Precipitation from alkaline solution necessitates masking the heavier elements to prevent their precipitation, and yields a fine precipitate which is difficult to filter. Increasing the acidity produces a coarser precipitate, and also reduces interference from other cations. Unfortunately, decomposition of the reagent increases with increase in acidity and temperature, and precipitation from a solution which is more than 0·1 M in mineral acid must be carried out at low temperature.

Precipitation at 0°C with sodium tetraphenylboron was used for the determination of potassium in sea water by Sporek (1956). He used a final acid concentration of 0·44 N HCl, and claimed a coefficient of variation of 0·6% for the method. The results for potassium in synthetic solutions were about 1% high because of interference by other cations.

C. SULPHATE

Methods of Determination. Few methods are available for the determination of sulphate, and of these, the most accurate is probably the classical gravimetric method of precipitation as barium sulphate. This is the method which has been applied almost exclusively since 1819 to the determination of sulphate in sea water, although some workers have introduced minor modifications of procedure to reduce the many interferences to which the method is subject.

It is not possible to purify barium sulphate by repeated precipitation because of its low solubility in reagents other than hot concentrated sulphuric acid, and co-precipitation must, therefore, be reduced to a minimum in one precipitation. The most serious cause of contamination of the precipitate is co-precipitation of calcium as calcium sulphate. The extent to which this can be reduced by carrying out the precipitation in the presence of hydrochloric acid is limited by the increased solubility of barium sulphate in the acid. Bather and Riley (1954) investigated the effect of hydrochloric acid concentration on the recovery of sulphate from a synthetic sea water. They obtained quantitative recovery by carrying out the precipitation in 0·06 N hydrochloric acid and confirmed that the precipitate contained only negligible amounts of calcium and strontium.

Co-precipitation of the alkali metals is less serious, and can be minimized by carrying out the precipitation in the presence of picric acid. Rapid addition of the barium chloride to the sample at $>90°C$ also helps to reduce co-precipitation.

The high precision of the barium sulphate method is certainly largely caused by a compensation of negative and positive errors, but it is still the most satisfactory method of determining sulphate in sea water.

D. STRONTIUM

Strontium occurs in sea water in concentrations which are significant in terms of salinity, but low enough to create difficulties in its determination. No colorimetric or gravimetric reagents are known with which strontium can be determined in the presence of large quantities of calcium and magnesium. In many of the determinations of the strontium content of sea water, co-precipitation with calcium as oxalate has been used to concentrate it, prior to its determination. The earlier workers used spectral methods, but these were subject to interference from other constituents of sea water. Miyake (1939b) converted the oxalates to nitrates, extracted the calcium with an alcohol–ether mixture and weighed the residual strontium nitrate. Kawasaki and Sugawara (1958) and Odum (1951a, 1957) used a flame photometric method for the determination of the strontium in the oxalate precipitate. All methods in which strontium is concentrated by oxalate precipitation are subject to error however, because it has been found that co-precipitation of strontium is seldom quantitative, and depends on the method of precipitation. The most satisfactory method of separating strontium from the other constituents of sea water is ion exchange (Carpenter 1957; R. Greenhalgh *et al.*, unpublished work), although it is time consuming.

When applied to the direct determination of strontium in sea water, flame photometric methods are subject to considerable interference from the other constituents of the water. Sodium, calcium and potassium enhance the emission of strontium, and sulphate, magnesium and chloride have a depressive effect. Chow and Thompson (1955a) overcame these interferences by using an "internal standard" technique, analogous to that which they used for calcium.

The use of radiochemical methods, either alone or in combination with one of the above-mentioned methods, has done much to establish the true concentration of strontium in sea water. Smales (1951) used a combination of radiochemistry and flame photometry in his preliminary work. Hummel and Smales (1956) used both isotope dilution and radioactivation techniques. In the isotope dilution method, a radioactive tracer, ^{89}Sr or ^{84}Sr, was added to a large sample of sea water and the calcium and strontium were then precipitated as carbonates. After separation from calcium as nitrate, the strontium was converted to sulphate and weighed. The activity of the tracer in the precipitate was measured, and a correction was applied for losses caused by incompleteness of precipitation. In the radioactivation method, the sea water sample was subjected to neutron irradiation, before chemical separation of strontium as sulphate. The radiation from ^{87}Sr or ^{89}Sr in the residue was counted and compared with that of a standard which had been similarly treated. The results obtained by the two methods were in good agreement, the isotope dilution method having a lower coefficient of variation ($\pm 1\%$) than the radioactivation method ($\pm 4\%$). Bowen (1956) also used a radioactivation technique in which strontium was separated by ion exchange.

E. BORON

Boron occurs in sea water as boric acid which, because of its low dissociation constant, is only *c.* 1% ionized at the pH of ocean water; it cannot therefore be titrated directly. However, certain organic polyhydroxy compounds, such as mannitol and glycerol, complex boric acid and convert it into a strong acid which can be titrated with alkali. The organic compound must contain *cis*-hydroxyl groups and it is thought that the monobasic acid which is formed is a cyclic ester of the type

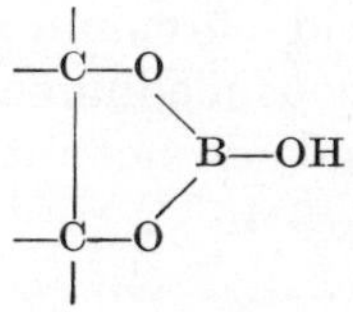

The titrimetric method has been used more than any other for the determination of boron in sea water. Standard acid is added, the solution is boiled to expel carbon dioxide, and excess of acid is titrated (to pH 7·0) with carbonate-free alkali. Mannitol is then added and the boric acid is titrated to pH 7·0. Most of the results shown in Table IX have been obtained by this method. In the earlier determinations, indicators were used to detect the end-point of the titration. More recently, Gast and Thompson (1958) and Hood and Noakes (1961) have found that potentiometric detection of the end-point (pH 7·0) is more satisfactory, the method having a coefficient of variation of *c.* ±1%.

In 1958, Gast and Thompson suggested the presence of *cis*-type polyhydroxy organic complexes of boron in sea water. Such compounds cause boric acid to be titrated together with the excess of acid and hence the usual mannitol method gives low results in their presence. In order to determine the total boron it is first necessary to destroy these organic compounds by vigorous oxidation with permanganate (Gast and Thompson, 1958) or peroxydisulphuric acid (Hood and Noakes, 1961).

Frederikson and Reynolds (1960) have used a spectrographic method for the determination of boron in sea water. High voltage spark excitation of the spectrum was accomplished using a rotating disc electrode system, and beryllium was used as internal standard. The method, which is simple, rapid and has a coefficient of variation of ±3%, does not appear to have been used by any other workers.

Barnes and Parker (1960) have described the fluorometric determination of boron in sea water with benzoin. The procedure involves a preliminary ion exchange resin treatment of the sample, and the coefficient of variation (±2%) is higher than that of the mannitol method.

One of the most sensitive tests for boron is the formation of a red dye by the reaction between boric acid and curcumin. This reaction forms the basis of numerous methods for the determination of boron in various materials. Usually an aqueous solution of the sample treated with hydrochloric and oxalic acids, and an alcoholic solution of curcumin is evaporated to dryness. Reaction between the boric acid and curcumin takes place only when all the water has been removed, and the method lacks reproducibility for a number of reasons. Greenhalgh and Riley (1962) have developed a method for the determination of boron in sea water in which the boric acid–curcumin complex is formed in a completely liquid and homogeneous medium consisting of phenol and acetic acid. Polyhydroxy compounds do not interfere, and the method has a coefficient of variation of $\pm$ 0·45%.

F. BROMINE

The determination of bromide in sea water is basically a problem of determining a small amount of bromide in the presence of a large quantity of chloride.

When the earlier determinations were carried out, the only available method consisted of precipitating the halides with silver nitrate, liberating the bromine by treating the precipitate with chlorine and determining the loss in weight caused by the conversion of silver bromide to silver chloride. The method requires a large sample of sea water, but, because of the low solubility of silver bromide, it is not necessary to precipitate all of the chloride. Dittmar (1884) achieved a satisfactory recovery of the bromide by twice precipitating *c.* 4% of the chloride and chlorinating the fractions separately.

Bromide can be oxidized to bromine with chromic acid or permanganate in acid solution, under which conditions only a trace of chloride is oxidized. This is the basis of a method developed by Evans (1931) for the determination of traces of bromide in chlorides, and it has been modified by some workers for the determination of bromide in sea water. The liberated bromine is distilled by passing a current of air through the solution and is then determined colorimetrically (Chelle and Denigès, 1912; Chelle, 1914; Miyake, 1939a) or volumetrically (Berglund, 1885; Winkler, 1916; Vasilev, 1937; Ratmanoff, 1937). Haslam and Moses (1950) showed that chromic acid alone was a more selective oxidizing agent than a mixture of chromic and sulphuric acids, but they rejected the method because of the long distillation time (7 hours) which was required for quantitative recovery of the bromine.

Oxidation of bromide to bromate, with hypochlorite in slightly acid solution, followed by treatment with potassium iodide and acid, and titration of the liberated iodine with standard thiosulphate is the basis of Kolthoff and Yutzy's (1937) method for the determination of bromide in chlorides. Excess of the oxidizing agent is destroyed with hydrogen peroxide or, preferably, sodium formate. This method, which has the advantage that one equivalent of bromine produces six equivalents of iodine, has been used successfully for the determination of bromide in sea water by Korpi and Thompson (1942), Haslam and Moses (1950) and Morris and Riley (1964). It has a coefficient of variation of $\pm 0{\cdot}2\%$ and appears to be the most satisfactory method available at present.

G. FLUORINE

The earliest determinations of fluoride in sea water were only semi-quantitative. Usually, the first step consisted of precipitating calcium fluoride or barium fluoride, and the methods then differed in their treatment of this precipitate. Forchhammer (1865) treated the precipitate with sulphuric acid and used the extent to which glass was etched by the liberated hydrogen fluoride as a measure of the fluoride content. Carnot (1896) precipitated fluoride with lime, converted the precipitate to silicon tetrafluoride, and finally weighed it as potassium fluorosilicate. Clausman and Gautier (1914) co-precipitated barium fluoride with barium sulphate. By means of an elaborate treatment of the precipitate in special apparatus, the fluoride was converted to lead fluoride and finally to lead sulphide, which was estimated colorimetrically. Taylor and Thompson (1933) investigated the precipitation techniques employed by the above-mentioned workers and concluded that fluoride was not quantitatively precipitated by any of the reagents used.

Most of the recent methods for the determination of fluoride in natural waters have been basically similar. Fluoride destroys the coloured complexes of certain metals, e.g. Al, Ti, Th, Zr, Fe, with organic dyes such as alizarin red S and thorin because of the tendency of these metals to form very stable complex fluorides. The resultant decrease in colour of the solution is a measure of the fluoride content. When applied to the determination of fluoride in sea water, these methods suffer from the disadvantage that chloride, and particularly sulphate cause severe interference by bleaching or precipitating the lake. This means that synthetic comparison standards containing the same amounts of chloride and sulphate as sea water must be used (Taylor and Thompson, 1933), or the reaction must be carried out in sulphuric acid medium (Anselm and Robinson, 1951). These modifications increase the time required for colour development and decrease the sensitivity of the method.

Recently, Belcher *et al.* (1959) have reported that the fluoride ion forms a blue complex with the red chelates of some rare earths and alizarin complexone (1,2-dihydroxyanthraquinonyl-3-methylamino-*N*,*N*-diacetic acid). Greenhalgh and Riley (1961) found that the lanthanum complex was the most sensitive towards fluoride. The spectrophotometric method which these workers developed for the determination of fluoride in sea water has a coefficient of variation of $0{\cdot}3\%$ and is by far the most satisfactory one available.

IV. Principal Determinations of Major Constituents

A. Calcium

Table II

Principal determinations of calcium in sea water

Reference	Ocean etc.	Method	Ca(g./kg.) for S = 35‰	Ca(g./kg.)/Cl‰
Dittmar (1884)	Various	1	0·4123	0·02128
Thompson and Wright (1930)	Pacific	1	0·4098	0·02115
Miyake (1939a)	Pacific	1	0·4107	0·02120
Matida and Yamauchi (1950)	Tokyo Bay	1	0·4127	0·02130
Fukai and Shiokawa (1955)	W. Pacific	1	0·4123	0·02128
	N. Pacific	1	0·4173	0·02154
Carpenter (1957)	Atlantic	1	0·4119	0·02126
Kirk and Moberg (1933)	Pacific	2	0·4136	0·02135
Gripenberg (1937)	Baltic	2	0·4177	0·02156
Chow and Thompson (1955b)	Pacific	3(a)	0·4241	0·02189
Kawasaki and Sugawara (1958)	W. Pacific	3(b)	0·4028	0·02079
	W. Pacific		0·3987	0·02058
	Indian		0·4067	0·02099
	Antarctic		0·4107	0·02120
Pate and Robinson (1958)	Pacific	4(a)	0·4134	0·02134
Voipio (1959)	Barents	4(a)	0·4039	0·02085
Carpenter (1957)	Atlantic	4(b)	0·4111	0·02122

Method 1. Precipitated as oxalate, weighed as oxide.
2. Precipitated as oxalate, titrated with permanganate.
3. (a) Direct flame photometric.
(b) Flame photometric after precipitation as oxalate.
4. (a) E.D.T.A. titration, direct.
(b) E.D.T.A. titration after ion exchange separation.

The results of the principal determinations of calcium are shown in Table II, and are grouped according to the method used.

The earlier results were obtained by the oxalate method. Comparison of these results with more recent ones is slightly complicated by the fact that accepted atomic weights have changed since the determinations were made. The earlier figures in Table II have been corrected by Fleming and Lyman (1940) for changes in atomic weight. A further correction has been applied for the strontium content of the precipitate (assuming a strontium/chlorinity ratio of 0·00042 for sea water). Where the calcium was weighed as oxide, usually after three precipitations, the results were in remarkably good agreement with one another. Apart from the results of Fukai and Shiokawa, they range from 0·02115 to 0·02130, the scatter was less than 0·5% about the mean. Gripenberg (1937), who carried out only one precipitation, and Kirk and Moberg (1933), who carried out two precipitations before determining the precipitate with permanganate, obtained higher results, presumably owing to contamination of the precipitate with other oxalates or oxalic acid.

Chow and Thompson (1955b), who applied a flame photometric method directly to sea water, obtained a calcium/chlorinity ratio of 0·02189, and found no significant variation with depth or position. This figure is unaffected by strontium but is considerably higher than the other figures in Table II. Kawasaki and Sugawara (1958) determined calcium by flame photometry, after precipitation as oxalate. Their figures are lower, and show considerably more variation than other figures obtained by oxalate precipitation.

Voipio (1959) obtained a mean calcium/chlorinity ratio of 0·02085 for the Barents Sea by direct E.D.T.A. titration, the ratio for deep waters being slightly higher than that for surface waters. Since no correction was made for a small negative error caused by adsorption of calcium ions on the magnesium hydroxide precipitate, Voipio's figures are probably low. Pate and Robinson (1958) also used direct E.D.T.A. titration, but they complexed most of the calcium before precipitating magnesium hydroxide. In this way they were able to reduce the negative adsorption error, and obtained an average calcium/chlorinity ratio of 0·02134 and found that the ratio did not vary significantly with depth. Probably the most elegant determination is that carried out by Carpenter (1957), who separated the calcium from magnesium and strontium before titrating with E.D.T.A. His figure of 0·02122 agrees well with that of Pate and Robinson, and confirms that Voipio's results were low.

The excellent agreement between the results obtained by the oxalate precipitation and E.D.T.A. methods for most of the major oceans suggests that calcium is a conservative element. Dittmar (1884), however, found that the differences between the mean values for deep waters and near surface waters were greater than his experimental error. He concluded that deep waters contained more calcium relative to total salt than did shallow waters. Voipio (1959) formed similar conclusions about the Barents Sea, although his deepest sample came from only 300 m. which is not deep by oceanic standards. This effect must be real for the Barents Sea, because although the negative error of this method rules out comparison of his results with those of other workers, it should not produce any bias with depth.

It is reasonable to expect that variations in the calcium content of sea water with position and depth might occur. Calcium carbonate is precipitated in large quantities in such areas as the Bahamas Bank under certain conditions (see Chapter 17, Volume 2). It is also extracted from surface waters in large quantities by animals with calcareous shells and when these animals die and fall to the bottom, some re-solution takes place. Calcium is a major constituent of river waters, and appreciable variations in calcium/chlorinity ratios can be expected in areas affected by land drainage. This has been amply demonstrated for the Baltic by Gripenberg (1937).

Whatever variations might be expected from these causes, the evidence to date suggests that deep waters contain only slightly more (<1%) calcium than surface waters, and that the average calcium/chlorinity ratio is 0·0213.

B. MAGNESIUM

TABLE III

Principal determinations of magnesium in sea water

Reference	*Ocean etc.*	*Method*	*Mg(g./kg.) for S = 35‰*	*Mg(g./kg.) / Cl‰*
Ellis and Matthews (1928)	E. Mediterranean	1	1·315	0·06785
	Gulf of Aden		1·320	0·06814
Dittmar (1884)	Various	2	1·318	0·06801
Thompson and Wright (1930)	N. Pacific	2	1·297	0·06695
Miyake (1939a)	N.E. Pacific	2	1·296	0·0669
	W. Pacific		1·310	0·0676
Matida and Yamauchi (1951)	Tokyo Bay	2	1·297	0·06697
Fukai and Shiokawa (1955)	W. Pacific	2	1·284	0·06627
	N. Pacific		1·285	0·06632
Voipio (1957)	Baltic	3(a)	1·297	0·06693
Voipio (1959)	Barents	3(a)	1·306	0·06742
Pate and Robinson (1961)	Pacific	3(a)	1·296	0·06689
F. Culkin (unpublished work)	Various	3(a)	1·294	0·06680
Carpenter (1957)	Atlantic	3(b)	1·292	0·0667

Method 1. Gravimetrically as Mg oxinate.
2. Gravimetrically as Mg pyrophosphate.
3. (a) Direct E.D.T.A. titration.
(b) E.D.T.A. titration after ion exchange separation.

With the exception of the determinations carried out by Ellis and Matthews (1928), who used the oxine method, which Thompson and Wright (1930) found to be inaccurate, all the major determinations of the magnesium content of sea water have been carried out either by precipitation of magnesium ammonium phosphate or by E.D.T.A. titration.

The earlier figures, obtained by the gravimetric method, show considerable variation, ranging from 0·06801 to 0·06627 for the magnesium/chlorinity ratio. This scatter gives the impression that magnesium is a non-conservative constituent of sea water, but it is probable that most of the variation is caused by analytical error. As with other elements, examination of the analyses of one worker for different oceans is more reliable than a comparison of analyses by different workers. Thus, Dittmar (1884) failed to find any variations in the ratio of magnesium/chlorinity which were significantly greater than the analytical errors. Voipio (1957, 1959) also found that the magnesium/chlorinity ratio was constant within the Baltic and Barents Seas, although the figures for the two seas were different. This finding for the Baltic is interesting in view of the fact that Gripenberg (1937) found significant variations in the calcium/chlorinity ratio for the same region. A simple explanation of this difference in behaviour between calcium and magnesium is that the fresh water sources of the Baltic are poor in magnesium and hence have little effect on its ratio to chlorinity, whereas the relatively higher influx of calcium produces regional variations.

Results obtained by E.D.T.A. titration methods agree much better with one another than do those obtained by gravimetry. The figures for the Baltic and Barents Seas, reported by Voipio, are probably a little high because his calcium figures, on which they depend, are low. The results obtained by Carpenter (1957), who used ion exchange separation, and by Pate and Robinson (1961) and F. Culkin (unpublished work), who used direct titration, are only slightly lower than the previously accepted best value obtained by Thompson and Wright (1930).

Since the number of samples which have been examined by E.D.T.A. titration is not very large, it would be premature to suggest that magnesium is a conservative element in sea water. Determinations of calcium and magnesium, preferably by independent methods, or after separation from each other, for more of the major oceans are desirable. At present it is only safe to say that the ratio of magnesium to chlorinity is *c.* 0·0668, and that there are insufficient data to estimate to what extent this ratio may vary in the oceans.

In attempting to assess the sodium results, it must be borne in mind

C. SODIUM

TABLE IV

Principal determinations of sodium in sea water

Reference	*Ocean etc.*	*Na(g./kg.) for S = 35‰*	*Na(g./kg.)/Cl‰*
Morton and Thorpe (1871)	Irish	10·797	0·5573
Schmidt (1878)	Baltic and White	10·725	0·5536
Schmelck (1882)	Atlantic	10·663	0·5544
Forsberg (1883)	Siberian	10·625	0·5484
Dittmar (1884)	Various	10·683	0·5514
Natterer (1892, 1893, 1894)	Mediterranean	10·288	0·5310
Kolotoff (1893)	Black	10·691	0·5518
Makin (1898)	Atlantic	10·609	0·5476
Schloesing (1906)	Atlantic and Mediterranean	10·710	0·5528
Wheeler (1910)	Atlantic	10·786	0·5567
Steiger (1910)	Atlantic	10·786	0·5567
Anderson and Thompson (1932)	Puget Sound	10·646	0·5495
Webb (1939)	Firth of Clyde	10·751	0·5549
Miyake (1939a)	W. Pacific	10·650	0·5497
Knapman and Robinson (1941)	Puget Sound	10·776	0·5562
	Pacific	10·751	0·5549
Fukai and Shiokawa (1955)	W. Pacific	10·774	0·5561
	N. Pacific	10·759	0·5553
F. Culkin (unpublished work)		10·762	0·5555

that often the samples were stored for some time in glass containers before analysis, and the possibility of contamination cannot be ignored. In addition, many of the figures were obtained by difference methods and hence are open to the criticism that they depend on the accuracy of other determinations. It is, therefore, not surprising that, even if Natterer's figure is neglected, the results range from 0·5476 to 0·5573 for the sodium/chlorinity ratio—a spread of $\pm 1\%$ about the mean.

In support of the sodium/chlorinity ratio of 0·5549 which he obtained by the zinc–uranyl acetate method, Webb (1939) computed a ratio of 0·5554 using accepted data on other ionic ratios and excess base. Using Knudsen's relationship between chlorinity and density, Webb concluded that the lower figures obtained by some workers were in error.

Using data on ionic ratios and alkalinity, Fleming and Lyman (1940) calculated sodium by difference and obtained a sodium/chlorinity ratio of 0·5556.

It may be concluded, therefore, that any differences which may occur in the relative concentration of sodium in sea water are probably smaller than the errors in the methods used for its determination.

D. POTASSIUM

TABLE V

Principal determinations of potassium in sea water

Reference	*Ocean etc.*	*K(g./kg.) for S = 35‰*	*K(g./kg.)/Cl‰*
Schmelck (1882)	N. Atlantic	0·3925	0·02026
Forsberg (1883)	Siberian	0·4088	0·0211
Dittmar (1884)	Various	0·3931	0·02029
Natterer (1892, 1893, 1894)	Mediterranean	0·3890	0·02008
Kolotoff (1893)	Black	0·4069	0·0210
Makin (1898)	Atlantic	0·3945	0·02036
Macallum (1903)	Atlantic	0·3923	0·02025
Schloesing (1906)	Atlantic and Mediterranean	0·3784	0·01953
Wheeler (1910)	Atlantic	0·5095	0·0263
Steiger (1910)	Atlantic	0·3852	0·01988
Anderson and Thompson (1932)	Puget Sound	0·3700	0·0191
Webb (1939)	Atlantic	0·3892	0·02009
Miyake (1939a)	Pacific	0·3700	0·0191
Fukai and Shiokawa (1955)	W. Pacific	0·4117	0·02125
	N. Pacific	0·4061	0·02096
Jentoft and Robinson (1956)	Atlantic and Pacific	0·3921	0·02023
Sporek (1956)	Copenhagen standard sea water	0·4033	0·0208
	English Channel	0·4010	0·0207

The figures for potassium show larger variations than those for sodium, despite the fact that all workers, except Webb (1939), Fukai and Shiokawa (1955), Jentoft and Robinson (1956) and Sporek (1956), used the chloroplatinate method. Wheeler's (1910) potassium/chlorinity ratio of 0·0263 for the Atlantic is so much higher than any of the others for the same ocean that it can be discarded. The low figures obtained by Miyake (1939a) and by Anderson and Thompson (1932) can also be rejected as they are comparatively low. If these extreme values are ignored, the figures range from 0·0195 to 0·0213. Dittmar (1884) found no significant variation in the potassium/chlorinity ratio with depth or position, and Jentoft and Robinson (1956) found no difference in the ratios for the Atlantic and Pacific. It is probable, therefore, that the differences in the ratio of potassium/chlorinity shown in Table V represent analytical errors rather than differences between the sea waters. Fleming and Lyman (1940) have suggested 0·0200, the weighted mean of all determinations up to 1930, as the best value of the ratio of potassium to chlorinity.

Potassium is more readily adsorbed or ion exchanged than sodium on particles of detritus, and particularly clay, and consequently becomes concentrated in bottom deposits. This has been put forward as a reason for the much lower concentration of potassium in sea water compared with that of sodium. Removal of potassium from sea water by living organisms will have the same effect, but it is unlikely that the regional variations caused by either of these processes could be detected by existing analytical methods.

E. Sulphate

Table VI

Principal determinations of sulphate in sea water

Reference	*Ocean etc.*	SO_4^{2-} (*g./kg.*) *for* $S = 35‰$	SO_4^{2-} (*g./kg.*) / *Cl‰*
Morton and Thorpe (1871)	Irish	2·707	0·1397
Dittmar (1884)	Various	2·689	0·1388
Anderson *et al.* (1927)	N. Pacific	2·705	0·1396
Johnson *et al.* (1931)	Atlantic	2·699	0·1393
	Pacific	2·701	0·1394
	Mediterranean	2·705	0·1396
	Red	2·703	0·1395
	Indian	2·710	0·1399
	Baltic	2·739	0·1414
Miyake (1939a)	W. Pacific	2·707	0·1397
Maeda *et al.* (1939)	Pacific and Atlantic	2·751	0·1420
Matida (1951)	Tokyo Bay	2·701	0·1394
Bather and Riley (1954)	Irish	2·710	0·1399
Fukai and Shiokawa (1955)	W. Pacific	2·710	0·1399
	N. Pacific	2·707	0·1397
J. P. Riley (unpublished work)	Various	2·712	0·1400

The sulphate results shown in Table VI are more or less comparable since they were all obtained by the gravimetric barium sulphate method. The sulphate/chlorinity ratios obtained up to 1927, which have been summarized by Anderson *et al.* (1927), range from 0·1363 to 0·1449 with a mean of 0·1395. In striking contrast, the results obtained since 1927, with two exceptions, fall within the range 0·1388 to 0·1400. The result obtained by Maeda *et al.* (1939) is so much higher than the others that it must be regarded with suspicion.

Johnson *et al.* (1931) studied the sulphate/chlorinity ratios of several of the major oceans and concluded that, with the exception of the Baltic, the ratio was remarkably constant.

A study of the sulphate content of the Irish Sea by Bather and Riley (1954) revealed no significant variation in the sulphate/chlorinity ratio with time of year, position or depth. The mean value of 0·1399 agrees well with that of 0·1397 obtained by Morton and Thorpe (1871) for the same region. In a recent investigation of more than three hundred samples from all the major oceans, J. P. Riley (unpublished work) has found that the sulphate/chlorinity ratio varies from 0·1395 to 0·1405, with a mean of 0·1400 $\pm$ 0·00025; most of this variation can be accounted for by analytical errors.

Although the sulphate/chlorinity ratio of open sea water may thus be regarded as constant, certain factors tend to cause regional variations. When sea water freezes, the ice contains proportionately more sulphate than chloride, and there is a consequent lowering of the ratio in the brine. Fukai and Shiokawa (1955) detected this effect in the N. Pacific where they found a decrease in the sulphate content of waters from the minimum temperature layer which had been affected by freezing. Surface waters were found to be relatively richer in sulphate, presumably because of dilution by melting ice. Lewis and Thompson (1950) found that the effect is a non-linear function of the rate of freezing and the temperature at which freezing takes place.

The sulphate/chlorinity ratio of river water is sometimes as much as a hundred times that of sea water, and some variations might be expected in regions where the two waters mix. Such effects are usually very small, however, because of the small quantities of dissolved solids in river water.

Sulphate-reducing bacteria have been found occasionally in sea water, and they occur in great abundance in marine sediments. Such bacteria are active only under anaerobic conditions, and are responsible for the very high concentrations of sulphide which have been found in the stagnant deep waters of the Black Sea (cf. Chapter 13). Certain parts of the Baltic are also sufficiently stagnant for sulphate reducing bacteria to exist. The production of 10 mg. of hydrogen sulphide/l., which is not uncommon in the Black Sea, would reduce the sulphate/chlorinity ratio by 0·0015. In the open sea, complete deoxygenation of the water is encountered only rarely. Sudden intense growths of phytoplankton do take place sometimes in regions of upwelling. Subsequent decomposition of this vegetable matter tends to produce anaerobic regions where sulphate reducing bacteria thrive. Generally speaking, however, biological reduction of sulphate in ocean water is not sufficiently widespread to have any significant effect on sulphate/chlorinity ratios.

F. STRONTIUM

TABLE VII

Principal determinations of strontium in sea water

Reference	*Ocean etc.*	*Sr(mg./kg.) for S = 35‰*	*Sr(mg./kg.) / Cl‰*
Thomas and Thompson (1932)	San Juan Channel	12·9	0·667
Miyake (1939b)	Pacific	13·7	0·707
Smales (1951)	Atlantic	[10 mg./l.]	—
Odum (1951a)	Atlantic	8·12	0·420
Chow and Thompson (1955a)	Arctic	8·22	0·424
	Atlantic	8·23	0·425
	Pacific	8·12	0·420
Hummel and Smales (1956)	Atlantic	8·1	—
Bowen (1956)	English Channel	8·3	0·437
Kawasaki and Sugawara (1958)	Antarctic	9·05	0·467
	Indian	8·62	0·445
	Pacific	8·0	0·413
F. Culkin (unpublished work)	Various	8·0	0·41

The first quantitative work on the occurrence of strontium in sea water was carried out by Desgrez and Meunier (1926) who found 13·5 mg./l. by arc spectrographic examination of English Channel water. Robinson and Thompson (1932) reported that Thomas and Thompson (1932) obtained a similar figure (13 mg./l. in water of chlorinity 19‰). Concentrations of 40 mg./l. (Ramage, 1933), 7 – 8 mg./l. (Noll, 1934) and 8 – 10 mg./l. (Vinogradov, 1938) were also obtained by spectographic methods. Miyake (1939b) determined strontium gravimetrically by the nitrate method and found 14 mg./l. for the Pacific.

On the basis of these earlier investigations, the accepted value for the concentration of strontium in sea water was 14 mg./l. until 1951. Smales (1951) used a combined radiochemical and flame photometric method and put forward a tentative figure of 10 mg./l. All subsequent determinations have confirmed that the concentration of strontium in sea water is *c.* 8 mg./l., and that the old figure was much too high.

Chow and Thompson (1955a) applied a flame photometric technique (using internal standards) directly to sea water and concluded that the strontium/chlorinity ratio did not vary with depth or position. F. Culkin's (unpublished work) results also suggest that the ratio is constant, but Kawasaki and Sugawara (1958) found considerable variations, especially for the Pacific, for which the ratio was found to vary from 0·35 to 0·502. Discharge of land water into the sea, preferential escape of strontium from the sea into the atmosphere, and formation of calcareous shells and deposits were some of the factors which these workers discussed as possible causes of these variations.

Much work has been carried out on the strontium content of marine animals. Thomas (cited in Robinson and Thompson, 1932) found that strontium and calcium occurred in a large number of types of shells in very nearly the same ratio as in sea water. The finding that an Antarctic radiolarian had a shell which was reported to consist almost entirely of strontium carbonate led Robinson and Thompson (1932) to suggest that temperature has an influence on the composition of calcareous shells. They pointed out that the solubility of calcium carbonate increases with decrease in temperature, whereas the converse is true for strontium carbonate, and in cold waters strontium might replace calcium as a building material for marine shells. Odum (1951b), however, concluded that temperature was not an important factor and, moreover, showed that strontium existed in radiolaria as celestite (strontium sulphate) and not strontianite as previously assumed (Odum, 1951a).

Chow and Thompson (1955c) also showed that many types of marine organisms had the same strontium/calcium ratios, irrespective of the water temperature of their environment. Swan (1956) has pointed out, however, that an inverse relationship between growth rate and strontium/calcium ratios could be inferred from the data of Chow and Thompson.

Variations in the strontium content of sea water caused by formation or dissolution of marine shells must, however, be only very minor and local. The bulk of the evidence suggests that strontium is a conservative constituent of sea water, and has a ratio to chlorinity of 0·42 to 0·43.

G. BORON

The earliest estimates of 9 mg./kg. for the concentration of boron in sea water (Agulhon and Bertrand, 1914; Goldschmidt and Peters, 1932)

TABLE VIII

Principal determinations of boron in sea water

Reference	*Ocean etc.*	*B(mg./kg.) for S = 35‰*	*B(mg./kg.) / Cl‰*
Buch (1933)	English Channel	4·80	0·248
	North	4·59	0·237
	Kattegat	4·51	0·233
	Kattegat	4·69	0·242
	Gulf of Finland	4·75	0·245
	Mean	4·67	0·241
Harding and Moberg (1933)	Pacific	4·65	0·239
Mahncke and Rakestraw (1935)	S. of Cape Cod (S. origin)	4·65	0·240
	S. of Cape Cod (N. origin)	4·94	0·255
Igelsrud *et al.* (1938)	N.E. Pacific	4·67	0·241
Miyake (1939b)	Tokyo Bay	4·30	0·222
Miyake and Sakurai (1952)		4·18–5·04	0·216–0·260
Gast and Thompson (1958)	N.E. Pacific	4·61	0·238
	Puget Sound	4·60	0·237
Barnes and Parker (1960)	Red	4·69	0·242
	Arctic	4·49	0·232
	English Channel (Selsey)	4·59	0·237
	English Channel (Portland)	4·73	0·244
Frederikson and Reynolds (1960)		4·65	0·240
Hood and Noakes (1961)	Gulf of Mexico	4·30	0·222
	Caribbean	4·30	0·222
R. Greenhalgh *et al.* (unpublished)	Various	4·45	0·230

are now considered to be too high. In more recent work, most determinations have been carried out by the mannitol method, and there has been a large measure of agreement between the results.

The titrimetric determinations made by Harding and Moberg (1933), Buch (1933), Gast and Thompson (1958), and Hood and Noakes (1961) and the fluorometric determination made by Barnes and Parker (1960) have all established that, within the limits of experimental error, boron occurs in sea water in a constant ratio to chlorinity. Igelsrud *et al.* (1938) also found that the ratio tended to be slightly higher in the oxygen minimum layer. They suggested that this might be caused by the degradation of a soluble organic borate complex in this zone, with liberation of a detectable quantity of uncomplexed boron.

Mahncke and Rakestraw (1935), however, found significant differences in the boron/chlorinity ratios for waters of northern and southern origin collected near Cape Cod. Miyake and Sakurai (1952) also found that the ratio was not constant, but varied with depth. They suggested that it could be used in the identification of water masses and in the tracing of ocean currents.

Gast and Thompson (1958) confirmed the existence of soluble organic complexes of boron. They found that total boron, determined after permanganate oxidation of the complexes, was about 1% higher than the inorganic boron, on average. Unfortunately, no indication of the precision of their method was given.

Hood and Noakes (1961) used peroxydisulphuric acid to break down the complexes and claimed a coefficient of variation of $\pm 1\%$ for their method. They found a definite correlation between inorganic boron and chlorinity, except in the oxygen minimum layer where the ratio was up to 6% higher for one station in the Gulf of Mexico. The lowest concentrations of complexed boron were found in the oxygen minimum zone, confirming that breakdown of organic boron complexes to the inorganic form is probably responsible for variations in the inorganic boron/chlorinity ratios.

H. BROMINE

The wide disagreement between the results of previous workers led Dittmar (1884) to include bromine in his analyses of sea water. Because of the large quantity of water which his method required, he had to analyse composite samples from different depths. He found that the ratio of bromide to chloride was the same for all depths. Dittmar also failed to detect any difference in the bromine content of a sample sea water taken off the Isle of Arran, a region which he thought might be affected by the extraction of bromine by seaweeds.

Berglund (1885) and Winkler (1916), both using methods involving oxidation of bromide and aspiration of the liberated bromine, obtained results in agreement with Dittmar's.

Chelle (1914), using a colorimetric method involving bromination of fuchsin, found that the bromide/chlorinity ratios for the Red Sea and Pacific Ocean were higher than those for the Atlantic and Indian Oceans and the Mediterranean. His results are the highest reported since 1851, and it must be concluded that his method was inaccurate.

Ratmanoff (1937) found that the bromine contents of the Bering Strait and Bering Sea were considerably higher than that of the Pacific, and he used his own and Vasil'ev's (1937) results to show that the bromide/chlorinity ratio decreased with decrease in latitude.

TABLE IX

Principal determinations of bromine in sea water

Reference	*Ocean etc.*	*Br(g./kg.) for S = 35‰*	*Br(g./kg.)/Cl‰*
von Bibra (1851)		0·354	0·0183
		0·312	0·0161
Morton and Thorpe (1871)	Irish	0·063	0·0033
Schmidt (1878)	Baltic and White	0·0475	0·00245
		0·0503	0·00260
Dittmar (1884)	Composite surface	0·0656	0·00339
	Composite medium depth	0·0660	0·00341
	Composite deep	0·0658	0·00340
	Off Arran	0·0658	0·00340
	Mean of all waters	0·0660	0·00341
Berglund (1885)	North	0·0638–0·0665	0·0033–0·00344
	Atlantic	0·0652–0·0660	0·00337–0·00341
	Gulf of Mexico	0·0660	0·00341
	Mediterranean	0·0663	0·00343
	Adriatic	0·0660	0·00341
	Kattegat	0·0652	0·00337
	Baltic	0·0611–0·0665	0·00316–0·00344
Makin (1898)	Atlantic	0·0628	0·00325
Chelle (1914)	Atlantic	0·074	0·0038
	Red	0·083	0·0043
	Mediterranean	0·074	0·0038
	Pacific	0·085	0·0044
	Japan	0·083	0·0043
	Indian	0·074	0·0038
Winkler (1916)		0·067	0·00347
Cameron (1922)	Departure Bay	0·0693	0·00358
Vasil'ev (1937)	Japan	0·0643	0·00333
Ratmanoff (1937)	Bering Sea and Strait	0·0668	0·00345
Miyake (1939a)	W. Pacific	0·0640	0·0033
Korpi and Thompson (1942)	N.E. Pacific	0·0673	0·00348
	Antarctic	0·0667	0·00345
	Bering	0·0670	0·00347
Matida and Yamauchi (1950)	Japan	0·0670	0·00347
Gibson and Haslam (1950)	Coasts of Great Britain	0·0684	0·00353
A. W. Morris and J. P. Riley (unpublished work)	Various	0·0673	0·00348

Miyake's (1939a) results for the W. Pacific agreed with those of Dittmar (1884), but he admitted that his method of determining bromine colorimetrically after oxidation with chlorine was not very accurate.

Probably the most reliable method is the one in which bromide is oxidized by hypochlorite to bromate which is then determined iodometrically. Using this method, Korpi and Thompson (1942) found that the bromide/chlorinity ratio was constant (0·00347) for all depths of the Bering Sea and the Pacific and Antarctic Oceans, and A. W. Morris and J. P. Riley (unpublished work) obtained a mean value of 0·00348 for all oceans. A slightly higher ratio of 0·00353 was obtained by Gibson and Haslam (1950), but this was the mean for a large number of in-shore waters collected around the coasts of Great Britain.

The form in which bromine exists in sea water has been investigated by Cameron (1922) who concluded that if any bromate is present it constitutes less than 1% of the total bromide.

It may be concluded, therefore, that bromine is a conservative constituent of sea water, and exists as bromide with a ratio to chlorinity of 0·00348.

I. FLUORINE

The presence of fluorine in sea water was first inferred in 1850 when Wilson detected it in boiler scale. Forchhammer (1865) reported *c.* 0·7 mg./l. in sea water. Carnot (1896) obtained a similar figure, but Carles (1907) found 12 mg./l. in a sample of water from the Bay of Archachon. Like Carles' values for the fluoride content of mineral waters, this figure is so much higher than any other reported that it can be ignored as unreliable. While studying the fluoride content of mineral waters, Clausman and Gautier (1914) examined three samples of sea water. Mediterranean samples from 10 m. and 1000 m. were found to contain 0·33 mg./l. and 0·30 mg./l. respectively, whereas an Atlantic sample contained 0·24 mg./l. All the above-mentioned results must be regarded as low because the precipitation methods with which they were obtained did not give quantitative recovery of fluoride.

The zirconium–alizarin method was used by Taylor and Thompson (1933) who found that the fluoride content of sea water was proportional to the chlorinity, and ranged from 1·0 to 1·4 mg./l. No seasonal variation was detected in the fluoride/chlorinity ratio which was found to be $7{\cdot}0 \times 10^{-5}$ for the Pacific Ocean and slightly lower for water contaminated by land drainage. Using the same method, Miyake (1939c) found 1·3 mg./l. for the N.W. Pacific, and Matida (1954) found 0·77 – 1·33 mg./kg. for surface and shallow waters of Tokyo Bay.

The only comprehensive investigation of the fluoride content of sea water reported so far was carried out by Greenhalgh and Riley (1963) who used the lanthanum–alizarin complexone method. They found no significant variation in the fluoride/chlorinity ratio with depth in the Pacific and Indian Oceans. The ratio was constant ($6 \cdot 7 \times 10^{-5}$) for surface waters generally, except for a few high salinity waters from the Mediterranean, for which the ratio averaged $6 \cdot 4 \times 10^{-5}$. It was suggested that this phenomenon might be associated with precipitation of calcium. Some deep waters from the China Sea and the Atlantic Ocean, however, were found to have abnormally high fluoride contents (cf. Table X).

TABLE X

*Variation of $F^-/Cl‰$ with depth in N. Atlantic at 60°12′N 39°42′W**

Depth (m.)	*F^-(mg./kg.)*	*$F^-/Cl‰ \times 10^5$*
0	1·284	6·67
1000	1·290	6·67
1800	1·329	6·86
2000	1·361	7·01
2250	1·483	7·65
2420	1·490	7·71

* J. P. Riley, unpublished work.

The fluoride/chlorinity ratios in the deep waters of certain restricted areas of the N. Atlantic ($9 \cdot 0 \times 10^{-5}$), Mid-Atlantic ($8 \cdot 0 \times 10^{-5}$) and S. Atlantic ($8 \cdot 4 \times 10^{-5}$) were considerably higher than the average for surface waters. Greenhalgh (1963) showed that these waters did not contain unusual amounts of aluminium, thus eliminating the possibility of the dissolution of the Greenland cryolite (Na_3AlF_6) deposits as a cause of the high fluoride values. A more likely cause was considered to be submarine volcanic activity on or around the Mid-Atlantic Ridge.

From the work of Greenhalgh and Riley it may be concluded that fluoride is not a conservative constituent of sea water. Its concentration, relative to chlorinity, is fairly constant for surface waters ($6 \cdot 7 \times 10^{-5}$), but some regional variations with depth do occur.

V. ISOTOPIC VARIATIONS

A. ISOTOPES OF HYDROGEN AND OXYGEN

Although the dissolved salts are regarded as the major constituents of sea water, the principal constituent is, of course, water. By far the most abundant isotopic components of water are ^{1}H and ^{16}O but other stable isotopes of these elements do occur in nature.

Naturally occurring hydrogen contains *c.* 0·0156% of its heavier isotope deuterium (^{2}H or D), and in water the equilibrium $D_2O + H_2O \rightleftharpoons 2HDO$ is always maintained. In addition to these species, water also contains $H_2{}^{18}O$, formed from a heavy oxygen isotope which has a natural abundance of only *c.* 0·2% of that of the ^{16}O isotope. The presence of these heavier isotopes has a marked effect on some of the physical properties of water, and consequently a number of studies have been made of the distribution of D and ^{18}O in the hydrosphere.

(1) *Determination and reporting of D and ^{18}O variations.* Probably the only satisfactory method of determining D and ^{18}O is by means of the mass spectrometer. Prior to deuterium determination, the water sample is converted to hydrogen gas by reaction with hot zinc or uranium metal (Friedman, 1953; Horibe and Kobayakawa, 1960). In the determination of ^{18}O, the water sample is equilibrated with CO_2 gas, which is then analysed mass spectrometrically (Epstein and Mayeda, 1953).

The presentation of results has given rise to considerable difficulties since absolute isotopic abundances cannot at present be determined with sufficient accuracy to be of use in studies of their natural variations. In much of the earlier work, deuterium analyses were reported relative to local tap water standards. This was unsatisfactory because differences occur in the deuterium content of tap waters from different sources. Horibe and Kobayakawa (1960) even found random seasonal variations in tap water from one source (Tokyo). For their work on $^{18}O/^{16}O$ ratios in natural waters, Epstein and Mayeda (1953) used the average of their isotopic data for a number of deep water samples taken from the Atlantic, Pacific and Indian Oceans as their standard. This was more satisfactory, but there was still no readily available standard. Craig (1961), who pointed out the desirability of a standard reference sample for both D and ^{18}O analyses, determined the D/H ratio for a set of ocean waters similar to the ones chosen by Epstein and Mayeda (1953) for their ^{18}O data. He suggested that both D and ^{18}O data could be specified simply in terms of the National Bureau of Standards isotopic reference samples No. 1 and No. 1-A. These are distilled water samples of large volume, intended for cross-check circulation to mass spectrometric laboratories. After consultation with various laboratories, Craig defined a "standard mean ocean water" (SMOW) in terms of the National Bureau of Standards reference sample 1 such that

$$D/H\ (SMOW) \equiv 1{\cdot}050\ D/H\ (NBS\text{-}1)$$

$$^{18}O/^{16}O\ (SMOW) \equiv 1.008\,^{18}O/^{16}O\ (NBS\text{-}1).$$

It was recommended that isotopic data for both isotopes be reported as per mille enrichments (δ) relative to the defined SMOW, so that

$$\delta = \left[\left(\frac{R_{\text{sample}}}{R_{\text{SMOW}}} - 1\right) \times 1000\right]$$

where the R values are the respective isotopic ratios. SMOW, as thus defined in terms of an actual water reference standard NBS-1, provides a convenient and consistent zero reference level for reporting enrichments.

Results. Most workers who have studied variations in isotopic ratios have included samples of natural waters from other sources as well as the ocean in an attempt to explain the sources of the variations. Quantitative comparison of these results is not possible because of the different standards used, but the conclusions which were reached can be compared.

(2) *Deuterium determinations.* Friedman (1953) found that ocean surface waters have widely different D/H ratios. Equatorial waters in general were found to be richer in the heavier isotope than high latitude waters. Deuterium was found to follow the same trend with depth as ^{18}O, viz. surface waters contained relatively more than deep waters. Friedman's figures showed that the ratio D/H varied from 6410 to 6536 for the Atlantic, Pacific and Indian Oceans, i.e. the values were much lower than the figure of 6880 found by Dole and Swartout (1939).

Horibe and Kobayakawa (1960) determined the deuterium content of waters from the Pacific, Antarctic and Indian Oceans. They confirmed Friedman's findings that high latitude waters tend to contain less deuterium than do equatorial waters, and that deep waters contain less than do surface waters.

(3) *Oxygen-18 Determinations.* Epstein and Mayeda (1953) examined the $^{18}O/^{16}O$ ratios of a number of fresh waters derived from snow, rain, lakes and rivers, as well as ocean waters. They found that the ^{18}O content of the fresh water sources varied considerably, and consequently the ^{18}O content of a sea water depended very much on the amount and source of its fresh water. Addition of fresh water from melted snow, for instance, decreased the ^{18}O content *c.* 3·5 times as much as addition of the same amount of water from the Mississippi River.

Some of the lower salinity ocean waters showed a correlation between salinity and ^{18}O content, but this was attributed to the fact that most of these samples had been collected in the Alaska and California currents where a lowering in salinity is caused mainly by water from melting snow and ice.

Deep samples generally were found to contain less ^{18}O than surface samples. The differences were greater than would have been expected from evaporation, according to the calculations of Epstein and Mayeda (*loc. cit.*). They suggested that, even in the equatorial regions, the deep water was diluted by sinking polar surface waters, with a consequent lowering of its ^{18}O content. Mediterranean surface and deep samples showed a much smaller difference, probably because the high evaporation rate causes less fractionation of the isotopes, which in turn implies that fresh water entering the ocean is less depleted in ^{18}O than in the other cases discussed.

(4) *Causes of isotopic variations in sea water*. The difference in masses of isotopes of the same element causes compounds of the isotopes to have different physical properties. This effect is most marked in the case for compounds of protium and deuterium where the isotopic mass difference is relatively large. The vapour pressure of HDO is lower than that of H_2O, which means that during evaporation of water, H_2O enters the vapour phase more readily than does HDO, with the result that deuterium becomes concentrated in the residual water. This undoubtedly accounts for the fact that surface waters, especially in equatorial regions where there is considerable evaporation, tend to be richer in deuterium than do deep waters.

In precipitation processes the situation is reversed, and the residual phase (vapour) becomes depleted in deuterium as HDO is preferentially precipitated. As the raindrops fall to the ground, further isotopic enrichment can take place owing to evaporation, especially in dry climatic conditions, as shown by Ehhalt *et al.* (1963).

The freezing of sea water can also have an effect on isotopic ratios. Friedman *et al.* (1961) made deuterium analyses of co-existing ice and Arctic sea water and found that the ice contained *c.* 2% more D than the water from which it was formed.

Much that has been said of deuterium applies to ^{18}O. The difference in atomic mass between the oxygen isotopes is relatively much smaller than that between the hydrogen isotopes, and therefore the effect on physical properties is also smaller. This means that in evaporation processes, differentiation of $H_2{}^{16}O$ and $H_2{}^{18}O$ will be less marked, although it still occurs. No work comparable to that of Friedman on the effect of freezing of sea water on D/H ratios appears to have been carried out for oxygen isotopes, although it is known that in snow the $^{18}O/^{16}O$ ratio is much lower than it is in rain or sea water.

In general, the two isotopic ratios are quite variable in sea water. Both heavier isotopes are concentrated in surface water relative to deep water, but for different reasons. Evaporation appears to be the main cause of concentration of deuterium at the surface, whereas dilution by sinking polar waters produces a decrease in the ^{18}O content of deep waters.

B. ISOTOPES OF SULPHUR

Naturally occurring sulphur contains four stable isotopes having atomic weights of 32, 33, 34 and 36. Of these four, the heaviest is by far the least abundant, and its occurrence in sea water does not appear to have been studied. Trofimov (1949) found the $^{32}S/^{33}S$ ratio to be 123·4 for sea water. His ratio of 21·74 for the $^{32}S/^{34}S$ ratio in sea water has been confirmed by Ault and Kulp (1959) for the Pacific and Atlantic, by Östlund (1959) for the Arctic, North and Baltic Seas, and by Fleming *et al.* (1953) for the Atlantic, Pacific and Arctic. This ratio is lower than that found in rain water, suggesting that the sulphate–sulphur in precipitation is not derived from sea water spray. Collins *et al.* (1949) suggested that the most important agents in isotopic fractionation of sulphur were sulphate-reducing bacteria. When sulphate is reduced to sulphide, the equilibrium

$$^{34}S^{2-} + {}^{32}SO_4^{2-} \rightleftharpoons {}^{32}S^{2-} + {}^{34}SO_4^{2-}$$

will tend to produce H_2S which is deficient in the heavier sulphur isotope. This will occur only in sediments or anaerobic water, and the effect on the isotope ratio will not alter the physical properties of sea water significantly. It is significant that the sulphate minerals occurring in marine evaporite deposits have the same $^{32}S/^{34}S$ ratio as the sulphate of sea water. Sulphide minerals associated with precipitation in marine anaerobic conditions have a rather higher ratio (as high as 23·3 for shales).

REFERENCES

Agulhon, H., and Bertrand, G. (1914). *Bull. Soc. chim. Fr.* **15**, 292.

Anderegg, G., Schwarzenbach, G., and Senn, H. (1957). *Helv. chim. acta*, **40**, 1886.

Anderson, T. G., Lang, J. W., and Thompson, T. G. (1927). *Publ. Puget Sd. Mar. (biol.) Sta.* **5**, 290.

Anderson, T. G., and Thompson, T. G. (1932). Quoted in Robinson and Thompson (1932).

Anselm, C. D., and Robinson, R. J. (1951). *J. Mar. Res.* **10**, 203.

Ault, W. U., and Kulp., J. L. (1959). *Geochim. et cosmoch. Acta*, **16**, 201.

Banewicz, J. J., and Kenner, C. T. (1952). *Analyt. Chem.* **24**, 1186.

Barnes, W. J., and Parker, C. A. (1960). *Analyst*, **85**, 828.

Bather, J. M., and Riley, J. P. (1954). *J. Cons. int. Explor. Mer*, **20**, 145.

Belcher, R., Close, R. A., and West, T. S. (1958). *Talanta*, **1**, 238.

Belcher, R., Leonard, M. A., and West, T. S. (1959). *Talanta*, **2**, 92.

Belcher, R., and Robinson, J. W. (1962). *Analyt. chim. acta*, **27**, 568.

Berglund, E. (1885). *Z. anal. Chem.* **24**, 184.

Bergman, T. O. (1779). "Physical and Chemical Essays" (translated from original Latin by E. Cullen). J. Murray, London. Published 1788.
Bertrand, G. (1922). *C.R. Acad. Sci., Paris,* **174**, 1251.
Bibra, von (1851). *Liebigs Ann.* **77**, 90.
Bowen, H. J. M. (1956). *J. Mar. biol. Ass. U.K.* **35**, 451.
Buch, K. (1933). *J. Cons. int. Explor. Mer,* **8**, 309.
Calamai (1848). *Z. prakt. Chem.* **5**, 235.
Cameron, A. T. (1922). *Contr. Canad. Biol.* **1**, N.S., 75.
Carles, P. (1907). *C.R. Acad. Sci., Paris,* **144**, 1240.
Carnot, A. (1896). *Ann. Min., Paris,* **10**, 175.
Carpenter, J. H. (1957). *Limnol. Oceanogr.* **2**, 271.
Carpenter, J. H., and Carritt, D. E. (1959). *Nat. Acad. Sci. nat. Res. Coun. Publ.* **600**, 67.
Chelle, L., and Denigès, H. (1912). *C.R. Acad. Sci., Paris,* **155**, 1010.
Chelle, L. (1914). *Bull. Inst. océanogr. Monaco, No.* **281**.
Chow, T. J., and Thompson, T. G. (1955a). *Analyt. Chem.* **27**, 18.
Chow, T. J., and Thompson, T. G. (1955b). *Analyt. Chem.* **27**, 910.
Chow, T. J., and Thompson, T. G. (1955c). *Deep Sea Res.* **3** (Suppl.), 20.
Clausman, P., and Gautier, A. (1914). *C.R. Acad. Sci., Paris,* **158**, 1631.
Collins, C. B., Macnamara, J., and Thode, H. G. (1949). *Canad. J. Res.* **27B**, 361.
Cox, R. A., Culkin, F., Greenhalgh, R., and Riley, J. P. (1962). *Nature, Lond.* **193**, 518.
Craig, H. (1961). *Science,* **133**, 1833.
Desgrez, M. A., and Meunier, J. (1926). *C.R. Acad. Sci., Paris,* **183**, 689.
Dittmar, W. (1884). *In* "The Voyage of H.M.S. *Challenger*" (J. Murray Ed.), Vol. 1. H.M. Stationery Office, London.
Dole, M., and Swartout, J. A. (1939). *J. Amer. chem. Soc.* **61**, 2025.
Ellis, B. A., and Matthews, D. J. (1928). *J. Cons. int. Explor. Mer,* **3**, 191.
Ehhalt, D., Knott, K., Nagel, J. F., and Vogel, J. C. (1963). *J. geophys. Res.* **68**, 3775.
Epperson, A. W. (1928). *J. Amer. chem. Soc.* **50**, 321.
Epstein, S., and Mayeda, T. (1953). *Geochim. et cosmoch. Acta,* **4**, 213.
Evans, B. S. (1931). *Analyst,* **56**, 590.
Fleming, R. H., and Lyman, J. (1940). *J. Mar. Res.* **3**, 134.
Fleming, W. H., Macnamara, J., and Thode, H. G. (1953). *Geochim. et cosmoch. Acta,* **3**, 235.
Forchhammer, G. (1865). *Phil. Trans.* **155**, 203.
Forsberg, K. (1883). *Vega Expedition Reports,* **2**, 376.
Frederikson, A. F., and Reynolds, R. C. (1960). *Oil Gas J.* **58**, 154.
Friedman, I. (1953). *Geochim. et cosmoch. Acta,* **4**, 89.
Friedman, I., Harris, J., and Schoen, B. (1961). *J. geophys. Res.* **66**, 1861.
Fukai, R., and Shiokawa, F. (1955). *Bull. chem. Soc. Japan,* **28**, 636.
Garrels, R. M., and Thompson, M. E. (1962). *Amer. J. Sci.,* **260**, 57.
Gast, J. A., and Thompson, T. G. (1958). *Analyt. Chem.* **30**, 1549.
Gibson, R. O., and Haslam, J. (1950). *Analyst,* **75**, 357.
Goldschmidt, V. M., and Peters, Cl. (1932). *Nachr. Ges. Wiss. Göttingen, Math. Phys. Kl. III,* **28**, 528.
Greenhalgh, R. (1963). Ph.D. Thesis, London University.
Greenhalgh, R., and Riley, J. P. (1961). *Analyt. chim. acta,* **25**, 179.
Greenhalgh, R., and Riley, J. P. (1962). *Analyst,* **87**, 970.

Greenhalgh, R., and Riley, J. P. (1963). *Nature, Lond.* **197**, 371.
Gripenberg, S. (1937). *J. Cons. int. Explor. Mer*, **12**, 293.
Gysling, H., and Schwarzenbach, G. (1949). *Helv. chim. acta*, **32**, 1314.
Harding, M. W., and Moberg, E. G. (1933). *Science*, **77**, 510.
Haslam, J., and Moses, G. (1950). *Analyst*, **75**, 343.
Headridge, J. B. (1961). "Photometric Titrations". Pergamon Press, Oxford.
Hildebrand, G. P., and Reilley, C. N. (1957). *Analyt. Chem.* **29**, 259.
Hood, D. W., and Noakes, J. E. (1961). *Deep Sea Res.* **8**, 121.
Horibe, Y., and Kobayakawa, M. (1960). *Geochim. et cosmoch. Acta*, **20**, 273.
Hummell, R. W., and Smales, A. A. (1956). *Analyst*, **81**, 110.
Igelsrud, I., Thompson, T. G., and Zwicker, B. M. G. (1938). *Amer. J. Sci.* **35**, 47.
Jensen, M. L., and Nakai, N. (1961). *Science*, **134**, 2103.
Jentoft, R. E., and Robinson, R. J. (1956). *J. Mar. Res.* **15**, 170.
Johnson, W. R., Thompson, T. G., and Wirth, H. E. (1931). *J. Cons. int. Explor. Mer*, **6**, 246.
Kampitsch, E., Schwarzenbach, G., and Steiner, R. (1945). *Helv. chim. acta*, **28**, 828.
Kawasaki, N., and Sugawara, K. (1958). *Rec. oceanogr. Wks Jap.* **2**, 1.
Kirk, P. L., and Moberg, E. G. (1933). *Industr. Engng. Chem.* (*Anal.*) **5**, 95.
Kitano, Y., and Tsubota, H. (1960). *Bull. chem. Soc. Japan*, **33**, 770.
Knapman, F. W., and Robinson, R. J. (1941). *J. Mar. Res.* **4**, 142.
Kobanov, V. V., and Skopintsev, B. A. (1958). *Trud. Morsk. Gidrofiz. Inst. U.S.S.R.* **13**, 130.
Kolotoff, S. (1893). *J. phys.-chim. russe. Soc.* **24**, 82.
Kolthoff, I. M., and Yutzy, H. (1937). *Industr. Engng. Chem.* (*Anal.*) **9**, 75.
Korpi, E., and Thompson, T. G. (1942). *J. Mar. Res.* **5**, 28.
Lewis, G. J., and Thompson, T. G. (1950). *J. Mar. Res.* **9**, 211.
Macallum, A. B. (1903). *J. Physiol.* **29**, 213.
Maeda, M., Nisikawa, T., Ogata, Y., and Okuno, T. (1939). *J. Soc. Chem. Ind., Japan*, **42**, Suppl. Bind. 71.
Mahncke, H. E., and Rakestraw, N. W. (1935). *Industr. Engng. Chem.* (*Anal.*) **7**, 425.
Makin, C. J. S. (1898). *Chem. News*, **77**, 171.
Marcet, A. (1819). *Phil. Trans.* **109**, 161.
Matida, Y. (1951). *Contrib. cent. Fish. Sta. Japan*, Nos. 138–143.
Matida, Y. (1954). *J. oceanogr. Soc., Japan*, **10**, 71.
Matida, Y., and Yamauchi, N. (1951). *Contrib. cent. Fish. Sta. Japan*, No. 139.
Maury, M. F. (1855). *In* "The Physical Geography of the Sea", p. 150. Sampson Low, Son & Co., London.
Miyake, Y. (1939a). *Bull. chem. Soc. Japan*, **14**, 29.
Miyake, Y. (1939b). *Bull. chem. Soc. Japan*, **14**, 55.
Miyake, Y. (1939c). *Bull. chem. Soc. Japan*, **14**, 467.
Miyake, Y., and Sakurai, S. (1952). *J. Mar. Met. Soc. Japan*, **30**, 1.
Molins, L. R., and Rial, J. R. B. (1962). *Boletin Inst. esp. Oceanog.* No. 111.
Morton, E. H., and Thorpe, T. E. (1871). *Ann. Chem. Pharm.* **158**, 123.
Natterer, K. (1892). *Mh. Chem.* **13**, 873, 897.
Natterer, K. (1893). *Mh. Chem.* **14**, 624.
Natterer, K. (1894). *Mh. Chem.* **15**, 530.
Noll, W. (1934). *Chem. d. Erde*, **8**, 507.
Odum, H. T. (1951a). *Science*, **114**, 211.

Odum, H. T. (1951b). *Science*, **114**, 407.
Odum, H. T. (1957). *Publ. Inst. Mar. Sci. Univ. Tex.* **4**, 22.
Östlund, G. (1959). *Tellus*, **11**, 478.
Pate, J. B., and Robinson, R. J. (1958). *J. Mar. Res.* **17**, 390.
Pate, J. B., and Robinson, R. J. (1961). *J. Mar. Res.* **19**, 12.
Pensar, G., Ringbom, A., and Wänninen, E. (1958). *Analyt. chim. acta*, **19**, 525.
Price, J. B., and Priddy, R. R. (1961). *Bull. Mar. Sci. Gulf Caribbean*, **11**, 198.
Ramage, H. (1933). *S. East Nat.* **38**, 54.
Ratmanoff, G. E. (1937). *U.S.S.R. Hydrolog. Publ.* **10**, 173.
Reilley, C. N., and Schmid, R. W. (1957). *Analyt. Chem.* **29**, 264.
Reilley, C. N., and Schmid, R. W. (1959). *Analyt. Chem.* **31**, 887.
Robinson, R. J., and Thompson, T. G. (1932). *Nat. Acad. Sci. nat. Res. Coun. Bull*, **85**, 174
Schloesing, T. (1906). *C.R. Acad. Sci., Paris*, **142**, 320.
Schmelck, L. (1882). *Norwegian N. Atl. Exped. 1876–8, Pt. 9*, 1.
Schmidt, C. (1878). *Bull. Acad. St. Petersberg*, **24**, 231.
Shtukovskaya, L. A., Skopintsev, B. A., and Vorob'eva, R. V. (1957). *Gidrokhim. Materialy*, **27**, 146.
Smales, A. A. (1951). *Analyst*, **76**, 348.
Sousa, A. de (1954). *Analyt. chim. acta*, **11**, 221.
Sporek, K. F. (1956). *Analyst*, **81**, 540.
Steiger, G. (1910). Quoted in Clarke, F. W. (1924).
Swan, E. F. (1956). *Deep Sea Res.* **4**, 71.
Taylor, H. J., and Thompson, T. G. (1933). *Industr. Engng. Chem.* (*Anal.*) **5**, 87.
Thomas, B. D., and Thompson, T. G. (1932). Cited in Robinson and Thompson (1932).
Thompson, T. G., and Wright, C. C. (1930). *J. Amer. chem. Soc.* **52**, 915.
Trofimov, A. (1949). *Dokl. Akad. Nauk S.S.S.R.* **66**, 181.
Vasil'ev, V. V. (1937). *J. appl. Chem.* (*U.S.S.R.*) **10**, 1296.
Vinogradov, A. P. (1938). *Usp. Khim*, **7**, 639.
Voipio, A. (1957). *Suom. Kemistilehti*, **B.30**, 84.
Voipio, A. (1959). *Suom. Kemistilehti*, **B.32**, 61.
Webb, D. A. (1939). *J. exp. Biol.* **16**, 178.
Wheeler, A. S. (1910). *J. Amer. chem. Soc.* **32**, 646.
Winkler, L. W. (1916). *Z. Allgem. Chem.* **29**, 68

Chapter 5

Minor Elements in Sea Water

EDWARD D. GOLDBERG
Department of Earth Sciences
University of California, San Diego
La Jolla, California, U.S.A.

I. Introduction 163
II. Chemical Speciation 166
III. Reactivities of Elements in the Oceans 171
IV. Residence Times 172
V. Degrees of Undersaturation 175
VI. Activity Coefficients 177
VII. Atmosphere – Hydrosphere Interactions 178
VIII. Biosphere – Hydrosphere Interactions 179
A. Effects on Sea-Water Composition 179
B. Anoxic Waters 181
C. Concentration by Organisms 183
D. Skeletal Materials 186
IX. Hydrosphere – Lithosphere Interactions 187
A. Reducing Environments 187
B. Oxidizing Environments 189
References 194

I. Introduction

Although the electrolyte solution sea water contains species of all of the elements, only fourteen (O, H, Cl, Na, Mg, S, Ca, K, Br, C, Sr, B, Si and F) have concentrations equal to, or greater than, one part per million by weight, and by and large these are geochemically the more unreactive elements (Table I). The remaining elements, which can be defined as the *minor* ones, exist in micromolar and sub-micromolar concentrations, and yet can be involved in some of the more important inorganic and biochemical reactions of the marine environment.

The last few years have witnessed a constantly increasing interest in the inorganic and physical chemistry of sea water, the aim of which is to recognize the chemical principles which might account for the curious behaviours of some of the elements in the marine environment. Possibly the most important advances have been concerned with the elucidation for sea-water systems of the various chemical species present and their thermodynamic properties. Thus, before proceeding to the consideration

TABLE I

Abundances of the elements in sea water and residence times

Element	*Concentration in mg./l.*	*Residence time (years)*
H	108,000	
He	0·000005	
Li	0·17	$2{\cdot}0 \times 10^7$
Be	0·0000006	150
B	4·6	
C	28	
N	0·5	
O	857,000	
F	1·3	
Ne	0·0001	
Na	10,500	$2{\cdot}6 \times 10^8$
Mg	1350	$4{\cdot}5 \times 10^7$
Al	0·01	100
Si	3·0	$8{\cdot}0 \times 10^3$
P	0·07	
S	885	
Cl	19,000	
A	0·6	
K	380	$1{\cdot}1 \times 10^7$
Ca	400	$8{\cdot}0 \times 10^6$
Sc	0·00004	$5{\cdot}6 \times 10^3$
Ti	0·001	160
V	0·002	$1{\cdot}0 \times 10^4$
Cr	0·00005	350
Mn	0·002	1400
Fe	0·01	140
Co	0·0001	$1{\cdot}8 \times 10^4$
Ni	0·002	$1{\cdot}8 \times 10^4$
Cu	0·003	$5{\cdot}0 \times 10^4$
Zn	0·01	$1{\cdot}8 \times 10^5$
Ga	0·00003	$1{\cdot}4 \times 10^3$
Ge	0·00006	$7{\cdot}0 \times 10^3$
As	0·003	
Se	0·0004	
Br	65	
Kr	0·0003	
Rb	0·12	$2{\cdot}7 \times 10^5$
Sr	8·0	$1{\cdot}9 \times 10^7$
Y	0·0003	$7{\cdot}5 \times 10^3$
Zr		
Nb	0·00001	300
Mo	0·01	$5{\cdot}0 \times 10^5$
Tc		
Ru		
Rh		

TABLE I—*continued*

Element	*Concentration in mg./l.*	*Residence time (years)*
Pd		
Ag	0·00004	$2·1 \times 10^6$
Cd	0·00011	$5·0 \times 10^5$
In	<0·02	
Sn	0·0008	$1·0 \times 10^5$
Sb	0·0005	$3·5 \times 10^5$
Te		
I	0·06	
Xe	0·0001	
Cs	0·0005	$4·0 \times 10^4$
Ba	0·03	$8·4 \times 10^4$
La	$1·2 \times 10^{-5}$	440
Ce	$5·2 \times 10^{-6}$	80
Pr	$2·6 \times 10^{-6}$	320
Nd	$9·2 \times 10^{-6}$	270
Pm		
Sm	$1·7 \times 10^{-6}$	180
Eu	$4·6 \times 10^{-7}$	300
Gd	$2·4 \times 10^{-6}$	260
Tb		
Dy	$2·9 \times 10^{-6}$	460
Ho	$8·8 \times 10^{-7}$	530
Er	$2·4 \times 10^{-6}$	690
Tm	$5·2 \times 10^{-7}$	1800
Yb	$2·0 \times 10^{-6}$	530
Lu	$4·8 \times 10^{-7}$	450
Hf		
Ta		
W	0·0001	10^3
Re		
Os		
Ir		
Pt		
Au	0·000004	$5·6 \times 10^5$
Hg	0·00003	$4·2 \times 10^4$
Tl	<0·00001	
Pb	0·00003	$2·0 \times 10^3$
Bi	0·00002	$4·5 \times 10^4$
Po		
At		
Rn	$0·6 \times 10^{-15}$	
Fr		
Ra	$1·0 \times 10^{-10}$	
Ac		
Th	0·00005	350
Pa	$2·0 \times 10^{-9}$	
U	0·003	$5·0 \times 10^5$

of the diverse reactions involving the minor elements in sea water, it is of paramount importance to consider the forms of the reacting species.

II. Chemical Speciation

Sea water can be considered to be a solution of electrolytes together with minor amounts of non-electrolytes. Following the scheme suggested by Robinson and Stokes (1959), the electrolytes can be divided into two main classes: non-associated and associated.

The *non-associated* electrolytes are presumed to exist in the forms of simple cations and anions, all of which are probably solvated. Substances are placed in this category if there is no evidence for strong covalent bonding or electrostatic association between oppositely charged ions. This class is typified by the alkali metal ions, Li^+, Na^+, K^+, Rb^+ and Cs^+.

The *associated* electrolytes are subdivided into two classes. The *weak electrolytes* include those solutes which can exist as undissociated (covalent) molecules in equilibrium with their ions. This group includes all those substances which in classical nomenclature are regarded as weak acids and all bases. Of the components of sea water, the typical example is boric acid. The experimental observation that boric acid exerts a detectable vapour pressure above sea water (Gast, 1959) indicates that uncharged molecules of this solute are present in solution. Future analyses of the equilibrium vapour phase in contact with sea water may reveal other members of this group.

A second group of associated electrolytes comprises *ion-pairs* which result from a purely electrostatic attraction between oppositely charged ions. For example, the association of chloride ion with cadmium ion gives the ion pair $CdCl^+$

$$Cd^{2+} + Cl^- = CdCl^+.$$

In addition to ion-pair aggregates, in which ideally the bonding is purely electrostatic, there exist dissolved complexes (in which covalent bonding may be important) which are formed by interaction between a cationic species (central atom or central group) and donor groups (ligands). The number of linkages attaching ligands to a central group is known as the *co-ordination number*. Complexes in which the ligand attaches itself to the central atom through two or more bonds are known as chelate or multidentate complexes.

The interactions between metallic cations and ligands in the sea-water system can be appropriately considered on the basis of the classification given by Schwarzenbach (1961) who subdivided central atoms or groups into three classes.

(a) Ions of metallic elements with a rare gas configuration (d^0 cations), e.g. those of the alkali metals, alkaline earths, Al, Sc, the lanthanides, Ti, Th, and Nb.

(b) Cations with an outer shell of 18 electrons (d^{10} cations). Characteristic of this group are Ag^+, Zn^{2+}, Ga^{3+}, and Sn^{4+}.

(c) Those transition metal cations in which the number of d electrons is greater than zero and less than ten (e.g. Mn^{2+}, Fe^{2+}, Co^{2+}, and Ni^{2+}).

The d^0 series is characterized by the fact that members form few complexes, these mainly with fluoride ion and ligands where oxygen is the donor atom. There is no evidence as yet that any ions of this group form complexes with the heavier halides. The stability of the complexes in general increases with increasing cationic charge, and for cations of the same charge, with decreasing ionic radius.

Garrels and Thompson (1962) have considered the distribution of the major cations of sea water as ion-pairs with sulphate, bicarbonate and carbonate in water of chlorinity 19‰ and pH 8·1 at 25°C and at one atmosphere pressure (Table II). The stability constants of the various species, as well as their activities, were taken from the literature or from experimental work. For the divalent cations, magnesium and calcium,

TABLE II

Distribution of major cations as ion-pairs with sulphate, carbonate and bicarbonate ions in sea water of chlorinity 19‰, pH 8·1 at 25°C and one atmosphere pressure (from Garrels and Thompson, 1962)

Ion	*Molality*	*Free Ion* (%)	*Sulphate ion-pair* (%)	*Bicarbonate ion-pair* (%)	*Carbonate ion-pair* (%)
Ca^{2+}	0·0104	91	8	1	0·2
Mg^{2+}	0·0540	87	11	1	0·3
Na^+	0·4752	99	1·2	0·01	—
K^+	0·0100	99	1	—	—

			Ca ion-pair (%)	*Mg ion-pair* (%)	*Na ion-pair* (%)	*K ion-pair* (%)
SO_4^{2-}	0·0284	54	3	21·5	21	0·5
HCO_3^-	0·00238	69	4	19	8	—
CO_3^{2-}	0·000269	9	7	67	17	—

the importance of the sulphate ion-pairs is evident, about 10% of each of these ions being complexed. Only 1% of the total potassium or sodium is complexed with sulphate. Since the stability constants of nearly all ion-pairs formed between divalent metal ions and sulphates are close to 200, it seems likely that about 10% of the strontium, barium and radium in sea water exists as the undissociated sulphates. Further, probably only 1% or so of the rubidium and caesium is present as the sulphate ion-pair.

For d^0 cations with an oxidation number of two or more, fluoride complexing competes with the uptake of hydroxyl ions. For such reactions, we have

$$K_F = \frac{a_{MF}}{a_M . a_F} \text{ and } K_{OH} = \frac{a_{MOH}}{a_M . a_{OH}}$$

where K_F and K_{OH} are the stability constants of the fluoride and hydroxyl ion-pairs with the metal M respectively; and the a terms refer to the thermodynamic activities of the appropriate species.

Sillén (1961) has pointed out that for the fluoride complex to predominate the following condition must be met:

$$\log K_F - \log K_{OH} > \log a_{OH} - \log a_F > -1{\cdot}6$$

since for the sea-water system $\log a_{OH} = -6$ and $\log a_F = \log(1{\cdot}2 \times 10^{-3}/19) - 0{\cdot}2 = -4{\cdot}4$, if the activity coefficient of a monovalent ion is taken as 0·7, or $\log \gamma_{\pm} = -0{\cdot}2$. From Table III, in which the stability constants for the alkaline earths complexes are given, it would appear that this requirement is met only by magnesium. It is important to note that neither the fluoride nor the hydroxyl complex is very strong.

TABLE III

Stability constants for complexes of the alkaline earth metals with fluoride and hydroxyl ions (Bjerrum et al., *1958). Temperatures between 18 and 25°C. Values at zero ionic strength, except for Be*

	log K_F	*log* K_{OH}
Be	4·29	10·28
Mg	1·82	2·1–2·6
Ca	<1·04	1·29–1·51
Sr	—	0·82–0·96
Ba	<0·15	0·64–0·85

In order to ascertain the significance of the magnesium fluoride complex, it is possible to write

$$\log K_F = \log c_{MgF^+} + \log \gamma_{\pm} - \log c_{Mg^{2+}} - \log \gamma_{\pm\pm} - \log c_{F^-} - \log \gamma_{\pm}.$$

$$1{\cdot}82 = \log x - \log (0{\cdot}0555 - x) - (-1{\cdot}0) - \log (6{\cdot}65 \times 10^{-5} - x)$$

where the activity coefficient of divalent ions in sea water is assumed to be 0·1. On solving one obtains:

$$x = c_{MgF^+} = 1{\cdot}5 \times 10^{-5}.$$

Thus, although a trivial amount of the magnesium is complexed as the fluoride, somewhat over 20% of the fluoride ion in sea water appears to be associated with the magnesium.

The observation made by Goldberg *et al.* (1963) that the heavier rare earths (Gd to Lu) in sea water are enriched relative to their lighter counterparts (La to Eu) with respect to sediments, may reflect one or both of two general effects: the enhanced stabilization of the heavier rare earths in solution by complex formation, or the greater reactivity of the lighter members of the lanthanides owing to their increased basicity with respect to incorporation in, or on, solid phases. Moeller (1963) has pointed out that the lanthanide ions are like other d^0 cations in that they engage in complex formation only with the strongest chelating agents. Thus, the stabilities of the rare earth–sulphate ion-pairs are all closely similar, whereas the stability constants of the EDTA chelates vary over four orders of magnitude. If the enrichment of the heavier lanthanides results from this factor, then complexing of these elements, whose concentrations are of the order of 10^{-10}–10^{-11} M, by soluble organic debris appears reasonable. However, scavenging of the dissolved rare earths by either biological or inorganic solids has been suggested by Goldberg *et al.* (1963) to account for the conveyance of these elements from surface to deeper waters and to the sediments. During such processes fractionation of the rare earths between soluble and solid phases may occur.

The univalent d^{10} metallic ions behave in a completely different way from their d^0 counterparts. For these ions the halide complexes increase in stability with increasing atomic weight of the ligand, viz.

	AgF	AgCl	AgBr	AgI
$\log K_1$	−0·3	3·0	4·3	8·1

The very high concentration or activity of chloride relative to that of bromide or iodide in sea water ($\log c_{Cl} = -0{\cdot}37$; $\log c_{Br} = -3{\cdot}1$; and $\log c_I = -6{\cdot}6$) indicates that the chloride complexes would always dominate inasmuch as the respective K_1 values do not increase proportionately sufficiently to compensate for the decrease in concentration.

The possible combinations of the ions of this group with chloride and hydroxyl ions may now be considered. For chloride ion to compete successfully with hydroxyl ion in the formation of a simple complex,

$$\log K_{\mathrm{Cl}} - \log K_{\mathrm{OH}} > \log a_{\mathrm{OH}} - \log a_{\mathrm{Cl}} = -6 - (-0{\cdot}37 - 0{\cdot}2) \approx -5{\cdot}4.$$

Table IV gives the values of ($\log K_{\mathrm{Cl}} - \log K_{\mathrm{OH}}$) and it is evident that chloride complexing takes place to a greater extent than hydroxyl complexing for Ag^+, Hg^{2+}, Cd^{2+} and Zn^{2+}, but not for Cu^{2+} and Pb^{2+}.

The significance of sulphate complexing with members of this group can be estimated by considering the inequality

$$\log K_{\mathrm{Cl}} - \log K_{\mathrm{SO_4}} > \log a_{\mathrm{SO_4}} - \log a_{\mathrm{Cl}} = -3{\cdot}55 - (-0{\cdot}57) \approx -3$$

which must be fulfilled if the chloride ion-pair is to predominate over the sulphate association. From Table IV it is clear that in no instance is the sulphate complex more important than the ion association between chloride and the d^{10} cation. It should be pointed out that the stability constants for processes leading to the uptake of the additional chloride may be high enough to result in significant contributions of such complexes as $AgCl_2^-$, $AgCl_3^{2-}$, $HgCl_3^-$ and $HgCl_4^{2-}$.

To evaluate the relative contributions of the free ion, chloride complex and sulphate complex to the distribution of a d^{10} ion, the example of cadmium may be cited. Let $x = c_{\mathrm{CdSO_4^0}}$; $y = c_{\mathrm{CdCl^+}}$; and $z = c_{\mathrm{Cd^{2+}}}$, and assume activity coefficients of 0·7, 0·1 and 1·0 for monovalent ions, divalent ions, and for uncharged dissolved species respectively. Then, since the cadmium concentration in sea water is 10^{-9} M

$$x + y + z = 10^{-9}$$

$$K_{\mathrm{SO_4^{2-}}} = 2 \times 10^2 = \frac{a_{\mathrm{CdSO_4^0}}}{a_{\mathrm{Cd^{2+}}} . a_{\mathrm{SO_4^{2-}}}} = \frac{x}{z(0{\cdot}1)\,(2{\cdot}84 \times 10^{-2})\,(0{\cdot}1)}$$

$$x = 5{\cdot}68 \times 10^{-2}\, z$$

and

$$K_{\mathrm{Cl^-}} = 10^2 = \frac{a_{\mathrm{CdCl^+}}}{a_{\mathrm{Cd^{2+}}} . a_{\mathrm{Cl^-}}} = \frac{0{\cdot}7y}{z(0{\cdot}1)\,(0{\cdot}7)\,(0{\cdot}536)}$$

$$y = 5{\cdot}36\, z.$$

Therefore

$$c_{\mathrm{CdSO_4^0}} = 0{\cdot}01 \times 10^{-9}\ \mathrm{M}$$

$$c_{\mathrm{CdCl^+}} = 0{\cdot}83 \times 10^{-9}\ \mathrm{M}$$

$$c_{\mathrm{Cd^{2+}}} = 0{\cdot}16 \times 10^{-9}\ \mathrm{M}$$

TABLE IV

Values of log K_{OH}, K_{SO_4} and K_{Cl} for some d^{10} ions (from Bjerrum et al., 1958). Corrected to zero ionic strength where the data are available

Species	*log* K_{OH}	*log* K_{Cl}	*log* K_{SO_4}	*log* K_{Cl} − *log* K_{OH}	*log* K_{Cl} − *log* K_{SO_4}	*log* $K_{SO_4^-}$ − *log* K_{Cl}
Ag^+	2·3	3·1	1·3	0·8	1·8	0·5
Cd^{2+}	5·5	2·0	2·3	−3·5	−0·3	5·8
Hg^{2+}	11·5	7·3	1·4	−4·2	5·9	5·6
Zn^{2+}	4·4	−0·5	2·3	−4·9	2·8	7·2
Cu^{2+}	6·3	0·4	2·3	−5·9	1·9	8·2
Pb^{2+}	7·8	1·5	—	−6·3	—	—

Thus, by far the most important species of cadmium in sea water is the ion-pair with chloride.

The transition metal cations with between 0 and 10 *d*-electrons have been the subject of extensive investigations with respect to complex formation. Out of such work has evolved the so-called "Irving-Williams" order which states that for almost every ligand the stability of its complexes increases in the order: $Mn^{2+} < Fe^{2+} < Co^{2+} < Ni^{2+} < Cu^{2+} > Zn^{2+}$.

This is especially significant for organic ligands as may be seen from Table V. This concept will be returned to when the enrichment of trace elements in marine organisms is considered.

TABLE V

Log stability constants for the "Irving-Williams" order

Chelating agent	Mn^{2+}	Fe^{2+}	Co^{2+}	Ni^{2+}	Cu^{2+}	Zn^{2+}
EDTA	14	14	16	18	19	16
Malonic acid	3·3	—	3·7	4·0	5·8	3·5
Ethylenediamine	2·7	4·3	5·9	7·9	10·5	6·0
Nitrilotriacetic acid	7·4	8·3	10·5	11·4	12·8	10·5
Cystine	4·1	6·2	9·3	—	—	8·84
8-Hydroxyquinoline	6·8	8·0	9·1	9·9	12·2	8·5

III. REACTIVITIES OF ELEMENTS IN THE OCEANS

An element may have a low concentration in ocean water for one or both of two reasons. It may be very reactive in the marine environment and, subsequent to its introduction to the oceans, be rapidly removed to the sediments. Secondly, it may occur only in very low concentrations in the source domains, crustal rocks or the gaseous emanations from the interior of the earth. Thus aluminium, although a minor element in ocean water, is one of the most abundant constituents of igneous rocks; its high reactivity in marine waters reduces its concentration there to very low values. Caesium has a low concentration in crustal rocks and similarly a very low concentration in the oceans.

Insight into the comparative behaviour of elements in sea water has been gained by considerations of the relative reactivities of the elements on the basis either of the average time which they spend in sea water before removal to the ocean bottom or of the degree of undersaturation of sea water with them.

IV. Residence Times

The concept of the world's oceans as a simple reservoir system for elements introduced into it during the major sedimentary cycle was initiated by Barth (1952). He assumed a steady state system in which the amount of an element entering per unit time was compensated by the settling out of an equivalent amount. The residence time of an element, τ, can then be defined as the average time which it remains in the sea water before removal by some precipitation process. Thus,

$$\tau = \frac{A}{(\mathrm{d}A/\mathrm{d}t)}$$

where A is the total amount of the element in suspension or solution in the oceans, and $\mathrm{d}A/\mathrm{d}t$ is the amount introduced or precipitating per unit time. It is further assumed that there is a complete mixing of the element in a time which is short compared with the residence time.

Materials are introduced into the oceans via two principle paths—the fall-out of substances from the atmosphere and the influx of river water. Some of the input is in the form of non-reactive solid phases, such as quartz and the feldspars which remain in sea water only a relatively short time before settling to the sea floor. Other substances brought into the marine hydrosphere are incorporated as dissolved chemical species which are normally considered to be components of the waters of the oceans.

Barth used the figure of $2{\cdot}73 \times 10^{15}$ g./year given by Clarke (1924) for the total amount of dissolved substances supplied by the rivers. The absolute concentrations of elements in this river load cannot be used directly to calculate $(\mathrm{d}A/\mathrm{d}t)$ since significant amounts of materials are cycled from the sea, via the atmosphere and rains to the rivers, and subsequently returned to the oceans. Conway (1943) has pointed out that nearly all of the chlorine and about 53% of the sodium in rivers is of cyclic origin. Barth used Conway's data to obtain the net transfer of material between the oceans and the continents, and was able to derive the residence times of five elements (Table VI). The calculations of Barth can be extended to include some minor elements by utilizing the data given by Durum and Haffty (1963) for the median river water compositions of the United States. It has been assumed that these values are representative of the world's rivers, and that no significant amounts of these elements are recycled.

TABLE VI

The residence times of elements in sea water as calculated by river input and sedimentation

Element	*Ocean Content* ($\times 10^{20}$ *g.*)	*Residence time (millions of years)* *River Input*	*Sedimentation*
Na (a)	147·8	210	260
Mg (a)	17·8	22	45
Ca (a)	5·6	1	8
K (a)	5·3	10	11
Sr (b)	0·11	10	19
Si (a)	0·052	0·035	0·01
Li (b)	0·0023	12	19
Rb (b)	0·00165	6·1	0·27
Ba (b)	0·00041	0·05	0·084
Al (b)	0·00014	0·0031	0·0001
Mo (b)	0·00014	2·15	0·5
Cu (b)	0·000041	0·043	0·05
Ni (b)	0·000027	0·015	0·018
Ag (b)	0·0000041	0·25	2·1
Pb (b)	0·00000041	0·00056	0·002

(a) From Barth (1952). The silicon oceanic content has been modified to be consistent with the value given in Table I. The residence time given by Barth was an upper limit.
(b) Calculated from Durum and Haffty (1963).

Goldberg and Arrhenius (1958) using independent data calculated the residence times on the basis of the rates of sedimentation presumed to take place in the marine environment. They obtained the value of 2×10^{15} g./year as the amount of weathered products and pyroclastics introduced into the oceans. Assuming the composition of such materials to be the same as that of crustal rocks, they computed a comparable set of residence times (Table I).

In spite of this drastically oversimplified model of the oceans, there is a remarkable agreement between the two sets of data. The values span a time range of six orders of magnitude—sodium with $2{\cdot}6 \times 10^8$ years to aluminium with 100 years. It should be noted that one further condition must be satisfied if these considerations are to be valid. Both A and $(\mathrm{d}A/\mathrm{d}t)$ should not change during a period of the order of 3–4 τ. Even for sodium, this assumption appears to be valid, since a 10^9 year age of the oceans is in agreement with present-day geological concepts.

Other elements with long residence times—the lower atomic number alkali metals and alkaline earths (excluding beryllium)—are characterized by the lack of reactivity of their aqueous ions in the marine environment. Further, the decrease in residence times of the alkali metals in passing from sodium to caesium reflects variations in their oceanic reactivities. It has been suggested by Grim (1952) that the primary reactions regulating the marine concentrations of the alkali metals involve ion-exchange equilibria with the clay minerals on the sea floor. For positively charged ions, retention on the clay phases increases with increasing ionic radius (i.e. decreasing hydrated radius). Thus, the sequential decrease in residence times with increasing atomic numbers is in accord with the known chemical behaviour of the alkali metals.

However, certain elements (Be, Al, Ti, Cr, Fe, Nb and Th) have residence times of less than 10^3 years, periods of the order of, or less than, the mixing times for oceanic water masses. These elements, in part, enter the oceans as particulate phases from the continents or from volcanic activity, in the form of clay minerals, feldspars, augite, volcanic glass etc., and rapidly settle to the sediment deposits. Further, some are also the reactants in such authigenic substances as the ferromanganese minerals, zeolites etc. Thus, their entry into the oceans as solids and/or their high chemical reactivity can account for their low residence times.

The absolute values of these residence times are somewhat tenuous as the assumption that there is a complete mixing of such elements in the oceanic pool is not satisfied. Nevertheless, certain deductions can be made with regard to their marine geochemistry. Ocean to ocean, as well as temporal, variations in their sea water concentrations are to be expected. The heterogeneity in the distribution of iron (Lewis and Goldberg, 1954), which exists in sea water primarily as solid phases, emphasizes this point. Also, Goldberg and Koide (1962) have pointed out that the thorium contents of authigenic minerals of the sediments, which presumably derived their content of this element from the overlying waters, vary as a function of geographic location. The thorium concentrations apparently reflect the amounts of river run-off introduced into oceanic areas. The short residence time of thorium in the oceans allows various water masses to maintain different concentrations of it.

Certain minor elements have intermediate values of residence times: manganese (7000 years), zinc (180,000 years), cobalt (18,000 years) and copper (65,000 years). The reactivities of some of these elements are clearly connected with the formation of the ferromanganese minerals which have a widespread distribution in all oceans. Menard and Shipek (1958) estimated that these minerals cover between 20 and 50% of the sea floor in the southwestern Pacific; their chemical composition, which is unlike that of any terrestrial counterpart, shows a suite of elements, including copper, nickel, cobalt, zinc, lead, thorium and the rare earths, as guests in the lattice (see Chapter 15, Volume 2).

V. Degrees of Undersaturation

Estimates of relative reactivities can also be made on the basis of the degrees of undersaturation of metallic ions and dissolved gases. The upper limit of concentration that a metal ion could attain might be expected to be regulated by the solubility of its least soluble compound, as determined by the anionic species present, provided that no other chemical reactions reduced its abundance. As a first approximation it might be deduced that those elements with the highest degrees of undersaturation would be the most reactive, whereas those near, or at, saturation would be essentially inert in the marine environment.

Krauskopf (1956) calculated the theoretical maximum concentrations of a group of metal ions by considering those compounds formed with the major anions from sea water, which existing data in the literature suggested, would result in the lowest amount of the metal ion in solution. His data, which took into account activity coefficients of the ions as described in the following section, are given in Table VII.

He further initiated an experimental attack on this problem. A solution of a given metal was added to sea water until a precipitate formed; adequate precautions were taken to maintain the pH of the sea water relatively constant (between 7·8 and 8·2, values characteristic of open ocean waters). The amount of the metal ion in solution was then determined, and this was assumed to represent an equilibrium value, or possibly an upper limit. Unfortunately, the solid phases were not fully characterized. An approach to equilibrium from the opposite direction was then made by adding some of the precipitate to a fresh sea water aliquot and measuring the metal ion concentration in the solution when it reached a constant value. Such concentrations should be at least minimal values. Where the two values did not coincide, it was presumed that the equilibrium figure must lie somewhere in between.

TABLE VII

Comparison of calculated and observed concentrations of metals at saturation with the observed concentrations in sea water. Adapted from Krauskopf (1956)

Metal	*Insoluble compound*	*Solubility product*	*Concentration of element in saturated solution (mg./l.)* Calculated	Measured	*Concentration of element in ocean water (mg./l.)*	*Ratio of measured saturated concentration of element to that in ocean water*	*Residence time (years)*
Pb	$PbCO_3$	$1·5 \times 10^{-13}$	0·01	0·3–0·7	0·00003	10,000–20,000	$2·0 \times 10^{3}$
Ni	$Ni(OH)_2$	$1·6 \times 10^{-16}$	150	20–450	0·002	10,000–225,000	$1·8 \times 10^{4}$
Co	$CoCO_3$	$8·0 \times 10^{-13}$	0·02	25–200	0·0005	50,000–400,000	$1·8 \times 10^{4}$
Cu	$CuCO_3$	$2·5 \times 10^{-10}$	5·7	0·4–0·8	0·003	133–266	$5·0 \times 10^{4}$
Ba	$BaSO_4$	$1·0 \times 10^{-10}$	0·03	0·11	0·03	3·7	$8·4 \times 10^{4}$
Zn	$ZnCO_3$	$2·0 \times 10^{-10}$	4·6	1·2–2·5	0·01	120–250	$1·8 \times 10^{5}$
Cd	$Cd(OH)Cl$	$3·2 \times 10^{-11}$	105	4–1000	0·0001	40,000–10,000,000	$5·0 \times 10^{5}$
Ca	$CaCO_3$	$5·0 \times 10^{-9}$	70	100–480	400·0	0·25–1·2	$8·0 \times 10^{6}$
Sr	$SrCO_3$	$3–16 \times 10^{-10}$	9–44	22	8·0	2·75	$1·9 \times 10^{7}$
Mg	$MgCO_3$	$1·0 \times 10^{-5}$	84,000	36,000	350·0	27	$4·5 \times 10^{7}$

There is much uncertainty associated with both the theoretical treatment and the measured values. Thus, the activity coefficients are at best rough figures, particularly since the effects of complex ion formation have been disregarded. It is not certain that the presumed solid phases are in fact those present. For example, silicates or phosphates of the metals may be precipitated; the maximum concentrations of most cations which can be attained in equilibrium with these precipitates may be lower than that found in the absence of silicate or phosphate. In addition, the solid phases may be basic carbonates rather than the simple compounds tabulated. Furthermore, both the experimental and the theoretical examinations related to temperatures of 18–25°C, i.e. higher than the average oceanic value.

However, certain uniformities in the results justify their usefulness. In spite of the number of assumptions used in obtaining both the measured and calculated values of maximum concentrations, there is often close agreement between them. The measured values as a rule are higher than the calculated ones. This result is to be expected, if the metal ions engage in complex or ion-pair formation as has been indicated earlier. An important deduction is that there is little doubt that the concentrations of the listed metal ions, excluding calcium, strontium, and possibly barium, are not controlled by solubility equilibria. Finally, there is a reasonable correlation between the residence times and the ratios of measured maximum concentration to the observed concentration in sea water. This ratio is an indicator of the degree of undersaturation.

Those chemical species which reached saturation in sea water by equilibration with the atmosphere include the dissolved rare gases and nitrogen. Their chemical passivity is reflected by the fact that they subsequently undergo little, if any, concentration changes in the water mass (see Chapter 6, Sections III, B and C). However, dissolved gaseous oxygen, which is highly reactive in the marine biochemical cycle and which enters ocean water from the atmosphere or from photosynthesis, is often found to be present in undersaturated concentrations and may even be absent from some waters (see Chapter 13).

These considerations of reactivity have been concerned with metallic ions which have been derived from rock-weathering processes and with the dissolved atmospheric gases. Those anionic species in the sea which have not been derived in these ways but which apparently originated from the outgassing of the earth, cannot be treated in the same way.

VI. Activity Coefficients

Various approaches have been used to estimate the thermodynamic activity coefficients of the dissolved ions in sea water. Although the values used by different investigators for a given ion vary slightly, for the purposes of most thermodynamic calculations the use of a figure within the published ranges appears to be adequate.

Sillén (1961) proposed as a very rough estimate for uni- and di-valent ions respectively, that $\gamma_1 = 0{\cdot}63$ ($\log \gamma_1 = -0{\cdot}2$) and $\gamma_2 = 0{\cdot}16$ ($\log \gamma_2 = -0{\cdot}8$). Krauskopf (1956) has suggested that in solutions of ionic strength 0·7 (about that of sea water) $\gamma_1 = 0{\cdot}7$ and $\gamma_2 = 0{\cdot}12$ (cf. Latimer, 1952).

Garrels and Thompson (1962) have provided the most extensive compilation of activity coefficients for sea-water species and their data are given in Table VIII. For the uncharged dissolved species they have utilized the value of 1·13 which falls within the range 1·1–1·2 given by Harned and Owen (1958) for solutions with an ionic strength similar to that of sea water. The activity coefficients of sodium and potassium were determined experimentally by using glass electrodes sensitive to each of these ions; the values for the remaining ions were calculated from literature data.

For the purpose of calculation in this chapter the following values of the activity coefficients will be used: for monovalent ions, $\gamma_1 = 0{\cdot}7$, for divalent ions, $\gamma_2 = 0{\cdot}1$ and for undissociated molecules $\gamma_0 = 1{\cdot}0$.

Table VIII

Activity coefficients for dissolved species in sea water (ionic strength, 0·70; chlorinity, 19‰; temperature 25°C). Data from Garrels and Thompson (1962)

Species (*dissolved molecules such as*)	*Activity coefficient*
$MgCO_3^0$, $CaCO_3^0$, $CaSO_4^0$, $NaHCO_3^0$	1·13
Na^+	0·76
HCO_3^-, $NaCO_3^-$, $NaSO_4^-$, $CaHCO_3^+$, $MgHCO_3^+$, KSO_4^-	0·68
K^+	0·64
Mg^{2+}	0·36
Ca^{2+}	0·28
SO_4^{2-}	0·12
CO_3^{2-}	0·20

VII. Atmosphere–Hydrosphere Interactions

The observations that marine rains contain relatively high concentrations of both boron and iodine are well-documented in the literature

(Gast, 1959; Duce *et al.*, 1963). The transfer of chemical species through the liquid–gas interface, although having no discernible effect on the chemical composition of sea water with regard to these elements, is nevertheless important both in their geochemical cycles and possibly in productivity problems in surface waters.

The evaporation of boric acid appears to be a straightforward distillation process. However, the vaporization of iodine depends on a photochemical reaction (Miyake and Tsunogai, 1963). Evaporation of sea water containing either iodide or iodate alone, or in combination, in the dark did not produce significant quantities of iodine in the vapour phase. However, under ultra-violet irradiation, free iodine was found to escape from the aqueous phase; this has been explained by the reaction

$$2\,I^- + \tfrac{1}{2}\,O_2 + H_2O + h\nu = I_2 + 2OH^-.$$

Irradiation with light of wavelength of up to 560 mμ was effective in promoting this reaction.

VIII. Biosphere–Hydrosphere Interactions

A. Effects on Sea-Water Composition

There are two general areas of interest concerned in the interactions of the marine biosphere with the minor elements present in sea water. The first involves the regulation of the abundances of certain dissolved and particulate species as a function of space and time by the life processes. The second concerns the much higher concentrations of heavy metals in the marine biosphere relative to those in the hydrosphere, and also the remarkable specificity of certain organisms for the uptake of particular minor elements.

The uptake at the surface and subsequent deep-water oxidative regeneration of such nutrient species as dissolved phosphate, nitrate, silicate etc., as well as of vitamins, results in the non-uniform oceanic distribution of these materials as a function of depth, location and time. Such processes are appropriately treated elsewhere in this volume (see Chapters 8–12). However, recent investigations have suggested that some highly reactive trace metals (which consequently have short residence times) may be conveyed from shallow to deeper waters as a result of life processes in which these metals are not directly involved. Chow and Goldberg (1960) found that barium concentrations in several oceanic water columns from the Pacific increased with depth (Table IX). It was suggested that during the bacterial oxidation of the detrital organic phases, high concentrations of sulphate ions were formed within the bio-mass, thereby causing precipitation of barium sulphate to occur.

The barytes crystals, incorporated in, and sinking with, the decomposing organic matter, may either dissolve in deeper waters during the destruction of their carbonaceous matrix or be incorporated in the sediments on the sea floor. Such a mechanism explains not only the depth profiles of barium, but also the observation that sediments below zones of high organic productivity are enriched with barium (Goldberg and Arrhenius, 1958; Turekian and Tausch, 1964). This was also true for deposits containing large numbers of diatom frustules (Goldberg, 1958). Dissolution of barium sulphate on the sea floor and the consequent return of barium to the overlying waters may contribute to the increased barium concentrations in deeper waters. However, Chow and Goldberg (1960) have pointed out that thermodynamic considerations suggest that saturation values of barium appear to be approached at great depths in the oceans.

TABLE IX

Barium concentrations as a function of depth at Downwind Station 8 (03° 14′ N.; 130° 31′ W.)

Depth (m.)	*Barium (μg./l.)*
0	12
517	16
1051	31
2167	29
3002	43
4392	63

Several other investigations have shown increases in minor element concentrations with depth. Hosohara *et al.* (1961) in studies of the abundance of mercury in the equatorial West Pacific (Lamapo Deep) gave values for surface concentrations of *c.* 0·1 μg./l., whereas at 3000 m. the values ranged between 0·15 and 0·27 μg./l. at four stations. Intermediate concentrations were found at 500 m. Goldberg *et al.* (1963) found that sea water off the coast of California contained concentrations of rare earths at a depth of 4000 m. four times as great as at the surface.

The radio-isotope lead-210, which enters the ocean subsequent to its production in the atmosphere by radon-222 decay, shows an increase in concentration with depth in the sea (Rama *et al.*, 1961). Using a simplified two-layer model of the oceans, consisting of an upper mixed layer and a deep reservoir, a residence time of lead in the upper mixed layer of less than two years was derived for steady state conditions.

For mercury, the rare earths and lead-210, two transport mechanisms, both involving the biosphere, are likely for the conveyance of these species from surface to deeper waters. The well-known fact that most heavy metals are markedly enriched in the marine biosphere (see below) has led to the suggestion that their source in the deep-water may be solid phases resulting from the death of organisms or from the formation of metabolic waste products, which fall through the sea-water column and undergo oxidation. Alternatively, dissolved species of these metals may be adsorbed onto the disorganized masses of organic matter and would result in a net transfer from the upper to the lower layers of the ocean. The relative importance of these two processes in relation to inorganic scavenging is, as yet, unknown.

The oceanic distribution of at least one minor metal in sea water today seems to be influenced measurably by the activities of man. The work of Tatsumoto and Patterson (1963) has indicated that the present rate of introduction of lead into the world's oceans is 27 times greater than it was in the Pleistocene period, and has suggested that this increased input results from the addition of industrial combustion products, primarily lead tetraethyl. These investigators have concluded that during pre-industrial times, $1 \cdot 1 \times 10^{10}$ g. of dissolved lead per year were added to the oceans in the northern hemisphere. Chemical denudation from resulting agriculture has increased this rate threefold during the past several centuries. However, in the last three decades the exhaust of lead from internal combustion engines in the United States has risen from zero to about $1 \cdot 6 \times 10^{11}$ g. of lead per year. Tatsumoto and Patterson have calculated that 3×10^{11} g. of lead are present in the atmosphere of the northern hemisphere and $3 \cdot 5 \times 10^{11}$ g. of lead per year are washed out of the atmosphere annually with $2 \cdot 5 \times 10^{11}$ g. per year falling from the atmosphere into the oceans.

In sea waters off the southern California coast, the surface values of dissolved lead ranged from 0·08–0·4 μg./l. (average 0·2 μg./l.), whereas the deep waters possessed almost uniform lead concentrations at a level of 0·03 μg./l. This pattern of distribution of a heavy metal is the inverse of that of barium, the rare earths, lead-210 and mercury. The high surface lead concentrations are attributed to an input of industrially produced lead. On the basis of a non-steady state model, the residence time of lead in the mixed layer is found to be seven years, in satisfying agreement with that of two years derived for lead-210.

B. ANOXIC WATERS

The distributions of minor metals in anoxic waters have been investigated in only limited ways. Nevertheless, studies so far undertaken appear to give results not unexpected on the basis of chemical considerations. Anoxic waters can be defined as those devoid of dissolved oxygen and containing hydrogen sulphide in varying amounts (F. A. Richards, personal communication, see Chapter 13). Such waters are a consequence of a high rate of organic productivity at the surface and of a consequent high rate of oxygen consumption in deeper waters. Anoxic conditions usually prevail in basins or fjords where the physical circulations of the bottom water masses are so restricted that replenishment of dissolved oxygen in the deeper waters does not take place.

Mokyevskaya (1961) analysed the total manganese content and various forms of manganese in both the oxygenated and anoxic zones of the Black Sea. The hydrogen sulphide waters contained high concentrations of dissolved manganese (0·1–0·45 mg./l.), presumably in the divalent form, whereas surface waters had abundances of the order of, or less than, 0·005 mg./l. Minor amounts of manganese were present in the hydrogen sulphide zone and were associated with organic suspensions and dissolved organic compounds.

F. A. Richards (personal communication) determined the dissolved copper and iron contents in the Cariaco Trench. The copper concentrations in the upper oxygen-containing layers were similar to those detected in the open ocean, whereas in the sulphide bearing waters, the copper content was too low to measure by conventional techniques. Soluble iron, on the other hand, had concentrations 1000-fold greater in the anoxic zones than in normal oceanic waters (see also Chapter 13 and Chapter 18 in Volume 2).

Nikolaev *et al.* (1960) found between 1·5 and 2·8 μg. of uranium/l. in Black Sea waters in both the upper oxygenated and lower anoxic waters with no obvious differences in abundance between the two zones. Kolyadin *et al.* (1960) investigated the mode of occurrence of uranium in these deep reducing waters where the hydrogen sulphide concentrations approach 8 ml./l. and they found that the dissolved metal was in the 6+ oxidation state. The authors suggested that this is in accord with the known chemistry of uranium. The reduction

$$U(OH)_4\,(s) + 3\,CO_3^{2-} \rightleftharpoons UO_2(CO_3)_3^{4-} + 2\,H_2O + 2e$$

requires a redox potential of $-0{\cdot}4$ to $-0{\cdot}5$ V for the concentrations at which uranium and carbonate concentrations are present in sea water, whereas the redox potentials in the Black Sea never fall below $-0{\cdot}2$ V. Thus, the observation of an almost constant distribution of uranium between the upper and lower levels is in agreement with the concept that uranium is not reduced in the anoxic zone.

The absence of copper in the reduced lower levels of the Cariaco Trench most probably results from the precipitation of cupric sulphide

($K_{s.p.} = 8 \times 10^{-36}$), whereas manganese with a much higher solubility product ($K_{s.p.} = 1 \times 10^{-11}$) can remain in solution in the anoxic waters of the Black Sea. The reduction of iron to the ferrous state and/or the complexing of dissolved iron by the large amounts of organic matter present in the reducing basins would account for the tremendous iron abundances. It seems reasonable that all of these metals are transferred from the surface to deeper waters by the fall-out of organic matter in the water column as discussed above.

C. CONCENTRATION BY ORGANISMS

The remarkable accumulation by members of the marine biosphere of many metals which exist only in very minor concentrations in sea water is amply recorded in the oceanographic literature. Several salient features of this phenomenon are noteworthy (summarized by Goldberg, 1957). Concentration factors as high as millions have been found, and it is not uncommon for transition metals to be enriched in animals of the sea 100- to 100,000-fold compared with their aqueous environment. The minor metals are retained by fairly strong chemical bonds by the organisms and are not readily eluted. Further, the ability to remove specifically a given minor metal from sea water is often a property belonging to a family of organisms or even to particular members of a family. The rather dramatic observation reported by Carlisle (1958) that certain tunicates concentrate vanadium, and others its vertical periodic table neighbour niobium and still others neither element, but that none can accumulate both elements, emphasizes the most unusual specificities that have been encountered in recent investigations. Levine (1961), working with the colonial ascidian *Eudistoma ritteri*, found that titanium was enriched over a million times compared with its sea-water concentration for samples dried at 110°C. Such results have led Nicholls *et al.* (1960) to suggest that "for any given chemical element there will be found at least one planktonic species capable of spectacularly concentrating it".

The problem of the selectivity in these enrichment processes has been reviewed by Bayer (1964). He pointed out that simple chelating or complexing agents bind a series of metal ions and form complexes having varying degrees of stability, and he suggested that specificity in nature requires macromolecules which provide both appropriate chemical groups and a proper geometry for the uptake of a given metallic species from sea water. He argued that the copper complexing ability of haemocyanin, the blue pigment responsible for the transport of oxygen in the blood of molluscs, is most probably a function of the macromolecular protein since the haemocyanin itself contains no low molecular weight complexing agents. Of further interest is his observation that the resynthesis of haemocyanin from apo-haemocyanin (up to 75% of the complexed copper can be removed from the blood pigment without destroying its complex forming ability) can be accomplished with cuprous but not with cupric salts! Finally, other metal ions cannot be introduced into the apo-haemocyanin without denaturing the complexing agent. On the basis of such considerations, Bayer was led to synthesize macromolecular chelating agents which possessed unusually high specificities. One, which contained the glyoxal bis-2-hydroxyanil grouping, concentrated copper and uranium from sea water, and another, containing glyoxal bis-2-mercaptoanil as the complexing group, selectively enriched gold from oceanic samples.

The general problem of the enrichment of metals in the marine biosphere has been treated by Goldberg (1957) who pointed out that in general, the relative concentration factors closely parallel the order of stability of the complexes formed by metal ions with ligands. The stability order given in Table X, is for the most part independent either of the nature of the ligand or of whether the metal atoms are attached to oxygen, nitrogen or sulphur, and is one commonly encountered in the chemical literature.

Concentration factors for these metals by whole organisms clearly indicate that the d^{10} cations are much more strongly enriched in the biomass than are the d^0 cations, and that with the d^{10} cations there is a correspondence between the enrichments and the "Irving-Williams" order. It should be emphasized that the tabulated concentration factors are not intercomparable from investigation to investigation (i.e. along the vertical rows of Table X), but that consideration should only be given to variations along horizontal lines. Differences in experimental techniques for the reduction of the sample (i.e. temperature of drying, washing procedures etc.), combined with problems of systematic analytical errors, make vertical intercomparisons difficult. Nevertheless, along horizontal rows the trend of enrichment factors does follow the elemental stability order in these examples. Over-riding the general picture is the phenomenon of selectivity. For example, fish have long been known to concentrate zinc (Vinogradov, 1953), and the marked accumulation of zinc in the whole organism, compared to other d^{10} minor metals, is evident from Table X.

The trace element concentrations in various parts of the organisms show great variations. The d^{10} cations in fish are found to be most abundant in the internal organs; a smaller degree of concentration occurs in the skeletal parts, the heart, flesh and skin. In studies of cadmium in the marine environment, Mullin and Riley (1956) found a

TABLE X

Concentration factors of elements in marine organisms in relation to the stability of complexes formed by metal ions with a variety of ligands. The only intercomparisons which are valid are along horizontal lines (see text). Elements arranged in "Irving Williams" order, see p. 171.

Organisms	*Metal* Cu	Ni	Pb	Co	Zn	Mn	Mg	Ca	Sr	Ba
Seaweeds (Black and Mitchell, 1952)	—	550	—	—	900	—	—	—	23	—
Benthic algae (Lowman, 1964)	—	2000–40,000	—	—	—	1000–30,000	—	—	—	—
Plankton (Nicholls *et al.*, 1960)	400–90,000	<20–8000	30–12,000	<100–16,000	—	—	—	—	—	—
Marine animals (F. G. Lowan, personal communication)	—	3000–70,000	—	—	—	2000–10,000	—	—	—	—
Anchovetta (Goldberg, 1962)	80	—	10,000	—	400	1000	0·1	7	8	20
Yellow fin tuna (Goldberg, 1962)	200	—	—	—	700	80	0·2	6	7	2
Skipjack tuna (Goldberg, 1962)	100	50	—	—	500	40	0·3	5	5	3
Sponges (Bowen and Sutton, 1951)	1400	420	—	50	—	—	—	0·07	3·5	—

low content of this element in the muscle and gonads while much higher values were found in the digestive glands and renal organs.

The selective and general accumulation processes are most probably initiated by complexing reactions, including chelation, taking place at exposed external surfaces of the organisms. Goldberg (1957), in a review of the work carried out using tracer techniques on the uptake of dissolved species from sea water, suggested that for organisms possessing mucus layers exposed to sea water, this part of the bio-mass acts as a surface adsorption sheet for either selective or general accumulation. A second form of assimilation involves the uptake from the oceans of particulate species, either of an organic or inorganic nature, with a subsequent retention or rejection of minor metals through metabolic processes. Iron, for example, exists in sea water as solid phases, most probably hydrated oxides with accompanying sorbed ions. The uptake of the iron particles would provide a mechanism for obtaining iron and/or the sorbed species.

D. SKELETAL MATERIALS

The solid phases secreted by marine organisms include calcium carbonate, silicon dioxide, calcium phosphate, iron oxides and strontium sulphate (a review of the types and characteristics of these materials has been given by Lowenstam, 1963). These biological mineral precipitates have been the subject of many paleontologically-based investigations since their mineralogical and chemical compositions may reflect some of the physico-chemical parameters of the environment in which deposition took place. By far the most extensive researches have centred about the calcareous phases, with emphasis on mineralogy, isotopic analyses of oxygen, and strontium and magnesium contents. The systematics of the major element distributions in calcium carbonates are now emerging, and set a pattern for investigations of minor elements in such materials. The salinity, chemistry, temperature of the water, crystal structure of the $CaCO_3$, or biochemistry of the species can control the calcium and magnesium contents (Lowenstam, 1963). Most studies of minor elements in calcium carbonates have not exercised controls over these variables, and hence all the factors regulating trace element distributions are not as yet clearly discernible (see, for example, the data compiled by Graf, 1960). The following examples are illustrative of the types of relationships between minor element concentrations and the parameters cited above.

Turekian and Armstrong (1960) found that barium is slightly enriched in clam and snail shells relative to sea water. The most important parameter controlling the concentration of barium appeared to be the generic association and clams showed a higher barium concentration than did the snails. Uranium showed no evident preference for the aragonitic, as opposed to the calcitic, structures although, like strontium, which does prefer the aragonitic mineralogy, it showed its highest concentrations in the aragonitic corals (Tatsumoto and Goldberg, 1959).

The other skeletal material on which minor element analyses have been performed is the fluorapatite of bathypelagic fish debris (Arrhenius *et al.*, 1957). The organic phases were separated from the inorganic phosphate matrix and it was found that the rare earth elements and most of the strontium and barium occurred in the apatite structure, whereas the organic phases contained zinc, tin, lead, titanium, copper and silver and much of the magnesium, aluminium, chromium and nickel. Samples from Recent and Tertiary strata did not show significant differences in these minor element concentrations. Analyses of fish bones taken from living organisms did not reveal high concentrations of these minor metals, and hence the accumulation at the surfaces of the sedimentary deposits must have taken place subsequent to the death of the organisms.

IX. Hydrosphere–Lithosphere Interactions

Those inorganic reactions in the oceans by which dissolved species in sea water form precipitates which are subsequently incorporated in the sediments may be characterized in the following ways. The elements whose concentrations in sea water are influenced or possibly controlled by such reactions, exist at micromolar or sub-micromolar concentrations, and possess short residence times. The rates of formation of the solid phases often have time constants of the order of geological time periods. Hundreds to hundreds of thousands of years pass before appreciable amounts of material accumulate. The nature of the precipitates is in part controlled by the physico-chemical characteristics of the water, which can determine the speciation of the reacting elements. Since Eh and pH are two of the prime variables which can define the nature of the aqueous system, it appears appropriate to consider such reactions with respect to the minor elements accommodated in the solid phases on the basis of the type of environment, i.e. oxidizing or reducing.

A. Reducing Environments

Manheim (1961) investigated the relationship between the type of depositional environment and the distribution of minor elements in the sediments of the Baltic Sea. A significant feature of this environment is the change in bottom character in which the overlying waters change from an oxygenated to an anoxic condition as the depth increases. Three classes of sediments were defined: (i) a coarse sand—gravel deposit near the coast with high Eh and mildly alkaline pH in both the sediments and the overlying waters; (ii) a green-grey organic mud which usually contains H_2S (the gyttja) and which lies close to the basin; (iii) black sapropels formed in the deep basin, under strongly reducing conditions, as indicated by Eh values. There is a clear difference between the two reducing environments, the sapropels and the gyttjas. The gyttja may, or may not, contain H_2S even though the water overlying it does contain free oxygen; the sapropel and its overlying water both contain H_2S.

Heavy metals such as Cu, Ag, U, Zn, and possibly Mo, tend to be concentrated in the sapropels—not in the most highly reducing centres of the basins, but at the peripheries of the stagnant basins. Ferromanganese minerals were found in the oxygenated sediments (see below) whereas in the deep anoxic basins a solid solution of calcium, magnesium and manganese carbonates occurred. Although the finding of tetravalent and divalent manganese in oxidizing and reducing zones respectively is satisfying, the occurrence of the maximal concentrations of heavy metals in the modestly reducing areas of the sapropels is surprising. Manheim suggested that the heavy metals are removed initially from the basin peripheries where reducing conditions prevail, and when the water mass reaches the deeper basin it is impoverished in these elements. Such a concept gains credence from the observations of a salinity stratification in the basins which may reflect the very sluggish movements of the bottom waters.

Of particular interest in this study is the enrichment of uranium in the reducing environments. The highest uranium content encountered attained a level of 130 p.p.m. and was in a black, homogenous clay. The organic-poor coastal sediments contained about 5 p.p.m. of uranium. The lowest values of the Eh measured in the Baltic Sea waters were about −0·2 V in the deepest parts of the basins. Since a redox potential of −0·4 to −0·5 V is necessary for the conversion of U VI to U IV, at pH values encountered in anoxic waters, in the absence of appreciable complex formation this reduction, if it occurs, must take place within the sedimentary column. It would be particularly useful to know its oxidation state in these deposits.

Studies on the inorganic deposition of marine apatites (often referred to as phosphorites) have been materially enhanced by investigations of the oxidation state of uranium within these minerals. Extensive deposits of apatite with the general formula $Ca_{10}(PO_4,CO_3)_6F_{2-3}$ occur off the coast of southern California where over an area of about 6000 square miles it occurs in the forms of nodules, slags, and oölites in depths ranging from 15 – *c.* 150 m. (Emery, 1960).

Krumbein and Garrels (1952) suggested that phosphorites form in environments which have a pH somewhat less than that of normal sea waters approaching or under anoxic conditions. The work of Altschuler *et al.* (1958) appears to bear out this concept. These workers found that in the phosphorites from off the coast of southern California, the tetravalent uranium attained levels of 55–74% of the total uranium. The reduced uranium occurred in the apatites where it substituted for calcium in the lattice, whereas the uranyl ions were apparently adsorbed at the surfaces of the apatite crystallites (Sheldon, 1959). Again, since nowhere in the water column off southern California does the Eh attain sufficiently negative values for the uranium reduction, the hypothesis that the formation of the tetravalent uranium, and most probably of the phosphorite itself, must take place within the reducing environment of the sediment appears valid.

The geological evidence on phosphorite formation bears out this theory. McKelvey *et al.* (1953) have indicated that phosphorites occur in the eastern oceanic coastal deposits above which upwelling waters give rise to high organic productivity. The subsequent accumulation of this organic matter on the sea bed can give rise both to reducing conditions, through the depletion of dissolved oxygen gas in the sea water, and by the decomposition of biologically produced phases to a supply of phosphate for mineral formation.

B. OXIDIZING ENVIRONMENTS

A comparison of the minor element concentrations in the slowly depositing open ocean sediments (oxidizing systems) either with the more rapidly accumulating near shore deposits or with the crustal rocks, from both of which the substances in question were presumably derived, has revealed some rather well-defined differences. Wedepohl (1960) has noted that elements such as Zn, Ni, Pb, Cu, Co, Mn and Mo increase in concentration along a section from the near-shore argillites to the pelagic sediments of the Atlantic. Furthermore, the Pacific open ocean sediments were richer in these elements than were their counterparts of the Atlantic (Table XI). However, the average abundances of Ti, V, Cr, Ga, Rb and Zr are much the same in the near-shore clays as in the pelagic solid phases. Finally, the elements La, Sc, Y and Ba are equally abundant in both pelagic clays from the Atlantic and in the near-shore sediments, but are more enriched in the deep-sea sediments of the Pacific.

TABLE XI

Ratios of concentrations of elements in deep-sea sediments to those in near shore deposits (from Wedepohl, 1960)

Element	Abundance ratio Atlantic Ocean	Pacific Ocean
Sc	1·0	2·0
Ti	1·0	1·0
V	1·0	1·0
Cr	0·9	0·8
Mn	4·0	11·0
Co	3·0	8·5
Ni	2·5	5·5
Cu	3·0	8·5
Zn	1·5	2·0
Ga	1·0	1·0
Rb	1·0	0·85
Y	1·0	5·5
Zr	0·8	1·0
Mo	9·0	46·0
Ba	1·0	5·5
La	1·0	1·5
Pb	2·0	2·5
Fe	1·2	1·5

Goldberg and Arrhenius (1958) compared the ratios of the average elemental concentrations in pelagic sediments of the Pacific to those of igneous rocks, and reported no fractionations for Si, Al, Fe, Ti, Mg, Ca, Na, K, Sr, Ga, V and Zr. However, Mn, Ni, Cu, Co, Pb, Mo, Yb, Y, La, Sc, Ba and B were enriched by factors of up to 30 in the deep-sea deposits. The analyses were performed on the gross samples, i.e. the interstitial solutions were considered as a part of the sedimentary domain. The estimates made by previous workers were based on sediments from which the sea salts had been washed out. However, it is very difficult by any washing procedure to remove solely the sea salts so as to leave unaffected the composition of the minor elements in the solid phases.

This unusual chemical composition of deep-sea sediments is undoubtedly related to the widespread dispersion of ferromanganese minerals on the sea floor. These authigenic deposits contain in high concentrations relative to sea water (Table XII) the same elements which are enriched in the open ocean sediments (see discussion above). Menard (1964) has discussed their distribution in the world's oceans where submarine photographs have shown that in some areas they may cover the entire sea floor. Although it is difficult to evaluate an average coverage in a given area because of the somewhat patchy distribution of the minerals, nevertheless, areas having large surface accumulations have been delineated. In the southwestern Pacific, a region with 26–46% of the bottom covered with manganese nodules has been found. A region in the North Pacific, centred at latitude 18–20° N., possesses a bottom coverage of nodules and slabs ranging between 10 and 100%. Menard (1964) has shown on the basis of coring and dredging the sea floor in combination with camera studies, that the nodules are as abundant below the sediment surface as on it.

TABLE XII

The average composition of ferromanganese nodules from the Pacific

Element	*Weight* (%)
Fe	14
Mn	19
Ni	0·4
Co	0·3
Cu	0·5
Ti	0·8
Zn	0·04
Pb	0·1
P	0·5
Si	0·7
Zr	0·006

The ferromanganese minerals occur as coatings on exposed rocks, and as slabs, nodules and micronodules. The micronodules range up to several mm. in diameter and are often found dispersed throughout unconsolidated sediments. The shapes of the nodules are often near spherical or reniform, and usually a nucleus of pumice, consolidated clay, zeolite, a shark's tooth or a whale earbone is evident.

In sea water, the mechanism of formation of the ferromanganese minerals, involves the transformation of iron and manganese at submicromolar concentration levels into the two main components of these solids at levels *c.* 15% on the average, and must therefore represent a powerful concentrating process. For a discussion of the other trace metals incorporated in these materials, it is advantageous to consider the approach of Goldberg and Arrhenius (1958) who directed attention to the problem of the oxidation of dissolved manganese in sea water. A simple and direct representation of the overall oxidation is given by:

$$Mn^{2+} + 2\,OH^- + 1/2\,O_2 = MnO_2 + H_2O$$

which appears to be thermodynamically possible in the marine environment. A reaction site may be provided by the accumulation on the sea floor of the oxides of iron, particles of which are ubiquitous in the oceans. Thus, one would expect, and in fact one finds, the primary occurrence of the ferromanganese minerals to be in areas where the bottom topography and/or the lack of supply of sedimentary components other than oxides of iron do not allow any appreciable accumulation of solid phases. In rapidly accumulating deposits, where the ferromanganese accretions do not occur, probably the burial of the iron oxide by other sediment components minimizes the formation of the minerals. Where the iron accumulation is more rapid than the mineral formation, the mineral goethite appears incorporated in the ferromanganese deposits.

The ferromanganese phases have rather extensive surfaces ranging from 6–190 m.^{2}/g. (Buser and Grütter, 1956) and also contain appreciable amounts of organic matter (Graham, 1959), which has not yet been characterized. The occurrence of this organic matter has stimulated a certain amount of discussion concerning a possible biological origin for these minerals. However, sorption of organic matter from sea water or the activities of benthic organisms or bacteria on the surfaces appears to offer a much more reasonable explanation for the mode of accumulation of these carbonaceous materials.

The rates of build-up of the ferromanganese nodules have been measured by two techniques in the author's laboratories—the ionium/thorium ratio (Goldberg and Koide, 1962) and the uranium-234/uranium-238 ratio (Thurber, 1962) methods—and have an indicated range of between $0 \cdot X$ and $0 \cdot 0X$ mm./10^3 years. Although accumulations are neither uniform nor continuous, it is of interest to note that on average, the minerals form by the addition of several atomic layers per day, if the distance between the δ-MnO_2 sheets is taken as 10Å. Thus, the submicromolar concentrations of the reacting species are complemented by very low reaction rates.

The processes responsible for the enrichment of minor elements in the components of deep-sea sediments, especially the ferromanganese minerals, will probably be found in extended investigations of the crystallography of the phases and of the sorption processes on the surfaces of the solids. However, with our present state of knowledge of elemental distributions, it appears worthwhile to seek out those species which might be diagnostic of the oxidizing or reducing characters of the sedimentary systems.

The ratio of the abundances of the two naturally occurring radioactive elements, thorium and uranium, in marine deposits may well serve as an environmental index. The Th/U ratio in crustal rocks is about 3·5, a value which also appears to be representative of common shales (Adams *et al.*, 1959). Thorium is concentrated in the ferromanganese minerals by a factor of up to ten times that of deep-sea sediments, and varies between 24 and 124 p.p.m. (Goldberg and Picciotto, 1955), whereas the uranium concentrations reach levels of 4–6 p.p.m. (Tatsumoto and Goldberg, 1959). In the pelagic deposits of the Atlantic, Indian and Pacific Oceans, the thorium concentrations range in value between 6 and 15 p.p.m. (on a calcium carbonate-free basis), whereas the uranium contents are normally only a few p.p.m. (Starik *et al.*, 1958; Goldberg and Koide, 1962; Goldberg and Koide, 1963; Goldberg *et al.*, 1964). A Th/U ratio somewhat greater than five for the ferromanganese minerals and deep-sea sediments seems reasonable.

However, phosphorites have Th/U ratios close to 0·02 (Adams *et al.*, 1959; Goldberg and Parker, 1960). The average uranium content of the Black Sea sediments is four to five times higher than that of deep-sea deposits, and the Th/U ratio in these anoxic deposits is about unity (Agamirov, 1963). Thus, high and low values of the Th/U ratio can be indicative of oxidizing and reducing environments respectively in the oceanic system. The uranium, which is stabilized in aerated marine waters as its carbonate complex, is most readily taken up in solid phases by reduction processes, most probably within the sedimentary phases. Thorium, the 4+ form of which is not subject to reduction in such environments, is preferentially removed from sea water in the pelagic areas and enriched in the ferromanganese minerals.

A second approach to the use of minor elements in defining the Eh of a marine system in which the inorganic precipitation of solid phases has occurred, involves the group of metals, cobalt, cerium and lead. The members of this trio, like manganese, can exist in sea water as dissolved species, but may be oxidized to insoluble oxides of higher oxidation states. The possibility that such reactions have taken place may be examined by a comparison of the abundances of these metals in sedimentary components, sea water and crustal rocks with those of elements which display similar chemical properties in lower oxidation states but which do not possess the property of being readily oxidized.

Cobalt has a concentration in sea water about 1/20 of that of nickel (Table I); the two elements occur in similar concentrations in crustal rocks. However, the ferromanganese phases contain about as much cobalt as nickel on the average (Table XII), with cobalt having a greater range of concentration relative to manganese than nickel. The oxidation potential necessary to transform cobalt from the 2+ to the solid CoOOH is slightly higher than that for the corresponding oxidation of manganese. On the basis of the known marine concentrations and the postulated activity coefficients of the various species in sea water, it appears that the reaction should not proceed. However, a combination of the cobalt and manganese oxidation reactions leading to the formation of a solid solution of the cobalt oxides in the manganese dioxide might give a thermodynamically possible reaction which would result in the preferential concentration of cobalt compared with nickel. However, it should be remembered that the oxidation potentials necessary for a particular process are influenced strongly by environmental conditions (e.g. pH and complex formation).

The remarkable enrichments of cerium compared to its periodic table neighbour lanthanum, which exhibits only the 3+ oxidation state, in ferromanganese minerals has been emphasized recently (Goldberg *et al.*, 1963; Pachadzhanov *et al.*, 1963). In the ferromanganese minerals, the Ce/La ratio ranges from 5–20, whereas in crustal rocks, the ratio has a value of three. The oxidation of cerium to the 4+ state as CeO_2 provides a mechanism for its enrichment. Since cerous is even more difficult to oxidize than is cobaltous, it is suggested that the reaction proceeds through the incorporation of ceric oxides into the manganese dioxide lattice, with the coupling of the manganous and cerous oxidation reactions, in the same manner as was suggested above for cobalt.

Although lead exists in crustal rocks at concentrations between 10 and 20 p.p.m., in the ferromanganese minerals it is enriched to average levels of 10,000 p.p.m. (Table XII). Sillén (1961) has indicated that PbO_2 would be the favoured form of this minor metal in oxygenated marine systems, provided that the proper conditions for the oxidation of the divalent dissolved lead exist. He suggested, for example, that the formation of a solid solution of lead dioxide with the manganese dioxide might account for the accumulation of lead in solid phases.

Thus, the observed simultaneous enrichments of lead, cobalt and cerium in open ocean deposits can indicate that the solid phases were formed under oxidizing conditions.

REFERENCES

Adams, J. A. S., Osmond, J. K., and Rogers, J. J. W. (1959). *Phys. Chem. Earth*, **3**, 298.

Agamirov, A. (1963). *Geochemistry*, 630.

Altschuler, Z. S., Clarke, R. S., and Young, E. J. (1958). *U.S. Geol. Surv. Prof. Paper 314D*, 1.

Arrhenius, G., Bramlette, M., and Picciotto, E. (1957). *Nature, Lond.* **180**, 85

Barth, T. W. (1952). "Theoretical Petrology." Wiley, New York.

Bayer, E. (1964). *Angew. Chem.* (*Int. Ed.*) **3**, 325.

Bjerrum, J., Schwarzenbach, G., and Sillén, L. G. (1958). "Stability Constants. Part II: Inorganic Ligands". The Chemical Society, London.
Black, W. A. P., and Mitchell, R. L. (1952). *J. Mar. biol. Ass. U.K.* **30**, 575.
Bowen, V. T., and Sutton, D. (1951). *J. Mar. Res.* **10**, 153.
Buser, W., and Grütter, A. (1956). *Schweiz. min. petrogr. Mitt.* **36**, 49.
Carlisle, D. B. (1958). *Nature, Lond.* **181**, 922.
Chow, T. J., and Goldberg, E. D. (1960). *Geochim. et cosmoch. Acta,* **20**, 192.
Clarke, F. W. (1924). *Bull. U.S. geol. Surv.* 770.
Conway, E. J. (1943). *Proc. R. Irish Acad.* **48B**, 161.
Duce, R. A., Wasson, J. T., Winchester, J. W., and Burns, F. (1963). *J. geophys. Res.* **68**, 3943.
Durum, W. H., and Haffty, J. (1963). *Geochim. et cosmoch. Acta,* **27**, 1.
Emery, K. O. (1960). "The Sea off Southern California", pp. 68–73. Wiley, New York.
Garrels, R. M., and Thompson, M. E. (1962). *Amer. J. Sci.* **206**, 57.
Gast, J. A. (1959). "Some Factors Influencing the Boron-Chlorinity Ratio of Sea Water". Unpublished thesis. University of Washington, Seattle.
Goldberg, E. D. (1957). *Mem. geol. Soc. Amer.* **67**, 345.
Goldberg, E. D. (1958). *J. Mar. Res.* **17**, 178.
Goldberg, E. D. (1962). *Limnol. Oceanogr.* **7**, 72.
Goldberg, E. D., and Arrhenius, G. O. S. (1958). *Geochim. et cosmoch. Acta,* **13**, 153.
Goldberg, E. D., and Koide, M. (1962). *Geochim. et cosmoch. Acta,* **26**, 417.
Goldberg, E. D., and Koide, M. (1963). *In* "Earth Sciences and Meteoritics" (J. Geiss and E. D. Goldberg, eds.), pp. 90–102. North Holland Publishing Co., Amsterdam.
Goldberg, E. D., Koide, M., Griffin, J., and Peterson, M. N. A. (1964). *In* "Isotopic and Cosmic Chemistry" (H. Craig, S. Miller and G. J. Wasserburg, eds.), pp. 211–232. North Holland Publishing Co., Amsterdam.
Goldberg, E. D., Koide, M., Schmitt, R. A., and Smith, R. H. (1963). *J. geophys. Res.* **68**, 4209.
Goldberg, E. D., and Parker, R. H. (1960). *Bull. geol. Soc. Amer.* **71**, 631.
Goldberg, E. D., and Picciotto, E. (1955). *Science,* **121**, 613.
Graf, D. L. (1960). *Circ. Ill. geol. Surv.* **301**, 71 pp.
Graham, J. (1959). *Science,* **129**, 1428.
Grim, R. W. (1952). "Clay Mineralogy". McGraw-Hill Book Co., New York.
Grütter, A., and Buser, W. (1957). *Chimia,* **11**, 132.
Harned, H. S., and Owen, B. B. (1958). "Physical Chemistry of Electrolyte Solutions". Reinhold Publishing Corp., New York.
Hosohara, K. (1961). *Nippon Kagaku Zasshi,* **82**, 1107.
Kolyadin, L. B., Nikolaeu, D. D., Grashchenko, S. M., Kuztetzou, Y. U., and Lazareu, K. F. (1960). *C.R. Acad. Sci. U.R.S.S.* **132**, 915.
Krauskopf, K. B. (1956). *Geochim. et cosmoch. Acta,* **9**, 1.
Krumbein, W. C., and Garrels, R. M. (1952). *J. Geol.* **60**, 1.
Latimer, W. M. (1952). "Oxidation Potentials." Prentice Hall, Inc., New York.
Levine, E. P. (1961). *Science,* **133**, 1352.
Lewis, G., and Goldberg, E. D. (1954). *J. Mar. Res.* **13**, 183.
Lowenstam, H. A. (1963). *In* "The Earth Sciences" (T. W. Donnelly, ed.), pp. 137–195. Rice University Press, Houston.
Manheim, F. T. (1961). *Geochim. et cosmoch. Acta,* **25**, 52.

McKelvey, V. E., Swanson, R. W., and Sheldon, R. P. (1963). *19th Int. Geol. Congr.* (*Algiers*) *C.R.* **11**, 45.
Menard, H. W. (1964). "Marine Geology of the Pacific." McGraw-Hill Book Co., New York.
Menard, H. W., and Shipek, C. J. (1958). *Nature, Lond.* **182**, 1156.
Miyake, Y., and Tsunogai, S. (1963). *J. geophys. Res.* **68**, 3989.
Moeller, Th. (1963). "The Chemistry of the Lanthanides." Reinhold Publishing Corp., New York.
Mokyevskaya, V. V. (1961). *C.R. Acad. Sci. U.R.S.S.* **137**, 1445.
Mullin, J. B., and Riley, J. P. (1956). *J. Mar. Res.* **15**, 103.
Nicholls, G. D., Curl, H., and Bowen, V. T. (1960). *Limnol. Oceanogr.* **5**, 472.
Nikolaev, D. S., Korn, O. P., Lasarev, K. F., Koliadin, L. B., Kusnetsov, J., and Graschenko, S. M. (1960). *C.R. Acad. Sci. U.R.S.S.* **132**, 1411.
Pachadzhanov, D. N., Bandurkin, G. A., Migdisov, A. A., and Girin, Yu. P. (1963). *Geochemistry*, 520.
Rama, Koide, M., and Goldberg, E. D. (1961). *Science*, **134**, 98.
Robinson, R. A., and Stokes, R. H. (1959). "Electrolyte Solutions." Butterworths Scientific Publications, London.
Schwarzenbach, G. (1961). *Adv. Inorg. Radiochem.* **3**, 257.
Sheldon, R. P. (1959). *Bull. U.S. Geol. Surv.* **1084b**, 83.
Sillén, L. G. (1961). *In* "Oceanography" (M. Sears, ed.), pp. 549–582. AAAS, Washington, D.C.
Starik, I. E., Kuznetsov, Y. V., Grashchenko, S. M., and Frenklikh, M. S. (1958). *Geochemistry*, 1.
Tatsumoto, M., and Goldberg, E. D. (1959). *Geochim. et cosmoch. Acta*, **17**, 201.
Tatsumoto, M., and Patterson, C. C. (1963). *In* "Earth Science and Meteoritics" (J. Geiss and E. D. Goldberg, eds.), pp. 74–89. North Holland Publishing Co., Amsterdam.
Turekian, K. K., and Tausch, E. H. (1964). *Nature, Lond.* **201**, 696.
Turekian, K. K., and Armstrong, R. L. (1960). *J. Mar. Res.* **18**, 133.
Thurber, D. L. (1962). *J. geophys. Res.* **67**, 4518.
Vinogradov, A. P. (1953). "The Elementary Composition of Marine Organisms." Sears Foundation Marine Research Memoir 2, New Haven, Connecticut.
Wedepohl, K. H. (1960). *Geochim. et cosmoch. Acta*, **18**, 200.

Chapter 6

Dissolved Gases Other than Carbon Dioxide*

FRANCIS A. RICHARDS

Department of Oceanography
University of Washington
Seattle, U.S.A.

I. Introduction 197
A. General 197
B. Studies of Dissolved Oxygen 199
C. Studies of Dissolved Nitrogen 200
D. Processes Leading to the Alteration of the Concentrations of Dissolved Gases 200
II. Physical Factors Controlling the Gas Content of Sea Water 201
A. General 201
B. The Solubility of Oxygen in Sea Water 204
C. The Solubility of Nitrogen and Argon in Sea Water 205
D. The Solubility of the Other Noble Gases in Sea Water 206
E. The State of Equilibrium at the Sea Surface 208
F. Exchange of Gas across the Sea Surface 211
III. Changes of Gas Concentration caused by *in situ* Processes 214
A. Oxygen 214
B. Nitrogen 217
C. The Noble Gases 217
IV. Advective and Diffusive Processes 218
V. Some Oceanographic Applications of Dissolved Gas Distributions 220
Acknowledgments 223
References 223

I. Introduction

A. General

A useful assumption in chemical oceanography is that at one time every parcel of water in the world's oceans was at the sea surface, and at that time became equilibrated, or nearly so, with the atmospheric gases. In the course of oceanic circulation, the dissolved gases are distributed throughout the entire water mass by the processes of advection and diffusion. Between the time the water leaves the surface layers and the time it returns to them, the concentrations of some of the dissolved gases, particularly oxygen, may be altered by *in situ* processes, which are primarily biological.

* Contribution No. 338 from the Department of Oceanography, University of Washington.

By far the most abundant of the atmospheric gases are nitrogen, oxygen, argon, and water vapour (Table I). There are traces of carbon dioxide, the noble gases, hydrogen, various hydrocarbons, and a variety of contaminants introduced by man's activities, but except for carbon dioxide, which is treated in Chapter 7, these minor atmospheric gases have received a minimum of attention from oceanographers. However, the distribution of, and processes altering, the concentration of dissolved oxygen have been studied by oceanographers since the beginning of the century, when a reasonably satisfactory analytical method for its determination was introduced to oceanography. Earlier attempts to study dissolved oxygen in the oceans, using less satisfactory methods, were made from about the middle of the last century (see Chapter 1).

TABLE I

Composition of the atmosphere (from Glueckauf, 1951)
A. Summary of non-variable components

Constituent	%	*p.p.m.*	*Atmospheres*
N_2	78·084 ± 0·004		0·7808
O_2	20·946 ± 0·002		0·2095
CO_2	0·033 ± 0·001		0·0003
Ar	0·934 ± 0·001		0·0093
Ne		18·18 ± 0·04	$1·82 \times 10^{-5}$
He		5·24 ± 0·004	$5·24 \times 10^{-6}$
Kr		1·14 ± 0·01	$1·14 \times 10^{-6}$
Xe		0·087 ± 0·001	$8·7 \times 10^{-8}$
H_2		0·5	5×10^{-7}
CH_4		2·0	2×10^{-6}
N_2O		0·5 ± 0·1	5×10^{-7}

B. Variable constituents of dry atmospheric air at the surface of the ground

Constituent	*Source*	*Amount*
O_3	Ultra-violet radiation	0–0·07 p.p.m. (summer) 0·02 (winter)
SO_2	Industrial	0–1 p.p.m.
NO_2	Industrial	0–0·02 p.p.m.
CH_2O	Biological or oxidation of CH_4	Uncertain
I_2	Industrial	Up to 10^{-4} g. m.$^{-3}$
NaCl	Sea spray	Order of 10^{-4} g. m.$^{-3}$
CO	Industrial	0 to trace
NH_3	Industrial	0 to trace

C. Isotopic composition of the main atmospheric gases

Element	*Atomic mass numbers and % of isotopic species*
H in H_2O	(1) 99·98% (2) 0·02%
He	(3) 1×10^{-4}% (4) 100%
C in CO_2	(12) 98·9% (13) 1·1% (14) $0{\cdot}95 \times 10^{-12}$ %
N	(14) 99·62% (15) 0·38%
O	(16) 99·757% (17) 0·039% (18) 0·204%
Ne	(20) 90·00% (21) 0·27% (22) 9·73%
Ar	(36) 0·307% (38) 0·061% (40) 99·632%

Figures in brackets represent the isotopic mass.

D. Reported variability of some air constituents

Constituent	*Locality*	%
O_2	Mediterranean Sea	20·94 ± 0·01
O_2	Atlantic Ocean	20·94 ± 0·01
O_2	Arctic Ocean	20·91 ± 0·01
O_2	Eastern Indian Ocean	Variable, mostly low
O_2	Montreal to Liverpool (7 samples)	20·950 average
O_2	Genoa to Boston (36 samples)	20·946
Ar	Ionian Sea (37°N., 15°E.)	0·936
Ar	Atlantic Ocean (37°N., 24°W.)	0·932
Ar	Atlantic Ocean (43°N., 22°W.)	0·949
Ar	Average of 11 places	0·9343 ± 0·0006
CO_2	World average, 1895	*c.* 292 p.p.m.
CO_2	World average, 1950	*c.* 330 p.p.m.

B. STUDIES OF DISSOLVED OXYGEN

Studies of oxygen in the ocean appear to have been begun by W. L. Carpenter, working aboard the *Porcupine*, as early as 1869. Oxygen determinations were attempted during the cruise of the *Challenger* (1873–6), but it was not until the introduction of the Winkler (1888) titrimetric method to the analysis of sea water that oceanographers began to accumulate enough data on dissolved oxygen to get a clear picture of its distribution in the ocean. The Winkler method was probably first used on an extensive oceanographic cruise on the *Planet*, during her circumnavigation of the globe in 1906 and 1907.

Various modifications of the Winkler determination have been, and are still, in use in oceanography, and although modern oceanographers have expressed some dissatisfaction with the method (Carritt, 1963), it still remains the basic reference method.

C. STUDIES OF DISSOLVED NITROGEN

There have been, until the introduction of gas chromatographic procedures, no reasonably satisfactory methods for the more or less routine determination of dissolved nitrogen. Nitrogen was determined by extracting the dissolved gases from solution and determining the nitrogen gasometrically after other components of the gas mixture were removed. In most procedures, no attempt was made to separate the nitrogen from the argon and other inert gases, and the analyses therefore returned a combination of inert gases, of which the nitrogen was by far the most abundant component. Rakestraw and Emmel (1937) removed the nitrogen by combining it with molten lithium to form the nitride, but the method is too tedious to be applied to the routine or semi-routine analysis of large numbers of samples. The existing gasometric analysis procedures suffer from this limitation, and in addition, probably lack sufficient precision to distinguish between significant departures in the nitrogen content from that expected from the solution of air and scatter resulting from analytical uncertainty. Consequently, our knowledge of the distribution of dissolved nitrogen in the oceans is in general and rather imprecise terms, but it is much better than that of the distribution of dissolved argon or of any of the other less abundant components of air except carbon dioxide.

D. PROCESSES LEADING TO THE ALTERATION OF THE CONCENTRATIONS OF DISSOLVED GASES

The processes which alter the concentrations of dissolved gases in a parcel of sea water are as follows.

(1) *Exchanges with the atmosphere across the sea surface.* These processes, limited by rates of diffusion and replacement of the surface waters, are confined to a thin upper layer, but they may include exchanges with bubbles which are carried below the sea surface under turbulent conditions (Kanwisher, 1963). Exchange across the surface will occur in both directions, but the direction of the net transport will depend on the sign of the difference in the partial pressure of the gas in solution and that in the gas phase; the rate of the net transport will

be proportional to the magnitude of this difference and to the area of contact between the two phases.

(2) In situ *processes*. These include the photosynthetic production of oxygen by marine plants and the respiratory consumption of oxygen by plants and animals, including bacteria. These processes are much more significant quantitatively than reactions of dissolved oxygen with dissolved inorganic substances such as Fe(II), Mn(II), sulphides etc., and the extracellular enzyme-catalysed consumption of oxygen. The concentration of dissolved nitrogen can be altered *in situ* by nitrogen fixation and, under special conditions, when all or almost all of the oxygen has been consumed from the water, by denitrification, during which free nitrogen is produced by the biochemical reduction of nitrate and nitrite ions. The fraction of the total nitrogen in any unit mass of water produced or consumed by these processes is small. We know of no *in situ* processes by which the concentration of argon is altered except by the radioactive decay of ^{40}K, a process which adds amounts of argon that are small, except over geological periods of time, compared with the total amounts present. ^{4}He is produced *in situ* by the radioactive decay of uranium and its daughter nuclides (see Section IIF and Revelle and Suess, 1963).

Advective and mixing processes will alter the local concentrations of any dissolved constituent of sea water in accordance with the principles of continuity and the theory of the distribution of variables in the ocean (Sverdrup *et al.*, 1942). Given sufficient data and an adequate knowledge of eddy diffusion coefficients, the equation for the distribution of variables could be solved so as to give the value of the non-conservative changes.

II. Physical Factors Controlling the Gas Content of Sea Water

A. General

Dissolved gases will exchange across the sea surface in proportion to the area of the interface and to the difference in partial pressures of the gas in the two phases. When the partial pressures become equal, the fugacities in the two phases are equal, and no net exchange will follow.

(1) *Composition of the atmosphere.* The composition of the atmosphere may be assumed, for most practical purposes, to be constant except for major changes in its water vapour pressure and smaller changes in its carbon dioxide content. Therefore, at the sea surface, the partial pressure of oxygen, nitrogen, or argon can be assumed to be a specific fraction of the atmospheric pressure when corrected for its content of water vapour. It is probably safe to assume that the air at the sea surface

is saturated with water vapour, and therefore that the partial pressures of the major gases are functions of the barometric pressure, the temperature, and the salinity of the water, because the latter two parameters control the equilibrium vapour pressure of sea water. However, the effect of variations of the salinity on the equilibrium vapour pressure can be neglected for most practical purposes. Compensation for this effect can be made by use of the equation:

$$p_s/p_0 = 1 - 0{\cdot}000969\, Cl‰,$$

where p_s and p_0 are respectively the vapour pressures of the sea water and of distilled water at the same temperature, and $Cl‰$ is the chlorinity (Witting, 1908). Within the range of salinities and temperatures in the open ocean, the vapour pressure of sea water will not vary by more than 0·02 Torr from that of distilled water. The approximate composition of the atmosphere is given in Table I.

(2) *Partial pressures of gases in solution.* The partial pressures of the gases dissolved in the water *at the sea surface* (1 atmosphere pressure) are not so simply estimated because the surface of the sea may not be in equilibrium with the atmosphere. An estimate of the partial pressures at the sea surface depends on a knowledge of the actual concentration of the particular gas and of its solubility. The first of these depends on an analytical determination, and the second on empirically derived knowledge of the solubility as a function of the temperature and the salinity of the water. The state of the art is at present such that neither of these can be determined with complete satisfaction, but good and useful approximations of both can be made for dissolved oxygen. The solubility of dissolved nitrogen is reasonably well known, although it has not received as much attention as has that of oxygen. Analytical methodology for dissolved nitrogen is lacking, and knowledge of the solubility of argon and analytical methods for its determination in sea water are both wanting. Benson and Parker (1961a) have reviewed the probable reliability of the various solubility data.

(3) *The solubility of atmospheric gases in sea water.* Equilibrium gas solubilities are expressed either in terms of the solubility coefficient, α, which applies to the solubility of a pure gas, or in terms of the quantities of the gases contained in a unit quantity of sea water which is in equilibrium with air under specified conditions. In either case, they are functions of the temperature and the salinity of the water.

Henry's Law states that at equilibrium, the solubility of a gas in solution will be directly proportional to its partial pressure in the gas phase. At equilibrium:

$$p = P$$

where p is the partial pressure in the gas phase and P is the partial pressure in the solution. Thus, at equilibrium the amount of gas in solution is proportional to its partial pressure in the atmosphere:

$$c = \alpha p$$

where c is the concentration of gas in solution, and α is a solubility coefficient which is characteristic of the gas, and which varies with the temperature and, for sea water, with the salinity. The coefficient α can be determined empirically by equilibrating the solvent (sea water) with the pure gas. The solubility can be computed from a knowledge of the partial pressure of the gas in the atmosphere and by assuming the validity of Henry's Law for the system. Historically, solubilities of gases in sea water have been determined by Fox (1909), Whipple and Whipple (1911), Rakestraw and Emmel (1938a), and by Truesdale *et al.* (1955) by the reverse method, i.e. by equilibrating sea water samples having various temperatures and salinities with air and determining the amounts of the gases so dissolved.

(4) *Measures of gas quantities.* The solubility of a gas in sea water having a given salinity and temperature, in equilibrium with air at a standard total pressure of 1 atmosphere (760 Torr) is usually designated by the chemical symbol for the gas with a prime, such as O_2', N_2', Ar′ etc. These values are reported in two ways in the literature: the equilibrium solubility corrected to a dry atmosphere having a total pressure of 760 Torr, and the solubility from an atmosphere saturated with water vapour at the temperature in question having a total pressure of 760 Torr. It is necessary to specify which system is used.

It is sometimes useful to use two primes, O_2'', N_2'', Ar″, to indicate saturation concentrations when the water is equilibrated with air under other than the above standard conditions, that is, when the total pressure is other than 760 Torr, when the gas phase is neither dry nor saturated with water vapour. The values of O_2' and O_2'' can be related by the equation

$$O_2'' = O_2' \left[\frac{\left(p - p_s \dfrac{h}{100} \right)}{760 - p_s} \right]$$

where p is the barometric pressure (Torr), h is the relative humidity in per cent, p_s is the aqueous tension of sea water at the temperature in question, and O_2' is the solubility from a water-saturated atmosphere (Richards and Corwin, 1956). $(p - p_s h/100)$ is the partial pressure of dry air in the atmosphere. It seems safe to assume that at the air–sea interface, h does not vary appreciably from 100%.

Oceanographers frequently refer to the "per cent saturation" of a gas in a sample:

$$\% \text{ saturation} = 100.\, O_2/O_2',$$

where O_2' is the solubility as defined above for water having the observed *in situ* temperature and salinity, and O_2 is the observed oxygen content. When this expression is applied to subsurface samples, the O_2' implies the oxygen content which the water would have if it were brought to the surface and equilibrated there with the air under standard conditions. The potential temperature should therefore be used, but this refinement is not always introduced into the calculation.

Another useful term is the oxygen consumption, or apparent oxygen utilization (A.O.U.) (Redfield, 1942) given by

$$\text{A.O.U.} = O_2' - O_2.$$

This is used to obtain approximate estimates of the changes in the oxygen concentration which have taken place since the water was equilibrated with the atmosphere.

B. THE SOLUBILITY OF OXYGEN IN SEA WATER

In 1909, Fox reported on a series of experiments which he had conducted at the University of Christiania (now Oslo) on the equilibrium solubility of oxygen, nitrogen and carbon dioxide in sea water (Fox, 1909). Whipple and Whipple, in the United States, had been working on the solubility of oxygen in sea water, and when Fox published his results, they found that their data were in essential agreement with those of Fox. They decided to suspend their experiments, because their results so well confirmed those of Fox, and published a note to that effect in the Journal of the American Chemical Society (Whipple and Whipple, 1911). The impact of these papers led to the conclusion that Fox's solubility data were definitive, and they were accepted as such, at least by oceanographers, for nearly 50 years. Later, Truesdale and his co-workers, working at Stevenage, England, noted anomalies in rates of oxygen solution, which they felt might be the result of inaccurate oxygen solubility data. They made new determinations of oxygen solubilities, which they published in 1955 (Truesdale *et al.*, 1955), that indicated that the solubility was appreciably lower than Fox's figures suggested. A controversy concerning the accuracy of the data arose and still awaits a definitive answer (see Steen, 1958; Elmore and Hayes, 1960; Benson and Parker, 1961a,b). Most of the new solubility data that have ensued have been for distilled water solutions rather than sea water, and the oceanographer therefore remains in a quandary. However, as Benson and Parker have pointed out, highly precise methods which permit the repetition of determinations with a minimum of scatter in the results, must be distinguished from highly accurate methods which may return larger random but smaller systematic errors.

Fox's solubility data can be fitted to the empirical equation

$$1000a = 10{\cdot}291 - 0{\cdot}2809t + 0{\cdot}006009t^2 - 0{\cdot}0000632t^3$$
$$- (0{\cdot}1161 - 0{\cdot}003922t + 0{\cdot}0000631t^2).Cl\text{‰},$$

in which $1000a$ is the solubility of oxygen in ml. at NTP in 1 litre of water equilibrated with *dry* air at a total pressure of 760 Torr, t is the temperature in °C and Cl‰ is the chlorinity. Kalle (1939) has published a nomogram based on Fox's equation. The data of Truesdale, Downing and Lowden can be fitted to the equation

$$1000a = 9{\cdot}9096 - 0{\cdot}2759t + 0{\cdot}005398t^2 - 0{\cdot}00004527t^3$$
$$- (0{\cdot}05896 - 0{\cdot}00179t + 0{\cdot}00002618t^2).S\text{‰}$$

where t has the above meaning, S‰ is the salinity of the water, and $1000a$ is the solubility expressed in the above units in water equilibrated with air *saturated with water vapour*. These solubilities can be derived from a nomogram given by Richards and Corwin (1956).

The Water Pollution Research Laboratory at Stevenage has recently reported (Department of Scientific and Industrial Research, 1964) that subsequent investigations at their laboratory indicate that the solubilities published by Truesdale *et al.* (1955) were in error because of systematic errors introduced by volatilization of iodine during the variant of the Winkler method which they used while making their solubility determinations. Workers in the same laboratory (Montgomery *et al.*, 1964) have used a modified Winkler method to determine the solubility of oxygen in pure water and sea water. The results for sea water are in fair agreement with those of Fox (1909) (see Appendix, Table XII).

C. THE SOLUBILITY OF NITROGEN AND ARGON IN SEA WATER

The experimental work by Fox included determinations of the solubility of "residual" gases, primarily nitrogen plus argon (see Appendix, Table XIII). These solubilities could be fitted to the equation

$$1000a = 18{\cdot}639 - 0{\cdot}4304t + 0{\cdot}007453t^2 - 0{\cdot}0000549t^3$$
$$- (0{\cdot}2172 - 0{\cdot}007187t + 0{\cdot}0000952t^2).Cl\text{‰},$$

where $1000a$ is the number of ml. of (N_2 + Ar) at S.T.P. absorbed by 1000 ml. of sea water from a free, dry atmosphere of 760 Torr pressure.

Cl‰ is the chlorinity, and *t* is the temperature. Rakestraw and Emmel (1938a,b) made new determinations of the solubility of nitrogen in sea water which they attempted to correct for the argon (and other inert gases) content of the mixture extracted from sea water equilibrated with air, by measuring the residue left after combining the nitrogen in the mixture with molten lithium. This also gave them estimates of the solubility of argon (plus the other inert gases) in sea water, but they claimed no great precision or accuracy for their results. Their data for the solubility of nitrogen are given by the equation

$$C_s = 18{\cdot}174 - 0{\cdot}4872t + 0{\cdot}01152t^2 - 0{\cdot}0000886t^3$$
$$- (0{\cdot}199 - 0{\cdot}00736t + 0{\cdot}000214t^2).Cl‰$$

where C_s is the solubility in ml./l. at standard temperature and pressure and *Cl*‰ is the chlorinity of the sample.

Benson and Parker (1961a) have given new data on the *ratios* of the solubilities of nitrogen, argon and oxygen in sea water, but they were careful to point out the uncertainties in the solubilities of the individual gases. For distilled water, they gave the least squares equation

$$N_2'/Ar' = 37{\cdot}90 + 0{\cdot}0590t,$$

where N_2'/Ar' is the ratio (volumetric or molecular) of the solubilities of nitrogen and argon in distilled water equilibrated with air at, or near, 760 Torr, at *t*°C. The analogous expression for the ratio of the solubilities of nitrogen to oxygen in distilled water is

$$N_2'/O_2' = 1{\cdot}800 + 0{\cdot}00533t.$$

Owing to inadequacies in their methods they were unable to determine accurately the effect of salinity on the N_2'/Ar' and N_2'/O_2' ratios in the range of salinities (32–36‰) which they investigated. Within this range, the ratios (by volume at S.T.P.) they found were given by the equations

$$N_2'/Ar' = 37{\cdot}48 + 0{\cdot}0522t \text{ (standard deviation } = 0{\cdot}175, c.\ 0{\cdot}46\%)$$

and

$$N_2'/O_2' = 1{\cdot}807 + 0{\cdot}00396t \text{ (standard deviation } = 0{\cdot}028, c.\ 1{\cdot}5\%)$$

for a salinity of 34·5‰.

D. The Solubility of the Other Noble Gases in Sea Water

König (1963) has reported data for the solubilities of He, Ne, Ar, Kr and Xe in sea water of 19·12‰ chlorinity (34·54‰ salinity), equilibrated with 760 Torr of the pure gas, and has calculated the solubilities of the gases in sea water equilibrated with air (Table II).

TABLE II

Solubilities of noble gases in sea water equilibrated with 760 Torr of each of the gases and calculated for equilibrium with air containing 5·24 × 10^{-6} atm of He, 1·82 × 10^{-5} atm of Ne, 9·32 × 10^{-3} atm of Ar, 1·14 × 10^{-6} atm of Kr, and 8·6 × 10^{-8} atm of Xe respectively (from König, 1963). The units are ml. (S.T.P.)/kg. water

Temperature (°C)	He		Ne		Ar		Kr		Xe	
	Pure	Air	Pure	Air	Pure	Air	Pure	Air	Pure	Air
0 ± 1	—	—	9·37	171 × 10^{-6}	—	—	71·5	81·5 × 10^{-6}	136	11·70 × 10^{-6}
1 ± 1	7·91	41·4 × 10^{-6}	—	—	38·5	0·359	—	—	—	—
5 ± 1	—	—	9·02	164 × 10^{-6}	35·3	0·329	63·9	72·8 × 10^{-6}	115	9·89 × 10^{-6}
10 ± 1	7·40	38·8 × 10^{-6}	8·67	158 × 10^{-6}	32·8	0·306	58·2	66·3 × 10^{-6}	103	8·86 × 10^{-6}
15 ± 0·5	6·95	36·4 × 10^{-6}	8·35	152 × 10^{-6}	30·2	0·282	51·6	58·5 × 10^{-6}	90·0	7·74 × 10^{-6}
17·5 ± 0·5	—	—	—	—	—	—	50·2	57·2 × 10^{-6}	—	—
20 ± 0·5	7·00	36·7 × 10^{-6}	8·20	149 × 10^{-6}	26·3	0·245	44·8	51·1 × 10^{-6}	80·0	6·88 × 10^{-6}
22·8 ± 0·5	—	—	—	—	—	—	44·2	50·4 × 10^{-6}	—	—
24 ± 0·5	—	—	—	—	—	—	42·9	48·9 × 10^{-6}	—	—
25 ± 0·5	—	—	8·07	147 × 10^{-6}	—	—	—	—	70·2	6·04 × 10^{-6}

König (1963) also compared Rakestraw and Emmel's data on argon solubilities with those which he determined, as well as with argon solubilities computed from the N′/Ar′ ratios determined by Benson and Parker (1961a,b) using nitrogen solubilities determined by Rakestraw and Emmel, and by Fox (1909). These comparisons are shown in Table III. Mazor *et al.* (1964) have commented that there are uncertainties in König's data, that other solubilities of argon in sea water have recently been reported in a paper by Douglas (1964), and that Douglas found values about 11% higher than König's. It is suggested that the data on the solubility of rare gases in sea water can now be accepted only provisionally.

TABLE III

Solubility of argon in sea water as a function of temperature (from König 1963). (In ml. at S.T.P./l.)

	König (1963)	*Rakestraw and Emmel (1938a)*	*Benson and Parker (1961a)*	
			(Fox, 1909)	*(Rakestraw and Emmel, 1938a)*
Chlorinity (‰)	19	19	20	20
Temperature (°C)				
0	(0·383)	—	0·375	0·379
5	0·338	0·354	0·335	0·334
10	0·314	0·314	0·306	0·301
15	0·289	0·283	0·282	0·272
20	0·251	0·256	0·262	0·248

E. THE STATE OF EQUILIBRIUM AT THE SEA SURFACE

For practical oceanographic purposes, the sea surface is represented by mixtures taken from the top metre of the water column. An excellent suite of data on the oxygen content of the sea surface (thus defined) of the South Atlantic Ocean is available from the *Meteor* expedition (Wüst, 1932; Wattenberg, 1933, 1938). These data have been examined in an attempt to estimate the degree of equilibrium between the sea surface and the atmosphere. The problem is complicated by the uncertainty in our knowledge of the equilibrium solubility of oxygen in sea water, referred to in Section IIB above, but the percentage saturation has been calculated using the solubility data of both Fox (1909) and Truesdale *et al.* (1955). In both sets of calculations the effects on the solubility of the relative humidity and of the total barometric pressure have been taken into consideration. Because the rate of response of the solution of gases to changes in barometric pressure is unknown, the data were examined using the barometric pressures observed when the water was sampled and 6 and 24 hours before the sampling, but this made inconsequential differences in the saturation values. The results are given in the histograms in Fig. 1, and are inconclusive, largely because of the uncertainties in the solubility data. However, it appears that a large fraction of the surface samples departs significantly from 100% saturation. The fact that a maximum number of samples appears to be 100% saturated if the solubility data of Trusedale *et al.* are used, whereas the corresponding maximum falls at 97% of saturation using Fox's solubility data, can neither be interpreted as giving the data of Truesdale *et al.* more credibility nor as indicating that true departures from 100% are scattered about a 100% median value. Plots of the above saturation values on charts of the ocean (see Richards and Corwin, 1956, Fig. 2) indicate a systematic distribution of the saturation values in which undersaturated surface waters appear to be associated with areas of

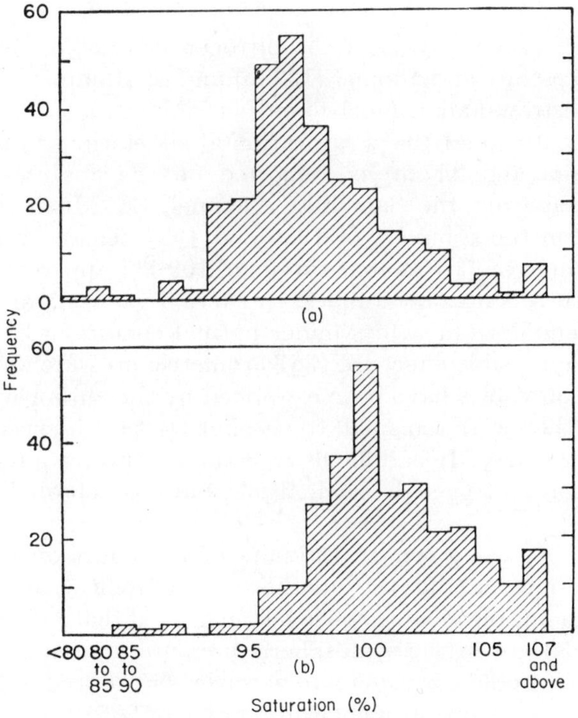

FIG. 1. The apparent percent saturation of oxygen in surface samples taken during the *Meteor* expedition. The solubilities were corrected for the barometric pressure observed at the time of sample collection. (a) Solubility data of Fox (1909); (b) solubility data of Truesdale *et al.* (1955).

divergence whereas oversaturated waters are associated with regions of convergence or high winds.

It is particularly difficult to evaluate the degree of equilibration of the surface layers with oxygen because biological processes may alter its concentration rapidly. These complications should be less severe for nitrogen, and presumably negligible when the rare gases are considered, but good, precise methods that are practical for the analysis of large numbers of sea water samples for these gases have been lacking. The introduction of gas chromatographic methods for dissolved gas analysis may assist in overcoming this problem.

Swinnerton and Sullivan (1962) and Swinnerton *et al.* (1962a,b) have described a rapid system for the gas chromatographic determination of oxygen (plus argon) and nitrogen in sea water. The method involves stripping the dissolved gases from the sample by bubbling the carrier gas, helium, through it. This, or similar methods, may soon improve our knowledge of the distribution of dissolved nitrogen in the oceans.

Among the best analyses of the nitrogen content of sea water are those of Rakestraw and Emmel (1938b) and of Hamm and Thompson (1941). Rakestraw and Emmel found surface nitrogen values of 100, 108, 101, 102, 101, and 104% saturation at six stations in the Atlantic Ocean. Hamm and Thompson reported on the analyses of surface samples taken from the San Juan Channel, San Juan Archipelago, Washington in the summers of 1936 and 1937. These samples varied from a minimum of 93·7 to a maximum of 109·1% apparent saturation. In both of these suites of samples, the actual gas contents were compared with equilibrium values under normal conditions of barometric pressure. The possible effects of the barometric pressure on the Hamm and Thompson values have been examined by the author who has concluded that they were too small to account for the observed variations in saturation values. It is difficult to estimate the accuracy of Hamm and Thompson's nitrogen determinations, but it was probably no better than ± 2%.

Carritt (1954) has given an estimate of the effect of varying total atmospheric pressure on the equilibrium solubilities of N_2 and O_2 (N_2'' and O_2'' in our terminology). He has suggested that in high latitudes, seasonal variations in the atmospheric pressure may amount to 19 to 22 Torr, which would correspond to variations in O_2'' and N_2'' of 0·34 and 0·65 ml./l., respectively, at a chlorinity value of 20‰ and 0°C.

The state of the initial equilibrium saturation in sub-surface samples is even more difficult to evaluate because the relationship between solubility and temperature is not linear. Thus, at constant gas pressure, a mixture of two water samples, each saturated with gas at a different temperature, will be oversaturated with the gas.

There are many observations in the oceanographic literature of surface oxygen concentrations of 110%, or 120%, or even higher saturation values which are attributable to *in situ* photosynthetic production of oxygen. The fixation of nitrogen by marine blue-green algae in the upper layers of parts of the ocean has been demonstrated (Dugdale *et al.*, 1961), and could account for values of nitrogen saturation of less than 100%.

At the present time, our knowledge of the state of equilibration of the surface layers of the ocean is uncertain. The thermodynamic equilibrium state will be approached, but the balance between the rate at which it is approached, the rate at which water is moved into and out of the surface layers, and the rates at which *in situ* processes alter the gas content of the water, cannot be precisely evaluated. These evaluations are some of the outstanding problems in chemical oceanography today.

F. EXCHANGE OF GAS ACROSS THE SEA SURFACE

The net rate of exchange of gas across the sea surface depends on the difference between the partial pressure of the gas in the atmosphere and its partial pressure in the surface water. In the gas phase, the partial pressure will be the product of the mole fraction of the gas and the total pressure. In the liquid phase at the surface, the partial pressure of the gas will be the ratio of its actual concentration to its equilibrium solubility, under the given conditions, multiplied by its partial pressure in the gas phase.

There are other factors in the exchange of gases across the sea surface. If we assume a thin layer at the sea surface such as Kanwisher (1963) proposed, the flux through this layer would be given in terms of his formulation

$$\text{Flux} = \frac{\Delta\,\text{concentration} \times \text{diffusivity}}{\text{thickness}}$$

where the concentration is the partial pressure multiplied by the solubility. For the atmospheric gases, the diffusivity will have a value of 2×10^{-5} cm./second (the coefficient of molecular diffusion). It can be assumed that generally the flux of gas to the film will take place by eddy diffusion and advection and that molecular diffusion will prevail in the film.

Kanwisher has pointed out that although the diffusion coefficients of the atmospheric gases are not very different, the flux through the surface film will vary widely from gas to gas, for a given partial pressure, because of the large variations in their solubilities. Thus, the flux for a given difference in partial pressures would be about twice as great for O_2 as for N_2, whereas for CO_2 it would be about 70 times that for N_2.

Kanwisher suggested that these surface films have counterparts in the ocean, but because turbulent motion will be imparted by the shear of the wind stress, the thickness of the films decreases rapidly beyond a certain critical wind speed. Laboratory experiments indicated that this speed was of the order of 2 m./second at a height of 10 cm. above the surface.

It is clear that large gradients in the oxygen concentrations can exist in the surface layer. Observations made by Dr. R. M. Pytkowicz, in which a dropping mercury electrode oxygen recorder (Føyn, 1955) was located in the surface waters, indicated large variations in the oxygen concentration of the uppermost metre.

The flux by eddy diffusion to the surface film will depend on the concentration gradient in the horizontal (dc/dx) and in the vertical (dc/dz) directions. These gradients will generally be larger for oxygen than for nitrogen (particularly dc/dz), as is evident from inspection of vertical distributions of dissolved oxygen and dissolved nitrogen, and values of dc/dz will generally be much larger than those of dc/dx, because the concentrations tend to change more rapidly with changes in depth than they do in the horizontal direction.

There have been few attempts to assess the exchange of a gas across the sea surface, but Redfield (1948) estimated the amount and rate of exchange across the surface in the Gulf of Maine for the period July, 1933 to September, 1934. His methods and results have been reviewed by Richards (1957). Redfield assumed that the amount of oxygen produced in, or consumed from the water column could be estimated from the changes in its phosphate content. These estimated *in situ* changes were then compared with the observed changes in the oxygen content. The differences were attributed to the passage of oxygen across the sea surface. He estimated that about 30×10^4 ml. of oxygen per $m.^2$ left the surface of the Gulf of Maine between the end of October and the latter part of March, and that a similar amount entered the surface during the rest of the year. He also estimated that the differences between the partial pressures of oxygen in the water and in the air varied between $+1{\cdot}4$ and $-2{\cdot}1 \times 10^{-2}$ atmospheres, where a positive value indicates that the partial pressure in the atmosphere was greater than that in the water.

The above discussion has implied exchange across a smooth sea surface, which, in nature, is the exception rather than the rule. Kanwisher

(1963) has examined some of the effects that sea state and wind may have on the exchange of gas between sea and air. Laboratory experiments indicated that exchange increased with wind velocity after a critical value of c. 2 m./second. At higher velocities, the exchange increased approximately with the square of the wind velocity. These results might be interpreted as indicating that increased wind speeds (a) decrease the effective thickness of the boundary film, and (b) increase the area of liquid surface.

In addition to the above effects, in certain sea states, bubbles will be produced and carried downwards, perhaps to an appreciable depth. The hydrostatic pressure increases by 0·1 atmosphere for every metre of depth in the sea, and thus the partial pressure of gas in such a bubble will be increased according to the equation given by Ramsey (1962)

$$p = g_A P_t$$

where p is the partial pressure of the gas in the bubble, g_A is the mole fraction of the gas in the bubble, and P_t is the total hydrostatic pressure at the depth in question. Thus, the partial pressure in the bubble will increase by approximately 10% for every metre the bubble is carried beneath the sea surface. Kanwisher (1963) has pointed out that bubbles of the size apt to be carried downwards to considerable depths (100–1000 μ in diameter) are excellent gas exchangers and can be assumed to be important in the exchange of gas across the sea surface.

Ramsey (1962) has considered the physics of the formation of bubbles in the water column from supersaturation concentrations of oxygen. Although he reports having observed oxygen saturations of up to 170% in the upper 3·2 m. off San Diego, California, the average depth at which bubble growth, at the expense of the dissolved oxygen, would occur was in the range of 0·56–1·09 m.

Because the biologically less active gases in sea water, nitrogen and the rare gases, do not depart from their surface saturation values as markedly as does oxygen, one would expect less net exchange of them across the sea surface. Exceptions would be expected where the circulation of the oceans brings relatively cold water, containing close to saturation concentrations of these gases, into the surface layers, there to be warmed with a consequent decrease in the solubility of the gas, resulting in a tendency for the gas to escape to the atmosphere.

Another exception would be expected to be helium. According to Revelle and Suess (1963), helium escapes from the atmosphere, but its terrestrial concentration is maintained by that which is produced by radioactive decay, primarily in the lithosphere including oceanic sediments and suboceanic rock. They postulated that the helium in sea water is at a stationary concentration; in other words, helium is escaping from the ocean, through the atmosphere, and into space at the same rate as it is being produced. Argon-40, although also produced by radioactive decay (of ^{40}K), accumulates in the hydrosphere and atmosphere, because it is not lost from the atmosphere. They estimated that it would have taken of the order of 10^{11} years to produce the amount of ^{40}Ar dissolved in the oceans by the radioactive decay of the ^{40}K present in sea water. This must be compared with the apparent radio-carbon age of deep-sea water which is of the order of 10^3 years.

III. Changes of Gas Concentration caused by *in situ* Processes

A. Oxygen

The chief *in situ* processes by which oxygen concentrations are altered are photosynthetic production and respiratory consumption. The former of these is strictly limited to the upper layers into which adequate light energy penetrates. In recent years, there have been extensive studies of the rates of photosynthetic production of carbon compounds in the world's oceans, and generally these rates can be approximately equated to *in situ* oxygen production; viz.

$$CO_2 + H_2O = CH_2O + O_2,$$

where the term CH_2O should be interpreted as meaning carbohydrate (see Chapter 12). Geographical variations in the rates of photosynthetic productivity have been reviewed by Ryther (1963). Although photosynthesis in the ocean is never limited by the availability of carbon, it may well be limited by the availability of nutrient ions, generally nitrogen and phosphorus compounds, and by light. The latter is severely limiting, and nowhere penetrates in sufficient intensity to support photosynthetic activity at depths greater than 100 or 200 m. At greater depths, all chemical and biochemical processes tend to consume oxygen. The processes of oxygen consumption are treated in Chapter 13, and need not be covered here.

Generally, photosynthetic rates are at a maximum somewhat below the sea surface, presumably because light intensities at the surface are greater than optimal. Below this subsurface maximum, photosynthetic rates decrease with depth until they become too small to measure. Oxygen is consumed by respiration, even in the upper layers where photosynthesis may be quite rapid. The depth down to which photosynthetic production of oxygen exceeds respiratory consumption is the photosynthetic zone, and the depth at which they just balance is called the compensation depth. This depth varies widely (depths from

0·8–100 m. .have been reported, according to Richards, 1957) in the oceans, depending largely on the transparency of the water. As a consequence, the compensation depth will be shallower where plankton populations are denser (see Chapter 12).

In situ production of oxygen by photosynthesis is accompanied by the uptake of inorganic carbon and nutrient ions generally in proportion to their ratios in the organisms. These changes have been discussed by Redfield (1934, 1942), Richards (1957), Redfield *et al.* (1963), and others. It has been somewhat difficult to relate these changes to photosynthetic production, but Stefánsson and Richards (1964) have demonstrated the concurrent uptake of phosphate, nitrate, and silicate ions and the production of oxygen off the Washington and Oregon coasts. In these waters, during parts of the year, oxygen is produced and these ions are taken up in approximately the same average ratios as those found in plankton organisms; C : N : P = 106 : 16 : 1 (Fleming, 1940). Frequently the oxygen concurrently produced (in approximately the ratio of one mole of oxygen produced to one mole of CO_2 photosynthetically fixed) is lost to the system by escape across the sea surface, but Stefánsson and Richards (1964) were able to demonstrate that under certain hydrographic regimes, the oxygen remained in the water and accumulated in quantities stoichiometrically related to the uptake of inorganic phosphate, thereby resulting in concentrations greater than 100% saturation. However, under other hydrographic regimes, oxygen was lost to the atmosphere and there was then no apparent relationship between the phosphate uptake and oxygen production at the time of the oceanographic sampling (see also Postma, 1964).

The converse of the above process, the *in situ* consumption of oxygen by respiration and the decomposition of organic matter, is also accompanied by regular changes in the carbon dioxide, inorganic nitrogen, and inorganic phosphate contents of the water. Because these processes continue and predominate in the strata below that of active exchange with the atmosphere, good agreement between the consumption of oxygen and the regeneration of inorganic nutrients is more evident than the converse changes.

Because of the quasi-stoichiometric relationships between biological oxygen changes, the formation and decomposition of organic compounds, and the uptake and release of nutrient elements, especially nitrogen and phosphorus, measurements or estimates of any one of these changes can be approximately converted to the concurrent changes in any of the others. Richards (1960) has demonstrated the possibility of estimating organic productivity from phosphate changes in the waters of the upper layers of the Cariaco Trench and in the Gulf of Cariaco. It should be possible to obtain from these estimates approximate values for the amounts of nitrogen compounds concurrently consumed from the water and the amounts of dissolved oxygen produced in it.

It is clear that respiratory consumption of oxygen (including that by bacteria) proceeds at all depths in the ocean, because at no depths are the waters sterile. The rates of oxygen consumption, on the other hand, clearly decrease to low values at great depth. Several estimates of these rates are available in the literature, including those by Sverdrup and Fleming (1941), Riley (1951), and Barnes and Collias (1958). The rates found by Barnes and Collias for the relatively shallow waters of Puget Sound are presented in Table I of Chapter 13.

Wyrtki (1962) has shown that the rates estimated by Riley (1951) tend to decrease exponentially with depth. The form of the relationship is given by the equation

$$R = R_0 e^{-\alpha z}$$

in which R and R_0 are the consumption rates at the depth z and at the surface, respectively. He estimated a value for R_0 of between 60 and 140×10^{-10} ml./l./sec., and found α to be $3–4 \times 10^{-5}$ per cm. The rate of consumption at any depth was given approximately by the relationship

$$R = 100 \times 10^{-10} e^{-3 \cdot 5 \times 10^{-5} z}.$$

Riley (1951) arrived at his estimates by evaluation of the equation for the distribution of variables, using oceanographic observations of salinity, temperature, and dissolved oxygen from the Atlantic Ocean. Sverdrup and Fleming (1941) used somewhat similar methods of physical oceanography to estimate rates of oxygen consumption off the coast of southern California. Their estimates decreased from 1·8 ml./l./year at 200 m. to 0·05 of these units at 700 m. They are somewhat higher than Riley's estimates (0·21 and 0·035 ml./l./year at 200 and 700 m., respectively), but Riley's figures are average values for the whole Atlantic Ocean, and the region to which the Sverdrup and Fleming estimates apply should be one of more intense biochemical activity than the Atlantic Ocean as a whole. The estimates by Barnes and Collias ranged from 1·8–7·3 ml./l./year and appear to be consistent with the other estimates. The consumption rate, R_0, that Wyrtki used is *c.* 0·32 ml./l./year, and is apparently lower because the waters which Barnes and Collias studied are much richer than are Atlantic surface waters as a whole. Other estimates of rates of oxygen consumption have been reviewed by Richards (1957).

B. NITROGEN

In situ changes in dissolved nitrogen can be reviewed only briefly, because they are less well known than are *in situ* changes of dissolved oxygen, and also because they are much less extensive. They are brought about by nitrogen fixation and by denitrification. The latter process appears to be confined to anoxic waters (see Chapter 13).

Nitrogen fixation by terrestrial organisms has long been known, and the phenomenon has been demonstrated to take place in lakes (Dugdale *et al.*, 1959; Dugdale and Dugdale, 1962). Redfield (1958) suggested that nitrogen fixation in the ocean may be important in maintaining the nitrogen–phosphorus balance in marine organisms, but it has been demonstrated only recently in the ocean. Dugdale *et al.* (1961) conducted experiments which demonstrated reasonably conclusively the ability of the blue-green alga, *Trichodesmium thiebautii*, to fix nitrogen. This organism occurs rather abundantly in tropical and semi-tropical waters during periods of thermal stability of the upper layers. During these periods, nutrient elements are stripped from the surface layers to very low concentrations, and Dugdale *et al.* reasoned that organisms thriving under such conditions might have the ability to supply their requirements for nitrogen compounds by fixation. Dugdale *et al.* have since reported (at the meeting of the American Society of Limnology and Oceanography, Miami Beach, Florida, 15–18 June, 1964) on observations which indicate that nitrogen fixation by *Trichodesmium* takes place over extensive areas of the Indian Ocean.

Investigations of nitrogen fixation have not yet demonstrated its importance to the nitrogen balance or to the distribution of dissolved nitrogen in the ocean. However, Benson and Parker (1961b), investigated the dissolved N_2/Ar ratios in a large number of samples from the North and South Atlantic Ocean, in an attempt to detect deviations in these ratios from those expected from the solution of air, on the premise that significant increases or decreases would indicate that *in situ* processes had altered the ratios. Their results indicated that the ratios were within 1% of those to be expected from the solution of air. The conclusion to be drawn is that nitrogen fixation or denitrification had not taken place in the samples they investigated to an extent greater than that needed to alter the ratio by 1%.

C. THE NOBLE GASES

According to König *et al.* (1964), the analysis of 11 samples of sea water from depths of up to 4274 m. in the Pacific Ocean and up to 3300 m. in the Indian Ocean indicated that the concentrations of Ar, Ne, and He were within 3% (considered to be the experimental error) of those to be expected from equilibration of the waters at the sea surface with air.

In an earlier section (IIF), the production of He and Ar by radioactive decay was discussed, but it should be pointed out that Revelle and Suess (1963) attributed most of this production to radioactive processes in the lithosphere rather than in the water itself. They estimated that in each litre the water of the world's oceans, about 30 atoms of He are produced per minute from the decay of uranium and its daughters, and about 50 atoms per minute of ^{40}Ar from the decay of ^{40}K. These can be considered to be *in situ* processes, and will have negligible effects so far as short term changes in the concentrations of these gases in sea water are concerned.

IV. Advective and Diffusive Processes

At depths greater than the photosynthetic zone, all *in situ* processes tend to decrease the oxygen content of sea water, and if advective and diffusive processes were not effective in ventilating the depths of the ocean, the deep waters would become anoxic and azoic. However, advection and diffusion are effective, and except for the stagnant environments discussed in Chapter 13, the depths of the ocean are well ventilated and always contain dissolved oxygen.

The vertical distribution of oxygen observed at a variety of locations in the world's oceans is shown in Fig. 2. In general, the vertical distribution of oxygen shows the following features.

(1) A surface layer, which is essentially in equilibrium with the sea surface. This layer is generally well mixed and has a uniform oxygen content down to the thermocline. The depth of this layer usually corresponds to the depth of the wind-mixed layer.

(2) In stable water columns, such as develop in the spring in the temperate zones, the oxygen frequently shows a subsurface maximum, generally in the upper 50 m.; this is associated with the photosynthetic maximum.

(3) At depths greater than the above layers, there is a decrease in the oxygen content as the depth increases. This reflects the consumption of dissolved oxygen by reaction with organic matter in the upper layers. The decrease with depth may be sharp or relatively gentle. In the east tropical North Pacific Ocean, oxygen concentrations decrease from 5 ml. to a few hundredths of a ml./l. in 100 m., but in the Antarctic Convergence, the oxygen decreases only from 7·5–4·0 ml./l. in about 450 m.

(4) Oxygen concentrations generally pass through a minimum value,

except in the great convergences. The minimum is usually at a depth of between 700 and 1000 m., but may vary widely.

(5) Concentrations gradually increase again at depths beneath the minimum, but the oxygen content of the deep ocean may be nearly constant for hundreds of metres of depth. This represents the ventilation of the deep water by the circulation of water which has sunk from the surface at high latitudes.

The balance between the *in situ* consumption of oxygen and the

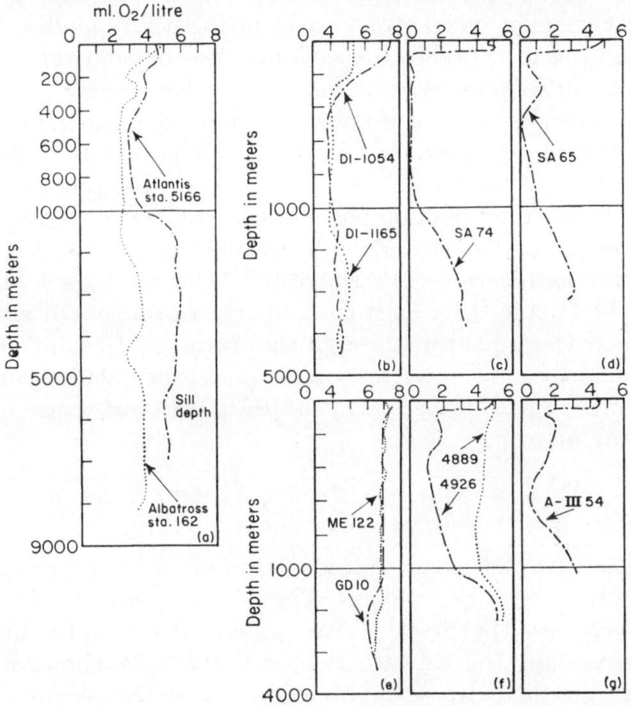

FIG. 2. Representative vertical distributions of oxygen. (a) *Albatross* station 162 in Mindinao deep. 23 January, 1948. 5°23′ N., 127°48′ E., and *Atlantis* station 5166 in the Brownson Deep, 24 February, 1954. 19°55′ N., 64°51′ W., (b) *Discovery* station 1054 in Antarctic Convergence. 3 December, 1932. 50°07·8′ S., 35°48·6′ W. *Discovery* station 1165 in Subantarctic Convergence, 24 February, 1932. 41°00′ S., 09°30′ E. (c) *Albatross* station 74, in eastern tropical Pacific. 22 September, 1947. 11°39′ N., 114°15′ W. (d) *Albatross* station 65 in eastern tropical Pacific, showing two minima. 16 September, 1947, 6°21′ N., 103°42′ W. (e) *Godthaab* station 10, 3 June, 1928, 56°56′ N., 51°17′ W., and *Meteor* station 122, 55°03′ N., 44°46′ W., 9 March, 1935, in North Polar Front. (f) *Atlantis* stations 4926 in tropical eastern Atlantic 28 February, 1952, 12°53′ N., 22°59′ W., and 4889 in south-eastern North Atlantic, 4 February, 1952, 29°52′ N., 30°12′ W. (g) *Albatross* III station 54, showing two minima and lowest concentrations observed in tropical eastern Atlantic. 16 February, 1952, 17°00·5′ N., 30°03′ W., from Richards (1957).

advective and diffusive renewal of oxygen leads to the occurrence of, and determines the position of, the oxygen minimum zone. Wyrtki (1962) has examined this balance, and has concluded that the minimum lies in the upper part of the layer of smallest advection, since rates of oxygen consumption decrease exponentially with depth.

Richards (1957) has reviewed other features of the oxygen minimum zone, and the discussion need not be repeated here. However, the concept that the amount of oxygen removed from a parcel of water is the stoichiometric consequence of the oxidation of organic matter must be emphasized. The tropical Pacific Ocean contains a thick layer having much lower oxygen concentrations than the corresponding regions of the Atlantic Ocean. These waters contain less oxygen than the Atlantic waters because more organic matter has decomposed in them—not because the circulation of the Pacific is more sluggish than that of the Atlantic. This can, in turn, be related to the generally higher level of nutrient concentrations which are brought into the photic zone of the Pacific Ocean, thereby making possible the production of larger quantities of organic matter to decompose below the photic zone. A. C. Redfield (personal communication) has suggested that the higher level of nutrients in the Pacific Ocean is a consequence of the loss of nutrient-rich deep Atlantic Ocean water, through the Antarctic Ocean to the Pacific Ocean, in exchange for nutrient-poorer, shallower Pacific water. Thus, a water balance is maintained, but the Pacific Ocean tends to act as a nutrient sink or trap.

V. Some Oceanographic Applications of Dissolved Gas Distributions

The earliest applications of knowledge of dissolved oxygen in the ocean were to biological problems. The discovery aboard the *Porcupine* and the *Challenger* that oxygen was present at all depths in the ocean went far in settling the Edward Forbes–C. Wyville Thompson controversy regarding the existence of an azoic zone in the ocean.

Biologists have long attempted to relate oxygen concentrations to densities of marine populations, but without much success. It appears that most marine organisms can tolerate reasonably large variations in oxygen tensions until some threshold limit is reached, and many organisms appear to be able to tolerate low oxygen concentrations for relatively long periods (Richards, 1957).

Although oxygen concentrations appear to have little relationship to the distributions of marine organisms, it has become evident that biological activities are the principle cause of changes in oxygen concentrations in the body of the ocean, and these changes can be used to reconstruct much of the biological history of a water mass. The work of Redfield has gone far towards elucidating many of these concepts, and his investigations of the relationship between oxygen consumption and the decomposition of organic matter have constituted an important step in biological and chemical oceanography.

The photosynthetic production of dissolved oxygen was for a long time the only practical method for estimating primary production in the ocean. It has generally been replaced by ^{14}C fixation methods for experimental determinations (see Chapter 12, Section VA), but observations of *in situ* changes in oxygen production or consumption are potentially useful in estimating these activities in oceanic areas. These observations are probably less useful for this purpose than observations of the changes in dissolved nutrients, because oxygen can readily pass through the sea surface.

The relationships between oxygen changes and biological activities tend to be gross, and will probably continue to be so in spite of refinements in methods, because the ecosystem is too complex to be described in terms of such simple parameters. However, other refinements can be added which may increase our understanding of biological effects in the oceans by orders of magnitude. Most biological processes are accompanied by isotope fractionation of one or more of the air gases. Rakestraw *et al.* (1951) demonstrated that $^{18}O/^{16}O$ ratios in the ocean vary in a manner which is clearly related to biological activity. The techniques of following isotope ratios in the dissolved gases vary in their precision, since they depend on the gas species and the state of the art, but it seems probable that intensive studies of variations in isotope ratios may supply a marked impetus to our understanding of biological events in the sea.

Photosynthesis is known to be accompanied by relatively large fractionations of carbon and oxygen isotopes, and it is suggested that studies of these effects could add much to our knowledge of the fate of photosynthetically produced oxygen, and this could, in turn, lead to an increased knowledge of advective and diffusive processes in the upper layers of the ocean.

Studies of the distribution of the dissolved gases in sea water are a valuable adjunct to the standard approaches of physical oceanography. Richards and Redfield (1954) used the oxygen content of waters of the Gulf Stream system to distinguish them from the waters of the adjacent Sargasso Sea. This was useful, because these two water masses are very similar in their temperature and salinity characteristics, although they differ markedly in their oxygen contents at a given σ_t value. Stefánsson and Richards (1964) have recently used similar methods to study the water masses off the Washington and Oregon coasts. They found that there was, during much of the year, a good statistical correlation between the oxygen content and the density of these waters, but there were local departures from the correlation which could be used to identify water masses, estimate biological changes in the oxygen content, and elucidate some of the processes of upwelling and the effects of the discharge of the Columbia River into the area.

In the above examples, the dissolved oxygen was used as a quasi-conservative property, and this was permissible because the rates of oxygen consumption in the deeper layers of the ocean are low. The rates of biological and biochemical changes in dissolved nitrogen are generally negligible, although no good estimates of the effect of nitrogen fixation on dissolved nitrogen concentrations are yet available. It was suggested by earlier oceanographers (see Buch, 1929) that dissolved nitrogen could be used as a conservative property of sea water, and that its concentration could be used to estimate the temperature the water had when it last left the sea surface. Buch recognized that the nitrogen determinations made up to that time were too greatly in error to be useful for this purpose, but new methods for nitrogen analysis make the suggestion again interesting.

If nitrogen can be accepted as being conservative in sea water, an accurate knowledge of its distribution would have other uses. In an earlier section the uncertainty of our knowledge of the state of equilibration of the sea surface with the atmosphere was discussed. Precise nitrogen determinations should enable one to assess the combined effect of the nearness to which equilibrium was approached and the actual partial pressure of the nitrogen in the atmosphere when a sea-water sample acquired its burden of air gases. Laboratory measurements could then be made to determine whether the other gases could be assumed to have approached saturation to the same degree as nitrogen. This information should then be useful, in conjunction with a detailed knowledge of gas distributions, in making better estimates of the value of the apparent oxygen consumption, A.O.U. The distribution of dissolved nitrogen itself, if known to a high degree of accuracy, should be a useful water mass property, particularly if it responds to seasonal changes in barometric pressure as Carritt (1954) has suggested.

Since argon concentrations are even less likely to have been altered by *in situ* processes than are those of dissolved nitrogen, precise determinations of its concentrations should be particularly valuable. Richards and Benson (1961) made use of argon as a conservative element when they assessed the quantities of nitrogen biologically produced in anoxic environments.

Some of the geochemical uses of the distributions of the rare gases in the ocean have been discussed by Revelle and Suess (1963) and are summarized in earlier sections.

Progress in oceanography demands an increased knowledge of the physical, chemical, biological, and geological processes which occur in the oceans, and also a better understanding of the distribution of oceanographic variables. The dissolved gases in sea water are such variables; information concerning their distributions and the processes involved is at present incomplete.

ACKNOWLEDGMENTS

The author would like to express his thanks to his colleagues at Woods Hole Oceanographic Institution and the University of Washington for fruitful discussions on this subject. He is particularly grateful for the technical assistance of M. Meller, to Edwyna von Gal for preparing the data for Fig. 1, and to E. E. Collias for fitting the Rakestraw and Emmel nitrogen and argon solubility values to the equations given in the text. Part of the work was supported by the Office of Naval Research Contract Nonr-477(37) Project RR 004-03-01 and National Science Foundation Grant NSF-G 19784.

REFERENCES

Barnes, C. A., and Collias, E. E. (1958). *J. Mar. Res.* **17**, 68.

Benson, B. B., and Parker, P. D. M. (1961a). *J. phys. Chem.* **65**, 1489.

Benson, B. B., and Parker, P. D. M. (1961b). *Deep Sea Res.* **7**, 237.

Buch, K. (1929). *J. Cons. int. Explor. Mer*, **4**, 162.

Carritt, D. E. (1954). *Deep Sea Res.* **2**, 59.

Carritt, D. E. (1963). *In* "The Sea, Ideas and Observations on Progress in the Study of the Seas" (M. N. Hill, ed.), Vol. II, pp. 109–123. Interscience Publishers, New York and London.

Department of Scientific and Industrial Research (1964). "Water Pollution Research, 1963". H.M.S.O., London.

Douglas, E. (1964). *J. Phys. Chem.* **68**, 169.

Dugdale, V. A., and Dugdale, R. C. (1962). *Limnol. Oceanogr.* **7**, 170.

Dugdale, R. C., Dugdale, V. A., Neess, J., and Goering, J. (1959). *Science*, **130**, 859.

Dugdale, R. C., Menzel, D. W., and Ryther, J. H. (1961). *Deep Sea Res.* **7**, 298.

Elmore, H. L., and Hayes, T. W. (1960). *Proc. Amer. Soc. civ. Engrs* **86**, No. SA4, 41.

Fleming, R. H. (1940). *Proc. Pacif. Sci. Congr.* **3**, 535.

Fox, C. J. J. (1909). *Trans. Faraday Soc.* **5**, 68.

Føyn, E. (1955). *Rep. Norweg. Fish. Invest.* **11**, 8 pp.

Glueckauf, E. (1951). *In* "Compendium of Meteorology" (T. F. Malone, ed.), pp. 3–10. American Meteorological Society, Boston.

Hamm, R. E., and Thompson, T. G. (1941). *J. Mar. Res.* **4**, 11.
Kalle, K. (1939). *Ann. Hydrogr., Berl.* **67**, 267.
Kanwisher, J. (1963). *Deep Sea Res.* **10**, 195.
König, H. (1963). *Z. Natur.* **18a**, 363.
König, H., Wänke, H., Bien, G. S., Rakestraw, N. W., and Suess, H. E. (1964). *Deep Sea Res.* **11**, 243.
Mazor, E., Wasserburg, G. J., and Craig, H. (1964). *Deep Sea Res.* In the press.
Montgomery, H. A. C., Thom, N. S., and Cockburn, A. (1964). *J. Appl. Chem.* **14**, 280.
Postma, H. (1964). *Neth. J. Sea Res.* **2**, 258.
Rakestraw, N. W., and Emmel, E. V. M. (1937). *Industr. Engng Chem.* **9**, 1.
Rakestraw, N. W., and Emmel, E. V. M. (1938a). *J. phys. Chem.* **42**, 1211.
Rakestraw, N. W., and Emmel, E. V. M. (1938b). *J. Mar. Res.* **1**, 207.
Rakestraw, N. W., Rudd, DeF. P., and Dole, M. (1951). *J. Amer. chem. Soc.* **73**, 2976.
Ramsey, W. L. (1962). *Limnol. Oceanogr.* **7**, 1.
Redfield, A. C. (1934). *In* "James Johnstone Memorial Volume" (R. J. Daniel, ed.), pp. 179–192. University Press, Liverpool.
Redfield, A. C. (1942). *Pap. phys. Oceanogr.* **9**, 22 pp.
Redfield, A. C. (1948). *J. Mar. Res.* **7**, 347.
Redfield, A. C. (1958). *Amer. Scient.* **46**, 205.
Redfield, A. C., Ketchum, B. H., and Richards, F. A. (1963). *In* "The Sea, Ideas and Observations on Progress in the Study of the Seas" (M. N. Hill, ed.), Vol. II, pp. 26–77. Interscience Publishers, New York and London.
Revelle, R., and Suess, H. E. (1963). *In* "The Sea, Ideas and Observations on Progress in the Study of the Seas" (M. N. Hill, ed.), Vol. I, pp. 313–321. Interscience Publishers, New York and London.
Richards, F. A. (1957). *In* "Treatise on Marine Ecology and Paleoecology" (J. W. Hedgpeth, ed.), Vol. I, pp. 185–238. Memoir 67, Geological Society of America, New York.
Richards, F. A. (1960). *Deep Sea Res.* **7**, 163.
Richards, F. A., and Benson, B. B. (1961). *Deep Sea Res.* **7**, 254.
Richards, F. A., and Corwin, N. (1956). *Limnol. Oceanogr.* **1**, 263.
Richards, F. A., and Redfield, A. C. (1954). *Deep Sea Res.* **2**, 182.
Riley, G. A. (1951). *Bull. Bingham oceanogr. Coll.* **13**, 1.
Ryther, J. H. (1963). *In* "The Sea, Ideas and Observations on Progress in the Study of the Seas" (M. N. Hill, ed.), Vol. II, pp. 347–380. Interscience Publishers, New York and London.
Steen, H. (1958). *Limnol. Oceanogr.* **3**, 423.
Stefánsson, U., and Richards, F. A. (1964). *Deep Sea Res.* **11**, 355.
Sverdrup, H. U., and Fleming, R. H. (1941). *Bull. Scripps Instn Oceanogr.* **4**, 261.
Sverdrup, H. U., Johnson, M. W., and Fleming, R. H. (1942). "The Oceans", 1087 pp. Prentice Hall, Englewood Cliffs, New Jersey.
Swinnerton, J. W., and Sullivan, J. P. (1962). NRL Report 5806, 13 pp. U.S. Naval Research Laboratory, Washington, D.C.
Swinnerton, J. W., Linnenbom, V. J., and Cheek, C. H. (1962a). *Analyt. Chem.* **34**, 483.
Swinnerton, J. W., Linnenbom, V. J., and Cheek, C. H. (1962b). *Analyt. Chem.* **34**, 1509.

Truesdale, G. A., Downing, A. L., and Lowden, G. F. (1955). *J. appl. Chem.* **5, 53.**
Wattenberg, H. (1933). *Wiss. Ergebn. dtsch. atlant. Exped. Meteor*, **7**, 333 pp.
Wattenberg, H. (1938). *Wiss. Ergebn. dtsch. atlant. Exped. Meteor*, **9**, 132 pp.
Whipple, G. C., and Whipple, M. C. (1911). *J. Amer. chem. Soc.* **33**, 362.
Winkler, L. W. (1888). *Ber. dtsch. chem. Ges.* **21**, 2843.
Witting, R. (1908). *Finnl. hydrogr.-biol. Untersuch. No. 2*, 246 pp.
Wüst, G. (1932). *Wiss. Ergebn. dtsch. atlant. Exped. Meteor*, **7**, 290 pp.
Wyrtki, K. (1962). *Deep Sea Res.* **9, 11.**

Chapter 7

The Dissolved Gases — Carbon Dioxide

GEOFFREY SKIRROW

Donnan Laboratories

The University, Liverpool, England

I. Introduction 227
II. Carbonate Equilibria in Sea Water 231
A. Dissociation of Water and the pH Concept 231
B. Alkalinity, Carbonate Alkalinity, Total Carbon Dioxide and P_{CO_2} 246
C. Solution of Carbon Dioxide and the Dissociation of Carbonic Acid 253
D. Ion Pair Formation in the Sea-Water Carbonate System 278
E. Relationship between Changes in Atmospheric Carbon Dioxide Pressure and the Total Carbon Content of the Sea 282
III. The Isotopes of Carbon 285
IV. Carbon Dioxide in the Atmosphere 288
A. Short-term Variations 288
B. Long-term Changes 294
C. Changes in Atmospheric Radio-carbon Activity 297
V. The Radio-carbon Activity of Sea Water 300
VI. Circulation of Carbon Dioxide between the Atmosphere and the Sea and within the Sea 305
VII. Transfer of Carbon Dioxide across a Gas–Liquid Interface 312
Acknowledgments 316
References 317

I. Introduction

The major portion of oceanic dissolved carbon occurs in the carbonate system in which equilibria between dissolved carbon dioxide, carbonic acid, bicarbonate ion, carbonate ion and hydrogen ion provide the principal pH regulating system in the ocean. This system is part of one of wider geochemical significance in which carbon is in circulation between the biosphere, the lithosphere, the atmosphere and the oceans. Carbon dioxide is produced by volcanic activity, the burning of fuels, forest fires, agricultural activity and respiratory processes, and passes into the atmosphere where it is available for use in biosynthetic and weathering processes, and where it is brought into contact with the oceans and enters the oceanic bicarbonate system. Although it is not correct to regard the oceanic system as being in complete regional equilibrium, nevertheless, some parts of it do approximate to such a state, and can be described in physico-chemical and thermodynamic terms. However, even those facets of the overall problem which can be so treated are far from simple, and the position of the carbonate equilibrium may vary from locality to locality and also with time in sympathy with seasonal and diurnal temperature changes, and with the oxidative production and biosynthetic utilization of dissolved carbon dioxide.

Carbon dioxide which enters the ocean participates in the equilibria indicated above which may be summarized by

$$H_2O + CO_2 \rightleftharpoons H_2CO_3 \rightleftharpoons HCO_3^- + H^+ \rightleftharpoons CO_3^{2-} + 2H^+.$$

Associated with each step in this series is an equilibrium constant which can be expressed in terms of the thermodynamic activities of the participating species. Since equilibrium constants are, in general, temperature and pressure dependent, a shift in the equilibrium position will occur when a water mass moves from a warm region to a cooler one or when it changes depth. The activity coefficients of many species, particularly ions, depend on the ionic strength of the medium and are, therefore, functions of salinity. Further and as yet incompletely resolved, complications arise from the possible formation of organic complexes or of ion-pair carbonates such as $NaCO_3^-$ and $MgCO_3^0$, and a complete description of the equilibria demands consideration of the conditions under which insoluble carbonates are precipitated. This important problem is discussed in Chapter 16, Volume 2 (see also Cloud, 1962, 1963; Ingerson, 1962).

Diurnal and seasonal variations in the carbonate system are caused by the removal of carbon dioxide by photosynthesis and by solar heating. Superimposed on these short-term effects is a probable gradual long-term increase in total oceanic carbon resulting from an increase in the total amount of atmospheric carbon dioxide owing to the increasing usage of fossil fuels. If, as has been suggested, the world is undergoing a progressive increase in its atmospheric carbon dioxide content, then it is a matter of both climatological and oceanographical interest to examine the extent to which this increase can be assimilated by the oceans. The climatological interest stems from the suggestion that an increase in atmospheric carbon dioxide content may, by virtue of the strong infra-red absorption by carbon dioxide, modify the earth's mean temperature (Plass, 1956, 1959, 1961; see also Kaplan, 1960), although this suggested importance of atmospheric carbon dioxide in influencing climatic variations has been questioned (Möller, 1963). If the oceans were well-mixed reservoirs and were able to maintain equilibrium with the atmosphere, much of the increase would be absorbed, but the present indications are that the oceans as a whole respond only slowly to changes in the atmospheric carbon dioxide content, mainly because of slow rates of mixing within the sea. Part of this transfer problem—the mixing of different water masses—involves purely physical processes, but the absorption or evolution of carbon dioxide by the sea surface requires a combination of physical factors (e.g. diffusion across the interface) and chemical processes. The chemical processes include hydration of dissolved carbon dioxide, dehydration of carbonic acid and the ionization of the carbonate and bicarbonate ion. The ionization processes are rapid, but the hydration and dehydration steps are relatively slow and may possibly, in certain circumstances, influence the transfer rate.

Approximate time scales for many of the transfer processes within the carbon system can be estimated by radio-carbon counting methods. Radio-carbon produced in the upper atmosphere mixes relatively quickly with the lower atmosphere, and participates together with inactive carbon in the geochemical carbon cycle. From a knowledge of the specific activities of carbon in the various natural reservoirs, and of the natural rate of production of radio-carbon, it is possible to gain some idea of reservoir turn-over and mixing times. In order to make estimates of this sort, information on the total carbon content of each reservoir is needed. Most recent inventories of the carbon system agree reasonably well with respect to the amounts allocated to the major reservoirs, although the amount contained in the marine biosphere is not accurately known. Table I lists some of the most recently suggested values.

TABLE I

Carbon content of natural exchange reservoirs

Carbon reservoir	*Carbon content ($g./cm.^2$ of earth's surface)*
Atmosphere	0·125
Humus	0·19
Biosphere (land)	0·06
Biosphere (marine)	0·002
Oceanic dissolved carbon (organic)	0·533
Oceanic inorganic carbon (above thermocline)	0·20
Oceanic inorganic carbon (below thermocline)	7·25

These values, which have been taken from Craig (1957), Arnold and Anderson (1957), Revelle and Suess (1957) and Fergusson (1958), are not necessarily the best estimates available, and others are to be found in the literature. Takahashi (1961), for example, has recently estimated the total atmospheric carbon dioxide to amount to 0·477 $g./cm.^2$ of the earth's surface (0·13 $g./cm.^2$ as carbon). The amount of inorganic carbon in the ocean is known reasonably accurately, and various estimates do not differ by more than a few per cent from those given in Table I. The carbon inventories of the biosphere and of humus are less precisely known, and estimates for the marine biosphere differ by orders of magnitude. In addition, the carbon dioxide found in the air occluded in sea ice amounts to 0·1–0·5 ml. at S.T.P./kg. (Miyake and Matsuo, 1963).

The division of the sea into a two-layered structure by the thermocline is a rather arbitrary way of summarizing a complex of interacting reservoirs. It is based on a simple concept of an upper layer 75–100 m. deep which interacts and exchanges only slowly with the rest of the sea. Although more detailed models have been proposed and are discussed later, this simple picture at least emphasizes that for many purposes it is not correct to regard the sea as a single well-mixed reservoir.

Figure 1 shows diagrammatically the approximate amounts in the

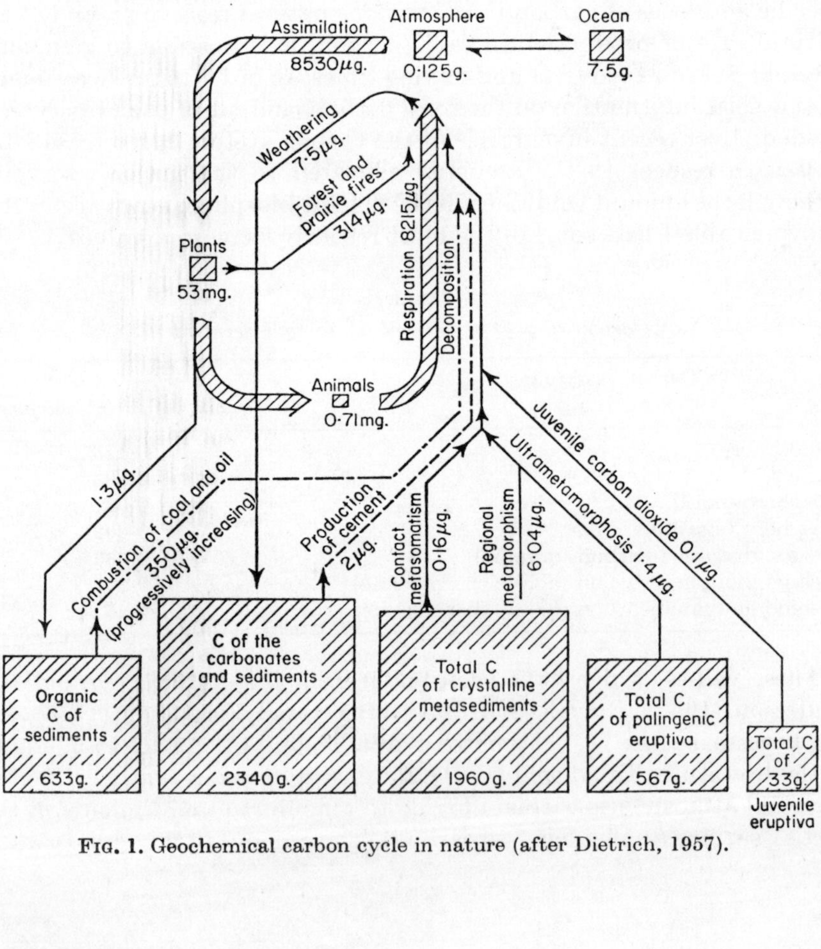

FIG. 1. Geochemical carbon cycle in nature (after Dietrich, 1957).

various natural carbon reservoirs, and the transfer routes between them. It is clear that the amount of carbon stored as carbonate rock, coal and oil is very much greater than that in the atmospheric or marine reservoirs. However, much of the carbon in these vast reserves must, at some time, have passed through the atmosphere.

II. Carbonate Equilibria in Sea Water

If equilibrium were to be attained in the atmosphere–sea water carbon system, a state of dynamic balance would exist between atmospheric carbon dioxide, dissolved carbon dioxide, carbonic acid, hydrogen ion, bicarbonate ion and carbonate ion. The main features of the relevant equilibria are now well established, although some details are still not fully explained, and it is not necessary to make an independent determination of each of the above quantities in order to characterize a water sample. By making use of tabulated values of equilibrium constants it is possible to obtain much of the desired information (at least in approximate form) from a few determinations—temperature, salinity, depth, alkalinity and pH—which are relatively easy to make on board ship. Present understanding of the carbonate system owes much to the early work of Buch and his collaborators who estimated values for the constants used when these computations are attempted. Although agreement between measured and computed values of the partial pressure of carbon dioxide in the water is not always good, Buch's treatment offers a convenient basis for describing the main features of the system. In the following sections, concepts relevant to the carbonate buffer system are discussed.

A. DISSOCIATION OF WATER AND THE pH CONCEPT

(1) *Dissociation of water.* Water is a very weak electrolyte and dissociates according to

$$H_2O \rightleftharpoons H^+ + OH^-.$$

More correctly, this equation should be written

$$2H_2O \rightleftharpoons H_3O^+ + OH^-$$

as H^+ ions as such do not exist in aqueous solution*, but form with water the solvated oxonium ion, H_3O^+. However, for simplicity, it is

* In this chapter, the symbols H^+ and c_H, and the term hydrogen ion are to be understood to refer to the oxonium ion H_3O^+.

customary to represent the equilibrium by the first of the above equations, for which the equilibrium constant is

$$K_{H_2O} = \frac{a_H \cdot a_{OH}}{a_{H_2O}} = c_H \cdot c_{OH} \frac{\gamma_H \cdot \gamma_{OH}}{a_{H_2O}}$$

$$= k_w \cdot \frac{\gamma_H \cdot \gamma_{OH}}{a_{H_2O}}$$

where a denotes the thermodynamic activities of the various species, γ the activity coefficients and k_w is the ionic product (strictly, concentration ionic product). For dilute solutions, the activity of water and the activity coefficients of H^+ and OH^- are each unity, and the product of the ionic activities (thermodynamic ionic product $k_{w(TD)}$), is identical with the concentration ionic product (k_w), i.e.

$$k_{w(TD)} = a_H \cdot a_{OH} \approx k_w = c_H \cdot c_{OH}.$$

It should be noted that the thermodynamic ionic product and the concentration ionic product can only be equated under these conditions. At 25°C, k_w is about 10^{-14}, and if pH is tentatively defined by

$$\text{pH} = -\log_{10} c_H \tag{1}$$

it follows that the pH at the neutral point where the concentrations of H^+ and OH^- ions are equal is 7. k_w is influenced by pressure (Table II) and is strongly dependent on temperature, the pH of neutrality falling from about 7·5 at 0°C to about 6·5 at 60°C.

The concentrations of the two ions in equilibrium are also affected by the presence of electrolytes since at higher ionic strengths the activity coefficients of the ions H^+ and OH^- and the activity of water are no longer unity. The thermodynamic dissociation constant of water is, of course, independent of salt concentration, and as $K_{H_2O} = k_w \cdot (\gamma_H \cdot \gamma_{OH})/a_{H_2O}$, changes in k_w are compensated by changes in the activity coefficient product $(\gamma_H \cdot \gamma_{OH})/a_{H_2O}$. The decrease of a_{H_2O} caused by added electrolyte will be less than the change in the ionic activity coefficients since water is a neutral species, and up to moderate concentrations at

TABLE II

Relative effect of pressure on the ionic product of water (Harned and Owen, 1958)*

Pressure (bars)	1	200	400	600	800	1000
$k_w/k_w(p=1)$	1	1·225	1·490	1·800	2·163	2·585

* Temperature, 15°C.

least, the activity coefficient product decreases with increasing salt concentration (see e.g. Harned and Owen, 1958).

(2) *The pH concept.* The definition of pH by equation (1) is essentially that given by Sørensen (1909) who attempted to set up a scale based on measurements of the E.M.F.'s of cells of the type

$$H_2(Pt) \mid soln(X) \,\vdots\, \text{salt bridge} \mid \text{reference electrode.}$$

Observed E.M.F.'s were interpreted in terms of theories current at that time, and were related to the concentration of hydrogen ion by the Nernst expression, i.e. the cell E.M.F. was expressed in terms of the hydrogen ion concentration, viz.

$$E = E^0 - \frac{RT}{F} \ln c_H \text{ (solution X)}$$

$$= E^0 + \frac{2{\cdot}303RT}{F} \text{ pH (solution X).}$$

The H^+ concentrations in the series of NaCl–HCl mixtures used to standardize the method were calculated on the basis of conductance measurements, and it was assumed that $c_H = c_{HCl} \cdot \alpha$, where $\alpha = \lambda/\lambda_\infty$, λ_∞ and λ being the equivalent conductivities at infinite dilution and at the experimental concentration respectively. An attempt was made to eliminate the liquid junction potential by applying an extrapolation procedure. These methods were applied to buffer solutions to establish what was intended to be a series of standards of known pH.

For several reasons the basis of the above procedure is now known to be unsound, and equation (1) is regarded as unacceptable. It is by no means certain that the extrapolation method did remove the liquid junction potential, the E.M.F. of a cell is more directly related to ionic activity than to concentration, and it is now realized that the hydrochloric acid solutions used as standards are, in fact, fully ionized. Because of these shortcomings, Sørensen's scale does not give pH values which have the interpretation originally intended, and several attempts have been made to provide alternative definitions which satisfy practical needs, and which allow at least a limited amount of theoretical interpretation (see Bates, 1954, 1959; Gold, 1956).

It would be advantageous if a scale of pH could be defined which rendered values obtained open to immediate interpretation in terms of some clearly understood physical or thermodynamic concept. If, for example, it were possible to measure individual ionic activities, then a useful definition of pH would be

$$pH = -\log_{10} a_H \tag{2}$$

where a_H is the hydrogen ion activity. Unfortunately, it is impossible to obtain individual ionic activities without making use of extra-thermodynamic assumptions, and it does not seem that there is hope of ever being able to do so. The difficulties which arise when an attempt is made to determine pH using the above definition can be seen from a consideration of typical cells which might be used in pH investigations; consider, for example, cell A without a liquid junction

$$H_2(Pt) \mid H^+ Cl^- \mid Hg_2Cl_2 \mid Hg \tag{A}$$

or cell B in which solution X of pH interest is separated from the reference electrode by a KCl bridge

$$\underset{}{H_2(Pt)} \mid \underset{soln(X)}{H^+ X^-} \mid \underset{soln(K)}{KCl} \mid Hg_2Cl_2 \mid Hg. \tag{B}$$

If the hydrogen gas is at 760 mm. pressure, the E.M.F. of the first cell is given by

$$E = E^0 - \frac{RT}{F} \ln a_{H(X)}.\, a_{Cl(X)}$$

whereas that of the second differs from this by a term which takes account of the liquid junction potential (E_l), and by the fact that the chloride activity refers to that in solution (K) rather than that in (X), viz.

$$E = E^0 - \frac{RT}{F} \ln a_{H(X)}.\, a_{Cl(K)} - E_l.$$

Both of these expressions embody quantities not directly accessible experimentally, such as a_{Cl} and the liquid junction potential, and only by introducing assumptions which, however reasonable, cannot be rigorously verified is it possible to obtain an estimate of a_H, and consequently the pH on the basis of equation (2).

These difficulties have led to the introduction of a scale agreed upon and recommended by several different authorities. This scale is operational, but it allows a limited amount of theoretical interpretation under favourable circumstances. It is defined in terms of two E.M.F. measurements on cells of the type

$$H_2(Pt) \mid soln\ (X) \mid KCl(aq) \mid \text{reference electrode}$$

in which in the first measurement X is the test solution, while in the second measurement it is a pH standard. Denoting the two E.M.F. readings by E_X and E_S respectively, the relationship between them and the two pH values is given by

$$pH(X) - pH(S) = \frac{E_X - E_S}{2{\cdot}303RT/F}\,. \tag{3}$$

Solution S, according to the recommendation of British Standard Specification 1649 (1950), is a 0·05 M solution of potassium hydrogen phthalate, the pH of which is given by

$$pH(S) = 4{\cdot}00 + \tfrac{1}{2}\left[\frac{t - 15}{100}\right]^2$$

where t is the temperature in °C. The National Bureau of Standards have recommended a series of standards of pH, one of which is identical with the British Standard, and for all practical purposes the two systems are in full agreement.

This operational definition of pH furnishes an agreed scale of acidity by means of which comparisons between different solutions can be made, and for many routine applications this is all that is required. However, interpretation of the measured values is not always straightforward. From the manner in which the scale is defined, it is evident that the pH determination, in addition to requiring a standard of known pH, rests on the determination of the difference of two E.M.F.'s. Thus, since the E.M.F. of a typical cell used in pH determination is given by equation (3) above, the difference pH(X) − pH(S) is given by

$$pH(X) - pH(S) = \frac{E_X - E_S}{2{\cdot}303RT/F}$$

$$= \log_{10}(a_H)_S - \log_{10}(a_H)_X + \frac{E_{l(X)} - E_{l(S)}}{2{\cdot}303RT/F}$$

where $E_{l(S)}$ and $E_{l(X)}$ refer to the liquid junction potentials when the standard and test solutions are used respectively. Thus, only if it were possible to assign an activity to the pH standard and to know the difference in the liquid junction potentials could the activity of the hydrogen ions in the test solution be determined.

The methods by which the pH values of standard buffers have been determined have been discussed by Bates (1954) whose monograph should be consulted for details, but it should be noted that in the NBS method, the standardization is achieved through appeal to non-thermodynamic assumptions. The existence of the liquid junction potential is a source of uncertainty in the interpretation of the measured pH, particularly when the pH difference between the test and standard is large, and when the two solutions differ considerably in ionic strength. Considerations of this sort have led to the suggestion that an alternative scale of pH which is based on measurements made on buffers made up in sea water may be useful to oceanographers (Smith and Hood, 1964).

Not all chemists are agreed on the legitimacy of attempting to identify single ionic activities since, it is argued, they have no physical significance and are meaningless. Although this is probably an extreme view, care is nevertheless necessary if equation (2) is used for interpreting pH measurements directly in terms of hydrogen ion activity. Bates, who places rather severe restrictions on the type of solution with which some limited interpretation is possible, states that "It is safe to say that no theoretical interpretation of measured pH values should be attempted unless the medium can be classified as a dilute aqueous solution of simple solutes. This requirement excludes all non-aqueous media, suspensions, colloids and aqueous solutions of ionic strength greater than 0·2. From this point of view the ideal solutions are those which match the standards of reference, namely aqueous solutions of buffers and simple salts with ionic strengths between 0·01 and 0·1."

Some of these remarks seem to be particularly pertinent to sea water which may have an ionic strength as high as 0·7. This does not mean that pH values cannot be obtained for such systems. It is perfectly possible to go through the operations and to obtain a pH value which may be reproducible, and of value for record or comparison purposes. However, it does mean that pH values obtained under these conditions cannot be interpreted in terms of activities without the danger of straining the basic theoretical framework.

It should be noted that many of the earlier measurements in the literature are reported in terms of the Sørensen scale (psH). The relationship between measurements on this scale and those on the more modern scale is given by

$$\mathrm{pH} = \mathrm{psH} + 0{\cdot}04.$$

For a more detailed discussion of the pH concept, reference should be made to Bates (1954, 1959), Gold (1956) and Feldman (1956).

(3) *Measurement of pH*. pH determination involves two separate measurements—one on the unknown and the other on a standard. Practical methods fall into two classes, electrometric methods using hydrogen, quinhydrone or glass electrodes, and indicator methods. Of these two classes, the former (particularly with the glass electrode) is nowadays generally favoured. Detailed description of practical methods are to be found in works by Clark (1928), Dole (1941), Kolthoff and Laitinen (1948), Bates (1954), Britton (1956) and Gold (1956).

(a) *Electrometric methods*. Although the hydrogen electrode is the ultimate standard on which all electrode potentials are based, it does not lend itself well to routine oceanographical measurements since the stream of hydrogen necessary for its operation purges carbon dioxide from the samples, and thereby raises the pH. Nevertheless, it has been used for certain specific measurements (Buch, 1938), and it is useful as a laboratory standard against which other electrodes may be checked. It consists of a platinum foil or wire covered with platinum black (produced by electrolysis of a solution of chloroplatinic acid to which lead acetate has been added) over which a stream of oxygen-free hydrogen is passed. As the potential is a function of hydrogen pressure, the gas should be at 760 mm. pressure or a correction must be applied. The electrode is subject to poisoning by phthalate, cyanide, sulphide, silver and mercury ions and by arsenic compounds.

The quinhydrone electrode is set up by dipping a bright platinum wire, previously heated in a flame, into the solution which is saturated with quinhydrone (equimolecular proportions of quinone and hydroquinone). Its use is generally confined to solutions of pH lower than 8·5 since above this it departs from simple behaviour owing to the tendency of hydroquinone to ionize as a weak acid. Even at lower pH values it is subject to a "salt error" because of the dependence of the ratio of the activity coefficients of the quinone and hydroquinone ions on the ionic strength, although the system can be modified to reduce this effect by saturating the solution with quinone as well as quinhydrone.

The general versatility of the glass electrode, particularly when used in conjunction with a pH meter, has made it the most widely used of all the pH electrodes for most purposes. The electrode consists of a thin envelope of conductive glass containing a buffer electrolyte into which an electrode (e.g. Ag : AgCl) dips. The whole constitutes a half-cell which is immersed in the test solution. The cell is completed by means of a suitable reference electrode (e.g. calomel). Since the glass electrode has a very high internal resistance, it cannot be used with a simple potentiometer, but requires a circuit capable of tolerating a high input resistance. Although the pH response of a glass electrode may occasionally agree with that of a hydrogen electrode (i.e. show the theoretical E.M.F. *vs* pH slope), this is not always so, and it is desirable to standardize it with at least two buffer solutions having pH values lying on each side of that of the unknown. Because of strain in the glass, the electrodes often show what is termed asymmetry potential. Thus, if similar electrolyte solutions are placed on each side of the pH sensitive membrane, a definite potential is shown. Since this potential may change with time, it is necessary to standardize the electrode at frequent intervals.

Most commercial glass electrodes are designed for operation at atmospheric pressure, but electrodes capable of working up to much higher pressures, and suitable for *in situ* oceanographical pH determination have been described (Disteche, 1959; Disteche and Dubuisson. 1960).

The buffer systems normally used for the calibration of glass electrodes (Bates, 1954) are usually solutions of low ionic strength. The difference in ionic strength between these solutions and sea water gives rise to a large change in junction potential when the calibration buffer is replaced by the sample. This deprives the operational definition of pH of much of the meaning which it might otherwise have. Smith and Hood (1964) have investigated the use of buffers made by dissolving tris-(hydroxymethyl)-aminomethane and its hydrochloride in sea water. By this means, they have established a sea-water pH scale (pH_X). The authors claim that measured pH_X values for sea water are better defined than when the conventional buffers are used, and that the drift of E.M.F. often observed when electrodes are transferred between solutions of greatly different ionic strength does not occur with this buffer system.

(b) *Indicator methods.* The essential feature of indicators is that their light-absorbing properties are a function of the pH of the medium. They may be either acids or bases; a simplified description based on a weak-acid indicator, HA, is as follows. When added to a buffered aqueous solution, the indicator will ionize according to

$$HA \rightleftharpoons H^+ + A^-.$$

The ion A^-, or a tautomer with which it is in equilibrium, and the unionized indicator each absorb in a different spectral region, and the observed colour will depend on the relative proportions of the two forms and will, therefore, be a function of pH. The thermodynamic dissocia tion constant K_{in} is given by

$$K_{in} = \frac{a_H \,.\, a_A}{a_{HA}} = a_H \cdot \frac{c_A \,.\, \gamma_A}{c_{HA} \cdot \gamma_{HA}}$$

so that

$$pH = -\log a_H = pK_{in} + \log \frac{c_A}{c_{HA}} + \log \frac{\gamma_A}{\gamma_{HA}}\,.$$

If K_{in} is known and estimates can be made of γ_{HA} and γ_A (the activity coefficients of the unionized indicator and the ion A^- respectively), the pH can be estimated from the ratio of the intensities of the colours of the two forms. In general, however, K_{in} is not known and estimates of γ_A are not readily made for solutions of high ionic strength. The most general procedure is to compare the colour of the test solution with those given by a series of buffers to which identical concentrations of indicator have been added. The colour matching may be carried out visually, or, with more precision, spectrophotometrically. Subject to a salt-error correction, the pH of the test solution is the same as that of the buffer having an identical colour.

The salt-error correction arises from the effect of the different ionic strengths of the buffer and test solutions on the activity coefficient terms. A temperature correction is also necessary if the test and the buffer are not at the same temperature; this arises from the temperature coefficient of K_{in}. Further temperature correction is necessary if the buffer is not at the temperature at which it was calibrated.

The indicator method is of limited use with turbid or coloured solutions, or when there is the possibility of chemical interaction with the indicator. It is also of doubtful value in certain poorly buffered natural waters in which dissociation of the indicator may influence the pH.

For specific applications of indicators to the determination of the pH of sea water see Buch (1929), Buch and Nynäs (1939) and Harvey (1955).

Sometimes it is inconvenient to measure the pH of a water mass at its *in situ* temperature. Adjustment of the measured pH can be made by assuming that at constant total carbon dioxide content the pH of the sample changes by about $-0{\cdot}01$ pH unit per °C rise in temperature (Buch and Nynäs, 1939; Harvey, 1955). As will be seen below, changes in pH arising purely from temperature changes are generally small compared with those produced by biological or oxidation processes which effect the carbonate system.

(4) *Buffer affect.* Aqueous solutions of certain electrolytes show a pH buffering effect in that additions of small amounts of strong acid or base cause only a small change of pH. This is in contrast to the change observed following similar additions to pure water when a marked pH shift occurs.

Buffering is shown by relatively concentrated solutions of strong acids and bases, and also by solutions of weak acids and bases in the presence of their salts. Thus, a solution of a weak acid (HA) will dissociate according to

$$\text{HA} \rightleftharpoons \text{H}^+ + \text{A}^-$$

and the pH of the solution will be given by

$$\text{pH} = \text{p}K_{\text{HA}} + \log \frac{c_{\text{A}}}{c_{\text{HA}}} + \log \frac{\gamma_{\text{A}}}{\gamma_{\text{HA}}} \tag{4}$$

where K_{HA} is the thermodynamic dissociation constant of the acid, and the c and γ terms refer respectively to the molar concentrations and activity coefficients of the various species. If the solution contains the fully ionized salt of the acid, and if K_{HA} is small, c_{A} and c_{HA} will be given approximately by the concentrations of the salt and acid respectively. Additions of small amounts of strong acid or base will be without appreciable effect on the last term of equation (4). The equilibrium dissociation of the acid will be maintained, and for an added increment Δc_H of hydrogen ion, the ratio c_A/c_{HA} will become $(c_A - \Delta c_H)/(c_{HA} + \Delta c_H)$. Thus, provided that Δc_H is small compared with c_A and c_{HA}, the pH shift will be small. Equation (4) underlines the fact that the pH of a buffered solution depends not only on the thermodynamic dissociation constant of the weak electrolyte and the ratio of c_A/c_{HA}, but also on the activity coefficients of the participating species. These, particularly those of the ions, are functions of the amount and type of electrolyte present, and consequently the pH values of buffered solutions depend on the salt content.

A measure of the ability of a solution to resist changes of pH on the addition of strong acid or base is given by the buffer capacity (Van Slyke, 1922) defined by

$$\beta = dc_b/d\text{pH}$$

where dc_b is an infinitesimal addition of strong base. If, for simplicity of discussion, the dissociation constant of the weak acid is written in terms of the concentrations of the various species, viz.

$$k_{HA} = c_H \cdot c_A/c_{HA}$$

and pH is defined by equation (1), then the buffer capacity of a solution in which the combined concentration of the weak acid and its conjugate base is c_t $(= c_A + c_{HA})$ is given by

$$\beta = 2{\cdot}303 \left[\frac{k_{HA} \cdot c_t \cdot c_H}{(k_{HA} + c_H)^2} + c_H + c_{OH} \right]$$

The last two terms in the bracket of this equation determine the buffer capacity at low and high pH values respectively. At intermediate pH values, these two contributions to the buffer capacity are insignificant compared with that arising from the weak acid term. If buffering associated with the H^+ and OH^- ions is ignored, the maximum buffer capacity can be shown to occur at a pH equal to pK_{HA}, and to be proportional to c_t.

Solutions of weak dibasic acids show two pH regions in which the buffer capacity is a maximum, and if the first and second dissociation constants of the acid H_2A are k_1 and k_2, the buffer capacity will be given by

$$\beta = 2{\cdot}303 \left[\frac{k_1 \cdot c_t \cdot c_H}{(k_1 + c_H)^2} + \frac{k_2 \cdot c_t \cdot c_H}{(k_2 + c_H)^2} + c_H + c_{OH} \right]$$

The buffering effect is shown during the potentiometric titration of weak acids with a strong base. With a weak monobasic acid, the buffer capacity is a maximum in the region of half-titration, where $c_A = c_{HA}$ and where the pH changes only slowly with addition of base, and is a minimum at the end-point. In a similar titration of a weak dibasic acid (H_2A), the two maxima in the buffer capacity-titration curve correspond to the points where $c_A = c_{HA}$ and $c_{HA} = c_{H_2A}$.

The buffer capacity as defined above is of great importance in many branches of chemistry, but in natural water bodies in which the buffering is caused by the carbonate system, it is usually gain or loss of carbon dioxide which is responsible for changes of pH. Thus, for such systems, a buffer capacity defined by

$$\beta' = \mathrm{d} \sum CO_2/\mathrm{dpH}$$

is the quantity of interest.

The pH of sea water is relatively constant, and only under exceptional circumstances falls outside the range 7·8–8·3. Such constancy is characteristic of the buffer action described above, the bulk of which can be attributed to the dissolved carbon dioxide and the carbonate and the bicarbonate ions (Buch, 1930; Buch *et al.*, 1932; Moberg *et al.*, 1934). The amounts of these vary somewhat with conditions and locality, but in the open ocean away from the effects of contamination or dilution, the combined carbonate and bicarbonate is in the order of 2·4 meq./l. Traces of organic acids and other substances potentially capable of giving rise to weak acids are known to occur in the sea, but their concentrations are small (see Table III, and also Chapter 11 and Slowey *et al.*, 1962), and only the amount of boric acid is sufficient to merit

TABLE III

Approximate concentrations of acid-forming substances in sea water at pH 8·3

	g. moles/l.
Bicarbonate	18×10^{-4}
Carbonate	$3{\cdot}5 \times 10^{-4}$
Borate	$4{\cdot}5 \times 10^{-4}$*
Phosphate	1×10^{-6}
Arsenite	4×10^{-8}
Sulphide	$0 - c.\ 2{\cdot}0 \times 10^{-5}$†
Organic acids	See Chapter 11, Table I

* Only 20% ionized at 10°C.

† Occurs only in anoxic waters (see Chapter 13).

consideration. Under anoxic conditions appreciable amounts of hydrogen sulphide are formed (Chapter 13), but the dissociation constants of H_2S are very small. Many other natural waters are also buffered by the carbonic acid system, although the contribution of the dissolved carbon dioxide to the observed pH is often outweighed by other factors; in extreme situations abnormal values such as 2, for certain volcanic lakes in which sulphuric acid occurs, or 10, for alkaline lakes, may be observed (see Hutchinson, 1957). Usually, the observed pH values of inland waters are much less extreme than this; if the alkalinity is low the waters will be poorly buffered. Contributions made to the pH by organic materials or substances other than carbonic acid often make the pH problems of the limnologist different from those of the oceanographer.

Despite the general buffering effect of the ocean carbonate system, some pH variations do occur within the sea. The range of values obtained is illustrated in Fig. 2(a) which is taken from a paper by Postma (1964). Although the bulk of the measurements fall within the expected range, there is a good deal of scatter, some of which may be caused by measurement errors. However, some of the variations are no doubt genuine (see the upper curve of Fig. 2(b), taken from Postma (1964) and compiled from the Russian *Ob* and *Vityaz* cruises). Some pH variations would be expected as a result of changes of carbon dioxide content caused by temperature shift or biological demand. Since the pH is largely determined by the bicarbonate buffer system, changes in pH of ocean water are intimately linked with changes in the partial pressure of dissolved carbon dioxide (see Sections II, B and C). In principle, the upper layers of the sea are in equilibrium with atmospheric carbon dioxide, the amount of which varies only slightly with locality and time over much of the sea. However, true air–sea equilibrium is rarely maintained, largely because of the restricted rate of passage of carbon dioxide across the sea surface.

One of the major causes of change in the amount of dissolved carbon dioxide and, therefore, of the pH is solar radiation. Its effect is twofold; the amount of photosynthetic activity and utilization of carbon dioxide is increased with an increase in the amount of solar radiation, and the temperature of the surface layers is increased.

In regions in which photosynthetic activity is high, undersaturation with respect to carbon dioxide will occur and the pH will rise (see e.g. Cooper, 1932). Effects of this sort are often seasonal or diurnal, and Park *et al.* (1958) have noted a marked diurnal effect in certain shallow Texas bays (see Fig. 3) The pH rises to its maximum value during the hours of daylight, and the maximum coincides with that of the dissolved oxygen.

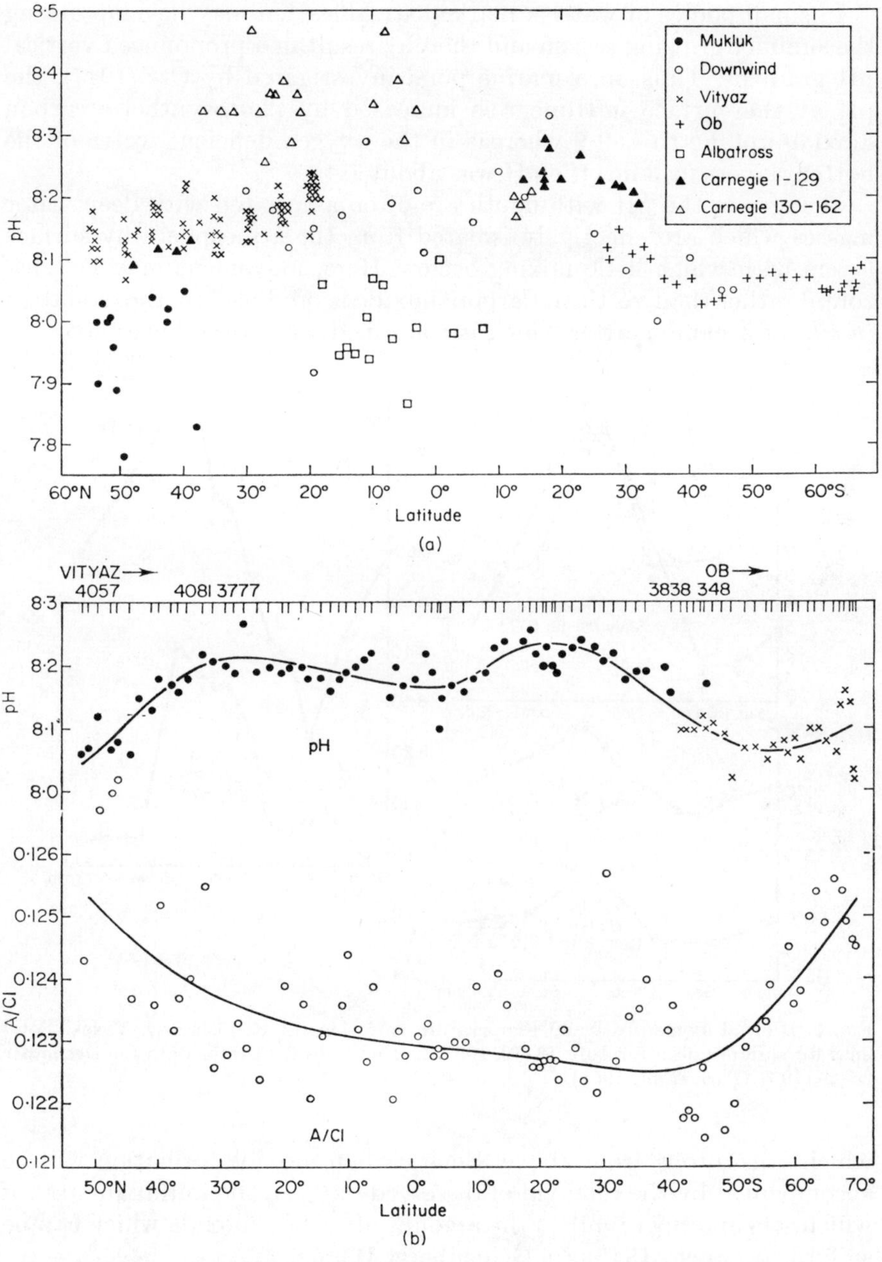

FIG. 2. (a) pH *in situ* of surface ocean water in the western Pacific Ocean between 110° and 150°W. (b) pH *in situ* and specific alkalinity of surface water in the Pacific (Postma, 1964).

In small bodies of water a marked stratification may develop during the summer growing season and thereby result in a pronounced vertical pH gradient. Thus, in a marine pond investigated by Orr (1947), the pH at the surface in June was increased by photosynthetic carbon dioxide uptake to *c.* 9·9 whereas in the oxygen-deficient water of the bottom layers (*c.* 3 m.) the pH was about 7·4.

Changes in the pH with depth are often associated with deep water masses which are effectively isolated from the atmosphere by surface layers with which little mixing occurs. Here, movement may be horizontal rather than vertical. Carbon dioxide is produced by the oxidation *in situ* of organic matter which is contained within the lower layer, or

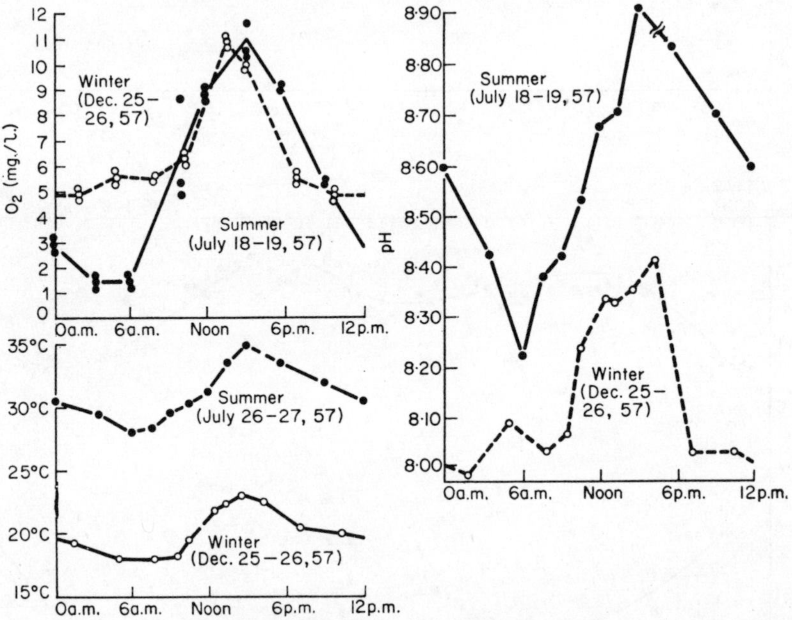

FIG. 3. Diurnal oxygen, pH and temperature variations at Redfish Bay, Texas. Solid lines for summer data for July 18–19, 1957. Dotted lines for winter data for December 25–26, 1957 (Park *et al.*, 1958).

which falls into it from above. Such carbon dioxide formation will be accompanied by the removal of dissolved oxygen, the initial amount of which sets an upper limit to the amount of carbon dioxide which can be so formed (Buch, 1945; see Gripenberg, 1960).

The change in pH (and in the partial pressure of dissolved carbon dioxide) which accompanies unit production of carbon dioxide depends on the alkalinity. This (see page 243) varies only slightly in most ocean waters, but in many lakes and inland waters, the alkalinity is low and a greater pH shift will accompany unit carbon dioxide production than would be observed in ocean water.

In many oceanic areas, vertical profiles show the existence of an oxygen-impoverished layer which corresponds with a region of high total carbon dioxide and low pH (see e.g. Harvey, 1955; Hood *et al.*, 1963). The detailed shape of the oxygen-depth or pH-depth profiles depends on the water circulation pattern, but low oxygen concentrations and relatively low pH tend to be characteristic of water masses which are poorly ventilated and which receive much decaying organic matter.

Changes of pH are sometimes used to follow losses of carbon dioxide during productivity studies (see Chapter 12). For work of this sort it is necessary to estimate the pH change associated with unit loss or gain of carbon dioxide; this may be accomplished by making use of theoretical relationships between pH, alkalinity and total carbon dioxide, or by preparing calibration plots of pH *vs* total carbon dioxide (or carbon dioxide change). The latter approach is the more desirable, but care is necessary in the choice of the method by which this is done (see Beyers and Odum, 1959, 1960; Verduin, 1960), and direct titration with a solution of carbon dioxide in distilled water has been recommended by Beyers and Odum.

Changes associated with the production or loss of carbon dioxide and oxygen can provide a useful basis for comment on circulation and transfer problems (Postma, 1964). The uptake of carbon dioxide in biosynthetic processes is accompanied by the removal of other nutrients (notably phosphate) from the water. Conversely, the removal of oxygen during the oxidation of dead organic matter is accompanied by the liberation of phosphate. Thus, in the deep Pacific, as water moves northwards, oxygen is consumed and phosphate liberated in the ratio

$$\frac{\Delta PO_4^{3-}(\mu g.\text{-at./l.})}{\Delta O_2(\text{ml./l.})} = 0{\cdot}27.$$

The ratio of phosphate to carbon dioxide liberated in this process is

$$\frac{\Delta PO_4^{3-}(\mu g.\text{-at./l.})}{\Delta CO_2(\text{ml./l.})} = 0{\cdot}33.$$

Temperature and depth (i.e. pressure) affect pH through their influence on the equilibrium constants of the carbonate system (see Section II, C(2)). Both the first and second dissociation constants of carbonic acid increase with an increase in temperature. In a mass of water isolated from carbon dioxide exchange with the atmosphere, an increase of temperature would, therefore, result in an increase in the hydrogen ion concentration; this effect is not very great since dpH/dt is only about $-0{\cdot}01$ pH unit/1°C rise. With surface waters, there is the possibility that carbon dioxide exchange with the atmosphere may occur and thereby reduce the magnitude of the pH change compared with that which would be observed if interchange were not possible. In practice, the observed pH shift will depend on how rapidly the temperature changes relative to the rate at which mixing and diffusion processes enable re-adjustments in the equilibrium position to be made. Rapid temperature changes are more likely to be noted in the surface layers of the ocean, where solar heating is most pronounced, than in deeper water masses.

Diurnal changes in the pH or in the partial pressure of carbon dioxide in sea water caused by solar heating have been observed by Takahashi (1961) and Ibert and Hood (1963). The time of the maximum in the carbon dioxide cycle arising from this cause is different from that produced by photosynthetic activity, and the minimum pH value or maximum in the partial pressure of carbon dioxide coincides with the daylight hours.

Superimposed on the above causes of pH variation are those produced by localized concentrations of pH influencing substances. Brackish water, estuarine, tidal and harbour waters and muds, and areas of local industrial contamination may show considerable deviations from the pH ranges expected in the open ocean. Park *et al.* (1958) have given a list of some of the reported diurnal pH ranges of a number of shallow water masses. Ranges amounting to as much as 1·5 pH units are occasionally noted. Ibert and Hood (1963) and Park *et al.* have observed that rather marked changes in pH may accompany heavy rainfall. During a series of observations made in Redfish Bay (Texas), Park *et al.* (1958) noted that a period of three hours rain caused the pH to fall from 8·41–8·03 while the temperature fell from 22·4–20·4°C. These changes were more rapid than those normally caused by a decrease in light intensity, but certainly cannot be accounted for by the temperature change.

B. ALKALINITY, CARBONATE ALKALINITY, TOTAL CARBON DIOXIDE AND P_{CO_2}

Sea water owes its alkaline properties to the presence of the ions, bicarbonate, carbonate and borate. In terms of the Brønsted-Lowry theory, these and the OH^- ion are to be regarded as bases because of their ability to accept protons (see e.g. Rakestraw, 1949). Water itself may be regarded as either an acid or a base; as an acid it is capable of donating a proton (and thereby yielding its conjugate base, OH^-), and as a base it will accept a proton to give its conjugate acid, the oxonium ion H_3O^+. The relatively alkaline pH of sea water is a consequence of reactions of the type

$$HCO_3^- + H_2O \rightleftharpoons H_2CO_3 + OH^- \tag{5}$$

in which the above concept of acid–base behaviour is displayed, and in which the base HCO_3^- accepts a proton from the acid H_2O to give the conjugate acid and base H_2CO_3 and OH^-.

Because of the above process, the analytical concentration of bicarbonate in an aqueous solution will be slightly greater than the equilibrium concentration, c_{HCO_3}. Thus, the analytical concentration determined by titration to the bicarbonate–carbonic acid equivalence point will be given by

$$c_{\text{anal}} = c_{HCO_3} + (c_{OH} - c_H). \tag{6}$$

In a sea-water sample, the alkalinity is the sum, in terms of equivalents per litre, of the analytical concentrations of the anions of carbonic and weaker acids. Since the only ions of importance in determining the alkalinity of sea water are those derived from carbonic and boric acids,

$$\text{alkalinity (eq./l.)} = c_{HCO_3} + 2c_{CO_3} + c_{H_2BO_3} + (c_{OH} - c_H) \tag{7}$$

where the c terms refer to the equilibrium concentrations of the various species in g. ions/l. For markedly anoxic waters containing traces of HS^-, S^{2-} and NH_4^+, it may be necessary to introduce additional terms into equation (7) (see Chapter 13).

In sea water of pH 8·3 at 10°C, the concentrations of bicarbonate, carbonate and borate ions are in the order of 18×10^{-4}, $3{\cdot}5 \times 10^{-4}$ and 10^{-4} g.moles/l. respectively, whereas $(c_{OH} - c_H)$ is about 2×10^{-6} g.moles/l. It is evidently a reasonable approximation to ignore the last term of the above expression, and to identify the alkalinity with the total number of equivalents of weak acid anions which exist as such in the solution. This approximation is not tenable if the pH falls outside the range 5·5–8·5.

The alkalinity may be estimated by treating the sample with a measured excess of strong acid, driving off the liberated carbon dioxide in a stream of CO_2-free air, and back-titrating the solution to a pH (6·0–7·0) at which the boric acid is negligibly ionized (Gripenberg, 1937), or by titrating the sample directly in the presence of the liberated carbon dioxide to a pH of 4·5–5·0 (Greenberg *et al.*, 1932). An alternative method has been described by Thompson and Anderson (1940) and subsequently refined by Anderson and Robinson (1946) and by

Bruneau *et al.* (1948). An accurately measured volume of hydrochloric acid (e.g. 25 ml. of 0·01 N) is added to 100 ml. of the sample and the pH of the mixture determined. From the pH it is possible to estimate the amount of acid equivalent to the alkalinity of the sample.

At the stoichiometric end-point, if the dissociation of the water is ignored, $c_{HCO_3} = c_H$. Addition of more than sufficient acid to reach this point introduces further hydrogen ions, a fraction of which react with bicarbonate ion, the remainder serving to depress the pH. If v ml. of acid of normality N is added to V ml. of the sample, the alkalinity can be shown to be given by

$$\text{Alkalinity} = \frac{1000N \, . \, v}{V} - (\text{H}^+ \text{ ion lost by reaction with } \text{HCO}_3^- + \text{H}^+ \text{ ion used in depressing the pH below the stoichiometric end-point})$$

$$\approx \frac{1000}{V}\left[N \, . \, v + (V + v)\left\{\frac{K'_{L1} \, . \, c_{H_2CO_3}}{a_H} - \frac{a_H}{\gamma_H}\right\}\right]$$

where K'_{L1} is the first apparent dissociation constant of carbonic acid, and the term a_H/γ_H corresponds to the concentration of hydrogen ions. Various methods have been proposed for realizing in practice the correction terms embodied in the above equation. The method used by Anderson and Robinson (1946) is less complete than that of Bruneau *et al.* (1948) since it fails to apply a bicarbonate correction. Bruneau *et al.* considered that some of the carbon dioxide produced by the reaction with the acid may be lost from the solution, and in the application of the bicarbonate correction they assumed a value for $c_{H_2CO_3}$ of about one-half of that corresponding to the total carbon dioxide content of the original sample.

Hood *et al.* (1963) used direct potentiometric titration for the estimation of alkalinity. A recording potentiometric method would probably offer the advantage of greater objectivity in the selection of the end-point.

Specific alkalinity is defined as

$$\text{Specific alkalinity} = \frac{\text{alkalinity} \times 10^3}{\text{chlorinity}} \qquad (8)$$

but has the disadvantage of using "mixed" units. It is often stated to be constant at 0·123 for most ocean waters. This relationship appears to be true for the bulk of Atlantic water (Wattenberg, 1933), although it increases slightly near the sea bed. However, considerable deviations from this value occur, and Koczy (1956) considers that the range falls within the limits 0·119–0·130, with an average for all waters of 0·126.

Circumstances which lead to the solution of calcium carbonate by bicarbonate-rich waters lead to high values of specific alkalinity. Values of the specific alkalinity as a function of latitude for surface waters of the Pacific (Soviet vessels *Ob* and *Vityaz*) are shown by the lower curve of Fig. 2(b) (Postma, 1964). In the low salinity water of the Baltic, the alkalinity/chlorinity ratio is rather higher than in the open ocean, and shows slight variations from year to year (Gripenberg, 1960).

Gripenberg has suggested that organic matter may contribute to the alkalinity, and in connection with an investigation of Baltic water, she has defined quantities Alk_{min} (minimum alkalinity) and Alk_x (surplus alkalinity). The minimum alkalinity is regarded as the sum of two parts, one equal to the alkalinity of that fraction of the water which derives from the Atlantic, and the other equal to the difference between the calcium equivalents of Baltic and Atlantic water, both parts being functions of chlorinity. Thus, since the calcium contents of the Baltic and its adjacent areas are given by

$$\text{Bottom Water, Baltic Ca meq./kg.} = 1{\cdot}130 + 1{\cdot}018\, Cl‰$$

$$\text{Surface Water, Baltic Ca meq./kg.} = 0{\cdot}413 + 1{\cdot}191\, Cl‰$$

$$\text{Gulf of Bothnia Ca meq./kg.} = 0{\cdot}095 + 1{\cdot}289\, Cl‰$$

and that of the open ocean is given by (Wattenberg, 1933)

$$\text{Ca meq./kg.} = 1{\cdot}078\, Cl‰$$

subtraction of $1{\cdot}078\, Cl‰$ from each of the three equations relating to the Baltic region, and addition to each of them of the expression for the alkalinity of ocean water ($1{\cdot}20$ meq./kg. $\times Cl‰$) gives

$$\text{Bottom Water, Baltic } Alk_{min} \text{ meq./kg.} = 1{\cdot}130 + 0{\cdot}060\, Cl‰$$

$$\text{Surface Water, Baltic } Alk_{min} \text{ meq./kg.} = 0{\cdot}413 + 0{\cdot}232\, Cl‰$$

$$\text{Gulf of Bothnia } Alk_{min} \text{ meq./kg.} = 0{\cdot}095 + 0{\cdot}331\, Cl‰$$

This procedure assumes that the increased alkalinity of the Baltic is caused by the solution of calcium carbonate. No account needs to be taken of dissolved magnesium carbonate since the $Mg/Cl‰$ ratio is the same in the Baltic as in the ocean. Gripenberg noted that the observed alkalinities, Alk_{obs}, were mostly greater than Alk_{min}, and the surplus alkalinity, Alk_x, was defined by her as the difference between the two. The amount of the surplus alkalinity was small ($< 0{\cdot}2$ meq./kg.), but was greater than the experimental error. Since it tended to be highest when the amount of organic matter in the water was greatest, she concluded that it was attributable to organic constituents of Baltic water.

Gripenberg considers that at least part of the effect arises from the complexing of the organic compounds with boric acid. It is well known that polyhydroxy compounds do complex with boric acid in such a way as to enhance the strength of the acid, and in fact the mannitol method of boric acid estimation rests on this property. Evidence for complex formation between the organic matter and boric acid in sea water has been given by Gast and Thompson (1958) who found that in the mannitol method of boron estimation, higher values were obtained if the sample was first oxidized with permanganate solution to destroy organic matter, and similar observations have been reported by Noakes and Hood (1961).

It would seem that there is scope for further investigation of the possible interaction between organic matter and boron in sea water. Both boron and organic matter occur in ocean water only in relatively low concentrations, and presumably only a fraction of the total dissolved organic matter is of a form suitable for complex formation. Thus, detectable amounts of the total boron can be complexed only if the formation constant for the complex is very high. It might also be noted that in titrimetric investigations of this sort, a high order of experimental precision is required if meaningful results are to be obtained. Proper appreciation of the form of the pH titration curve is necessary, since the end-point must be selected with care.

The carbonate alkalinity ($c_{HCO_3} + 2c_{CO_3}$) can be estimated by subtracting the borate ion contribution from the alkalinity. The boron concentration in sea water has been discussed in Chapter 4, and from the values listed it appears that a working relationship between the boron content and chlorinity is

$$\frac{[B]\text{g./kg.}}{Cl\text{‰}} = 2{\cdot}3 \times 10^{-4}. \tag{9}$$

This ratio is applicable to most open ocean water, although Gripenberg (1960) considers that for the Baltic to be somewhat greater. Except at high pH values, only a small proportion of the total boron is ionized, and the concentration of borate ions must be estimated by making use of the apparent ionization constant of boric acid in sea water (e.g. Buch, 1933a, b, c; Lyman, 1956).

For the ionization

$$H_3BO_3 \rightleftharpoons H^+ + H_2BO_3^- \tag{10}$$

the apparent ionization constant is

$$K'_B = \frac{a_H \cdot c_{H_2BO_3}}{c_{H_3BO_3}}$$

and since

$$c_{H_2BO_3} + c_{H_3BO_3} = c_{\Sigma B}$$

the concentration of borate ions is given by

$$c_{H_2BO_3} = \frac{K'_B \cdot c_{\Sigma B}}{(K'_B + a_H)}$$

when $c_{\Sigma B}$ is the total concentration of boron in moles/l.

Thus, denoting the carbonate alkalinity by CA

$$\begin{aligned} CA &= c_{HCO_3} + 2c_{CO_3} \\ &= \text{alkalinity} - \frac{K'_B \cdot c_{\Sigma B}}{(K'_B + a_H)} - (c_{OH} - c_H) \end{aligned} \quad (11)$$

or,

$$CA = \text{alkalinity} - \frac{K'_B \cdot c_{\Sigma B}}{(K'_B + a_H)} \quad (12)$$

between pH 5·5 and 8·5.

Tables of the apparent dissociation constant of boric acid (K'_B) have been compiled by Buch (1933a) and are reproduced in works by Harvey (1955) and by Barnes (1959). These values are based on the Sørensen pH scale. More recent estimates of K'_B have been given by Lyman (1956) and are shown in the Appendix Table III. Lyman's estimates are based on the NBS pH scale. At high pH values, some dissociation of the $H_2BO_3^-$ ion may occur, and thus cause a further contribution to the alkalinity. The Appendix Table IV (Lyman, 1956) shows that the second apparent dissociation constant is small, and therefore the contribution to the alkalinity of HBO_3^{2-} ions will be small over the natural range of pH values of sea water. As might be expected, the third apparent dissociation constant of boric acid is too small to have any appreciable effect on the alkalinity of sea water (see Lyman, 1956).

Total alkalinity is, of course, greater than the carbonate alkalinity, but the difference between the two decreases as the pH falls and the degree of ionization of the very weak boric acid becomes small.

The total carbon dioxide, $\sum CO_2$, comprises dissolved carbon dioxide as such, H_2CO_3, HCO_3^- and CO_3^{2-}, although as shown in a subsequent section (II, C) the bulk of the quantity is made up of the HCO_3^- and CO_3^{2-} contributions. $\sum CO_2$ can be estimated by measuring the amount of carbon dioxide liberated when the sample is treated with hydrochloric or phosphoric acid (e.g. Buch *et al.*, 1932). The estimation can be made using a Van Slyke apparatus (Van Slyke and Neill, 1924a,b; Van Slyke and Sendroy, 1927; Strickland and Parsons, 1960), although modern laboratory vacuum systems (in which the liberated carbon dioxide can be separated from water and permanent gases by the use of suitable refrigerants and subsequently estimated by the measurement of its pressure at known volume and temperature) will recommend themselves to many chemists. A Conway microdiffusion method has been described by Saruhashi (1953). It seems likely that gas chromatographic methods could be devised.

The partial pressure of carbon dioxide in sea water, P_{CO_2}, is that pressure of carbon dioxide above the liquid with which the sample would be in equilibrium. It can be measured by shaking the sample in contact with a bubble of air and estimating the carbon dioxide content of the air (Krogh, 1910; Buch, 1939a,b), or by circulating air which has been brought into intimate contact with the water sample in an equilibrator (and subsequently dried) through an infra-red gas analyser (Smith, 1953; Kanwisher, 1960; Takahashi, 1961; Ibert and Hood, 1963). Ibert and Hood have described a sampling apparatus for ship-borne use in which the sea water was pumped through a series of six equilibrators. The effluent gas, which was in carbon dioxide equilibrium with the water, was passed to an infra-red differential analyser. Since water vapour also gave a response in the analyser, the gas was dried by passage over silica gel and magnesium perchlorate. A reproducibility of $\pm$ 2 p.p.m. was possible. Temperature control in the equilibrator is important, particularly when the ship's laboratory temperature differs considerably from that of the water being sampled.

The partial pressure of carbon dioxide in the atmosphere, p_{CO_2}, can be determined by conventional gas analysis of the air sample (Fonselius *et al.*, 1955) or by infra-red analysis of the dried sample (Fonselius *et al.*, 1956; Takahashi, 1961; Ibert and Hood, 1963).

It is customary to express the carbon dioxide content of a gas sample in units of atmospheres or of p.p.m. by volume. If the total atmospheric pressure is 1 atm, then the *in situ* partial pressure expressed in atm $\times\ 10^6$ is numerically the same as the concentration expressed in p.p.m. by volume. When the atmospheric pressure deviates appreciably from 760 mm. (for example in high altitude air sampling, in connection with investigations of the partial pressure of carbon dioxide of high altitude lakes, or in variations of atmospheric pressure at sea level) the numerical values of the two systems of units differ unless the partial pressure is expressed in terms of the sample reduced to S.T.P.

When reporting the carbon dioxide concentrations of an air sample, it is important to indicate whether the quoted values refer to the gas as sampled or to the dry gas. It is usually assumed that air sampled at the sea surface or air emergent from an air–sea equilibrator is saturated with water vapour. Removal of 23 mm. of water vapour from a sample of air having a carbon dioxide concentration of 320 p.p.m. by volume will increase the concentration by about 9 p.p.m.

Reference should be made to other analytical investigations of carbon dioxide and bicarbonate containing solutions. Many of these investigations, although not specifically designed for oceanographical use, may possibly be adaptable for this purpose (Jennings and Osborn, 1957; Loomis, 1958; Loveland *et al.*, 1959; Underwood, 1961; Lodge *et al.*, 1962; Pobiner, 1962; Underwood and Lyman, 1962).

C. SOLUTION OF CARBON DIOXIDE AND THE DISSOCIATION OF CARBONIC ACID

When carbon dioxide is brought into contact with an aqueous solution the following equilibria are established

$$CO_2(\text{gas}) \rightleftharpoons CO_2(\text{solution}) \quad (13)$$

$$CO_2(\text{solution}) + H_2O \rightleftharpoons H_2CO_3 \quad (14)$$

$$H_2CO_3 \rightleftharpoons HCO_3^- + H^+ \quad (15)$$

$$HCO_3^- \rightleftharpoons CO_3^{2-} + H^+. \quad (16)$$

The ionization reactions (15) and (16) are extremely rapid, but the hydration of dissolved carbon dioxide and the dehydration of carbonic acid are relatively slow, as also is the process

$$CO_2 + OH^- \rightleftharpoons HCO_3^- \quad (17)$$

which is important only at higher pH values (i.e. > 9). The slow hydration and dehydration reactions are possibly contributory factors in determining the exchange rate of carbon dioxide between the atmosphere and the sea under certain conditions (see also Section VII). Their industrial and physiological importance have stimulated many investigations (see papers by Faurholt, 1924; Roughton and Booth, 1937; Mills and Urey, 1940; Roughton, 1941; Berg and Patterson, 1953; Wissburn *et al.*, 1954; Pinsent *et al.*, 1956; Edsall and Wyman, 1958; Ljunggren and Lamm, 1958; Sirs, 1958; Danckwerts and Melkersson, 1962, for details and earlier references). Both reactions are subject to catalysis by a wide range of substances including selenite, sulphite, chromate and phosphate, but particularly by the enzyme carbonic anhydrase which plays an essential role in physiological processes involving carbon dioxide.

The main kinetic features of the CO_2–H_2O and CO_2–OH^- reactions may be summarized as follows.

(i) The reaction $CO_2 + H_2O \rightarrow H_2CO_3$ is first order with respect to the dissolved carbon dioxide as such, and has a first order velocity constant (k_i) of *c.* 0·03 sec.$^{-1}$ at 25°C. The half-life of the carbon dioxide under these conditions is thus several minutes. The apparent activation energy for this reaction is about 19 kcal./mole at 0°C, but probably falls to about 11 kcal./mole at 40°C (Kern, 1960).

(ii) The process $OH^- + CO_2 \rightarrow HCO_3^-$ obeys the second order rate law $-dc_{CO_2}/dt = k_{ii} \cdot c_{CO_2} \cdot c_{OH}$, and at 25°C the velocity constant (k_{ii}) has a value of about 8500 l./mole/sec. The activation energy is about 13·2 kcal./mole. Process (ii) is insignificant compared with process (i) at pH values $< 7{\cdot}5$, although the relative importance of (ii) probably increases somewhat with temperature (Kern, 1960).

(iii) The dehydration reaction $H_2CO_3 \rightarrow H_2O + CO_2$ is first order and has a velocity constant (k_{iii}) of about 20 sec.$^{-1}$ at 25°C. Its activation energy is 16·5 kcal./mole.

(iv) The reaction $HCO_3^- \rightarrow CO_2 + OH^-$ has a first order velocity constant (k_{iv}) at 25°C of 2×10^{-4} sec.$^{-1}$.

From processes (i) and (iii) it is possible to estimate the equilibrium ratio of carbon dioxide as such to carbonic acid. This is given by the ratio of the velocity constants, k_{iii}/k_i, and has a value of about 670 at 25°C. These general features appear to be reasonably well established, although a number of details, and some discrepancies between the results of different workers, remain to be resolved.

Processes (i) and (iii) are subject to base catalysis; when this effect is important, the generalized expression for the net rate of loss of dissolved carbon dioxide as such is given by

$$-\frac{dc_{CO_2}}{dt} = k_i(1 + l \,.\, c_b)c_{CO_2} - k_{iii}(1 + l \,.\, c_b)c_{H_2CO_3}$$

where c_b is the concentration of catalytically active base and l is the catalytic constant. The value of l varies widely between different substances, and has a value of 8 for phosphate, 150 for borate and 10,000 for arsenite. Nitrogenous bases are also catalytically active. The enzyme carbonic anhydrase is a catalyst *par excellence*, and about 0·1 p.p.m. of it will produce a pronounced effect.

Possible catalytic effects should be borne in mind when applying the results of kinetic investigations of the rates of hydration of dissolved carbon dioxide obtained from solutions in which such effects were absent to solutions of high ionic strength and which may contain catalytically active substances. No direct measurements of the rate of hydration of carbon dioxide or the rate of dehydration of H_2CO_3 appear to have been made for solutions approximating in composition to that of natural sea water.

Equations (13) to (16) give an over-simplified picture of the equilibria which exist in the sea-water carbonate system since they take no account of the existence of ion-pairs which may be formed between carbonate and bicarbonate ions and certain metal cations (Garrels *et al.*, 1961) nor of carbamino-carboxylic acid formation which may account for 3×10^{-6} moles CO_2/l. (Smith *et al.*, 1960). Evidence has been obtained from the kinetics of the reaction of carbon dioxide with water, (Koefoed and Engel, 1961) that still other species may occur, but their existence remains to be confirmed, and there seems little doubt that at the present time the bulk of limnological and oceanographical interest in equilibrium aspects of the water–carbon dioxide system lies in the entities given in the above equations and possible complexes.

(1) *Solution of carbon dioxide.* Although when carbon dioxide dissolves in pure water all the above reactions occur with the result that the pH of the solution falls to the acid side, most of the carbon dioxide entering the solution remains as dissolved gas. Bicarbonate ion formation is small below pH 6, and carbonate becomes significant only when the pH exceeds about 8. Figure 4 shows the way in which the proportions of the ions HCO_3^-, CO_3^{2-} and the dissolved carbon dioxide depend on the pH. The term dissolved carbon dioxide implies the dissolved carbon dioxide as such plus unionized carbonic acid.

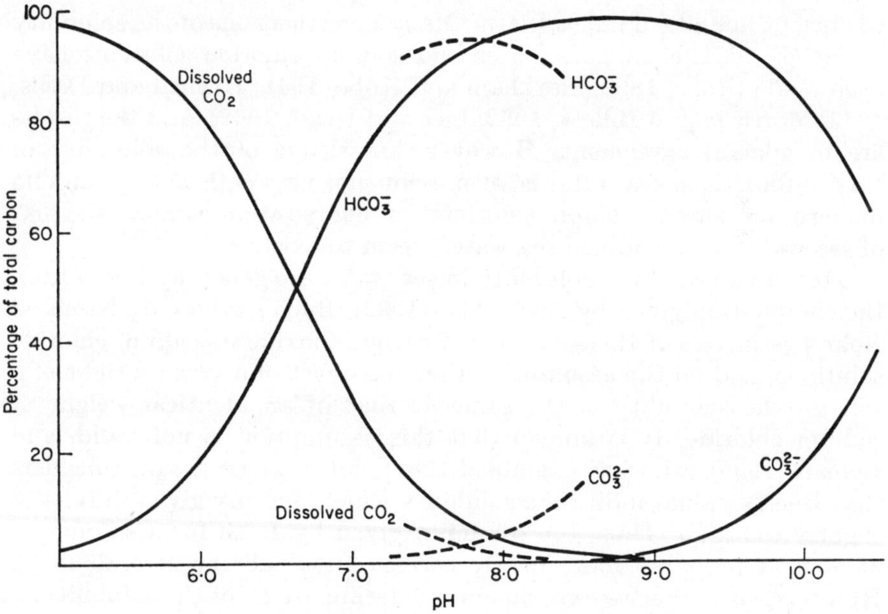

FIG. 4. Percentages of dissolved carbon dioxide, bicarbonate ion and carbonate ion as a function of pH (data plotted from tables by Saruhashi, 1955). ———, fresh water; - - - - - -, sea water ($Cl = 20$‰), both at 0°C.

It is noted that at 0°C, in fresh water the optimum proportion of bicarbonate is attained at a pH of *c.* 8·5, at which value almost 98% of the total carbon of the carbonate system exists as bicarbonate. The effect of increasing salinity is to shift the curves for all three components towards lower pH values.

Investigations of the kinetics and mechanism of the CO_2–H_2O system have shown that, in general, the carbonic acid amounts to only a fraction of a percent of the dissolved $CO_2 + H_2CO_3$. That is, the equilibrium constant of process (14) defined by

$$K_0 = \frac{a_{H_2CO_3}}{a_{H_2O} \cdot a_{CO_2}}$$

ranges from about 2×10^{-3} at 5°C to about $3{\cdot}2 \times 10^{-3}$ at 38°C (see above references). Thus, in pure water, carbon dioxide dissolves in accordance with Henry's Law over a wide range of pressures, i.e.

$$c_{CO_2} = p_{CO_2} \cdot \alpha$$

where c_{CO_2} is the concentration of dissolved gas in moles/l., p_{CO_2} is the applied pressure of carbon dioxide in atmospheres and α is the solubility coefficient in units of moles/l./atm. Many investigations of the solubility of carbon dioxide in pure water and sodium chloride solutions have been made (Bohr, 1899; Markham and Kobe, 1941; Harned and Davis, 1943; Morrison and Billett, 1952; Cox and Head, 1962), and the results are in general agreement. However, knowledge of the solubility of carbon dioxide in sea water is, at present, not very satisfactory, and no modern measurements on solutions of composition similar to that of sea water (e.g. acidified sea water) seem to exist.

Quoted values of the solubility in sea water are generally taken from the compilation given by Buch *et al.* (1932). Buch's tables are based on Bohr's estimates of the solubility of carbon dioxide in sodium chloride solutions, and on the assumption that the effect of a given weight of a salt on the solubility is the same as that of an identical weight of sodium chloride. It is known that this assumption is not valid, and Lyman (1956), who has examined this point in some detail, considers that Buch's values indicate a solubility which, for any given salinity is slightly too high. Thus, the solubility given by Buch for a salinity of 36·00‰ is probably more nearly correct for a salinity of *c.* 35·00‰. However, since precise experimental determinations of the solubility in sea water are not available, the values given by Buch are given in Table V of the Appendix.

It is seen from this Table that the solubility decreases with an increase of both temperature and salinity. The solubility of carbon dioxide is considerably greater than that of oxygen and nitrogen, and as a consequence, when the three gases dissolve in an aqueous solvent to equilibrium with their atmospheric partial pressures, there is an enrichment with respect to the more soluble gases. Thus, whereas the atmospheric partial pressures of N_2, O_2 and CO_2 are in the ratio *c.* 2400: 630 : 1 (see Table I of Chapter 6), the ratio of the dissolved equilibrium concentrations are *c.* 28 : 19 : 1. The precise value depends on the temperature. The observed dissolved concentration ratio will be altered if the equilibrium is disturbed by, for example, the oxidation of organic matter or the utilization of carbon dioxide in photosynthesis.

Under alkaline conditions, Henry's Law apparently breaks down in the sense that there is no longer proportionality between the applied pressure of carbon dioxide and the total amount of carbon in the solution because of appreciable bicarbonate and carbonate ion formation. In sea water, for example, only about 1% of the total dissolved inorganic carbon exists as dissolved carbon dioxide. If, however, account is taken only of the dissolved carbon dioxide as such, Henry's Law is still obeyed.

(2) *Dissociation of carbonic acid in water*. The true first ionization constant of carbonic acid, defined by

$$K_f = \frac{c_{\mathrm{H}} \cdot c_{\mathrm{HCO_3}}}{c_{\mathrm{H_2CO_3}}} \tag{18}$$

in which $c_{\mathrm{H_2CO_3}}$ is the concentration of carbonic acid as such has been found to have a value of about 2×10^{-4} at ordinary temperatures (Mills and Urey, 1940; Roughton, 1941; Berg and Patterson, 1953); various estimates differ too widely to allow a precise assignment of a value, but it is evident that carbonic acid is stronger than is generally realized. The illusion of low strength stems from the usage made of the more practical constant K'_f when

$$K'_f = \frac{c_{\mathrm{H}} \cdot c_{\mathrm{HCO_3}}}{(c'_{\mathrm{CO_2}} + c_{\mathrm{H_2CO_3}})}$$

when $c'_{\mathrm{CO_2}}$ is the concentration of dissolved carbon dioxide as such.

When dealing with the thermodynamics of the dissociation of aqueous carbonic acid it is often found convenient to combine equations (14) and (15) and to write the first stage of the ionization as

$$\mathrm{CO_2(dissolved)} + \mathrm{H_2O(liquid)} \rightleftharpoons \mathrm{HCO_3^-(aq.)} + \mathrm{H^+(aq.)} \tag{19}$$

and the first and second thermodynamic ionization constants can be written

$$K_1 = \frac{a_H \cdot a_{HCO_3}}{a_{H_2O} \cdot a_{CO_2}} \tag{20}$$

and

$$K_2 = \frac{a_H \cdot a_{CO_3}}{a_{HCO_3}} \tag{21}$$

where the a terms refer to the activities of the appropriate species. These thermodynamic dissociation constants are of great fundamental importance not only because of their intrinsic chemical interest, but also because of the role which this system plays in respiratory processes. As a consequence, considerable effort has been devoted to their determination. Particularly careful measurements have been made by Shedlovsky and McInnes (1935), Harned and Scholes (1941), Harned and Davis (1943) and Nasanen (1946, 1947). Some of these workers' results are shown in Table IV.

TABLE IV

First and second thermodynamic dissociation constants of carbonic acid

	$K_1 \times 10^7$			$K_2 \times 10^{11}$
t(°C)	*(1)*	*(2)*	*(3)*	*(4)*
0	2·647	—	2·613	2·36
10	3·430	—	—	3·24
15	3·802	3·79	3·722	3·71
25	4·452	4·48	4·310	4·69
35	4·914	4·89	—	5·62

K_1 refers to the process $CO_2(aq.) + H_2O \rightleftharpoons H^+(aq.) + HCO_3^-(aq.)$.
K_2 refers to the process $HCO_3^-(aq.) \rightleftharpoons H^+(aq.) + CO_3^{2-}(aq.)$.
(1) Harned and Davis (1943).
(2) Nasanen (1947).
(3) Shedlovsky and McInnes (1935).
(4) Harned and Scholes (1941).

Although thermodynamic constants have theoretical advantages and are independent of the salt concentration, they do have certain limitations. In particular, they lack immediate practical utility since not all the activity terms are readily accessible to experimental measurement.

The requirements of the practical oceanographer are better met by using apparent constants such as

$$K_1' = \frac{a_H \cdot c_{HCO_3}}{a_{H_2O} \cdot a_{CO_2}} \tag{22}$$

and

$$K_2' = \frac{a_H \cdot c_{CO_3}}{c_{HCO_3}} \tag{23}$$

in which the activities of the bicarbonate and carbonate ions used in the thermodynamic constants are replaced by molar concentration terms (Buch *et al.*, 1932). Because apparent constants are "mixed" in the sense that they contain both activity and concentration terms, they are not independent of the salinity, although at zero salinity the numerical value of an apparent constant is identical with that of the parent thermodynamic constant. If they are to be useful over a range of sea-water compositions it is necessary to know the way in which they change with salinity as well as with temperature. Detailed investigations of the apparent constants of carbonic acid have been made by Buch *et al.* (1932), Moberg *et al.* (1934), Buch (1933a,b, 1938, 1951) and Lyman (1956).

At the present time, two sets of apparent constants are in general use. One of these is the compilation given by Buch (1951) as the result of a long series of investigations (see above references). Buch's values are based on the definitions given by equations (22) and (23), and his tables have been widely reproduced (Harvey, 1955; Saruhashi, 1955; Barnes, 1959). The other set of values has been prepared by Lyman (1956). Since Lyman's values are based on a definition of the first constant which differs slightly from that used by Buch, in the following account his first apparent constant is given the symbol K'_{L1}. It is related to K_1' by

$$K'_{L1} = K_1'(p_s/p_0) \cdot (\alpha_0/\alpha_s) \tag{24}$$

where p_0 is the vapour pressure of pure water and p_s is that of the sea-water sample. α_0 and α_s are the solubility coefficients of carbon dioxide in pure water and the sample respectively. The second apparent constant used by both Buch and Lyman is based on the definition given by equation (23) above. Constants obtained by Hindman (1943) are sometimes used (see e.g. Cloud, 1963).

It must be recognized that apparent constants do not have the same fundamental significance as do thermodynamic constants, and that they are devices designed to facilitate the estimation of the properties of the carbonate system of a water sample on the basis of two measurements (pH and alkalinity). As stated by equation (22) it is implied that the hydrogen ion activity of the sea water must be known before the constants can be used. The impossibility of knowing individual ionic activities and the uncertainty of the pH relationship (equation (2)) at high ionic strengths have been referred to earlier (Section II, A(2)). However, since the apparent constants are determined on the assumption of the validity of equation (2), correction for this uncertainty is, as it were, built into the tabulated values.

Buch's values for the first apparent constant were obtained by making use of the expression

$$K_1' = \text{CA} \,.\, \frac{a_{\text{H}}}{P_{\text{CO}_2} \,.\, \alpha_0 \,.\, \left[1 + \dfrac{2K_2'}{a_{\text{H}}}\right] a_{\text{H}_2\text{O}}} \tag{25}$$

which can be obtained by re-arrangement of equation (37) below. This expression can be re-written in logarithmic form

$$\text{p}K_1' = \text{psH} - \log \frac{\text{CA}}{P_{\text{CO}_2} \,.\, \alpha_0 \,.\, \left[1 + \dfrac{2K_2'}{a_{\text{H}}}\right] a_{\text{H}_2\text{O}}} \,. \tag{26}$$

The term psH rather than pH is used in order to emphasize that Buch's determinations were based on the Sørensen pH scale. psH was determined by means of the quinhydrone electrode, CA by an alkalinity titration, and P_{CO_2} by gas analysis. Thus, by employing a pH region in which $2K_2'/a_{\text{H}}$ was very small, approximate values for K_2' would suffice to enable equation (26) to be used to estimate values for K_1'.

Values for the second apparent constant were obtained by Buch and his co-workers by using solutions of relatively high pH in order that appreciable dissociation of the bicarbonate occurred, and by relating the second constant to the total carbon dioxide (see equation (39) below). The earlier determinations made by Buch *et al.* (1932) were in error because at the time of these measurements the importance of the borate contribution to the alkalinity was not fully appreciated, but later publications by Buch (1933a,b,c, 1938, 1945, 1951) have corrected and refined the earlier estimates.

Lyman has criticized the determination of the first dissociation constant made by Buch *et al.* on the grounds that the procedure employed required three separate determinations, each on a separate sample (i.e. pH, alkalinity and P_{CO_2}). An alternative approach by Lyman (1956) made use of the fact that during the titration of a weak dibasic acid, the pH is a function of the two dissociation constants. His experiments were designed to allow correction to be made for the boric acid in sea water. Three titrations were performed on each water sample:

(i) titration with HCl to an end-point slightly beyond that corresponding to the stoichiometric equivalent of a solution of carbon dioxide in the presence of a mixture of salts;

(ii) the solution resulting from titration (i) was freed from carbon dioxide and was titrated with CO_2-free NaOH solution to a pH of about 10;

(iii) the solution resulting from titration (ii) was treated with sufficient HCl to neutralize the excess NaOH, mannitol added, and titration (ii) repeated. The pH throughout each titration was measured by means of a glass electrode, and from titrations (ii) and (iii) the total boron content of the sample and the apparent dissociation constants of boric acid could be determined, as well as the correction which needed to be applied to titration (i) in order to allow the first and second dissociation constants of carbonic acid to be estimated.

Values for K'_{L1} and K'_2 have been given by Lyman (1956) for a wide range of chlorinities and temperatures, and are shown in Tables VI and VII of the Appendix. These values are based on the NBS pH scale.

Table V (Lyman, 1956) gives a comparison of Buch's and Lyman's estimates for the first and second apparent constants. Buch's original values for the first constant have been adjusted by equation (24) to

Table V

Comparison of Buch's and Lyman's values for the first and second apparent dissociation constants of carbonic acid

First apparent constant (as pK values)				
Temperature (°C)	*0*	*10*	*20*	*30*
Buch *et al.* (1932)	6·18	6·09	6·02	5·96
Lyman (1956)	6·16	6·09	6·03	5·99

Second apparent constant (as pK values)				
Temperature (°C)	*0*	*10*	*20*	*30*
Buch (1951)	9·33	9·18	9·08	9·00
Lyman (1956)	9·42	9·30	9·19	9·06

make them consistent with Lyman's definition, and a pH correction factor has been applied to Buch's first and second constants so as to base them on the NBS scale. Good agreement is noted between the values obtained by both workers for the second constant. The first constants also agree reasonably well, although strict comparison is not possible since the exact values for the solubility of carbon dioxide in sea water which are needed for the application of equation (24) are not known.

(3) *Use of the apparent constants.* In the derivation of the first and second apparent constants as defined by Buch, and in their use in the calculation of the properties of the sea-water carbonate system (see below), estimates are needed for a_H, a_{H_2O} and a_{CO_2}. The activity of the hydrogen ion is calculated from the pH using equation (2). The standard state for water in aqueous solutions is taken as the pure solvent, and the activity of water in a sea-water sample is the ratio of its vapour pressure (p_s) to that of pure water (p_0), i.e.

$$a_{H_2O} = \frac{p_s}{p_0}. \tag{27}$$

It is related to the chlorinity by

$$a_{H_2O} = 1 - 9{\cdot}69 \times 10^{-4}\, Cl‰ \tag{28}$$

(Harvey, 1955), although other relationships are available (Arons and Kientzler, 1954; Robinson, 1954). It is apparent that the activity of the water in most ocean waters never falls much below 0·98.

The expression for the activity of the dissolved carbon dioxide deserves special comment since the a_{CO_2} term of equations (20) and (22) refers to the dissolved carbon dioxide as such (designated below by c'_{CO_2}) rather than the total dissolved carbon dioxide which comprises the sum of the concentrations of the dissolved carbon dioxide as such plus the carbonic acid, and which is denoted by c_{CO_2}.

The chemical potential of the dissolved carbon dioxide as such is given by

$$\mu_{CO_2} = \mu^0_{CO_2} + RT \ln a_{CO_2} \tag{29}$$

$$= \mu^0_{CO_2} + RT \ln \gamma_{CO_2} \cdot c'_{CO_2}$$

where μ^0 is the standard chemical potential of the dissolved carbon dioxide as such (i.e. its chemical potential when its activity is unity). In view of the equilibrium

$$CO_2 + H_2O \rightleftharpoons H_2CO_3$$

it is evident that over the rather low concentration range of dissolved gas, the concentration of carbonic acid is proportional to c'_{CO_2}, i.e.

$$c'_{CO_2}/c_{H_2CO_3} = K_r.$$

Thus, since

$$c_{CO_2} = c'_{CO_2} + c_{H_2CO_3}$$

then

$$c_{CO_2} = c'_{CO_2}(1 + 1/K_r)$$

and the expression for the chemical potential of the dissolved carbon dioxide as such becomes

$$\begin{aligned}\mu_{CO_2} &= \mu^0_{CO_2} + RT \ln \gamma_{CO_2} \cdot c_{CO_2}/(1 + 1/K_r) \\ &= \mu^0_{CO_2} + RT \ln \gamma_{CO_2} \cdot c_{CO_2} - RT \ln (1 + 1/K_r) \\ &= \mu^{0'}_{CO_2} + RT \ln \gamma_{CO_2} \cdot c_{CO_2} .\end{aligned}$$

It might be noted in passing that since K_r is quite large, $\mu^{0'}_{CO_2}$ does not differ significantly from $\mu^0_{CO_2}$. Since the solution of carbon dioxide obeys Henry's Law over all carbon dioxide concentrations of interest, for each particular salinity it is possible to write

$$\mu_{CO_2} = \mu^{0'}_{CO_2} + RT \ln c_{CO_2}$$

i.e. the activity coefficient is unity over the carbon dioxide concentration range at fixed salinity. The value of $\mu^{0'}_{CO_2}$ will depend on the salinity, but since in two solutions of carbon dioxide—one in pure water and the other in a saline solution—subjected to the same applied carbon dioxide pressure, the chemical potentials of the dissolved carbon dioxide are the same,

$$\mu^{*'}_{CO_2} + RT \ln c_{CO_2(O)} = \mu^{0'}_{CO_2} + RT \ln c_{CO_2(S)}$$

where $\mu^{*'}_{CO_2}$ is the standard chemical potential corresponding to $\mu^{0'}_{CO_2}$ but referring to a solution in pure water, and $c_{CO_2(O)}$ and $c_{CO_2(S)}$ are the concentrations of dissolved carbon dioxide plus H_2CO_3 in pure water and the saline solution respectively.

Thus,

$$\begin{aligned}\mu^{0'}_{CO_2} &= \mu^{*'}_{CO_2} + RT \ln c_{CO_2(O)}/c_{CO_2(S)} \\ &= \mu^{*'}_{CO_2} + RT \ln \alpha_O/\alpha_S\end{aligned}$$

where α_O and α_S are the solubility coefficients of carbon dioxide in pure water and in the saline solution respectively. Therefore, the chemical potential of dissolved carbon dioxide as such in a solution of any particular salinity is given by

$$\mu_{CO_2} = \mu^{*'}_{CO_2} + RT \ln \alpha_O/\alpha_S + RT \ln c_{CO_2}. \tag{30}$$

Writing the chemical potentials of the hydrogen ion, the bicarbonate ion and water as

$$\mu_{H} = \mu^{0}_{H} + RT \ln a_{H}$$
$$\mu_{HCO_3} = \mu^{0}_{HCO_3} + RT \ln a_{HCO_3}$$
$$\mu_{H_2O} = \mu^{0}_{H_2O} + RT \ln a_{H_2O}$$

and remembering that at equilibrium for reaction (19)

$$\sum \mu \text{ reactants} - \sum \mu \text{ products} = 0$$

it follows that

$$K_1 = \exp\left[(\mu^{*\prime}_{CO_2} + \mu^{0}_{H_2O} - \mu^{0}_{H} - \mu^{0}_{HCO_3})/RT\right] = \frac{a_H \cdot a_{HCO_3}}{\frac{\alpha_0}{\alpha_s} \cdot c_{CO_2} \cdot a_{H_2O}}$$

and that

$$K_1' = K_1/\gamma_{HCO_3} = \frac{a_H \cdot c_{HCO_3}}{\frac{\alpha_0}{\alpha_s} \cdot c_{CO_2} \cdot a_{H_2O}}. \qquad (31)$$

The quantity $(\alpha_0/\alpha_s)c_{CO_2}$ is thus to be identified with the a_{CO_2} of equation (22), and for this reason (α_0/α_s) is often designated the activity coefficient of the dissolved carbon dioxide. Since the term activity coefficient is usually reserved for application to systems which do not show ideal behaviour, it is more satisfactory in this instance, where the solution of carbon dioxide obeys Henry's Law, to regard it as a factor which accounts for the change in properties of the solvent with change of salinity. However, on both view-points the algebraic expression for the apparent constant is the same.

By suitably combining the expression for the carbonate alkalinity (equation (11)) and that for the solubility of carbon dioxide with those for the first and second apparent constants, it is possible to derive equations (32) to (41) below, from which the concentrations of bicarbonate ion, carbonate ion, dissolved carbon dioxide, total carbon dioxide and the partial pressure of carbon dioxide in the water can be calculated (Buch, 1951; Harvey, 1955).

(a) *Bicarbonate ion concentration* (c_{HCO_3}). Elimination of the carbonate ion concentration between the second apparent constant (equation (23)) and the carbonate alkalinity (equation (11)) enables the bicarbonate ion concentration to be calculated

$$c_{HCO_3} = \mathrm{CA} \cdot \left[\frac{1}{1 + \frac{2K_2'}{a_H}}\right] \qquad (32)$$

(b) *Carbonate ion concentration* (c_{CO_3}). Combination of equation (32) and equation (11) gives

$$c_{CO_3} = CA \, . \left[\frac{K_2'}{a_H + 2K_2'} \right] . \tag{33}$$

(c) *Concentration of dissolved carbon dioxide* (c_{CO_2}). Substitution of c_{HCO_3} from equation (32) in equation (31) gives on re-arrangement

$$c_{CO_2} = CA \, . \, \frac{a_H}{K_1' \frac{\alpha_0}{\alpha_s} \left[1 + \frac{2K_2'}{a_H} \right] a_{H_2O}} \tag{34}$$

or, in terms of Lyman's first apparent constant,

$$c_{CO_2} = CA \, . \, \frac{a_H}{K_{L1}' \left[1 + \frac{2K_2'}{a_H} \right]} . \tag{35}$$

(d) *Partial pressure of carbon dioxide in the water* (P_{CO_2}). The partial pressure of carbon dioxide in the water is related to c_{CO_2} by the equation

$$c_{CO_2} = P_{CO_2} \, . \, \alpha_s . \tag{36}$$

Combination of equations (34) and (36) gives

$$P_{CO_2} = CA \, . \, \frac{a_H}{K_1' \, . \, \alpha_0 \left[1 + \frac{2K_2'}{a_H} \right] a_{H_2O}} . \tag{37}$$

In terms of Lyman's first apparent constant, P_{CO_2} is given by

$$P_{CO_2} = CA \, . \, \frac{a_H}{K_{L1}' \, . \, \alpha_s \, . \left[1 + \frac{2K_2'}{a_H} \right]} . \tag{38}$$

(e) *Total carbon dioxide* ($\sum CO_2$). The total carbon dioxide content of the water is defined by

$$\sum CO_2 = c_{HCO_3} + c_{CO_3} + c_{CO_2} .$$

Substituting in this expression equations (32), (33), and (35) gives on re-arrangement

$$\sum CO_2 = CA \, . \left[\frac{1 + \frac{K_2'}{a_H} + \frac{\alpha_s \, . \, a_H}{K_1' \, . \, \alpha_0 \, . \, a_{H_2O}}}{1 + \frac{2K_2'}{a_H}} \right] . \tag{39}$$

If Lyman's first apparent constant is to be used, equation (39) becomes

$$\sum CO_2 = CA \,.\left[\frac{1 + \dfrac{K_2'}{a_H} + \dfrac{a_H}{K_{L1}'}}{1 + \dfrac{2K_2'}{a_H}}\right]. \tag{40}$$

Since c_{CO_2} is very much less than $(c_{HCO_3} + c_{CO_3})$, its contribution to total carbon dioxide is sometimes ignored, and equation (39) becomes

$$\sum CO_2 = CA \,.\left[\frac{a_H + K_2'}{a_H + 2K_2'}\right]. \tag{41}$$

The tabulated constants are based on determinations made at surface pressure, and if the intention is to apply the calculations to deep water samples, then corrections must be applied (Buch and Gripenberg, 1932).

When a sample is brought up from a great depth, the equilibrium position of the carbonate system shifts because of the pressure dependence of the equilibrium constants (see Owen and Brinkley, 1941; Ellis, 1959). This alters the pH and the distribution of carbon between the various species, but is without effect on the alkalinity, the carbonate alkalinity and the total carbon except in so far as the total volume of the sample alters. The pH, alkalinity and carbonate alkalinity of the sample are estimated at surface pressure and adjustments made to the *in situ* conditions as follows. If the ratio $\sum CO_2/CA$ is denoted by Q, and substituted in equation (39), then on re-arranging one obtains

$$a_H^2 - K_1' \frac{\alpha_0}{\alpha_s} a_{H_2O} \,.\, (Q-1)a_H - K_1' \,.\, K_2' \,.\, a_{H_2O} \,.\, \frac{\alpha_0}{\alpha_s}(2Q-1) = 0 \tag{42}$$

which allows the pH to be estimated for various depths if the dependence of K_1' and K_2' on pressure are known. Buch and Gripenberg (1932) used Brander's (1932) measurements on the carbon dioxide system to estimate the change of K_1' with pressure, and obtained

$$\Delta\, pK_1' = -0{\cdot}48 \times 10^{-4}\Delta z \tag{43}$$

where Δz is the depth in metres. The corresponding dependence of K_2' on pressure was not investigated by Brander, but it was assumed that as most weak acids behave similarly, it was possible to estimate the pressure dependence from measurements on acetic acid. The relationship is given by

$$\Delta pK_2' = -0{\cdot}18 \times 10^{-4}\Delta z. \tag{44}$$

Table VI shows the dependence of pH on depth. The change is a function of surface pH; at low surface pH values, ΔpH tends to $\Delta pK_1'$, and at high pH values it tends to $\Delta pK_2'$. The alkalinity and carbonate alkalinity at depth are obtained by multiplying the surface values by the ratio of the density of sea water at depth to that at the surface (Table VII), and from these and the corrected *in situ* estimates of pH, K_1' and K_2' the properties at depth can be derived by means of equations (32) to (41).

TABLE VI

Change of pH with depth (Buch and Gripenberg, 1932)

Surface pH		7·5	7·6	7·7	7·8	7·9
pH decrease (ΔpH) for 1000 m. increase in depth		0·035	0·031	0·028	0·025	0·023
Surface pH	8·0	8·1	8·2	8·3	8·4	8·5
pH decrease (ΔpH) for 1000 m. increase in depth	0·022	0·021	0·020	0·020	0·020	0·020

TABLE VII

Depth/surface density ratios (Buch and Gripenberg, 1932)

Depth (m.)	2200	4700	7300	9800
Depth/surface density ratio	1·010	1·020	1·030	1·040

Pytkowicz (1963) has pointed out that the Buch and Gripenberg estimates of the effect of pressure on the carbonate equilibria are based on constants obtained in pure water, and the results adapted for saline solutions. He also pointed out that the changes predicted by Buch and Gripenberg's treatment (decrease of pH with depth) involve the implicit assumption that there is no gain or loss of carbon dioxide, and no solution or precipitation of calcium carbonate when the water mass changes depth. Such gain or loss of material may be important if a mass of surface water which is in equilibrium with the atmosphere is submerged and brought into contact with bottom sediments containing calcium carbonate. Pytkowicz has estimated that if the effect of pressure on the dissociation constants of carbonic acid and on the apparent solubility product of calcium carbonate (Revelle and Fairbridge, 1957) is considered, the pH of a water mass of surface pH 7·89 will increase as calcium carbonate enters solution when the submerged water is brought into contact with bottom sediments. If this water mass is subsequently raised to the surface, it will be undersaturated with respect to atmospheric carbon dioxide and have a pH higher than that of the original mass before submergence. Uptake of carbon dioxide from the atmosphere will follow, but the final pH will still be higher than that of the original surface water mass. Further cycling may ensue until a limiting pH is reached. At this point, unless calcium carbonate precipitates at the surface, no further solution of bottom sediments will occur when the mass is submerged as the water will be saturated with respect to calcium carbonate at the depth of submergence. Pytkowicz considers that these processes may have some bearing on the control of the pH of surface water.

The above procedures assume that the carbonate alkalinity can be obtained by applying the borate correction to the observed alkalinity. This conclusion itself is subject to some uncertainty (Section II,B), and for waters which contain organic matter, humus or appreciable quantities of weak acids other than carbonic, methods of estimation of c_{HCO_3}, c_{CO_3} and c_{CO_2} via alkalinity may be in error, and Buch (1951) has recommended that under these circumstances the estimations should be made via the experimental determination of $\sum CO_2$. From the previous relationships it can be shown that

$$c_{CO_2} = \frac{\sum CO_2}{\dfrac{\alpha_0}{\alpha_s} \cdot a_{H_2O} \cdot \left[\dfrac{K_1'}{a_H} + \dfrac{K_1' \cdot K_2'}{a_H^2}\right] + 1} \cdot \tag{45}$$

In waters of low salinity this equation simplifies to

$$c_{CO_2} = \frac{\sum CO_2}{\dfrac{K_1'}{a_H} + \dfrac{K_1' \cdot K_2'}{a_H^2} + 1} \cdot \tag{46}$$

The concentration of bicarbonate ion is given by

$$c_{HCO_3} = c_{CO_2} \cdot \frac{K_1'}{a_H} \tag{47}$$

and the concentration of carbonate ion by

$$c_{CO_3} = c_{CO_2} \cdot \frac{K_1' \cdot K_2'}{a_H^2} \cdot \tag{48}$$

The partial pressure of CO_2 is estimated using equation (36), the value for c_{CO_2} being obtained by equation (45) or (46).

Saruhashi (1955) also favours estimation of the components of the carbonate system via the experimental determination of $\sum CO_2$, and has prepared tables for both fresh water and sea water from which the proportions of dissolved carbon dioxide, bicarbonate ion and carbonate ion can be read off when the temperature, salinity and total carbon dioxide content are known (Appendix Tables VIII and IX). Although direct comparison between predicted and field values for every component of the carbonate system is not possible because of the lack of suitable analytical methods, it is possible to examine P_{CO_2} values.

As the P_{CO_2} of sea water is very sensitive towards pH and temperature, its indirect estimation by equation (37) or (38) demands high precision in the measurement of these values. Some of the literature reports of poor agreement between experimental and predicted P_{CO_2} values may be caused by failure to appreciate this point, to the use of pH scales which are inconsistent with the particular tabulations of dissociation constants used in the calculations, or possibly to imperfectly understood inadequacies in the basis of the derivation of the equations and constants. Exact agreement would not be expected if, because of slow reactions in the carbonate system, disequilibrium existed within the sea. Although the rate of hydration of dissolved carbon dioxide (and the reverse process) is slow, its rate is high enough to make it appear unlikely that disequilibrium arising from this cause is significant. A further possible cause of divergence between the observed and predicted values is the formation of complexes between the dissolved carbon dioxide and organic matter.

The differences between observed and predicted P_{CO_2} values is often quite considerable. Hanya and Ishiwatari (1961) found that during their laboratory studies of gas–liquid carbon dioxide exchange it was necessary to adjust the observed pH values in order to obtain agreement. Hood and Ibert (1960) detected serious discrepancies between observed and predicted values. They measured the P_{CO_2} of sea water by determining the equilibrium partial pressure of carbon dioxide in the gas phase above the water in a closed vessel, and found this to be the same as that reached under a canopy on the sea surface. The values found from Buch's equations differed considerably from the experimental estimates. In Batabano Bay, where carbonate precipitation was occurring, the measured values were almost 80% higher than those predicted. Figure 5 (Hood, 1963) shows the observed and calculated P_{CO_2} values and the measured p_{CO_2} values for a Batabano Bay station. Similar observations have been reported for laboratory studies of algal cultures by Supernaw (see Hood, 1963). In other areas (e.g. the Atlantic 31°29·5′N., 66°52·5′W.) the measured values were lower than the predictions, and the differences, although considerable, were less than those shown in Batabano Bay, and appeared to decrease with depth. The reason for these discrepancies is not clear, but it is evident that further investigations of the carbonate equilibria are needed. However, it is evident that when precise values of P_{CO_2} are required, it is desirable to make direct measurements rather than to place reliance on calculated values.

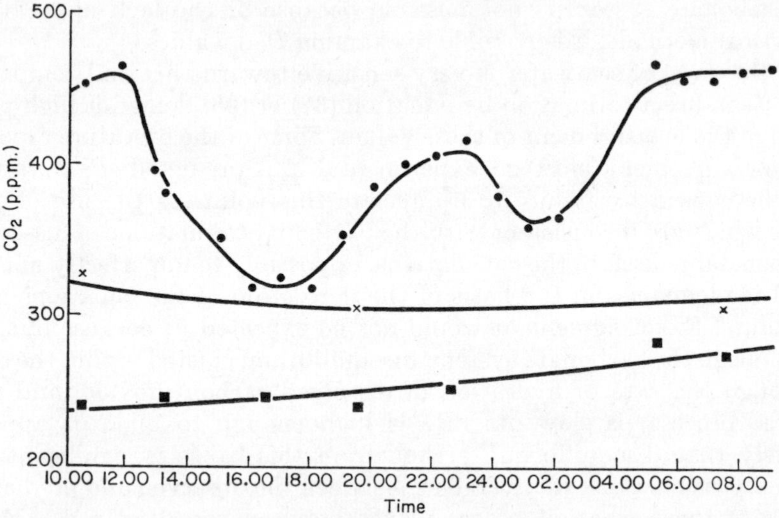

FIG. 5. Diurnal cycle of carbon dioxide partial pressures in Batabano Bay (Hood, 1963). —•—•—, measured carbon dioxide partial pressure of the water (P_{CO_2}); —×—×—, measured carbon dioxide in the air (p_{CO_2}); —■—■—, calculated carbon dioxide partial pressure of the water (Buch *et al.*, 1932).

It should be noted that, in general, the P_{CO_2} of sea water differs from the partial pressure of carbon dioxide in the atmosphere above the sea (p_{CO_2}) (Buch, 1942; Cooper, 1932, 1956; Takahashi, 1961; Ibert and Hood, 1963; Waterman, 1964); only in restricted areas is equilibrium between the atmosphere and the sea attained, and then only in a transient sense. Cooper (1956) has emphasized that with a suitable pattern of water movement (e.g. sinking in Polar regions and upwelling in the tropics), there must be a net transfer of carbon dioxide via solution in the cooler seas of high latitudes and evolution in the warmer tropical seas.

Takahashi's detailed investigations of the P_{CO_2} of Atlantic water have confirmed that there is a marked dependence of P_{CO_2} on latitude; at higher latitudes the relative carbon dioxide pressures of the water and the atmosphere are such that a net transport from the atmosphere to the sea must occur. In warmer regions (the equator to *c.* 40°S.), the net transport is in the opposite direction. Takahashi's results (Fig. 6) show the P_{CO_2} maximum to occur just south of the equator; this is probably because the observations were made during the summer months of the southern hemisphere, and the maximum would be expected to move northwards during the summer months of the northern hemisphere.

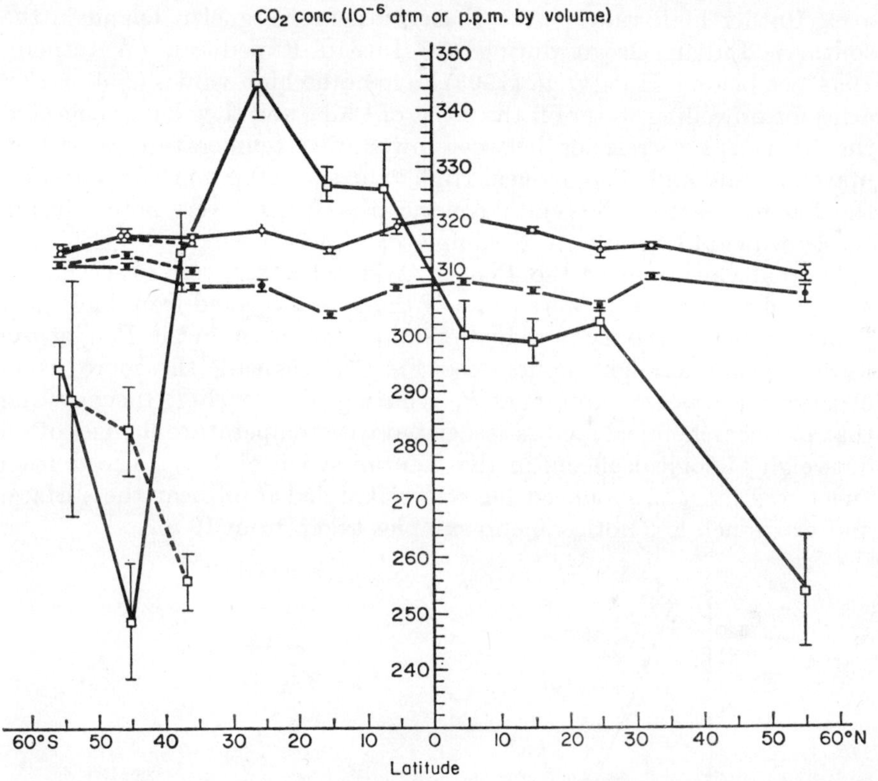

FIG. 6. Partial pressures of carbon dioxide in the surface water and the carbon dioxide concentrations in the air over the North and South Atlantic Oceans (Takahashi, 1961). ———, measurements in the Western Hemisphere; - - - - - -, measurements in the Eastern Hemisphere; —o—o—, carbon dioxide concentration in the atmosphere (p.p.m. by volume); —•—•—, partial pressure of carbon dioxide in atmosphere saturated by water at surface temperature (10^{-6} atm); —□—□—, partial pressure of carbon dioxide in sea water (10^{-6} atm).

In the more southerly latitudes examined by Takahashi, rather wide fluctuations in the measured P_{CO_2} values were observed. The precise reason for these is not clear, but several factors may contribute. Temperature changes of the water are probably important, although the range of observed P_{CO_2} values is larger than can be accounted for by temperature changes alone. Mixing of different water masses, change of pH, biological effects and the precipitation or solution of calcium carbonate will all influence P_{CO_2}. Other workers have noted high values for the P_{CO_2} of Antarctic water. Miyake and Matsuo (1963) quote measurements by Torii which indicate that the partial pressure of carbon dioxide in the water in this region may be as high as 6×10^{-4} atm. Rather high values were observed near Kerguelen Island in the southern Indian Ocean during the Lusiad Expedition (Waterman, 1964, see below). Hood *et al.* (1963) have noted high values (*c.* 6×10^{-4} atm) for upwelling water off the coast of Chile, and they have suggested that there is a correlation between low surface temperature caused by upwelling and high P_{CO_2} values. High values (*c.*440 p.p.m.) for water off the Pacific coast of the South American continent were noted during the Downwind Expedition (Keeling *et al.*, 1965).

Diurnal variations of the P_{CO_2} of Atlantic water were observed by Takahashi; only rarely were these of the type expected from biological photosynthetic activity (a reduction and a minimum in the P_{CO_2} of the water during daylight hours—see Fig. 7). Instead, the more usual observation was a maximum of P_{CO_2} during the day (Fig. 8) suggesting that physico-chemical factors associated with temperature changes often outweigh biological effects in the determination of P_{CO_2}. The diurnal effect was most pronounced for water sampled from near the surface, and was much less noticeable in samples taken from 10 m.

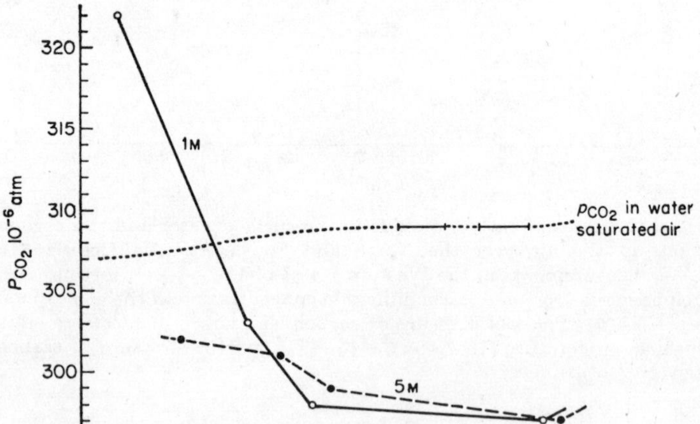

FIG. 7. Diurnal variation of the partial pressure of carbon dioxide in the sea 15°30′N., 40°33′W. (November 26, 1957) (Takahashi, 1961).

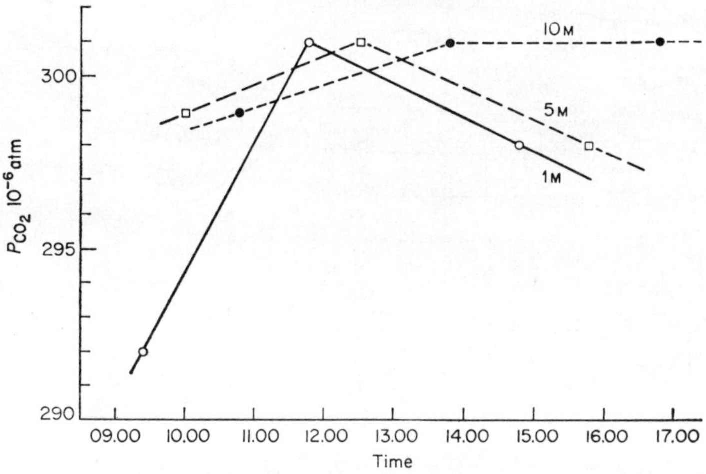

FIG. 8. Diurnal variations of the partial pressure of carbon dioxide in the sea 14°26′N., 38°52′W. (November 27, 1957) (Takahashi, 1961). The partial pressure of carbon dioxide in air above water = 309 × 10^{-6} atm (in water-saturated air).

A preliminary report of the Lusiad Expedition (1962–3) (Waterman, 1964) has given details of the measurement of atmospheric p_{CO_2}, the partial pressure of carbon dioxide of the sea water and sea-water temperature for tracks in the Pacific, Indian and Atlantic Oceans. The tracks and summaries of the relevant observations are indicated in Figs. 9 to 11. These data emphasize the general inequality of the partial pressures of the carbon dioxide in the atmosphere and in the water, and show the wide range of values for the sea as compared with the atmosphere. The Pacific traverse shows the partial pressure of carbon dioxide in the water to be generally less than that in the atmosphere, and to reach a minimum of about 252 p.p.m. at 175°E. At this point, the difference in the partial pressures of carbon dioxide between the atmosphere and the sea is about 68 p.p.m. The change in P_{CO_2} along the traverse is too large to be accounted for solely by the change in temperature.

The relationship between the partial pressure of carbon dioxide in the atmosphere and that in the sea for several areas in the Gulf of Mexico has been investigated by Ibert and Hood (1963). Again, the two partial pressures were generally unequal, and in a shallow, highly productive carbonate basin, a pronounced photosynthetic diurnal cycle with a maximum at 24.00 hours, a minimum at 14.00 hours, and a range of more than 70 p.p.m. was observed. Relatively large fluctuations in P_{CO_2} were reported for samples taken in the Mississippi delta where fresh water

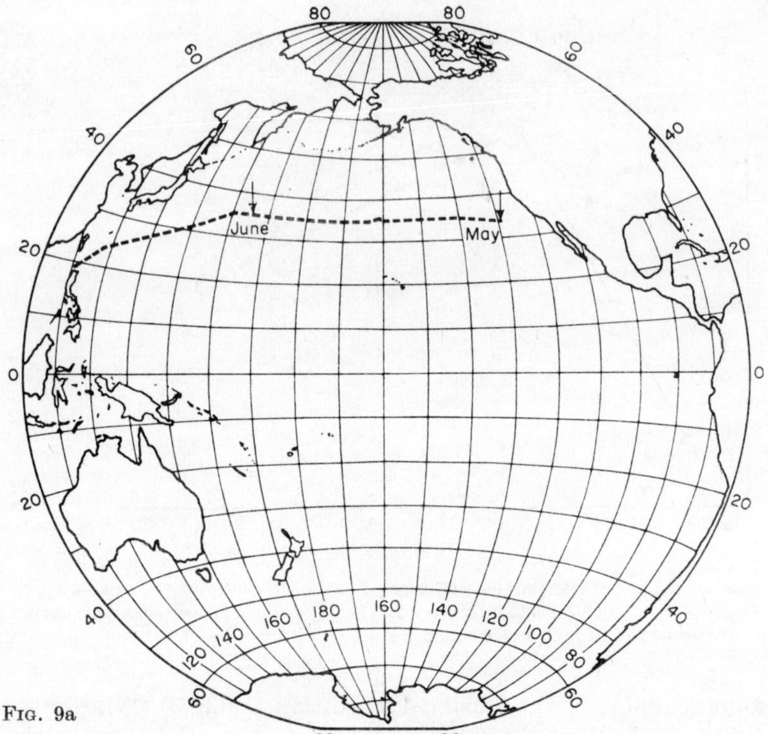

FIG. 9a

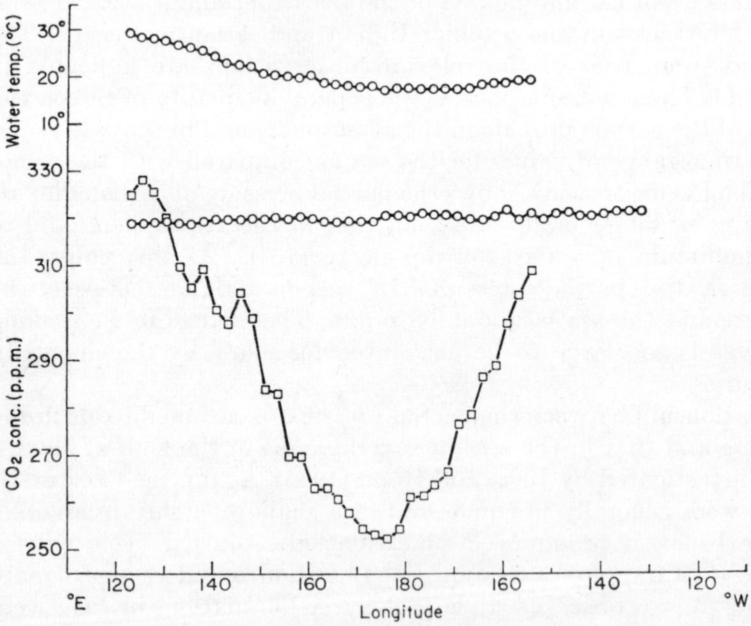

FIG. 9b. See legend on facing page

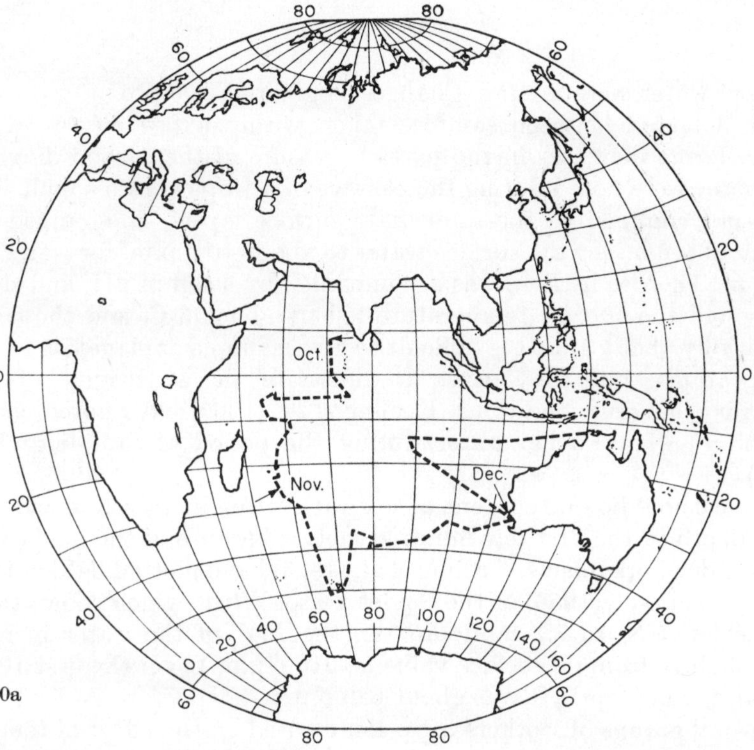

FIG. 10a

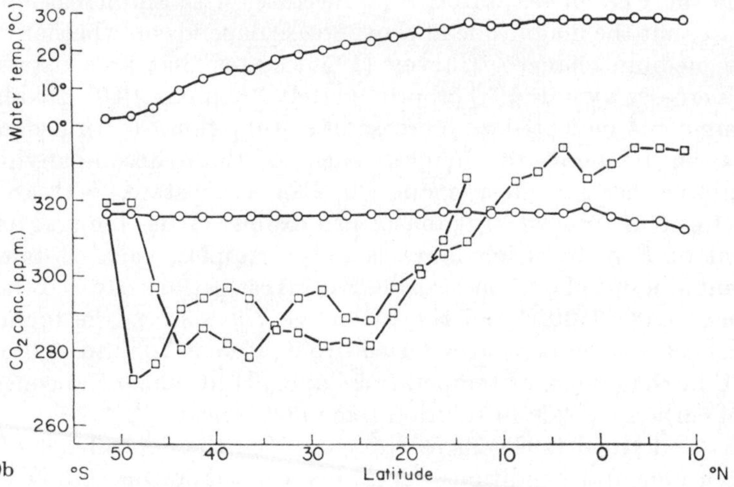

FIG. 10b

FIG. 9. (a) Track of RV *Argo*, Lusiad Expedition, May 18 to June 9, 1962 (Waterman, 1964). (b) Average partial pressure of carbon dioxide in sea water (□) and in the atmosphere near the sea surface (○) *vs* longitude along the track shown in (a) (Waterman, 1964).

FIG. 10. (a) Track of RV *Argo*, Lusiad Expedition, October 6 to December 21, 1962 (Waterman, 1964). (b) Average partial pressure of carbon dioxide in the sea (□) and in the atmosphere near the sea surface (○) *vs* latitude along the track shown in (a) (Waterman, 1964).

and sea water were mixing. Changes as great as 200 p.p.m. were sometimes detected between samples taken within a few metres of each other. Large increases in the partial pressure of the carbon dioxide in the water were noted during the passage of a tropical rain squall. These were not confined to the immediate surface layers, and ranged from about 70 p.p.m. for the surface water to about 40 p.p.m. for water from *c*. 70 m. The rise in P_{CO_2} was accompanied by a fall in pH, and during the squall the surface temperature fell by about 3°C, and the surface salinity by about 3‰. It is difficult to give a simple explanation of these observations, but they serve to illustrate the sensitivity of P_{CO_2} towards changing conditions. Keeling *et al.* (1965) have noted changes in the P_{CO_2} of Pacific water during the period of sampling (10–20 min.).

The carbon dioxide content of sea water often increases considerably with depth in the first few hundred metres because of the oxidation of sinking dead organisms. Keeling *et al.* (1965) have plotted data obtained during Cruise 26 (1958) of the Soviet vessel *Vityaz* which shows that in the region 7°N. to 7°S. and 175°W., the P_{CO_2} of the water increased only slightly from its surface value (*c*. 310 p.p.m.) over the first 100 m., but at 450 m. depth, it was about 620 p.p.m.

Several groups of workers have commented on the effect of temperature on the P_{CO_2} of sea water. P_{CO_2} increases with an increase of temperature, but the magnitude of the increase depends on whether the pH of the medium changes. Harvey (1955) states that at constant pH, P_{CO_2} increases by *c*. 1%/°C (approximately 3·2 p.p.m./°C); Takahashi's measurements indicated an increase of *c*. 4·0 p.p.m./°C. In general, for short-term temperature changes, such as those accompanying the diurnal solar heating effect, changes in P_{CO_2} at constant $\sum CO_2$ are more likely to be of interest. Takahashi has examined the temperature coefficient of P_{CO_2} by using four sea-water samples, each containing a different amount of carbon dioxide. His investigations covered the temperature range 3–30°C and the pH range 7·9–8·3, and he found that "the effect of temperature varied from $12 \cdot 5 \times 10^{-6}$ to $8 \cdot 0 \times 10^{-6}$ atm/°C in that range of temperatures and pH at which the concentration of carbon dioxide in solution remained constant".

Ibert and Hood (1963) have also examined the dependence of P_{CO_2} on environmental conditions. They made a laboratory study of sea-water samples adjusted at 20°C to pH values of 8·0, 8·2 and 8·4 (carbonate alkalinities of 1·4, 2·4 and 3·4 meq./l. respectively) over the temperature range of 5–20°C, and found $d(P_{CO_2})/dt$ values of 7·0, 4·4 and 0·3 p.p.m./°C respectively. Field data from the Caribbean area were in reasonable agreement with the laboratory measurements.

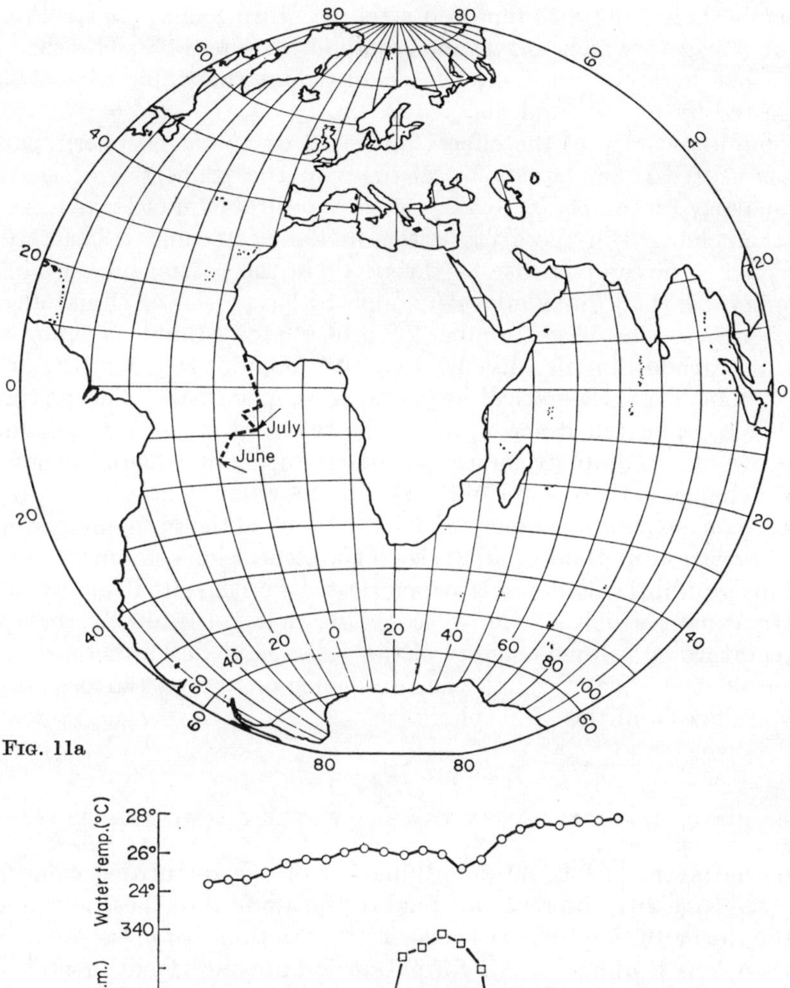

FIG. 11a

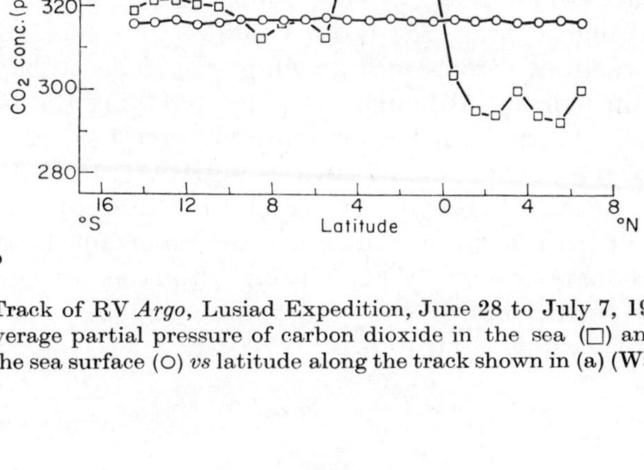

FIG. 11b

FIG. 11. (a) Track of RV *Argo*, Lusiad Expedition, June 28 to July 7, 1963 (Waterman, 1964). (b) Average partial pressure of carbon dioxide in the sea (□) and in the atmosphere near the sea surface (○) *vs* latitude along the track shown in (a) (Waterman, 1964).

Kanwisher (1960) has reported a temperature coefficient for P_{CO_2} of about 4·5%/°C, which corresponds to about 14 p.p.m./°C. This seems to be rather high for most sea waters, as does the value of *c.* 7%/°C suggested by Revelle and Suess (1957).

Some knowledge of the effects produced on the atmospheric partial pressure of carbon dioxide by changes in the properties of the sea, particularly by temperature variations, is desirable in order to examine whether changes in these properties in the past could reasonably be expected to have given rise to variations in the partial pressure of atmospheric carbon dioxide large enough to have been of climatological consequence. The large number of interrelated equilibria embracing gaseous carbon dioxide, dissolved carbon dioxide, H_2CO_3, HCO_3^-, CO_3^{2-} and H^+ makes a theoretical assessment far from easy. The picture is further complicated if account is to be taken of changes in alkalinity which would accompany solution or precipitation of calcium carbonate, or of a change in the volume of the sea (by for example, transfer of water to land-locked ice). Eriksson (1963) has examined the problem by applying a variation method of analysis to the expressions summarizing the various equilibria. Eriksson's paper should be consulted for details of the treatment which led him to conclude that it is unlikely that past temperature or volume changes of the ocean have had any appreciable climatological effect through their influence on the partial pressure of carbon dioxide in the atmosphere.

D. Ion Pair Formation in the Sea-Water Carbonate System

In the methods for the determination of the quantities c_{HCO_3} and c_{CO_3} (see page 264), no regard is paid to the amount of these ions bound in the form of ion pair associates. Interaction between oppositely charged ions leading to pair formation is an established phenomenon (Robinson and Stokes, 1959; Davies, 1962), and is favoured by small ionic size and high charge. Sea water contains ions such as Ca^{2+} and Mg^{2+} which are known from work on simpler systems to be capable of participating in pairing. Although ion pairs involving the bicarbonate and carbonate ions have not been extensively investigated, Greenwald (1941) has given evidence for the entities $MgHCO_3^+$, $CaHCO_3^+$, $MgCO_3^0$ and $CaCO_3^0$. Recently, Garrels *et al.* (1961) have investigated the possibility that pair formation may be an important feature of the sea-water carbonate system by attempting estimates of the individual ionic activity coefficients of the bicarbonate and carbonate ions in the presence of various electrolytes.

From the first thermodynamic dissociation constant of carbon dioxide in water*

$$K_1 = \frac{a_H \cdot a_{HCO_3}}{a_{H_2O} \cdot a_{CO_2}}$$

$$= \frac{a_H \cdot c_{HCO_3} \cdot \gamma_{HCO_3}}{a_{H_2O} \cdot a_{CO_2}}$$

an expression for the bicarbonate activity coefficient, γ_{HCO_3} is obtained.

$$\gamma_{HCO_3} = \frac{K_1 \cdot a_{CO_2} \cdot a_{H_2O}}{a_H \cdot c_{HCO_3}} \, . \qquad (49)$$

Numerical values can readily be assigned to a_{CO_2} and a_{H_2O}, and c_{HCO_3} can be controlled by addition of a suitable amount of bicarbonate. Thus, provided it is possible to use pH as defined by equation (2) as a measure of hydrogen ion activity, this expression allows the bicarbonate ion activity coefficient to be estimated from laboratory experiments. Similarly, from the expression for the second thermodynamic dissociation constant, the ratio

$$\frac{\gamma_{CO_3}}{\gamma_{HCO_3}} = \frac{K_2 \cdot c_{HCO_3}}{a_H \cdot c_{CO_3}} \qquad (50)$$

is obtained and allows estimation of γ_{CO_3} by insertion of the activity coefficient of the bicarbonate ion estimated as above.

Garrels *et al.* (1961) prepared solutions of known bicarbonate and carbonate content in which a_{CO_2} was controlled, and measured the pH values after the addition of various amounts of electrolyte. The results were expressed as plots of carbonate and bicarbonate ion activity coefficients as functions of ionic strength (Figs. 12 and 13).

Electrolytes examined included NaCl, $MgCl_2$ and $MgCl_2$–NaCl in 1 : 10 molar ratio (that of sea water). Figures 12 and 13 also include data for $NaHCO_3$, $KHCO_3$, Na_2CO_3 and K_2CO_3 (from Walker *et al.*, 1927), for sea water (Sverdrup *et al.*, 1942) and for a Debye-Hückel calculation. The agreement between the activity coefficients of the bicarbonate and the carbonate ions on the basis of the Debye-Hückel formula and those observed experimentally in $KHCO_3$ and K_2CO_3 solutions suggests that complex or pair formation does not contribute to the observed lowering of activity coefficients in these solutions. The greater depression of

* Garrels *et al.* (1961) used the constant for the equilibrium CO_2(gas) + H_2O(liquid) = H^+(aq) + HCO_3^-(aq), and took the activity of gaseous carbon dioxide as the pressure of the gas in atmospheres. Also, they expressed concentration in terms of the molarity scale (g. molecules/kg. solvent).

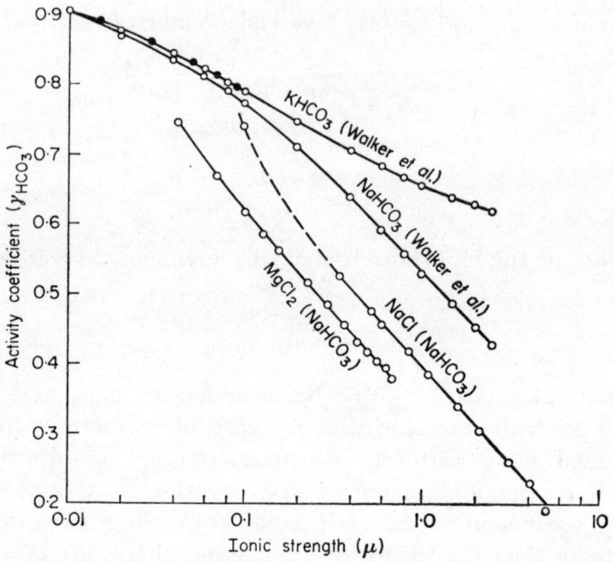

FIG. 12. The activity coefficient of the bicarbonate ion *vs* ionic strength for various solutions (Garrels *et al.*, 1961). —•—•—, according to the Debye-Hückel equation; —o—o—, experimental values.

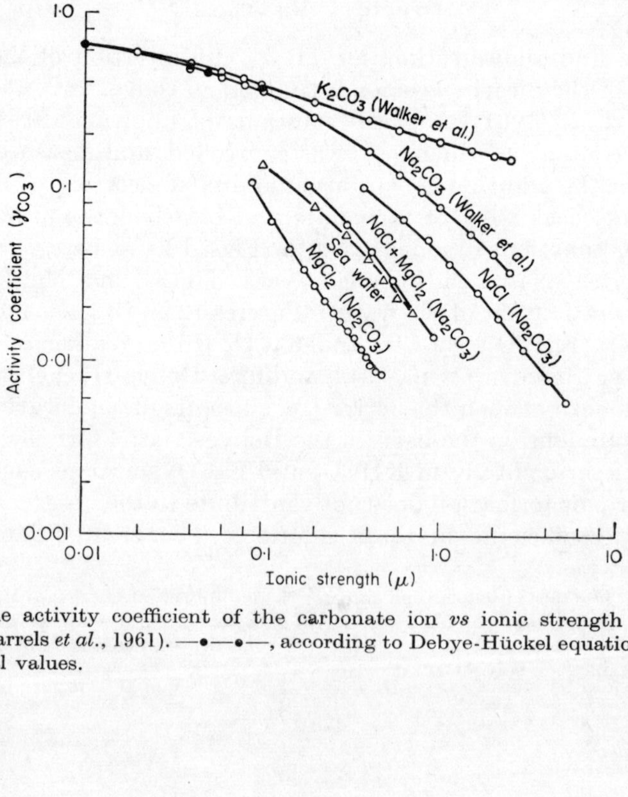

FIG. 13. The activity coefficient of the carbonate ion *vs* ionic strength for various solutions (Garrels *et al.*, 1961). —•—•—, according to Debye-Hückel equation. —o—o—, experimental values.

γ_{HCO_3} and γ_{CO_3} in the presence of sodium and magnesium ions is indicative of ion association in these solutions. Garrels *et al.* (1961) explain their results in terms of the entities $NaCO_3^-$ and $MgCO_3^0$ for which they estimated dissociation constants at 25°C of $5{\cdot}4 \times 10^{-2}$ and $4{\cdot}0 \times 10^{-4}$ respectively. Rather weaker associations were also thought to occur involving the bicarbonate ion, but the complexes were not characterized. The values of the dissociation constants for the sodium and magnesium carbonate complexes indicate that in sea water only about 9% of the total carbonate exists as unassociated ion.

Other workers have previously commented on the low values obtained for the activity coefficients of the bicarbonate and carbonate ions in sea water (Buch *et al.*, 1932; Moberg *et al.*, 1934), and the possibility of complex formation has been suggested (Greenberg and Moberg, 1932), but prior to the work of Garrels *et al.* (1961) no detailed examination had been attempted.

Complex or ion pair formation in sea water is, of course, not confined to interactions involving carbonate and bicarbonate, and Garrels and Thompson (1962) have extended the above considerations in an attempt to set up a model for sea water based on the assumption that only the ions Ca^{2+}, Mg^{2+}, Na^+, K^+, SO_4^{2-}, CO_3^{2-} and HCO_3^- participate in complex formation. Their approach, which rests heavily on the concept of individual ionic activities, led to the values listed in Table VIII. It should be noted that these estimates apply to a total pressure of one atmosphere.

TABLE VIII

*Distribution of major dissolved species in representative sea water of chlorinity 19‰, pH 8·1, at 25°C and one atmosphere total pressure**

Ion	*Molality (total)*	*% Free ion*	*% Me-SO₄ pair*	*%Me-HCO₃ pair*	*%Me-CO₃ pair*
Ca^{2+}	0·0104	91	8	1	0·2
Mg^{2+}	0·0540	87	11	1	0·3
Na^+	0·4752	99	1·2	0·01	—
K^+	0·0100	99	1	—	—

Ion	*Molality (total)*	*% Free ion*	*% Ca-anion pair*	*% Mg-anion pair*	*% Na-anion pair*	*% K-anion pair*
SO_4^{2-}	0·0284	54	3	21·5	21	0·5
HCO_3^-	0·00238	69	4	19	8	—
CO_3^{2-}	0·000269	9	7	67	17	—

*Garrels and Thompson, 1962.

It is seen that of the cations, Mg^{2+} is most strongly involved in complex formation, whereas very little of the potassium participates. Of the anions, most of the carbonate, about one-third of the bicarbonate and one-half of the sulphate form ion pairs. The usage made of the concept of individual ionic activities, the mean salt method and the Debye-Hückel extrapolations are features of the method which will not appeal to thermodynamic purists, but it must be remembered that the numerous ionic species present in sea water and the high ionic strength demand that some compromise be made if any progress is to be made in understanding the detailed ionic chemistry. The investigator must use whatever tools are to hand, and Garrels *et al.* (1961) have succeeded in setting up a model which is reasonable in the relative proportions of the different ions allocated to the pairs, and which should serve as a useful working model for future investigations.

E. Relationship between Changes in Atmospheric Carbon Dioxide Pressure and the Total Carbon Content of the Sea

The increasing use of fossil fuels over the last hundred years has meant that relatively large amounts of radio-carbon-free carbon dioxide are being injected into the atmosphere. The amounts produced in this way are considerable in comparison with the rate at which juvenile carbon dioxide is being produced, and an important question is whether the sea is capable of assimilating an appreciable proportion of the atmospheric increase. Even if it is, there remains the possibility of a time lag between the generation of an atmospheric increase and its equilibrium uptake by the oceans. On the assumption of a two-layer ocean system (an upper layer of 75–100 m. depth and a deep layer comprising the rest of the ocean), a time lag would be expected if mixing between the upper and the lower layer was slow.

The equilibrium effect following the imposition on the atmospheric carbon dioxide (partial pressure $P_{CO_2} = p_{CO_2}$) of a small increase δP_{CO_2} can be estimated (Bolin and Eriksson, 1959) by using the equations given earlier (Section II, C(3)). The concentration of dissolved carbon dioxide will increase in order to re-establish equilibrium, but since dissolved carbon dioxide is in equilibrium with bicarbonate, carbonate and hydrogen ions, the solution process will be accompanied by a decrease in pH. That is, the aqueous system will react so as to oppose the change, and although the total carbon in the sea will increase, the fractional increase $\delta \sum CO_2 / \sum CO_2$ will be less than the fractional increase in the partial pressure, $\delta P_{CO_2}/P_{CO_2}$.

The equations given in an earlier section can be adapted to give the following approximate forms for $\sum CO_2$ and the carbonate alkalinity (CA)

$$\sum CO_2 = \left[1 + \frac{K_1'}{c_H} + \frac{K_1' \cdot K_2'}{c_H^2}\right] c_{CO_2} \tag{51}$$

$$CA = \left[\frac{K_1'}{c_H} + \frac{2K_1' \cdot K_2'}{c_H^2}\right] c_{CO_2} \tag{52}$$

and at equilibrium the partial pressure of carbon dioxide is given by

$$P_{CO_2} = c_{CO_2}/\alpha \tag{53}$$

where α is the solubility constant.

A small change in $\sum CO_2$ results from the change δP_{CO_2}, and is given by

$$\delta \sum CO_2 = \left(\frac{\partial \sum CO_2}{\partial c_H}\right)_{c_{CO_2}} \delta c_H + \left(\frac{\partial \sum CO_2}{\partial c_{CO_2}}\right)_{c_H} \delta c_{CO_2} \tag{54}$$

and the fractional change is

$$\frac{\delta \sum CO_2}{\sum CO_2} = \frac{\delta c_{CO_2}}{c_{CO_2}} + \frac{\frac{\partial}{\partial c_H}\left(1 + \frac{K_1'}{c_H} + \frac{K_1' \cdot K_2'}{c_H^2}\right)}{1 + \frac{K_1'}{c_H} + \frac{K_1' \cdot K_2'}{c_H^2}} \delta c_H . \tag{55}$$

Similar expressions for the fractional changes of CA and P_{CO_2} are

$$\frac{\delta CA}{CA} = \frac{\delta c_{CO_2}}{c_{CO_2}} + \frac{\frac{\partial}{\partial c_H}\left(\frac{K_1'}{c_H} + \frac{2K_1' \cdot K_2'}{c_H^2}\right)}{\frac{K_1'}{c_H} + \frac{2K_1' \cdot K_2'}{c_H^2}} \delta c_H \tag{56}$$

and

$$\frac{\delta P_{CO_2}}{P_{CO_2}} = \frac{\delta c_{CO_2}}{c_{CO_2}} . \tag{57}$$

CA, the carbonate alkalinity, must remain constant over relatively short periods of time (i.e. $\delta CA = 0$), and using this fact in combination with equations (55), (56) and (57) and the assumption that $K_1'/c_H = 143$ and $K_1'K_2'/c_H^2 = 18$, the relationship

$$\frac{\delta P_{CO_2}}{P_{CO_2}} = 12{\cdot}5 \, \frac{\delta \sum CO_2}{\sum CO_2} \tag{58}$$

is obtained (Bolin and Eriksson, 1959).

Equation (58), which expresses quantitatively the concept outlined above, shows that a relatively large fractional increase in P_{CO_2} is balanced by a much smaller fractional increase in the total carbon of the water. If it is supposed that a 10% increase in P_{CO_2} occurs, and that equilibrium is established with the upper layer only, then because this layer contains about as much carbon as does the atmosphere, the atmospheric increase would be balanced when about a tenth of it has passed into the sea. If equilibrium were to be established between the atmosphere and the entire sea (which contains about 50 times as much carbon as the atmosphere), then approximately four-fifths of the increase would pass into the sea and the final increase in the partial pressure of carbon dioxide in the atmosphere would be about 2%. Thus, although the potential capacity of the sea to buffer an increase of atmospheric carbon dioxide is appreciable, if the rate of establishment of equilibrium between the surface layer and the deep layer is slow with a time scale in the order of centuries, then even quite rapid equilibration between the atmosphere and the upper layer will not appreciably reduce the atmospheric increase.

The relationship between P_{CO_2} and $\sum CO_2$ has been demonstrated experimentally on a laboratory scale by Kanwisher (1960). He circulated sea water and air through an equilibrator in a counter-current closed system, and after drying the emergent air, monitored its carbon dioxide content with an infra-red gas analyser. He was able to show that a change in P_{CO_2} of about 10% was produced by a change in $\sum CO_2$ of about 0·6%, a result in reasonable agreement with the predictions of Bolin and Eriksson. The change in partial pressure of carbon dioxide of sea water on the addition or removal of carbon dioxide is shown in Fig. 14. Also included in Fig. 14 is a plot of the change in partial pressure of carbon dioxide over distilled water on the addition of carbon dioxide (Kanwisher, 1963a,b). The difference in slope of the two plots illustrates the importance of alkalinity in controlling the amount of carbon dioxide gained or lost by natural waters in response to changes in the applied pressure of carbon dioxide.

Over very long periods of time it is unlikely that the assumption that $\delta CA = 0$ is justified since calcium carbonate solution or deposition may accompany changes in the total oceanic carbon. Bolin and Eriksson have extended their treatment to consider the effect that this would have on the equilibrium attained after very long periods, and concluded that a 1% increase in $\sum CO_2$ obtained from the atmosphere would be balanced by a 4·2% increase in the atmospheric pressure of carbon dioxide. This means that of any instantaneous increase of atmospheric carbon dioxide, ultimately 11/12 would pass into the sea.

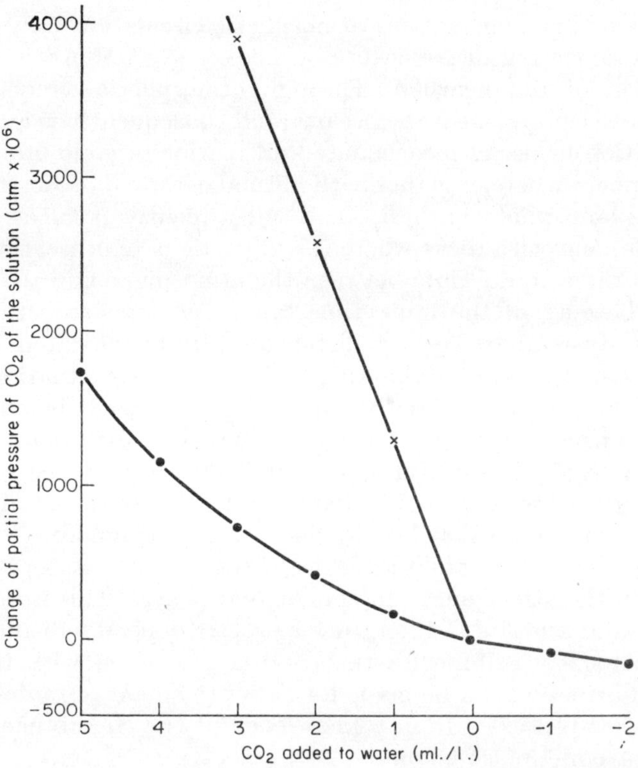

FIG. 14. Change of partial pressure of carbon dioxide as a function of change of total carbon dioxide for sea water and fresh water (Kanwisher, 1963a,b). —×—×—, distilled water; —•—•—, sea water, $S = 31‰$.

III. THE ISOTOPES OF CARBON

Carbon occurs in several different isotopic forms, but only the naturally occurring isotopes of masses 12, 13 and 14 are relevant to the present discussion. Of these, those of masses 12 and 13 are the most abundant, and in atmospheric carbon dioxide they together account for all but about 10^{-10}% of the total carbon. Both are stable isotopes. Carbon-14 or radio-carbon as it is often termed, never occurs naturally in anything other than trace amounts. It is radioactive having a half-life of about 5600 years, and decays by beta emission. It is produced in the upper atmosphere by the interaction of cosmic ray neutrons with nitrogen; the carbon-14 atoms so produced react readily with oxygen or ozone to give carbon dioxide (possibly via CO—see e.g. MacKay *et al.*, 1963).

The short time constant for atmospheric circulation compared with that for radioactive decay ensures a more or less uniform atmospheric distribution of radio-carbon. Through atmospheric circulation it is brought into contact with the oceans (and consequently is available for incorporation in ocean bicarbonate and marine organic life), with the land biosphere where, together with normal carbon dioxide, it is assimilated in plant photosynthesis (and subsequently is incorporated in animal life) and with rocks where it will participate in weathering.

True isotopic equilibrium between the atmosphere and the sea is not attained because of the finite time taken for transfer between these reservoirs. Instead (if certain disturbing effects indicated below are ignored), something approximating to a steady state situation exists in which the rate of entry of active carbon into the sea is balanced by its rate of loss from the sea by transfer in the reverse direction, plus decay in the sea. Even if true isotopic equilibrium were to be established between atmospheric carbon dioxide and the carbon of an aqueous solution, in the sense that loss by decay could be ignored, it would be found that the relative abundance of the three carbon isotopes would be different in the atmospheric and condensed phases. This is because the equilibria and kinetics of the transfer process depend on isotopic mass, and consequently isotope fractionation is associated with the carbon transfer. Correction can be made for the enrichment of radio-carbon in such transfer processes from a knowledge of the simultaneous enrichment of carbon-13 (see below).

The use of radio-carbon methods for archaeological dating is well established (Libby, 1955), and depends on the fact that following death, the carbon of a plant is isolated from the general circulation system, and its specific activity falls in accordance with radioactive decay laws. Thus, from a knowledge of the specific activity of a sample and the half-life of radio-carbon, the time which has elapsed since death can be estimated. The method assumes that over the period of applicability, atmospheric activity has remained constant. This implies that the atmospheric cosmic ray flux and nitrogen pressure and the concentration of ^{14}C-free carbon dioxide have remained constant, and although it has been suggested (Crowe, 1958) that in the past relatively large variations in atmospheric activity may have occurred, it seems more likely that at least until relatively recent times only minor fluctuations have occurred (Munnich *et al.*, 1958; Broecker *et al.*, 1959).

Two factors arising from human activity have more recently contributed to the disturbance of the constancy of the specific activity of atmospheric carbon dioxide. One is the progressive dilution of the atmospheric activity by "dead" carbon dioxide derived from the burning of fossil fuels which, by virtue of their age, contain negligible amounts of carbon-14. Their use has increased almost exponentially since the second half of the 19th century, and the resultant atmospheric carbon-14 dilution effect has been observed in tree ring samples taken from wood grown in relatively recent times; such samples show a lower activity than those grown in the last century when allowance is made for the difference in real age. The other disturbing factor is the production of carbon-14 by nuclear explosions which has caused the atmospheric radio-carbon content to rise sharply since about 1954; this has completely outweighed the dilution effect. Both these effects have been used to estimate the time scales for the transfer of carbon between the natural reservoirs.

One further effect which may in the past have imposed fluctuations on an otherwise constant atmospheric radio-carbon content is the possible catastrophic overturning of the deep sea as a result of prolonged periods of cold (Worthington, 1954; Broecker *et al.*, 1958; de Vries, 1958). Deep water brought to the surface in this way would be out of isotopic equilibrium with the atmosphere, and the subsequent readjustment would cause a lowering of atmospheric specific activity.

Carbon-14 measurements are applicable to problems involving oceanic carbonate and carbon interchange between the natural reservoirs. Relative to surface water, the deep ocean may show an appreciable reduction in activity, but measurements are frequently open to ambiguous interpretation. As Cooper (1956) has pointed out, the complicated pattern of oceanic circulation and the possible mixing of waters of different origins and different specific activities makes it difficult to equate "radio-carbon age" with "real age" of a water mass.

When radio-carbon data are used in the examination of the distribution and transfer of carbon between the natural reservoirs, attention must be given to the effects of isotope fractionation. Correction can be made for carbon-14 fractionation from a knowledge of the fractionation factor for carbon-13. This last quantity is defined by

$$\alpha_f = R_t/R_s$$

where R_t and R_s are the $^{13}C/^{12}C$ ratios in the two compounds t and s respectively (Craig, 1954). In order to provide a working scale of reference, R_s is a standard of known isotopic composition (Craig, 1957), and the composition of the test sample is expressed as the enrichment per mille ($\delta^{13}C$) relative to the standard, i.e.

$$\delta^{13}C = \left[\frac{R_t - R_s}{R_s}\right] \times 1000.$$

In an isotope fractionation process, the carbon-14 fractionation factor should be the square of the corresponding carbon-13 fractionation factor. Both factors are close to unity, and consequently the permillage enrichment of carbon-14 should be almost twice that of the carbon-13 for the same source material. A $\delta^{14}C$ value can be defined in terms of the radio-carbon activities of the sample and a standard (e.g. the NBS oxalic acid standard), and combination of the $\delta^{13}C$ and $\delta^{14}C$ values by means of equation (3) of Chapter 22 gives a $\Delta^{14}C$ value. The scale of $\Delta^{14}C$ values defined in this way provides a convenient framework within which natural radio-carbon activities can be discussed (see Chapter 22, Volume 2, and Broecker *et al.*, 1959).

IV. Carbon Dioxide in the Atmosphere

Atmospheric carbon dioxide occurs to the extent of about 320 p.p.m., and relative to the Chicago standard (Craig, 1957) shows a $\delta^{13}C$ enrichment of about $-7‰$. These should be taken as rough values only, since atmospheric carbon dioxide exchanges between various carbon reservoirs, particularly the oceans and the biosphere, and is produced in industrial areas. It consequently shows some variation in its atmospheric concentration with time and region in accordance with local and regional supply and demand. The intimate connection between atmospheric carbon dioxide and the carbon of the sea demands that some consideration be given to the concentration of atmospheric carbon dioxide and its possible short- and long-term changes, to possible variations in its isotopic composition, and to the rate of exchange of carbon dioxide between the atmosphere and the sea.

A. Short-term Variations

Diurnal variations about a mean value are an established phenomenon over most land areas, the range depending on the locality of the station, the season, prevailing wind and weather conditions, the height of sampling above ground level and the type of vegetation.

Excellent illustrations of the daily range at ground level have been given by Keeling (1958, 1961) for stations which include woodlands, grasslands and mountain and coastal areas (e.g. Figs. 15 and 16). His data show that the daily range can be as large as 100 p.p.m., but is strongly influenced by plant cover and is reduced when the air is derived directly from over the sea or is taken over relatively barren ground. In conjunction with $\delta^{13}C$ data (also given in Figs. 15 and 16) these results indicate that near-ground air samples taken during the night or in the early hours are particularly affected by the local environment (see also Lieth, 1963).

In regions of fairly dense vegetation cover, the atmospheric carbon dioxide content increased from about 320 p.p.m. at midday to almost

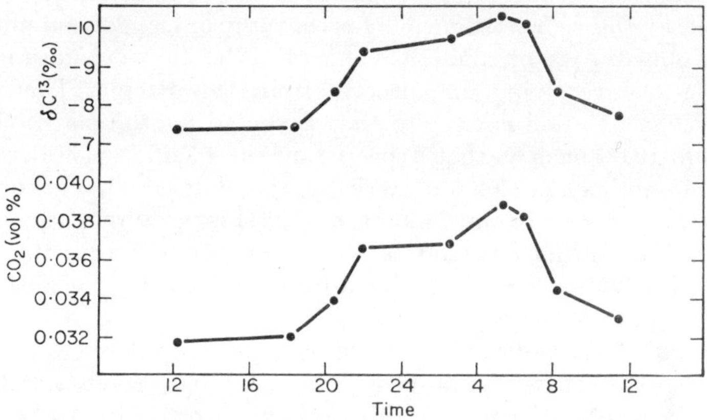

FIG. 15. Diurnal variation in the concentration and isotopic ratios of atmospheric carbon dioxide in a coastal redwood forest of California, 18–19 May, 1955, Big Sur St. Pk. (Keeling, 1958).

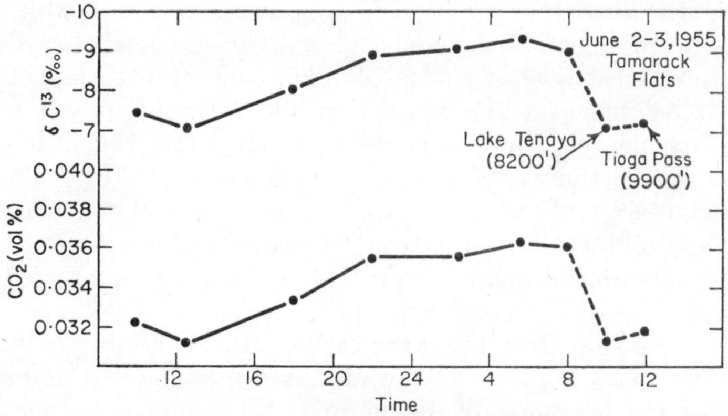

FIG. 16. Diurnal variation in the concentration and isotopic ratios of atmospheric carbon dioxide in a forest of the Sierra Nevada, 2–3 June, 1955, Tamarack Flats (Keeling, 1958).

400 p.p.m. during the night. This increase is, of course, related to plant respiratory processes. Variations followed a similar pattern in different types of plant cover, but were most intense in regions of rain forest, less marked over grassland, and small over barren land, mountains and the sea. The similarity between the plots for the total carbon dioxide

content and for $\delta^{13}C$ suggests that the two effects have a common origin in the addition to the atmosphere of carbon dioxide containing carbon of $\delta^{13}C$ value ranging from -21 to $-25‰$, which Keeling points out is in the range generally found for the carbon of terrestrial plants. Turbulence arising during the day is presumed to cause mixing with the "free" atmosphere, and to be responsible for the more or less general minimum value of 309–317 p.p.m. and $\delta^{13}C$ value of $-7‰$ shown at most stations including one receiving air directly from the Pacific. This carbon dioxide concentration is close to that expected for the maritime polar air (Buch, 1948) such as that expected off the Pacific coast at the time of this investigation. The pattern of the diurnal change observed depends on the season, and Takahashi (1961) for example has observed a peak value during dark hours and a minimum during the day in autumn, but little diurnal change during early spring because of the absence of photosynthesis.

Since daily variations are the result of plant respiratory processes, it is to be expected that their range will become progressively smaller with increase in height of sampling. Measurements made by Huber (1951) suggest that the local effect is still felt at 20 m., and Chapman *et al.* (1954) have been able to detect it at heights of 150 m.

Bischof (1960) has analysed some of the results obtained for the years 1955–9 in the Scandinavian investigation of the carbon dioxide content of the atmosphere, and has noted seasonal variations. He has also commented on the change of carbon dioxide content with heights up to 3 km. At heights of 3 km. the carbon dioxide content was found to change considerably from time to time (Bischof, 1962), but above 1000 m. was in the range 308–320 p.p.m. (average 314 p.p.m.). The range and average increased towards the ground, and a seasonal variation with a minimum in summer and a maximum in winter was noted. Many of the variations noted at ground level were absent or reduced at height.

Huber (1952) has shown that the carbon dioxide produced during the hours of darkness is progressively dispersed into the first 100 m. or so during the first few hours of the morning, since plots of carbon dioxide concentration as a function of the time of sampling for various altitudes showed that as the altitude of sampling was increased, the peak concentration was reduced and occurred at progressively later times (see Lieth, 1963).

Diurnal and seasonal atmospheric carbon dioxide changes appear to be much less pronounced over the open sea than those noted over land areas, and it seems clear from recent measurements (see e.g. Bolin and Keeling, 1963; Keeling *et al.*, 1965) that the range of variations with latitude is much smaller than the earlier measurements made by Buch (1934, 1939a,b, 1942) had indicated.

Bolin and Keeling (1963) have summarized the results of a large number of atmospheric carbon dioxide determinations made over the Pacific Ocean. Their observations include measurements made by continuous infra-red gas analysers sited at Barrow (Alaska), Mauna Loa (Hawaii), Little America (Antarctica) and the South Pole. In addition, numerous samples were taken on aircraft flights at various altitudes, and samples were taken at surface level on the Monsoon, Downwind and other cruises. The whole survey covered the period from 1957 to 1962, and was designed to investigate seasonal and latitudinal variations of the atmospheric carbon dioxide content. Some data processing was necessary in order to present the results in a form useful for giving a comprehensive picture. Measurements taken at Mauna Loa for example indicated an increase in the total atmospheric carbon dioxide content of 0·06 p.p.m./month (Fig. 17), and this increase had to be taken into account in order to adjust values taken in different years to a common

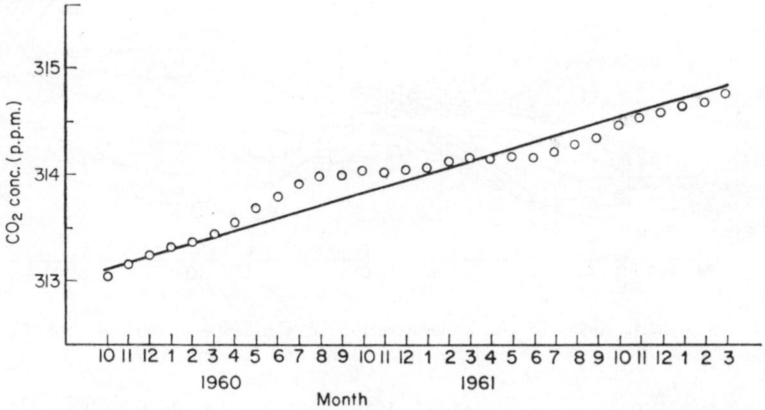

FIG. 17. Annual average concentration of carbon dioxide at Mauna Loa Observatory, Hawaii (Bolin and Keeling, 1963). —o—o—, average carbon dioxide concentration during the preceding year; ———, assuming rate of increase of 0·06 p.p.m. per month.

datum. Smoothed curves showing the seasonal carbon dioxide content as a function of latitude for each month are given in Fig. 18. Figures 19(a) to (d) show the concentration of atmospheric carbon dioxide at various altitudes and latitudes as a function of the month of the year.

Bolin and Keeling explained their observations in terms of a model in which there is a natural source of carbon dioxide in the tropics and an industrial source in the mid-latitudes of the northern hemisphere, and sinks in the polar regions. Their analysis led to the time-averaged sources and sinks indicated in Fig. 20. It was concluded that about $1{\cdot}5 \times 10^{10}$ tons of carbon dioxide are assimilated by the vegetation of the northern hemisphere north of 45°N. during the summer months.

Bolin and Keeling's treatment of their observations is based on the assumption that the exchange coefficient is independent of latitude.

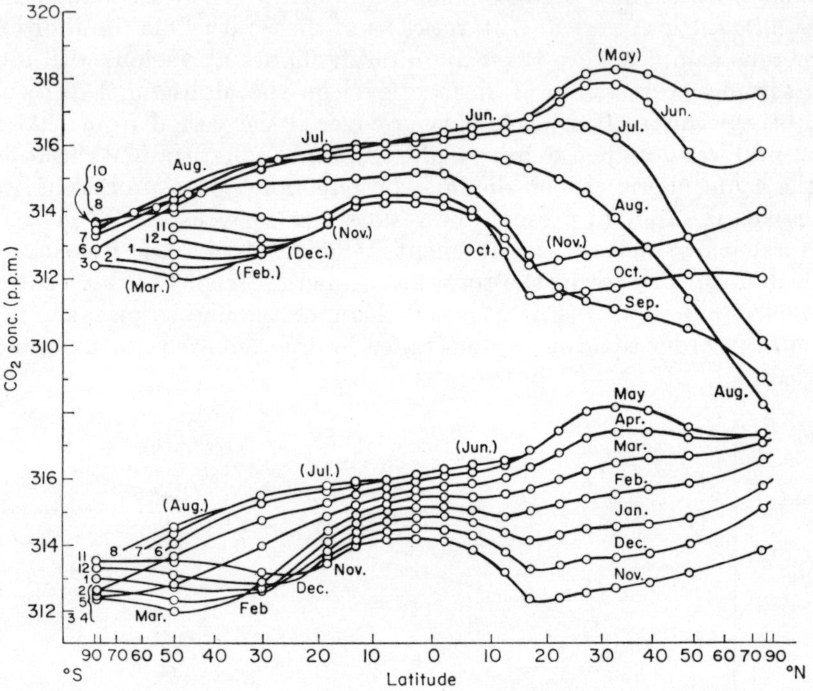

FIG. 18. Smoothed curves of the concentration of atmospheric carbon dioxide as a function of latitude for each calender month (Bolin and Keeling, 1963).

This assumption is not strictly valid since it is probable that the coefficient is less at low than at high latitudes. Junge (1962, 1963) has attempted to examine inter-hemisphere mixing by using a model which regards each hemisphere as well mixed, and separated from the other by an equatorial zone which acts as a barrier to latitudinal exchange. He has pointed out that any atmospheric constituent which has a source or sink or an annual variation restricted to one hemisphere can be used to give estimates of the exchange time between the two hemispheres. Approaches which have been made make use of the hemispherical differences of (i) the industrial dilution (Suess) effect, (ii) the rise in the concentration of bomb-produced $^{14}CO_2$, (iii) the annual variations in the carbon dioxide contents, or (iv) the rise in the concentration of tritiated methane. Although estimates obtained in this way are rather imprecise, Junge has concluded that the exchange time is about two years.

Postma (1964) has made estimates of the global regions in which carbon dioxide transfer from the sea to the atmosphere occurs. His estimates are based on considerations of the relationships between

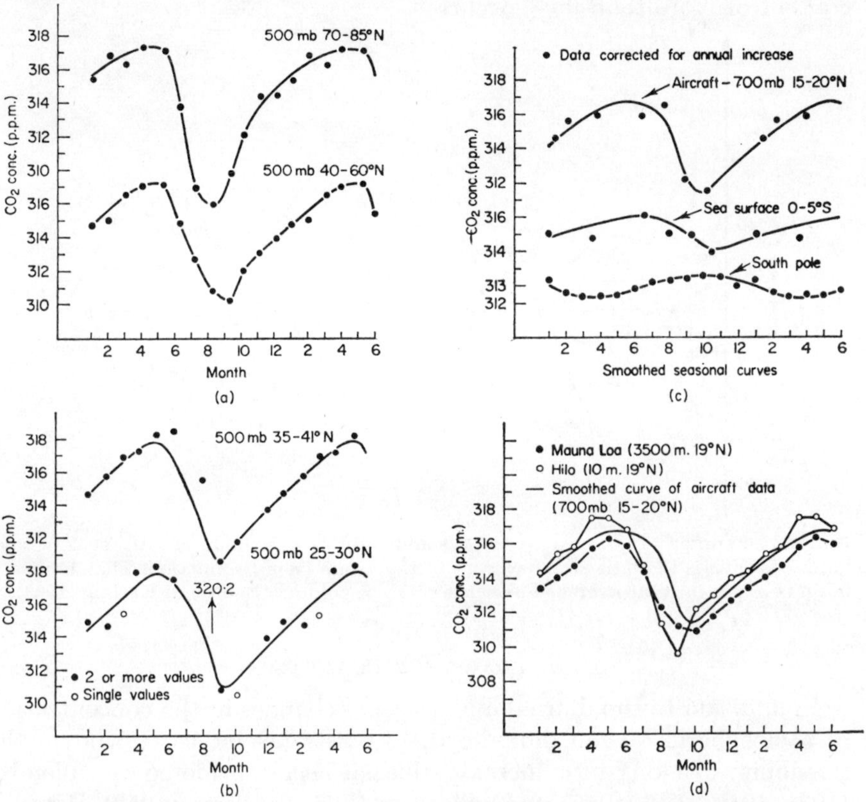

FIG. 19. (a) to (d). The concentration of carbon dioxide at various altitudes and latitudes as a function of the month of the year. January to June (months 1–6) are plotted twice to show the seasonal pattern more fully (Bolin and Keeling, 1963).

oxygen consumption and phosphate and carbon dioxide liberation in the sea. The relationships were established by reference to Pacific deep water, and deviations in the surface layers were attributed to oxygen and carbon dioxide exchange with the atmosphere. The main escape regions of carbon dioxide from the sea were thought to lie in a broad equatorial belt, and in somewhat narrower belts at high latitudes.

Postma considers that the current increase in total atmospheric carbon dioxide is causing a progressive decrease in the area of the escape regions, and he has suggested that an increase of 25% in the total atmospheric carbon dioxide content, which may well occur in the course of the next few decades, may eliminate the high latitude escape regions. The more extensive equatorial belt would diminish during this period, but would not disappear until an increase in atmospheric carbon dioxide content of more than 60% occurred.

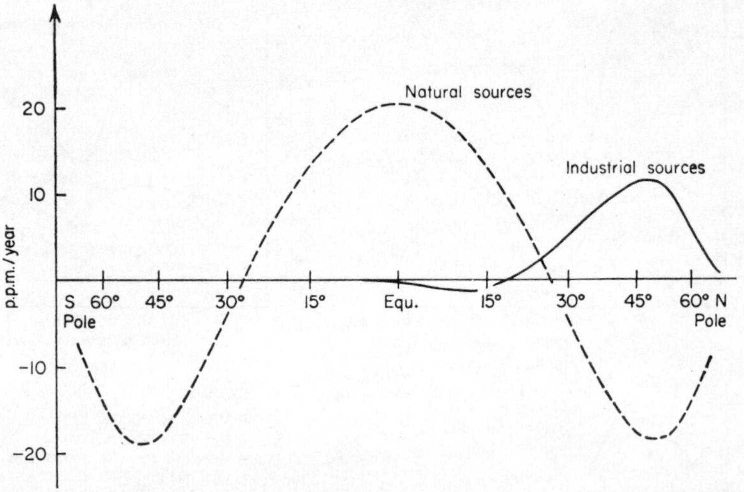

FIG. 20. Computed time-averaged sources and sinks as a function of latitude. The solid horizontal line is the latitudinal average of the natural source function. The abscissa is in units of p.p.m./year averaged over the total air column (Bolin and Keeling, 1963).

B. LONG-TERM CHANGES

In addition to the diurnal and seasonal changes in the concentration of atmospheric carbon dioxide, there exists (as indicated above) the possibility of a long-term increase; this has been considered by Callendar (1938, 1940, 1949, 1957, 1958), Slocum (1955) and Bray (1959). Detailed consideration of the trend over the last century is hampered by lack of generally accepted data covering a sufficient time span, particularly as many older measurements are often isolated observations taken with no particular regard to local conditions, time or the method of sampling.

Callendar (1940) has examined in detail published atmospheric carbon dioxide analyses for the period between 1860 and the 1930's, and on the basis of certain criteria has rejected those estimates which he considered to be unreliable indicators of any possible change.* He

* Callendar's criteria were: (i) period mean values differing by 10% or more from the average; (ii) samples taken in the neighbourhood of towns; (iii) isolated measurements; (iv) measurements taken for special (e.g. biological) purposes.

concluded that an increase of about 10% had occurred over the period considered. Since the accelerated accumulation of atmospheric carbon dioxide effectively started when the rapid consumption of fossil fuels was commencing, he attributed the increase to that cause. Although Callendar's main conclusions are not universally accepted (Slocum, 1955), Bray has also examined atmospheric carbon dioxide data for a similar period and has concluded that the period from 1907 to 1956 had a higher mean than the period 1859 to 1906. Only by making precise observations of future changes in the atmospheric carbon dioxide content will it be possible to check the validity of the suggested increase and to comment with confidence on its origin. However, careful measurements over a relatively short time span do support the suggestion of a gradual increase (see Fig. 17).

There are several possible causes of a long-term increase in atmospheric carbon dioxide. These include the use of coal and oil as fuels, changes in the biosphere and soil resulting from deforestation or changes in agricultural practice, a gradual increase in the mean temperature of the sea and an increase in the rate at which juvenile carbon dioxide is produced.

Although the reliability of the older carbon dioxide data makes it unwise to be too dogmatic about the reality of the suggested rise in atmospheric carbon dioxide over the last century, it seems evident that in the absence of compensating changes through other factors (e.g. rapid uptake by the oceans), the burning of fossil fuels should have resulted in a considerable increase, and most commentators favour this as a probable origin of the supposed effects.

Brannon *et al.* (1957) have made estimates of the cumulative total of carbon dioxide released to the atmosphere by the burning of fuels since about 1860. Their estimates, based on the sum of the contributions from coal and oil, show that if all the carbon dioxide produced in this way had remained in the atmosphere, the accumulation from this cause would have been exponential from 1860 to 1910; after 1910 the rate of accumulation deviates slightly from the exponential form. In 1957 the cumulative total given by these estimates was some 14% of the total carbon dioxide in the atmosphere.

Table IX, taken from Revelle and Suess (1957), gives details of United Nations estimates of fossil fuel combustion over the period 1860 to 1949, together with forecasts to the end of the first decade of the next century. The figures are in reasonable agreement with the conclusions of Brannon *et al.* (1957), and the forecast for the years 2000 to 2009 suggests that over this decade the carbon dioxide produced from artificial sources may be as much as 20% of the present atmospheric total. According to Arnold and Anderson (1957), industrial carbon dioxide production has increased exponentially with a half-time of seventeen years up to 1910, and of forty years since that date.

TABLE IX

*Carbon dioxide added to atmosphere by consumption of fossil fuels**

	Decade	*Average amount added per decade (measured)*		*Cumulative total added*			
		10^{18} *g.*	*% Atm* CO_2	10^{18} *g.*	*% Atm* CO_2	10^{18} *g.*	*% Atm* CO_2
	1860–69	0·0054	0·23	0·0054	0·23		
	1870–79	0·0085	0·36	0·0139	0·59		
	1880–89	0·0128	0·54	0·0267	1·13		
	1890–99	0·0185	0·79	0·0452	1·92		
	1900–09	0·0299	1·27	0·0751	3·19		
	1910–19	0·0405	1·72	0·1156	4·91		
	1920–29	0·0470	2·00	0·1626	6·91		
	1930–39	0·0497	2·11	0·2123	9·02		
	1940–49	0·0636	2·71	0·2759	11·73		
Predicted†	1950–59	0·091	3·9	0·367	15·6	0·367‡	15·6‡
Predicted†	1960–69	0·128	5·4	0·495	21·0	0·458‡	19·5‡
Predicted†	1970–79	0·176	7·5	0·671	28·5	0·549‡	23·4‡
Predicted†	1980–89	0·247	10·5	0·918	39·0	0·640‡	27·2‡
Predicted†	1990–99	0·340	14·5	1·258	53·5	0·731‡	31·1‡
Predicted†	2000–09	0·472	20·0	1·730	73·5	0·822‡	35·0‡

* Revelle and Suess (1957).

† Assuming fossil fuels are used to meet future requirements of fuel and power as estimated by UN (1955).

‡ Assuming fossil fuel consumption remains constant at estimated 1955 rate.

Fergusson (1958) has combined these periods and assumed a half-time of 28 years for the period up to 1954. He represented the rate of industrial carbon dioxide production by $Ce^{\mu t}$ where μ is 0·025 year^{-1}, and taking $t = 0$ at 1850, obtained for C a value of 0·024% year^{-1}. In a similar expression used by Bolin and Eriksson (1959) and derived by them from the United Nations data given by Revelle and Suess, μ is 0·029 year^{-1}.

Hutchinson (1954) has suggested that changes in the soil and the biosphere over the last century have been responsible for the injection of considerable quantities of carbon dioxide into the atmosphere. Deforestation and conversion of land to agricultural use has been pronounced over the last century, and is thought by him to have reduced the soil content by an amount equivalent to about 4% of the carbon dioxide now in the atmosphere.

An increase in the mean temperature of the sea would cause an increase in P_{CO_2} of the water and also a rise in mean sea level. However, Revelle and Suess (1957) consider that the increase in average sea temperature over the last century is probably not more than 0·05°C; this would be insufficient to cause an appreciable increase in the partial pressure of carbon dioxide.

It seems unlikely that an appreciable proportion of any increase in atmospheric carbon dioxide over the last century is attributable to an enhanced rate of production of juvenile carbon dioxide, since the average rate at which juvenile carbon dioxide is at present being produced is a good deal less than the rate of generation of fossil fuel carbon dioxide (see Revelle and Suess, 1957).

C. CHANGES IN ATMOSPHERIC RADIO-CARBON ACTIVITY

In the pre-bomb era (i.e. before about 1954), addition of ^{14}C-free carbon dioxide to the atmosphere on the scale indicated by the industrial effect should have caused a detectable reduction of atmospheric carbon specific activity unless isotopic equilibrium was rapidly established between the atmosphere and the larger carbon reservoir of the ocean. This would lead to a reduction in the specific activity of plants which had obtained their carbon since the beginning of the century, compared with that which would have been observed in the absence of fossil carbon dioxide production (Hutchinson, 1954). This effect (the Suess, industrial or dilution effect) noticed in 1953 by Suess has subsequently been confirmed by a number of investigators (Suess, 1955; Hayes *et al.*, 1955; Brannon *et al.*, 1957; Fergusson, 1958), and depletions of radiocarbon in modern wood amounting to as much as 4% have been reported. To some extent the magnitude of the effect seems to be geographical, tending to be smaller in the southern than in the northern hemisphere. Possibly this is associated with the heavier concentration of industry in the northern hemisphere and the finite rate of atmospheric mixing, and results obtained from rings sampled from trees growing near industrial areas may show a higher effect (Fig. 21). Fergusson has examined rings from trees grown in New Zealand, Peru and North America. His results (Fig. 22) show an increasing impoverishment in the activity of all samples over the last 100 years and indicate an average Suess effect of about 2%. Also included in Fig. 22 is a plot of the amount of fossil fuel produced over this period, and the similarity of shape of the two curves supports the supposed origin of the dilution effect.

Although the reduction of the specific activity of atmospheric carbon over the last century, as judged from tree ring data, is about what would be expected from the dilution of naturally occurring ^{14}C with industrial "dead" carbon, the picture is complicated by the possibility that the total change observed may be caused by the superposition on the industrial effect of a natural effect of indeterminate size. Evidence for small changes of some 1 or 2% in the activity of atmospheric carbon

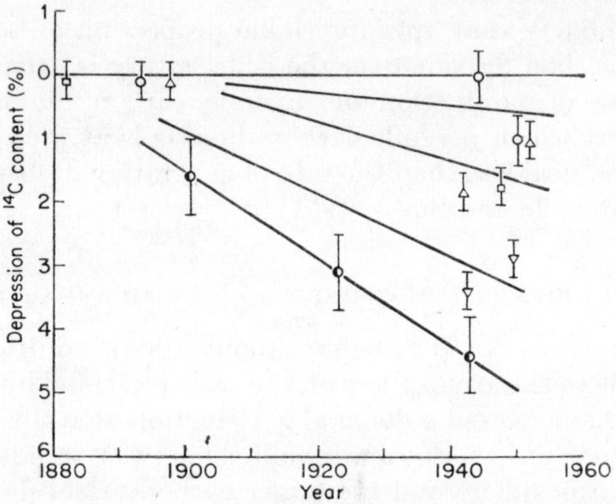

FIG. 21. Depression of ^{14}C count as a function of year of growth for wood grown in several areas (Arnold and Anderson, 1957). —o—o—, Peru; —Δ—Δ—, Yosemite, California; —□—□—, Alaska; —▽—▽—, Massachusetts; —◐—◐—, Wisconsin.

dioxide over the last few centuries has been obtained by de Vries (1958) from an examination of tree and other samples. Although there appears to be a difference between American and European specimens in their absolute activity (Fig. 23), they show similar trends.

De Vries has discussed his observations in terms of carbon dioxide exchange between various carbon reservoirs using a model for the sea similar to that described by Craig (1957) (see Section VI). He has suggested that since the fluctuations in atmospheric activity are unlikely to occur purely from atmosphere–surface water interaction, deep water must participate. In prolonged periods of very cold weather, the surface of the ocean may become cold enough to exchange with deep water of low carbon activity which, on undergoing isotopic exchange with the atmosphere, will cause a reduction in atmospheric carbon activity (Broecker *et al.*, 1958). As a guide to past temperature variations, de Vries has included with his activity plots a glacial advance and retreat curve. The sympathy between the two plots is quite marked up to the last glacial retreat, and the lack of atmospheric radio-carbon increase from this time onwards may have been caused by the overwhelming effect of fossil fuel carbon dioxide.

The development and testing of nuclear weapons caused a marked increase in atmospheric radio-carbon activity. The increase has been monitored both by atmospheric sampling and by examination of plant leaves, although the latter method can be unreliable as there is occasionally a time-lag between the rise in the atmospheric $^{14}C/^{12}C$ ratio and that in certain plant leaves (Broecker and Olson, 1960).

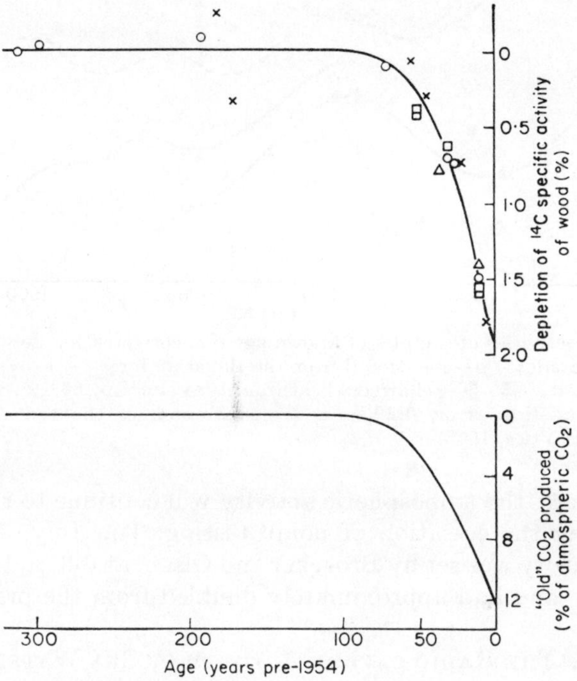

FIG. 22. The total amount of carbon dioxide produced by the burning of fossil fuels and the resulting depletion of the ^{14}C specific activity of wood over the period ending 1954. Samples: —□—□—, Wisconsin; —Δ—Δ—, Peru; —o—o—, Wellington, N.Z.; —×—×—, Christchurch, N.Z. (Fergusson, 1958).

Broecker and Olson (1960) have summarized much of the data up to 1959 for tropospheric $\Delta^{14}C$ values. Although there was a distinct difference between the two hemispheres, both showed $\Delta^{14}C$ values which rose rapidly in a roughly parallel way. Besides the difference between the northern and southern hemispheres, other regional and seasonal variations have been noted (Willis, 1960; Tauber, 1960; Munnich and Vogel, 1959). Munnich and Vogel (1958), for example, reported an increase of about 3·2% $year^{-1}$ in the activity of European plants but only about half this rate of increase in South African plants (see also Broecker and Walton, 1959).

Since much of the activity arising from nuclear explosions has been injected into the stratosphere, which mixes only relatively slowly with

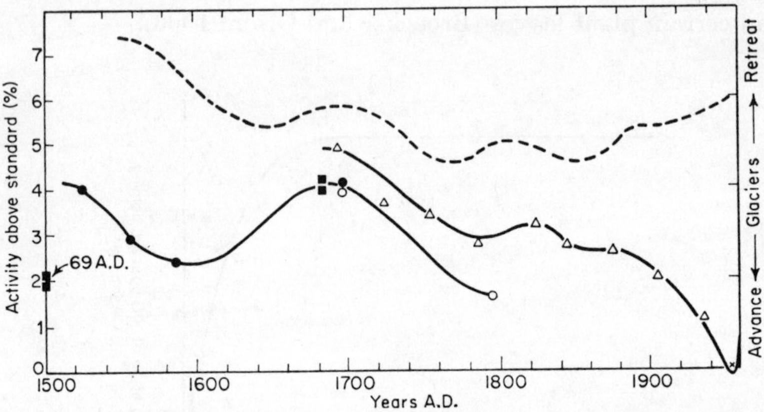

Fig. 23. Initial activities of samples of known age (i.e. corrected for decay). —Δ—Δ—, tree A from Colorado; —o—o—, tree B from the Bavarian forest; —•—•—, tree C from the Spessart forest; —▪—▪—, charred wheat from 1684 and A.D. 69; ×, recent calibration sample. Statistical error, 0·2%. - - - - - -, extension of the continental glaciers (schematical) (de Vries, 1958).

the troposphere, the tropospheric activity will continue to rise for some time following the cessation of bomb testing. The July, 1959, tropospheric inventory was set by Broecker and Olson, at $6{\cdot}3 \times 10^{27}$ atoms of radio-carbon, and had approximately doubled from the previous year.

V. The Radio-carbon Activity of Sea Water

When oceanic circulation removes surface water from contact with the atmosphere, the radio-carbon activity of the water will decrease, but many factors combine to complicate the relationship between the activity and other features of oceanographical interest, such as the time which has elapsed since submergence. Mixing between adjacent water masses of different origins and activities will give rise to water of intermediate activity. Upwelling sub-surface water of low activity may sink again before being at the surface long enough to reach isotopic equilibrium with the atmosphere. Decaying organic matter in equilibrium with surface water of relatively high radio-carbon content may fall into older deep water, and on oxidation, ultimately give bicarbonate which will alter the apparent age of the lower layers. There is also the possibility (usually regarded as unimportant) of interaction and isotope exchange between sub-surface water and sea-bed carbonate. Further difficulties of interpretation stem from the fact that the atmosphere–ocean system does not, at the present time, constitute in the strictest sense a steady state system with regard to carbon-14 activity owing to the progressive injections into the atmosphere of low-activity fossil fuel carbon dioxide and high-activity carbon dioxide rising from nuclear tests.

Since 1955 there has been a marked increase in the number of investigations of the radio-carbon content of both surface and deep waters, and it is possible to recognize within collected sets of results patterns which are consistent with standard data on ocean circulation. Results obtained from carbon-14 measurements can be used to trace the origin and course of water masses and to estimate their mixing times (see Chapter 2).

The $^{13}C/^{12}C$ ratio in marine carbonates is approximately 2·5% higher than that in land plants (Craig, 1953, 1954). It would be expected that because of isotope fractionation, if equilibrium between the ocean surface and the atmosphere was complete, the carbon-14 activity of marine materials should be some 5% higher than that of land plants. Early measurements by Suess (1953, 1955) showed that the activities of these two classes were the same, and that when isotope fractionation was taken into account, the Atlantic samples examined by him showed an apparent age of 400 years relative to the atmosphere. A similar age was deduced from the measurements made by Rafter (1955) on samples from near New Zealand when correction was made for the standard used by him (Revelle and Suess, 1957; Rafter and Fergusson, 1958a,b).

Examinations of samples of surface and sub-surface Pacific water have been made by Rafter (1955), Rafter and Fergusson (1958a,b), Brodie and Burling (1958), Garner (1958), Burling and Garner (1959) and Bien *et al.* (1960, 1963a,b). Figures 24 to 26, taken from Bien *et al.* (1963a,b), indicate the locality of a number of stations and the carbon-14 activities expressed as $\Delta^{14}C$ values.

Surface samples show some scatter, probably caused by the uptake of nuclear bomb produced $^{14}CO_2$, but tend to centre around a $\Delta^{14}C$ value of about −50‰ except in latitudes farther south than 60°S. where low values of about −170‰ were observed. Similar low values for surface water in the Antarctic south of the Atlantic have been reported by Broecker *et al.* (1960), and probably are characteristic of outcrops of the deep sea.

Pacific deep-water samples examined by Bien *et al.*, are in the range −190 to −250‰ and show a progressive increase in apparent age from south to north corresponding to a water movement of about 0·06 cm./second in a northerly direction. The apparent ^{14}C age is probably

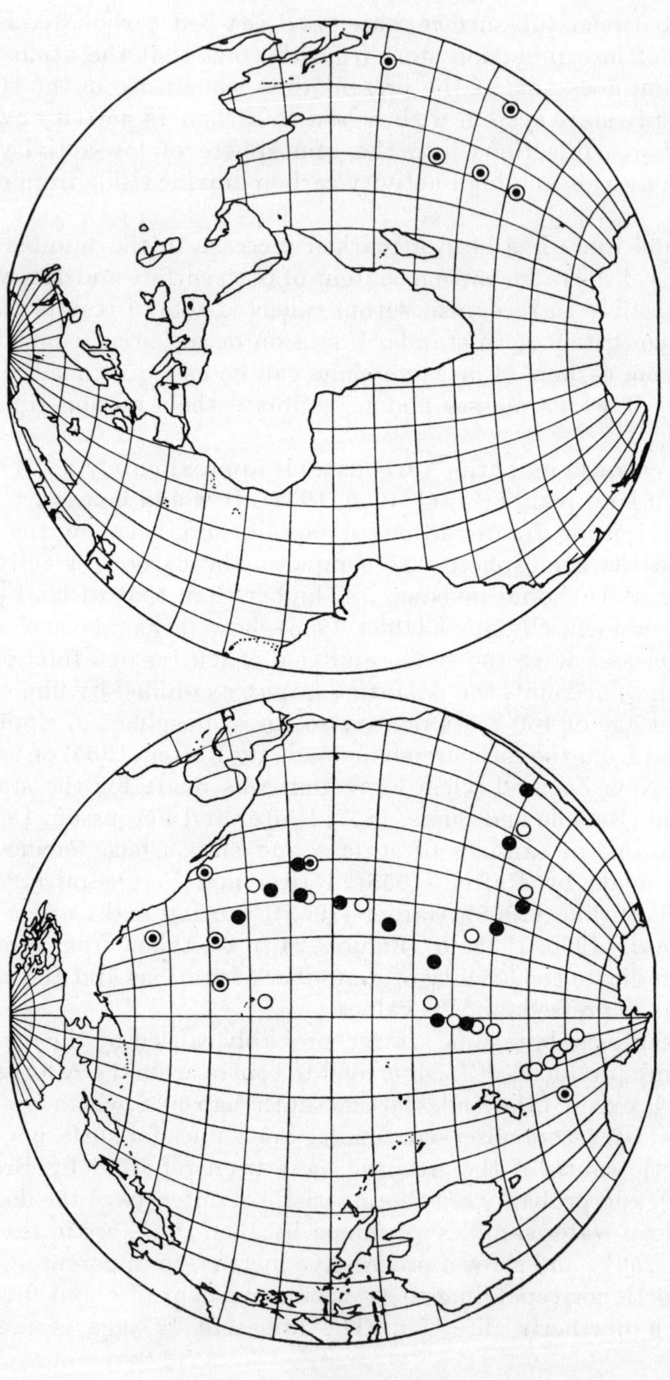

FIG. 24. Location of radio-carbon sample stations. Open circles denote surface samples, solid circles denote deep water samples, and concentric circles denote profiles (Bien *et al.*, 1963a,b).

affected by oxidation of particulate organic matter relatively rich in radio-carbon falling from above, since the northward movement is accompanied by a decrease in dissolved oxygen. The activity decreased sharply with depth down to about 1000 m.; below this point the trend was less pronounced.

Indian Ocean deep water is slightly more active (i.e. younger) than the Pacific deep water of similar latitude, as would be expected if the Weddell Sea were the source of both waters. A decrease in activity from west to east across the Indian Ocean corresponds to a water movement of 0·03 cm./second. At a position 10°S. and at a depth of 400 m., water of high apparent age was detected, and was probably representative of an

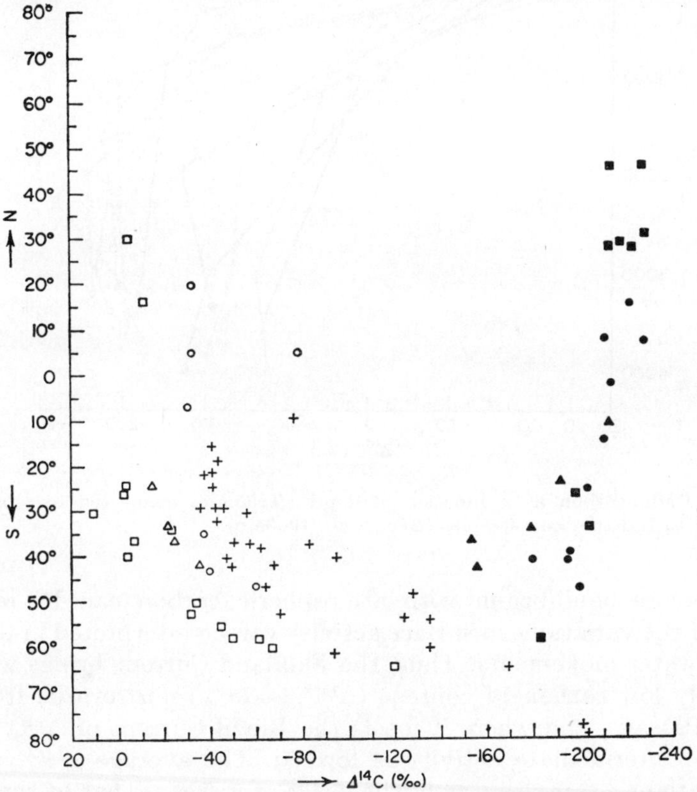

FIG. 25. Latitudinal variation of the ^{14}C content of surface and deep (3500 m.) water. Circles denote La Jolla measurements collected during the IGY; squares and triangles denote Pacific and Indian Ocean samples respectively, collected in 1960 and 1961 and measured at La Jolla. Open circles, squares or triangles refer to surface samples; solid triangles, squares and circles refer to water from a depth of about 3500 m. Crosses refer to New Zealand measurements on surface samples collected before 1958 (Bien *et al.*, 1963a,b).

influx of Northwest Intermediate Indian Ocean Water which, since it does not mix well, shows a relatively low activity.

A detailed study of the pattern of radio-carbon distribution in the Atlantic has been made by Broecker *et al.* (1960) who examined 135 samples comprising deep and surface waters taken from the major water masses over the period 1955–7. The lowest surface activity ($\Delta^{14}C$, $-120‰$) was found in a sample taken south of the Antarctic Convergence where water of low activity rising from below probably sinks

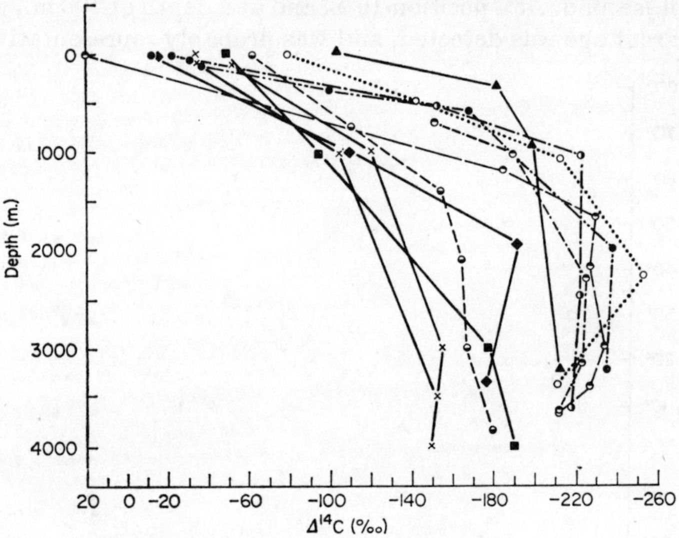

FIG. 26. Radio-carbon as a function of depth. Triangles connected by continuous lines refer to Indian ocean samples (Bien *et al.*, 1963a,b).

again before equilibrium with atmospheric carbon can be reached. Many of the variations in surface activity can be interpreted in terms of known water movements. Thus, the Falkland Current brings water of relatively low carbon-14 content ($\Delta^{14}C$, $-78‰$) northwards from the Drake Passage, and where it meets the Brazil Current of $\Delta^{14}C$ $-39‰$ water of intermediate activity is formed. The average $\Delta^{14}C$ value of South Atlantic samples was in the order of $-57‰$. That of the North Atlantic samples was about $-49‰$, and the Caribbean, which draws water from both the North and South Atlantic, has an intermediate value. Higher northern latitudes (60°–80°N.) have been investigated by Fonselius and Ostlund (1959) who obtained $\Delta^{14}C$ values of about $-35‰$.

The core of each of the deep-water masses of the Atlantic examined by Broecker *et al.* (1960), showed a standard deviation in $\Delta^{14}C$ values only slightly greater than the measurement error, and since clear differences were observed between the various masses, circulation within each mass is probably rapid relative to residence time within the mass. Deep water originating in the high latitudes of the southern hemisphere has a lower carbon-14 content than that originating in the high latitudes of the northern hemisphere, and bottom water of the Eastern Atlantic Basin shows a $\Delta^{14}C$ value of about 27‰ lower than that of the Western Basin. The lowest activity found in the Atlantic system ($\Delta^{14}C$, -144‰) was that of Antarctic bottom water taken from the South Western Atlantic. An important contrast between the Atlantic and Pacific Oceans is the smaller difference between surface and deep-water activities in the former.

Broecker (1963) has summarized much of the evidence relating to ocean surface waters, and has recognized three categories of fossil marine organisms with respect to $^{14}C/^{12}C$ ratios: (i) those forming within the warm surface waters ($t > 5°C$) where the $\Delta^{14}C$ is approximately -55‰; (ii) those forming in colder regions where deep water of low ^{14}C content outcrops (fauna from these areas will show anomalously high "ages" unless proper account is taken of the ^{14}C content of the environment); and (iii) those forming in brackish coastal waters or near the discharge of large rivers where the amount of inactive carbon derived from the land may be high.

A few estimates of $\Delta^{14}C$ values for surface and near-surface waters from under the Antarctic pack-ice showed abnormally low $\Delta^{14}C$ values of about -350‰. If confirmed, these values would suggest that this water experiences only very slow vertical and horizontal exchange with neighbouring water masses.

VI. Circulation of Carbon Dioxide between the Atmosphere and the Sea and within the Sea

Although many aspects of the equilibria of the oceanic carbonate system are reasonably well understood, rate factors associated with the distribution and circulation of carbon between the various natural reservoirs are less clearly defined. A knowledge of the residence times of carbon in these reservoirs is desirable because of its bearing on such problems as the probable accumulation of carbon dioxide in the atmosphere, and its obvious value in conjunction with conventional oceanographical data to supplement information on circulation and exchange between water masses. These problems of circulation and distribution are potentially capable of at least a partial solution through appeal to radio-isotope tracer methods. The differences between the fractionation-corrected specific activities of the atmosphere, ocean surface water and deep water are caused by the finite time taken for inter-reservoir transfer, but one of the principal difficulties in obtaining a solution is that of setting up a model of the atmosphere–ocean system which is sufficiently realistic to give useful results, but yet which is simple enough to allow theoretical treatment. No solution can be any more exact than the model on which it is based, and in this connection a vital assumption usually made is that apart from recent alterations in the system caused by industrial and bomb effects, a steady state prevails in the sense that the specific activities in each reservoir are invariant with time. It is not possible at present to prove that this hypothesis is correct, but the alternative—non-steady state mixing resulting from periods of relatively rapid oceanic mixing alternating with periods of readjustment towards equilibrium—does not lend itself well to theoretical treatment, although it has been considered in a general way by Broecker *et al.* (1960) (see also Broecker, 1962).

Clearly it is desirable to use a model in which the sea is sub-divided in such a way as to reproduce at least to a limited extent the main features of the various water masses. Craig (1957, 1958) has discussed the distribution of radio-carbon using a model in which the ocean is separated at the thermocline into two layers across which mixing takes place only slowly (Fig. 27(a)). In a full discussion, account is also taken of the participation of the biosphere and humus in the carbon cycle. If mixing between the atmosphere and the sea was infinitely rapid compared with the decay time of carbon-14, the specific activity of the carbon, when corrected for fractionation, would be the same everywhere. Imposition of a mixing barrier between the upper layer and the atmosphere would result in a "piling up" of radio-carbon in the atmosphere, and a further barrier between the upper and deep layers would result in a "piling up" of carbon-14 in the surface layer. The difference in the activities of the two layers is a function of the mixing time, and Craig's analysis and assumptions showed that relative to the hypothetical case of infinitely rapid exchange, the activity of the atmosphere is increased by about 0·74% for each year's residence in the atmosphere, and the activity of both the atmosphere and the mixed layer are increased by about 1·2% for each 100 years residence time in the deep sea. The difference between the activities of the atmosphere and the surface layer is almost independent of the residence time in the deep water for times of up to several thousand years, and corresponds to an atmospheric residence time of about seven years. These calculations were consistent with a deep-sea residence time of up to 2500 years, but did not give precise information on this point, although Craig considered it to lie between 500 and 1000 years.

An alternative calculation with this model by Craig (1957), employing the steady state balance between the rate of introduction of

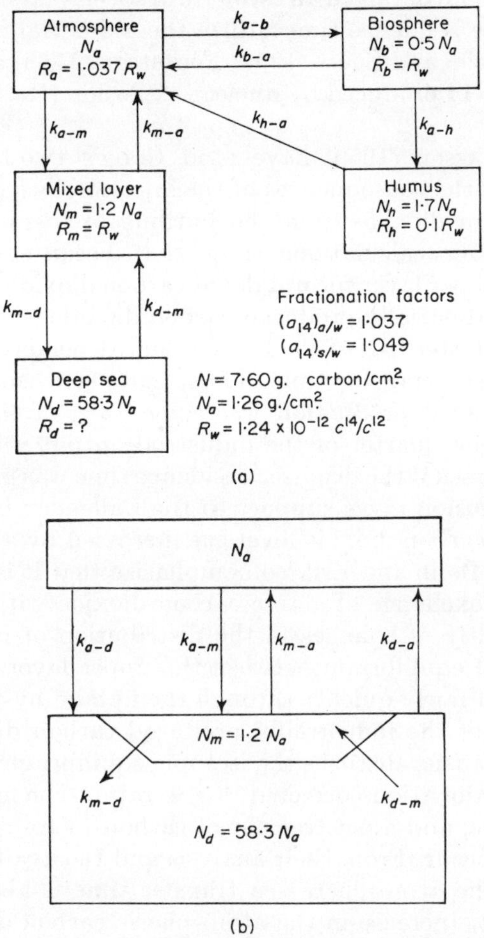

FIG. 27. (a) Carbon reservoirs in the "chain" model of the carbon dioxide exchange system (Craig, 1957). (b) Carbon reservoirs in the "cyclic" model of the carbon dioxide exchange system (Craig, 1958). N, N_a, N_m, N_d, N_b and N_h are, respectively, the total amounts of carbon (g./cm.^{2} of the earth's surface) in the system, the atmosphere, the mixed layer, the deep layer, the biosphere and the humus. The symbol R_i indicates the $^{14}C/^{12}C$ ratio in reservoir i, and R_w is the ratio for standard wood. $(\alpha_{14})_{a/w}$ and $(\alpha_{14})_{s/w}$ are the fractionation factors for ^{14}C relative to wood for the atmosphere and marine carbonates respectively. The k terms are the velocity constants for the appropriate transfer processes.

radio-carbon to the atmosphere by natural production plus evolution from natural reservoirs (the sea, humus and biosphere), and its loss from the atmosphere by decay and physical removal, gave a similar residence time of about six years. Steady state calculations used by Arnold and Anderson (1957) gave a somewhat higher value of about 14 years, while Broecker *et al.* (1960) obtained an estimate of seven years. Other methods of calculating the residence time utilize the industrial effect (Brannon *et al.*, 1957; Revelle and Suess, 1957; Fergusson, 1958) and the rate of uptake of carbon-14 produced by nuclear explosion (Rafter and Fergusson, 1958a,b).

Bolin and Eriksson (1959) have used Craig's two-layer model to examine in detail the consequences of the injection into the atmosphere of inactive carbon dioxide from the burning of fossil fuels (see also Bolin, 1959). Their calculations show that because of the limited capacity of the upper layer for uptake of carbon dioxide (Section II,E), the fraction of industrially produced carbon dioxide which has entered the sea since the latter half of the last century depends only slightly on the atmosphere–sea transfer time. Thus, in 1954 when the estimated cumulative industrial production was some 13% of the atmospheric total, less than one quarter of the industrial output would have been transferred to the sea if the deep-sea residence time was greater than 500 years. This conclusion gives support to the Callendar hypothesis that the atmospheric carbon dioxide level has increased by about 10% over the last century. Bolin and Eriksson emphasize that it is not correct to assume that the exchange of active carbon dioxide with the sea takes place independently of changes in the distribution of inactive carbon dioxide. Chemical equilibrium between the upper layer and the atmosphere is attained fairly quickly through the uptake by the sea of only a small fraction of the industrially produced carbon dioxide (Section II,E), but this transfer disturbs the isotopic equilibrium by causing the surface layer fractionation-corrected $^{14}C/^{12}C$ ratio to be higher than that of the atmosphere, and a net transfer of carbon-14 from the sea to the atmosphere will occur. From their analysis and the available data they concluded that the atmosphere–sea transfer time is about five years, and that a 7–10% increase in the atmospheric carbon dioxide content caused by fossil fuel combustion is compatible with a Suess effect of 3–4%.

Although these estimates of the mean atmospheric residence time of carbon dioxide cover a range of values from about three to twenty years, they do bracket the most probable value more precisely than earlier estimates which covered the range from several hours (Dingle, 1954) to about a thousand years (Plass, 1956). Some uncertainty must exist owing to imperfections in the assumptions underlying the models and calculations (Revelle and Suess, 1962), but a value of less than a decade is probably of the right order, and is not inconsistent with an experimental determination of the invasion coefficient of carbon dioxide into a stirred solution (Bohr, 1899).

This mean atmospheric residence time of about seven years is at best an average value for the ocean as a whole, and the real rate of entry of carbon dioxide into the sea will vary with region, as it is a function of temperature, turbulence and the partial pressure of carbon dioxide in the atmosphere. An analysis of factors which determine the comparatively long atmospheric residence time of carbon dioxide has been made by Bolin (1960). A rate determining factor may well be the slowness of passage of carbon dioxide by molecular diffusion through a boundary layer at the surface of the sea, and Bolin also considers that slow hydration of dissolved carbon dioxide is important.

Recognition is given to possible regional variations of the atmosphere–sea transfer rate in one variant of Broecker's world-wide "box" model (see below) in which a higher rate of entry of carbon dioxide to the sea is supposed to prevail in the stormy southern seas than exists in lower latitudes (Broecker, 1962; Kanwisher, 1963a,b). A lower limit to the rate of entry would be that obtained for the carbon dioxide invasion of an unstirred solution of sea water, quoted by Broecker (1962) as 6 moles/m.^{2}/year compared with a rate of 20 moles/m.^{2}/year corresponding to a mean residence time of seven years.

The simple two-layer model is defective since it makes no provision for direct entry of atmospheric carbon dioxide to the deep layer as might occur when the thermocline is absent (as for example in Antarctic regions), and since because of its simplicity it is not suited for dealing with mixing between the major deep-water masses. A basis for discussion of mixing phenomena when direct entry of carbon dioxide to the deep layer is important is provided by Craig's (1958, 1963) cyclic or outcrop model (Fig. 27(b)). In this, the deep layer is supposed to be in exchange contact with the atmosphere in outcrop regions north of 45°N. and south of 45°S., and between these latitudes the two layers are separated by the thermocline. Carbon enters the deep sea from the upper layer by mixing across the thermocline and directly from the atmosphere in the outcrop regions. By taking the ratio of the atmosphere contact areas of the upper and deep layers as 3 : 1, Craig (1963) estimated the residence time of carbon dioxide in the atmosphere to be 15 $\pm$ 5 years, although the estimate was rather more sensitive to the age assumed for the deep-sea carbon than was the corresponding estimate made from the chain model.

Comparison of the properties of the outcrop system with those of models in which the sea is treated either as a single or as a two-layer system illustrates the difficulties associated with the concept of residence time in the deep sea as judged from radio-carbon data, and shows that its interpretation is markedly dependent on the choice of model. In the outcrop model, the total residence time for carbon in the deep layer (τ_t) is defined by $1/\tau_t = 1/\tau_a + 1/\tau_u$, where τ_a and τ_u are the residence times with respect to loss to the atmosphere and upper layer respectively, i.e. it is the residence time with respect to all modes of physical loss. Obviously τ_u is greater than τ_t, that is, the age of the deep-water relative to that of the mixed layer as judged by the radio-carbon evidence is less than the "real" age implied by the oceanographical flushing time of the deep sea. The outcrop model is intermediate between the uniform sea (single-layer) model and the two-layer model. In a hypothetical single-layer system, in which the radio-carbon concentration is uniform throughout the sea, the radio-carbon residence time in the sea refers only to the transfer to the atmosphere. In the two-stage model, the residence time in the deep layer refers to the mixing time with the upper layer, and the residence time for carbon and water are the same.

Broecker *et al.* (1960) have examined the circulation of carbon dioxide on the basis of a world-wide steady state circulation model which is a good deal more specific in regard to geographical areas and water masses than are the above models. The factors underlying selection of a suitable model have been discussed by Broecker (1962). The system shown in Fig. 28 was based on ocean radio-carbon data obtained by a number of different workers from a wide range of surface and sub-surface water masses (Rafter and Fergusson, 1958a,b; Fonselius and Ostlund, 1959; Burling and Garner, 1959; Bien *et al.*, 1960; Broecker *et al.*, 1960). Within the ocean there is a flux of water from the Antarctic via the South and North Atlantic surfaces and a return at depth via the Atlantic Deep Water. The Pacific and Indian Ocean waters are combined into one system in which the surface and sub-surface waters are separated by the thermocline across which negligible mixing is supposed to occur. Both the surface and deep waters are in exchange with the vertically mixed Antarctic reservoir through which the Pacific and Atlantic waters communicate. Together with the atmosphere, the entire system comprises eight reservoirs, and the stationary state condition can be described by a set of eight equations, one for each reservoir, in which the rate of gain of carbon-14 by each reservoir by transfer from outside is balanced by the rate of loss by decay, plus transfer in the reverse direction. In making calculations using this model, corrections were made for the disturbance of the steady state by the introduction to the surface layers of inactive carbon dioxide from industrial production and of active carbon from nuclear explosions.* The residence times of carbon and of dissolved solids in each of the water masses are indicated in Fig. 28. These times differ, since the carbon dioxide is subject to transport through the atmosphere as well as by direct mixing of the different masses.

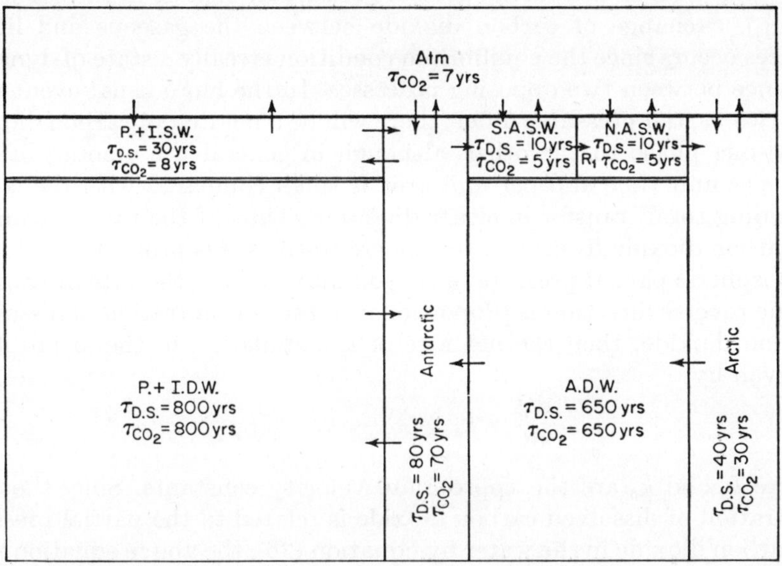

FIG. 28. Model of the steady-state cycle in the atmosphere and oceans. The arrows indicate possible modes of transfer between reservoirs. The carbon dioxide and dissolved solid (DS) mean residence times differ because significant quantities of carbon dioxide are transferred through the atmosphere (after Broecker *et al.*, 1960).

The validity of conclusions drawn from this model depends not only on the existence of a steady state but also on the following assumptions: (i) that the activity of the indicated source region of each mass is characteristic of the water added to the mass; (ii) that mixing between adjacent masses does not affect their respective activities; (iii) that solution of sea-bed carbonates or oxidation of organic material produces only a negligible change in activity.

Although simplifications are necessarily built into this model, which probably does not provide a unique interpretation of world-wide ocean bicarbonate radio-carbon data, it does illustrate the potentialities of

* In later discussions Broecker (1963) suggests that correction for industrial carbon dioxide can be ignored.

radio-isotope tracer methods in this field by the way in which it brings together and correlates the results of a number of separate investigations, and indicates mixing times which appear reasonable.

VII. Transfer of Carbon Dioxide across a Gas–Liquid Interface

Even when the partial pressure of carbon dioxide in the atmosphere (p_{CO_2}) is equal to the partial pressure of carbon dioxide in the sea water (P_{CO_2}), exchange of carbon dioxide between the gaseous and liquid phases occurs since the equilibrium condition is really a state of dynamic balance between two opposing processes. In the more usual event that the two partial pressures differ, there will be a net flux of carbon dioxide from one phase to the other, although in general the amount of this flux per unit time and per unit area is small compared with the corresponding total transfer in either direction. Thus, if the rate of transfer of carbon dioxide from the atmosphere to the sea is proportional to the atmospheric partial pressure of carbon dioxide, and the rate of transfer in the reverse direction is proportional to the concentration of dissolved carbon dioxide, then the net rate of accumulation in the atmosphere is given by

$$\frac{dQ}{dt} = k_s \,.\, c_{CO_2} - k_a \,.\, p_{CO_2} \tag{59}$$

where k_s and k_a are the appropriate velocity constants. Since the concentration of dissolved carbon dioxide is related to the partial pressure of carbon dioxide in the water by equation (36), the above equation may be written

$$\frac{dQ}{dt} = k_s \,.\, \alpha \,.\, P_{CO_2} - k_a \,.\, p_{CO_2} \,. \tag{60}$$

At equilibrium, when there is no net transfer across the interface, dQ/dt is zero, and consequently $k_s \,.\, \alpha \,.\, P_{CO_2} = k_a \,.\, p_{CO_2}$. Since at equilibrium P_{CO_2} is equal to p_{CO_2}, it follows that $k_a = k_s\alpha$, and equation (60) becomes

$$\frac{dQ}{dt} = k_a \left(P_{CO_2} - p_{CO_2} \right) \tag{61}$$

which is an expression of the Bohr (1899) equation used in studies of gas–liquid invasion and evasion rates.

The simplest model to which the above considerations can be applied assumes that the interface is a single plane surface, that each phase is characterized by a unique value for the carbon dioxide concentration (i.e. no concentration gradient) and that the exit rate from each phase is proportional to the concentration of carbon dioxide in that phase.

However, it is unlikely that under most conditions the sea surface even approximates to a single surface separating two well-mixed phases. Usually, and certainly under relatively quiescent conditions, it is more likely that two boundary layers exist, one in the gaseous phase and the other in the liquid phase; passage of molecules through these layers probably occurs by the relatively slow process of molecular diffusion. The boundary on the liquid side of the interface is probably more important in determining the transfer rate than is that on the gaseous side; its thickness, which may amount to 0·1 mm., is a function of the amount of turbulence in the water. Bolin (1960) has pointed out that the short atmospheric residence time of water compared with that of carbon dioxide (water, *c.* 10 days; CO_2, *c.* 5–10 years) indicates that it is slow transport in the aqueous boundary layer rather than in the atmospheric layer which is the rate-determining factor in carbon dioxide uptake by the sea.

In the region of the interface, when the partial pressure of carbon dioxide in the atmosphere exceeds the P_{CO_2} of the water, the net rates of downward transfer through the gaseous and the liquid boundary layers will be given by $-D_g(dCO_2/dz)_g$ and $-D_l(dCO_2/dz)_l$, where the subscripts g and l refer to the gaseous and liquid phases respectively, D_g and D_l being the appropriate diffusion coefficients. For steady state conditions, these two rates will be equal. The diffusion constant in the gaseous phase will be in the order of 10^{-2} cm.^{2}/second, whereas that in the liquid phase will be in the order of 3×10^{-5} cm.^{2}/second, and thus it is evident that in most circumstances the rate controlling factor is the rate of diffusion through the liquid phase. Diffusion through the gaseous boundary layer could be an important limiting factor only if the layer were some 10^3 times thicker than the liquid boundary layer. As turbulence increases, the thickness of the film at the liquid surface will be reduced and will no longer be constant with time, and the film may be alternately renewed and removed by eddy diffusion.

The situation with respect to carbon dioxide is more complex than with other atmospheric gases since carbon dioxide participates in the series of equilibria indicated in Section II,C. The attainment of equilibrium is not instantaneous, and there is the possibility that the finite rate of hydration of the dissolved carbon dioxide to give carbonic acid (and the reverse process) may be a factor in controlling the rate of entry of carbon dioxide to the water (Bolin, 1960).

Bolin has examined the steady state passage of carbon dioxide through the boundary layer, and has considered the effect of slow hydration. Two extreme situations can be visualized, (i) diffusion slow compared with the rate of hydration; (ii) diffusion rapid compared with the rate of hydration. If hydration is rapid, equilibrium will be rapidly established at the surface, and the downwards transport through the layer will be caused by the migration of all the species of the carbonate equilibria (HCO_3^-, CO_3^{2-}, H_2CO_3 and dissolved CO_2) down their respective concentration gradients. If hydration is slow compared to the time taken for diffusion through the boundary layer, the transport would be essentially that caused by the migration of dissolved carbon dioxide only down its concentration gradient. By making use of reasonable values for the velocity constant for the hydration process (*c.* 3×10^{-2}/second) and the diffusion coefficient in water (2×10^{-5} cm.^{2}/second), Bolin was able to show that his treatment indicated that the total flux was within 10% of that for the limiting case (i) above if the thickness of the boundary layer was greater than 20 cm., and was within 10% of the limiting flux for case (ii) above if the thickness of the layer was less than 130μ. Bolin considered that the mean atmospheric residence time of carbon dioxide is consistent with a thickness of the boundary layer of about 35μ. This is the thickness obtained by Wyman *et al.* (1952) from studies of the diffusion of oxygen and nitrogen into sea water.

Several workers have examined the rate of transfer of carbon dioxide through liquid interfaces on a laboratory scale (e.g. Hanya and Ishiwatari, 1961; Sugiura *et al.*, 1963; Kanwisher, 1963a,b), and by controlled field experiments (Sugiura *et al.*, 1963). Laboratory investigations can be made by measuring the rate at which equilibrium is approached between a stirred liquid and a circulating gas by following the rate of loss or gain of carbon dioxide by the gas phase. Radio-carbon is useful as a tracer in this respect since, with it, it is possible to follow the transfer process while the system is virtually at chemical equilibrium (Sugiura *et al.*, 1963).

Most workers agree that factors on the liquid side of the interface are mainly responsible for limiting the rate of transfer, and in general they find that increased turbulence leads to an enhanced transfer rate. Under natural conditions, wind is the main agent which causes turbulence and Kanwisher has carried out wind tunnel experiments in which it was possible to show that below a rather low threshold velocity the rate of transfer changed little with the air speed. Above this critical velocity the rate of transfer was roughly proportional to the square of the wind speed. Hanya and Ishiwatari (1961) have examined the rate of transfer in an apparatus in which air and sea water could be arranged to flow with controlled velocities in a counter-current system. They found that the coefficient k_a of equation (61) tended to increase with increasing rate of flow, although they detected a slight difference between the values for sorption and desorption. Their experiments also suggested that the value of k_a increased when the partial pressure of carbon dioxide in the air was close to the P_{CO_2} of the water.

Kanwisher (1963a,b) has examined the effect of additions of catalysts to the aqueous phase on the rate of passage of carbon dioxide from the gas phase to a stirred aqueous solution. Addition of carbonic anhydrase (which catalyses the hydration process) was almost without effect on the rate at which carbon dioxide entered a stirred solution which had been slightly impoverished in dissolved carbon dioxide. This observation suggests that in his experiments the rate of hydration was not an important factor in the determination of the rate of entry. This sort of experiment (i.e. varying the rate of hydration by the addition of a catalyst) is probably the most direct way of testing the suggestion that hydration controls the air–sea transfer rate, and further investigations using boundary layers of different thicknesses would be of value.

Values of invasion and evasion rates reported in the literature often differ widely, probably because of the strong influence of turbulence which cannot be controlled in a reproducible manner. Sugiura *et al.* (1963) have investigated the rates of transfer from the gas phase to a stirred solution, and have shown that the transfer rate was independent of the salinity. They estimated the rate at hypothetical zero turbulence by extrapolating the results obtained at various stirrer speeds to zero rate of stirring. Table X, taken from the paper by Sugiura *et al.*, summarizes field measurements made by monitoring the carbon dioxide content of air circulating through canopies floating on the sea surface, together with the results of previous workers.

The low values correspond to measurements made in static systems, the intermediate values to measurements made under conditions of mild turbulence and the high values to measurements made under

TABLE X

*Rate constants for the transfer of carbon dioxide through the air–sea interface (mg./cm.²/minute/atm)**

Field experiment 1 (invasion)	1·91
Field experiment 2 (invasion)	0·91
Field experiment (evasion)	1·38
Bohr (1899)	0·26
Becker (1924)	0·033
Miyake and Hamanda (see Sugiura *et al.*, 1963)	0·020
Guyer and Tobler (1934)	6·6
Sugiura *et al.* (1963)	0·32

* Sugiura *et al.* (1963).

conditions of vigorous turbulence. Field measurements fall between the values found for static systems and for mild turbulence in laboratory experiments. It might be noted that an atmospheric residence time for carbon dioxide of seven years would correspond to an entry rate of about 20 moles/m.^{2}/year or $1{\cdot}77 \times 10^{-4}$ mg./cm.^{2}/minute. For an atmospheric partial pressure of carbon dioxide of 320×10^{-6} atm, this entry rate would be satisfied by a rate constant of 0·55 mg./cm.^{2}/minute/atm.

An estimate of the rate of evolution of carbon dioxide by the sea surface of the equatorial Pacific Ocean has been attempted by Keeling (1965). He made use of the analysis of the distribution of atmospheric carbon dioxide over the Pacific Ocean given by Bolin and Keeling (1963) in which they concluded that between latitudes 30°N. and 30°S., carbon dioxide is lost from the sea to the atmosphere at a rate equivalent to 10 p.p.m./year of the atmosphere confined within these latitude limits. This figure corresponds to a rate of passage of 13 p.p.m./year if the evolved carbon dioxide were confined to the atmosphere immediately above the sea within this belt, and is equivalent to an average equatorial net flux of $13 \times 10^{-6} \times N$ moles of CO_2/cm.^{2}/year, where N is the total number of moles of atmospheric gas above 1 cm.^{2} of the earth's surface. Keeling combined this estimate with that for the average difference in the carbon dioxide partial pressures of the atmosphere and surface water in Pacific equatorial regions. Results obtained on the Monsoon and Downwind Expeditions (see Keeling *et al.*, 1965) indicated that the carbon dioxide partial pressure of equatorial ocean surface water is independent of the time of year, and that the average difference between the atmospheric and surface water partial pressures of carbon dioxide is about 25 p.p.m. By inserting this estimate for the partial pressure difference and the above estimate for the net flux through unit area of the surface into equation (61), Keeling (1965) obtained an evasion rate of *c.* 1·5 mg.CO_2/cm.^{2}/minute/atm. Although this value corresponds to an atmospheric residence time (*c.* two years) less than the average for the seas as a whole (*c.* seven years) and is, therefore, appreciably less than the value that might be expected for this region where exchange should be less vigorous than the average, it is at least in the same order of magnitude as the time indicated by radio-carbon methods, and indicates that, in general terms, the basic assumptions underlying Bolin and Keeling's estimate of the rate of evolution of carbon dioxide by the equatorial Pacific Ocean are correct.

ACKNOWLEDGMENTS

The author would like to express his thanks to Dr. Stina Gripenberg and to Dr. D. W. Hood for their many helpful comments. He would also like to thank Dr. John Lyman for the loan of his thesis, and Drs. C. D. Keeling and L. S. Waterman for providing an advance copy of the Report of the Lusiad Expedition.

REFERENCES

Anderson, D. H., and Robinson, R. J. (1946). *Industr. Engng Chem.* (*Anal.*) **18**, 767.

Arnold, J. R., and Anderson, E. C. (1957). *Tellus*, **9**, 28.

Arons, A. B., and Kientzler, C. F. (1954). *Trans. Amer. geophys. Un.* **35**, 722.

Barnes, H. (1959). "Apparatus and methods of oceanography," 341 pp. Interscience Publishers Inc., New York.

Bates, R. G. (1954). "Electrometric pH determinations", 331 pp. Wiley, New York.

Bates, R. G. (1959). *In* "Treatise on Analytical Chemistry" (I. M. Kolthoff, P. J. Elving and E. B. Sandell, eds.), Vol. I. Interscience Publishers Inc., New York.

Becker, H. G. (1924). *Ind. Engng Chem.* **16**, 1220.

Berg, D., and Patterson, A. (1953). *J. Amer. chem. Soc.* **75**, 5197.

Beyers, R. J., and Odum, H. T. (1959). *Limnol. Oceanogr.* **4**, 449.

Beyers, R. J., and Odum, H. T. (1960). *Limnol. Oceanogr.* **5**, 228.

Bien, G. S., Rakestraw, N. W., and Suess, H. E. (1960). *Tellus*, **12**, 436.

Bien, G. S., Rakestraw, N. W., and Suess, H. E. (1963a). *Bull. Inst. océanogr. Monaco*, **61**, No. 1278.

Bien, G. S., Rakestraw, N. W., and Suess, H. E. (1963b). Nuclear Geophysics. *Nat. Acad. Sci., Nat. Res. Coun. Publ.* **1075**, 152.

Bischof, W. (1960). *Tellus*, **12**, 216.

Bischof, W. (1962). *Tellus*, **14**, 87.

Bohr, C. (1899). *Ann. Phys., Lpz.* **68**, 500.

Bolin, B. (1959). *Proc. Nat. Acad. Sci., Wash.* **45**, 1663.

Bolin, B. (1960). *Tellus*, **12**, 274.

Bolin, B., and Eriksson, E. (1959). *In* "Rossby Memorial Volume", pp. 130–142. Rockefeller Institute Press, New York.

Bolin, B., and Keeling, C. D. (1963). *J. geophys. Res.* **68**, 3899.

Brander, E. (1932). *Comment. phys.-math., Helsingf.* **VI**, 8.

Brannon, H. R., Daughtry, A. C., Perry, D., Whitaker, W. W., and Williams, M. (1957). *Trans. Amer. geophys. Un.* **38**, 643.

Bray, D. (1959). *Tellus*, **11**, 220.

Britton, H. T. S. (1956). "Hydrogen Ions", 476 pp., 4th ed., Vol. I. Van Nostrand, Princeton, New Jersey.

Brodie, J. W., and Burling, R. W. (1958). *Nature, Lond.* **181**, 107.

Broecker, W. S. (1963). *In* "The Sea" (M. N. Hill, ed.), Vol. II, pp. 88–108. Interscience Publishers Inc., New York.

Broecker, W. S. (1963). Nuclear Geophysics. *Nat. Acad. Sci., Nat. Res. Coun. Publ.* **1075**, 138.

Broecker, W. S., and Walton, A. (1959). *Science*, **130**, 309.

Broecker, W. S., and Olson, E. A. (1960). *Science*, **132**, 712.

Broecker, W. S., Olson, E. A., and Bird, J. (1959). *Nature, Lond.* **183**, 1582.

Broecker, W. S., Gerard, R., Ewing, M., and Heezen, B. C. (1960). *J. geophys. Res.* **65**, 2903.

Broecker, W. S., Ewing, M., Gerard, R., Heezen, B. C., and Kulp, J. L. (1958). *Nat. Acad. Sci., Nat. Res. Coun. Publ.* **572**, 118.
Bruneau, L., Jerlov, N. G., and Koczy, F. F. (1948). *Report of the Swedish Deep Sea Expedition 1947-8*, **3**, 99.
Buch, K. (1929). *Meeresforch. No. 61*
Buch, K. (1930). *Rapp. Cons. Explor. Mer*, **67**, 51.
Buch, K. (1933a). *J. Cons. int. Explor. Mer*, **8**, 309.
Buch, K. (1933b). *Nature, Lond.* **131**, 688.
Buch, K. (1933c). *Rapp. Cons. Explor. Mer*, **85**, 71.
Buch, K. (1938). *Acta Acad. åbo.* **11**, No. 5.
Buch, K. (1939a). *Acta Acad. åbo.* **11**, No. 9.
Buch, K. (1939b). *Acta Acad. åbo.* **11**, No. 12.
Buch, K. (1942). *Ann. Hydrogr., Berl., Jahrg.* **70**, 193.
Buch, K. (1945). *Fennia*, **68**, 208 pp.
Buch, K. (1948). *Eripainos Geophys.* **3**, 63.
Buch, K. (1951). *Meeresforch. No. 151.*
Buch, K., and Gripenberg, S. (1932). *J. Cons. int. Explor. Mer*, **7**, 233.
Buch, K., and Nynäs, O. (1939). *Acta Acad. åbo.* **12**, No. 3.
Buch, K., Harvey, H. W., Wattenberg, H., and Gripenberg, S. (1932). *Rapp. Cons. Explor. Mer*, **79**, 1.
Burling, R. W., and Garner, D. M. (1959). *N.Z. J. Geol. Geophys.* **2**, 799.
Callendar, G. S. (1938). *Quart. J. R. met. Soc.* **64**, 223.
Callendar, G. S. (1940). *Quart. J. R. met. Soc.* **66**, 395.
Callendar, G. S. (1949). *Weather*, **4**, 310.
Callendar, G. S. (1957). *Tellus*, **9**, 421.
Callendar, G. S. (1958). *Tellus*, **10**, 243.
Chapman, H. W., Gleason, L. S., and Loomis, W. E. (1954). *Plant Physiol.* **29**, 500.
Clark, M. W. (1928). "The Determination of Hydrogen Ions", 717 pp., 3rd ed. Williams and Wilkins, Baltimore.
Cloud, P. E. (1962). *Geochim. et cosmoch. Acta*, **26**, 867.
Cloud, P. E. (1963). *Prof. Pap. U.S. geol. Surv. 350.*
Cooper, L. H. N. (1932). *J. Mar. biol. Ass. U.K.* **18**, 729.
Cooper, L. H. N. (1956). *J. Mar. biol. Ass. U.K.* **35**, 341.
Cox, J. D., and Head, A. J. (1962). *Trans. Faraday Soc.* **58**, 1839.
Craig, H. (1953). *Geochim. et cosmoch. Acta*, **3**, 53.
Craig, H. (1954). *J. Geol.* **62**, 115.
Craig, H. (1957). *Tellus*, **9**, 1.
Craig, H. (1958). *Proc. Second Int. Conf. Peaceful uses of Atomic Energy*, **18**, 358.
Craig, H. (1963). *In* "Earth Science and Meteoritics" (J. Geiss and E. D. Goldberg, eds.), pp. 103–114. North Holland Publishing Co., Amsterdam.
Crowe, C. (1958). *Nature, Lond.* **182**, 470.
Danckwerts, P. V., and Melkersson, K. H. (1962). *Trans. Faraday Soc.* **58**, 1832.
Davies, C. W. (1962). "Ion Association", 190 pp. Butterworths, London.
de Vries, H. (1958). *Proc. Acad. Sci. Amst.* **61B**, 1.
Dietrich, G. (1957). "General Oceanography", 588 pp. Interscience Publishers Inc., New York.
Dingle, H. N. (1954). *Tellus*, **6**, 342.
Disteche, A. (1959). *Rev. sci. Instrum.* **30**, 474.
Disteche, A., and Dubuisson, M. (1960). *Bull. Inst. océanogr. Monaco*, **57**, No.11 74.

Dole, M. (1941). "The Glass Electrode", 332 pp. Wiley, New York.
Edsall, J. T., and Wyman, J. (1958). "Biophysical Chemistry", 699 pp., Vol. I. Academic Press, New York.
Ellis, A. J. (1959). *J. Chem. Soc.* 3689.
Eriksson, E. (1963). *J. geophys. Res.* **68**, 3871.
Faurholt, C. (1924). *J. Chim. phys.* **21**, 400.
Feldman, I. (1956). *Analyt. Chem.* **28**, 1859.
Fergusson, G. J. (1958). *Proc. roy. Soc.* **A 243**, 561.
Fonselius, S., and Ostlund, G. (1959). *Tellus,* **11**, 77.
Fonselius, S., Koroleff, F., and Buch, K. (1955). *Tellus,* **7**, 258.
Fonselius, S., Koroleff, F., and Wärme, K. E. (1956). *Tellus,* **8**, 176.
Garner, D. M. (1958). *Nature, Lond.* **182**, 466.
Garrels, R. M., and Thompson, M. E. (1962). *Amer. J. Sci.* **260**, 57.
Garrels, R. M., Thompson, M. E., and Siever, R. (1961). *Amer. J. Sci.* **259**, 24.
Gast, J. A., and Thompson, T. G. (1958). *Analyt. Chem.* **30**, 1549.
Gold, V. (1956). "pH Measurements", 125 pp. Methuen, London.
Greenberg, G., and Moberg, E. G. (1932). *Nat. Acad. Sci., Nat. Res. Coun. Bull.* **No. 89**, 73.
Greenberg, G., Moberg, E. G., and Allen, E. (1932). *Industr. Engng Chem. (Anal.)* **4**, 309.
Greenwald, I. (1941). *J. biol. Chem.* **141**, 789.
Gripenberg, S. (1937). *5th Hydrol. Conf. Helsingfors. Comm.* **10B**
Gripenberg, S. (1960). *J. Cons. int. Explor. Mer,* **26**, 5.
Guyer, A., and Tobler, B. (1934). *Helv. Chim. Acta,* **17**, 550.
Hanya, T., and Ishiwatari, R. (1961). *J. Oceanogr. Soc. Japan,* **17**, 28.
Harned, H. S., and Davis, R. (1943). *J. Amer. chem. Soc.* **65**, 2030.
Harned, H. S., and Owen, B. B. (1958). "The Physical Chemistry of Electrolyte Solutions", 803 pp., 3rd. ed. Reinhold, New York.
Harned, H. S., and Scholes, S. R. (1941). *J. Amer. chem. Soc.* **63**, 1706.
Harvey, H. W. (1955). "The Chemistry and Fertility of Sea Waters". Cambridge University Press, Cambridge.
Hayes, F. N., Anderson, E. C., and Arnold, J. R. (1955). *Proc. Int. Conf. Peaceful Uses of Atomic Energy, Geneva,* **14**, 188.
Hindman, J. C. (1943). Thesis, University of California. Now in proof, p. 12.
Hood, D. W. (1963). Great Lakes Research Division, University of Michigan. Publ. **10**, 91.
Hood, D. W., and Ibert, E. R. (1960). *Oceanographical Congress, Helsinki. Comm. K* 6.
Hood, D. W., Berkshire, D., Adams, R., and Supernaw, I. (1963). A & M College of Texas, Technical Report, Ref. 63-3D.
Huber, B. (1951). *Arch. Met., Wien,* **4**, 54.
Huber, B. (1952). *Forstwiss. Zbl.* **71**, 372.
Hutchinson, G. E. (1954). *In* "The Earth as a Planet" (G. Kuiper, ed.). Chicago University Press, Chicago.
Hutchinson, G. E. (1957). "A Treatise on Limnology", Vol. I, 1015 pp. Wiley & Sons, New York.
Ibert, E. R., and Hood, D. W. (1963). A & M College of Texas. Technical report. Ref. 63-9-T.
Ingerson, E. (1962). *Geochim. et cosmoch. Acta,* **26**, 815.
Jennings, P. P., and Osborn, E. M. (1957). *Analyst,* **82**, 671.

Junge, C. E. (1962). *Tellus*, **14**, 242.
Junge, C. E. (1963). *J. geophys. Res.* **68**, 3849.
Kanwisher, J. (1960). *Tellus*, **12**, 209.
Kanwisher, J. (1963a). *J. geophys. Res.* **68**, 3921.
Kanwisher, J. (1963b). *Deep Sea Res.* **10**, 195.
Kaplan, L. D. (1960). *Tellus*, **12**, 204.
Keeling, C. D. (1958). *Geochim. et cosmoch. Acta*, **13**, 322.
Keeling, C. D. (1961). *Geochim. et cosmoch. Acta*, **24**, 277.
Keeling, C. D. (1965). *J. Geophys. Res.* In the press.
Keeling, C. D., Rakestraw, N. W., and Waterman, L. S. (1965). *J. geophys. Res.* In the press.
Kern, D. M. (1960). *J. chem. Educ.* **37**, 14.
Koczy, F. F. (1956). *Deep Sea Res.* **3**, 279.
Koefoed, J., and Engel, K. (1961). *Acta chem. scand.* **15**, 1319.
Kolthoff, I. M., and Laitinen, H. A. (1948). "pH and Electrotitrations", 190 pp, 2nd ed. John Wiley & Sons, New York.
Krogh, A. (1910). *Skand. Arch. Physiol.* **23**, 224.
Lieth, H. (1963). *J. geophys. Res.* **68**, 3887.
Libby, W. F. (1955). "Radioactive Dating", 175 pp., 2nd ed. University of Chicago Press, Chicago.
Ljunggren, S., and Lamm, O. (1958). *Acta chem. scand.* **12**, 1834.
Lodge, J. P., Frank, E. R., and Ferguson, J. (1962). *Analyt. Chem.* **34**, 702.
Loomis, W. F. (1958). *Analyt. Chem.* **30**, 1825.
Loveland, J. W., Adams, R. W., King, H. H., Nowak, F. A., and Cali, L. J. (1959). *Analyt. Chem.* **31**, 1008.
Lyman, J. (1956). Thesis, University of California.
MacKay, C., Pandow, M., and Wolfgang, R. (1963). *J. geophys. Res.* **68**, 3929.
Markham, A. E., and Kobe, K. A. (1941). *Chem. Rev.* **28**, 519.
Mills, A., and Urey, H. C. (1940). *J. Amer. chem. Soc.* **62**, 1019.
Miyake, Y., and Hamada, A. (1960). *Oceanographical Congress, Helsinki Comm. K 13.*
Miyake, Y., and Matsuo, S. (1963). *Pap. Met. Geophys., Tokyo*, **14**, 120.
Moberg, E. G., Greenberg, D. M., Revelle, R., and Allen, E. C. (1934). *Bull. Scripps. Instn. Oceanogr.* **3**, 231.
Möller, F. (1963). *J. geophys. Res.* **68**, 3877.
Morrison, T. J., and Billett, F. (1952). *J. Chem. Soc.* 3819.
Munnich, K. O., Ostlund, G., and de Vries, H. (1958). *Nature, Lond.* **182**, 1432.
Munnich, K. O., and Vogel, J. C. (1958). *Naturwissenschaften*, **48**, 327.
Munnich, K. O., and Vogel, J. C. (1959). Paper presented at Radiocarbon Dating Conference, Gröningen, Netherlands.
Nasanen, R. (1946). *Acta chem. fenn.* **19B**, 90.
Nasanen, R. (1947). *Acta chem. scand.* **1**, 204.
Noakes, J. E., and Hood, D. W. (1961). *Deep Sea Res.* **8**, 121.
Orr, A. P. (1947). *Proc. roy. Soc. Edinb.* **B63**, 3.
Owen, B. B., and Brinkley, S. R. (1941). *Chem. Rev.* **29**, 461.
Park, K., Hood, D. W., and Odum, H. T. (1958). *Institute of marine Sci.* **5**, 47.
Pinsent, B. R. W., Pearson, L., and Roughton, F. J. W. (1956). *Trans. Faraday Soc.* **52**, 1512.
Plass, G. N. (1956). *Tellus*, **8**, 140.
Plass, G. N. (1959). *Sci. Amer.* **201**, 41.

Plass, G. N. (1961). *Tellus*, **13**, 296.
Pobiner, H. (1962). *Analyt. Chem.* **34**, 878.
Postma, H. (1964). *Netherlands J. Sea Research*, **2**, 258.
Pytkowicz, R. M. (1963). *Deep Sea Res.* **10**, 633.
Rafter, T. A. (1955). *N.Z. J. Sci. Tech. Bull.* **37**, 20.
Rafter, T. A., and Fergusson, G. J. (1958a). *N.Z. J. Sci. Tech. Bull.* **38**, 871.
Rafter, T. A., and Fergusson, G. J. (1958b). *Proc. 2nd U.N. Intern. Conf. Peaceful Uses of Atomic Energy*, **18**, 526.
Rakestraw, N. W. (1949). *J. Mar. Res.* **8**, 14.
Revelle, R., and Fairbridge, R. (1957). *In* "Treatise on Marine Ecology and Paleoecology", Vol. I (J. W. Hedgpeth, ed.). *Geol. Soc. Amer. Mem.* **67**, 239.
Revelle, R., and Suess, H. E. (1957). *Tellus*, **9**, 18.
Revelle, R., and Suess, H. E. (1962). *In* "The Sea" (M. N. Hill, ed.). Interscience Publishing Inc., New York.
Robinson, R. A. (1954). *J. mar. biol. Ass. U.K.* **33**, 448.
Robinson, R. A., and Stokes, R. H. (1959). "Electrolyte Solutions", 2nd ed. Butterworths, London.
Roughton, F. J. W. (1941). *J. Amer. Chem. Soc.* **63**, 2930.
Roughton, F. J. W., and Booth, U. H. (1937). *Chemy Ind.* **56**, 1053.
Saruhashi, K. (1953). *Pap. Met. Geophys., Tokyo*, **3**, 202.
Saruhashi, K. (1955). *Pap. Met. Geophys., Tokyo*, **6**, 38.
Shedlovsky, T., and McInnes, D. A. (1935). *J. Amer. Chem. Soc.* **57**, 1705.
Sirs, J. A. (1958). *Trans. Faraday Soc.* **54**, 201, 207.
Slocum, G. (1955). *Monthly Weather Rev. Oct.* 225.
Slowey, J. E., Jeffrey, L. M., and Hood, D. W. (1962). *Geochim. et Cosmoch. Acta*, **26**, 607.
Smith, J. B., Tatsumoto, M., and Hood, D. W. (1960). *Limnol. Oceanogr.* **5**, 425.
Smith, V. N. (1953). *Instruments*, **26**, 421.
Smith, W. H., and Hood, D. W. (1964). *J. Oceanogr. Soc. Japan.* In the press.
Sørensen, S. P. L. (1909). *Biochem. Zbl.* **21**, 131, 201.
Strickland, J. D. H., and Parsons, T. R. (1960). *Bull. Fish. Res. Bd. Can.* **125**, 185.
Suess, H. E. (1953). *Nuclear Processes in Geologic Settings. Nat. Acad. Sci. Nat. Res. Coun. Pub.* 52.
Suess, H. E. (1955). *Science*, **122**, 415.
Sugiura, Y., Ibert, E. I., and Hood, D. W. (1963). *J. Mar. Res.* **21**, 11.
Sverdrup, H. U., Johnson, M. W., and Fleming, R. H. (1942). "The Oceans", 1087 pp. Prentice Hall, New York.
Takahashi, T. (1961). *J. geophys. Res.* **66**, 477.
Tauber, H. (1960). *Science*, **131**, 921.
Thompson, T. G., and Anderson, D. H. (1940). *J. Mar. Res.* **3**, 224.
Underwood, A. L. (1961). *Analyt. Chem.* **33**, 955.
Underwood, A. L., and Lyman, H. H. (1962). *Analyt. Chem.* **34**, 692.
Van Slyke, D. D. (1922). *J. Biol. Chem.* **52**, 525.
Van Slyke, D. D., and Neill, J. M. (1924). *J. Biol. Chem.* **61**, 523.
Van Slyke, D. D., and Sendroy, J. (1927). *J. Biol. Chem.* **73**, 127.
Verduin, J. (1960). *Limnol. Oceanogr.* **5**, 228.
Walker, A. C., Bray, V. B., and Johnston, J. (1927). *J. Amer. chem. Soc.* **49**, 1235.
Waterman, L. S. (1964). Preliminary Report on Lusiad Expedition. *Scripps Institution of Oceanography.*

Wattenberg, H. (1933). *Wiss. Ergebn. 'Meteor'*, **8**.
Willis, E. H. (1960). *Nature, Lond.* **185**, 552.
Wissburn, K. F., French, D. M., and Patterson, A. (1954). *J. phys. Chem.* **58**, 693.
Worthington, L. V. (1954). *Deep Sea Res.* **1**, 244.
Wyman, J., Scholander, P. F., Edwards, G. A., and Irving, L. (1952). *J. Mar. Res.* **11**, 47.

Chapter 8

Phosphorus

F. A. J. ARMSTRONG

Marine Biological Association
Plymouth, Devon, England

I. Introduction 323
II. Occurrence 323
A. Dissolved Phosphorus 324
B. Suspended Phosphorus 324
C. Analytical Considerations 324
III. Abundance of Various Forms 326
IV. Seasonal Variations 329
A. Coastal Waters 329
B. Diurnal Changes 334
C. The Oceans 335
V. Vertical and Horizontal Distribution of Phosphorus in the Sea 335
A. The North Sea 335
B. The Arctic 337
C. The Atlantic Ocean 342
D. The Antarctic Ocean 352
E. The Pacific Ocean 355
F. The Indian Ocean 358
VI. Comparison of the Major Oceans 359
VII. Nitrogen: Phosphorus Ratio 360
References 362

I. Introduction

The average phosphorus content of sea water has been estimated (Redfield, 1958) at 70–75 μg.P/l., although in the uppermost 100 m. or so it is usually a good deal less. It is in this layer that the element is taken up by phytoplankton and enters the food chain in the sea, and is thereby concentrated by a factor of 10^7–10^8, since plant tissues may contain 2–3% of their dry weight of the element (Ketchum and Redfield, 1949). This phosphorus is returned to the water, possibly well distant from the place and depth at which it was abstracted, on the death and decomposition of the plants or of organisms which have eaten them. The processes of removal and return of phosphorus, combined with physical movement of the water, bring about seasonal and geographical changes in the distribution of the element.

II. Occurrence

Phosphorus may exist in sea water as dissolved phosphate, dissolved organic phosphorus compounds, insoluble and adsorbed phosphates in suspension, and organically combined phosphorus in suspension. These forms may be estimated separately or together by suitable analytical methods (Strickland and Austin, 1960), and are considered below.

A. DISSOLVED PHOSPHORUS

Dissolved phosphate is almost certainly present as orthophosphate, as $H_2PO_4^-$ and HPO_4^{2-} ions, with only negligible amounts of PO_4^{3-} ion or free phosphoric acid (Cooper, 1948). There is no evidence of the presence of pyrophosphate or of polyphosphates in sea water, though the latter have been found in some seaweeds (Stich, 1953).

The nature of the dissolved organic phosphorus compounds can only be guessed at, but if they are derived from decomposing organic material, they may be expected to include phospho- and nucleoproteins, phospholipids, and sugar phosphates, and their breakdown products. No less than six (unidentified) organic phosphorus fractions were detected by paper chromatography in an extract from a medium in which phytoplankton had grown and died (Watt and Hayes, 1963).

B. SUSPENDED PHOSPHORUS

Although many phosphates are, in principle, sufficiently insoluble to be found in suspension in sea water, only those of calcium and ferric iron are likely to be measured in detectable amounts. Dietz *et al.* (1942) have suggested that sea water is saturated with respect to tricalcium phosphate, and a value of $10^{-32 \cdot 5}$ for the solubility product (Holt *et al.*, 1925) supports this. A figure of $10^{-31 \cdot 5}$ for that of ferric phosphate has been suggested by Cooper (1948).

Some phosphate may be adsorbed on detrital material, and this may be desorbed on contact with the strongly acid reagents used for phosphate determination; in addition, this treatment may cause suspended calcium or ferric phosphates to dissolve. Unfiltered water may show higher phosphate estimates than filtered water. Furthermore, with unfiltered water increase with the time of contact with the reagent is frequently noted.

Insoluble combined organic phosphorus is that contained in living or dead organisms or in detritus.

C. ANALYTICAL CONSIDERATIONS

Phosphate in sea water is usually determined by the Dénigès reaction in which the sample is allowed to react with molybdate in strongly acid (0·2–1·1 N) conditions leading to the formation of phosphomolybdate which is reduced to a molybdenum blue and determined absorptiometrically. This reaction determines orthophosphate only, and if other phosphorus compounds are to be determined, they must be converted to orthophosphate. The Dénigès reaction is, however, unsatisfactory in several ways. No well-defined chemical entity is formed, so that colour intensity is rather dependent on experimental conditions. In many variations of the method, the intensity of colour is markedly depressed by the salts in the sea water, an effect not appreciated in early work when samples were compared with standards made with distilled water. Correction factors ranging from about 1·10–1·36 (depending on the method) have been found (Cooper, 1938a).

Many methods are subject to interference by arsenate, which is returned as phosphate (see also Chapter 20, Volume 2). The error introduced is uncertain because the arsenate content of sea water is unknown. Of recent determinations (early ones now appear improbably high), only those of Gorgy *et al.* (1948) give the proportion (8–16%) as arsenate in the total (15–38 μg.As/l.), for East Pacific water. Later figures of total arsenic content are almost an order of magnitude lower, viz. 1–3 μg.As/l. in the English Channel (Armstrong and Harvey, 1950), 3–6 μg.As/l. in the West Pacific (Ishibashi *et al.*, 1951), 1·8–3·6 μg.As/l. in the English Channel and East Atlantic (Smales and Pate, 1952) and 2–3 μg.As/l. in the English Channel (Portmann and Riley, 1964). If these values are typical of the oceans, the error in phosphate determinations from this cause would only be a few per cent, at least at levels above about 20 μg.P/l. even if all arsenic were present as arsenate.

It appears improbable that the strongly acid molybdate reagent can hydrolyse appreciable amounts of organic compounds in the cold and liberate phosphate (Murphy and Riley, 1958). With a rigorous enough digestion, however, organic compounds can be broken down completely and total phosphorus determined (Kreps and Osadchih, 1933; Kalle, 1935; Cooper, 1937a; Redfield *et al.*, 1937; Harvey, 1948; Hansen and Robinson, 1953). The last three methods eliminate interference by arsenic.

Inorganic phosphate and total phosphorus determinations on a sea-water sample before and after filtration will allow the estimation of the four forms of phosphorus listed above. The technique is discussed by Strickland and Parsons (1960). Organically combined phosphorus in suspension, which is found as a small difference between two estimations, is the least accurately determined, and better estimates may be obtained by digestion of the matter filtered from a large volume of water (Redfield, *et al.*, 1937; Ketchum, *et al.*, 1958; Vaccaro, 1963).

III. Abundance of Various Forms of Phosphorus

Early studies showed that organically combined phosphorus in sea water was present in rather smaller amounts than inorganic phosphate (Kreps and Osadchih, 1933; Kalle, 1935, 1937). Redfield *et al.* (1937) studied the changes during a year at a position in the Gulf of Maine (depth 238 m.). The inorganic, organic and particulate phosphorus results are given for various depths in Table I. The table shows a fairly constant total phosphorus content in the whole water column through the year, with little change in the amount in the particulate form. Total phosphorus increases with depth, whereas organic phosphorus decreases.

Three years' observations of total and inorganic phosphorus for a single station 20 miles offshore in the English Channel are given by Armstrong and Harvey (1950). In winter, inorganic phosphate is 85–90% of the total, but decreases to 45% or so in summer. The distribution with depth is shown in Fig. 1(a) and (b).

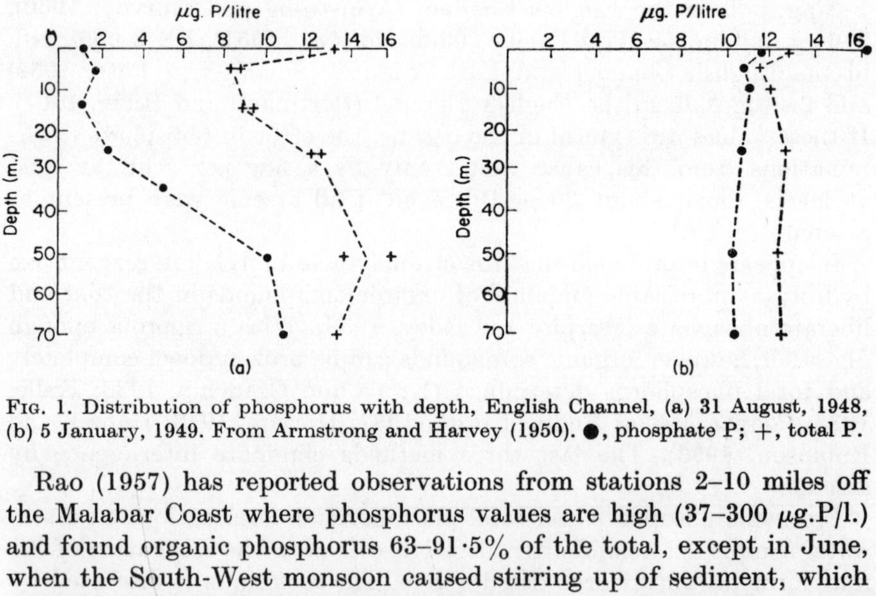

Fig. 1. Distribution of phosphorus with depth, English Channel, (a) 31 August, 1948, (b) 5 January, 1949. From Armstrong and Harvey (1950). ●, phosphate P; +, total P.

Rao (1957) has reported observations from stations 2–10 miles off the Malabar Coast where phosphorus values are high (37–300 μg.P/l.) and found organic phosphorus 63–91·5% of the total, except in June, when the South-West monsoon caused stirring up of sediment, which resulted in relatively less organic phosphorus in solution. Kawamoto (1955) found that organic phosphorus exceeding by 10 times or more the inorganic phosphate concentration in the surface water occurred in deep water in the Kuroshio region of the Western Pacific. Phosphate was less than 30% of the total in the uppermost 100 m., but below

TABLE 1

*Vertical distribution of phosphorus in the Gulf of Maine**

	18 May 1935				20 August 1935				8 November 1935				20 February 1936				14 May 1936			
	Total μg./l.	*Inorganic*	*Organic*	*Particulate*	*Total μg./l.*	*Inorganic*	*Organic*	*Particulate*	*Total μg./l.*	*Inorganic*	*Organic*	*Particulate*	*Total μg./l.*	*Inorganic*	*Organic*	*Particulate*	*Total μg./l.*	*Inorganic*	*Organic*	*Particulate*
Depth m.		%	%	%		%	%	%		%	%	%		%	%	%		%	%	%
0– 60	1·5	12	1·5	3·0	2·1	13	6·2	1·8	1·9	11	5·0	1·7	2·2	20	1·4	0·9	1·7	11	2·3	2·0
60–120	2·2	22	0·4	1·2	2·3	17	5·4	1·0	2·7	19	5·4	0·9	2·3	19	2·7	0·9	3·1	21	6·0	1·2
120–180	2·5	26	0·2	0·8	2·7	22	3·2	0·6	3·0	22	5·0	1·4	2·4	21	3·3	0·6	3·0	25	2·5	0·4
180–240	3·1	32	0·0	0·9	3·0	26	3·7	0·6	3·1	21	6·4	1·4	2·9	29	0·6	0·7	3·2	26	1·7	1·0
TOTAL	9·3	92	2·1	5·9	10·1	78	18·5	4·0	10·7	73	21·8	5·4	9·8	89	8·0	3·1	11·0	83	12·5	4·6

* Redfield *et al.* (1937), percentage of phosphorus in various forms.

500 m. the proportion exceeded 80%. There was little organic phosphorus below 600 or 700 m. This decrease in the organic phosphorus content of the deeper water was also observed in the Equatorial Atlantic (Ketchum *et al.*, 1955). Nearly 50% of the surface samples contained more than 8 μg. organic P/l. (about 50% of the total phosphorus), but this proportion decreased with depth so that below 1000 m. there was little, if any, organic phosphorus. This is in agreement with the author's observations for the deep water in the Eastern North Atlantic (Fig. 2)..

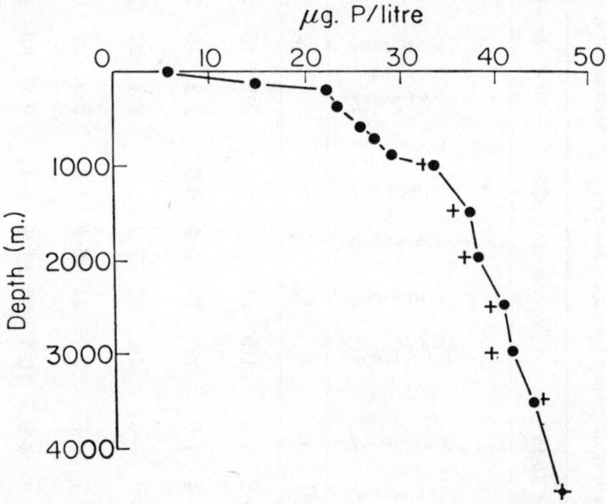

Fig. 2. Distribution of phosphorus with depth, Eastern North Atlantic (46°28'N., 8°06'W.) September, 1952. From Harvey (1955). ●, phosphate P; +, total P.

The distribution of organic phosphorus in the South-East Indian and South-West Pacific Oceans has been reported (Rochford, 1963). Analytically significant concentrations (3–12 μg.P/l.) were found at most depths on three meridional sections. Above 200 m., maximal values of 6–12 μg.P/l. were found along the boundary between the South Equatorial Current and the Equatorial Divergence, and within the boundary of the counter current and the North Equatorial Current. South of 30–40°S. another maximum was found which could not be associated with water movements. The maxima at *c.* 1000 m. found in vertical profiles coincided in depth at 41% of the stations with the salinity minimum indicative of Antarctic Intermediate Water.

In contrast with these waters, where organic phosphorus is present in such small concentrations that reliable estimation is very difficult, are those of the North-East Pacific (Strickland and Austin, 1960). In this area, the highest phosphorus concentrations found in the open ocean occur (even exceeding 120–130 μg.P/l.), of which a substantial fraction, down to all depths, exists as organically combined phosphorus. In the shallow coastal waters off Vancouver Island, the inorganic phosphate varied from 80–95% of the total in winter, and decreased to 25–35% in summer. At the same time, particulate phosphorus rose from less than 5% to 25%. In oceanic waters, particulate phosphorus was appreciable only in the upper 50–100 m. where, in summer, it amounted to about a quarter of the total.

IV. Seasonal Variations

A. COASTAL WATERS

(1) *Temperate latitudes.* Apart from some inconclusive results obtained with insufficiently sensitive methods of analysis (Matthews, 1916, 1917; Brandt, 1920), the study of the phosphorus cycle in sea water was initiated by Atkins in a series of papers starting in 1932. Atkins applied the Dénigès method, and measured phosphate concentrations at standard depths at positions in the English Channel, particularly at one station about 20 miles offshore from Plymouth, with a depth of about 70 m. Maximum values of about 20 μg.P/l. were found early in the year when plant growth was at its minimum because of lack of daylight. In the spring, the increased duration and intensity of daylight caused increased growth of plants, and phosphate concentrations were found to decrease throughout the water column so long as wave and wind action maintained the vertical mixing characteristic of winter conditions. As summer advanced, solar warming of the upper layers set up stable layering with a thermocline at 20–30 m. which restricted vertical mixing. In the warm, well illuminated epithalassa, plant growth could proceed vigorously until phosphate fell to undetectable levels. In the lower part of the hypothalassa, plants tended to die from lack of light, and phosphate was returned to the water. At the end of summer, cooling and mixing ended the layering, and as daylight decreased with the advance of winter, phosphate increased throughout the now homogeneous water column. Typical records are shown in Figs. 3, 4 and 5.

Similar patterns of events have been found to be general in temperate latitudes in the north (e.g. Buch, 1932; Thompson and Robinson, 1934; Bigelow *et al.*, 1940; Riley and Conover, 1956) and also in the south (Dakin and Colefax, 1935), to mention only a few of these studies. In detail, however, different regions vary in the time of year at which the maximum phosphate utilization occurs, and even in the same region, events do not necessarily exactly repeat themselves from year to year.

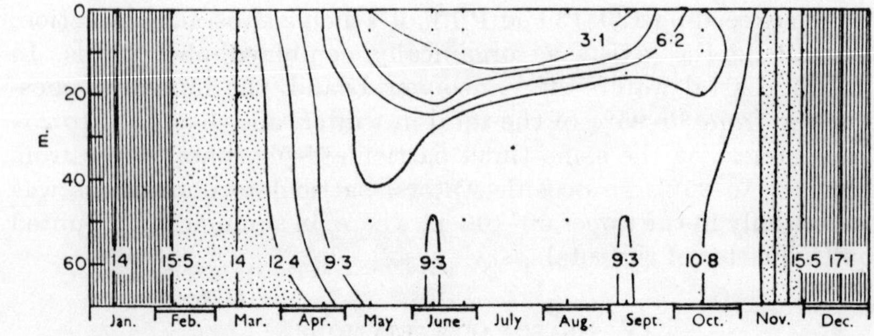

FIG. 3. Vertical distribution of phosphate, for the English Channel (50°04′N., 4°22′W.) 1959 (in μg.P/l.). From Armstrong and Butler (1960b).

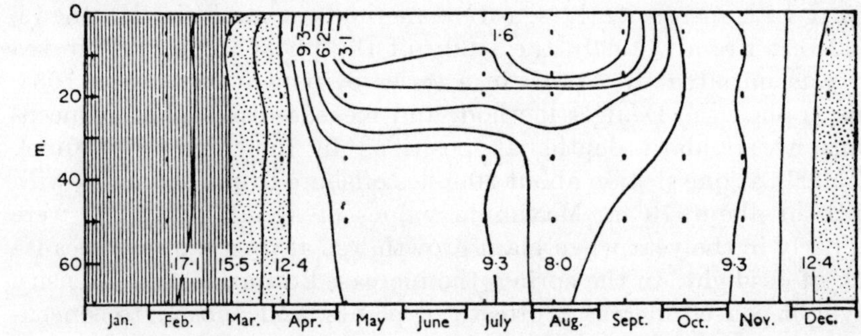

FIG. 4. Vertical distribution of phosphate, for the English Channel (50°04′N., 4°22′W.) 1960 (in μg.P/l.). From Armstrong and Butler (1962).

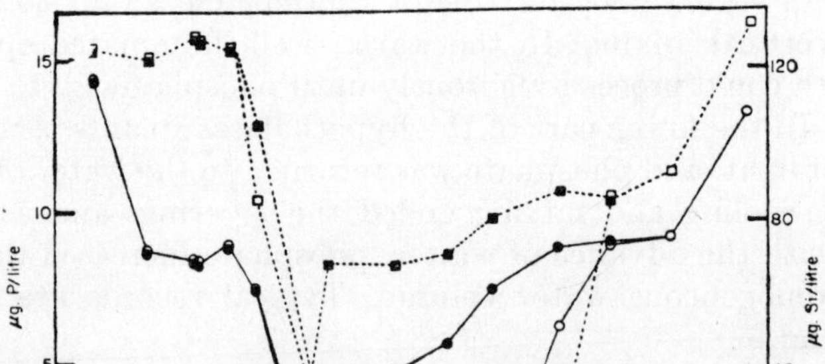

FIG. 5. Seasonal change of phosphate, together with that of silicate at 10 and 50 m., for the English Channel (50°04′N., 4°22′W.) 1954. From Armstrong (1955). – – – –, phosphate; □, 10 m.; ■, 50 m.; ———, silicate; ○, 10 m.; ●, 50 m.

(2) *The tropics*. Probably the first observations are those of Orr (1933) off the NE. Australian coast in the neighbourhood of the Great Barrier Reef. At 33 m. depth, there was little variation from a mean concentration of about 5 μg.P/l., values rarely exceeded 8 and only once or twice approached zero. Orr concluded that there was no seasonal variation, and considered that since phosphate in the local river was negligible, the supply of phosphate was maintained by bacterial decomposition of bottom muds, two samples of which contained respectively 3 and 90 mg.P/kg.

Similarly, Delsman (1939) found in the Java Sea, in about 50 m. of water, only 2–3 μg.P/l. at the surface and 3–4 μg.P/l. in deeper water, and concluded that variations were small.

In the Singapore Straits, almost on the Equator and in very shallow water, higher values and larger fluctuations were found by Tham (1953), as shown in Fig. 6. He concluded that in these waters there was a perceptible annual cycle of phosphorus which correlated with rainfall.

Newell (1959) noted that in the waters off Zanzibar, at a depth of

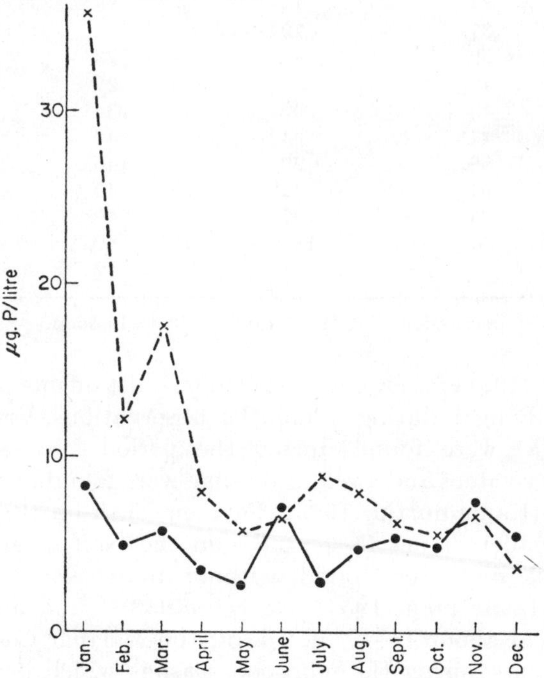

FIG. 6. Monthly average phosphate concentrations in Singapore Straits 1948–49. From Tham (1953). – – – –, 1948; ——, 1949. By permission of H.M. Stationery Office, London.

500 m., a seasonal change in phosphate concentration occurred which reached a maximum of up to 34 μg.P/l. in the period from June to September, with a minimum in October to January, just after the commencement of the North-East monsoon. The vertical distribution is shown in Table II.

TABLE II

*Phosphate and total phosphorus (μg.P/l.) in East African waters**

Depth (m.)	*June 1956 Between Lathom Island and Zanzibar*		*October 1956 Station "Z" 40 km. E. of Zanzibar Island 6°30′S., 40°0′E.*	
	Inorganic	*Total*	*Inorganic*	*Total*
0	19	—	11	—
20	22	—	11	—
40	25	62	12	111
60	28	62	16	124
80	31	62	19	124
100	33	65	22	124
120	33	70	25	124
140	34	74	28	127
160	37	79	28	130
180	39	83	28	136
200	40	88	29	140
240	43	99	34	146
280	47	94	40	152
320	48	90	40	—
360	51	92	40	—
400	53	107	40	—
450	57	136	40	—
500	—	—	42	—

* Newell (1959), by permission of H.M. Stationery Office, London.

Jayaraman (1951) found seasonal changes just offshore from Madras, in the Bay of Bengal, during 17 months' observations. Very high values (15–90 μg.P/l.) were found during the period May to September. Rather smaller values and a reduced range were found in a continuation of this work (Ramamurthi, 1953). Farther south on the east coast of India still smaller values (5–9 μg.P/l. in the Gulf of Manar and 4–8 μg.P/l. in Palk Bay) were found, without any seasonal cycle (Chacko *et al.*, 1954; Jayaraman, 1954). George (1953) found a normal range of phosphate of about 11–17 μg.P/l. off the Malabar Coast with some very high values during the monsoon seasons which he attributed to stirring up of bottom muds; storm water from the land probably also contributed. During the post-monsoon period he observed variations in phosphate content related to outbursts of phytoplankton, but although there were heavy growths of diatoms during the monsoon period no decrease of phosphate could be observed.

Seshappa and Jayaraman (1956) have studied the phosphate content of mud banks along the Malabar Coast, and they (1957) have summarized processes along the coasts of India as follows.

(i) During the South-West monsoon (June to August) both phosphate and plankton contents are very high; at the commencement of the monsoon there is a heavy mortality of bottom animals. The sea is rough throughout the period.

(ii) During the post-monsoon period the sea is calm. Resettlement of bottom animals takes place and there are minor peaks in both phytoplankton and phosphate values, the fluctuations being erratic.

(iii) During the subsequent pre-monsoon months the sea is calm, and both phytoplankton and phosphate values are low. The bottom deposits contain relatively large amounts of adsorbed and interstitial phosphate; the latter is low during the monsoon, and shows erratic fluctuations. These fluctuations are ascribed to the alternation of periods of growth, consuming phosphate, with periods of decay and of grazing by zooplankton with release of phosphate. It is probable that these processes for release of phosphate are more rapid in the tropics than in temperate waters because of the higher temperatures.

The interchange of phosphorus between bottom deposits and the water above may be important in shallow waters, but it has not received much attention, probably because of experimental difficulties. In fresh-water lakes, the liberation of phosphate from mud is controlled by the state of oxygenation of the water next to the bottom; this is subject to seasonal change. High oxygen content in winter produces an oxidized layer which hinders removal of phosphate (Mortimer, 1941, 1942).

Estuarine muds may contain 0·04–0·18% phosphorus (Moore, 1930), or about 10^5 times the concentration in the overlying water, so that the release of only a small fraction of the total would greatly affect the concentration of dissolved phosphorus.

Rochford (1951) studied the interchange between sea water and silt in estuaries in Australia. He estimated total phosphorus in the silt and in the interstitial and adsorbed fractions, and found that the adsorbed phosphorus showed a seasonal cycle following that of the dissolved phosphorus. He found also that there was an increase in phosphate in the bottom water when the oxygen content decreased.

Carritt and Goodgal (1954), found from laboratory experiments that uptake of phosphate by suspended sediments was reversible, and that desorption was favoured by increased pH. They suggest that in nature this effect may be important as a means of increasing the dissolved phosphate available to plants where the suspended matter in rivers meets more alkaline water in estuaries, and also, possibly, when suspended matter from near the bottom is brought by water movement, into upper layers which have been rendered more alkaline by the growth of plants.

Experiments by Jitts (1959) on the uptake of phosphate by river mud using isotope dilution showed that adsorption falls off rapidly below pH 3 and above pH 9. The adsorptive capacity of the mud ranged up to about 3 mg.P/g., and increased as the ratio of iron to organic matter increased.

In the western hemisphere, observations in the Caribbean Sea (Rakestraw and Smith, 1937) showed that outflow of surface waters through the shallow Straits of Florida leads to accumulation of regenerated phosphate in the Caribbean Sea and Gulf of Mexico.

The distribution of phosphorus in the Gulf of Venezuela has been investigated during a hydrographic survey (Redfield, 1955). Seasonal observations were not made, but high concentrations (up to 30 μg.P/l.) were found in the deep water of the Gulf, and at all levels in the innermost part. These were about four times those found in the open Caribbean. It was shown that prevailing winds move the surface water of the Gulf offshore, and this impoverished surface water is replaced by richer water from lower levels outside. This effect is reinforced by sinking and decomposition of organic matter inside the Gulf.

B. DIURNAL CHANGES

Variations of phosphate concentrations with the time of day were noted in Chesapeake Bay (Newcombe and Lang, 1939). Quite large changes were found; phosphate at the surface decreasing slowly after sunrise from 13 μg.P/l. to a minimum value of 3 μg.P/l. at 15.00 hours. In the afternoon, values rose, fell again, and then after sunset increased rapidly to maximum values at 02.00 hours. Newcombe and Brust (1940) found similar effects.

In some Texas bays, Bruce and Hood (1959) observed smaller diurnal changes. For example, phosphate decreased from 2·5 μg.P/l. to minimum values of about 1·5 μg.P/l. at 18.00 hours on a day of bright sunshine. Total and particulate phosphorus showed only slight variations. The results are consistent with the findings of Pratt (1950) that the rate of regeneration of phosphate is proportional to the quantity of phytoplankton crop, and that the rate of utilization is proportional to the growth rate.

In contrast to these observations are some made off Cananéia (Occhipinti *et al.*, 1961). Changes were small, with maximum values of about 1·2 μg.P/l. at noon, and minimum values of about 0·6 μg.P/l. at 13.00–17.00 hours. These observations in the tropics were in very shallow water and at high light intensities.

C. THE OCEANS

It is accepted that seasonal changes in the open ocean in temperate latitudes are similar to those outlined for shallower water, although they will be restricted to the uppermost 50–100 m. This has been demonstrated at three stations from 30–200 miles off the east coast of North America which were worked seven times in one year. It was shown that phosphate above the thermocline was used in summer and regenerated in winter (Rakestraw and Carritt, 1948).

V. Vertical and Horizontal Distribution of Phosphorus in the Sea

Apart from a few observations of total and particulate phosphorus mentioned above, observations in the ocean are of inorganic phosphate only, the other determinations being difficult on board ship. Very large areas of the sea have as yet no observations, but more details will become available as the results of international oceanographic surveys, such as those of the International Geophysical Year and the International Indian Ocean Expedition, are published.

A. THE NORTH SEA

Phosphate in this important fishing region has received attention for many years. Summer conditions were investigated by Atkins (1923), Seiwell (1935), Thomsen (1931) and Graham (1938) among others. Seiwell found the surface in the central area was almost depleted (0·4 –1·8 μg.P/l. of phosphate in July, apart from high values in the mouth of the Elbe), and suggested that the shallowness of the water may allow photosynthesis to proceed at all depths. The much higher values (up to 20 μg.P/l.) which were found between the Faroe and Shetland Isles, were attributed to mixing of surface and deep waters. South of Iceland, where values from 3·5–12 μg.P/l. were found, there was an inverse relationship between phosphate content and stability of the water column; higher values were noted in deep off-shore water. Thomsen (1931) found nutrients to be highest to the north-west and north of the North Sea, and suggested that the inflow of Atlantic water between the

Orkneys and Shetlands carried in phosphate and nitrate, and that mixing occurred on the shelf in this area.

Graham (1938) worked stations across some of the nearer fishing grounds on a section 60–70 miles offshore and roughly parallel to the east coast of England (Fig. 7). In May, the density structure caused little

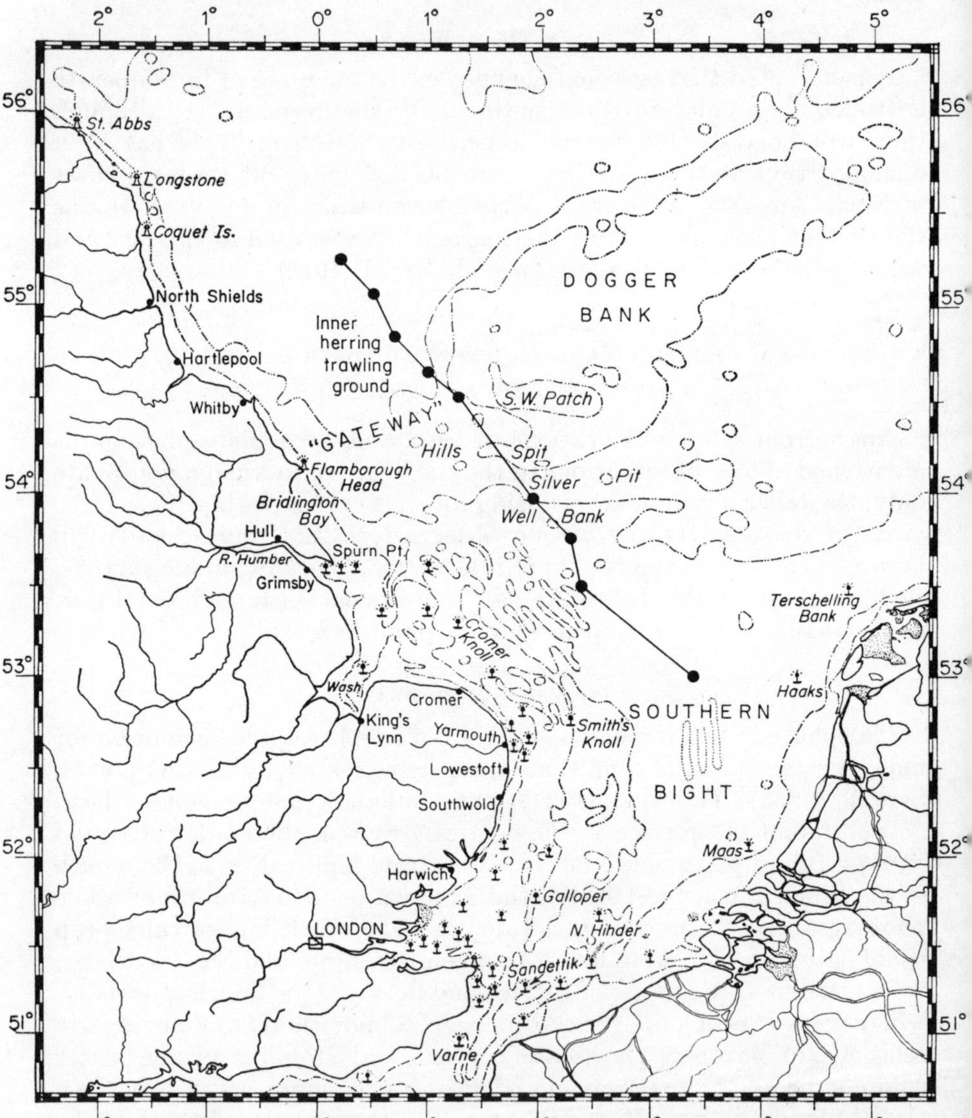

FIG. 7. Fishing grounds and stations in the North Sea. From Graham (1938), by permission of H.M. Stationery Office, London.

hindrance to vertical mixing as there was no marked thermocline. Surface waters over the Inner Herring Trawling Ground to the north were exhausted of phosphate (concentrations were given only in arbitrary units) although the lower layers contained a supply which extended upward to the surface over the Dogger Bank. In June, two large masses of phosphate-rich water were found, one extending from surface to bottom in the South-West Dogger Swirl near the South-West Patch, and the other in the Inner Herring Trawler Ground; between these areas, phosphate concentrations were low. In July, there was some thermal stratification, which was marked to the north-west of the Hills ground, but which decreased to the south-east. Higher phosphate values were found below the thermocline and extended to the Swirl region. Similar conditions with, however, high values at the surface at one place were found in August, and this distribution was repeated in the August of 1936, two years later, when high phosphate values were found in the same position, which suggested that the water of high phosphate concentration had a local origin. Its most likely source was the deep water of the Silver pit. This may also be an additional source of phosphate for the Dogger. There appears to be a relationship between phosphate concentrations and production of diatoms and, therefore, the abundance and size of herring. The spring growth of diatoms in the Clyde, and its effect on phosphate concentrations, is discussed by Marshall and Orr (1927, 1930). Their observation that a specified diatom count was attained by very nearly the same date in a number of years suggests that the total light flux is the controlling factor.

Changes in phosphate concentrations during 1951–3 in the Fladen Ground in the northern North Sea have been used to estimate organic production in this area (Steele, 1956). Temperature observations and the use of diffusivity equations (Sverdrup *et al.*, 1942) allowed the change in phosphate concentration caused by biological activity in each 10 m. layer to be estimated independently of changes caused by water mixing. After discussion of some of the assumptions and necessary corrections, Steele concluded that the zone of phosphate utilization sinks during spring and summer to a depth of *c.* 30–40 m., and that variations from year to year may be related to the varying incidence of high winds.

B. THE ARCTIC

Both seasonal and geographical variations in Arctic waters were shown in the observations of Kreps and Verjbinskaya (1930, 1932). Ten cruises in the Barents Sea were made in 1929–31, stations at 30-mile intervals being worked from the Murmansk coast to the ice edge, along the meridians 33°30′E. and 38°E. The depth of the water was generally less than 300 m. The region is affected by the easterly drift of Atlantic water (the North Cape current) round North Cape and into the Barents Sea, where it divides into north- and south-lying streams. Less Atlantic water reaches the 38°E. meridian than that at 33°30′E.

When the observations were started in May 1929, this inflow of Atlantic water had diminished, and cold Arctic water had advanced to the south. On both meridians, phosphate was high in the deep water (more than 33 μg.P/l.). It was lower and rather variable in the upper layers owing to consumption by phytoplankton. Both sections showed a band of water with a rather uniform vertical distribution; at 33°30′E. it spread from about 70–71°N., but at 38°E., it was narrower at the 70° parallel. These sections apparently represented an intruded east-west band of water which, at least at the surface, was of high phosphorus content but of low phytoplankton content. Temperature and other observations showed it to be Atlantic water, and the authors suggested that the spring growth of plants and the concomitant fall in the phosphate content of the water was delayed because of species differences in the phytoplankton. In spite of the higher nutrient levels in the Atlantic water, the phytoplankton was later in responding to light increases in the early spring than was the Arctic water phytoplankton, which they considered might be composed of species adapted to low light intensities. Vertical upwelling of nutrients from bottom water over submarine banks may be seen in the sections (Figs. 8 and 9).

In September, summer conditions allowed longer sections, to 79° and 76°N., to be worked. There were then no detectable differences in phosphate content between the Arctic and Atlantic water; the upper layers, with less than 10 μg.P/l. being depleted in phosphate. There was, however, a greater growth of phytoplankton in the Atlantic water, perhaps because that in the Arctic had grown earlier and had been consumed by zooplankton. The influence of the bottom topography on the distribution of phosphate in the lower layers could still be seen.

In December, with phytoplankton production brought to a standstill by the polar night, utilization of phosphate in the upper layers had ceased, and strong convectional mixing was making its vertical distribution more uniform. Vertical uniformity was attained by the end of March, 1930, and the beginning of April. No sign of the onset of vernal conditions was detected, and no phytoplankton and little zooplankton were found. In these conditions of winter maximal nutrient concentrations, there were differences in the concentrations at different parts

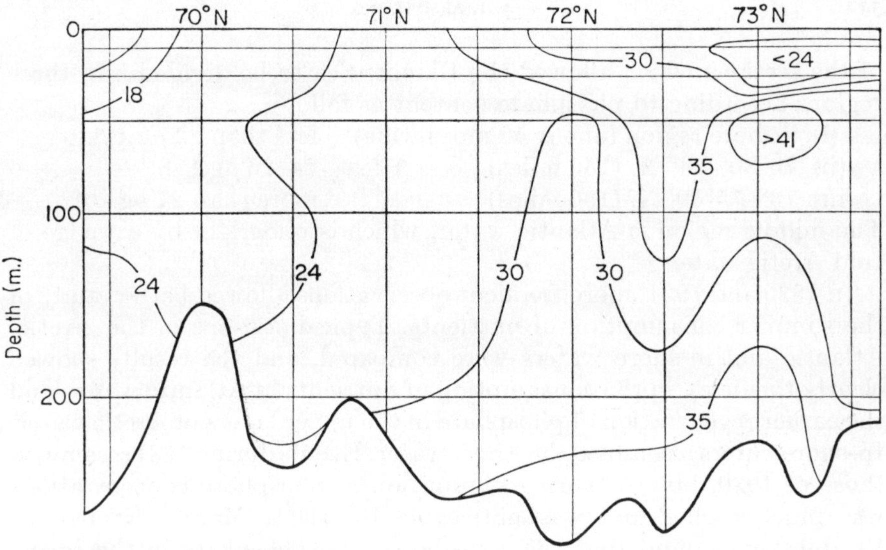

FIG. 8. Phosphate concentrations in Barents Sea (33°30′E.), May 1929 (in μg.P/l.). From Kreps and Verjbinskaya (1930).

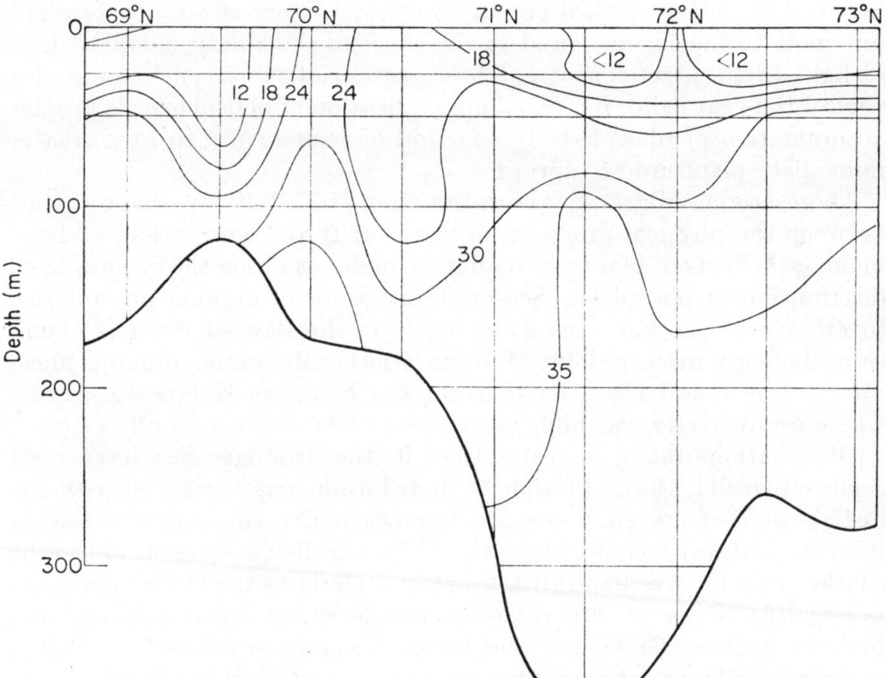

FIG. 9. Phosphate concentrations in Barents Sea (38°E.), May 1929 (in μg.P/l.). From Kreps and Verjbinskaya (1930).

of the sections which allowed the Barents Sea to be divided into three regions according to phosphate content as follows:

(i)	inshore region (about 50 miles wide)	less than 22 μg.P/l.,
(ii)	70°30′–73° N. (150 miles)	24–26 μg.P/l.,
(iii)	73°–75°30′ N. (180 miles)	more than 27 μg.P/l.

The middle region is Atlantic water, which is underlain by a wedge of cold Arctic water.

In 1930 and 1931, more frequent observations allowed better study of the summer consumption of nutrients. Typical stations in the Arctic, Atlantic and in-shore waters were compared, and the results showed clearly the delay in the consumption of nutrients in Atlantic water, and the earlier regeneration of phosphate in the upper layers of Arctic water. In-shore waters resembled the Arctic water. Events during 1931 resembled those of 1930, but the spring diminution in phosphate concentrations was much earlier and was noticeable by 19–21 March. Kreps and Verjbinskaya found that the consumption of phosphate in the spring and summer, when calculated for the whole water column, was greater than that found in the English Channel (Atkins, 1926). This they ascribed to better vertical circulation (which more effectively brought nutrients to the surface), and to the effect of the almost complete lack of light in the Arctic winter. The cessation of photosynthesis in this part of the year led to increased mineralization of phosphorus, a greater accumulation of phosphate by the end of winter, and so to a greater immediate plant crop in spring.

These studies illustrate the interaction, in relatively shallow seas, between the physical processes of water drift and convection, and the biological effects of plants and animals, in determining the geographical distribution of phosphate. Seasonal variations of organic phosphorus, together with those of organic nitrogen, in the Barents Sea have more recently been discussed by Minkina (1961). Maximum organic phosphorus was found near the shore of the Kaminsk-Kolgnevsk region, where productivity was high. A seasonal variation was found.

Observations made in June 1955 in the Irminger Sea have been reported (Kalle, 1957). Phosphate distribution was compared with the distributions of oxygen, fluorescence and turbidity, three sections having been worked: (a), roughly along the 60°N. parallel, from Scotland to the southern tip of Greenland; (b), south of Iceland to the Greenland coast at about 64°N., 38°W., and (c) parallel to the eastern coast of Greenland, midway between Greenland and Iceland. The most interesting finding was a marked layer of maximum phosphate content rising from a depth of 1200 m. in the east to above 300 m. in the west. This layer, with phosphate exceeding 37 μg.P/l., was in places thicker than 1000 m., and the inclination of its upper boundary followed that of the isopycnic lines. The phosphate distribution indicated a greater amount of low-phosphate Gulf Stream water to the east of the area, contrasting with the higher phosphate concentrations in the surface waters of the Irminger Sea itself in the west.

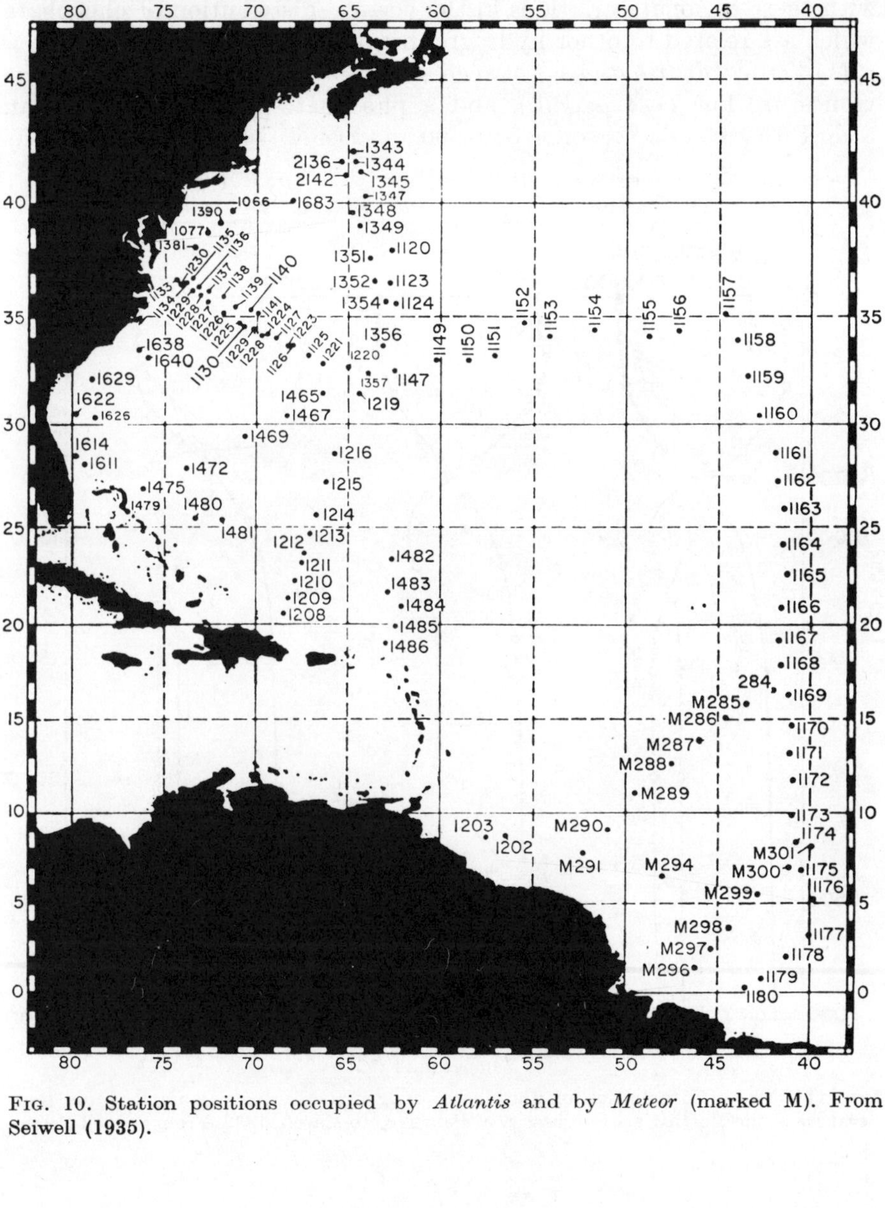

FIG. 10. Station positions occupied by *Atlantis* and by *Meteor* (marked M). From Seiwell (1935).

C. THE ATLANTIC OCEAN

Seiwell (1935) has given a comprehensive account of observations made mostly in 1932–33 in the North Atlantic from the research vessel *Atlantis*. Figure 10 shows the station numbers and positions, and includes some occupied by *Meteor*, which were also taken into consideration. The descriptive part of this account is concerned primarily with those regional variations in the vertical distribution of phosphate which are related to other hydrographic observations.

(1) *Vertical distributions*. Between about 35°W. and Bermuda, surface values are low (1–4 μg.P/l.), and a phosphate-poor layer (less than 5 μg.P/l.) extends to some 350–500 m., below which concentrations

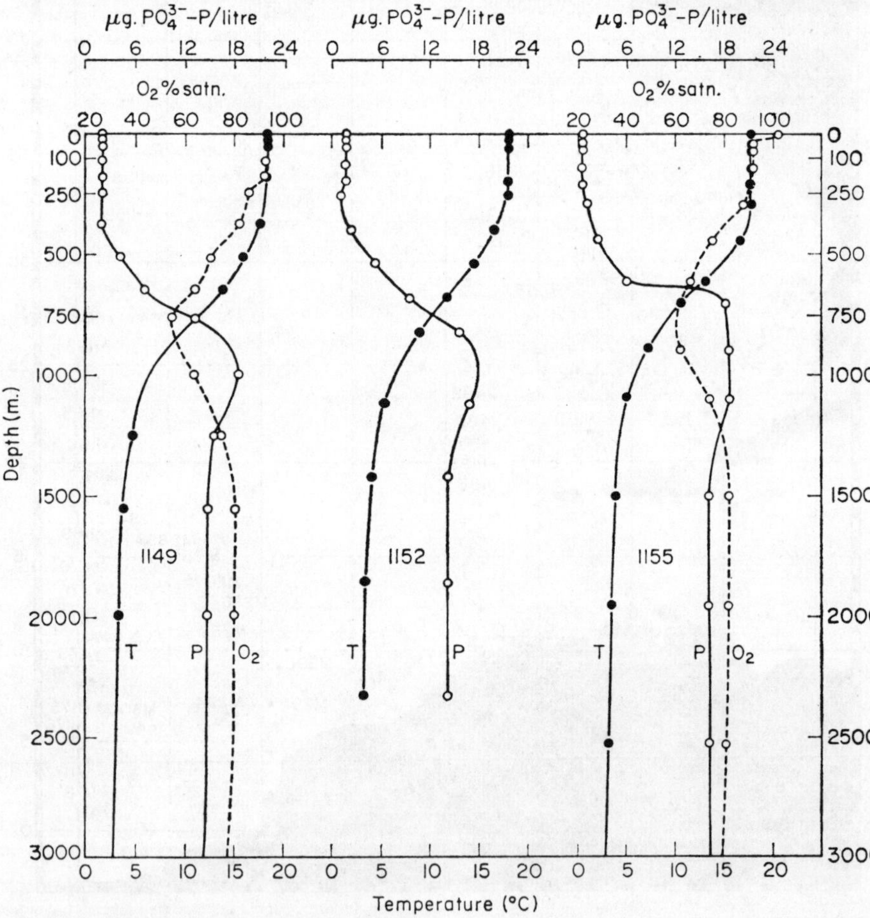

FIG. 11. Vertical distribution of phosphate, oxygen (percentage saturation) and temperature in the North-East Sargasso Sea, February to March, 1932. From Seiwell (1935).

increase to a maximum of about 40 μg.P/l. at depths of 800–1000 m. At greater depths, there is a slight decrease in phosphate. Throughout the section, significant increases in phosphate concentrations begin at about the level of the thermocline (Fig. 11).

In the mid-Atlantic, along the 40° meridian, between 35°N. and the equator, the distribution of phosphate varied from north to south. North of 26°N., surface concentrations were 0–5 μg.P/l., and the phosphate-poor layer extended to some 400 m. At lower latitudes it decreased in thickness, being only 25 m. near 8°N., and increased again to 100 m. in depth near the equator. Below this impoverished layer, phosphate increased rapidly with depth, and northwards of 27°N. the maximum of 40 μg.P/l. was at 900–1000 m. South of this latitude, it increased, reaching values of 80 μg.P/l., and in general the level at which the gradient of phosphate content increased was higher, occurring at depths of 300–400 m. The deeper water had a more uniform phosphate content (*c.* 40 μg.P/l. at 2000 m.), and in the south, the phosphate maximum was very clearly defined, as is shown in Fig. 12. In the northern part of this section, phosphate contents below this maximum tend to be low, possibly because of the intrusion of Mediterranean water of low phosphate content.

The vertical distribution in the upper part of the water column for the section between Bermuda and Haiti is shown in Fig. 13. Surface concentrations were around 1–4 μg.P/l., and low values of phosphate extended downward to 250–300 m. and were similar to those of the same latitudes farther east. Maximum phosphate concentrations were at 800–1000 m., the higher maximum values being found in the north; below these depths, the phosphorus content decreased somewhat with depth to values of 34–39 μg.P/l. at 2000 m.

On the section between Bermuda and Chesapeake Bay, several series of stations were worked in the period December 1931 to April 1932, and some variations of phosphate concentration were observed which may have been caused partly by water movement, and partly by seasonal biological events. The section crossed the American Coastwise Convergence about 400 miles to the west of Bermuda, and showed abrupt transitions in phosphate concentration and other properties of the water. East of the convergence, phosphate concentrations at the surface were 5–6 μg.P/l. in December, and decreased to 0–4 μg.P/l. in the succeeding spring. To the west, the seasonal variation *appeared* to be more marked, changing from 11–15 μg.P/l. in February to about 3 μg.P/l. in April. However, since the convergence was displaced to the west at this time, changes in phosphate concentration were not necessarily caused by biological action alone.

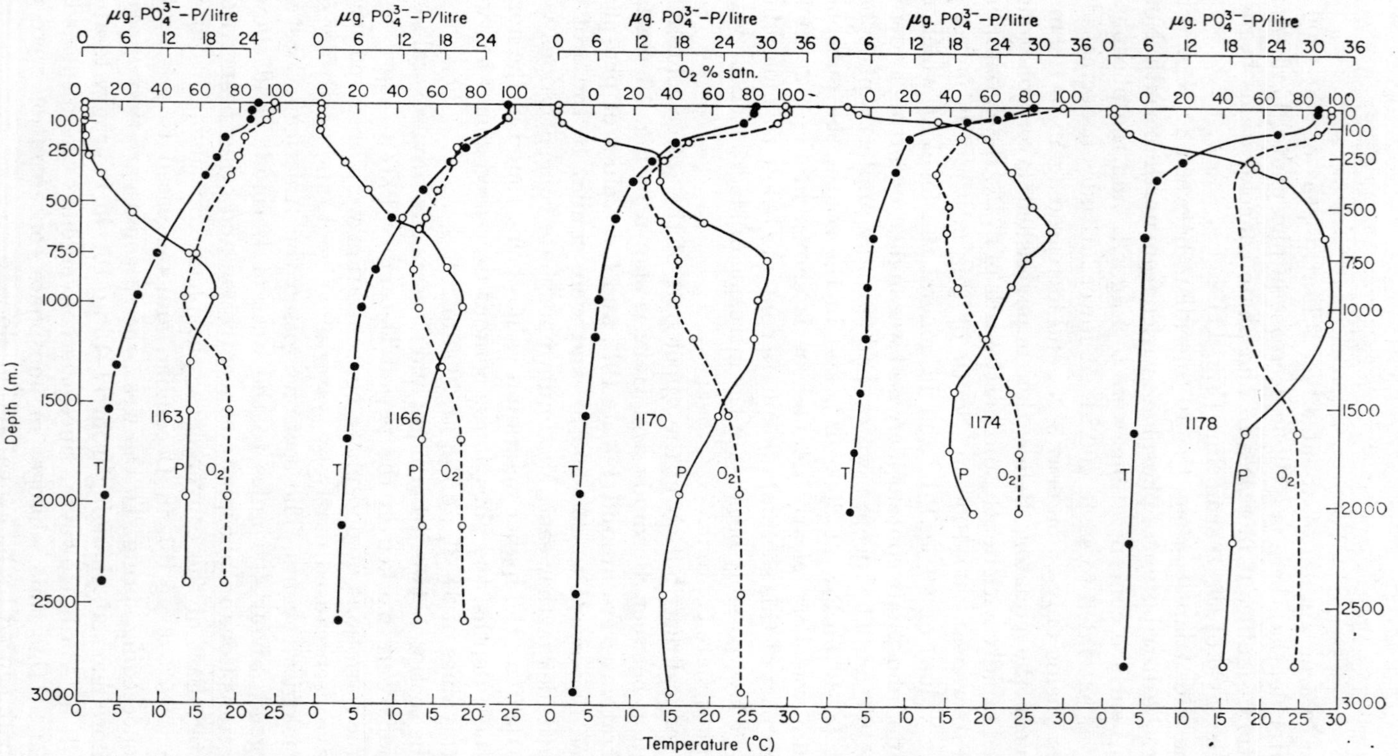

FIG. 12. Vertical distribution of phosphate, oxygen (percentage saturation) and temperature in mid-Atlantic along 40°W. meridian, March 1932. From Seiwell (1935).

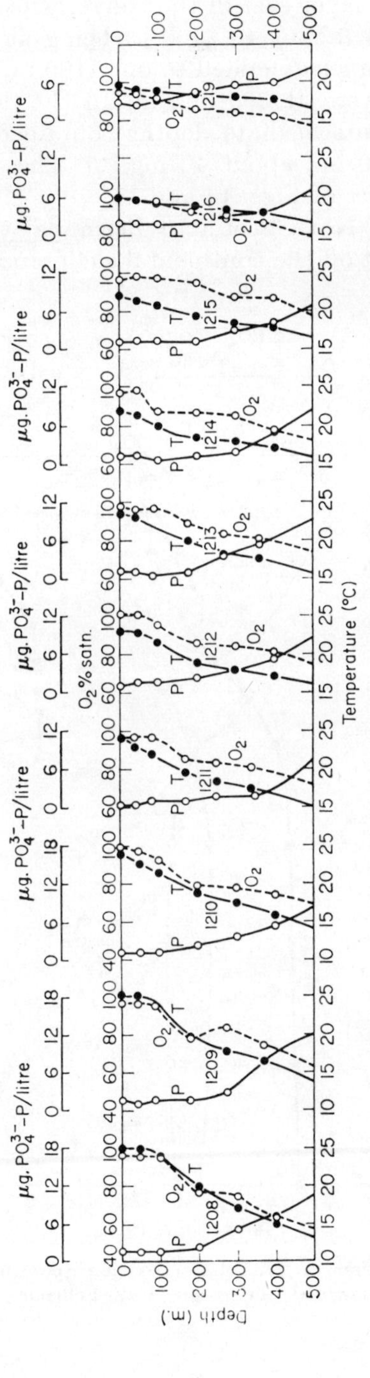

FIG. 13. Vertical distribution of phosphate, oxygen (percentage saturation) and temperature in the upper 500 m. between Haiti and Bermuda, April 1932. From Seiwell (1935).

The phosphate-poor layer east of the convergence extended to 400–500 m., the maximum (about 40 μg.P/l.) being at 1000 m., whereas to the west, the poor layer extended to only 100 m. or less, and maximum concentrations were at 300–500 m. In the deeper water in all parts of the section, intermediate depths contained about 45 μg.P/l. and those around 2000 m., about 40 mg.P/l. These distributions are illustrated in the profiles of Figs. 14 and 15.

In the section from Nova Scotia to Bermuda, three convergences were observed, one just off the continental shelf, and others at 200 and

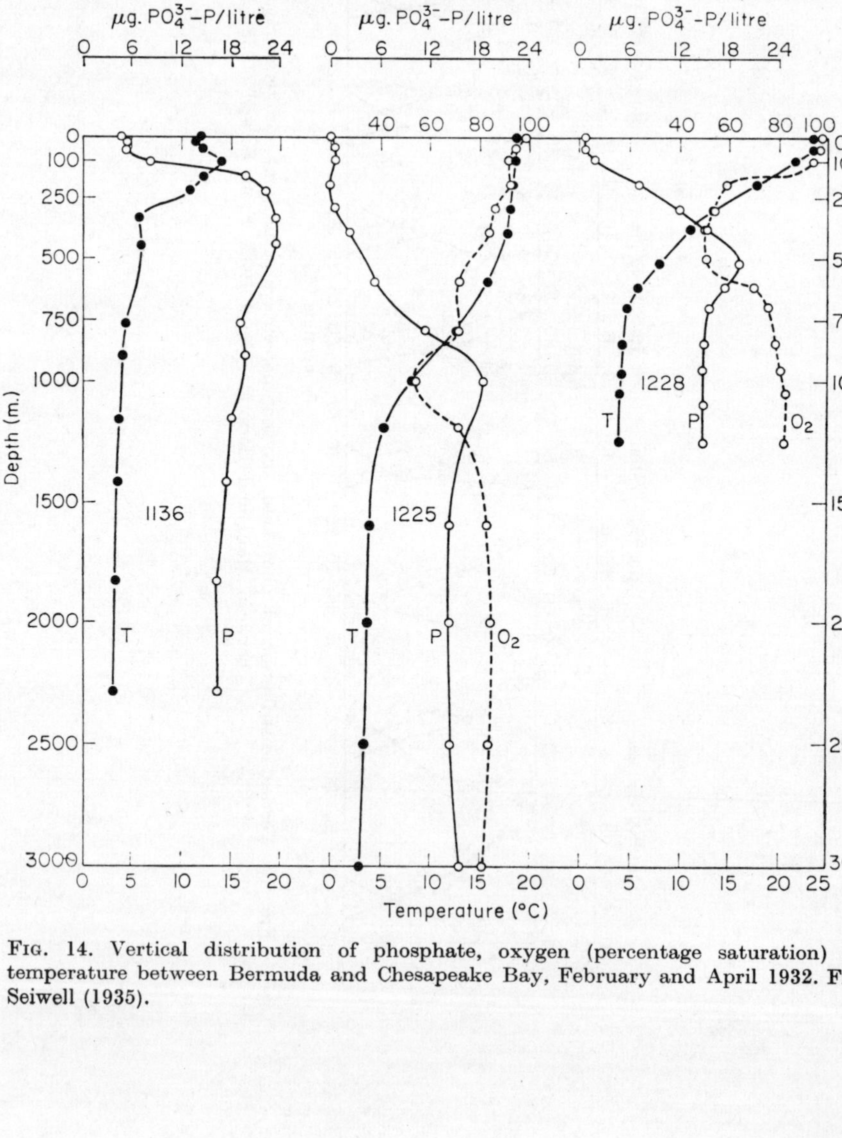

FIG. 14. Vertical distribution of phosphate, oxygen (percentage saturation) and temperature between Bermuda and Chesapeake Bay, February and April 1932. From Seiwell (1935).

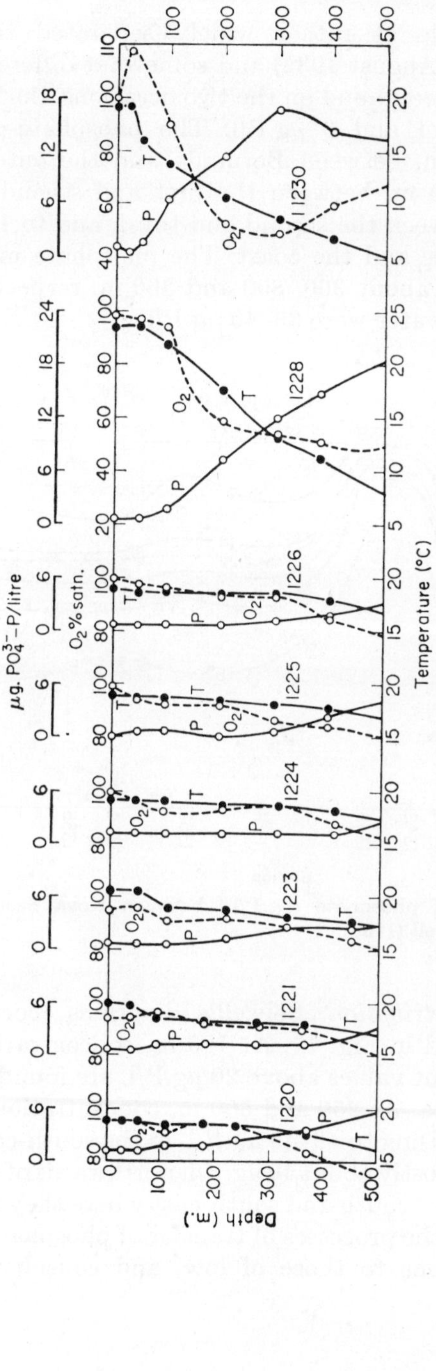

FIG. 15. Vertical distribution of phosphate, oxygen (percentage saturation) and temperature between Bermuda and Chesapeake Bay, April 1932. From Seiwell (1935).

300 m. offshore. They are clearly demonstrated by the phosphate distribution (Fig. 16).

Cruises were made at rather widely separated times (between November 1931 and August 1932) and somewhat different positions of these convergences were found on the two occasions. Surface concentrations were between 1 and 9 μg.P/l. The phosphate-poor layer extended to 200–400 m. between Bermuda and the outermost convergence, to only 30–40 m. between the first and second convergences, to about 340 m. between the second and third, and to 100 m. between the third convergence and the coast. The phosphate maxima were at 900–1000 m., and at about 300, 800 and 300 m. respectively. Concentrations in the deep water were 35–40 μg.P/l.

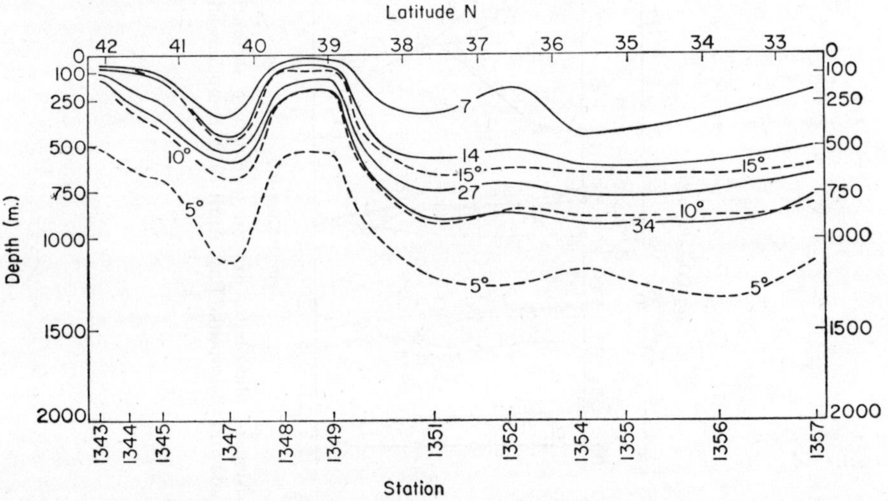

Fig. 16. Distribution of phosphate (μg.P/l.) between Nova Scotia and Bermuda, August, 1932. From Seiwell (1935).

(2) *Horizontal distribution.* Seiwell's diagrams (corrected for salt error) are reproduced in Fig. 17. At 100 m. concentrations are mostly less than 5 μg.P/l., but values above 20 μg.P/l. are found in the extreme south-east of the area. At 250 and 500 m. concentrations are generally less than 13 μg.P/l. rising to 40–45 μg.P/l. in the south-east. At 1000 m., concentrations are mostly about 40 μg.P/l. northwards of the 25°N. parallel, and increase to the south and south-east where they reach 80 μg.P/l.

Seiwell examined the processes of transfer of phosphate from domains of high concentrations to those of low, and considered particularly vertical transport by eddy conduction, which he considered to be the main factor in vertical exchanges of nutrients in the open ocean.

The important horizontal layer through which phosphate is transported upward into the regions where it can be utilized by plants (the

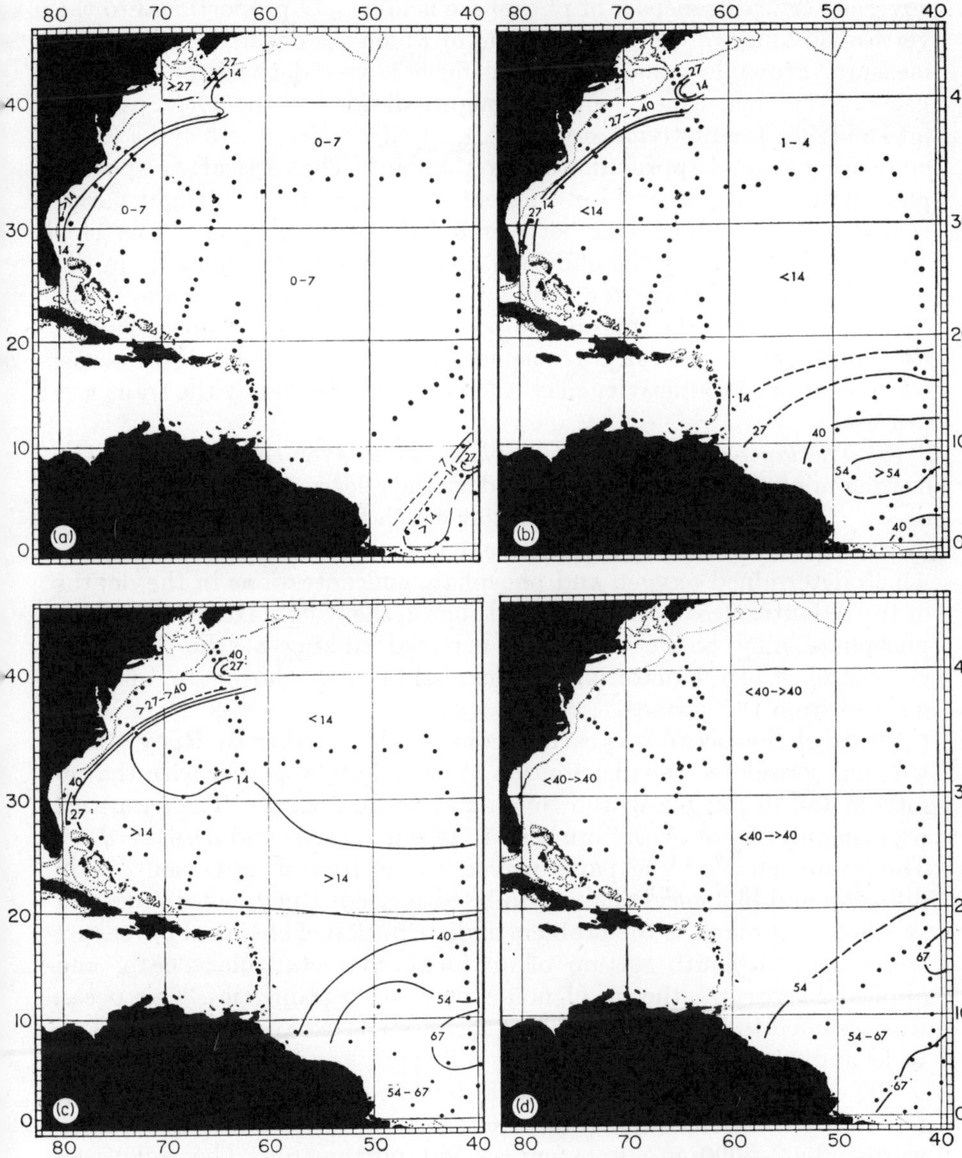

FIG. 17. (a)–(d) Horizontal distribution of phosphate (μg.P/l.) at 0, 250, 500 and 1000 m, respectively in the North West Atlantic. From Seiwell (1935).

transport layer), is the region between the beginning of the thermocline and the level at which phosphate no longer increases significantly with increasing depth. It is marked by downwardly increasing phosphate concentration and density. The coefficient of eddy conductivity, which governs vertical transport of phosphate is inversely proportional to the vertical stability of the water column of which the density gradient is a measure. Probable values of this coefficient were derived for different parts of the area, which could be roughly divided at the 30°N. parallel into a higher conductivity area (30 c.g.s. units) to the north and a lower one (10 c.g.s. units approximately) to the south. The vertical transport of phosphate was calculated for different regions, and only a slight variation was found, values lying between 3 and 6×10^{-2}g.P/cm.^{2}/year. This was brought about by the effects of greater phosphate gradients in the south opposing the greater conductivity to the north. However, in the north, where the phosphate-poor layer is much deeper than in the south, the vertically transported phosphate appears in lower concentration. Productivity may, therefore, be affected by the transport process.

(3) *The layer of maximum phosphate.* This layer coincides with the oxygen minimum layer and is caused by the release of phosphate during decay and oxidation of organic matter which fall from the upper layers (Seiwell and Seiwell, 1934). Redfield (1942) examined the processes which determined oxygen and phosphate concentrations in the depths of the Atlantic Ocean and suggested that in the North Atlantic the deep phosphate may be derived from material sinking at the Antarctic Convergence. In equatorial regions, additional phosphate may be derived from the surface.

Study of this ocean was carried considerably further by Riley (1951) who has considered the distribution of phosphate, together with that of nitrate and of oxygen in the North and South Atlantic. The data used were mainly those of the cruises of *Atlantis*, *Dana* and *Discovery II*. This study enabled the parts played by horizontal movement, eddy diffusion and biological changes to be estimated. Riley first discussed a synthetic approach to the problem of distribution of chemical properties along a north–south section of an idealized rectangular ocean, and applied Sverdrup's theory of non-conservative properties. This ocean was assumed to be 12,000 km. long, between 54°N. and 54°S., 2500 km. wide and 5000 m. deep, with the following vertical velocity profile: 0–200 m., 1·2 cm./second northwards; 200–800 m., 0·13 cm./second northwards; 800–2000 m. zero; 2000–3500 m., 0·24 cm./second southwards; 3500–5000 m., 0·03 cm./second northwards. The quantities transported by these currents across the equatorial Atlantic have been estimated by Sverdrup *et al.* (1952). At the ends of this ocean, the currents flow into north and south polar seas in which it is assumed that turbulence maintains constant conditions. In the south polar seas, nitrate-nitrogen and phosphate-phosphorus concentrations are assumed to be 520 and 78 μg./l. respectively. The distribution within the ocean under steady state conditions is to be calculated. The rate of horizontal eddy diffusion is assumed to be 10^8 g./cm^2/second everywhere, that of vertical eddy diffusion is assumed to vary with depth and latitude, between 1 and 50 g./cm.^{2}/second. Other assumptions required for calculation are that the rate of nitrogen regeneration is 140×10^{-8} μg./l./second at the surface, decreasing to zero at 1400 m., and that of phosphorus regeneration is 37×10^{-8} μg./l./second at the surface, decreasing to zero at 800 m.

The mathematical treatment of this model requires some further assumptions about boundary conditions. It is then possible to derive models for distribution of phosphorus and nitrogen along the north–south profile of the ocean. These distributions show some of the major features of those actually observed—for example, minimum values at the surface, the mid-depth maximum and the decrease in nutrients from north to south—although the simplifying nature of some of the assumptions limits the utility of the model. Rather more realistic results were obtained when more sophisticated assumptions were made, although Riley holds that the value of experimenting with such mathematical models is not so much the supplying of solutions to problems of distribution as that of clarifying the effects of the various processes and their mutual interactions. He has also shown that the processes of derivation of physical coefficients and biological rates from the observations, although usually regarded as difficult, are possible, and may be done in stages. For this, the area of the North and South Atlantic Oceans is sub-divided into some 70 areas with centres 1000 km. apart; observations from 660 stations were used to provide mean values which were allocated to the area centres. Current velocities, relative to certain arbitrary reference surfaces, were then calculated from temperature and salinity observations by conventional hydrographic methods, and from these the velocities of northward and southward movements of water across selected latitudes, and at different density levels, were calculated in several stages. From the estimated water movements, rates of horizontal and vertical eddy diffusion were computed; these allowed estimates to be made of biological rates of change at the various σ_t surfaces. Finally, these results were applied to obtain values for oxygen consumption and phosphate regeneration between 200 m. and the bottom, and it was shown that these values corresponded to about one-tenth of the production of organic matter at the surface, and that biological rates of change increased with increasing depth.

This important paper represents a considerable advance in the provision of methods whereby quantitative description of the biological processes in the sea may be made without extrapolation from laboratory experiments.

D. THE ANTARCTIC OCEAN

Ruud (1930) made hydrographic observations which included phosphate determinations at a number of stations in the Weddell Sea, and showed the abundance of nutrients in this area. He recognized three types of vertical distribution of phosphate; the first two, which showed sharp or gradual increases from about 68–85 μg.P/l. at 150–200 m., occurred in the north-eastern part of the area, and the third type, which showed a sharp increase from 40–60 μg.P/l. nearer the surface (at 10–40 m.) was found in the south-west region.

Extensive cruises by R.R.S. *Discovery II* in the Antarctic between 1929 and 1935 allowed the collection of observations of phosphate and silicate which have been reported by Clowes (1938). The water movements in this area are the key to the understanding of the vertical and horizontal distribution of the nutrients, and are discussed by Deacon

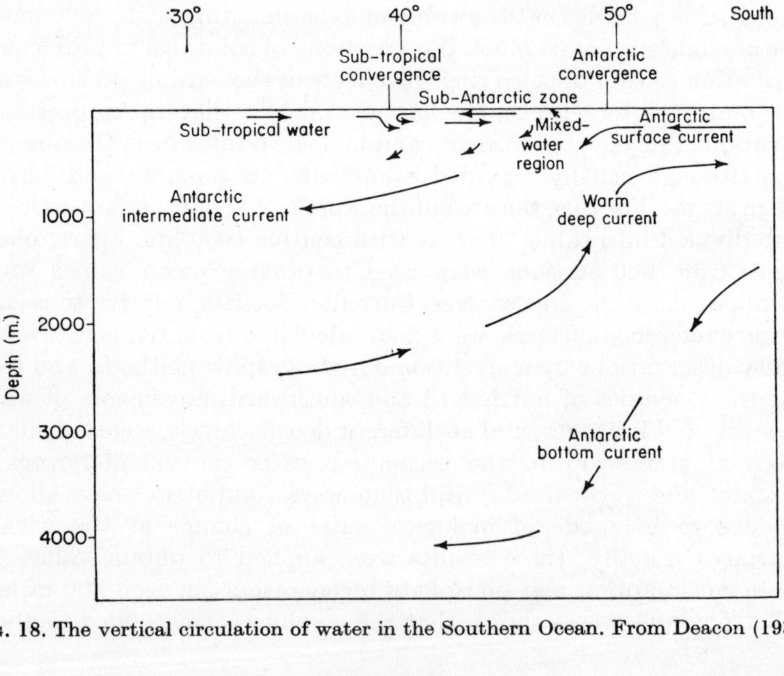

FIG. 18. The vertical circulation of water in the Southern Ocean. From Deacon (1937a).

entering the Antarctic region. Higher concentrations of up to 90 μg.P/l. at the 1000–2000 m. level are found more to the south between 38° and 45°S.; they are derived from the rich plankton of the north-going Antarctic Surface Water which sinks through the Sub-Antarctic Current. At the lower boundary of this current, mixing transfers remineralized phosphate to the south-going, warm deep water. The vertical distribution is shown in Fig. 20.

An interesting consequence of this mechanism of northward transport of phosphate can be traced in from the ice-edge to the equator (Fig. 21). The Antarctic Intermediate Current is characterized by a salinity minimum (Deacon, 1933) and it can thereby be located at any latitude along the section. As this current flows northwards, mixing should cause a slow and steady decrease of phosphate concentration from the region where the current starts and is charged with its burden of phosphate. Such a steady falling off is not in fact found, but rather a series of maxima and minima in step with changes in silicate content (Fig. 21). The explanation suggested by Clowes is that the alternation of seasons, by causing a yearly cycle of phosphate in the sub-Antarctic and Antarctic regions, produces a fluctuation in the input of phosphate to the Antarctic Intermediate Current. It appears (Fig. 22) that maximum decomposition of organic matter supplying phosphate occurs on this section at or to the south of station 1165 (40°S.).

The vertical distribution in this section is shown in Fig. 22, the high concentrations at 1000 m. and at the bottom at the station at 41°S. being well marked. Phosphate in the bottom water decreases to the north of this position, possibly because the Atlantic–Indian Cross-ridge at this latitude acts as a barrier to northward flow.

Seasonal variations were observed in surface water round South Georgia, in the Scotia Sea and at the ice edge, but the spacing of the observations made it difficult to draw firm conclusions. Minimum phosphate following the spring outburst of phytoplankton appeared to occur in October to November in the sub-Antarctic regions, and in November to December in the Antarctic. The occasional high phosphate values in the Antarctic Intermediate Current (e.g. the values at 41°S. in the western part of the South Atlantic, north of the Rio Grande Ridge) considerably exceed those in the warm deep water below. This suggests rapid regeneration of inorganic phosphate from decomposing plankton.

E. THE PACIFIC OCEAN

Extensive observations in this ocean were made on the cruises of *Dana* (Thomsen, 1937) and *Carnegie* (Graham and Moberg, 1944).

Parts of the north-eastern region have been described by Strickland and Austin (1960) and the equatorial region by Wooster and Cromwell (1958). The north-western region of the Kuroshio Current is described by Kawamoto (1955).

Generally, the northern part of the ocean has high surface values of phosphate; Graham and Moberg reported 11–66 μg.P/l. for much of the area north of 40°N. This they ascribe to the strong vertical mixing which takes place in this stormy region. Surface concentrations are affected by a current which appears to originate in the Bering Sea, moves southwards and is diverted eastward by the North Pacific West Wind Drift (Japan Current) and then carried southward by the California Current.

The surface layer of low phosphate content (less than about 50 μg.P/l.) is generally less than 200 m. in thickness. The depth of the layer of maximum phosphate (75–90 μg.P/l.) is about 300 m. in the central part (166°E.) of this zone, and is deeper (1000–2000 m.) in the eastern and western parts.

In the central region of the ocean between 40°N. and 10°N., and away from the coasts, lower surface concentrations of less than 5 μg.P/l. were found, and extended down to 100 m. or more. This appears to be caused by the great vertical (density) stability of the water column and the absence of strong currents. Higher surface values which exceeded 60 μg.P/l. and which were attributable to upwelling, were found off the Californian coast. The transition zone in these latitudes was unusually thick, and in some places between Guam and Japan was as deep as 1000 m. The concentrations in the transition zone (35–48 μg.P/l.) were less than in more northerly regions.

South of 10°N., intermediate values of phosphate were found in the surface layers. However, higher concentrations (exceeding 20 μg.P/l.) were noted off the coast of Peru and were caused (Guenther, 1936) by upwelling and by the transport of nutrients in the Peru Current. This region shows some variability, probably owing to the irregularity of the currents and of the consequent vertical movements. The part of this area extending from the west coasts of Central and South America to about 130°W. was described by Wooster and Cromwell (1958), who collected data from a co-operative survey in 1952 in which four survey vessels took part (Fig. 23). West of the Galapagos Islands some upwelling along the equator caused high values of phosphate toward the surface, that is, more than 30 μg.P/l. at 100 m. A very marked transition was found at about 18°N. 124°W., phosphate concentrations increasing suddenly on passing from north to south. This change appeared to be related to a change in the temperature–salinity relationship and

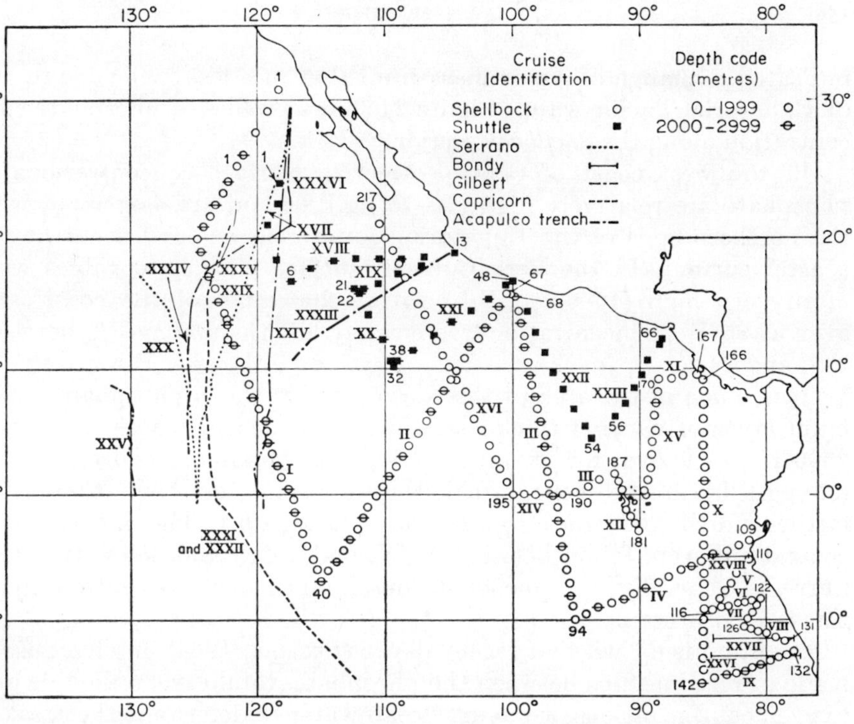

FIG. 23. Station positions in the North-East Pacific, showing profile numbers. From Wooster and Cromwell (1958).

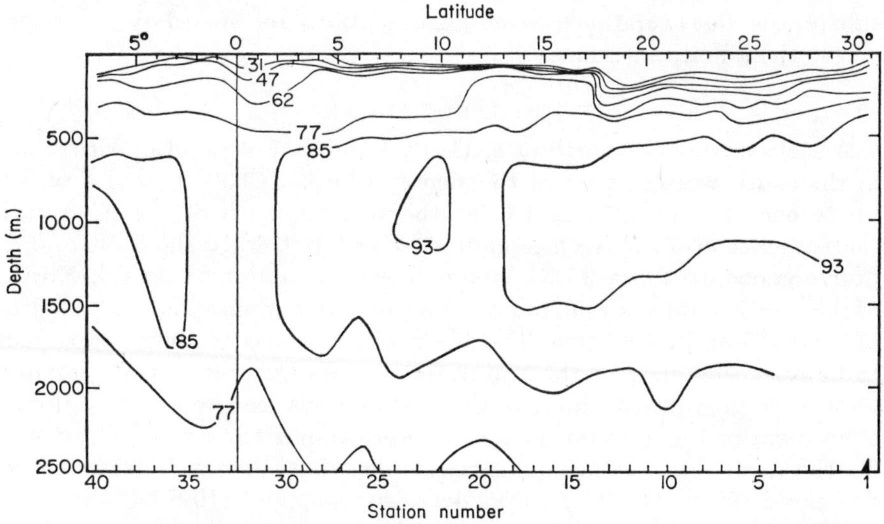

FIG. 24. Distribution of phosphate along *Shellback* profile 1. May to June, 1952 (as μg.P/l.). From Wooster and Cromwell (1958).

indicated a change of water mass from that of a less saline one to one of Equatorial Pacific water. Figure 24 shows profiles of phosphate concentration along the *Shellback* section.

Off the west coast of Central America, surface concentrations of phosphate are relatively high (10–15 μg.P/l.) and are associated with the northern edge of the Equatorial Countercurrent and a northward coastal current. In the Peru Current, surface phosphate values were often very high (45–60 μg.P/l.) within 200 miles of the coast, and even at 85°W., concentrations were rarely less than 30 μg.P/l. between 3°S. and 14°S.

Over much of the area of this survey, there was a phosphate maximum layer of varying thickness which extended in places from 500–2500 m., as shown for example in Fig. 24. Maximum concentrations (85 μg.P/l.) in the central region were rather less than they were farther north and south (more than 90 μg.P/l.). The highest concentrations were found between 10°N. and 15°N. and below the Peru Current. Generally, the phosphate maximum layer was centred about the 4–5°C isotherms.

Attention is drawn to a recent paper (Postma, 1964) in which use is made of relationships between the phosphate, total carbon dioxide and oxygen concentrations of Pacific Ocean water to determine the areas of regional exchange of oxygen and carbon dioxide between the surface layers of the ocean and the atmosphere. Thus, in the deep water, simple linear relationships exist between the concentrations of these three substances, but in the surface layers, deviations are caused by exchange across the air–sea interface.

F. THE INDIAN OCEAN

A section from the north of Enderby Land to Freemantle, Australia, in the south-western part of this ocean (Clowes, 1938) showed a phosphate content of 50–71 μg.P/l. at the surface south of the Antarctic Convergence at 52°S. Surface values fell slowly toward the Subtropical Convergence at about 39°S., where there was about 28 μg.P/l. North of this, surface waters were more or less depleted of phosphate to a depth of several hundred metres. The Kerguelen-Gaussberg Ridge appeared to hamper the effect of the Antarctic Bottom Current in transporting phosphate northward, since, to the south of this feature, fairly uniform concentrations of around 60 μg.P/l. were found from top to bottom. To the north of it, concentrations of phosphate in the bottom water decreased from 80–70 μg.P/l. between 53° and 40°S. Continuing northward, the phosphate concentration at the bottom increased to values approaching 90 μg.P/l. This appeared to be caused by the presence of North Indian Ocean Deep Water, which has a very high phosphate concentration, as was shown by the *Dana* cruises (Thomsen, 1937) and the more recent ones of the *William Scoresby*. Typical values for the western part of this ocean are shown in Table III, taken from a *Discovery* Station list (*Discovery* Report, Vol. 26).

TABLE III

*Phosphate concentration, Western Indian Ocean**

Station	*Depth* *m.*	*Phosphate* *μg.P/l.*	*Station*	*Depth* *m.*	*Phosphate* *μg.P/l.*
WS 976	0	18·0	WS 1056	0	1·2
From	10	18·0		10	1·2
22°50′ S.	20	15·5	28°41′ S.	20	1·2
11°38′ E.	30	18·0	13°25′ E.	30	1·2
to	50	18·0		50	5·3
22°48′ S.	75	28·5		75	6·2
11°35′ E.	100	33·5		100	6·2
	150	38·5		150	25·5
	200	47·5		200	28·5
	300	52·0		300	43·0
	400	52·0		400	57·5
	500	73·0		600	57·5
	700	69·0		820	69·0
	900	73·0		1030	69·0
	1100	73·0		1240	69·0
	1400	65·0		1540	59·5
	1450	57·0		2060	49·5
	2420	48·0		2580	49·5
	2920	52·0			
	3098	85·0 (Filtered from ooze)			

* *William Scoresby*, 1950, from *Discovery Rep.* **26**, 221, 245.

Some total and particulate phosphorus observations from the South-East Indian Ocean are referred to above (Rochford, 1963).

This ocean will become more fully understood as the observations of the International Indian Ocean Expedition become available.

VI. Comparison of the Major Oceans

The vertical distribution of phosphate shows a common pattern in the deep oceans, and four layers may be recognized. These are:

(a) a thin surface layer of fairly uniform and often low phosphate concentration;

(b) a transition region in which the concentration increases rapidly;

(c) a region, usually lying between 500 and about 2000 m., of maximum concentration, most marked, however, in the Atlantic Ocean;
(d) a lower layer, which extends to the bottom, in which phosphate concentration may remain fairly constant, or decrease slightly with increasing depth.

In equatorial regions, the surface layer may be deep and of low phosphate content because of the high consumption by plants. On lines of divergence, however, a continuing supply of phosphate-rich water from below keeps this layer thin. In higher latitudes it may become less marked, and in regions of turbulence it may be absent.

The depth of the transition region and the concentration gradient of phosphate in it are governed by events in the layers above and below so that, for example, it is most marked where high consumption of phosphate in the surface layer overlies rich water at a fairly high level, as in the tropical Pacific and Indian Oceans.

It is in the deep water of the major oceans that phosphate concentrations differ most. In the Pacific and Indian Oceans, phosphate concentrations are highest north of the equator, with values exceeding 100 μg.P/l., and with no very clearly marked maximum layer so that phosphate concentrations do not alter greatly below 1000–1500 m.

In Antarctic regions, these oceans and also the Atlantic show somewhat similar characteristics, although movement of bottom water, and the northward movement of phosphate-rich Atlantic Intermediate Water, may depend on the configuration of the bottom. The Atlantic Ocean, however, is different since it is open at the north to the Arctic Ocean; this feature influences the deep-water circulation and the overall distribution of phosphate. A clearly marked maximum of about 60 μg.P/l. reaches northward from the Antarctic, and is centred on depths of about 1000 m. Below this depth, phosphate diminishes from south to north. Exchange of bottom water across the Equator is considerable in the Atlantic, since the water moves south, and is replaced by surface water of low phosphate content and the North Atlantic is, therefore, poorer in phosphate at all depths than the South. It is also poorer in this respect than the Pacific and Indian Oceans in which there is little deep-water movement across the Equator.

VII. Nitrogen: Phosphorus Ratio

Harvey (1926) drew attention to the fact that in the English Channel, phosphate and nitrate were used up simultaneously in constant proportion by the phytoplankton. Redfield (1934) showed that this was general for other bodies of water, and found that nitrogen : phosphorus ratios in the sea in a number of oceans and at all depths, and also in plankton, were generally close to 15 : 1 by atoms. (Redfield's ratios have been corrected for salt error.) Cooper (1937b, 1938b) showed that although this ratio held in many areas, some variations occurred, anomalously high ratios being found in the Mediterranean, and where deep Mediterranean water intruded into the North Atlantic; low ratios occurred in certain waters of the South Atlantic. Cooper (1937a) and Fleming (1940) showed that analyses of plankton give N : P ratios close to 15 : 1 or 16 : 1 (by atoms).

There are, nevertheless, a good many observations which show that constancy of N : P ratio in the sea has exceptions, particularly in restricted waters or near the surface. Phifer and Thompson (1937) found ratios of 11·9 : 1 to 13·9 : 1 (Mean 12·5 : 1) in the San Juan Channel. Riley and Conover (1956), showed that the ratio in the waters of Long Island Sound varied from zero to 8 : 1. Megia and Lao (1955) evaluated the results of some thousands of analyses of water to depths of 2000 m. in the Philippines and adjacent areas, and they found values close to a mean of 13·5 : 1 but which varied slightly with depth. Ketchum *et al.* (1958) found the ratio in New England Coastal surface waters varied from zero in summer to values of between 5 : 1 and 10 : 1 in winter. On the continental shelf they were from 1·2 : 1 to 7·1 : 1, and in deep water from 11·1 : 1 to 14·6 : 1.

Armstrong and Butler (1960a) found ratios in the English Channel of 19 : 1 in summer and 10·5 : 1 in winter. Similarly, Stefansson and Richards (1963) for water off the Washington and Oregon coasts, found, in surface waters clear of the efflux of the Columbia River, ratios of 5 : 1 in winter, which decreased in spring. In subsurface waters the ratio was 13 : 1 and at the deepest stations 15·8 : 1.

Nearly all the ratios given above are calculated from the concentrations of nitrate and phosphate found, although Ketchum *et al.* give one series of ratios from "total nitrogen" and "total phosphorus" at intermediate depths on the continental shelf. These varied from 14·9 : 1 to 30·2 : 1.

The ratios of N : P utilized in spring have been found in many instances to be higher than those in the water, and this has resulted in some phosphorus remaining after the nitrogen has been exhausted. Ketchum *et al.* found the ratio of nutrients utilized to be 15 : 1, and Stefansson and Richards observed that it was 16 : 1. Armstrong and Butler found the ratio to be as high as 24 : 1.

Analyses of suspended matter filtered from sea water from the North Atlantic (Menzel and Ryther, 1964) have shown N : P ratios of 4·7 : 1 in January, and about 12·5 : 1 in April. Unpublished analyses by the writer, for a section across the English Channel have given results varying rather widely about a mean of 10·5 : 1 during several seasons. At 15°45′N., 53°08′E. in the Arabian Sea, N : P ratios in the suspended matter were 14 : 1 at 0–20 m., 20 : 1 at 20–400 m. and 26 : 1 at 400–2400 m.

REFERENCES

Armstrong, F. A. J. (1955). *J. Mar. biol. Ass. U.K.* **34**, 223.
Armstrong, F. A. J., and Butler, E. I. (1960a). *J. Mar. biol. Ass. U.K.* **39**, 525.
Armstrong, F. A. J., and Butler, E. I. (1960a). *J. Mar. biol. Ass. U.K.* **39**, 299.
Armstrong, F. A. J., and Butler, E. I. (1962). *J. Mar. biol. Ass. U.K.* **42**, 253
Armstrong, F. A. J., and Harvey, H. W. (1950). *J. Mar. biol. Ass. U.K.* **29**, 145.
Atkins, W. R. G. (1923). *J. Mar. biol. Ass. U.K.* **13**, 119.
Atkins, W. R. G. (1926). *J. Cons. int. Explor. Mer*, **1**, 197.
Bigelow, H. B., Lillick, L. C., and Sears, M. (1940). *Trans Amer. phil. Soc.* **31**, 149.
Brandt, K. (1920). *Wiss. Meeresuntersuch. Abt. Kiel*, **18**, 185.
Bruce, H. E., and Hood, D. W. (1959). *Univ. Texas Inst. Mar. Sci.* **6**, 135.
Buch, K. (1932). *Meerentutkimuslait Julk Nr.* **86**, 1.
Carritt, D. E., and Goodgal, S. (1954). *Deep Sea Res.* **1**, 224.
Chacko, P. I., Valsan, A. P., and Pillai, C. M. (1954). Contrib. from Marine and Fishing Biology Station Krusadai, Government Press, Madras, p. 1.
Clowes, A. J. (1938). '*Discovery*' *Rep.* **19**, 1.
Cooper, L. H. N. (1937a). *J. Mar. biol. Ass. U.K.* **21**, 673.
Cooper, L. H. N. (1937b). *J. Mar. biol. Ass. U.K.* **22**, 177.
Cooper, L. H. N. (1938a). *J. Mar. biol. Ass. U.K.* **23**, 171.
Cooper, L. H. N. (1938b). *J. Mar. biol. Ass. U.K.* **23**, 179.
Cooper, L. H. N. (1948). *J. Mar. biol. Ass. U.K.* **27**, 314.
Dakin, W. J., and Colefax, A. N. (1935). *Proc. Linn. Soc. N.S.W.* **60**, 303.
Deacon, G. E. R. (1933). '*Discovery*' *Rep.* **7**, 173.
Deacon, G. E. R. (1937a). '*Discovery*' *Rep.* **15**, 1.
Deacon, G. E. R. (1937b). '*Discovery*' *Rep.* **15**, 125.
Delsman, H. C. (1939). *Treubia*, **17**, 139.
Dietz, R. S., Emery, K. O., and Shepard, F. C. (1942). *Bull. geol. Soc. Amer.* **53**, 815.
Fleming, R. H. (1940). *Proc. 6th Pacif. Sci. Congr. Calif. 1939*, **3**, 535.
George, P. C. (1953). *J. zool. Soc. India*, **5**, 76.
Gorgy, S., Rakestraw, N. W., and Fox, D. L. (1948). *J. Mar. Res.* **7**, 22.
Graham, M. (1938). *Fish. Invest. Lond. Ser. II*, **16** (**3**).
Graham, H. W., and Moberg, E. G. (1944). *Publ. Carneg. Instn*, **562**, 1.
Guenther, G. R. (1936). '*Discovery*' *Rep.* **13**, 107.
Hansen, A. L., and Robinson, R. J. (1953). *J. Mar. Res.* **12**. 31.
Harvey, H. W. (1926). *J. Mar. biol. Ass. U.K.* **14**, 71.
Harvey, H. W. (1948). *J. Mar. biol. Ass. U.K.* **27**, 337.
Harvey, H. W. (1955). The Chemistry and fertility of Sea Waters. Cambridge Univ. Press, 234 pp.
Holt, L. E., LaMer, V. K., and Chown, H. B. (1925). *J. biol. Chem.* **64**, 509.
Ishibashi, M., Shigematsu, T., Nakagawa, Y., and Ishibashi, Y. (1951). *Bull. Inst. chem. Res. Kyoto Univ.* **24**, 68.

Jayaraman, R. (1951). *Proc. Indian Acad. Sci.* **33**, 92.
Jayaraman, R. (1954). *Indian J. Fish.* **1**, 345.
Jitts, H. R. (1959). *Aust. J. Mar. Freshw. Res.* **10**, 7.
Kalle, K. (1935). *Ann. Hydrogr., Berl.* **63**, 195.
Kalle, K. (1937). *Ann. Hydrogr., Berl.* **65**, 1.
Kalle, K. (1957). *Ber. dtsch. Komm. Meeresforsch*, **14**, 313.
Kawamoto, T. (1955). *J. Oceanogr. Soc. Japan*, **11**, 59.
Ketchum, B. H. (1954). *Ann. Rev. Pl. Physiol.* **5**, 55.
Ketchum, B. H., and Redfield, A. C. (1949). *J. Cell. comp. Physiol.* **33**, 281.
Ketchum, B. H., Corwin, N., and Keen, D. J. (1955). *Deep Sea Res.* **2**, 172.
Ketchum, B. H., Vaccaro, R. F., and Corwin, N. (1958). *J. Mar. Res.* **17**, 282.
Kreps, E., and Verjbinskaya, N. (1930). *J. Cons. int. Explor. Mer*, **5**, 327.
Kreps, E., and Verjbinskaya, N. (1932). *J. Cons. int. Explor. Mer*, **7**, 25.
Kreps, E., and Osadchih, M. (1933). *Int. Rev. Hydrobiol.* **29**, 221.
Marshall, S. M., and Orr, A. P. (1927). *J. Mar. biol. Ass. U.K.* **14**, 837
Marshall, S. M., and Orr, A. P. (1930). *J. Mar. biol. Ass. U.K.* **16**, 853.
Matthews, D. J. (1916). *J. Mar. biol. Ass. U.K.* **11**, 122.
Matthews, D. J. (1917). *J. Mar. biol. Ass. U.K.* **11**, 251.
Megia, T. G., and Lao, R. G. (1955). *Philippine J. Fish.* **3**, 55.
Menzel, D. W., and Ryther, J. H. (1964). *Limnol. Oceanogr.* **9**, 179.
Minkina, A. L. (1961). *Okeanolog. Issled. (Moscow Akad. Nauk, S.S.S.R.)* **3**, 162.
Moore, H. B. (1930). *J. Mar. biol. Ass. U.K.* **16**, 595.
Mortimer, C. H. (1941). *J. Ecol.* **29**, 280.
Mortimer, C. H. (1942). *J. Ecol.* **30**, 147.
Murphy, J., and Riley, J. P. (1958). *J. Mar. biol. Ass. U.K.* **37**, 9.
Newcombe, C. L., and Brust, A. F. (1940). *J. Mar. Res.* **3**, 76.
Newcombe, C. L., and Lang, A. G. (1939). *Proc. Amer. phil. Soc.* **81**, 393.
Newell, B. S. (1959). *Fish. Publ., Lond.* **12**, 18.
Occhipinti, A. G., Magliocca, A., and Teixera, C. (1961). *Bol. Inst. Oceanogr. S. Paulo*, **11**, 17.
Orr, A. P. (1933). Physical and chemical conditions in the sea in the neighbourhood of the Great Barrier Reef. Great Barrier Reef Expedition Reports 1928–9, No. 2 (3) London.
Phifer, L. D., and Thompson, T. G. (1937). *J. Mar. Res.* **1**, 34.
Portmann, J. E., and Riley, J. P. (1964). *Analyt. Chim. Acta.* In the press.
Postma, H. (1964). *Netherl. J. Sea. Res.* **2**, 258.
Pratt, D. M. (1950). *J. Mar. Res.* **9**, 29.
Rakestraw, N. W., and Carritt, D. E. (1948). *J. Mar. Res.* **7**, 362.
Rakestraw, N. W., and Smith, H. P. (1937). *Bull. Bingham oceanogr. Coll.* **6**, 1.
Ramamurthi, S. (1953). *J. Madras Univ.* **B 23**, 148.
Rao, S. V. S. (1957). *Proc. Indian Acad. Sci.* **B 45**, 77.
Redfield, A. C. (1934). *In* "James Johnstone Memorial Volume (R. J. Daniel, ed.), p. 176. University of Liverpool Press, Liverpool.
Redfield, A. C. (1942). *Pap. phys. Oceanogr.* **9**, 1.
Redfield, A. C. (1955). *Deep Sea Res. (Supplement)* **3**, 115.
Redfield, A. C. (1958). *Amer. Scient.* **46**, 205.
Redfield, A. C., Smith, H. P., and Ketchum, B. H. (1937). *Biol. Bull. Woods Hole*, **73**, 421.
Riley, G. A. (1951). *Bull. Bingham oceanogr. Coll.* **13**, 1.
Riley, G. A., and Conover, S. A. M. (1956). *Bull. Bingham oceanogr. Coll.* **15**, 47.

Rochford, D. J. (1951). *Aust. J. Mar. Freshw. Res.* **2**, 1.
Rochford, D. J. (1963). *Aust. J. Mar. Freshw. Res.* **14**, 119.
Ruud, J. T. (1930). *J. Cons. int. Explor. Mer*, **5**, 347.
Seiwell, H. R. (1935). *Pap. phys. Oceanogr.* **3** (**4**), 1.
Seiwell, H. R., and Seiwell, G. E. (1934). *Ann. Hydrogr., Berl.* **62**, 7.
Seshappa, G., and Jayaraman, R. (1956). *Proc. Indian Acad. Sci.* **B43**, 287.
Seshappa, G., and Jayaraman, R. (1957). *Proc. Indian Acad. Sci.* **B46**, 110.
Smales, A. A., and Pate, B. D. (1952). *Analyst*, **77**, 188.
Steele, J. H. (1956). *J. Mar. biol. Ass. U.K.* **35**, 1.
Stefansson, U., and Richards, F. A. (1963). *Limnol. Oceanogr.* **8**, 394.
Stich, H. (1953). *Z. Naturf.* **8**, 36.
Strickland, J. D. H., and Austin, K. H. (1960). *J. Fish. Res. Bd. Can.* **17**, 337.
Strickland, V. D. H., and Parsons, T. R. (1960). *Bull. Fish. Res. Bd. Canada* **125**, 185 pp.
Sverdrup, H. U., Johnson, M. W., and Fleming, R. H. (1942). "The Oceans, their physics, chemistry and general biology." 1087 pp. Prentice Hall, New York.
Tham, A. K. (1953). *Fish. Publ. Lond.* **1** (**4**), 65.
Thompson, T. G., and Robinson, R. J. (1934). *Proc. 5th Pacif. Sci. Congr.* pp. 2101–7.
Thomsen, H. (1931). *Rep. Danish oceanogr. Exped. Medit.* **3** (**6**), 1.
Thomsen, H. (1937). '*Dana*' *Rep.* **12,** 1.
Vaccaro, R. F. (1963). *J. Mar. Res.*'**21**, 284.
Watt, W. D., and Hayes, F. R. (1963). *Limnol. Oceanogr.* **8**, 276.
Wooster, W. S., and Cromwell, T. (1958). *Bull. Scripps. Instn. Oceanogr.* **7**, 169.

Chapter 9

Inorganic Nitrogen in Sea Water*

Ralph F. Vaccaro

Woods Hole Oceanographic Institution, Woods Hole Massachusetts, U.S.A.

I. Introduction 365
II. Geochemical Evolution of Inorganic Nitrogen in the Hydrosphere 366
A. Prebiogenic Oceanic Nitrogen 366
B. Nitrogen and the Primary Biosphere 367
III. Inorganic Nitrogen Exchanges across the Sea Surface 367
A. Molecular Nitrogen 367
B. Inorganic Nitrogen in Oceanic Rain 369
IV. Biological Factors Affecting the Distribution of Molecular and Combined Nitrogen 370
A. Nitrogen Fixation 370
B. Denitrification 372
C. Nitrogen Assimilation by Marine Phytoplankton 373
D. Ammonia and the Decomposition of Organic Nitrogen 376
E. Oxidation of Inorganic Nitrogen 377
V. Combined Inorganic Nitrogen in the Sea 383
A. Coastal Areas within Continental Slopes 383
B. The Deep Ocean 390
C. Mediterranean and Adjacent Seas 398
VI. Qualitative and Quantitative Aspects 401
References 404

I. Introduction

An unceasing and balanced exchange of nitrogen between resident organisms and their environment is essential for the continuity of life in the sea. For a given oceanic location, the many and diverse forms of nitrogen present depend upon a variety of assimilative and regenerative transformations which have long intrigued the marine ecologist. Collectively, these processes bear a fundamental relationship to all marine protein synthesis, and, along with comparable conversions for carbon, phosphorus and sulfur, often play a decisive role in controlling fertility.

Despite the chemical stability of molecular nitrogen, nitrogen in the sea readily responds to enzymatic activity and hence can occur in any one of nine different oxidation states (-3 to $+5$). Reduced nitrogen

* Contribution No. 1473 from the Woods Hole Oceanographic Institution. This research was in part supported by the National Science Foundation Grant, NSF 544, and by the U.S. Atomic Energy Commission under contract AT (30 – 1) – 1918.

(−3 to −1) occurs as ammonia and as organic nitrogen compounds in dissolved or particulate form. These highly reduced substances, which are commonly the end-products of nitrogen assimilation by marine plants or bacteria, represent about 35% of the total combined nitrogen in the oceans. However, molecular or gaseous nitrogen(0) is the most abundant form in the sea, and amounts to some 20 times the entire amount of combined nitrogen. The more highly oxidized forms of marine nitrogen are nitrite (+3) and nitrate (+5), the latter comprising about 65% of that in combined form. As the terminal oxidation state for combined nitrogen in the sea, nitrate-nitrogen can originate only from oxidative processes which presumably occur at a rather constant rate for the oceans as a whole. Since the major fraction of nitrogen in the combined state is highly oxidized, it would appear that over extended periods the amount of inorganic nitrogen which is oxidized to nitrate equals or exceeds the quantity reduced.

Normally, inorganic nitrogen in combined form is utilized by marine phytoplankton and certain bacteria for primary production. However, following bloom conditions it is not uncommon for algal populations to remove the available nitrogen quantitatively from the photosynthetic layer of the sea. Thereafter, the resumption of protein synthesis depends upon fixation or on the enrichment of the surface water with available nitrogen by various mixing phenomena, in either vertical or horizontal directions. Atmospheric precipitation, continental drainage, and the migrations of marine animals which excrete nitrogen compounds all play important roles in the supply and distribution of nitrogen.

II. Geochemical Evolution of Inorganic Nitrogen in the Hydrosphere

A. Prebiogenic Oceanic Nitrogen

Modern geochemical theory postulates that most of the nitrogen in the hydrosphere was originally evolved as free nitrogen and ammonia from the primitive Earth as a result of radiogenic heating by elements such as uranium, thorium and potassium. According to Vinogradov (1959), the minimum temperature required to melt the earth's interior would have been achieved some 5×10^9 years ago. During the subsequent 100 million years, the primordial ocean is presumed to have evolved (see Chapter 20, Volume 2). Apparently, these phenomena occurred during an intensively volcanic period when immense quantities of steam and other volatiles, including molecular- and ammoniacal-nitrogen, escaped from the earth's mantle. Throughout geological time the major fraction of juvenile nitrogen is believed to have entered the atmosphere from recurrent vulcanism. Other nitrogen contributions from rock weathering and meteoritic activity have also played important roles in determining the chemical characteristics of ancient seas and ultimately of life itself. It is commonly believed that the prebiogenic ocean was anoxic and of almost neutral pH, and that most of the combined nitrogen was in the reduced chemical form of ammonia. Presumably, even at this stage some oxygen was provided by the photochemical dissociation of water vapor at high altitudes.

B. NITROGEN AND THE PRIMARY BIOSPHERE

The appearance and evolution of life on our planet depended on the earlier synthesis of complex nitrogen-containing molecules which are vital constituents of all living material. The final molecular combinations are believed to have developed from the catalytic synthesis of organic nitrogen compounds by the interaction of ammonia with other chemical precursors such as CH_4, CO, CO_2, PO_4^{3-}, H_2S, HCl and HF. Some of the earliest recognized fossils, including the blue-green algae of Rhodesian limestone, originated in shallow continental seas some 2×10^9 years ago. The photosynthetic mode of existence of these organisms provides striking evidence that the biological oxidation of the atmosphere and ocean began at an extremely early date. Hutchinson (1944) has considered this matter in his discussion of the evolution of the nitrogen cycle, and has concluded that life, atmospheric oxygen and the ocean appear to have developed concurrently.

As a gaseous equilibrium between an oxygenated atmosphere and the oceans was approached, many of the less stable chemical compounds in the sea were subjected to both chemical and biological oxidations. Thus, sulphides became sulphates and ammonia was converted to nitrate and, at the same time, the concentration of molecular nitrogen probably increased both in the atmosphere and in the sea. The chemical compositions of neither the atmosphere nor the ocean appear to have undergone appreciable change since the late Proterozoic and Cambrian periods (Rubey, 1951).

III. Inorganic Nitrogen Exchanges across the Sea Surface

A. MOLECULAR NITROGEN

Like other atmospheric gases, molecular nitrogen interchanges between the atmosphere and the sea surface and the system may reach equilibrium in accordance with Henry's Law. In the sea, the dissolved atmospheric gases are distributed by the various oceanic circulatory and mixing processes. The noble gases, such as argon, whose subsequent distribution depends solely on mixing phenomena, because they are conservative, can be used for studying the participation of dissolved nitrogen in chemical and biological reactions.

Basic data on the solubility of nitrogen in sea water as given first by Fox (1909) and later by Rakestraw and Emmel (1938), have initiated considerable controversy. The latter authors believed that the oceans are normally saturated with molecular nitrogen, and recent improved measuring techniques as employed by Benson and Parker (1961) appear to substantiate this view. On this basis, most of the molecular nitrogen would appear to behave conservatively by remaining outside the marine biochemical cycle. However, specialized marine environments wherein molecular nitrogen is either utilized or liberated are becoming better known.

Naudé (1929) was the first worker to show that two naturally occurring non-radioactive isotopes of nitrogen, ^{14}N and ^{15}N, exist. In addition, there are four radio-isotopes (masses 12, 13, 16, and 17) which are short lived and presumably do not occur in nature. Marine biologists and geologists have shown considerable interest in the manner of distribution of the two stable nitrogen isotopes in the sea, since most of the atmospheric and sedimentary nitrogen has, at some time, passed through the oceanic biochemical cycle. The next few paragraphs summarize some important aspects of this subject.

Chapman and Brodia (1956) reviewed various reported values for the ratio $^{15}N : (^{15}N + ^{14}N)$ for atmospheric nitrogen. They considered that the ground level atomic ratio is constant at about 0·00365. According to Dole *et al.* (1954) this ratio is valid up to altitudes as high as 51·6 km. Hoering (1955) and Dole *et al.* (1954) have given isotopic evidence which appears to substantiate the contention of Rakestraw and Emmel (1938) that the molecular nitrogen content of the oceans is close to that which would be expected at equilibrium with the atmosphere.

It is not yet finally established to what extent biologically induced transformations involve a selectivity towards one or other of the nitrogen isotopes. Hoering (1955) has studied the relative abundance of nitrogen isotopes in plant proteins with inconclusive results, although he was able to show that in many animal proteins accumulation of the heavier isotope was favored. Among the substances showing ^{15}N enrichment were the proteins of certain seaweeds, clams and grains. J. J. Goering (personal communication) also found a slight ^{15}N enrichment in nitrogen derived from fish and various plants and from lake water. However, the influence of biological systems on nitrogen isotope separations in the natural processes of ammonification, nitrification, denitrification and transamination remains uncertain. An exception was reported by Hoering and Ford (1960) who found that the ratio for the rates of nitrogen isotope fixation by *Azotobacter* approach unity, thus indicating no measurable isotopic effect. The remaining nitrogen transformation processes await a more thorough evaluation.

Unlike nitrogen, comparable isotopic measurements with respect to dissolved oxygen, $^{16}O : ^{18}O$, for Pacific Ocean water (Dole *et al.*, 1954) indicate significant enrichment with the heavier isotope at all sub-surface depths with the degree of enrichment varying inversely with the dissolved oxygen content. This observation is consistent with the general rule of *mass effect* wherein the lighter isotope is selectively removed during biological assimilation. For practical purposes, the differential behavior of molecular oxygen as compared to nitrogen appears to re-emphasize the more conservative role of the latter in the sea.

B. Inorganic Nitrogen in Oceanic Rain

Although nitrogen occurs in the atmosphere predominantly in the molecular form, it also occurs in organic compounds and as ammonia, nitrite, nitrate and nitrogen oxides. The various sources of these constituents include the ocean itself, land masses, industrial contamination, and volcanic emanations. It is frequently postulated that atmospheric, electrical and photochemical activity have led to the fixation of nitrogen as ammonia and nitrate; these are therefore found in rain water. This subject has been reviewed by Hutchinson (1954), who concluded that atmospheric ammonia is derived from the decomposition of organic matter and is largely adsorbed onto dust particles. He suggested that only minor quantities of nitrate originate from lightning fixation compared to those derived from the oxidation of ammonia in the atmosphere. If this is the case, both the ammonia and nitrate of the atmosphere must correspond with stages of a cycle, and are not products of non-biological fixation. Isotopic fractionation studies by Hoering (1957) on the ammonia- and nitrate-nitrogen in rain further substantiate the hypothesis that nitrate in rain water arises from the oxidation of ammonia rather than from the fixation of atmospheric nitrogen.

The available data on the ammonia and nitrate content of rainwater at specific maritime locations over extended periods demonstrate wide variations in combined nitrogen (Carroll, 1962). Analyses from northern Europe compiled by the Svenska Geofysika Förenigen (1955, 1956) for 62 locations over a 30-month period give concentration ranges of 0–6·8 mg. NH_3–N/l. and 0–0·36 mg. NO_3^-–N/l. The corresponding mean concentrations were 0·31 and 0·056 mg./l. respectively. Other analyses from the continental United States reported by Junge and Werby (1958) failed to show a positive correlation between the concentrations of ammonia or nitrate in rain water and the distance from the sea. Over the entire country, the average ammonia content was 0·32 mg. N/l. compared with 0·52 mg. N/l. for nitrate.

Evidence given by Menzel and Spaeth (1962) indicates that the ammonia-nitrogen content in the upper 100 m. of the water column off the oceanic island of Bermuda is greater than that of nitrate-nitrogen. These authors also showed a positive correlation between the amount of surface ammonia and that provided by the accumulated rainfall of the five days prior to each sea water measurement. The average ammonia concentration given by 14 rain water analyses over a ten-month period was 0·069 mg. N/l. Unfortunately, comparable information on the nitrate in rain *vs* the surface nitrate concentrations are not included. However, for the sea surface, these data appear to stress the importance of complementary inorganic nitrogen contributions from rainfall in addition to the more familiar *in situ* sources of oceanic ammonia.

IV. Biological Factors Affecting the Distribution of Molecular and Combined Nitrogen

A. Nitrogen Fixation

In agriculture, farmers have traditionally encouraged the activities of nitrogen-fixing bacteria such as the nodular symbiotic *Azotobacter* and the anaerobic *Claustridia*. More recently, microbiologists have emphasized that nitrogen is fixed by a number of photosynthetic algae (Farren, 1932) and bacteria (Kamen and Gest, 1949). However, fixation in the marine environment has received less attention, and the extent of biological nitrogen fixation in the sea awaits more detailed evaluation.

Prolonged phytoplankton growth within the photic layer of stratified tropical and semi-tropical waters often reduces the available supply of combined nitrogen to negligible or trace amounts. Under these conditions, marine plants capable of utilizing molecular nitrogen would enjoy an appreciable ecological advantage. In addition, it is generally recognized that terrestrial and fresh-water bacteria, algae, yeasts and moulds known to fix nitrogen have true marine counterparts (Benecke and Kentner, 1903; Issatchenko, 1914; Waksman *et al.*, 1933a). However, a review of nitrogen fixation by Nutman (1959) describes the fixation process as endothermic, requiring rather large sources of chemical energy. Presumably carbohydrate provides the necessary hydrogen source which, according to Waksman and Starkey (1931), approximates to 1 g.

of sugar for each 10 mg. of nitrogen fixed. Neither the total concentration of available carbon in sea water (Plunkett and Rakestraw, 1955; Duursma, 1961) nor the rate of carbon fixation by marine photosynthesis (Riley, 1949; Steemann-Nielson, 1952; Ryther, 1959, 1960) appears adequate to support the carbohydrate requirement for nitrogen fixation over extended periods. Table I shows the unit carbohydrate requirements of a variety of organisms capable of fixing atmospheric nitrogen.

TABLE I

*Nitrogen fixation by pure cultures of various micro-organisms**

Species	*Nitrogen fixed (mg.)*
Azotobacter chroococcum	17
Azotobacter vinlandii	16
Pseudomondas sp.	17
Nocardia calcarea	4
Nocardia cellulans	4
Saccharomyces sp.	4
Rhodotorula sp.	4

* In mg. of nitrogen per g. of mannitol dissimilated (after Nutman, 1959).

The difficulty in allocating sufficient chemical energy to support heterotrophic nitrogen fixation in the sea has led to the proposal that photosynthetic organisms such as the blue-green *Calothrix* and *Trichodesmium* (Allen, 1963) may be important for this function. Accordingly, Fogg and Than-Tun (1958) proposed that in these algae, nitrogen fixation is intimately associated with photosynthesis since elementary nitrogen acts as an alternative acceptor to carbon for hydrogen released photochemically from water. This concept has more appeal for the marine ecologist than does heterotrophic fixation since it substitutes solar radiation for the unrealistic carbohydrate requirement.

Newer and more realistic analyses based on the uptake of ^{15}N promise to provide the necessary sensitivity for more definitive experiments on marine nitrogen fixation by obviating interference from unrealistic highly enriched cultures. Recently, the ^{15}N technique of Neess *et al.* (1962) has been used to measure nitrogen fixation in samples of lake water (Dugdale and Dugdale, 1962). Dugdale *et al.* (1961) have used the same technique to study the apparent fixation of nitrogen during daylight hours in the surface waters of the Sargasso Sea by the mixed phytoplankton in which the blue-green alga, *Trichodesmium* was predominant. No comparable nitrogen fixation was measured in the absence of light.

According to Allen (1963), *Trichodesmium* can occur in massive blooms which may cover hundreds of square km. in the Sargasso Sea, the Indian Ocean and off the Great Barrier Reef of Australia. However, confirmatory evidence on the nitrogen-fixing characteristic of these organisms awaits the development of more successful pure culture techniques for experimental purposes.

B. DENITRIFICATION

The accepted definition of denitrification for our purposes is the biological reduction of nitrate or nitrite to either nitrous oxide or free nitrogen. A wide variety of facultative anaerobic bacteria within the sea and its sediments is involved in reactions of this type which, according to Brisou and Vargues (1961), combine proteolysis with nitrate reduction. However, in a stricter sense, these organisms can be classified as either heterotrophs or autotrophs according to whether they use an organic or inorganic energy source (Cooper, 1937; Baas Becking, 1959). When an organic substrate supplies the hydrogen there is a fall in free energy, but if the hydrogen is derived from an inorganic source, e.g. H_2, H_2S, etc., there is a fall in free energy only in the absence of excess oxygen. Thus at 25°C:

$$2NO_3^- + 2H^+ \longrightarrow N_2O + 2O_2 + H_2O$$

$$\Delta G^\circ = +21 \text{ kcal.}$$

$$4NO_3^- + 3CH_4 \longrightarrow 2N_2 + 3CO_2 + 6H_2O$$

$$\Delta G^\circ = -475 \text{ kcal.}$$

$$8NO_2^- + C_6H_{12}O_6 \longrightarrow 4N_2 + 4CO_3^{2-} + 2CO_2 + 6H_2O$$

$$\Delta G^\circ = -728 \text{ kcal.}$$

In the early days of marine microbiology, considerable attention was devoted to the precipitation of calcium carbonate in the sea, a process which may be accompanied by the reduction of nitrate to free nitrogen. Also, geologists concerned with the origin of carbonates in marine sediments have shown interest in this subject. Among the former, Drew (1911, 1913) proposed that specialized calcium bacteria bring about carbonate precipitation in tropical areas through the reduction of nitrate to molecular nitrogen, the reduction presumably being brought about by the oxidation of sedimentary organic matter which is converted to carbonate. However, more recent evidence, summarized by Bavendamm (1931, 1932) and ZoBell (1946), recognizes that although microbial carbonate precipitation is important in the sediments of tropical seas, its occurrence depends upon the complementary activities of many, rather than a few, bacterial species. Under these conditions, an association between carbonate precipitation and denitrification would be a fortuitous rather than a specific aspect of the denitrification process.

Denitrification is also known to occur in the water column as well as in the sediments, provided that the dissolved oxygen concentration is below a critical level. This situation may occur at intermediate depths over broad areas of the Pacific and Indian Oceans where the dissolved oxygen content between 150 and 500 m. approaches zero. According to Brandhorst (1959), at these sites there is an enrichment with nitrite at the expense of nitrate when the dissolved oxygen concentrations decrease to less than 0·15 ml./l. Comparable observations have been made in the Arabian Sea (Gilson, 1937), the Cariaco Trench (Richards and Vaccaro, 1956), the Black Sea (Kriss, 1949), and for the oxygen-depleted sediments of marine basins off the coast of California (Rittenberg *et al.*, 1955). Among these studies, the Cariaco Trench has been especially well investigated; the original observations on anaerobic nitrate depletion were later confirmed by Richards and Benson (1961) who used the nitrogen : argon and the nitrogen isotope ratios to estimate the degree of enrichment of the anoxic water with free nitrogen of biogenic origin.

The relative importance of N_2O and N_2 formation during denitrification remains obscure. However, the subject could well have important geochemical implications bearing upon the ultimate balance of nitrogenous gases in the atmosphere. Thus, Hutchinson (1944, 1954) stressed the role of denitrification in restoring free nitrogen to the atmosphere, and estimated that at the contemporary rate of terrestrial fixation, nitrogen from the atmosphere would be exhausted in about 2×10^8 years. Infra-red absorption studies by Adel (1939, 1941) have confirmed that considerable quantities of nitrous oxide occur in the earth's atmosphere. Besides denitrification, the decomposition of unstable hyponitrite, a commonly recognized intermediate produced during the oxidation of ammonia in soil, has also been mentioned as an alternative or additional source of atmospheric nitrous oxide. Regardless of its origin, nitrous oxide is believed to be decomposed at high altitudes by ultra-violet radiation (Berthelot and Gaudechon, 1910).

C. NITROGEN ASSIMILATION BY MARINE PHYTOPLANKTON

The fundamental nitrogen requirements for life in the sea depend upon the conversion of inorganic nitrogen into cellular amino acids by the photosynthetic plankton algae. Protein for each higher trophic level depends upon alternate polymerization and proteolysis of these same primary amino acids. At the phytoplankton level, organic synthesis is highly selective, and consequently for a given location, cellular material often has a rather consistent elementary composition in a statistical sense (Vinogradov, 1953; Redfield, 1934).

TABLE II

*Elementary composition of some aquatic plants**

Organism and Reference	*No. of analyses*	*C*	*H*	*O*	*N*	*P*
Diatoms	18	50·54	10·21	28·83	7·0	1·55
Brandt and Raben (1920)		±3·90	±1·90	±7·10	±3·7	±0·88
Ketchum and Redfield (1949)						
Vinogradov† (1953)						
Peridineans	7	48·12	7·50	33·85	10·40	0·80
Brandt and Raben (1920)		±1·90	±1·70	±3·90	±1·40	±0·36
Chlorophyceae	18	54·55	7·54	31·21	7·7	2·94
Ketchum and Redfield (1949)		±1·30	±0·50	±4·35	±0·56	±0·83
Macroscopic algae	6	46·80	7·62	41·65	5·0	0·44
Brandt and Raben (1920)		±1·40	±1·90	±3·30	±2·40	±0·37
Zostera	3	46·11	7·12	43·26	3·4	0·50
Brandt and Raben (1920)		±0·60	±1·10	±1·70	±1·6	±0·26

* As a percentage on an ash-free dry weight basis, together with standard deviations.
† N and P values only.
From Ryther (1956).

Ryther (1956) has reviewed this subject with respect to the carbon, hydrogen, oxygen, nitrogen and phosphorus contents of some important aquatic plants. Table II illustrates the similarity of the elementary combinations which ultimately provide the fats, proteins and organic acids necessary for growth and reproduction of these organisms. Given normal amounts of available carbon, nitrogen and phosphorus, plus other essential nutrients, the observed atomic ratios for assimilation of these elements closely approximate to 100:16:1 respectively (Fleming, 1940; Redfield, 1958; see also Chapter 12, Section IV, B).

In the absence of light, nitrogen assimilation by phytoplankton cannot persist for extended periods. Thus, primary nitrogen assimilation is essentially an autotrophic phenomenon limited to the photic layer (the upper 150 m. or less). Commonly, this layer is isolated physically by temperature and salinity from the deep, but potentially eutrophic, water wherein regenerative biochemical transformations are dominant. While the concept of a consistent elementary composition appears to have a general application in nature, the nitrogen and phosphorus contents of algal populations under experimental conditions in the laboratory are more dependent on the concentrations of these elements in the culture media. On this basis, the cellular nitrogen content of *Chlorella pyrenoidosa* can vary from 3–9% of the ash-free dry weight (Ketchum, 1939a,b; Ketchum and Redfield, 1949; Spoehr and Milner, 1949). Also, Ketchum (1939a) showed that prolonged nitrogen deprivation depressed the N:P ratio to 2·33:1, whereas for cells deficient only in phosphorus, the ratio increased to 31:1. However, even with sub-optimal concentrations of nitrogen available, more than 90% of the total cellular nitrogen of *Scenedesmus obliquus* was shown to be associated with protein (Thomas and Krauss, 1955).

All marine algae capable of utilizing inorganic nitrogen can use ammonia, and most of the common forms attain comparable rates of growth with ammonia, nitrite, or nitrate. With all three nitrogen sources available simultaneously, ammonia is often used preferentially, possibly because of a toxic effect on the enzymatic reduction of nitrate (Ludwig, 1938; Schuler *et al.*, 1953). In 1954, Syrett demonstrated that starved *Chlorella* cells given equivalent amounts of nitrogen recovered more quickly with ammonia than with nitrate. Furthermore, ammonia is often the only form of available nitrogen measurable in the surface layer of the sea during certain seasons of the year.

There appear to be four preliminary stages in the assimilation of nitrogen from nitrate by algae; the overall process for each step is given by (cf. Nicholas, 1959):

$$NO_3^- + 2H^+ + 2e \longrightarrow NO_2^- + H_2O$$

$$2NO_2^- + 4H^+ + 4e \longrightarrow N_2O_2^{2-} + 2H_2O$$

$$N_2O_2^{2-} + 6H^+ + 4e \longrightarrow 2NH_2OH$$

$$NH_2OH + 2H^+ + 2e \longrightarrow NH_3 + H_2O.$$

It is not uncommon for significant amounts of extracellular nitrite to accumulate during nitrate reduction, especially at low light intensities (Orr, 1926; ZoBell, 1935; Vaccaro and Ryther, 1960). This type of behavior will be examined later in relation to the seasonal appearances of high nitrite concentrations in coastal waters and to nitrite accumulations at, or near, the depth of maximum chlorophyll concentration in the open sea.

The reduction of nitrite and nitrate to ammonia requires a free energy increase. Thus, for the same energy increase, algal cells can assimilate more nitrogen from ammonia than from nitrite or nitrate. When nitrate is reduced in the dark, the oxidation of previously formed reduced chemical substances provides two "extra" moles of carbon dioxide for each mole of nitrogen reduced. Accordingly, Syrett, 1962, has proposed the following stoichiometric reaction:

$$2(CH_2O) + NO_3^- + 2H^+ \longrightarrow NH_4^+ + 2CO_2 + H_2O.$$

In the light, approximately two "extra" moles of oxygen rather than carbon dioxide are evolved for each mole of ammonia produced (Kessler, 1955; Bongers, 1958):

$$NO_3^- + 2H_2O \longrightarrow NH_3 + OH^- + 2O_2$$

and

$$NO_2^- + 2H_2O \longrightarrow NH_3 + OH^- + 1{\cdot}5O_2.$$

During photosynthesis, the fixation of one mole of carbon dioxide to carbohydrate increases the free energy by 112 kcal. In nitrate reduction, nitrogen replaces carbon dioxide as the hydrogen acceptor, but the energy changes are only 73 kcal. for nitrate and 54 kcal. for nitrite. Therefore, in terms of moles of oxygen evolved, the kinetic efficiency of nitrate reduction is only about 30% of that realized by the photosynthetic reduction of carbon dioxide.

Guillard (1963) studied the capacity of various marine algae to use organic nitrogen sources, and concluded that, with minor exceptions, growth was less favorable than with nitrate. Among the heterotrophs, the inshore forms from estuaries and coastal waters, often exposed to significant organic pollution, tended to thrive better with urea and uric acid, whereas the heterotrophic oceanic species grew best with glutamine. As was concluded by Harvey (1940), there is as yet no firm basis for believing that naturally occurring organic nitrogen in sea water provides a significant source of algal nitrogen.

D. AMMONIA AND THE DECOMPOSITION OF ORGANIC NITROGEN

Just as organic synthesis in the sea is dominated by phytoplankton, the destructive phases of the organic cycle are primarily regulated by the marine bacteria. Most of the latter are intensely proteolytic (Waksman and Carey, 1935), being able to attack an impressive variety of nitrogen linkages (Ostroff and Henry, 1939). For their material and metabolic requirements, these organisms convert plant and animal remains and excretions into simpler chemical entities such as carbon dioxide, hydrogen sulphide and ammonia. Their overall effectiveness ensures that the residual concentrations of dissolved organic nitrogen and carbon are maintained at relatively low and quite uniform concentrations (Krogh, 1934; Plunkett and Rakestraw, 1955). The combined organic nitrogen and carbon content of these highly resistant substances rarely exceeds 3 g./m.^{3} in the water column of the open ocean (see Chapter 11).

Besides bacterial proteolysis, zooplankton excretions also provide both ammonia and phosphate. When released in the photic layer, such material provides available nitrogen and phosphorus for phytoplankton either directly, or after bacterial or autolytic decomposition. Harvey *et al.* (1935) noted that the amount of grazing by zooplankton in the waters off Plymouth, England, was related more directly to the quantity of food available than to nutritional requirements. Redfield and Keys (1938) have related seasonal ammonia changes in the Gulf of Maine to the *in situ* changes in zooplankton abundance. Similarly, Harris (1959) concluded that ammonia from animal excretions in Long Island Sound provides over half the daily nitrogen requirement of the resident phytoplankton. J. R. Beers (personal communication) has also attributed the significant ammonia and phosphate enrichment in the surface waters adjacent to Bermuda to excretions of the carnivorous arrow-worm *Sagitta hispida*. Such observations emphasize that some ammonia may be injected directly into the water, thus eliminating the prolonged oxidative stage of the cycle wherein nitrogen is returned to the sea as nitrate.

E. OXIDATION OF INORGANIC NITROGEN

The oxidation of ammonia to nitrite or nitrate, commonly referred to as nitrification, is a subject of much ecological uncertainty, especially with regard to the biological and chemical mechanisms involved. Clearly, ammonia and its oxidation provides a valuable source of both nitrogen and potential energy for the fixation of carbon dioxide by propitious and autotrophic micro-organisms. However, to date, views of the role of these oxidations in the deep ocean are largely based on indirect evidence from oceanographic studies of the distribution and relationship between nitrate and phosphate and the oxygen deficit within the water column. The results of these investigations have repeatedly suggested the occurrence of *in situ* nitrification within the water column (Wattenberg, 1929; Seiwell, 1935; Riley, 1951), and have demonstrated that nitrate represents the terminal oxidation state for de-aminated nitrogen and is the most abundant form of combined nitrogen in the sea.

Until recently, the biological oxidation of ammonia and nitrite was attributed to only a limited number of marine species of bacteria, presumably oceanic counterparts of the more familiar terrestrial *Nitrosomonas* and *Nitrobacter* (ZoBell, 1946). Also, the recorded distribution of these organisms in the sea indicated that their habitat was limited to coastal waters and to sediments (Waksman *et al.*, 1933a,b; Carey, 1938). Watson (1962, 1963) and Murray and Watson (1963) have described a marine nitrifier *Nitrosocystis oceanus* Watson, which is capable of oxidizing ammonia to nitrite, and which appears to be physiologically similar to, but morphologically distinct from, previously described marine *Nitrosomonas*. Significantly, *Nitrosocystis* has been successfully isolated from oceanic depths as great as 1000 m. suggesting that this species is a distinct representative of the pelagic nitrifying micro-organisms which are active within the water column.

When suitably activated, the oxidations of ammonia to nitrite and of nitrite to nitrate proceed with a free energy decrease, possibly via hydroxylamine and hyponitrite as short lived intermediates (Cooper, 1937). Thus, at 25°C:

$$NH_4^+ + OH^- + 1{\cdot}5\ O_2 \longrightarrow H^+ + NO_2^- + 2H_2O$$

$$\Delta G^\circ = -59{\cdot}4 \text{ kcal.}$$

and for nitrite oxidation:

$$NO_2^- + 0{\cdot}5\ O_2 \longrightarrow NO_3^-$$

$$\Delta G^\circ = -18{\cdot}0 \text{ kcal.}$$

In nature, however, the mineralization of nitrogen proceeds simultaneously with oxidation of the organic matter, so that the total quantity of oxygen ultimately associated with the production of a mole of nitrate from amino-nitrogen depends largely on the oxygen:carbon ratio of the organic precursor.

The biological oxidation of ammonia to nitrite, and finally to nitrate, in sea water has also been confirmed in the series of laboratory experiments by von Brand and colleagues (von Brand *et al.*, 1937, 1939; von Brand and Rakestraw, 1940, 1941; von Brand *et al.*, 1942) and by Rakestraw and von Brand (1947). Their studies will be considered in some detail since they provide important descriptions (Sverdrup *et al.*, 1942; Barnes, 1957; Raymont, 1963) of regenerative aspects of the marine nitrogen cycle under controlled conditions. The experimental procedure consisted of adding various kinds of plankton and other organic substances to sea water derived from various sources. The culture vessels used were glass carboys which were stored in the dark, usually at room temperature, and samples were periodically removed for analysis for ammonia-, nitrite-, and nitrate-nitrogen throughout an extended period of decomposition.

The period of predominance of any one form of nitrogen varied between experiments, but decomposition usually proceeded most rapidly in the earliest (ammonia) stage (Fig. 1(a)). The initial appearance of nitrite usually coincided with the attainment of an ammonia maximum, and the appearance of nitrate was often delayed until the ammonia was reduced to small or negligible concentrations. The overall rate of decomposition was more dependent on the nature and source of the planktonic enrichment material than on its concentration, and once vigorous ammonia oxidation was in progress, no inhibitory effects were induced by additional organic enrichment. As many as three complete cycles were completed during a single experiment by interrupting the periods of illumination with prolonged dark periods in which nitrification was allowed to go to completion (Fig. 1(b)). The actual time required for the completion of each of the three regenerative periods varied markedly, being consecutively 3·5, 7·0, and 2·5 months.

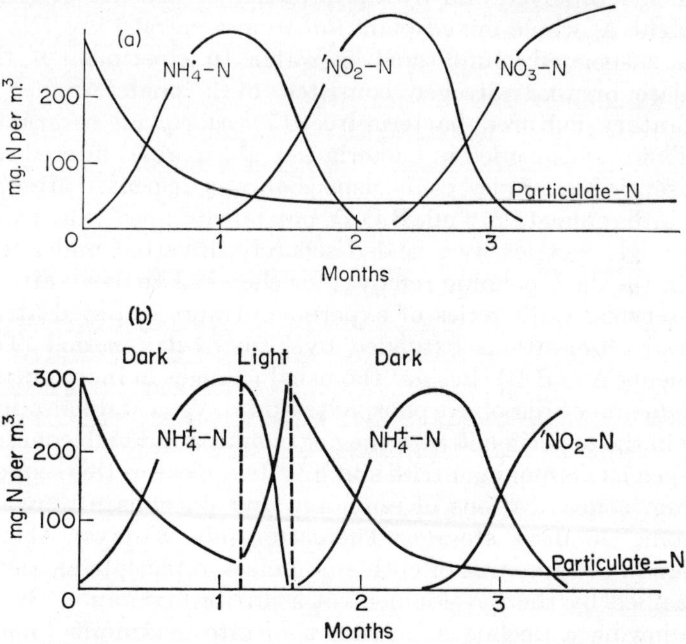

Fig. 1. (a) Production of NH_3–N and its subsequent conversion to nitrite and nitrate in sea water containing diatoms stored in the dark. (b) Production of NH_3–N in sea water containing diatoms stored in the dark, its utilization during a period of illumination and its subsequent reformation and conversion to nitrite in the dark (Harvey, 1960, reproduced by permission of Cambridge University Press).

The source of the sea water used in these experiments influenced the results in an important manner. Sea water from in-shore locations or from the upper 800 m. in the open sea usually supported complete nitrification, whereas in that taken from greater depths (1200 m.) the processes generally ceased after ammonia production. The responsible nitrifying flora was evidently associated and introduced with the plankton enrichment, since substitution of autoclaved sea-water substrates had little or no deleterious effect on the nitrification process. The authors suggested that the very deep waters used in their studies may have lacked a growth factor essential for the development of effective concentrations of nitrifying bacteria which, in any case, consistently resisted isolation by sub-culturing techniques.

To demonstrate the importance of marine bacteria in the nitrification process, observations similar to those of von Brand *et al.* (1937, 1939, 1942) have been made by the author. These studies were conducted with Vineyard Sound surface water collected from a single source in late August. Variations were introduced by the type of organic nitrogen enrichment employed and by the presence or absence of bacteria. In experiment A, whole mixed plankton from a surface tow (No. 20 mesh net) was suspended in unfiltered sea water. In experiment B, the added particulate organic nitrogen consisted of a predetermined number of laboratory cultured bacteria-free *Phaeodactylum tricornutum* cells (10^4 cells/ml.) suspended in bacteria-free Millipore®† filtered sea water. Using aseptic techniques, each suspension was dispensed after adequate mixing into replicate 270 ml. B.O.D. bottles equipped with ground glass stoppers. The bottles were sealed securely, inverted under water and stored in the dark pending removal for chemical analyses after selected time intervals. Both series of experiments were initiated at the same time, and observations extended over an 87-day period (Table III, experiments A and B). Besides the usual changes in inorganic nitrogen, measurements of dissolved phosphate and oxygen were also included.

Only in the experiment with the net plankton did high concentrations of nitrogen as ammonia, nitrite and nitrate appear in this sequence. The maximum concentrations of ammonia and phosphate were measured after some 30 days storage. Thereafter (30–48 days), there was a reduction in the amounts of both ammonia and phosphate, and this was accompanied by the development of a nitrite maximum. By the 72nd day, following a decline in nitrite, a nitrate maximum (oxidation of nitrite) appeared. Thereafter, no significant change in nitrate occurred up to the final observation at 87 days. The indicated changes in dissolved

† Registered trade mark, Millipore Filter Corporation.

oxygen showed a continual decrease throughout, with the total fraction removed amounting to about 35% of the initial concentration.

In the *Phaeodactylum* experiment, nitrogen regeneration ceased after the production of ammonia and phosphate. The time required for the appearance of maximum phosphate was 30 days, whereas that of ammonia required an additional 18 days. About 75% of the initial dissolved oxygen concentration remained at the termination of the experiment. Presumably, the absence of a nitrifying bacterial population in this experiment precluded the oxidation of ammonia to nitrite and nitrate, and the liberation of ammonia and phosphate was apparently brought about by cellular autolytic processes rather than by bacterial ammonification.

TABLE III

Regenerative changes in inorganic nitrogen and phosphorus vs *oxygen decrease accompanying the decomposition of marine plankton in Vineyard Sound Sea Water*

Time	NH_4^+–N		NO_2^-–N		NO_3^-–N		PO_4^{3-}–P		O_2–O	
(days)	*µg./l.*	*µg-at./l.*	*µg./l.*	*µg.at./l.*	*µg./l.*	*µg-at./l.*	*µg./l.*	*µg-at./l.*	*µg./l.*	*µg-at/l.*
					Experiment A *Enrichment, Whole Plankton Tow*					
0	0·0	0·00	0·0	0·00	0·0	0·00	25·7	0·83	0·00	0
7	10·4	0·74	0·0	0·00	1·7	0·12	24·5	0·79	−0·95	−59
17	39·3	2·81	0·0	0·00	0·0	0·00	26·0	0·84	−1·09	−68
30	42·0	3·00	0·8	0·06	0·3	0·02	32·5	1·05	−1·28	−80
48	0·0	0·00	53·5	3·83	0·4	0·03	30·4	0·98	−1·86	−116
72	3·1	0·22	—	—	54·5	3·90	30·6	0·99	−2·12	−132
87	2·0	0·14	0·7	0·05	55·1	3·95	32·2	1·04	−2·50	−156
					Experiment B *Enrichment,* Phaeodactylum tricornutum					
0	0·0	0·00	0·0	0·00	1·2	0·08	14·9	0·48	0·00	0
7	0·0	0·00	0·0	0·00	4·6	0·33	12·1	0·39	−0·29	−18
17	8·8	0·63	0·0	0·00	3·5	0·25	16·1	0·52	−0·48	−30
30	49·8	3·56	0·0	0·00	2·5	0·18	17·4	0·56	−0·58	−36
48	70·0	5·00	2·7	0·19	0·0	0·00	14·9	0·48	−0·77	−48
72	33·5	2·39	1·7	0·12	—	—	17·4	0·56	−0·87	−54
87	19·9	1·42	0·7	0·05	1·5	0·11	17·4	0·56	−1·44	−90

These data provide a basis for the comparison of the extent of oxygen utilization with the successive appearances of ammonia, nitrite and nitrate. Table IV summarizes this information in the form of atomic quotients ($\Delta O:\Delta N$) computed for the time intervals between zero and 17, 30, 48 and 87 days of storage. As shown, almost one half of the total oxygen uptake coincided with the liberation of ammonia (0–30 days), presumably owing to the simultaneous oxidation of organic matter. The balance of the oxygen utilization was about equally divided between that required for the oxidation of ammonia and of nitrite. The oxidative ratios for nitrogen show that the appearance of each atom of ammonia-nitrogen was accompanied by the utilization of about 25 atoms of oxygen, and each nitrite and nitrate atom formed required a total expenditure of 20 and 38 atoms of oxygen respectively. A comparable direct derivation of phosphorus oxidative ratios ($\Delta O{:}\Delta P$) from these data is not possible owing to erratic phosphate fluctuations which were probably associated with excessive internal recycling. Since the initial particulate N:P ratio was not determined, the $\Delta O{:}\Delta P$ ratio cannot be obtained directly. However, as a first approximation, average values for the particulate N:P ratio for typical August plankton populations within off-shore New England waters (Vaccaro, 1963) have been utilized in the construction of Table IV. The summer relationship ($\Delta O{:}\Delta N = 12$) is preferable since, at the time of the experiment, the available nitrogen in Vineyard Sound was at a seasonal minimum.

TABLE IV

Oxidative ratios for nitrogen and phosphorus from the atomic relationships between oxygen utilized and the ammonia, nitrite and nitrate produced during the decomposition of marine plankton in sea water

	Nitrogen change expressed as µg.-at./l.			
Storage (days)	NH_4^+–N	NO_2^-–N	NO_3^-–N	ΣN
0–17	2·81	0·00	0·00	2·81
0–30	3·00	0·06	0·02	3·08
0–48	0·00	3·83	0·03	3·86
0–87	0·14	0·05	3·95	4·14
	Oxygen Utilized			
	µg.-at./l.	*% Total*		
0–17	68	44		
0–30	80	51		
0–48	116	74		
0–87	156	100		
	Oxidative Ratios			
	$\Delta O{:}\Delta N$	$\Delta O{:}\Delta PO_4^{3-}$–P (*)	$\Delta O{:}\Delta PO_4^{3-}$–P (†)	
0–17	−24	−288	−384	
0–30	−26	−312	−414	
0–48	−30	—	—	
0–87	−38	—	—	

* Estimated from, $(\Delta O{:}\Delta N) = 12:1$.
† Estimated from, $(\Delta O{:}\Delta N) = 16:1$.

Processes other than the biological oxidation of ammonia have also been cited as possibly contributing to the formation of oxidized nitrogen in the sea. According to Cooper (1937), bacteria-free sea waters from the English Channel and Plymouth Sound, after ten days storage, showed decreases in the ammonia concentration without a compensating increase in nitrite. He concluded that the ammonia liberated from the organic matter was probably converted into hyponitrite by some unidentified but non-bacterial process. Kreps (1934) suggested that non-viable catalyzing "agents", probably of bacterial origin, can induce the oxidation of ammonia as well as other oxidations in the sea. Other investigators have described the oxidation of ammonia under the influence of solar ultra-violet radiation (ZoBell, 1933; Rakestraw and Hollaender, 1936) or by accelerated chemical reactions at an air–water interface (Cooper, 1937; Vaccaro, 1962). However, Hamilton (1964) has examined the role of photochemical processes in the marine inorganic nitrogen cycle and has concluded that the photochemical oxidation of ammonia is of no significance, whereas the photochemical reduction of nitrate is possible but of doubtful significance. Since most of these processes depend on conditions at the surface, their overall role in the nitrogen economy of the sea is presumably small.

V. Combined Inorganic Nitrogen in the Sea

In a strict sense, soluble nutrient elements in the sea are reserves temporarily outside the living biochemical cycle. The amounts of non-conservative elements found at a specific time and location will, therefore, largely reflect inherent inequalities in the rates of exchange among the reciprocal reactions of the overall nutrient cycle. More comprehensive measurements of multiple combinations of these elements are unlikely to provide an accurate description of the proportions which enter or leave the life cycles, since excessive residuals of one element in terms of another are not unusual. A more realistic approach to *in situ* nutrient cycles is provided by suitably controlled multiple sampling during periods of minimum non-biological interference. In this way, studies on the dispersal of nutrient elements in the sea can be put on a more quantitative footing.

A. Coastal Areas within Continental Slopes

(1) *Nitrate-nitrogen.* The main fluctuations in dissolved inorganic nitrogen in temperate surface waters as exemplified by the western North Atlantic, are mostly seasonal, especially within the confines of broad coastal areas overlying continental shelves. Typically, inorganic nitrogen concentrations decrease as a result of phytoplankton activity,

but are replenished by decomposition of organic matter *in situ* and by vertical mixing which returns available nitrogen to the surface layer. In early summer, solar heating at the surface leads to a stable density distribution in which warm mixed water extends to a depth of about 25 m. Completion of this process is closely attended by a quantitative removal of nitrate from the upper mixed layer because of an accompanying assimilation of nitrogen by phytoplankton populations. Where the water column exceeds 50 m., a sharp, positive gradient with respect to both density and nitrate is commonly observed between 25 and 50 m. Usually, the maximum summer concentrations of nitrate occur just above the sediment layer where, depending on the depth of water, nitrate-nitrogen concentrations may vary from 70–210 μg./l.

During winter, surface cooling and wind-mixing markedly decrease density stratification and result in high nitrate concentrations with a rather uniform vertical distribution. On the Atlantic continental shelf, the nitrate increase in waters above a depth of 50 m. is accompanied by a decrease in nitrate concentration below 50 m., indicating that much of the observed change may be caused by mixing within the water column rather than by a regeneration of nutrient *in situ*. By January, the shelf waters are homogeneous with respect to density, and nutrient concentration, and at the same time, nitrate-nitrogen concentrations of about 140 μg./l. extend from the surface down to a depth of about 100 m. or less. These conditions persist until the outbreak of the spring phytoplankton bloom which terminates with a return to early summer conditions. Figures 2 and 3 show the distribution of density (σ_t) and nitrate across the continental shelf south of Long Island as observed by Ketchum *et al.* (1958) at five different times during the year.

Seasonal studies by other investigators on the distribution of nitrate in temperate waters at a variety of comparable locations show marked similarities to the seasonal sequence outlined above. Along the northeast coast of the United States, corresponding seasonal changes presumably extend from the Gulf of Maine (Rakestraw, 1936; Bigelow *et al.*, 1940) to Long Island Sound (Riley and Conover, 1956; Harris and Riley, 1956) and southward to Cape Hatteras. Similar events have been described for the English Channel by Harvey (1928) and by Cooper (1933). Nitrate is often depleted to negligible or trace amounts during summer, and this fact is of importance to the marine ecologist because it implies that shortage of nitrogen limits plant growth during much of the year.

The distribution of nitrate is quite different in coastal waters located over narrow continental shelves where the prevailing off-shore winds cause upwelling of nutrient-rich deep waters. This situation is typical of western continental boundaries such as the northwest coast of

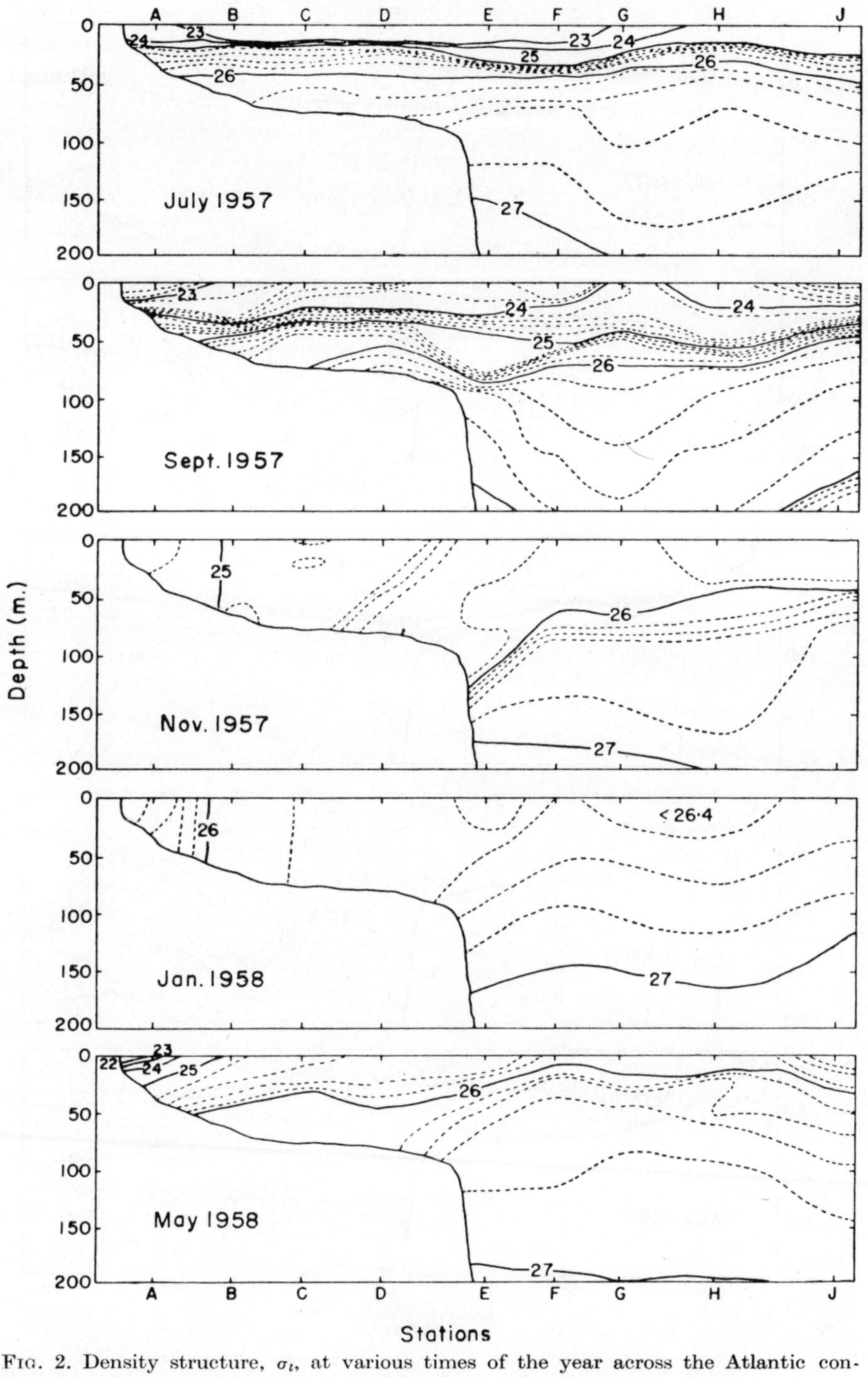

FIG. 2. Density structure, σ_t, at various times of the year across the Atlantic continental shelf south of Montauk, Long Island. From Ketchum *et al.* (1958).

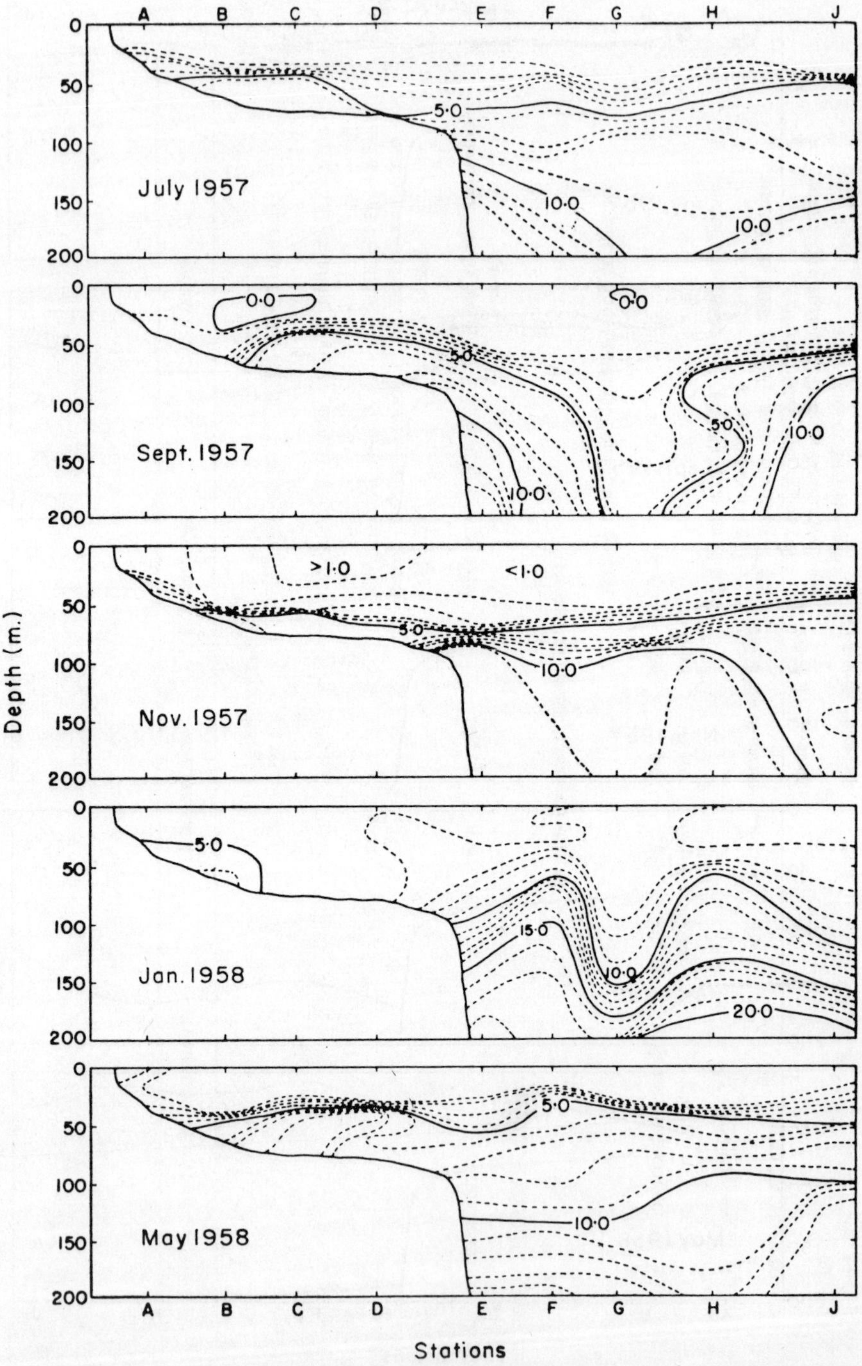

FIG. 3. Variations in nitrate-nitrogen, μg.–at./l., at various times of the year across the Atlantic continental shelf south of Montauk, Long Island (1 μg.–at./l. = 14 μg. NO_3–N/l.). From Ketchum *et al.* (1958).

North America and the west coasts of Africa and South America, all of which are characterized by persistently high surface nitrate concentrations. For example, Phifer and Thompson (1937) report a mean summer nitrate-nitrogen concentration of 210 μg./l. and 350 μg./l. during winter in the San Juan Channel of the northeastern Pacific. Similarly, off the West African coast, surface nitrate-nitrogen concentrations of 210 μg./l. are not unusual (R. F. Vaccaro, unpublished results). Clearly, the high nitrate concentrations of these areas exclude the possibility that primary production is nitrogen-limited.

(2) *Nitrite- and ammonia-nitrogen.* In addition to nitrate, ammonia and nitrite also provide available sources of nitrogen for phytoplankton populations, and are consequently subject to characteristic geographical and seasonal changes in the sea. Near the surface, ammonia occurs at significantly higher concentrations than does nitrite, a matter of interest in view of the physiological evidence favoring ammonia as the preferred nitrogen source for many species of phytoplankton. In addition, as previously stated, ammonia provides a convenient linkage for short-circuiting the more prolonged and extensive nitrogen transformations accompanying *in situ* nitrification. For these reasons, the nutrient role of ammonia in the sea is doubtless of considerably more importance than that of nitrite.

The seasonal ammonia variations in the English Channel have been described by Cooper (1933) as largely surface phenomena with the highest concentrations (up to 28 μg. NH_3–N/l.) occurring during the summer and autumn months. Similar seasonal changes were reported for the Gulf of Maine by Redfield and Keys (1938) who, as noted above, correlated high ammonia concentrations with *in situ* increases in the zooplankton populations. In Long Island Sound, Harris (1959) observed that the ammonia fraction in spring and summer was generally at least equal to the combined amounts of nitrate and nitrite. He concluded that ammonia, about half of which he attributed to the excretions of planktonic animals, provides about 75% of the daily nitrogen requirement for Long Island Sound phytoplankton. Earlier, Harvey *et al.* (1935) observed that grazing by zooplankton was related more directly to the quantity of food available than to the actual nutrient requirements of these animals. Much of the soluble ammonia and phosphate excreted near the surface during impulsive, luxury feeding (Marshall and Orr, 1961) is probably reassimilated rather than removed to subphotic depths by the sinking of plant cells. Consequently, at a given location intermittent nitrogenous animal excretions favor a rapid recycling of plant nutrients and probably enhance short-term primary production to an important extent.

Figure 4 shows the changes in the concentrations of ammonia, total available nitrogen, and phosphorus observed by Vaccaro (1963) during August and January in the water column off New England. For

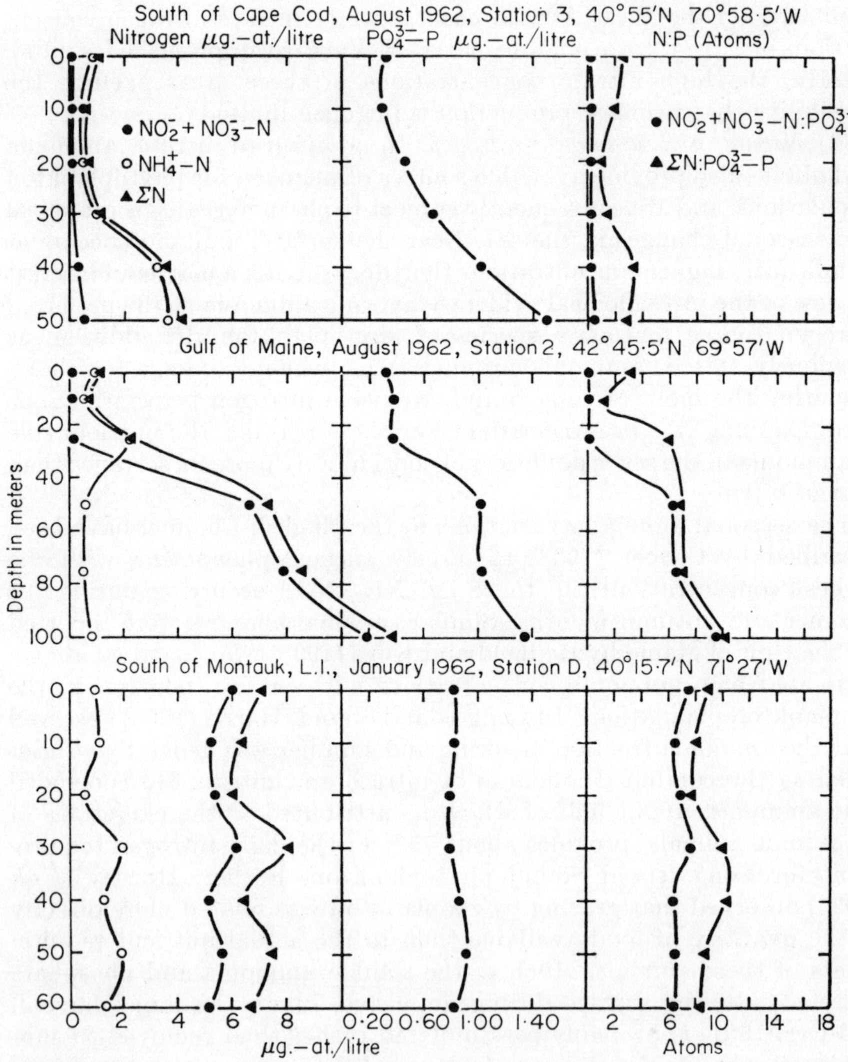

FIG. 4. Vertical distribution of available nitrogen species and PO_4^{3-}–P (in μg.–at./l.) and N:P atomic ratios during August and January, 1962 (1 μg.–at. NO_3^-–N/l. = 14 μg. NO_3–N/l. and 1 μg.–at. PO_4^{3-}–P/l. = 31 μg. PO_4^{3-}–P/l.). From Vaccaro (1963).

comparison, the ratios of the concentration of available nitrogen : phosphorus (atoms) are given with, and without, the inclusion of ammonia-nitrogen. At each of the August locations, ammonia is consistently the more abundant form of nitrogen within the upper 30 m. At greater depths, nitrate exceeds ammonia-nitrogen in the deeper parts of the Gulf of Maine where relatively large nitrate-nitrogen concentrations (more than 140 μg./l.) apparently persist throughout the year. South of Cape Cod, the maximum ammonia concentration occurred close to the bottom, suggesting that important nitrogen contributions arise from ammonification within the sediments. As shown, unless ammonia is taken into account, the apparent available nitrogen for August for the upper 30 m. both north and south of Cape Cod would be approximately zero.

In January, following the breakdown of the summer thermocline, homogeneity with respect to plant nutrients is once more established in the waters south of Cape Cod. High concentrations of nitrate are the rule throughout the water column; ammonia concentrations, although irregular, remain somewhat similar to those observed in summer. Throughout the winter months there is sufficient nitrate present to relegate ammonia to a minor fraction of the total available nitrogen, and the nutrient influence of ammonia is at a seasonal minimum.

The amount of nitrite in the sea is small compared with that of nitrate and also somewhat less than that of ammonia. Nitrite can be derived either from the oxidation of ammonia or the reduction of nitrate, and can be removed from solution during nitrogen assimilation by phytoplankton, although less readily than is ammonia (Ludwig, 1938; Schuler *et al.*, 1953). The relative instability of nitrite has been useful in the recognition of sites of nitrogen transformation. Because its production and utilization often occur at different depths, striking vertical changes in nitrite are not uncommon (Orr, 1926; Atkins, 1930; Braarud and Klem, 1931).

Within temperate coastal waters there is usually a progressive increase in nitrite during late summer and early winter (Harvey, 1926; Cooper, 1933). Later, a decrease in nitrite is accompanied by an increase in nitrate until, at mid-winter, only trace amounts of nitrogen remain as nitrite (Soot-Ryen, 1932; Verjbinskaya, 1932). At the same time, it is not unusual for significant nitrite concentrations to appear periodically in the bottom water, presumably as a result of oxidation of organic nitrogen within the sediments.

The excretion of extracellular nitrite by phytoplankton as observed by Orr (1926), ZoBell (1935) and Kessler (1957, 1959) may also influence the distribution of nitrite within the surface layers of natural waters (Rakestraw, 1936; Hutchinson, 1957; Vaccaro and Ryther, 1960). Such behavior is especially intense when nitrogen-deficient cells under reduced light intensity are abruptly confronted with an excess of available nitrate. In coastal waters off New England, optimal conditions for nitrite excretion by phytoplankton appear to coincide with the December nitrate maximum. At present, evidence which associates the nitrite in coastal areas with the nitrogen metabolism of marine phytoplankton populations is as strong as that supporting the more generally accepted theory which attributes this role to nitrifying or denitrifying bacteria.

B. THE DEEP OCEAN

(1) *Nitrite-nitrogen.* At oceanic locations remote from land, small but measurable concentrations of nitrite appear to be uniformly distributed throughout the upper photic layer. However, near the compensation depth for photosynthesis, nitrite and nitrate frequently increase rapidly, and the greatest nitrite concentrations, commonly appear between 75 and 125 m. At greater depths, at least in the Atlantic Ocean, increasing amounts of both nitrate and phosphate are unaccompanied by high nitrite concentrations and essentially all the inorganic combined nitrogen occurs as nitrate. The contrast between nitrite concentrations at the two extremes of the photic zone seems to be most pronounced in oligotrophic waters of tropical and semi-tropical areas which are characterized by a strong permanent thermocline. The most commonly accepted explanation for both the seasonal distribution of nitrite in coastal waters and the high nitrite associated with the photic zone in the open sea associates the intermediate appearance of nitrite with the formation of nitrate. It should be noted that nitrite accumulation can be facilitated by a favorable balance between density structure, nitrogen availability, and phytoplankton activity near the surface (Rakestraw, 1936; Riley *et al.*, 1949).

Figure 5 demonstrates the similarity between the horizontal nitrite distribution at the surface and that at 300 m. observed in early Spring 1952, from R.V. *Atlantis* for broad areas of the tropical Atlantic. The higher nitrite concentration at the surface off the West African coast is caused presumably by the upwelling of nitrite-rich water from an intermediate layer above 300 m. This possibility is supported by Fig. 6 which shows high nitrite reserves over broad but well defined areas centered near the 100 m. level.

Unlike nitrite in the Atlantic, that in the Pacific and Indian Oceans follows a more varied and singular pattern of distribution since the highest nitrite concentrations frequently occur far below the thermocline at intermediate depths (100–800 m.) In addition, the usual but

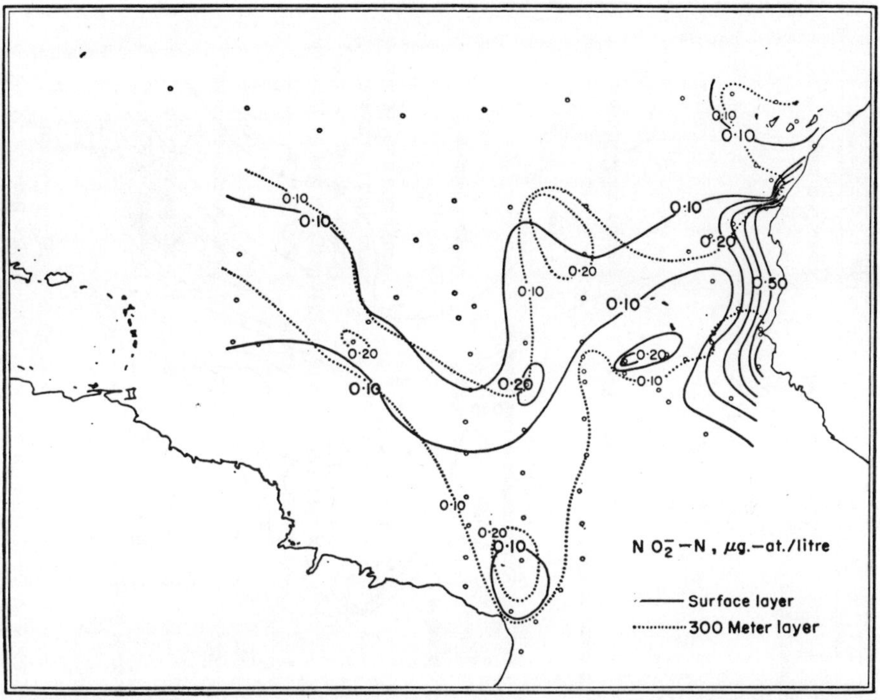

FIG. 5. Horizontal distribution of nitrite-nitrogen, μg.–at./l., at the surface and 300 m. in the tropical Atlantic Ocean (1 μg.–at.NO_2^-–N/l. = 14 μg. NO_2^-–N/l.). From R. F. Vaccaro (unpublished results).

smaller nitrite accumulation near the compensation depth for photosynthesis is typically present. The deeper nitrite maximum appears to be caused, at least in part, by hydrographic circumstances which introduce conditions favorable for the reduction of nitrate (Gilson, 1937; Brandhorst, 1959). Specifically, these include high surface productivity at the eastern oceanic boundaries induced by upwelling, and a more sluggish water circulation and less re-aeration than found in the Atlantic. These circumstances, combined with the sinking of organic particles from the surface and their subsequent decomposition, intensify the degree of oxygen depletion at intermediate depths in the Pacific and Indian Oceans, and allow high nitrite concentrations to develop where the dissolved oxygen concentration is reduced to 0·15 ml./l. or less.

(2) *Nitrate-nitrogen.*

(a) *Distribution and transport.* Soluble constituents of sea water which do not participate in the marine biochemical cycle are transported by advection, and are exchanged between adjacent water masses by eddy

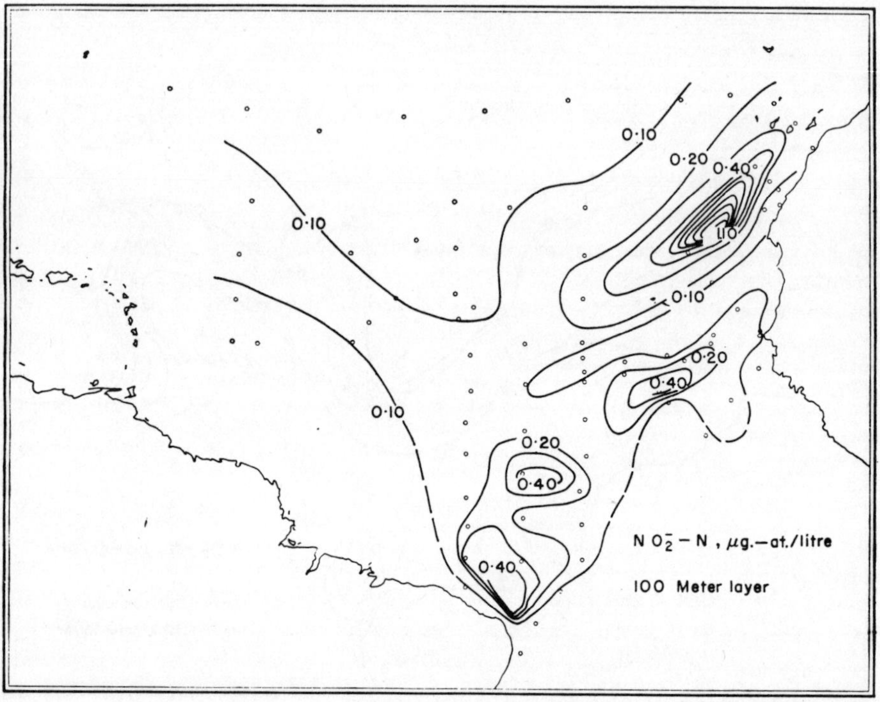

FIG. 6. Horizontal distribution of nitrite-nitrogen, μg.–at./l., at a depth of 100 m. in the tropical Atlantic Ocean (1 μg.–at.NO_2^-–N/l. = 14 μg.NO_2^-–N/l.). From R. F. Vaccaro (unpublished results).

diffusion. However, elements which become incorporated into the life cycle are further dispersed by the migratory and gravitational movements of marine organisms and their residues. Distinctions between these modes of transport correspond to definitions of conservative as opposed to the non-conservative behavior of constituents of sea water. Clearly, the latter description is characteristic of nitrate, since its distribution is dependent upon both hydrodynamic and biological processes. In general, the similarity between the spatial distributions of nitrate and phosphate as well as organic derivatives throughout the world's oceans emphasizes the high degree of uniformity of the processes controlling their oceanic distribution.

Iselin (1939) has traced the conservative properties which identify discrete water masses at various depths in the Atlantic to their ultimate origin at or near the surface at successively higher latitudes. The cold surface-orientated waters of higher latitudes become virtually saturated with oxygen because of their association with the atmosphere, and following winter periods of surface cooling, their density rises sufficiently to cause sinking. On this basis, the intermediate water layer ($\sigma_t = 27 \cdot 0$–$27 \cdot 5$) in the equatorial region which centers around a depth of 800 m. and extends northward to about 15°N., is contiguous with the Sub-antarctic area of surface convergence located between 42° and 50°S. After convergence, this relatively fresh but cold dense Antarctic water is transported horizontally towards the equator by isentropic movement along surfaces of equal potential density. This water mass is referred to as the Antarctic Intermediate Water; it penetrates the three oceans adjacent to the Antarctic, i.e. the Atlantic, Pacific and Indian Oceans, in a characteristic manner. Nutrient residuals associated with oxygen saturation during its residence at the surface have been designated *preformed nitrate and phosphate* by Redfield (1942), and like salinity, they can be expected to behave conservatively following descent below the photic layer.

Unlike the synthetic processes, biological regenerative ones are independent of light; hence, an increase in organic derivatives can be expected to accompany respiratory removal of dissolved oxygen after convergence and during isentropic flow. In this connection, the *apparent oxygen utilization* (A.O.U.) is a useful parameter which can be estimated from the difference between the calculated oxygen saturation value (standard wet atmosphere) at the *in situ* temperature and salinity and the actual oxygen concentration measured. This quantity can provide a basis for estimating subsequent and more localized nitrate contributions, provided that an accurate definition of the previously discussed equivalence between oxygen utilization and nitrate production is available. Nitrate added in this manner was termed *nitrate of oxidative origin* by Redfield (1942) who has examined the biological transformations accompanying the horizontal transport of some major oceanic water masses. For a comprehensive treatment of the dynamic balance between the physical and biological processes which ultimately determine the distribution of conservative and non-conservative properties in the sea, the reader is referred to Riley (1951), Redfield *et al.* (1963) and McGill (1964).

As in coastal waters, organic synthesis in the open ocean is limited to the illuminated surface layer, whereas the complementary regenerative processes extend throughout the water column. Following sinking, the decomposition of nitrogenous particulate material regenerates inorganic nitrogen as nitrate which accumulates in sizeable concentrations below the compensation depth. In the Atlantic, a sharply defined nitrate maximum occurs at moderate depths (350–800 m.) and coincides with a density level of σ_t 27·3. At comparable depths in the Pacific and

TABLE V

Data Sources for temperature, salinity and nitrate measurements at various locations and depths used to construct Fig. 7

Source	*Ocean area*	*Station*	*Date*	*Lat.*	*Long.*	*Sonic depth (m.)*	*Observed depths (m.)*
Fourth Japanese Antarctic Expedition, Fukase (1961)	Southern Antarctic	1	11–I–'60	66°42′ S.	38°06′ E.	4400	1–1500
		3	13–I–'60	66°14′ S.	37°44′ E.	1360	1–1194
		5	21–I–'60	66°38′ S.	45°15′ E.	3300	1–1247
		8	1–II–'60	67°53′ S.	40°33′ E.	2300	1–1439
Pacific Oceanography Group Fisheries Research Board, Canada Ocean Weather Station P.	Northeast Pacific	3	8–IX–'56	50°00′ N.	145°00′ W.	2000	1–1155
		4	15–IX–'56	50°00′ N.	145°00′ W.	2000	1–1190
		5	26–IX–'56	50°00′ N.	145°00′ W.	2000	1–1076
		7	27–IX–'56	50°00′ N.	145°00′ W.	2000	1–1113
Woods Hole Oceanographic Institution, R.V. *Chain*, Cruise 25	Equatorial Atlantic	494	23–IV–'62	11°00′ N.	57°30′ W.	3920	1–1881
		495	24–IV–'62	13°00′ N.	57°30′ W.	4930	1–1383
		496	24–IV–'62	15°00′ N.	56°30′ W.	5200	1–1463
International Indian Ocean Expedition R.V. *Anton Bruun*, Cruise A	Arabian Sea	6	27–II–'63	14°36′ N.	56°00′ E.	2610	1–1359
		7	28–II–'63	15°04′ N.	58°00′ E.	4453	1–1365
		8	1–III–'63	15°35′ N.	60°00′ E.	4453	1–1396
		9	1–III–'63	16°03′ N.	62°00′ E.	3920	1–1393
		10	2–III–'63	16°33′ N.	63°57′ E.	3675	1–1394

Indian Oceans, considerably higher nitrate concentrations are the rule with the layer of maximum concentration being less well defined because of differences in the deep-water circulation. At still greater depths in all three oceans, the water remains quite rich in nitrate which is rather homogeneously distributed to the bottom.

From the data sources listed in Table V, a comparison of nitrate concentrations at various locations within the major oceans has been prepared. The presentation is necessarily incomplete largely because of a scarcity of deep nitrate observations especially from the Pacific. Collectively, however, the data provide some basis for evaluating the factors controlling the overall distribution of nitrate below 150 m. Observations at shallower depths have been purposely omitted in order to minimize the effects of short-term seasonal variations. These data are summarized in Fig. 7 in which the base plot compares temperature with salinity (*TS*); the lines of constant density are also included. For a given (*TS*) relationship the corresponding nitrate concentrations are provided by the height of the superimposed vertical plane from which intercomparisons of all the data within the three-dimensional perspective is possible.

Throughout the Indian Ocean and within the southern and central regions of the Antarctic and Pacific Oceans, the nitrate concentrations below 2000 m. also depend in an important way on nitrate preformed in Antarctic regions. According to Sverdrup *et al.* (1942) the characteristic (*TS*) relationships of these deep waters are continuous with those of the Antarctic circumpolar sea and more southerly locations off Antarctica itself. Initially rich in nitrate, these waters converge at high latitudes and then undergo little attenuation as far north as the equator.

Figure 7 shows that preformed nitrate in the southernmost Antarctic Ocean is uniformly distributed at a concentration of *c.* 280 μg. NO_3^-–N/l. down to 1500 m. At this depth, the water temperature is $-0{\cdot}30$°C and the salinity 34·65‰. These measurements are very similar to those found below 4000 m. in the adjacent deep basins of the three neighboring oceans; this suggests that there is a minimum basic amount of nitrate at these great depths which is transported by horizontal advection. However, at intermediate depths the adjacent oceans show significant departures from the 280 μg. NO_3^-–N/l. level of the Antarctic. Presumably this is caused by long-term biological activity which accompanies the relatively sluggish circulation of the Indian Ocean (cf. the Atlantic Ocean), or by a dilution effect resulting from admixture with nitrate-poor deep water of Arctic origin within the Atlantic (Redfield, 1963).

Unfortunately, the data shown for the Pacific are from a northern location, and are independent of the Antarctic since they relate to a

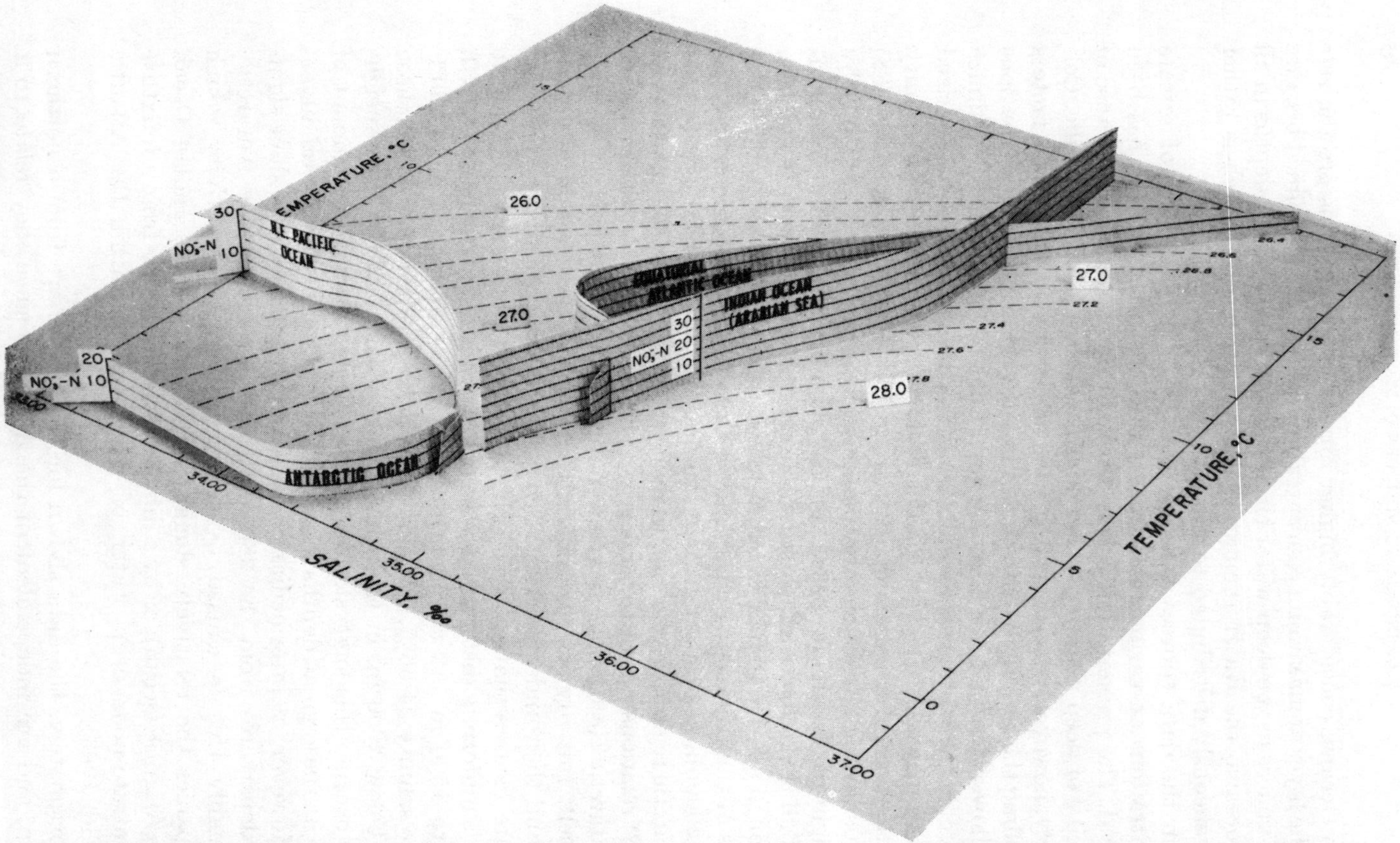

Fig. 7. Observed patterns of vertical nitrate-nitrogen distribution, μg.–at./l., in the major oceans. Base plot includes temperature °C (ordinate), salinity ‰ (abscissa) and connected density isopleths, σ_t. The corresponding nitrate-nitrogen concentrations appear in relief (see Table V; 1 μg.–at./l. = 14 μg./l.).

water column entirely dominated by North Pacific Sub-arctic water where the temperature range was 2·5–4·5°C, and the accompanying salinities were 33·0–34·2‰ with a core level centered near 350 m. There is no evidence for the presence of Pacific Intermediate, Pacific Central or Pacific Equatorial Waters which become very pronounced at more southerly locations. As shown, the corresponding nitrate-nitrogen concentrations are high, although the variations are not large, since the concentrations range only from 420–530 μg./l. between 150 and 1200 m.

The data for the equatorial Atlantic include sub-surface, (TS) relationships which show densities lying between σ_t 26·3 and 27·0 characteristic of South Atlantic Central Water having a nitrate content which varies between 140 and 280 μg. NO_3^-–N/l. This gradual increase in nitrate with depth depends on the degree of mixing with deeper water richer in nitrate. At greater depths and within the σ_t range 27·0–27·5, the previously described Antarctic Intermediate Water becomes prominent. This layer contains the maximum nitrate concentration which amounts to about 450 μg. NO_3^-–N/l. at these latitudes, and corresponds to a σ_t of 27·3. Between σ_t 27·5 and the deepest observations at about σ_t 27·8, the influence of colder more saline North Atlantic Deep Water relatively poor in nitrate is pronounced. The sharp contrast with respect to nitrate between this layer and the overlying Antarctic Intermediate layers accentuates the sharply defined nitrate maximum which is characteristic of the Central Atlantic. The remainder of the water column is presumably occupied by Middle and Lower North Atlantic Deep Water overlying a mixture of North Atlantic and Antarctic Bottom waters. The latter two water masses combine to provide a maximum density of *c.* σ_t 27·9 in the deepest water.

In the Arabian Sea at the northern extremity of the Indian Ocean, the (TS) relationships between 150 and 800 m. indicate a density structure which is largely temperature dependent since the salinities remain constant at about 35·6‰. Water of this description (Arabian Sea Water) appears to have a minimum temperature of about 11°C and a nitrate-nitrogen content which ranges from 370 to 420 μg./l. Directly below it is a layer which is more characteristic of the Red Sea with respect to salinity, and which is rich in nitrate; according to Defant (1961), it can be recognized by its temperature of 9°C and its salinity of 35·5‰. The core layer of this water is centered at *c.* 900 m. where the accompanying nitrate-nitrogen content exceeds 560 μg./l. This contrasts with the maximum concentration (920 μg./l.) reported for the open Pacific Ocean off Central America (Scripps Institution of Oceanography SIO Reference 61–18, 1961). The Arabian Sea location is of special interest since it corresponds with an area of vigorous upwelling during the period of the summer monsoon in July and August when the winds are off-shore. At such times, high concentrations of both nitrate and phosphate occur throughout the photic layer and lead to an extremely rapid rate of primary production. The observations were made during February and March, when less than 0·50 ml. of dissolved O_2/l. was present between 200 and 1000 m. The situation described is that which would be expected if periods having a rapid rate of decomposition and regeneration in the sub-surface layer alternate with periods of high surface production.

C. MEDITERRANEAN AND ADJACENT SEAS

In the seas adjacent to the oceans, the vertical distribution of combined nitrogen like that of other organic derivatives often differs markedly from that observed in the oceans themselves. The following details are of particular significance with respect to the ultimate internal distribution of the elements of biochemical importance.

1. Dissolved constituents outside the sill.
2. Sill topography.
3. Degree of isolation, rate of renewal.
4. Degree of vertical stability.
   a. Temperature changes with depth.
   b. Salinity changes with depth.
5. Surface productivity.

The Mediterranean, Caribbean and Black Seas and the Norwegian fjords encompass a wide variety of situations ranging from nitrate impoverishment to extreme nitrogen accumulation. Of particular importance in the Mediterranean is the relatively shallow sill (300 m.) which limits entrance to nitrate-depleted Atlantic water from the upper 100 m. since greater depths over the sill are occupied by a westward moving counter current from the Mediterranean. An important characteristic of this sea is the excess of evaporation over precipitation and run-off which causes above-average salinities throughout. The maximum salinity (up to 39·5‰) is found in the counter current itself which extends to the eastern boundary at depths of 100–500 m. According to Nielsen (1912), Thomsen (1931) and Wüst (1959), water of this salinity can originate only in the Levantine Basin following alternate periods of high summer evaporation and winter cooling. In late October 1961, hydrographic and chemical stations occupied by R.V. *Chain* between Rhodes and Scarpanto on the northern Levantine boundary, showed a maximum salinity of 39·2‰ accompanied by a nitrate-nitrogen content of less than 7·0 μg./l. within the upper 100 m. The ultimate departure of this water type from the Mediterranean via the counter current through the Straits of Gibraltar has a far reaching effect in the North Atlantic and causes discernible, high salinities and low nitrate concentrations between 1000 and 1500 m. as far west as Bermuda. The deepest waters of the Mediterranean which are formed in the northerly appendages are quite homogeneous with respect to salinity and temperature. They are also relatively low in nitrate, although there appears to be sufficient local regeneration to cause a discernible nitrate maximum (about 77 μg. NO_3^-–N/l.) at 1000 m. off Corfu. Correspondingly high concentrations of dissolved oxygen persist at all depths (Pollak, 1951; McGill, 1961) even down to depths of about 4000 m. thereby emphasizing the relatively rapid rate of water renewal.

Unlike the Mediterranean, the Caribbean and Gulf of Mexico System is composed of several basins with direct connections to the adjacent Atlantic across sills of great depth. An investigation of these sills by Wüst (1963) indicated variations in depth ranging from 1400–2200 m.; these permit the intrusion of cold, nutrient-rich, Atlantic water into the various Caribbean components. Atlantic water entering the Venezuelan Basin across the southern extremity of the Antilles Arc includes a prominent tongue of low salinity water between 600 and 800 m. This layer is an extension of the Antarctic Intermediate Water, previously described for the equatorial Atlantic, and also persists at about 800 m. throughout much of the southern and central Caribbean. It both provides and accounts for the vertical nitrate and phosphate maxima as well as the oxygen minimum described by Rakestraw and Smith (1937), and it can be detected as far west as the Yucatan Basin (Wüst, 1963).

The major transport within the Caribbean system is towards the Gulf of Mexico from which water shallower than 800 m. returns to the Atlantic Gulf Stream System via the Straits of Florida (Riley, 1951). The depth difference between the entrance and exit sills would allow an accumulation of regenerated nitrogen and phosphorus of local origin within the deeper Caribbean waters. Such a retention probably occurs, although there is insufficient data to demonstrate it convincingly. According to Wüst, oxygen concentrations at the core of the oxygen minimum within the Venezuelan and Columbian Basins are similar to those encountered in the North American Basin of the Atlantic (less than 3 ml./l. in both areas). Recognition of the degree of retention is further obscured by the intrusion of significant quantities of low nutrient Sargasso Water of relatively high oxygen content through the Windward Passage (Parr, 1937). The anoxic Cariaco Trench off the northern coast of Venezuela is an important exception to the otherwise well aerated

Caribbean, but the former appears to be a special phenomenon of limited scope (Richards and Vaccaro, 1956).

The actual vertical nitrate distribution in the Caribbean as recorded by Rakestraw and Smith (1937) indicates typically low concentrations at the surface which rapidly increase with depth. Between 400 and 600 m. the concentration of nitrate-nitrogen ranges from 280 μg./l. in the north to 420 μg./l. near the South American coast. At 800 m., there is a nitrate maximum where nitrate-nitrogen concentrations exceed 420 μg./l. Below 1000 m., the nitrate concentration remains relatively constant at about 350 μg. NO_3^-–N/l. down to the bottom.

Unlike the Mediterranean and Caribbean Seas, where there is an excess of oxygen available for nitrogen regeneration, other marine environments exhibit a more critical relationship between dissolved oxygen and the biological oxygen demand. For example, limited but sizeable areas of the Pacific and Indian Oceans provide an intermediate condition in this respect, since free oxygen is often limited to trace quantities. Redfield (1958) and Redfield *et al.* (1963) have suggested that the past histories of these marginal and micro-aerobic waters may well have included intermittent anoxic periods which had special biogeochemical consequences. More extreme degrees of anoxia occur in the deeper waters of the Black Sea, the Cariaco Trench, Norwegian fjords and other locations where organic enrichment is accompanied by extremely slow rates of deep water renewal (see also Chapter 13). Under these conditions, nitrogen regeneration takes an unusual course and leads to the formation of NH_3, N_2, N_2O together with H_2S, since chemically bound oxygen as NO_3^-, NO_2^-, SO_4^{2-} and CO_2 provide the respiratory hydrogen acceptors. Usually, anoxic regeneration does not attain the same degree of completion which characterizes aerobic conversions, and the accompanying accumulation of partially oxidized organic material can provide substantially different elementary proportions of organic derivatives.

The Black Sea is a two-layered system in which minimal vertical mixing occurs as a result of a pronounced salinity gradient between an upper fresh layer and a deep zone of relatively high salt content. Unlike the Mediterranean, where surface evaporation exceeds the combined contributions of precipitation and run-off by about 30%, the fresh water budget of the Black Sea is essentially in balance (Carter, 1956). The surface layer, comprising about 12% of the total water volume, extends to a depth of 150 m. and contains measurable quantities of dissolved oxygen and low concentrations of organic matter and plant nutrients (Gololobov, 1949). The deeper layer, which is anoxic to the bottom (2000 m.) has a high organic content and is extremely rich in regenerated nitrogen and phosphorus. The occurrence of large quantities of hydrogen sulfide throughout the lower layer is a consequence of the uninterrupted reduction of sulfate over an extended period. According to Gololobov (1949), the minimum age of the sequestered water, based on a continuity of measured phosphate exchange, is about 5600 years.

Probably the most important function of the anaerobic bacteria responsible for nitrogen transformations in areas such as the Black Sea is to limit the loss of potential nutrients to the sediments. The detailed chemistry of these processes is largely governed by thermodynamic considerations such as the free energy requirements of the individual reactions and the activities of the participating species; kinetic factors may also be of importance. According to McKinney and Conway (1957), the oxygen required for these processes is removed selectively in the following order of precedence.

$$O_2 > NO_3^- > SO_4^{2-} > CO_3^{2-}$$

Kriss (1949) reported that neither nitrate nor nitrite is present in the anaerobic layer of the Black Sea. Instead, the inorganic nitrogen comprises ammonia and molecular nitrogen; the latter is at supersaturated concentrations. Large accumulations of dissolved organic matter occur immediately above the bottom.

VI. Qualitative and Quantitative Aspects

Figure 8 prepared by Vaccaro and Ryther (1960), summarizes the complementary roles of bacteria, plants and animals with respect to the various oxidations and reductions of inorganic nitrogen within the sea. The blocked and circled numerals identify changes induced by plants and bacteria respectively, and the triangle-enclosed numbers indicate animal contributions to the cycle. The bacteria will be seen to bring about the greatest variety of reactions in the marine nitrogen cycle, especially in its regenerative aspects. However, phytoplankton as well as bacteria can assimilate nitrogen from ammonia, nitrite, and nitrate for use in protein synthesis during primary production. The animal contributions to the inorganic nitrogen cycle are limited to the excretion of ammonia or its organic precursors. The intermediate formation of nitrite in a wide variety of transformations which include nitrification, denitrification and nitrate reduction either by bacteria or plants is emphasized.

A rigorous analysis of the quantitative exchanges within the marine nitrogen cycle is not yet possible because of many uncertainties regarding the *in situ* exchange rates and owing to the unknown extent of terrestrial, atmospheric and sedimentary participation in the cycle.

However, within recent years, the application of more sensitive and reliable measuring techniques has provided a firmer basis for the evaluation of the relationship between available nitrogen reserves and the annual requirements needed to sustain the marine biochemical cycle. The true significance of such estimates, in terms of the entire cycle, depends on how closely nitrogen and other nutrient exchanges in the sea approximate to steady state conditions in which the addition of combined nitrogen to the oceans is just balanced by losses to the sediments. Under such conditions, the annual nitrogen requirements for primary production would be compensated by an equivalent amount of regenerated nitrogen. Thus, given an average oceanic water column

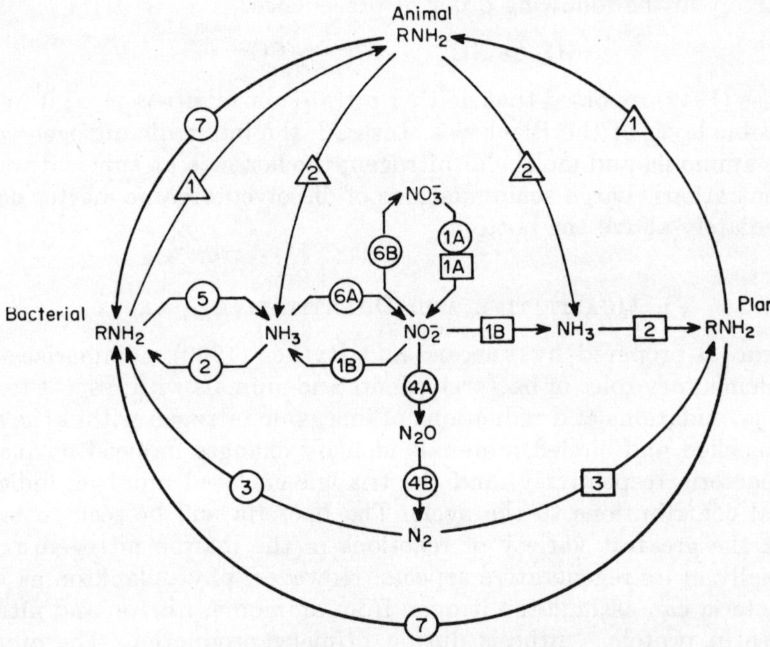

Fig. 8. The physiological origins of nitrite and other forms of inorganic nitrogen in the sea.

BACTERIAL (numbers in circles)

Nitrate reduction	1A 1B	Denitrification	4A 4B
Nitrogen assimilation	2	Ammonification	5
Nitrogen fixation	3	Nitrification	6A 6B

Heterotrophic decomposition 7

PLANT (numbers in squares)		ANIMAL (numbers in triangles)	
Nitrate reduction	1A 1B	Grazing	1
Nitrogen assimilation	2	Excretion	2
Nitrogen fixation	3		

containing 25 mg. NO_3^-–N/m.[2] (Sverdrup *et al.* 1942) and an average annual productivity of 150 g.C/m.[2]/year (Ryther, 1959), the nitrate reserve of the oceans would be adequate to support between 60 and 100 years of primary production. The lesser estimate ignores nitrogen re-cycled as ammonia, whereas the larger one assumes that 35% of the nitrogen turn-over is in the form of ammonia rather than as nitrate (Vaccaro, 1962).

TABLE VI

*Nitrogen budget of the earth**

	Nitrogen (metric tonnes $\times 10^9$)	%	*Source*
I Global distribution			
Atmosphere	3,860,000	97·5	Rankama and Sahama (1950)
Sedimentary rock	77,000	2·0	
Hydrosphere	23,000	0·5	
Total	3,960,000		
II Oceanic reserves			
Molecular as N_2	22,000		
Combined as:			
Nitrate	570	62·00	Sverdrup *et al.* (1942)
Ammonia plus nitrite	10	0·01	Krogh (1934)
Dissolved organic	340	37·00	
Total combined	920	99·01	
III In standing crop of organisms	1		Vinogradov (1953) Hutchinson (1944)
IV Annual oceanic requirement Equivalent to 18% of 150 g. Carbon/m²/yr.	10		Emery *et al.* (1955)
V External sources of fixed nitrogen			
River-flow	0·019	24·5	Clarke (1924)
Oceanic rain	0·059	75·5	Hutchinson (1944)
Total external sources	0·078		
VI Annually deposited in sediments	0·009		Trask (1939) Emery and Rittenberg (1952)

* After Emery *et al* (1955).

The source material compiled and utilized by Emery *et al.* (1955) still remains the best available analysis of the quantitative exchanges affecting the distribution of nitrogen in the sea. Within the reservations noted by the authors, the data emphasize the probable existence of a steady state condition with respect to marine nitrogen. Their estimates appear in Table VI and are summarized below.

1. The available nitrogen in the sea amounts to 100 times the annual requirement for primary production.

2. The nitrogen requirements for primary production exceed the annual contributions of combined nitrogen from river drainage and precipitation by about 125-fold.

3. The estimated renewal time for the amount of fixed nitrogen present in the oceans is about 12,000 years.

4. The total fixed nitrogen contribution from non-marine sources far exceeds the amount lost to the sediments. Hence, assuming steady state conditions, the difference can be attributed to denitrification.

5. Without denitrification, the annual deposition of nitrogen in the sediments would deplete the atmosphere of molecular nitrogen after about 4×10^8 years.

There is a distinct possibility that the above summary may require important modifications if there is a significant nitrogen contribution from oceanic nitrogen fixation. However, as a first approximation, the budget presented here assumes that essentially all nitrogen fixation is terrigenous or atmospheric, and that such nitrogen enters the sea via continental drainage and rainfall. Future investigations may well demonstrate that significant amounts of combined nitrogen are also provided by *in situ* fixation.

REFERENCES

Adel, A. (1939). *Astrophys. J.* **90**, 627.
Adel, A. (1941). *Astrophys. J.* **93**, 509 [Abstract].
Allen, M. B. (1963). *In* "Symposium on Marine Microbiology" (C. H. Oppenheimer, ed.), p. 85. Charles C. Thomas, Springfield, Illinois, U.S.A.
Atkins, W. R. G. (1930). *J. Mar. biol. Ass. U.K.* **16**, 515.
Baas Becking, L. G. M. (1959). *Mem. N.Z. oceanogr. Inst.* **3**, 48.
Barnes, H. (1957). *Mem. geol. Soc. Amer.* **67**, 297.
Bavendamm, W. (1931). *Ber. dtsch. bot. Ges.* **49**, 282.
Bavendamm, W. (1932). *Arch. Mikrobiol.* **3**, 205.
Benecke, W., and Kentner, J. (1903). *Ber. dtsch. bot. Ges.* **21**, 333.
Benson, B. B., and Parker, P. D. M. (1961). *Deep Sea Res.* **7**, 237.
Berthelot, D., and Gaudechon, H. (1910). *C.R. Acad. Sci., Paris*, **150**, 1517.
Bigelow, H. B., Lillick, L. C., and Sears, M. (1940). *Trans. Amer. phil. Soc.* **31**, 149.
Bongers, L. H. J. (1958). *Neth. J. Agric. Sci.* **6**, 79.

Braarud, T., and Klem, A. (1931). *Hvalråd. Skr.* **1**, 1.
Brandhorst, W. (1959). *J. Cons. int. Explor. Mer.* **25**, 3.
Brandt, K., and Raben, E. (1920). *Wiss. Meeresuntersuch. abt. Kiel* **19**, 175.
Brisou, J., and Vargues, N. (1961). *In* "Symposium on marine microbiology" (C. H. Oppenheimer, ed.). Charles C. Thomas, Springfield, Illinois, U.S.A.
Carey, C. L. (1938). *J. Mar. Res.* **1**, 291.
Carroll, D. (1962). *Wat. Supply Pap. Wash. 1535-G.*
Carter, D. B. (1956). *Publ. Clim.* **9**, 121.
Chapman, M. W., and Brodia, H. P. (1956). *Circ. U.S. nat. Bur. Stand.* **575**, 5.
Clarke, F. W. (1924). *Bull. U.S. geol. Surv.* **770**, 841.
Cooper, L. H. N. (1933). *J. Mar. biol. Ass. U.K.* **18**, 677.
Cooper, L. H. N. (1937). *J. Mar. biol. Ass. U.K.* **22**, 183.
Defant, A. (1961). "Physical Oceanography", Vol. I. MacMillan Co., New York.
Dole, M., Lane, G. A., Rudd, D. P. and Zaukelies, D. A. (1954). *Geochim. et cosmoch. Acta,* **6**, 65.
Drew, G. H. (1911). *J. Mar. biol. Ass. U.K.* **9**, 142.
Drew, G. H. (1913). *J. Mar. biol. Ass. U.K.* **9**, 479.
Dugdale, R. C., Menzel, D. W., and Ryther, J. H. (1961). *Deep Sea Res.* **7**, 297.
Dugdale, V. A., and Dugdale, R. C. (1962). *Limnol. Oceanogr.* **7**, 170.
Duursma, E. K. (1961). *Neth. J. Sea Res.* **1**, 1.
Emery, K. O., and Rittenberg, S. C. (1952). *Bull. Amer. Ass. Petrol. Geol.* **36**, 735.
Emery, K. O., Orr, W. L., and Rittenberg, S. C. (1955). "Nutrient budgets in the ocean", p. 299. University of Southern California Press, Los Angeles, California, U.S.A.
Farren, G. P. (1932). *Rapp. Cons. Explor. Mer.* **77**, 60.
Fleming, R. H. (1940). *Proc. Sixth Pacif. Sci. Congr. California* **1939**, **3**, 535.
Fogg, G. E., and Than-Tun (1958). *Biochim. biophys. Acta,* **30**, 209.
Fox, C. J. J. (1909). *Trans. Faraday Soc.* **5**, 68.
Fukase, S. (1961). *Antarctic Rec.* **13**, 1091.
Gilson, H. C. (1937). "The nitrogen cycle". *The John Murray Expedition, Sci. Rep.* **2**, 15.
Gololobov, Y. K. (1949). *C.R. Acad. Sci. U.R.S.S.* **66**, 451.
Guillard, R. R. L. (1963). *In* "Symposium on Marine Microbiology" (C. H. Oppenheimer, ed.), p. 93. Charles C. Thomas, Springfield, Illinois, U.S.A.
Hamilton, R. D. (1964). *Limnol. Oceanogr.* **9**, 107.
Harris, E. (1959). *Bull. Bingham oceanogr. Coll.* **17**, 31.
Harris, E., and Riley, G. A. (1956). *Bull. Bingham. oceanogr. Coll.* **15**, 315.
Harvey, H. W. (1926). *J. Mar. biol. Ass. U.K.* **14**, 71.
Harvey, H. W. (1928). *J. Mar. biol. Ass. U.K.* **15**, 183.
Harvey, H. W. (1940). *J. Mar. biol. Ass. U.K.* **24**, 115.
Harvey, H. W. (1960). "The Chemistry and Fertility of Sea Waters," p. 240. Cambridge University Press, Cambridge, England.
Harvey, H. W., Cooper, L. H. N., Lebour, M. V., and Russell, F. S. (1935). *J. Mar. biol. Ass. U.K.* **20**, 407.
Hoering, T. (1955). *Science,* **122**, 1233.
Hoering, T. (1957). *Geochim. et cosmoch. Acta,* **12**, 97.
Hoering, T., and Ford, H. T. (1960). *J. Amer. chem. Soc.* **82**, 376.
Hutchinson, G. E. (1944). *Amer. Scient.* **32**, 178.
Hutchinson, G. E. (1954). *In* "The solar system," Vol. II. "The earth as a planet" (G. P. Kuiper, ed.). University of Chicago Press, Chicago, Illinois, U.S.A.

Hutchinson, G. E. (1957). "A Treatise on Limnology," Vol. I. Wiley, New York.
Iselin, C. O'D. (1939). *Trans. Amer. geophys. Un.* **20,** 414.
Issatchenko, B. L. (1914). "Investigations on the Bacteria of the Glacial Arctic Ocean". Monograph. V. Th. Kirshbauma, Petrograd.
Junge, C. E., and Werby, R. T. (1958). *J. Met.* **15,** 417.
Kamen, M. D., and Gest, H. (1949). *Science,* **109,** 560.
Kessler, E. (1955). *Nature, Lond.* **176,** 1069.
Kessler, E. (1957). *Planta,* **49,** 505.
Kessler, E. (1959). *Symp. Soc. exp. Biol.* **13,** 87.
Ketchum, B. H. (1939a). *J. cell. comp. Physiol.* **13,** 378.
Ketchum, B. H. (1939b). *Amer. J. Bot.* **26,** 399.
Ketchum, B. H., and Redfield, A. C. (1949). *J. cell. comp. Physiol.* **33,** 281.
Ketchum, B. H., Vaccaro, R. F., and Corwin, N. (1958). *J. Mar. Res.* **17,** 282.
Kreps, E. (1934). *In* "James Johnstone Memorial Volume" (R. J. Daniel, ed.), p. 193. University Press, Liverpool, England.
Kriss, A. E. (1949). *Příroda, Brno,* **6,** 35.
Krogh, A. (1934). *Ecol. Monogr.* **4,** 430.
Ludwig, C. A. (1938). *Amer. J. Bot.* **25,** 448.
McGill, D. A. (1961). *Deep Sea Res.* **8,** 259.
McGill, D. A. (1964). "The distribution of phosphorus and oxygen in the Atlantic Ocean as observed during the I.G.Y., 1957–1958." Pergamon Press, Oxford, England.
McKinney, R. E., and Conway, R. A. (1957). *Sewage industr. Wastes,* **29,** 1097.
Marshall, S. M., and Orr, A. P. (1961). *J. Mar. biol. Ass. U.K.* **41,** 463.
Menzel, D. W., and Spaeth, J. P. (1962). *Limnol. Oceanogr.* **7,** 159.
Murray, R. G. E., and Watson, S. W. (1963). *Nature, Lond.* **197,** 211.
Naudé, S. M. (1929). *Phys. Rev. Ser.* **2, 34,** 1498.
Neess, J. C., Dugdale, R. C., Dugdale, V. A., and Goering, J. J. (1962). *Limnol. Oceanogr.* **7,** 163.
Nicholas, D. J. D. (1959). *Symp. Soc. exp. Biol.* **13,** 1.
Nielsen, J. N. (1912). *Rep. Danish oceanogr. Exped. Medit. 1908–1910,* **1,** 76.
Nutman, P. S. (1959). *Symp. Soc. exp. Biol.* **13,** 42.
Orr, A. P. (1926). *J. Mar. biol. Ass. U.K.* **14,** 55.
Ostroff, R., and Henry, B. S. (1939). *J. cell. comp. Physiol.* **13, 353.**
Parr, A. E. (1937). *Bull. Bingham oceanogr. Coll.* **5,** 110.
Phifer, L. D., and Thompson, T. G. (1937). *J. Mar. Res.* **1,** 34.
Plunkett, M. A., and Rakestraw, N. W. (1955). *Deep Sea Res.,* Suppl. to Vol. 3, 12.
Pollak, M. J. (1951). *J. Mar. Res.* **10,** 128.
Rakestraw, N. W. (1936). *Biol. Bull. Woods Hole,* **71,** 133.
Rakestraw, N. W., and Emmel, V. M. (1938). *J. Phys. Chem.* **42,** 1211.
Rakestraw, N. W., and Hollaender, A. (1936). *Science,* **84,** 442.
Rakestraw, N. W., and Smith, H. P. (1937). *Bull. Bingham oceanogr. Coll.* **6,** 1.
Rakestraw, N. W., and von Brand, T. (1947). *Biol. Bull. Woods Hole,* **92,** 110.
Rankama, K., and Sahama, T. G. (1950). "Geochemistry." University of Chicago Press, Illinois, U.S.A.
Raymont, J. E. G. (1963). "Plankton and Productivity in the Oceans." MacMillan Co., New York.
Redfield, A. C. (1934). *In* "James Johnstone Memorial Volume" (R. J. Daniel, ed.), p. 176. University Press, Liverpool, England.

Redfield, A. C. (1942). *Pap. phys. Oceanogr.* **9**, 22.
Redfield, A. C. (1958). *Amer. Scient.* **46**, 205.
Redfield, A. C., and Keys, A. B. (1938). *Biol. Bull. Woods Hole*, **74**, 83.
Redfield, A. C., Ketchum, B. H., and Richards, F. A. (1963). *In* "The Sea," (M. N. Hill, ed.), Vol. 2, p. 26. Wiley, New York.
Richards, F. A., and Benson, B. B. (1961). *Deep Sea Res.* **7**, 254.
Richards, F. A., and Vaccaro, R. F. (1956). *Deep Sea Res.* **3**, 214.
Riley, G. A. (1944). *Amer. Scient.* **32**, 129.
Riley, G. A. (1951). *Bull. Binghan oceanogr. Coll.* **13**, 1.
Riley, G. A., and Conover, S. A. M. (1956). *Bull. Bingham oceanogr. Coll.* **15**, 47.
Riley, G. A., Stommel, H., and Bumpus, D. F. (1949). *Bull Bingham oceanogr. Coll.* **12**, 1.
Rittenberg, C. S., Emery, K. O., and Orr, W. L. (1955). *Deep Sea Res.* **3**, 23.
Rubey, W. W. (1951). *Bull. geol. Soc. Amer.* **62**, 1111.
Ryther, J. H. (1956). *Limnol. Oceanogr.* **1**, 72.
Ryther, J. H. (1959). *Science*, **130**, 602.
Ryther, J. H. (1960). *In* "The Ecology of Algae" (C. A. Tryon, Jr., and R. T. Hortman, eds.), p. 72. *Special Publication* 2, Pymatuning Laboratory of Field Biology, University of Pittsburgh, Pennsylvania, U.S.A.
Schuler, J. F., Diller, V. M., and Kersten, H. J. (1953). *Plant Physiol.* **28**, 299.
Scripps Institution of Oceanography (1961). Progress report, STEP-I. Expedition, 15 Sept.–14 Dec. 1960. Part II. "Biological, Photometric and Additional chemical Data" SOD Ref. 61–18.
Seiwell, H. R. (1935). *Pap. phys. Oceanogr.* **3**, 1.
Soot-Ryen, T. (1932). *J. Cons. int. Explor. Mer*, **7**, 246.
Spoehr, H. A., and Milner, H. W. (1949). *Plant. Physiol.* **24**, 120.
Steeman-Nielsen, E. (1952). *J. Cons. int. Explor. Mer*, **18**, 117.
Svenska Geofysika Förenigen (1955, 1956). *Tellus*, **7**, 136, 266, 395, 522; **8**, 112, 283, 411, 513.
Sverdrup, H. U., Johnson, M. W., and Fleming, R. H. (1942). "The Oceans, their physics, chemistry and general biology." Prentice-Hall, New York.
Syrett, P. J. (1954). *In* "Autotrophic Micro-organisms" (B. A. Fry and J. L. Peel, eds.), p. 126. Cambridge University Press, Cambridge, England.
Syrett, P. J. (1962). *In* "Physiology and Biochemistry of Algae" (R. A. Lewin, ed.), p. 171. Academic Press, New York.
Thomas, W. H., and Krauss, R. W. (1955). *Plant Physiol.* **30**, 113.
Thomsen, H. (1931). *Rep. Danish oceanogr. Exped. Medit. 1908–1910*, 3, 1.
Trask, P. D. (1939). *In* "Recent Marine Sediments" (P. D. Trask, ed.), p. 428. Thomas Murby, Tulsa, Texas, U.S.A.
Vaccaro, R. F. (1962). *J. Cons. int. Explor. Mer.* **27**, 3.
Vaccaro, R. F. (1963). *J. Mar. Res.* **21**, 284.
Vaccaro, R. F., and Ryther, J. H. (1960). *J. Cons. int. Explor. Mer.* **25**, 260.
Verjbinskaya, N. (1932). *J. Cons. int. Explor. Mer.* **7**, 47.
Vinogradov, A. P. (1953). *Mem. Sears Found. Mar. Res.* **2**. 647 pp. Yale University Press, New Haven, Conn., U.S.A.
Vinogradov, A. P. (1959). *In* "The Origin of Life on the Earth" (A. I. Oparin, A. G. Pasynskii, A. E. Braunshtein, and T. E. Pavlovskaya, eds.), Vol. I, p. 23. Pergamon Press, New York.
von Brand, T., and Rakestraw, N. W. (1940). *Biol. Bull. Woods Hole*, **79**, 231.
von Brand, T., and Rakestraw, N. W. (1941). *Biol. Bull. Woods Hole*, **81**, 63.

von Brand, T., Rakestraw, N. W., and Renn, C. E. (1937). *Biol. Bull. Woods Hole*, **72**, 165.
von Brand, T., Rakestraw, N. W., and Renn, C. E. (1939). *Biol. Bull. Woods Hole*, **77**, 285.
von Brand, T., Rakestraw, N. W., and Zabor, J. W. (1942). *Biol. Bull. Woods Hole*, **83**, 273.
Waksman, S. A., and Carey, C. L. (1935). *J. Bact.* **29**, 531.
Waksman, S. A., and Starkey, R. L. (1931). "The Soil and the Microbe". John Wiley, New York.
Waksman, S. A., Hotchkiss, M., and Carey, C. L. (1933a). *Biol. Bull. Woods Hole*, **65**, **137**.
Waksman, S. A., Reuszer, H. W., Carey, C. L., Hotchkiss, M., and Renn, C. E. (1933b). *Biol. Bull. Woods Hole*, **64**, 183.
Watson, S. W. (1962). *Abstr. 8th Congr. int. Microbiol., Montreal, 1962.*
Watson, S. W. (1963). *In* "Symposium on Marine Microbiology" (C. H. Oppenheimer, ed.), p. 73. Charles C. Thomas, Springfield, Illinois, U.S.A.
Wattenberg, H. (1929). *Rapp. Proc. Verb.* **531**, 90.
Wüst, G. (1959). *Ann. Ist. Univ. Nav. Napoli*, **28**, 3.
Wüst, G. (1963). *Deep Sea Res.* **10**, 165.
ZoBell, C. E. (1933). *Science*, **77**, 27.
ZoBell, C. E. (1935). *Proc. Nat. Acad. Sci., Wash.* **21**, 517.
ZoBell, C. E. (1946). "Marine Microbiology." Chronica Botanica, Waltham, Massachusetts, U.S.A.

Chapter 10

Silicon

F. A. J. ARMSTRONG

Marine Biological Association
Plymouth, Devon, England

I. Introduction 409
II. Occurrence 410
A. Dissolved Silicate 410
B. Suspended Silicon 412
III. Changes in Silicate Concentration 415
A. Uptake by Plants 415
B. Re-solution of Diatom Frustules 415
C. Seasonal Changes 417
IV. Distribution of Silicon in the Sea 421
A. Coastal Regions 421
B. The Atlantic Ocean 423
C. The Pacific Ocean 426
D. The Indian Ocean 427
E. The Antarctic Ocean 428
F. Comparison of the Major Oceans 429
References 430

I. Introduction

The silicon content of sea water is affected by geochemical and biological events, and shows a wider range of variation than that of any other element. Silicate may vary from undetectable concentrations at the surface of the oceans or in shallow seas, to some 4000 μg. Si/l. in the deep water of the Pacific and Indian Oceans, where mean values may be 2000–3000 μg. Si/l. In the Atlantic Ocean, concentrations are about one half of these figures. Silicon dioxide is known to be an essential part of the solid structure of silicoflagellates, of diatoms, and of some radiolarians and sponges. The inorganic matter in diatoms may contain over 60% of SiO_2, and a vigorous growth of these plants may completely exhaust supplies of dissolved silicate in the water in a matter of days.

The silicon in the oceans is derived mainly from the silicates of the earth's crust, which, it has been suggested (Sillén, 1961) may have determined the ionic composition of sea water, soluble silicates having exerted a buffering action, and insoluble ones having acted through their ion-exchange capacity.

II. Occurrence

Silicon in sea water is found in solution as silicate ions, and also in suspension as silicon dioxide, both free and in diatoms and other living organisms, and as clay minerals. Organosilicon compounds have not, so far, been detected in sea water.

A. Dissolved Silicate

Sea water is undersaturated with respect to silica, although since reported values for its solubility are somewhat inconsistent, it is not possible to say by how much. The apparent solubility of quartz in water is influenced by the fact that on its surface, after breaking or grinding, there exists an amorphous mono-layer of silicic acid *c.* 0·03 μ thick which is considerably more soluble in water than the bulk of the material (see also Chapter 16, Volume 2, Section II, H). Determinations of the solubility of amorphous silica in sea water (Alexander *et al.*, 1954) gave values between 100 and 140 mg. SiO_2/l.; no dependence of solubility on hydrogen ion concentration could be detected below pH 9. These values agree with those found by Krauskopf (1956) using more elaborate methods, in which the solubilities were measured when equilibrium had been approached from both under- and over-saturation. At 22–27°C, a solubility of 100–110 mg. SiO_2/l. was found for silica gel in contact with sea water. When excess of dissolved silicate was added, SiO_2 was precipitated, leaving 100–140 mg. SiO_2/l. in solution. The solubility in fresh water was found to be temperature dependent (70 mg. SiO_2/l. at 0°C and 120 mg. SiO_2/l. at 25°C), and it is likely therefore that in sea water, it also increases with temperature. Precipitation of silica in the presence of magnesium salts was seen to occur. Krauskopf considered that the relative undersaturation of sea water with silica was attributable to:

(i) rapid circulation, which brings silicon to the surface where it can be utilized by plants;
(ii) removal by sponges;
(iii) formation of authigenic silicates;
(iv) precipitation by aluminium hydroxide.

Iler (1955) noted a decrease in solubility of amorphous silica in the presence of aluminium or magnesium ions, and suggested that the formation of a protective surface layer, possibly of the metal silicate, decreased the rate of solution. Goto (1956) found that colloidal silica was precipitated by traces of aluminium salts, and Okamoto *et al.* (1957) showed that the precipitating action of aluminium ions was also effective with dissolved silicate, although larger amounts of the metal were required. These authors showed that the solubility of amorphous silica increased with temperature and pH. Hydroxyl ions were found to cause polymerization or depolymerization of colloidal silica, depending on whether the initial dissolved silica concentration was high or low. Sea water was found to be a good depolymerizer, since it is undersaturated.

It seems probable that most of the dissolved silica in sea water is present as orthosilicic acid $Si(OH)_4$. The first dissociation constant of this acid has been given as 5×10^{-10} (Chatterjee, 1939) and 1×10^{-10} (Alexander *et al.*, 1954). Ingri (1959) determined the first and second dissociation constants in 0·5 M sodium chloride solution at 25°C, and found values of $3{\cdot}90 \times 10^{-10}$ and $1{\cdot}95 \times 10^{-13}$ respectively, which indicates that at pH 8·2, about 5% is ionized.

In nearly all the experimental work on the solubility of silica, and in all recent measurements of silica concentrations in sea water, the reaction of silicate with an acid molybdate reagent is used as the basis of a colorimetric method of analysis (see Chapter 21, Volume 2, Section IX, F(3)). Reaction with molybdate has been accepted as the criterion of solubility, and it is often assumed that only molecularly dispersed silica can so react. Nevertheless, Tourky and Bangham (1936) found that, in river water, a fraction of the reactive silica could be removed by ultrafiltration, and so also could part of the silica in a solution of sodium metasilicate. Chatterjee (1939) postulated that ionized silicate (mostly $SiO(OH)_3^-$) existed in equilibrium with a larger amount of inert material, this equilibrium being quickly restored if the ions were removed. Tourky and Bangham's findings are explicable if it is assumed that ultrafiltration removed this inert material which would otherwise have interacted with molybdate via the ionized form.

Chow and Robinson (1953) found that the silicate in solutions in which the ratio $SiO_2 : Na_2O$ was unity or less would react completely with molybdate. At higher ratios and at high concentrations, only part of the silica would react, although on being diluted and allowed to stand, up to 97% would do so. Solutions with $SiO_2 : Na_2O$ ratios of up to 3·78 were added to sea water to give silicate concentrations of up to 83 mg. SiO_2/l. On standing, the fraction of the silica which reacted with molybdate steadily increased until all would do so after one week. Colloidal silica added to synthetic or natural sea water gradually became reactive, and could be determined colorimetrically after standing for one or two weeks. This did not occur with distilled water.

It may be concluded that the colorimetric molybdate reaction determines ionized and molecularly dispersed silica, together with the less highly polymerized part of the colloidal material. It is relevant that all the silicon determined by the molybdate reaction can be utilized by some diatoms (Lewin, 1955).

B. SUSPENDED SILICON

Suspended matter, even in clear off-shore water, may amount to a few milligrams per litre. About half of it is inorganic, and contains a good deal of silicon; the quantity of inorganic material varies with location. Armstrong and Atkins (1950) found 0·5–2·0 mg./l. in the English Channel, and Armstrong (1958a,b) detected 0·05–1·0 mg./l. in the East North Atlantic. Goldberg *et al.* (1952) reported up to 5 mg./l. in the Eastern Pacific; Lisitsin (1961) found 2–4 mg./l. in the Bering Sea and Kuriles, 7–10 mg./l. in the Baltic, and 0·39–2·2 mg./l. in the Indian Ocean. Analyses of inorganic matter from the English Channel have shown silica contents of 17–55% SiO_2; these represented 37–410 μg. Si/l. in the water, quantities exceeding that in solution. In the deep water of the East North Atlantic, the suspended matter contained 8–80% SiO_2; this represents concentrations of 19–61 μg. Si/l., the exact amount varying with depth. At the surface, suspended silicon exceeded that in solution, but at depths greater than about 100 m. it represented only 2–10% of it. This suspended matter contained iron and aluminium and some calcium carbonate; there was an excess of silicon over that required for the common aluminosilicates (Armstrong, 1958a,b).

Examination of suspended material from the English Channel showed clay, quartz, cordierite, kaolinite, sponge spicules, mica (sericite), feldspar (orthoclase) and tourmaline (Atkins *et al.*, 1954). Lowndes (1955) identified kaolinite, microcline and nosean in the suspended matter from the same area. Gunnerson and Emery (1962) found that water from the San Pedro basin, California, contained feldspar, quartz, biotite, and small amounts of hornblende, muscovite and opaque minerals. Lisitsin (1961) reported kaolinite, mica and palygorskite in the surface water of the southern Indian Ocean. Ishii and Ishikawa, (1964) have, by means of X-ray diffraction, tentatively identified feldspar, quartz, clayey chlorite and illite in suspended matter in sea water off the east coast of Japan and in the east China Sea.

These minerals, and others, the presence of which in suspension can be inferred from their occurrence in bottom deposits, may enter the sea from the land, in river water, or from the atmosphere. It has been estimated (Clarke, 1924) that rivers bring into the sea some $3{\cdot}2 \times 10^8$ tons of suspended matter per year; this material contains much quartz, mica, feldspar, augite and hornblende. The aluminosilicates are likely to have been degraded by the partial leaching out of alkali and alkaline earth metals; on entering the sea they will rapidly adsorb potassium and magnesium; this will tend to restore their crystal structures (see Chapter 15, Volume 2, Section II, B). Bien *et al.* (1958) have reported observations which suggested that the silica content of clays already in the water may be enriched by the precipitation of soluble silica brought in by river water. However, this process cannot be of general occurrence, since Maeda (1952), Voipio (1961), Stefansson and Richards (1963) and others have shown linear relationships between the chlorinity and the silicate content of sea water near river mouths.

Suspended matter from the atmosphere may originate from wind-borne dust, or volcanic ash or from cosmic dust. Radczewski (1937a) identified quartz, clay, feldspar, green hornblende, biotite, tourmaline, garnet, epidote and zircon from the Sahara Desert in dust falls well away from land. He also (1937b) found that of the non-calcareous deposits of the Cape Verde Basin, up to 45% was volcanic augite from the volcanoes of the region. Quartz particles of 1–20 μ diameter in Pacific Ocean sediments were apparently of atmospheric origin (Rex and Goldberg, 1958).

Murray and Renard (1891) found small siliceous spherules, with inclusions of titaniferous magnetite, in bottom samples taken during the *Challenger* expedition; they concluded that these were extra-terrestrial particles probably thrown off on passage of meteorites through the earth's atmosphere. This conclusion was supported by Bruun *et al.* (1955) who found that many magnetic particles from the floor of the Pacific Ocean consisted of a siliceous ground-mass with inclusions of magnetite. These were thought to be of cosmic origin, from the trails of stony meteorites. However, they form only a small fraction of the cosmic dust known to fall on the earth (see also Chapter 15, Volume 2, Section II, D).

Clay minerals in suspension may affect the concentrations of dissolved organic and trace inorganic constituents owing to their adsorptive and ion-exchange properties. In shallow water, such effects may be appreciable (Carritt and Goodgal, 1954). Little is known of the nature of such suspended clays in the oceans, or of their composition, although their size distribution has been studied; Lisitsin (1961) for example, found that in the Indian Ocean over 70%, and in places up to 99%, of the particles were less than 10 μ in diameter. Particles of this size must sink very slowly and will remain in contact with the water for long periods, and so may be considered fairly reactive. It should not be assumed that clays found in sediments have necessarily existed in that form in the water.

The clay minerals of the ocean sediments appear closely to reflect their source areas. The commonest of these minerals appear to be illite, montmorillonite, chlorite, and kaolinite. Illite, which is concentrated, for example in the mid-Pacific sediments, does not appear to have been formed in the oceans (see Weaver, 1958; Hurley *et al.*, 1959), and Griffin and Goldberg (1963) consider it to be of aeolian origin from the arid regions of the world; together with chlorite it appears to be the most stable clay mineral under marine conditions. Montmorillonite, which is particularly abundant in the South Pacific and in certain near-shore areas, appears to be formed from volcanic ash (Correns, 1937; Ross and Hendricks, 1945; Griffin and Goldberg, 1963); it undergoes slow diagenetic changes in contact with sea water (Millot, 1942). There is evidence that kaolinite, which occurs mainly in near-shore sediments, originates from the land and is unstable in suspension, but its presence in the sediments suggests that it only slowly undergoes diagenesis. Chlorite occurs in considerable amounts in the sediments of certain areas such as the North Pacific (Griffin and Goldberg, 1963), although in other regions it is scarce (Murray and Harrison, 1956; Zen, 1959). It appears to have a continental origin.

Some or all of these minerals presumably will exist suspended in sea water. It is probable that their concentrations in the water column will be very low (10^{-5} g./l. or less) since the rate of sedimentation in the oceans is very low; this will render their detection very difficult.

Although it appears likely that the bulk of the clay minerals constituting marine sediments are detrital in origin, this does not preclude the authigenic formation of moderate amounts of clay minerals which may be difficult to distinguish from detrital minerals. Of the minerals which are probably authigenic, glauconite and phillipsite (Goldberg, 1961) are the most important (see Chapters 15 and 16 of Volume 2). Arrhenius (1954) has put forward the following arguments for authigenic formation of aluminosilicates.

(i) It has been shown that in the weathering of feldspars, the parent mineral breaks down into ions in solution, from which the secondary minerals are formed by recombination (Correns and Engelhardt, 1938). It is to be expected that some aluminium and silicon dissolved from continental rocks will be carried to the sea.

(ii) In the sea, dissolved silicate is present in excess, and enough aluminium is estimated to be present to account for the production of ten times the amount of solid material present in suspension.

(iii) An aluminosilicate sol is believed to be formed, which aggregates into particles which can settle out.

(iv) Support is given by the results of Grim *et al.* (1949) on the properties of Pacific clay minerals, and by the demonstration of the presence of authigenic illite, unrelated in composition to the surrounding detritus, in some permeable sandy horizons of Tyrrhenian Sea deposits (Norin, 1953).

Zen (1959) suggested that some aluminosilicates are formed by devitrification of volcanic glasses. However, Rex and Goldberg (1958) suggested that wind transport is a sufficient explanation, at least in the North Pacific.

In addition to the silicate minerals in suspension, there is also present in sea water a considerable amount of suspended silica of biogenic origin, amounting sometimes to 100 μg. Si/l. Diatom frustules for example can sink through deep water and form sediments of diatomaceous ooze. This material is opaline (Rogall, 1939). Goldberg (1958) determined opal in two cores from the Bering Sea and concluded (partly because of its covariance with barium) that it was of biogenic origin. There was no evidence of inorganic precipitation of opal (see also Chapter 15, Volume 2, Section II, A).

III. Changes in Silicate Concentration

A. Uptake by Plants

The mechanism of biological uptake of silicon has been studied mostly with fresh-water diatoms. The experimental difficulties of avoiding contact with glass and of providing non-toxic transparent containers is considerable. Jørgensen (1952, 1953, 1957) found that the growth rates of *Nitzschia palea* in fresh water and of *Bacillaria paradoxa* (measured by cell counting and by the chlorophyll content) in salt water increased in proportion to the silicate content of the medium as long as this exceeded 30–40 μg. Si/l. It decreased at lower concentrations even if other nutrients were provided. Silicate assimilation was inhibited in the later stages of growth, even in a flowing culture medium.

Lewin (1954) noted that washed cells of *Navicula pelliculosa* took up silicate more slowly than did unwashed ones, but that the uptake rate was restored by addition of sulphate or bivalent sulphur. Cadmium ion inhibited uptake without inhibiting respiration. In later experiments (1955), she showed that of the various soluble and insoluble silicates tested, only compounds having $K_2O:SiO_2$ ratios in the range 2 : 1 to 1 : 3·3 could be utilized. Silicon assimilation was shown to be linked with aerobic respiration, and thus respiratory inhibitors also inhibited uptake of silicon.

B. Re-solution of Diatom Frustules

As mentioned above, deep water has usually less than 10% of its total silicon in the suspended matter, and it follows that if this were to dissolve, only small changes in dissolved silicate contents would occur. However, at the surface of the oceans and in shallow seas there may be much more suspended than dissolved silica, and much of this is present as the frustules of living or recently dead diatoms. Re-solution of this material appears to be a vital part of the nutrient cycle in the sea. Cooper (1933a,b) compared estimates of minimum spring production of phytoplankton from changes in chemical constituents in the English Channel. Estimates from production of oxygen, and from consumption of carbon dioxide, phosphate and nitrate agreed well, being from 1000–1600 tons wet weight phytoplankton per km.^{2}, but that from silicon consumption was less than one-tenth of this amount (110 tons/km.^{2}). The discrepancy could be explained if siliceous plants comprised only one-tenth of the phytoplankton, but it was suggested that more probably the silicon passed through the life-cycle ten times in one season. Since little silicon is required by the herbivores, the element will be excreted unused, and if this excreted silicon is readily re-dissolved it becomes available again for growth of diatoms. This assumption was disputed by Atkins (1945), who considered that the phytoplankton crop contained a larger population of non-siliceous organisms than had previously been believed, and that the number of diatoms was small. The evidence for both views is circumstantial, and sometimes contradictory, as for example, in experimental studies of the rate of solution of diatoms, an understanding of which is crucial. Cooper (1952) has discussed earlier contradictory findings (Rogall, 1939; Conger, 1941; Atkins, 1945) and has suggested that although the silica in diatom frustules appears to be similar in chemical constitution to hydrated silica dispersed in water, it is necessary to postulate the existence of some sort of surface skin or coating to explain examples of low solubility. It seems to be difficult to obtain consistent results from one experiment to the next. For example, Jørgensen (1955) found that with the (freshwater) diatom *Nitzchia linearis*, in one experiment 0·3%, 5·0% and 17·2% of the silicon dissolved in 6 days at pH values of 5, 8 and 10, respectively. In another experiment at pH 10, only 5% dissolved in 6 days, whereas in a duplicate run, 17·2% had dissolved in 1 hour's time. The smaller marine diatom *Thalassiosira nana* was appreciably more soluble. Jørgensen was unable to confirm the observations made by King and Davidson (1933) which suggested that some enzymatic action might be involved. These authors had found the silica of boiled suspensions of diatoms to be less rapidly soluble than that of suspensions preserved with chloroform, an effect which they claimed to have been able to repeat several times.

The solubility of the frustules of *Navicula pelliculosa* has been investigated by Lewin (1961). Very little silica dissolved from living cells over a wide range of temperatures and pH values, but every treatment which killed the diatoms increased the rate of solution. With diatoms from which organic matter had been removed by digestion with nitric acid, silica dissolved most rapidly at high temperatures and high pH values. In solutions buffered at pH 9·0–9·3, silica dissolved in 3–4 weeks, saturation being reached at concentrations of 80–86 mg. SiO_2/l. The rate of solution was slower in sea water at a pH value of somewhat less than 8. The dissolution of killed cells was retarded by what appeared to be a cationic factor, the effect of which was diminished in the presence of chelating agents. At pH 8 it was found that aluminium, beryllium, iron, gallium, gadolinium and yttrium could combine in some way with the cell material and retard the dissolution of silica. In one experiment, 1% of aluminium was taken up; this metal was as effective with killed but intact cells as it was with acid-treated cells. It is worth noting that Rogall (1939) found about 1·5% Al_2O_3 or Fe_2O_3 associated with the silica of diatoms, with however, no titanium or alkaline earth metals.

Lewin further found that the silica in fossil diatoms was essentially insoluble, and that digestion with acid made only a small fraction reactive. However, the factor conferring insolubility could be removed by treatment with complexing agents. Lewin concluded that adsorbed cations were responsible for the low solubility of silica of dead diatoms, and that of living cells some additional mechanism was operative.

Direct evidence of solution of diatoms in the sea is scarce, although Wiseman and Hendey (1953) have commented on the eroded appearance of the remains of the very large diatom *Ethmodiscus rex* found in red clay deposits in the Marianas Trench.

C. SEASONAL CHANGES

Raben (1905a,b, 1910, 1914) found evidence for a seasonal variation of silicate in the North Sea and Baltic, with a correspondence between silicate concentration and the abundance of phytoplankton. He found that the concentrations varied between 190 and 510 μg. Si/l. in the late spring and summer. Brandt (1919), using a most laborious gravimetric method, found concentrations in the Baltic which decreased from about 420 μg. Si/l. in February to 280 μg. Si/l. in April and May, and increased to 420 μg. Si/l. again in June.

The application of a simple colorimetric method (Diénert and Wandenbulcke, 1923) by Atkins (1923) allowed more frequent observations to be made (Atkins, 1926, 1928, 1930, 1953). A great many were made at one position (International Hydrographic Station E1. 50°02′N. 4°22′W.) in the English Channel where the depth of water was about 70 m., and where the uppermost illuminated layer was a substantial part of the whole water column. Biological changes therefore produced noticeable effects. Seasonal changes in silicate were found to resemble those in phosphate content (Chapter 8), a winter maximum being followed by a marked depletion in spring corresponding to the utilization of silicate as the phytoplankton population increased. During the summer, phytoplankton growth lessened, and silicate was regenerated to some extent. There was often a second increase in the phytoplankton population in the autumn. The observations have been continued by others at Plymouth (Cooper, 1933a,b; Armstrong, 1954, 1955, 1957, 1958a,b; Armstrong and Butler, 1959, 1960a,b, 1962a,b) and have been compared with changes in other nutrient salts. Changes in silicate are generally greater, and more irregular, than those of phosphate. The winter maximum took place usually in late January or in February, and in some years it tended to come about a month later than that for phosphorus. Winter values varied from 70 to nearly 200 μg. Si/l., and were higher in the 1920's than in recent years. Minimum values occurred at various times during the summer from April to early August, and in the uppermost 20 m. of water there was present at times only 1 or 2 μg. Si/l. The decrease in silicate was usually most rapid in April and May. The seasonal change at two depths is shown in Fig. 5, Chapter 8, together with that of phosphorus. Two contrasting patterns of vertical distribution are shown in Fig. 1.

It is possible that this station, which is only 22 miles offshore from Plymouth, may be somewhat affected by coastal influences and may not be typical of the English Channel (Armstrong and Butler, 1962b).

However, long runs of observations at other places in the sea are rare. Thompson and Johnson (1930) have published results for just over twelve months' determinations at the Puget Sound Biological Station. A reduction in silicate was found in May and June, but this was only to about 70–80% of the winter maximum figure, which was around 1400 μg. Si/l.

Phelps (1937) made observations over three years in Monterey Bay, California, finding continuous and irregular variations at all levels down to the maximum depth of about 900 m. He considered that changes in the upper 100 m. were caused partly by biological effects and partly by upwelling of deep water of higher silicate content in spring and summer, combined with effects of water movements in the autumn and winter. The physical effects were demonstrated most clearly by the correlation between temperature and silicate content.

Phifer and Thompson (1937) have given hydrographic and nutrient observations in 10 m. of water for the San Juan Channel, for a period of five years. In the last three years of this period, phytoplankton counts were also carried out. Mean monthly figures for silicate were given, and

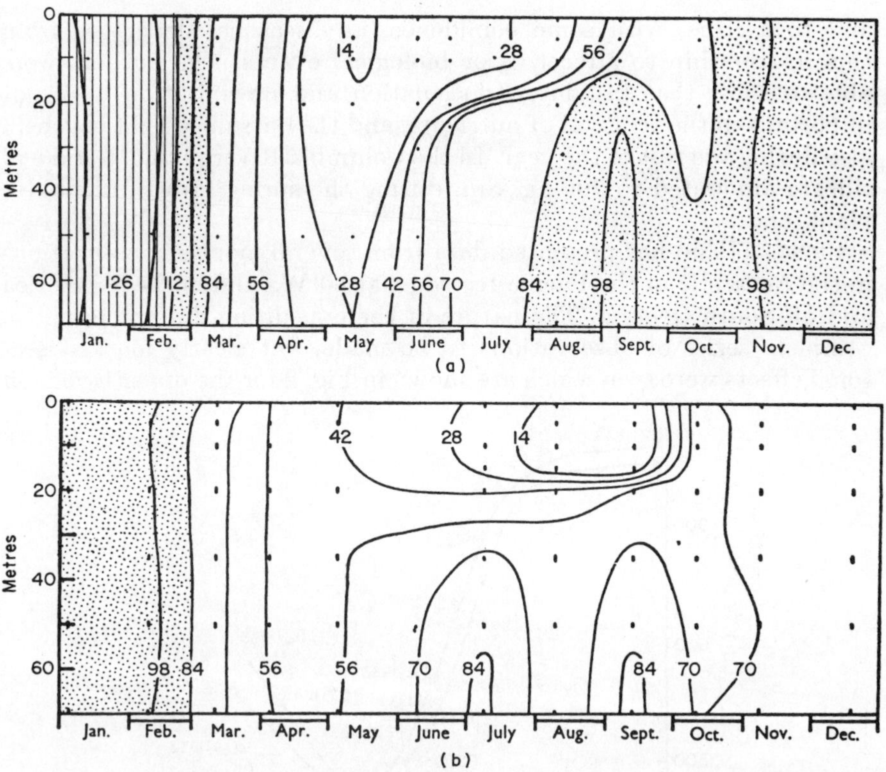

FIG. 1. Vertical distribution of silicate (in μg. Si/l.) for the English Channel (50°04′N., 4°22′W.). (a) 1959; (b) 1960, from Armstrong and Butler (1960b, 1962a).

a seasonal effect was observed. Winter maximum values of from 145–190 μg. Si/l. occurred, with summer minima from 106–123 μg. Si/l. Over the five years of the observations, the mean decrease in silicate amounted to 26% of that available. The fact that neither silicate nor any other nutrient showed complete utilization in the summer was ascribed by the authors to the influx of upwelled water into the Strait of Juan de Fuca, and to enrichment of the sea water by leaching processes occurring at the bottom in this area of marked turbulence.

Seasonal changes in silicate concentration were observed as part of a study of nutrient distributions of the plume of the Columbia River off the Washington and Oregon coasts (Stefansson and Richards, 1963). Outside the river plume, surface salinities and silicate concentrations were relatively uniform in winter, and the silicate content of the river water itself showed comparatively small variations. It was possible to show that the silicate content decreased linearly with increasing salinity, and to ascribe, with some confidence, any seasonal departure from this relationship to advective or biological events. The authors were able to show that in winter, flocculation and precipitation were not important in the removal of nutrients, and they assumed that this held also during the rest of the year. In the Columbia River plume, uptake of silicate amounted to 700 μg. Si/l. during the spring growth of phytoplankton.

Clowes (1938) has presented data from several months' observations in the Antarctic at the Pacific Ice-edge, at 80°W., and in the Scotia Sea and off South Georgia. The nature of the expedition did not permit a complete series of observations to be made, but clearly marked seasonal effects were seen which are shown in Fig. 2 for the upper layers off

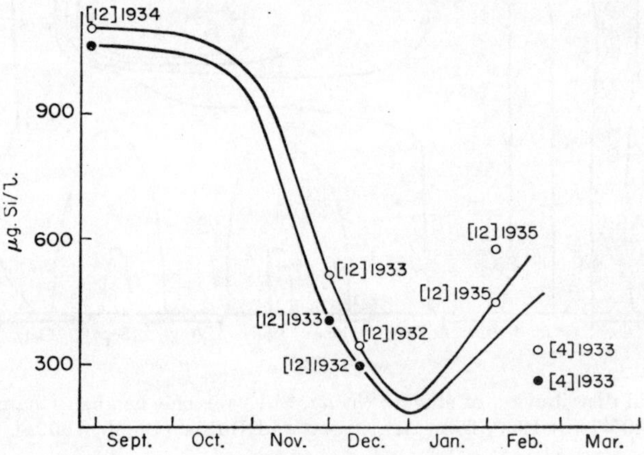

FIG. 2. Possible curves for the seasonal variations in silicate of the surface and of the 0–100 m. layer near South Georgia (from Clowes, 1938). Figures in brackets are numbers of observations. Open circles, 0–100 m. average; closed circles, surface.

South Georgia. The parts of the curves representing October and November, and the position of the minimum in early January are speculative; if they are correct, the spring consumption of silicate is over 80% of that available earlier. Estimates of the amount remaining for the other regions were: South Scotia Sea about 50%, North Scotia Sea approaching 80%, 80°W. about 45%. Averaged observations for the Antarctic are shown in Fig. 3.

The maxima and minima of silicate imposed by seasonal changes of silicate near the ice-edge and regeneration in the water of the northward flowing Antarctic Intermediate current are shown in Fig. 21, Chapter 8.

Two years of observations in the Gulf of Manar, and in Palk Bay (Southern India) are given by Jayaraman (1954). Both areas showed changes from 140–420 μg. Si/l., with a few higher or lower figures. The changes appeared to be caused mainly by the monsoon-driven current systems, the South West monsoon bringing in oceanic water of low silicate content in the period May–October, whereas coastal water with high silicate predominates in the area during the North East monsoon, from November–April. In June–October 1952, unusually high silicate concentrations were observed, and the association of these and other nutrients with concurrently observed swarms of *Noctiluca* has been discussed by Prasad and Jayaraman (1954).

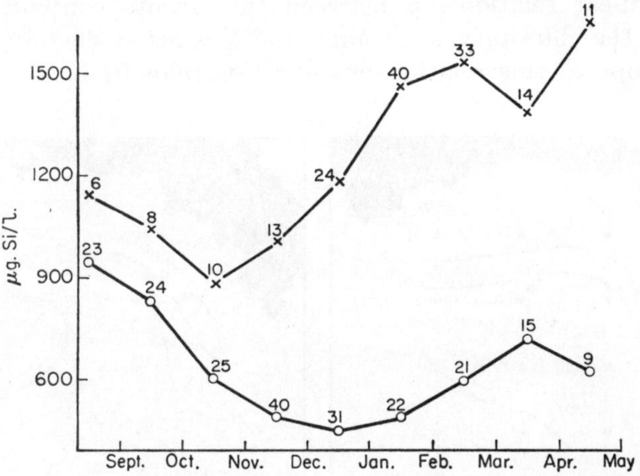

FIG. 3. The average silicate concentrations in the southern and northern regions in the Antarctic zone of the Southern Ocean (from Clowes, 1938). Figures against points are numbers of observations. —○—○—, northern region; —×—×—, southern region.

Observations off Waltair in the Bay of Bengal (Ganapati *et al.*, 1956) showed increases of silicate content at all depths from October–May. These have been attributed to a reduction in the volume of fresh water runoff, its replacement by Bay of Bengal water and by water upwelling from deeper layers. This incoming water maintains high silicate concentrations in spite of some consumption by phytoplankton. In March 1956 for example, silicate was 210 μg. Si/l. at the surface and 770 μg. Si/l. at 60 m.

IV. DISTRIBUTION OF SILICON IN THE SEA

A. COASTAL REGIONS

Coastal waters have, in general, quite high silicon contents, since they are affected by land drainage; as a consequence, off-shore silicate values can give useful information about the discharge of river water. Thus, for example, Voipio (1961) has shown a decrease in surface silicate contents from north to south in the Gulf of Bothnia and the Baltic which can be ascribed to the high silicate content of 1400–2800 μg. Si/l. of the larger rivers entering the Gulf of Bothnia. In this Gulf, surface silicate concentrations are typically 840–1680 μg Si/l., decreasing southward to 560–840 μg. Si/l. in the Bothnian Sea, and 280–560 μg. Si/l. in the Baltic Sea itself. Concentrations near the bottom, where the depth of water exceeded about 100 m., were about twice the surface values except in the more northerly parts, where they were about the same. A linear relationship between the silicate content of surface water and the chlorinity was found, and this fitted the data very well, but the slope in this relationship was less than that calculated from

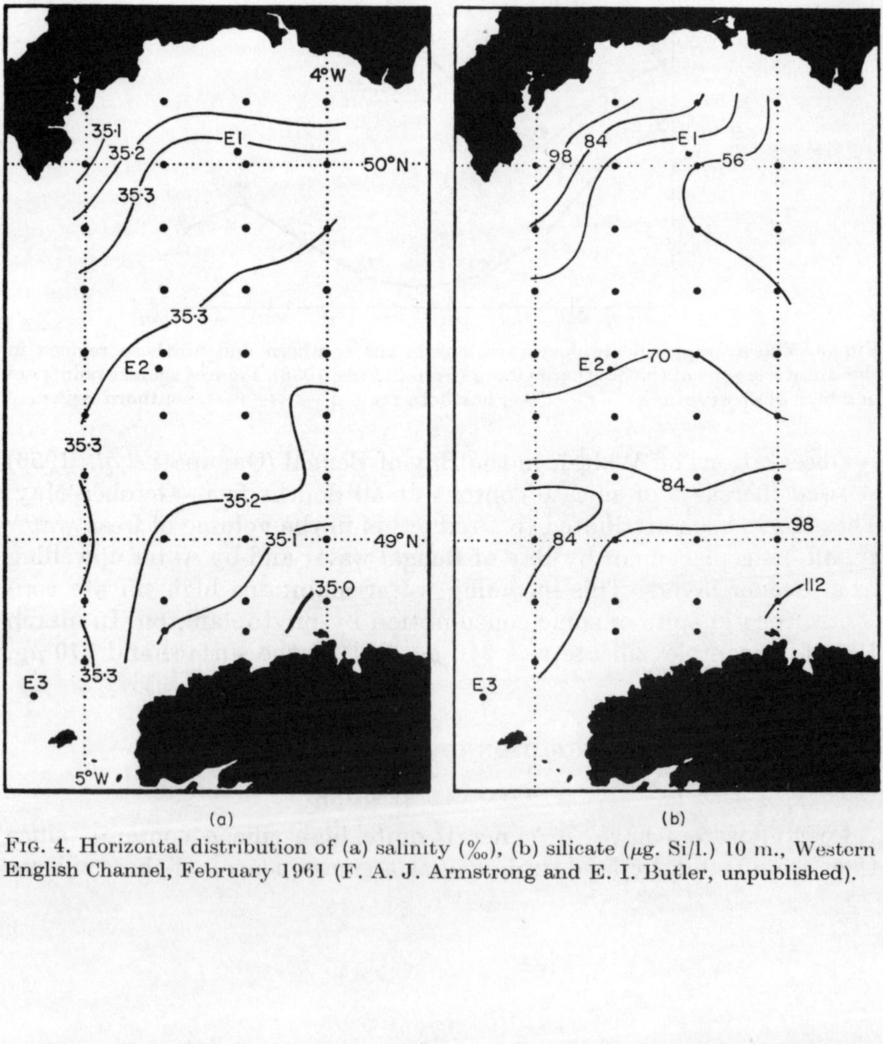

FIG. 4. Horizontal distribution of (a) salinity (‰), (b) silicate (μg. Si/l.) 10 m., Western English Channel, February 1961 (F. A. J. Armstrong and E. I. Butler, unpublished).

simple admixture of river water and silicate-poor ocean water, so that at any chlorinity there was a deficiency of silicate. In the Baltic, this deficiency amounts to about 950 μg. Si/l. in the water above the halocline, and may be ascribed to removal by primary production. The annual rate of removal may be calculated if the renewal time of this layer is known. Calculation shows that this is about 25 years, which gives an annual consumption of silicate of about 40 μg. Si/l. per year. That for the Gulf of Bothnia is, by similar calculation, about 30 μg. Si/l. per year.

Cooper (1933a,b) found a silicate consumption of 80–170 μg. Si/l. per year in the English Channel. Somewhat lower values have been found in recent years. The horizontal distribution in winter is shown in Fig. 4, and should be compared with that of salinity; near the coast, the correspondence between high silicate content and low salinity is evident. The vertical distribution resembles that of phosphate and can be seen in Fig. 1.

For the Pacific Ocean, Moberg (1928) reported concentrations at depths down to 150 m. at a position 10 miles offshore from La Jolla; these increased from 210 μg. Si/l. at the surface to about 780 μg. Si/l. at the bottom. Bigelow and Leslie (1940) reported concentrations at 31 stations in Monterey Bay, in July, 1938. They found silicate to increase from surface to bottom; averages over the whole area were from 240 μg. Si/l. at 0 m. to about 2000 μg. Si/l. at 600 m. They showed that high silicate values in this coastal region could be the result of upwelling of richer bottom water. The later observations by Phelps (1937) in this area have been referred to above.

B. THE ATLANTIC OCEAN

There are not enough observations for this ocean to allow the horizontal distribution of silicate to be charted in the same way as that of phosphate. Data for the northern part of this ocean are notably lacking.

In the eastern North Atlantic, records from single stations show a rapid increase of silicate with depth. Atkins and Harvey (1925) examined water from stations between Portugal and the Azores, and found 100 μg. Si/l. at the surface, increasing to about 500 μg. Si/l. at 3000 m. Armstrong (1951) reported observations to a depth of 4000 m. (Fig. 5). The increase between 2000 and 4000 m., and the non-linear relationship between silicate and other properties of the water between 1200 and 2000 m., have been discussed by Cooper (1952).

Clowes (1938) has described the vertical distribution of silicate in the eastern South Atlantic on a section from about 69°S. 14°E. to 4°N. 13°W., roughly parallel to the west coast of Africa. The southernmost part is shown as Section 3 in Fig. 19, Chapter 8, and the silicate distribution is shown in Fig. 6. Surface values of this nutrient are seen to be high (800–1300 μg. Si/l.) south of the Antarctic Convergence (at about 49°S.), and much lower (48–160 μg. Si/l.) to the north of it. Silicate is seen to move northward from the Antarctic in two ways: by sinking at the

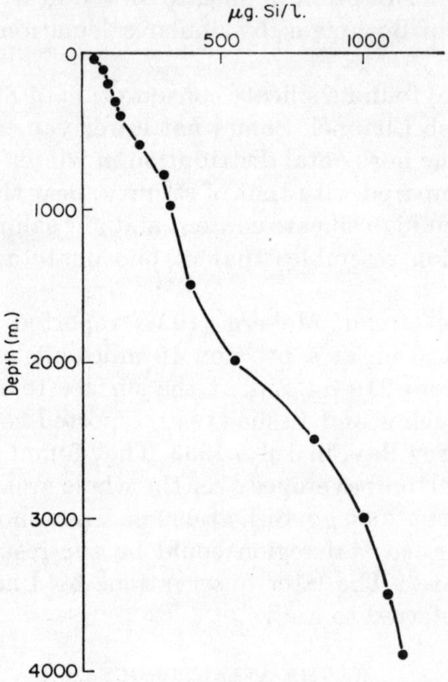

FIG. 5. Distribution of silicate with depth in the North Atlantic (47°24′N., 7°52′W.) (*Discovery* Report, 28).

Antarctic Convergence, and by entrainment with the Antarctic Bottom Water. The silicate content of the Antarctic water is over 2400 μg. Si/l., but movement appears to be hampered by the Atlantic–Indian cross-ridge at about 53°S. 14°E. North of the ridge, the southward moving warm Deep Water returns silicate to the Antarctic and rises near the Antarctic Convergence.

Richards (1958) has reported silicate, with other observations, for Western North Atlantic stations in the Antilles Arc region and in the Woods Hole–Bermuda area. Some silicate profiles are shown in Fig. 7; marked differences at different places are seen. At most stations there were silicate maxima at depths of 800–1200 m., and minima at 1400–1800 m., below which silicate again increased. No minima occurred in

(1933, 1937a,b). Components of motion toward and away from the Antarctic continent are shown schematically in Fig. 18. A shallow layer of surface water around the continent moves northward and sinks at the Antarctic Convergence. Somewhat north of this convergence is a region in which mixing of the Antarctic surface water with surface and deeper

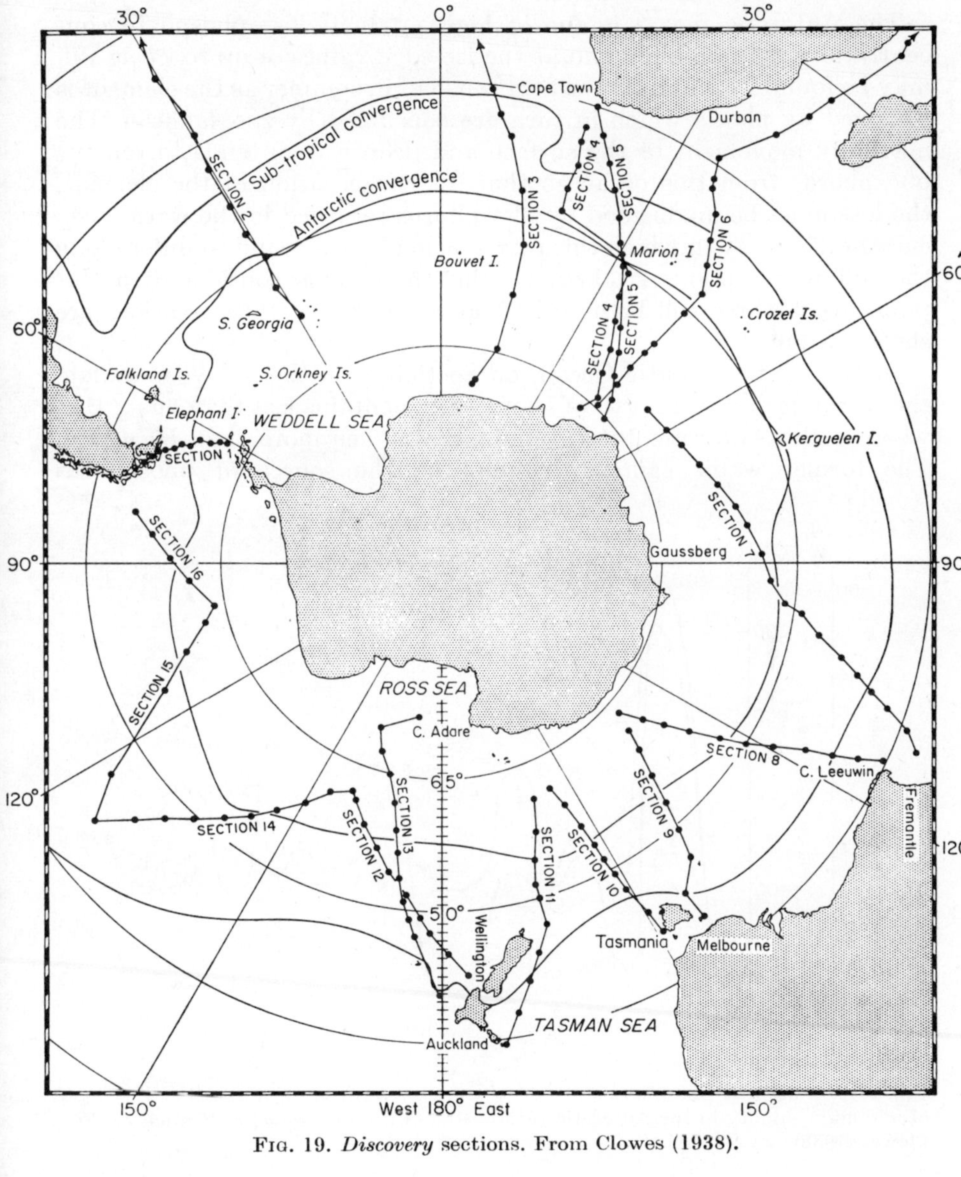

FIG. 19. *Discovery* sections. From Clowes (1938).

water of the sub-Antarctic zone (between about 40° and 50°S.) forms the Antarctic Intermediate Current which continues to move northward. At of a depth of *c.* 2000 m. in the sub-Antarctic zone there is a south-going current of warmer water; just south of the Antarctic Convergence this rises to rather less than 1000 m. Below it is the Antarctic Bottom Water, which moves towards the north.

The Antarctic region is one of high productivity; phosphate concentrations may be high, and at the ice edge values of up to 77 μg.P/l. may be found in winter. Concentrations fall in summer as the element is removed by plants which in turn are consumed by zooplankton. The northerly movement of the surface and deep waters tends to remove phosphorus from this ocean so that for conservation of the element, the loss must be made good by phosphorus returned in the warm deep current. The way in which this return is made was found to differ along the different sections studied. Of the 16 sections considered in this publication, three will be used here as illustrations; their locations are shown in Fig. 19.

In the South Atlantic Ocean, on Section 2 on the 30°W. meridian, the warm deep water north of about 35°S. contains less than 60 μg.P/l., whereas the Antarctic Intermediate Water has more than 80 μg.P/l. The former water cannot therefore be the source of phosphorus

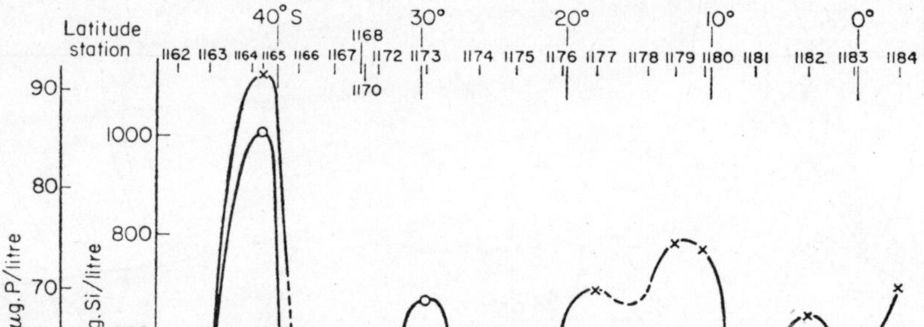

Fig. 21. Variation of phosphate and silicate contents from south to north at the depth of minimum salinity in the Antarctic Intermediate Current, *Discovery* Section 3. From Clowes (1938). ×, phosphate; ○, silicate.

the Venezuelan Basin, where the silicate content remained more or less constant below about 800 m. Concentrations were much higher in the anoxic Cariaco Trench than in other parts of the region (see Chapter 13). Comparisons with the other observations indicated that silicate con-

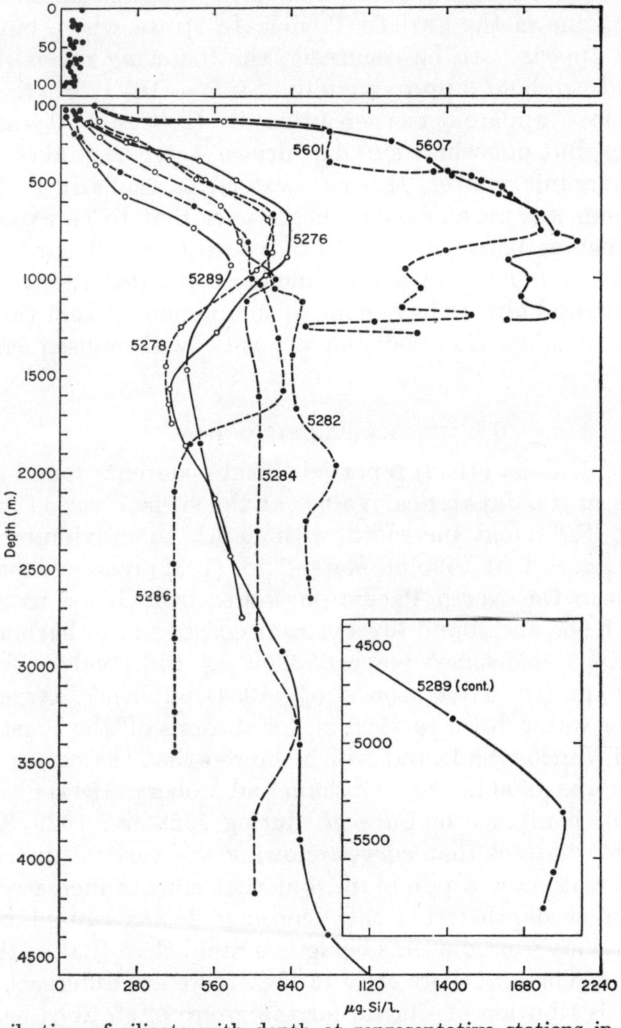

FIG. 7. Distribution of silicate with depth at representative stations in the Antilles Arc region (from Richards, 1958).
5601, 5607, Cariaco Trench.
5282, 5284, Venezuelan Basin.
5276, 5278, Atlantic Ocean, east of Windward Islands.
5289, Puerto Rico Trench.
5286, Jungfern Passage region.

centration is directly related to biological regeneration in some strata, and to hydrographic factors in others.

Direct relationships between silicate and dissolved oxygen and nutrient elements were found in the intermediate layers; these relationships were different in the cold water in the Venezuelan Basin, and in the anoxic zone in the Cariaco Trench. In strata where biological regeneration appeared to be occurring, the following ratios (by atoms) were found to hold approximately, $Si:P = 16:1$, $Si:N = 1:1$. A further ratio – "apparent oxygen utilized": $Si = 16{\cdot}9 : 1$ – also held in some places, but not where sinking surface water had taken with it its burden of organic matter. In some waters, particularly in the deeper regions, there was an excess of silicate over that to be expected from the relationships between it and other nutrients. Richards suggested that re-solution of silica may continue in deep water after regeneration of phosphate and nitrate have gone to completion, or that there may be re-solution of silica from bottom deposits, with subsequent upward diffusion.

C. THE PACIFIC OCEAN

Suda and Hidaka (1932) reported silicate contents found in 1929 at 77 stations in the Japan Sea. Values at the surface varied from about 25–350 μg. Si/l., and increased with depth to maximum values of 1200–1600 μg. Si/l. at 1500 m. Matsudaira (1932) recorded surface concentrations in the North Pacific on course from Kobe to San Diego to Hilo to Kobe and found (except near coasts and in harbours, where concentrations sometimes reached 3200 μg. Si/l.) values from 46 to about 310 μg. Si/l. Thompson *et al.* (1934) published average silicon contents for water down to 2000 m. for stations off the coast of Washington and Vancouver Island which showed that the concentration at the bottom was 4900 μg. Si/l. Graham and Moberg (1944) discussed the observations made on the *Carnegie* during 1928 and 1929. They drew attention to the fact that curves showing the vertical distribution of silicate did not show a minimum, but that silicate increased continuously to great depths; this they considered was caused by the resolution of silica from diatoms being less rapid than that of the organic constituents. These authors showed that there are differences between the mean distribution of silicate for the group of stations between 20° and 34°N. and that for the group between 20°N. and 14°S. At the southern stations, the increase of silicate with depth was more gradual below 300 m., and the quantities present were lower than those in the northern ones. At 1000 m. the concentrations were respectively 1760 and 3000 μg. Si/l., and at 3000 m., 3100 and 3750 μg. Si/l.

Wooster and Cromwell (1958) also found that in the eastern Equatorial Pacific silicate does not show a maximum at intermediate depths in the same way as does phosphate. This is shown in Fig. 8 which may be compared with the vertical distribution of phosphate on the same section (see Chapter 8, Fig. 24). Another difference is that the sharp increase in phosphate concentration north of 18°N. on this section is not paralleled by a similar increase in silicate. There are, however, many similarities between phosphate and silicate distributions, e.g. both are high off the west coast of Central America.

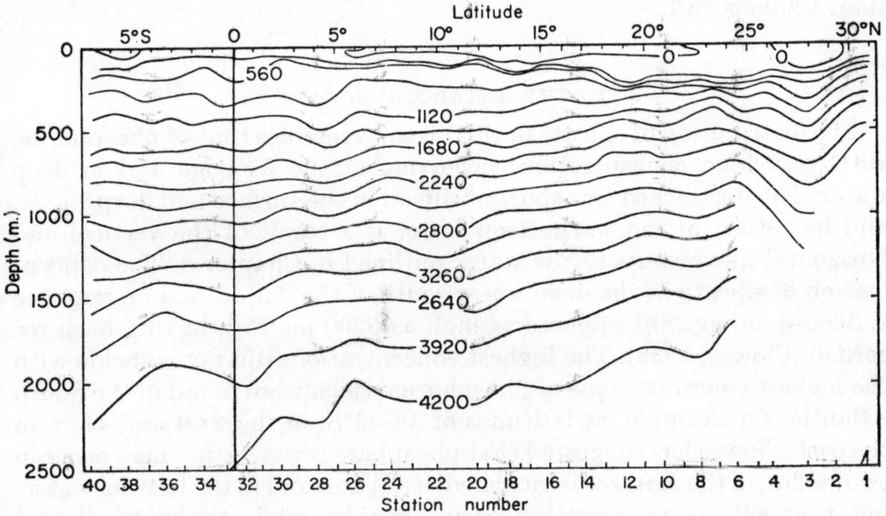

FIG. 8. Distribution of silicate (μg. Si/l.) along *Shellback* Profile 1. May–June, 1952 (from Wooster and Cromwell, 1958).

In the Peru Current, surface silicate is high near the coast (although over most of the eastern tropical Pacific it was below the limit of detection of the method), and farther west where the south equatorial current appears as a continuation of the Peru Current, silicate contents of up to 140 μg. Si/l. were found.

D. THE INDIAN OCEAN

The vertical distribution of silicate along a section from 65°S. to 11°N., starting at 34°E., and in its northernmost part passing between Madagascar and the east coast of Africa is shown in Fig. 9 (Clowes, 1938).

South of the Antarctic Convergence, surface silicate concentrations are very high at 500–1500 μg. Si/l.; they decrease and remain low (less than 150 μg. Si/l.) to the north. In the Antarctic, silicate is very high at all depths, generally exceeding about 3000 μg. Si/l. Northward movement of the bottom water appears to be hindered by the Prince Edward–Crozet Ridge, but even so, the concentrations at 15°S. are much higher (2400 μg. Si/l.) than in the Atlantic (900 μg. Si/l.). The vertical distribution in this ocean, as in the Pacific, shows no minimum.

In the northern Indian Ocean, silicate concentrations below about 1500 m. may exceed 3000 μg. Si/l., and at the bottom may be greater than 4000 μg. Si/l.

E. THE ANTARCTIC OCEAN

The distribution of silicate in this ocean resembles that of phosphorus, with very high concentrations occurring at the ice-edge and in deep water. The northward transport of silicon in the surface and deep water, and its return in the warm deep water is a result of the vertical and horizontal movements of the water outlined in Chapter 8. The concentration of silicate in the deep water south of the Antarctic Convergence is 3600–4800 μg. Si/l.; figures as high as 5200 μg. Si/l. having been recorded (Clowes, 1938). The highest concentrations did not coincide with the highest concentrations of phosphorus which were found in the South Atlantic; for example, at latitudes of 38–43°S. in the west and 41°S. in the east. Clowes has suggested that phosphate regeneration may occur in two stages, in the Antarctic Intermediate Water and in the bottom water, but that silicate regeneration occurs rapidly while the water is still within the Antarctic zone. There is little deposition of diatom remains in this region, and in fact these are found on the bottom only near the Antarctic Convergence where it is supposed that large numbers of diatoms are carried downward rapidly and are deposited before they have completely dissolved. These observations support others that suggest that solution of silica is rapid, and they are difficult to reconcile with experiments which indicate that it is slow. Clowes suggested that diatoms eaten and excreted by zooplankton are perhaps more rapidly dissolved.

In the sub-Antarctic, the winter surface concentrations were about 120 μg. Si/l. in the northern part of the region, and some 400 μg. Si/l. in the south. In summer, they were respectively 0–120 μg. Si/l. and 0–180 μg. Si/l. Where zero silicate concentrations (or at least, amounts not detectable) were found, it is reasonable to suppose that lack of silicate may have limited plant growth to some extent.

F. COMPARISON OF THE MAJOR OCEANS

The most striking differences between the oceans are seen in vertical profiles of silicon content. Some representative data are shown graphically in Fig. 10. On this scale, changes in the uppermost 100 m. cannot be shown clearly, although in general, appreciable amounts of silicon, adequate for phytoplankton growth, are found. Exceptions may be found in parts of the Antarctic as noted above, and in the Woods Hole–Bermuda area of the North Atlantic (Richards, 1958).

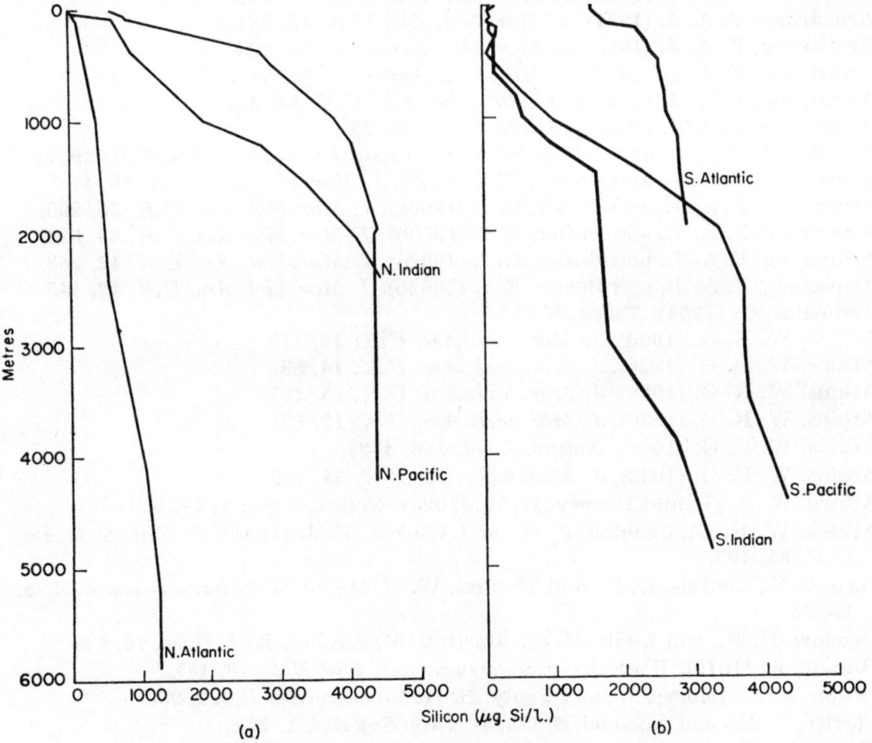

FIG. 10. Distribution of silicate with depth in the oceans.

(a)	N. Atlantic Ocean	43°23′N.	19°23′W.
	N. Pacific Ocean	49°59′N.	145°02′W.
	N. Indian Ocean	15°45′N.	53°08′E.
(b)	S. Atlantic Ocean	58°01′S.	33°03′W.
	S. Pacific Ocean	45°28′S.	152°41′W.
	S. Indian Ocean	38°08′S.	38°17′E.

The North Pacific and North Indian Oceans are notable for the high silicate content of their deep waters. This is undoubtedly caused by the restricted exchange of bottom water at the equator, compared with the freer movement and consequent dispersion of silicate which occurs in the Atlantic. Our understanding of water movements and of mixing processes in the Atlantic Ocean would be greatly increased by more knowledge of the silicate content of Arctic waters.

REFERENCES

Alexander, G. B., Heston, W. M., and Iler, R. K. (1954). *J. Phys. Chem.* **58**, 453.
Armstrong, F. A. J. (1951). *J. Mar. biol. Ass. U.K.* **30**, 149.
Armstrong, F. A. J. (1954). *J. Mar. biol. Ass. U.K.* **33**, 381.
Armstrong, F. A. J. (1955). *J. Mar. biol. Ass. U.K.* **34**, 223.
Armstrong, F. A. J. (1957). *J. Mar. biol. Ass. U.K.* **36**, 317.
Armstrong, F. A. J. (1958a). *J. Mar. biol. Ass. U.K.* **37**, 371.
Armstrong, F. A. J. (1958b). *J. Mar. Res.* **17**, 23.
Armstrong, F. A. J., and Atkins, W. R. G. (1950). *J. Mar. biol. Ass. U.K.* **29**, 139.
Armstrong, F. A. J., and Butler, E. I. (1959). *J. Mar. biol. Ass. U.K.* **38**, 41.
Armstrong, F. A. J., and Butler, E. I. (1960a). *J. Mar. biol. Ass. U.K.* **39**, 299.
Armstrong, F. A. J., and Butler, E. I. (1960b). *J. Mar. biol. Ass. U.K.* **39**, 525.
Armstrong, F. A. J., and Butler, E. I. (1962a). *J. Mar. biol. Ass. U.K.* **42**, 253.
Armstrong, F. A. J., and Butler, E. I. (1962b). *J. Mar. biol. Ass. U.K.* **42**, 445.
Arrhenius, G. (1954). *Tellus*, **6**, 215.
Atkins, W. R. G. (1923). *J. Mar. biol. Ass. U.K.* **13**, 119.
Atkins, W. R. G. (1926). *J. Mar. biol. Ass. U.K.* **14**, 89.
Atkins, W. R. G. (1928). *J. Mar. biol. Ass. U.K.* **15**, 191.
Atkins, W. R. G. (1930). *J. Mar. biol. Ass. U.K.* **16**, 821.
Atkins, W. R. G. (1945). *Nature, Lond.* **156**, 446.
Atkins, W. R. G. (1953). *J. Mar. biol. Ass. U.K.* **31**, 489.
Atkins, W. R. G., and Harvey, H. W. (1925). *Nature, Lond.* **116**, 784.
Atkins, W. R. G., Jenkins, P. G., and Warren, F. J. (1954). *J. Mar. biol. Ass. U.K.* **33**, 497.
Bien, G. S., Contois, D. E., and Thomas, W. H. (1958). *Geochim. et cosmoch. Acta,* **14**, 35.
Bigelow, H. B., and Leslie, M. (1940). *Bull. Mus. comp. Zool. Harv.* **70**, 429.
Brandt, K. (1919). *Wiss. Meeresuntersuch. Abt. Kiel N.F.* **18**, 187.
Bruun, A. F., Langer, E., and Pauly, H. (1955). *Deep Sea Res.* **2**, 230.
Carritt, D. E., and Goodgal, S. (1954). *Deep Sea Res.* **1**, 224.
Chatterjee, B. (1939). *J. Indian chem. Soc.* **14**, 589.
Chow, D. T. W., and Robinson, R. J. (1953). *Analyt. Chem.* **25**, 646.
Clarke, F. W. (1924). "The Data of Geochemistry", 841 pp. *Bull. U.S. Geol. Survey*, 770.
Clowes, A. J. (1938). '*Discovery*' *Rep.* **19**, 1.
Conger, P. S. (1941). *Univ. Wisconsin Symposium Hydrobiol.* 395.
Cooper, L. H. N. (1933a). *J. Mar. biol. Ass. U.K.* **18**, 677.
Cooper, L. H. N. (1933b). *J. Mar. biol. Ass. U.K.* **19**, 55.
Cooper, L. H. N. (1952). *J. Mar. biol. Ass. U.K.* **30**, 511.
Correns, C. (1937). *Wiss. Erg. dtsch. atlant. Exped. 'Meteor'*, **3**, 135.
Correns, C., and Engelhardt, W. (1938). *Chem. d. Erde*, **12**, 1.

Diénert, F., and Wandenbulcke, F. (1923). *C.R. Acad. Sci., Paris*, **176**, 1478.
Discovery Report (1957). **28**, 299.
Ganapati, P. N., La Fond, E. C., and Bhavanarayana, P. V. (1956). *Proc. Indian Acad. Sci.* **44B**, 68.
Goldberg, E. D. (1958). *J. Mar. Res.* **17**, 178.
Goldberg, E. D. (1961). *Phys. & Chem. Earth*, **4**, 281.
Goldberg, E. D., Baker, M., and Fox, D. L. (1952). *J. Mar. Res.* **11**, 194.
Goto, K. (1956). *Bull. chem. Soc. Japan*, **29**, 740.
Graham, H. W., and Moberg, E. G. (1944). Carnegie Institution of Washington, Publication 562.
Griffin, J. J., and Goldberg, E. D. (1963). *In* "The Sea" (M. N. Hill, ed.), pp. 728–41. Interscience, New York.
Grim, R. E., Dietz, R. S., and Bradley, W. F. (1949). *Bull. geol. Soc. Amer.* **60**, 1785.
Gunnerson, C. G., and Emery, K. O. (1962). *Limnol. Oceanogr.* **7**, **14**.
Hurley, P. M., Hart, S. M., Pinson, W. H., and Fairbairn, H. W. (1959). *Bull. geol. Soc. Amer.* **70**, 1622.
Iler, R. K. (1955). "Colloid chemistry of silica and silicates", 324 pp. Cornell University Press, Ithaca, N.Y.
Ingri, N. (1959). *Acta chem. scand.* **13**, 758.
Ishii, J., and Ishikawa, T. (1964). *In* "Studies in Oceanography" (Hidaka commemorative volume), p. 288. Tokyo.
Jayaraman, R. (1954). *Indian J. Fish.* **1**, 315.
Jørgensen, E. G. (1952). *Physiol. Plant.* **5**, 161.
Jørgensen, E. G. (1953). *Physiol. Plant.* **6**, 301.
Jørgensen, E. G. (1955). *Physiol. Plant.* **8**, 846.
Jørgensen, E. G. (1957). *Dansk. bot. Ark.* **18**, 1.
King, E. J., and Davidson, V. (1933). *Biochem. J.* **27**, 1015.
Krauskopf, K. B. (1956). *Geochim. et cosmoch. Acta*, **10**, 1.
Lewin, J. C. (1954). *J. Gen. Physiol.* **37**, 589.
Lewin, J. C. (1955). *J. Gen. Physiol.* **39**, 1.
Lewin, J. C. (1961). *Geochim. et cosmoch. Acta*, **21**, 182.
Lisitsin, A. P. (1961). *Okeanol. Issled.* (*Moscow Akad. Nauk, S.S.S.R.*) **3**, 52.
Lowndes, A. G. (1955). *Sci. Progr.* **171**, 434.
Maeda, H. (1952). *Publ. Seto mar. biol. Lab.* **2**, 249.
Matsudaira, Y. (1932). *J. Oceanogr. Soc. Japan*, **3**, 667.
Millot, G. (1942). *Géol. appl.* **2**, 1.
Moberg, E. G. (1928). *Proc. Nat. Acad. Sci., Wash.* **14**, 511.
Murray, H. H., and Harrison, J. L. (1956). *J. sediment. Petrol.* **26**, 363.
Murray, J., and Renard, A. F. (1891). Voyage of H.M.S. *Challenger*, 1872–76. Report on the Deep Sea Deposits. H.M.S.O., London.
Norin, E. (1953). *Bull. geol. Inst. Univ. Upfala*, **34**, 279.
Okamoto, G., Okura, T., and Goto, K. (1957). *Geochim. et cosmoch. Acta*, **12**, 123.
Phelps, A. (1937). *Trans. Amer. Phil. Soc.* **29**, 153.
Phifer, L. D., and Thompson, T. G. (1937). *J. Mar. Res.* **1**, 34.
Prasad, R. R., and Jayaraman, R. (1954). *Proc. Indian Acad. Sci.* **40B**, 49.
Raben, E. (1905a). *Wiss. Meeresuntersuch. Abt. Kiel N.F.* **8**, 99.
Raben, E. (1905b). *Wiss. Meeresuntersuch. Abt. Kiel N.F.* **8**, 279.
Raben, E. (1910). *Wiss. Meeresuntersuch. Abt. Kiel N.F.* **11**, 303.
Raben, E. (1914). *Wiss. Meeresuntersuch. Abt. Kiel N.F.* **16**, 207.

Radczewski, O. E. (1937a). *In* "Recent Marine Sediments" (Trask *et al.*, eds), p. 496. American Association Petroleum Geologists, Thomas Murby & Co., London.
Radczewski, O. E. (1937b). *Wiss. Erg. dtsch. atlant. Exped. 'Meteor'*, **3**, 262.
Rex, R. W., and Goldberg, E. D. (1958). *Tellus*, **10**, 153.
Richards, F. A. (1958). *J. Mar. Res.* **17**, 449.
Ross, C. S., and Hendricks, S. B. (1945). *U.S. Geol. Surv. Prof. Paper* **205***B*, 77 pp.
Rogall, E. (1939). *Planta*, **29**, 279.
Sillén, L. G. (1961). *In* "Oceanography" (M. Sears, ed.), p. 549. A.A.A.S., Washington, U.S.A.
Stefansson, U., and Richards, F. A. (1963). *Limnol. Oceanogr.* **8**, 394.
Suda, K., and Hidaka, K. (1932). *J. Oceanogr. Soc. Japan*, **3**, 291.
Thompson, T. G., and Johnson, M. W. (1930). *Publ. Puget Sd Mar.* (*biol.*) *Sta*, **7**, 345.
Thompson, T. G., Thomas, B. D., and Barnes, C. A. (1934). "James Johnstone Memorial Volume," p. 203. Liverpool University Press, Liverpool, England.
Tourky, A. R., and Bangham, D. H. (1936). *Nature, Lond.* **138**, 587.
Voipio, A. (1961). *Ann. Acad. Sci. Fenn. Ser. A II Chemica.* 14 pp.
Weaver, C. E. (1958). *Bull. Amer. Ass. Petrol. Geol.* **42**, 254.
Wiseman, J. D. H., and Hendey, N. I. (1953). *Deep Sea Res.* **1**, 47.
Wooster, W. S., and Cromwell, T. (1958). *Bull. Scripps Instn Oceanogr.* **7**, 169.
Zen, E. A. (1959). *Amer. J. Sci.* **257**, 29.

Chapter 11

The Dissolved Organic Constituents of Sea Water

E. K. DUURSMA

Nederlands Instituut voor Onderzoek der Zee
Den Helder, Holland

I. Introduction ... 433
II. Historical ... 434
III. Chemical Compounds possibly formed by the Breakdown of Organic Matter 437
A. Organic Compounds free from Nitrogen (other than Lipids) ... 438
B. Nitrogenous Compounds ... 440
C. Lipids ... 443
D. Complex Substances including Humic Acids ... 443
IV. Identity of the dissolved Organic Compounds found in Sea Water ... 446
A. Specific Substances Found ... 446
B. Elementary Composition ... 456
V. The Geographical Distribution and Seasonal Variation of Dissolved Organic Matter in the Sea ... 462
VI. Some Additional Remarks ... 470
A. Hydrocarbons ... 470
B. Solubility of Organic Compounds ... 472
References ... 473

I. Introduction

Of the subjects covered by marine chemistry, that of dissolved organic matter is one of the least well-known. Although in recent years there has been an increasing number of publications on the subject, it must be admitted that this particular field of research is still at a preliminary stage. We cannot give a complete picture of the molecular compositions of the organic compounds in sea water, and little quantitative information is available. Other major problems concern the distribution of dissolved organic matter in the sea. What are the variations in the organic contents of various water masses, do these contents change with depth and with season, and what is the effect of changes in the concentration of dissolved organic matter on those living organisms which may affect its distribution? At present, the answers to these complex problems are inconclusive and, furthermore, there is little information available concerning the distribution of specific compounds in the sea.

In spite of these shortcomings, further research on oceanic dissolved organic matter is being undertaken on an ever increasing scale in many marine institutes. Of particular interest to biologists and biochemists is the production of organic matter by living organisms. From many laboratory experiments, it is apparent that certain organic compounds may substantially influence the growth and mortality of micro-organisms, and, although it is not yet certain to what extent these biologically active compounds occur in the sea, their importance has been widely recognized (Provasoli, 1963), and this has greatly stimulated the gathering of further information.

The low concentrations of organic materials in sea water make their chemical analysis very complicated. The compounds are difficult to isolate because they occur at extreme dilution in a medium which contains many other compounds in far greater concentrations. Direct application of conventional macro-analytical methods is therefore out of the question because the quantities are below the lower limits of sensitivity of most test reactions, and because also the presence of relatively large quantities of inorganic salts often greatly reduces the sensitivity of test reactions. It is not practicable to isolate the organic constituents by evaporation or freezing, since any advantage so gained is offset by an increase in salt concentration. Other approaches, such as solvent extraction and adsorption, although presenting considerable difficulties, give promising results. However, dissolved compounds, though organic in composition, are often preferentially retained by the aqueous phase in extraction or adsorption procedures. Moreover, the recoveries of various dissolved compounds may differ, so that it is difficult to obtain a quantitative picture from the results.

Another difficulty is that the extremely small quantities of dissolved organic compounds being determined make it imperative that the reagents used for isolation or sorption must not be contaminated by organic matter in quantities approaching those of the organic compounds in the sample. It is known, for example, that even the purest commercial chlorides of iron and aluminium which are currently available are not free from organic matter, and complex decomposition products are formed in many purified organic solvents on storage.

These are only a few of the difficulties encountered in the analytical chemistry of these compounds which, although present in only minute quantities, play an important part in the cycle of living and dead matter in the sea. The analytical chemistry of organic compounds in sea water is discussed in Chapter 21, Volume 2, Section XI.

II. Historical

Natterer was one of the first investigators to publish data (1892–94) on dissolved organic matter in the sea. During cruises with the research vessel *Polar* in the eastern Mediterranean, it occurred to him that the dry weights of some sea-water samples exceeded the weight expected from the sum of the separate mineral compounds. He noted that on prolonged heating, the dried samples gave evidence of charring and the odour of acrolein (a compound which is also produced on heating fats). Natterer removed calcium and magnesium from concentrated sea water by precipitation with Na_2HPO_4, evaporated the filtrate, and extracted the organic compounds from the residue with ethyl alcohol. He found the brown syrupy liquor that remained after evaporation of the alcohol possessed the following properties: flammability in air, vapours smelling of palmitic and stearic acids evolved on heating, and a melting range for the sublimate of 62–98°C (palmitic acid m.p. 64°C, stearic acid m.p. 69°C). Again, a smell of acrolein could be detected; distillation *in vacuo* gave a colourless distillate smelling of impure glycerol. Organic nitrogen could not be detected by Lassaigne's test.

As a result of this work he found *c.* 2 mg. of dissolved organic matter per litre in the surface water of the open sea, and 10–20 mg./l. in water taken from near the coast of Greece. He attributed the dissolved organic matter in sea water to saponification of the fats of dead organisms.

Although Natterer failed to find organic nitrogen compounds in the evaporation residues, he did find them dissolved in the sea water itself. By application of the albuminoid–nitrogen method of Wanklyn *et al.* (1867), he found contents of 40–103 μg. albuminoid–N/l.; the amounts in surface water showed more variability than those from greater depths.

This first publication aroused much interest, not so much because dissolved organic matter had been found in the sea (it was known to occur in fresh water), but because the amount was high compared with that in suspension.

Further research on dissolved organic matter was stimulated chiefly by biologists, who were interested in its possible role as food for living organisms. In the theories of Pütter (1909), which followed this line of thought, it was considered that many marine animals, including some species equipped to take particulate food, utilized dissolved organic matter as their main diet, and that their large requirements of dissolved organic matter were met from extra-cellular excretion products of phytoplankton produced during the bloom. Although the carbon analysis method developed by Pütter was definitely unreliable when applied to sea water (Raben, 1910), Pütter continued to defend his view of the part played by dissolved organic substances in the physiological processes of marine animals.

The methods used by Pütter, as well as by his critics, to arrive at a rapid determination of the organic content of sea water were based on the ability of organic matter to reduce permanganate, and also on bacterial oxidation of dissolved organic matter by the dissolved oxygen. At the time of this work, these methods were far from reliable. The investigations of Moore *et al.* (1912) gave results which were more acceptable from a physiological standpoint. The final blow to Pütter's theories was dealt by Krogh and his collaborators (Krogh, 1930, 1931, 1934a,b; Krogh and Lange, 1930; Krogh and Keys, 1934), who developed an improved method for the determination of organic carbon, and who used a slightly modified adaptation of an earlier method for the determination of organic nitrogen (Krogh and Keys, 1934). Pütter's experiments were repeated, and the results left few grounds for the belief that dissolved organic matter was the main source of food for living organisms. Moreover, excretion of organic matter by living phytoplankton cells was shown to be unlikely, since Krogh's experiments demonstrated that excretion must be less than 10% (the accuracy of his experiments) of all organic matter formed.

In the following years, further attempts were made to determine the extent and distribution of dissolved substances in the sea. Kalle (1933), and Seiwell and Seiwell (1934) found that in addition to organic nitrogen compounds, dissolved organic phosphorus was present. Kalle (*loc cit*) attributed the specific light absorption in the blue part of the spectrum to organic compounds in solution ("Yellow substances"). Kreps (1934) demonstrated that some of the organic substances were biologically active compounds, such as enzymes or organic catalysts.

By 1940 the position was as follows: it was known that dissolved organic matter occurred everywhere in the sea, and that the experimental results agreed with the assumption that it was derived from living organisms via a long chain of decomposition processes. This explained its supposedly uniform distribution. This theory was based on Krogh's (1934a,b) assumption that the bulk of dissolved organic substances would be present in the sea in constant amount, regardless of the depth and the time of year. Besides these quantities of very "old" substances resistant to any sort of decomposition by organisms, minor amounts of biologically active matter were also expected to be present. No progress of any importance towards the identity of these materials had been made since Natterer's work in 1892.

The starting point for later research was the investigation of these stable end-products and the way in which they were formed. An important approach to the identification of the missing links in the decomposition chain in the sea, was the examination of the better known breakdown processes of terrestrial organic matter. It was known that amino acids and carbohydrates were important water soluble breakdown products, and attempts to detect these substances in sea water yielded positive results. As a result of these findings, the hitherto assumed stability of the so-called end-products was soon seriously questioned, and the similarity between terrestrial decomposition products and those in the sea became more and more apparent (see Chapter 12 and Vallentyne, 1957). The modern starting point for investigations on the chemical nature of the dissolved organic compounds in the sea is based on this analogy, and this means that, in principle, proteins and carbohydrates are basic "life" substances in the sea as well as on land.

In the sea the phytoplankton organisms constitute the bulk of living marine organic matter, although less than 10% is eventually converted into animal tissue. But sometimes dead organisms constitute a large proportion of the total particulate matter in the sea (cf. Hagmeier, 1962) and, therefore, dissolved organic substances seem to originate from dead phytoplankton and detritus, rather than from excretions from living cells (Duursma, 1961).

This principle of predicting breakdown products by considering the substances from which they originate will be followed for the remainder of this chapter. The approach is, of course, rather hypothetical, but it gives a framework within which the likely course of events can be discussed. Increased knowledge of the composition of the stock of living organism and of dead particulate matter in the sea will stimulate the analytical chemistry of dissolved organic substances.

III. Chemical Compounds possibly formed by the Breakdown of Organic Matter

When considering the formation of oceanic dissolved organic matter, one important self-evident hypothesis is that the less stable compounds will stand a better chance of occurring in relatively larger quantities in the sea than on land. This is because decomposition processes, which are chiefly brought about by the activities of heterotrophic micro-organisms, are likely to be more effective in the vicinity of solid surfaces, such as those provided by particulate matter. Thus, the amounts of many of the primary dissolved organic substances are probably related in an inverse way to the quantity of particulate matter in suspension, although this has yet to be established. Compounds which will be quickly decomposed in coastal waters containing much particulate matter, may occur in relatively large quantities in the deep seas where both particulate matter and bacteria are practically absent. Freshly formed organic material will become diluted by diffusion and will be transferred from its site of formation (e.g. from the euphotic zone to the deeper layers). As the degree of dilution increases, the importance of enzyme catalyzed decomposition processes relative to chemical hydrolysis will decrease, since the former process requires enzyme–substrate contact whereas the latter process proceeds via the reaction between the organic matter and the surrounding medium of water, hydrogen and hydroxyl ions.

Although these points do not favour a close analogy between the decomposition processes of organic material on land and in the sea, it is worthwhile examining the basic terrestrial organic life substances and their main breakdown products. The following discussion has been prepared by using texts dealing with decomposition processes on land (Tracey, 1948; Mallette *et al.*, 1960; Scheffer and Ulrich, 1960), and with organic chemistry (Fieser and Fieser, 1956; Karrer, 1959). It is obviously not possible to discuss all the organic substances which may occur in the sea, but at least such an approach enables the main principles to be recognized.

Four main groups of organic compounds will be considered:

(A) nitrogen-free organic matter;
(B) nitrogeneous substances;
(C) fat-like substances (lipids); and
(D) complex substances resulting from the groups (A) to (C) (humic substances).

A. ORGANIC COMPOUNDS FREE FROM NITROGEN (OTHER THAN LIPIDS)

Carbohydrates comprise the most important examples of this class, and their decomposition products will now be considered. All polymeric carbohydrates yield reducing sugars as hydrolysates; thus cellulose, which is a condensation product of glucose, gives glucose; and hemicelluloses, which occur mainly in the cell membrane together with pectins, give pentoses and hexoses. Polyuronic acids (compounds of hemicellulose + pectin) may be relatively quickly decomposed, so that the chances of detecting them will be small. Starch is readily hydrolysed by enzymes to yield glucose. There are considerable differences between the rates of decomposition of the polysaccharides and of lignin, as is shown in Fig. 1 (Schmalfuss, 1940).

The monosaccharides, which are the simplest carbohydrates, are usually decomposed, via a phosphorylized stage, by catalytic processes, in which amino compounds may play a role, to C_2–, C_3–, and C_4–compounds. All these substances and their breakdown products—

mostly carboxylic acids—may be expected to occur in the sea. The number of organic acids produced depends on the micro-flora present (Waksman, 1952; Rippel-Baldes, 1955).

Excretion of extracellular products by healthy cells may play a role in the sea (Fogg, 1958; Lewin, 1956). Carbohydrates are the most likely compounds to be excreted, and large amounts are formed by moribund cells of phytoplankton (Guillard and Wangersky, 1958). If the sugars which are already known to occur in the sea can be considered as representative, they are mainly pentoses (Collier *et al.*, 1953).

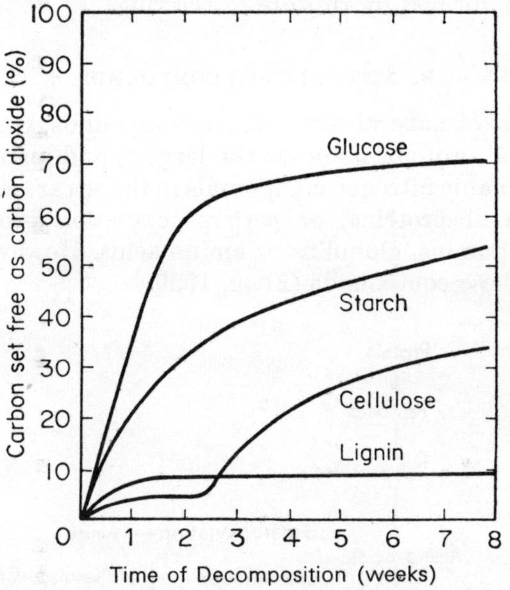

FIG. 1. Rate of decomposition of some important vegetable organic substances, Schmalfuss (1940).

2·5 g. carbon as glucose, starch, cellulose and lignin were mixed with 1 kg. sterile quartz sand and 150 ml. tap water. As the nitrogen source, glycine was added until C:N = 30:1; 0·5 g. KH_2PO_4 yielded the phosphorus, and 1 g. $CaCO_3$ was added as buffer. Inoculation was made with 25 ml. of fresh soil suspension. Reproduced by courtesy of Verlag Chemie GMBH, Weinheim, Germany.

Our knowledge of nitrogen-free compounds of aromatic structure is much more limited. It is known that numerous low-molecular weight phenols and quinones are liberated from dead vegetable matter, and these may be utilized by fungi in the building of molecules with a greater number of aromatic rings. These compounds are important in vitamin and antibiotic formation.

Terrestrial organic substances and their breakdown products might be expected to be found in the sea (Bader, 1956): Lignin is an example; it is an amorphous three-dimensional polymer derived from various unsaturated aromatic alcohols (Freudenberg, 1956). During its decomposition there is a loss of water and a decrease in the number of double bonds. All lignin-decomposing fungi produce extracellular phenol oxidases which catalyse the introduction of hydroxyl groups to the phenol ring. As a result, we may expect the formation of polyhydroxy carboxylic acids, or their derivatives, such as ferulic acid (3-methoxy-4-hydroxy-cinnamic acid), together with aliphatic acids containing six C-atoms or less, formed by rupture of the ring.

B. NITROGENOUS COMPOUNDS

Of the groups of natural nitrogenous compounds, that of the vegetable and animal proteins is by far the largest, and many, if not all, of the dissolved organic nitrogen compounds in the sea are either the breakdown products of proteins, or perhaps excretion products such as extracellular albumins, globulins or amino acids. However, little is yet known about these compounds (Fogg, 1959).

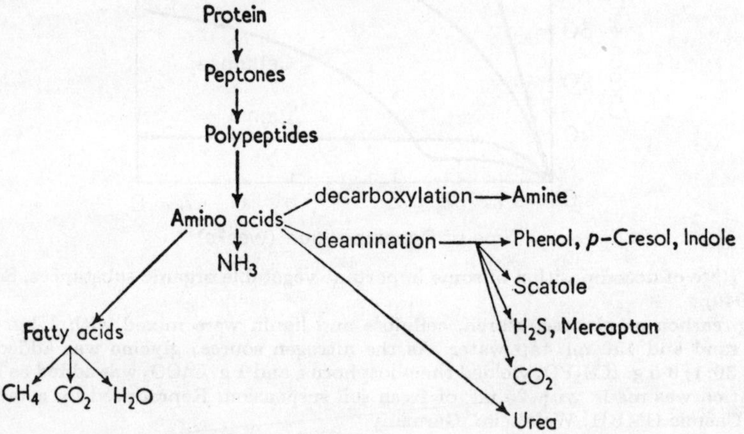

In nature, the breakdown of proteins is easily brought about by bacterial and fungal action. A simple scheme has been given by Rippel-Baldes (1955) for the generalized processes.

Proteolytic enzymes (proteases) cause the first two hydrolytic conversions, and deaminases catalyse the NH_3– and fatty acid conversions. Mercaptan is formed only under anaerobic conditions. A more detailed scheme for the breakdown of amino acids has been given by Waksman (1952):

Proteins can be broken down by micro-organisms through one or more of the following processes.

1. *Hydrolytic deamination.* The hydrolysis of an α-amino acid may result in the formation of α-hydroxy-carboxylic acids and ammonia; of an alcohol, CO_2 and ammonia; or of an aldehyde, lower acid and ammonia, as shown by the general equations:

$$R.CH.(NH_2).COOH + H_2O \rightarrow R.CHOH.COOH + NH_3$$
$$R.CH.(NH_2).COOH + H_2O \rightarrow R.CH_2OH + CO_2 + NH_3$$
$$R.CH.(NH_2).COOH + H_2O \rightarrow R.CHO + H.COOH + NH_3$$

2. *Decarboxylation with amine formation.*

$$R.CH.(NH_2).COOH \rightarrow R.CH_2.NH_2 + CO_2$$
$$R.CH_2.NH_2 + H_2O \rightarrow R.CH_2.OH + NH_3$$

3. *Reductive deamination.*

$$R.CH.(NH_2).COOH + H_2 \rightarrow R.CH_2.COOH + NH_3$$

or

$$R.CH.(NH_2).COOH + H_2 \rightarrow R.CH_3 + NH_3 + CO_2$$

4. *Ammonia formation without reduction.* Anaerobic bacteria may produce ammonia from amino acids, without reduction.

$$R.CH_2.CH.(NH_2).COOH \rightarrow R.CH{:}CH.COOH + NH_3$$

5. *Oxidative deamination.*

$$R.CH.(NH_2).COOH + O_2 \rightarrow R.COOH + NH_3 + CO_2$$

This process is brought about by aerobic organisms, and especially by fungi. As examples of this reaction are the transformation of leucine into *iso*-valeric acid ($R = (CH_3)_2.CH.CH_2-$), and of glutamic acid into succinic acid ($R = COOH.CH_2CH_2-$).

Thus, the decomposition of one amino acid may proceed through oxidative deamination, decarboxylation or reduction. Although an organism may bring about one series of the above reactions under a given set of conditions, different products may be obtained via the same organism under different conditions.

The formation of indole and its derivatives by decomposition processes is probably of biological importance. Many of these compounds have growth stimulating properties, and are plant hormones; they have a molecular structure based on 3-substituted indole viz.

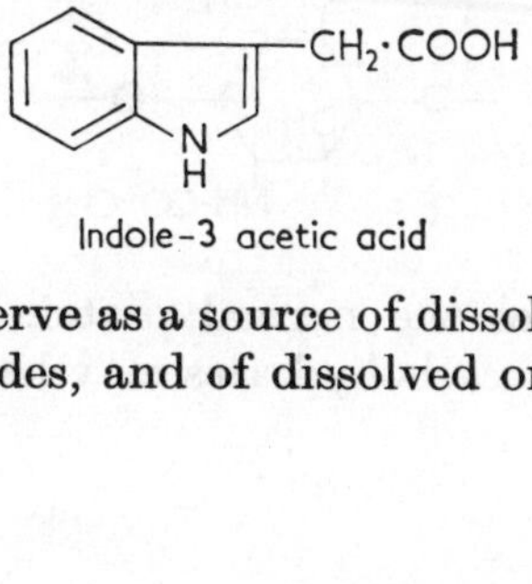

Indole-3 acetic acid

The nucleoproteins serve as a source of dissolved organic phosphorus in the form of nucleotides, and of dissolved organic molecules rich in nitrogen. They are broken down under alkaline conditions according to the following sequence of reactions (Mallette *et al.*, 1960).

Nucleoproteins
↓ (proteinase)
Proteins + Nucleic acids
↓ (nuclease)
Nucleotides
↓ (phosphatase)
Phosphoric acid + Nucleosides
↓
Ribose or deoxyribose + Purine- or Pyrimidine-base.

The most abundant purines are adenine and guanine:

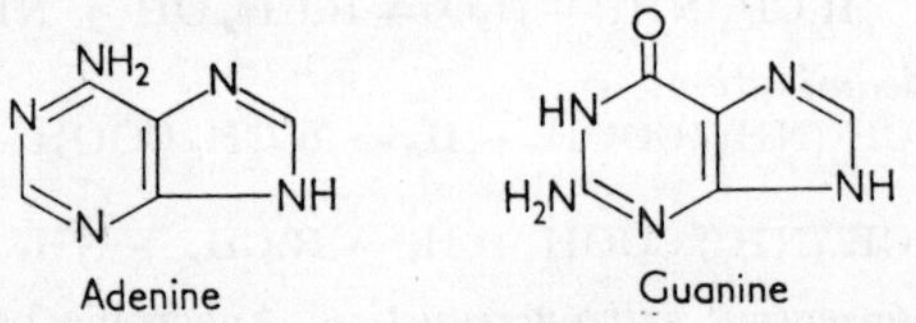

Skeletal nitrogenous substances such as the cartilage of vertebrates, and the chitin of invertebrates (especially that of crustaceans), are likely to be resistant to decomposition. It is unlikely that these substances will decompose in the surface waters of the sea, but during the protracted period spent in falling to the bottom, some decomposition will occur. Cartilage is a mucoprotein conjugated with chondroitin sulphuric acid

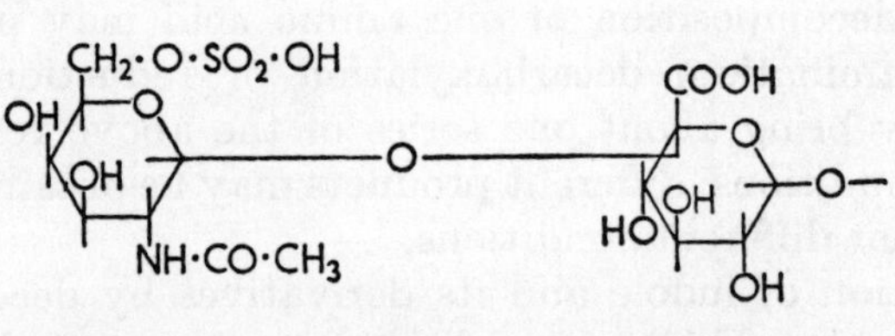

and chitin contains polymerized *N*-acetylglucosamine groups

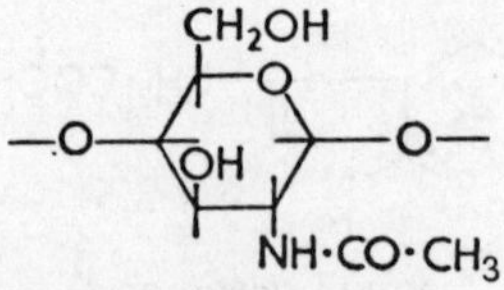

in its molecule. Besides the normal protein breakdown substances, glucosamines and other carbohydrates may be produced.

As is shown later, nitrogenous compounds may be important in the formation of humic substances. The humic acids, unlike total humus, always contain nitrogen.

Protein decomposition does not necessarily follow a straight path from complex molecules to simple ones, since micro-organisms responsible for some of the decompositions resynthesize protein from the decomposition fragments. In the sea, this protein regeneration by bacteria takes place to the extent of 30–40%, calculated as carbon (ZoBell, 1954). However, as the amino acid composition of the bacterially produced protein is almost identical with that of plankton-protein, this process will have little effect on the resultant dissolved organic matter, except that some special decomposition products may be formed.

C. LIPIDS

Members of this group, although often unrelated chemically, frequently give similar breakdown products. Lipids are generally esters of fatty acids and many in addition contain combined phosphoric acid. On hydrolysis they yield the parent acids, together with glycerol, aliphatic alcohols, carbohydrates, nitrogenous bases, or sterols. Usually the total amount of lipid is low compared with the other "life" substances, but because of their greater resistance to decomposition, lipids are more likely to be detectable.

D. COMPLEX SUBSTANCES INCLUDING HUMIC ACIDS

In terrestrial breakdown processes, a group of complex organic substances, collected under the name of humus is formed. As might be expected, an analogous group of compounds probably occurs in the sea, but owing to their complex nature little is known about their composition. Even so, a few principles concerning their formation can be discussed.

Three groups of compounds may be considered as starting points for the production of humic substances:

(1) phenol and quinones;
(2) breakdown products of proteins, possibly in combination with quinones or phenols;
(3) breakdown products of carbohydrates.

Humic substances are formed through the agency of enzymes rather than that of bacteria. An important role is played by reactions with strongly polar double bonds. Moreover, in the first two syntheses mentioned above, there is a tendency to de-aromatization of hydroquinone and phenol, owing to shifting of the charge in the molecule.

The transitions, which are mainly brought about by auto- or enzymatic oxidation reactions, take place under oxidizing and weakly alkaline conditions. Usually nitrogen is incorporated into the products as can be seen in the following examples:

1. Phenols, quinones with added nitrogen.

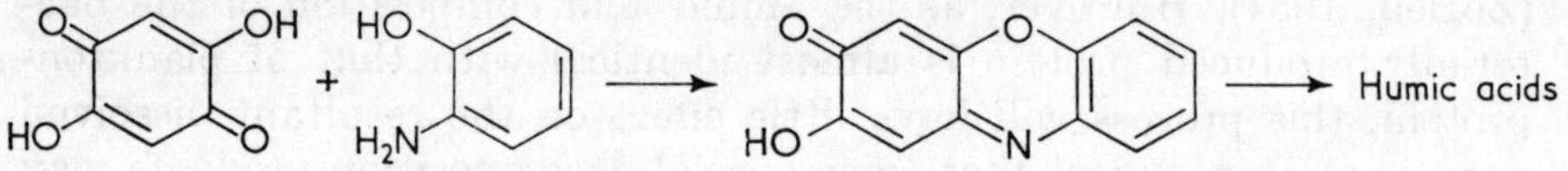

2. Melanin formation (protein fragments; e.g. from tyrosine).

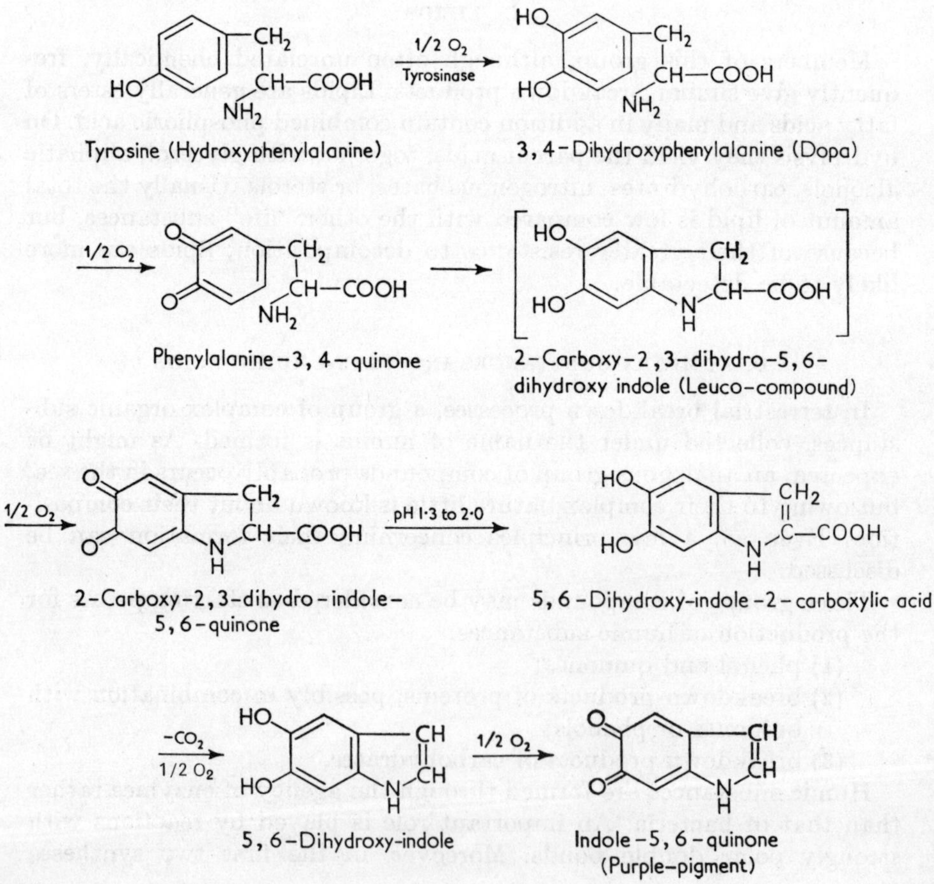

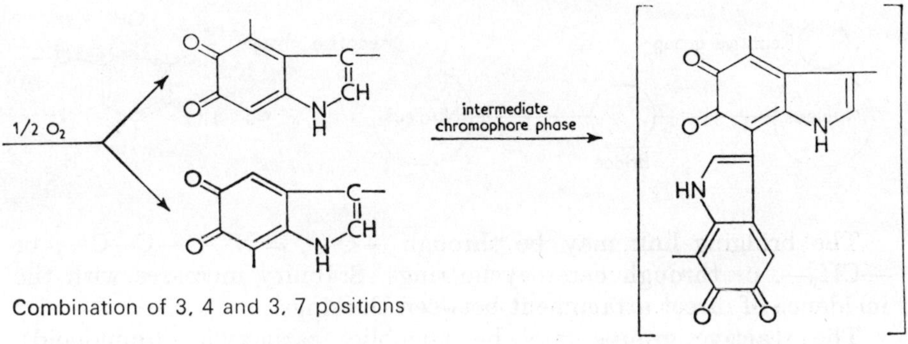

Combination of 3, 4 and 3, 7 positions

Melanin

3. Carbohydrates with amino-compounds (Maillard or melanin reaction).

Amino-compounds may act as catalysts in the splitting up of sugars (the reverse reaction of an aldol condensation). This initial degradation is followed by reactions involving phenols (and/or quinones), and amino-compounds (indicated by sequence 2 above). The reaction, which is initiated by addition of the amino-compound RNH_2 to the aldehyde form of the aldose, may be summarized by:

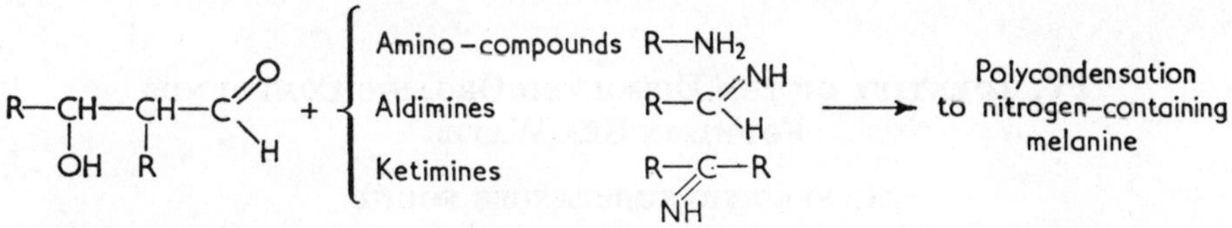

The final products include compounds with iso-cyclic and hetero-cyclic rings with five and six C-atoms, such as quinones, furfurol and their amino- and imino-derivatives; pyroles, imidazoles, pyridines and pyrazine are typical examples. In addition, compounds containing aldehyde groups may also be formed. Enders and his collaborators (Enders and Theis, 1938; Enders *et al.*, 1948) have stated that, with respect to equivalent weight, elementary composition and functional groups, humic acids from the soil strongly resemble the melanin formed *in vitro* from glucose and glycocol.

It is probable that both humic matter and similar complex materials, soluble in sea water, consist of three-dimensional poly-anions containing aromatic nuclei with certain reactive groups, linked through a kind of atomic bridge (Thiele and Kettner, 1953).

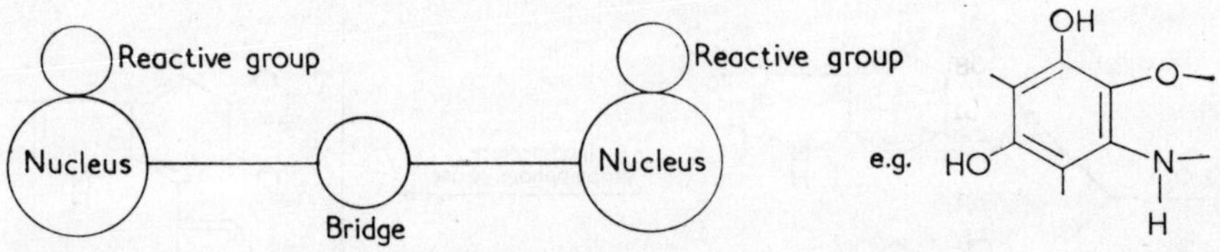

The bridging link may be through —O—, = N—, —C—C—, or —CH_2—, or through heterocyclic rings. Stability increases with the incidence of direct attachment between the rings.

The reactive groups may be phenolic, carboxylic, (quinonoid), methoxyl, carbonyl or amino-nitrogen (total up to 70% of N). The nitrogen does not occur as free —NH_2 groups, but rather as

$$\diagdown\!-\!C\!-\!N\!-\!R.$$

The properties of these three-dimensional poly-anions are such that they permit the addition of H^+ ions or —COO^- groups. Combination with cations may take place by complex or chelate formation. Much hydrogen-bonding also occurs in the molecule.

The elementary composition of the humic matter usually approximates to 50–60% C; 2·8–5·2% H; 5–10% N and about 35% O.

IV. Identity of the Dissolved Organic Compounds Found in Sea Water

A. Specific Substances Found

It is satisfying to find reasonable agreement between the prognoses of the possible organic compounds which might be formed from the breakdown of metabolites in the sea, and those known to occur. A wide range of amino acids and organic acids is found, and various carbohydrates have been reported, often in relatively large concentrations. More complicated compounds known collectively as "Yellow substances" or "Gelbstoffe" occur, and are analogous to the humic breakdown products found on land. Vitamins, such as B_{12}, biotin and thiamin have also been found, and plant hormones (auxins) are said to be present in biologically active quantities.

Until recently, there has been little evidence to indicate how environmental conditions influence the amounts of the organic substances present in sea water. These conditions include: salinity and pH of the water, amounts of phyto- and zooplankton present, season, temperature, depth and the history of the water. Consequently, Table I which summarizes most of the available information should be interpreted in a general way only.

Sugar-like compounds occur mostly in surface waters, where there is a high photosynthetic productivity, and it seems likely that many of them are extra-cellular excretion products of more or less healthy phytoplankton blooms (Lewin, 1956; Guillard and Wangersky, 1958; Fogg, 1958). However, since many carbohydrates can be formed by the decomposition of metabolites (Section III, C), it is unlikely that excretion is their sole means of origin.

The most abundant of the amino acids are aspartic acid, threonine, serine, glutamic acid, alanine, valine, iso-leucine and leucine (Hood, 1963). They probably occur in the free state to the extent of about 30% of the total organic nitrogen compounds, and are mostly present condensed as polypeptides. The spectrum of the combined amino acids seems to be different from that of the free constituents; an analogous difference has also been found to occur between the spectrum of amino acids of surface plankton and that of any other of the amino acids found in the sea (free dissolved, combined dissolved, particulate dead, and buried in sediments). The dissolved free amino acids are more uniformly distributed than the combined. The concentration of the latter decreases sharply with increase in depth in the euphotic zone, and this decrease is accompanied by changes in the relative amounts of the more basic amino acids. Below the bottom of the euphotic zone, the concentration of these combined amino acids remains almost constant (Degens *et al.*, 1964).

The observed distribution of the fatty acids is difficult to explain. Data do not yet indicate where they are formed, because no clear concentration maximum is found in water layers of high plankton and detritus content. Fogg and Nalewajko (1963) believe that glycollic acid is formed by excretion from phytoplankton when conditions are favourable for photosynthesis. Dissolved glycollate may represent a reservoir of potential energy for the maintenance of life when photosynthesis is not possible.

The biologically active compounds contribute much less to the total mass of dissolved organics, but their influence on life can be great. "Good" and "bad" waters in relation to the growth of diatoms were found in the open seas by Johnston (1963) (cf. Fig. 2). Recent experiments with sea water, and laboratory investigations with cultures, have greatly increased our knowledge about the organic regulation of the growth of marine organisms. An extensive review of the factors affecting the growth of phytoplankton has recently been presented by Provasoli

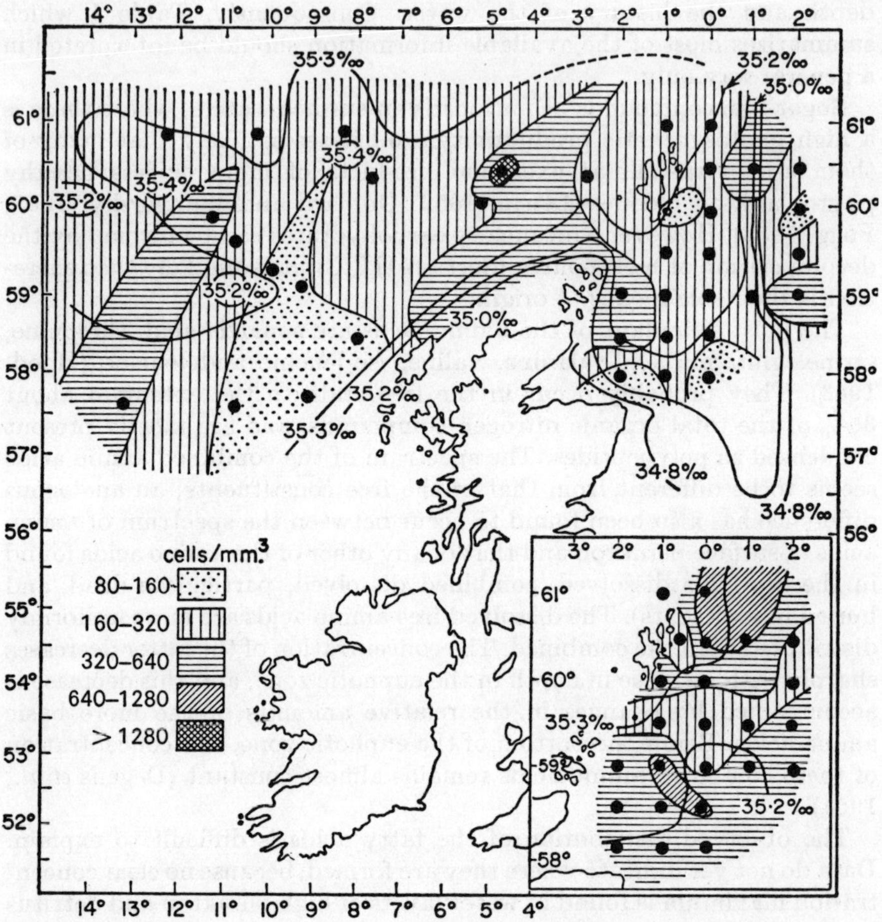

FIG. 2. Bio-assay activity of unialgal *Skeletonema*. North Sea and Faroe Islands, July–August 1956, surface samples (inset: bottom samples). The increase in number of cells/mm.³ is a measure of the amount of biologically active substances present in the water. The assay experiments were carried out on water samples enriched with inorganic nutrients, Johnston (1963). Reproduced by courtesy of H.M. Stationery Office, London, and the Council of the Marine Biological Association of the United Kingdom.

(1963). Table I summarizes our present knowledge of vitamin B_{12} compounds, and Fig. 3 illustrates the indole-like auxin activity in the sea (Bentley, 1960).

The name humic or "Yellow substance" gives no indication of the class of compounds involved. If, however, it is assumed that humic substances are formed as outlined above, then the observed fluorescence and absorption of light at several wavelengths agrees reasonably with the assumption that compounds with the same polar groups and double bond systems (melanoidines) as the terrestrial group of humic substances are present. Theoretical examination of the fluorescence and absorption spectra might yield useful information about the compounds present, but so far only Kalle (1962) has carried out analyses of these substances for both fresh water and sea water.

The average molecular weight of lake-water humic material was about 450 (Shapiro, 1957), but its infra-red spectrum did not confirm

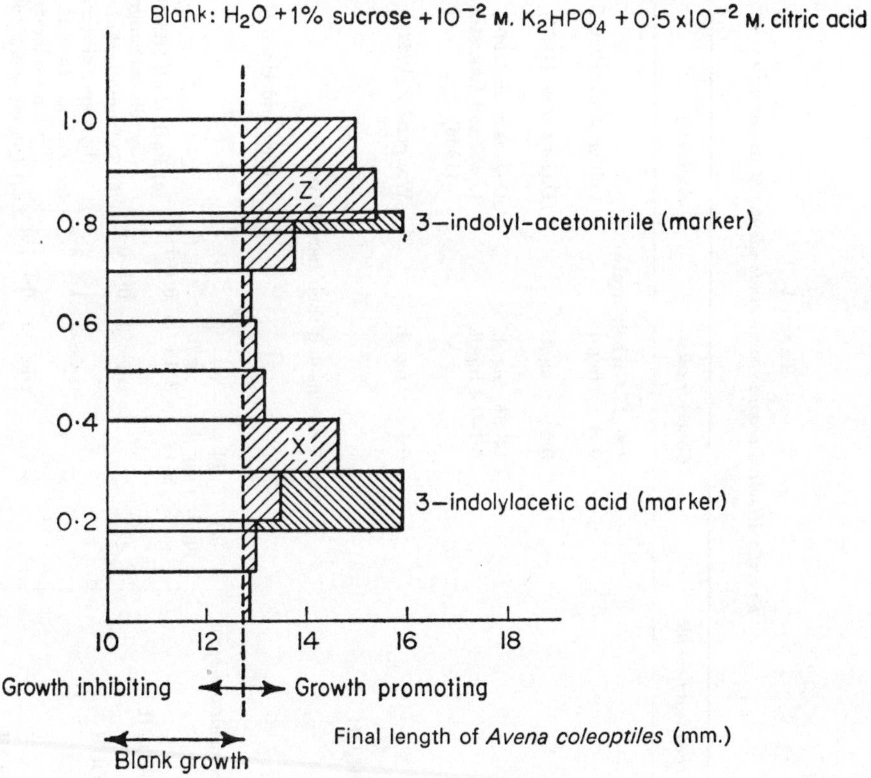

Fig. 3. Chloroform-ether extracts of sea water, separated by chromatography using *iso*-propanol-ammonia, and measured for auxin activity by measurement of cell-enlargement of 10 mm. lengths of *Avena coleoptiles* after 24 h. The bio-assay activities of the sea water extracts were comparable with those of phytoplankton auxin extracts. The main zones X and Z point to compounds closely related to indole. Concentration of X is about 0·06 μg./l. and of Z 0·315 μg./l. (Bentley, 1960). Reproduced by courtesy of H.M. Stationery Office, London, and the Council of the Marine Biological Association of the United Kingdom.

TABLE I

Specific dissolved organic compounds identified in sea water

Name of compound and chemical formula	*Concentration*				*Author(s)*	*Locality*
	I. Carbohydrates					
Pentoses $C_5H_{10}O_5$	0–8 mg./l.				Collier *et al.* (1950, 1956)	Gulf of Mexico
Pentoses $C_5H_{10}O_5$	0·5 μg./l.				Degens *et al.* (1964)	Pacific off California
Hexoses	14–36 μg./l.				Degens *et al.* (1964)	Pacific off California
Rhamnosides $C_6H_{12}O_5$ Rhamnosides	0·1–0·4 mg./l.				Lewis and Rakestraw (1955)	Pacific Ocean coast U.S.A.
Dehydroascorbic acid COCOCOCHCH(OH)CH_2OH (ring –O–)	0·1 mg./l.				Wangersky (1952)	Gulf of Mexico inshore water
	II. Proteins and their derivatives					
Peptides C:N ratio = 13·8:1					Jeffrey and Hood (1958)	Gulf of Mexico
Polypeptides and polycondensates of:	(*a*) μg./l.	(*b*) μg./l.	(*c*) μg./l.	(*d*) μg./l.		
Glutamic acid COOH$(CH_2)_2$CH(NH_2)COOH		8–13	8–13	0·1–1·8	(*a*) Park *et al.* (1962) (by ion-exchange)	Gulf of Mexico
Lysine $NH_2(CH_2)_4$CH(NH_2)COOH	<1	?	trace–3	0·1–0·9	(*b*) Tatsumoto *et al.* (1961) (by paper chromatography)	Gulf of Mexico
Glycine NH_2CH_2COOH		—	trace–3	1·2–3·7	(*c*) Tatsumoto *et al.* (1961) (by ion-exchange)	Gulf of Mexico
Aspartic acid COOHCH_2CH(NH_2)COOH		3–8	trace–3	0·1–1·0	(*d*) Degens *et al.* (1964)	Pacific off California

Serine $CH_2OHCH(NH_2)COOH$		?	trace–3	1·8–5·6
Alanine $CH_3CH(NH_2)COOH$		3–8	trace–3	0·7–3·1
Leucine $(CH_3)_2CHCH_2CH(NH_2)COOH$	0·5–1	8–13	trace–3	0·9–3·8
Valine $(CH_3)_2CHCH(NH_2)COOH$		trace–3	trace–3	0·1–1·7
Cystine $[SCH_2CH(NH_2)COOH]_2$		trace–3	—	0·0–3·8
Iso-leucine $CH_3CH_2CH(CH_3)CH(NH_2)COOH$		8–13	trace–3	—
Leucine $CH_3CH_2CH_2CH_2CH(NH_2)COOH$		—	—	0·9–3·8
Ornithine $NH_2(CH_2)_3CH(NH_2)COOH$		—	trace–3	0·2–2·4
Methionine sulphoxide $CH_3S(:O)CH_2CH_2CH(NH_2)COOH$				—
Threonine $CH_3CHOHCH(NH_2)COOH$		—	3–8	0·3–1·3
Tyrosine $HOC_6H_4CH_2CH(NH_2)COOH$	<0·5	—	trace–3	tr.–0·5
Phenylalanine $C_6H_5CH_2CH(NH_2)COOH$		—	—	0·1–0·9
Histidine $C_3H_3N_2CH_2CH(NH_2)COOH$		?	trace–3	tr.–2·4
Arginine $NH_2C(:NH)NH(CH_2)_3CH(NH_2)COOH$		?	trace–3	0·1–0·6
Proline C_4H_8NCOOH		?	—	0·3–1·4
Methionine $CH_3SCH_2CH_2CH(NH_2)COOH$		—	trace–3	tr.–0·4
Tryptophan $C_8H_6NCH_2CH(NH_2)COOH$		—	trace–3	—
Glucosamine $C_6H_{13}NO_5$		—	trace–3	—

(*a*) Park *et al*. (1962) (by ion-exchange) Gulf of Mexico
(*b*) Tatsumoto *et al*. (1961) (by paper chromatography) Gulf of Mexico
(*c*) Tatsumoto *et al*. (1961) (by ion-exchange) Gulf of Mexico
(*d*) Degens *et al*. (1964) Pacific off California

TABLE I—*continued*

Name of compound and chemical formula	*Concentration*		*Author(s)*	*Locality*
Free amino acids	(*e*)	(*f*) μg./l.		
Cystine $[SCH_2CH(NH_2)COOH]_2$	det	—	(*e*) Palmork (1963a), b	Norwegian coastal water
Lysine $NH_2(CH_2)_4CH(NH_2)COOH$	det.	0·2– 3·1	(*f*) Degens *et al.* (1964)	Pacific off California
Histidine $C_3H_3N_2CH_2CH(NH_2)COOH$	det.	0·5– 1·7		
Arginine $NH_2C(:NH)NH(CH_2)_3CH(NH_2)COOH$	det.	0·0		
Serine $CH_2OHCH(NH_2)COOH$	det.	2·3–28·4		
Aspartic acid $COOHCH_2CH(NH_2)COOH$	det.	tr.–9·6		
Glycine NH_2CH_2COOH	det.	tr. –37·6		
Hydroxyproline $C_4H_7N(OH)COOH$	det.	tr. – 2·8		
Glutamic acid $COOH(CH_2)_2CH(NH_2)COOH$	det.	1·4– 6·8		
Threonine $CH_3CHOHCH(NH_2)COOH$	det.	2·8–11·8		
α-Alanine $CH_3CH(NH_2)COOH$	det.			
Proline C_4H_8NCOOH	det.	0·0		
Tyrosine $HOC_6H_4CH_2CH(NH_2)COOH$	det.	tr. –5·0		
Tryptophan $C_8H_6NCH_2CH(NH_2)COOH$	det.	—		
Methionine $CH_3SCH_2CH_2CH(NH_2)COOH$	det.	—		
Valine	det.	0·3– 2·7		

$C_6H_5CH_2CH(NH_2)COOH$					
Iso-leucine $CH_3CH_2CH(CH_3)CH(NH_2)COOH$	det.	—			
Leucine $(CH_3)_2CHCH_2CH(NH_2)COOH$	det.	0·5– 5·5			
Free compounds					
Uracil NHCONHCOCH:CH (ring)		det.		Belser (1959, 1963)	Pacific coast near La Jolla
Iso-leucine $CH_3CH_2CH(CH_3)CH(NH_2)COOH$		det.			
Methionine $CH_3SCH_2CH_2CH(NH_2)COOH$		det.			
Histidine $C_3H_3N_2CH_2CH(NH_2)COOH$		det.			
Adenine $C_5H_3N_4NH_2$		det.			
Peptone		det.			
Threonine $CH_3CHOHCH(NH_2)COOH$		det.			
Tryptophan $C_8H_6NCH_2CH(NH_2)COOH$		det.			
Glycine NH_2CH_2COOH		det.			
Purine $C_5H_4N_4$		det.			
Urea CH_4ON_2		det.		Degens *et al.* (1964)	Pacific off California
III. Aliphatic Carboxylic and Hydroxy-carboxylic acids					
	mg./l. (0–200 m.)	mg./l. (200–600 m.)	mg./l. (>600 m.)		
Lauric acid $CH_3(CH_2)_{10}COOH$	0·01–0·32	0·01–0·28	0–0·28	Slowey *et al.* (1962)	Coastal waters of Gulf of Mexico
Myristic acid $CH_3(CH_2)_{12}COOH$	0·01–0·10	0·01–0·05	0–0·07		
Myristoleic acid $CH_3(CH_2)_3CH:CH(CH_2)_7COOH$	traces–0·02	0·01–0·03	0–0·05		

TABLE I—*continued*

Name of compound and chemical formula	*Concentration*			*Author(s)*	*Locality*
Palmitic acid	0·01–0·17	0·03–0·42	0–0·38		
$CH_3(CH_2)_{14}COOH$					
Palmitoleic acid	0·02–0·16	0·02–0·16	0–0·21		
$CH_3(CH_2)_5CH{:}CH(CH_2)_7COOH$					
Stearic acid	0·04–0·09	0·02–0·13	0–0·10		
$CH_3(CH_2)_{16}COOH$					
Oleic acid	0·01	0·02	0		
$CH_3(CH_2)_7CH{:}CH(CH_2)_7COOH$					
Linoleic acid	0·01	0·01	0		
$CH_3(CH_2)_4CH{:}CHCH_2CH{:}CH(CH_2)_7COOH$					
		mg./l. (1000–2500 m.)			
Fatty acids with:					
12 C-atoms		0·0003–0·02		Williams (1961)	Pacific Ocean coastal water
14 C ,,		0·0004–0·043			
16 C ,,		0·0027–0·0209			
16 C ,, + 1 double bond		0·0003–0·003			
18 C ,,		0·0037–0·0222			
18 C ,, + 1 double bond		0·0083			
18 C ,, + 2 double bonds		0·0000–0·0029			
20 C ,,		traces–0·0081			
22 C ,,		traces–0·0014			
		mg./l.			
Acetic acid		<1·0		Koyama and Thompson (1959)	Pacific Ocean
CH_3COOH					
Lactic acid					
$CH_3CH(OH)COOH$					
Glycollic acid					
$HOCH_2COOH$					
Malic acid		0·28		Creac'h (1955)	Atlantic coastal water
$HOOCCH(OH)CH_2COOH$					
Citric acid		0·14			
$HOOCCH_2C(OH)(COOH)CH_2COOH$					

Carotenoids and brownish waxy or fatty matter	2·5	Johnston (1955)	North Sea
		Wilson and Armstrong (1955)	English Channel
IV. Biologically active compounds (see also Provasoli, 1963)			
Organic Fe compound(s)		Harvey (1925)	Deep sea water
Vitamin B_{12} (Cobalamine) $C_{63}H_{90}O_{14}N_{14}PCo$	3·4–1·6 mμg./l.	Vishniak and Riley (1961)	Long Island Sound
Vitamin B_{12}	0·2 mμg./l. (summer) 2·0 mμg./l. (winter)	Cowey (1956)	Oceanic surface-water
Vitamin B_{12}	0·2–5·0 mμg./l.	Daisley and Fisher (1958)	
Vitamin B_{12}	0–2·6 mμg./l.	Kashiwada *et al.* (1957)	North Pacific Ocean
Vitamin B_{12}	0–0·03 mμg./l.	Menzel and Spaeth (1962)	Sargasso Sea 0—05 m.
Thiamine (Vit. B_1) $C_{12}H_{18}ON_4SCl_2$	0–20 mμg./l.	Cowey (1956)	Surface water, possibly from land drainage
Plant hormones (auxins)	3·41 mμg./l.	Bentley (1960)	North Sea near Scotland
V. Humic acids			
"Gelbstoffe" (Yellow substances) Melanoidin-like		Kalle (1949, 1962)	Coastal waters
		Jerlov (1955)	
		Armstrong and Boalch (1961a,b)	
VI. Phenolic compounds			
p-Hydroxy-benzoic acid HOC_6H_5COOH	1–3 μg./l.	Degens *et al.* (1964)	Pacific off California
Vanillic acid $CH_3(HO)C_6H_3COOH$	1–3 μg./l.		
Syringic acid $(CH_3O)_2(HO)C_6H_2COOH$	1–3 μg./l.		
VII. Hydrocarbons			
Pristane; (2, 6, 10, 14-tetramethylpentadecane)	tr.	Blumer *et al.* (1963)	Cape Cod Bay

tr. = trace
det. = detected
— = not detected
? = possibly present

the phenolic structure expected from the prognosis. These findings are not in agreement with those of Kalle (1962). The light absorption measurements at wavelengths of 387–665 mμ showed that phenol-humic acids, which have a brownish colour, seem to be mostly of terrestrial origin, whereas in the sea, purely yellow carbohydrate-humic acids (melanoidines) predominated. Kalle explained the occurrence of melanoidine by the fact that in the sea, humus formation takes place under circumstances of great dilution, and thus leads to the formation of compounds of low molecular weight which exist only in the initial stages of terrestrial humus production. The by-products of melanoidine formation are probably similar to the products of reaction between two methylglyoxal molecules (by double aldol condensation) which yields six-membered ring structures, and may very well be fluorescent. Therefore, the fluorescent substances may represent groups of organic matter different from the yellow substances.

A discussion about the identity of specific dissolved organic compounds would not be complete without referring to the principles of the analytical methods used. As can be seen from Fig. 3, the auxin compounds X and Z seem to be identical with the marker compounds of the determination, namely 3-indolylacetic acid and 3-indolylacetonitrile. This underlines the fact that many reactions used in organic analysis are not specific for particular compounds, but may be given by all compounds having a particular molecular grouping, for example, the anthrone or *N*-ethylcarbazole methods used for pentoses.

When compounds are isolated from sea water, a good many manipulations are necessary before they can be identified by paper-, ion exchange- or gas-chromatography, and it is possible that degradation may occur during these processes. This is especially likely when compounds which are stable in a dilute, almost sterile, marine environment are concentrated by evaporation, adsorption or extraction.

In spite of the difficulty of preventing contamination from chemicals and during manipulation, the results so far obtained provide confidence that the substances found do indeed occur in sea water, and have not arisen from contamination during the analysis. Methods for the determination of the organic components of sea water are reviewed in Chapter 21, Volume 2, Section XI.

B. Elementary Composition

So far we have considered only the possibility of the occurrence of certain specified compounds in the sea, as well as the distribution of the main elements carbon, nitrogen and phosphorus. We must now discuss whether these data are in agreement. If, for instance, we find 1·5 mg.

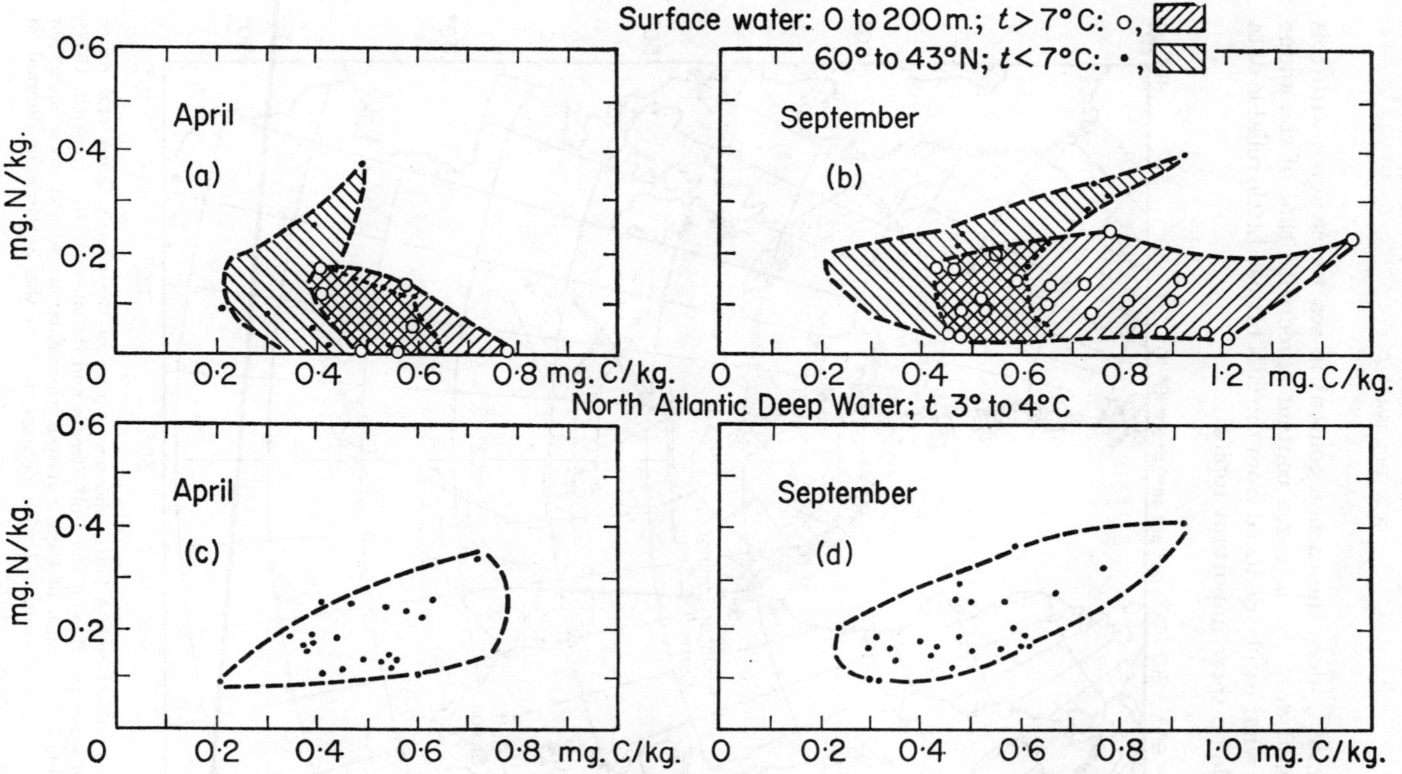

FIG. 4. Carbon–nitrogen relationships for the dissolved organic matter in the North Atlantic section mentioned in Figs. 5, 9(a) and 9(b). (a) and (b) show the results for surface samples between 0 and 200 m. The data for the colder surface waters ($t < 7°$ C) are shown by points, those for the warmer surface waters by circles. (c) and (d) show the data for North Atlantic deep water (cf. Figs. 9(a) and (b)). Reproduced with permission from the Netherlands Journal of Sea Research (Duursma, 1961).

carbohydrate/l., does the carbon content in sea water agree with this analysis, and does the nitrogen content agree with that of the amino acids, and what can be deduced from the C/N ratios? Little reliable data is available on these important topics.

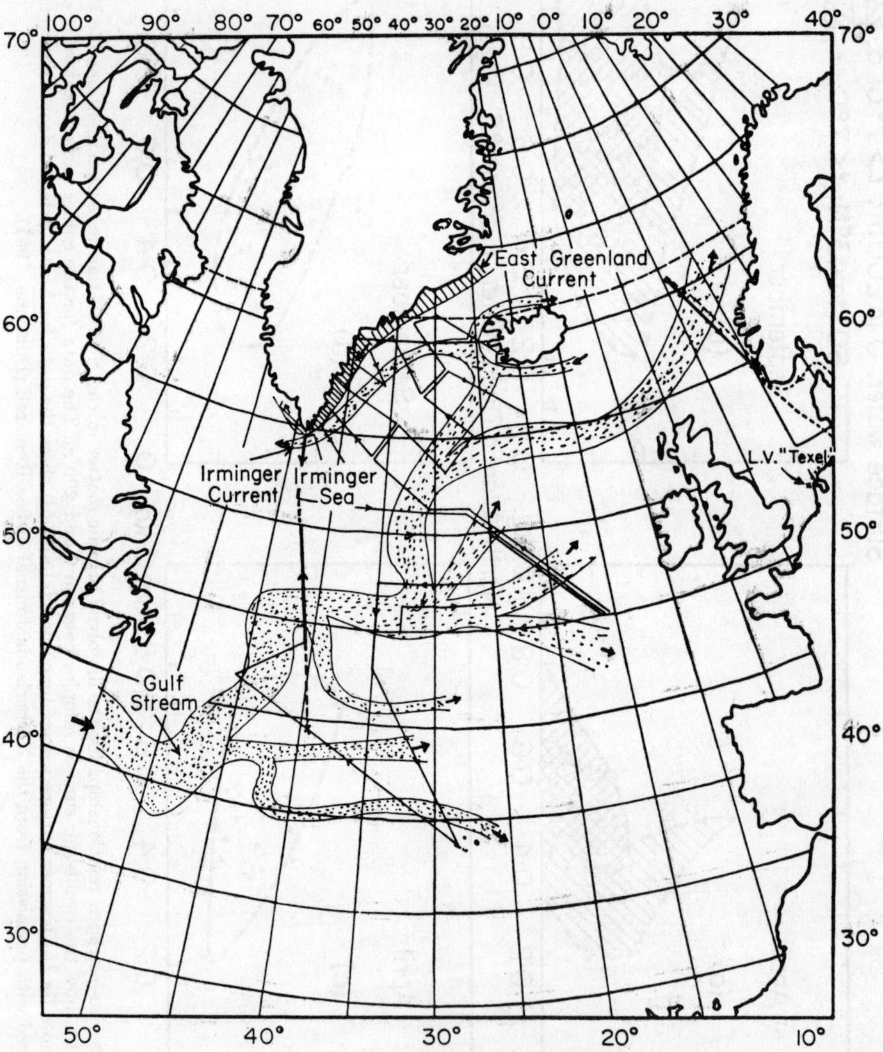

FIG. 5. Map of the North Atlantic showing cruises of the vessels *Gauss* and *Anton Dohrn* in 1958 and the flow of the Gulf Stream and its offshoots (see Dietrich, 1957). The heavy black line south of Cape Farewell represents the section investigated for organic components. Reproduced with permission from the Netherlands Journal of Sea Research (Duursma, 1961).

Most published water analyses for C and N have not been accompanied by attempts to identify the specific organic compounds present. However, it is possible to draw some conclusions about them from the C, N, and P concentrations and their ratios.

By means of the C and N contents and their ratios, it is possible to distinguish three water groups in the Atlantic between 43° and 60°N. (Figs. 4 and 5). In the surface layers of the warmer southern water, the C/N ratio tends to be very high (C contents are high, and N contents are low). This indicates a sequence of events which favours compounds rich in carbon, owing either to accelerated formation of these compounds, or to more intensive breakdown of nitrogenous compounds under the influence of the high water temperatures. Both possibilities are likely to occur. Thus, in these waters, carbohydrates, such as pentoses, or fatty and related acids (glycollic acid? Fogg and Nalewajko, 1963) may be the chief compounds.

The two remaining groups have C/N ratios lying between 2·5:1 and 6·5:1 (by weight). In the cold surface waters the relative amounts of the two elements show a seasonal variation, whereas the C and N contents of the corresponding deep waters show only slight variation.

It is surprising to find such low C/N ratios. By analogy with the organic matter in sediments, in which C/N ratios increase with age from 6:1 (the mean value for living matter) to 12:1 (Bader, 1955), one might expect high C/N ratios at great depths, where "old" water may occur. Since the "old" North Atlantic Deep Water (Fig. 9) and the water of its source area (the Irminger Sea, from which the deep Atlantic derives part of its winter supply) have identical C/N ratios, it is not possible to correlate the C/N ratio of this water with age.

There is, however, some indication that the C/N ratio may increase with time, if the C/N ratios of 6:1 to 8:1 for the deep water of the Norwegian Basin (60°N., 0°E.) are representative of "really old" water values. Probably it depends on what is considered "old" water, and on what is the initial value of the C/N ratio. The Norwegian deep water is assumed to be "old" since even in winter the warm surface water of the Gulf Stream does not mix with it. "New" water must be derived from the deep water of the Polar regions where convection occurs in winter. C/N ratios of between 2:1 and 3:1 have been found for "young" deep water of the Polar region south of Greenland (Duursma, 1961). The rapid flow of this bottom water southward over the Greenland–Iceland Ridge makes the conclusion acceptable (Dietrich, 1957).

Probably the C/N ratio of dissolved marine matter varies with time in contrast to the situation in overlaid terrestrial deposits. It is likely that at first a rather sharp decrease occurs in a matter of days or weeks.

Only after decades or perhaps centuries is this followed by an increase to or above the original value of the living matter. The high carbon contents of the dissolved organic matter (0·6–0·8 mg. C/kg.) relative to that of the sub-Arctic Bottom Water (0·3–0·4 mg. C/kg.), is in favour of this hypothesis. The increase in carbon, rather than the decrease in nitrogen, may be explained as a net gain from the breakdown of sinking organisms, and from decomposition of the dissolved products themselves.

These processes are, however, so slow that it is feasible to use dissolved organic carbon as a conservative property for water masses. Variation in the dissolved organic carbon concentration of deep water can be explained better by mixing of water masses rather than by "ageing" (cf. Fig. 6).

The relationship between the high C/N values and the type of organic compounds present in the water has already been mentioned. It remains to consider low C/N ratios. If, for instance, a C/N ratio of 2·5:1 (by weight) is found, this corresponds to an atomic ratio of 2·15:1. Such low ratios suggest that the waters contain only small amounts of non-nitrogenous compounds, and that the organic nitrogen compounds must have low C/N ratios. Many well-known metabolic products such as urea (C/N = 0·5:1), uracil (C/N = 2:1), creatinine (C/N = 1:1), guanidine (C/N = 1:1) and glycine (C/N = 2:1) may, therefore, be present. Of these compounds, only urea has so far been detected in sea water (Degens *et al.*, 1964). Comparison of the amino–nitrogen, calculated from the total amounts of amino acids present in the sea water (see Table I), with the dissolved organic carbon gives C/N ratios much higher than those normally observed. This suggests that appreciable amounts of other types of nitrogenous compounds are present or that non-nitrogenous organic compounds predominate.

There is yet another factor that must be borne in mind. The accuracy of a ratio is less than that of the individual analyses from which it is derived. With those samples of low C/N values in which the N contents are high rather than the C contents low, the error in the C/N ratio is less than that of samples of high C/N values. With the latter, the N contents are often extremely low. For instance, a sample which contains 0·5 ± 0·03 mg. C/kg. and 0·25 ± 0·06 mg. N/kg. gives a C/N ratio in the range 1·5:1 to 2·8:1. With a sample having a high C/N ratio containing 1·25 ± 0·03 mg. C/kg. and 0·10 ± 0·06 mg. N/kg., the limits of the ratios are 7·6:1 and 32:1. The above values for the precision of carbon and nitrogen determinations are those observed by Duursma (1961).

The C/P ratios of the marine dissolved organic matter are of much less

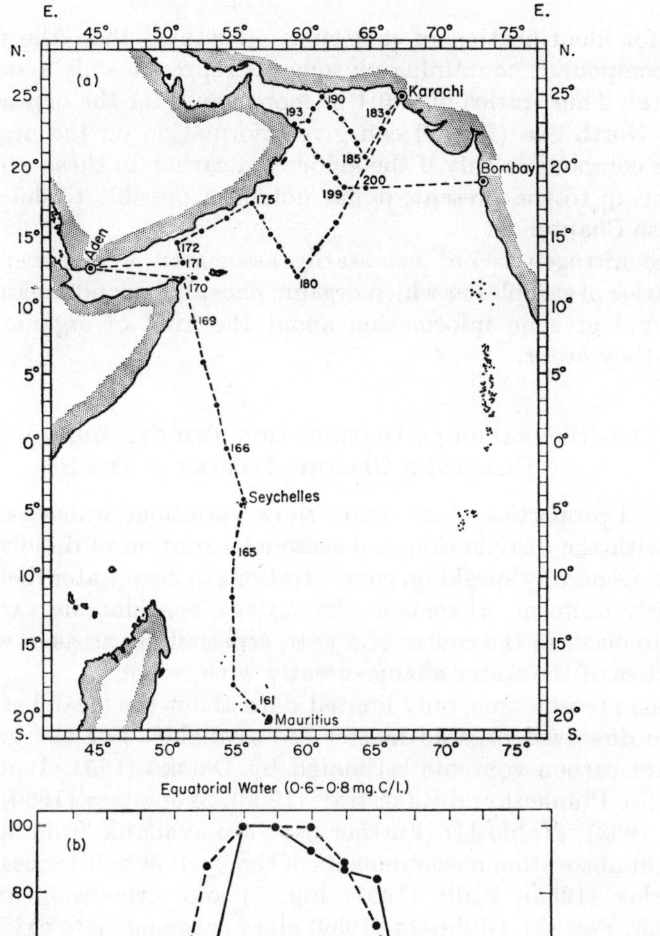

FIG. 6. (a) Track of the R.V. *Anton Bruun* (Cruise 4A, October–November, 1963) showing station locations (Menzel, 1964). (b) Water-mass dilution calculations, based on salinity (solid lines) and dissolved organic carbon (broken lines) at the σ_t level 27·0 (depth 400–800 m.) in the Western Indian Ocean. D is the percentage of the Equatorial Water present at different stations (Menzel, 1964). Reproduced with permission from Deep Sea Research.

value for identification of molecular structures than the C/N ratios, since compounds containing phosphorus represent only a small part of the total. Thus, ratios of 250:1 or more found for the organic material in the North Sea (Fig. 9) can give information on the organic phosphorus compounds only if the amount of carbon in these compounds is known; up to the present, it has not been possible to determine this (see also Chapter 8).

Since nitrogen is not necessarily associated with phosphorus, the N/P ratios of samples in which organic phosphorus and organic nitrogen are found give no information about the kind of organic matter in which they occur.

V. The Geographical Distribution and Seasonal Variation of Dissolved Organic Matter in the Sea

Several properties of sea water show variations which can be correlated with the distribution and seasonal variation of dissolved organic matter. Generally speaking, concentrations in deep water (below 400 m.) are fairly uniform, whereas at the surface, considerable variations are likely to occur in the course of a year, especially in areas in which other properties of the water change greatly with season.

At the present time, only limited data about the local distribution of oceanic dissolved organic matter are available. For the greater part they are carbon contents estimated by Datsko (1951, 1955a,b), Kay (1954a,b), Plunkett and Rakestraw (1955), Skopintsev (1960), Duursma (1961, 1963); (Table II). Further data are available from fluorescence and light absorption measurements of the "yellow substances", as given by Jerlov (1955), Kalle (1957; Fig. 7) and Armstrong and Boalch (1961a,b; Fig. 12). Gillbricht (1959) gives permanganate oxidation data for water from the North Atlantic and the North Sea which, in general, agree with the other results (see Fig. 8). Since in these oxidations it is unlikely that organic carbon is oxidized completely to carbon dioxide, it is difficult to draw a valid conclusion from them (cf. Chapter 20, Volume 2).

Judging from the data available, it would seem that the vertical distribution of organic matter follows a definite pattern. Generally, there is more organic matter in the surface layers than in deeper water, and below a depth of 400–600 m. the concentration variations become very slight. Distribution in the upper layer depends somewhat on the season, and probably also on water movements. For instance, coastal waters near Greenland are poorer in winter and richer in summer than the average deep water of the North Atlantic. In the Gulf Stream at the

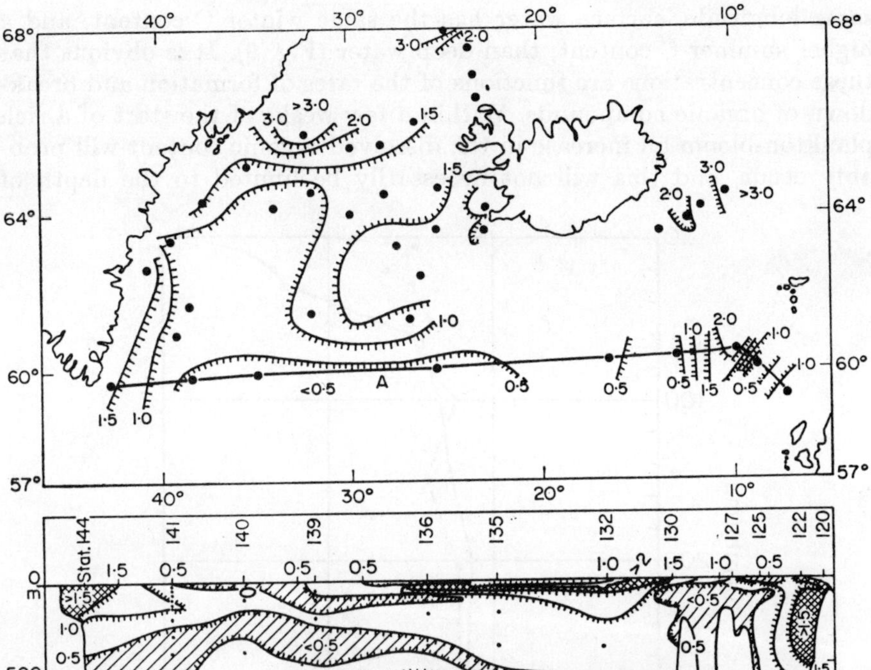

Fig. 7. Fluorescence (mFl) of sea water, caused by organic substances in the surface water of the North Atlantic between Greenland and Scotland (upper) and in the section A of this region (lower), Kalle (1957). Reproduced with permission from Berichte der Deutschen Wissenschaftlichen Kommission fur Meeresforschung.

same longitude, surface water has the same winter C content, and a higher summer C content, than deep water (Fig. 9). It is obvious that these concentrations are functions of the rates of formation and breakdown of organic compounds. Within a few weeks of the start of a rich plankton bloom an increase in the dissolved organic content will probably occur, and this will not necessarily be limited to the depth of

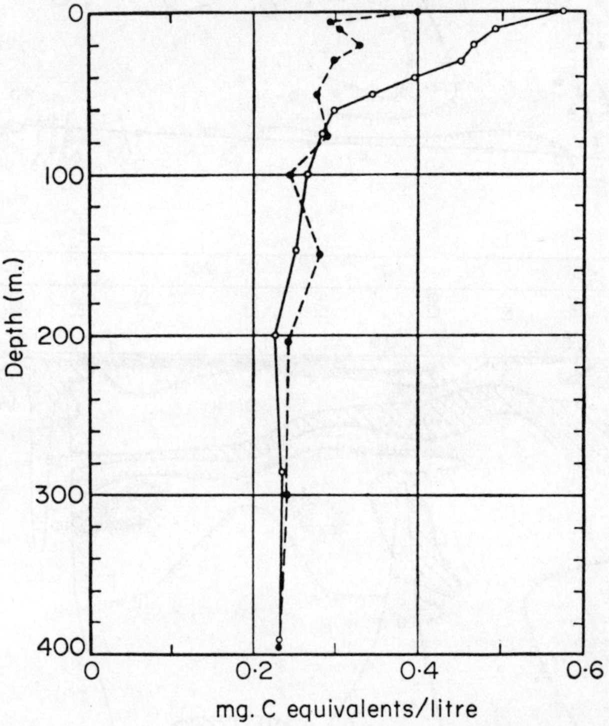

FIG. 8. Average values for the vertical distribution of easily oxidizable dissolved organic matter, expressed in carbon equivalents, in the Irminger Sea in March–April (●—●) and August–September (O—O) 1958 (Gillbricht, 1959). Reproduced with permission from Deutsche Hydrographischen Zeitschrift.

optimal photosynthesis in the euphotic zone, but may occur above or below it. Breakdown processes take some time, and the normal turbulence and diffusion in the surface layers will mix the waters sufficiently to equalize the concentrations.

The organic contents of deeper water of the North Atlantic, like those of the Indian Ocean (see Fig. 6), may often be explained if the origin of the water is known. Water masses south of Greenland can be characterized by their C and N contents (Fig. 9). It appears that deeper waters receive only small supplies of fresh organic matter (e.g. by sinking dead organisms). The rate of further breakdown of this dissolved organic matter is slow. Since the deep waters of the regions mentioned (North Atlantic Deep Water and sub-Arctic Bottom Water) obtain a fresh supply of water only during winter—by convection of cold surface water of the Irminger Sea and Arctic Sea respectively—they have the same carbon content as is found in the source areas in winter.

TABLE II

Bearings of stations shown in Figs. 9(a) and 9(b)

Figure	*Station*	*Date*		*Position*	
9(a)	A.D. 2336	April	11, 1958	59°28′ N.	43°54′ W.
	A.D. 2337		11	59°09′ N.	43°50′ W.
	A.D. 2338		11	58°47′ N.	43°50′ W.
	A.D. 2343		13	56°26′ N.	43°25′ W.
	Gauss 56		15	44°04′ N.	39°26′ W.
	Gauss 83		22	49°11′ N.	40°30′ W.
	Gauss 87		24	51°07′ N.	40°54′ W.
	Gauss 91		25	52°47′ N.	42°18′ W.
	Gauss 95		26	54°59′ N.	42°25′ W.
9(b)	A.D. 2621	September	15, 1958	59°41′ N.	43°56′ W.
	A.D. 2623		15	59°28′ N.	43°54′ W.
	A.D. 2624		15	59°08′ N.	43°57′ W.
	A.D. 2625		16	58°46′ N.	43°50′ W.
	A.D. 2626		16	58°25′ N.	43°47′ W.
	A.D. 2627		16	58°06′ N.	43°42′ W.
	A.D. 2628		16	57°44′ N.	43°38′ W.
	Gauss 169	August	24, 1958	43°55′ N.	38°57′ W.
	Gauss 218	September	8	49°35′ N.	40°05′ W.
	Gauss 220		8	50°09′ N.	40°28′ W.
	Gauss 222		9	50°51′ N.	40°48′ W.
	Gauss 224		9	51°30′ N.	41°14′ W.
	Gauss 226		10	52°06′ N.	41°45′ W.
	Gauss 231		12	54°25′ N.	42°49′ W.

The deep water of the Pacific is older than that of the Atlantic, and it is not known quantitatively whether its carbon content increases in the same way as that of the Norwegian Basin (p. 459). It does seem, however, from the data obtained by Plunkett and Rakestraw (1955), that the deep water of the Pacific contains more dissolved carbon than is present in the deep waters of the Atlantic. Thus, in this region, deep water (below 1000 m.) may be supplied with dissolved organic matter which results from the breakdown of sinking dead organisms.

As noted above, the accumulation of high concentrations of dissolved organic matter in the surface waters is a process requiring some weeks.

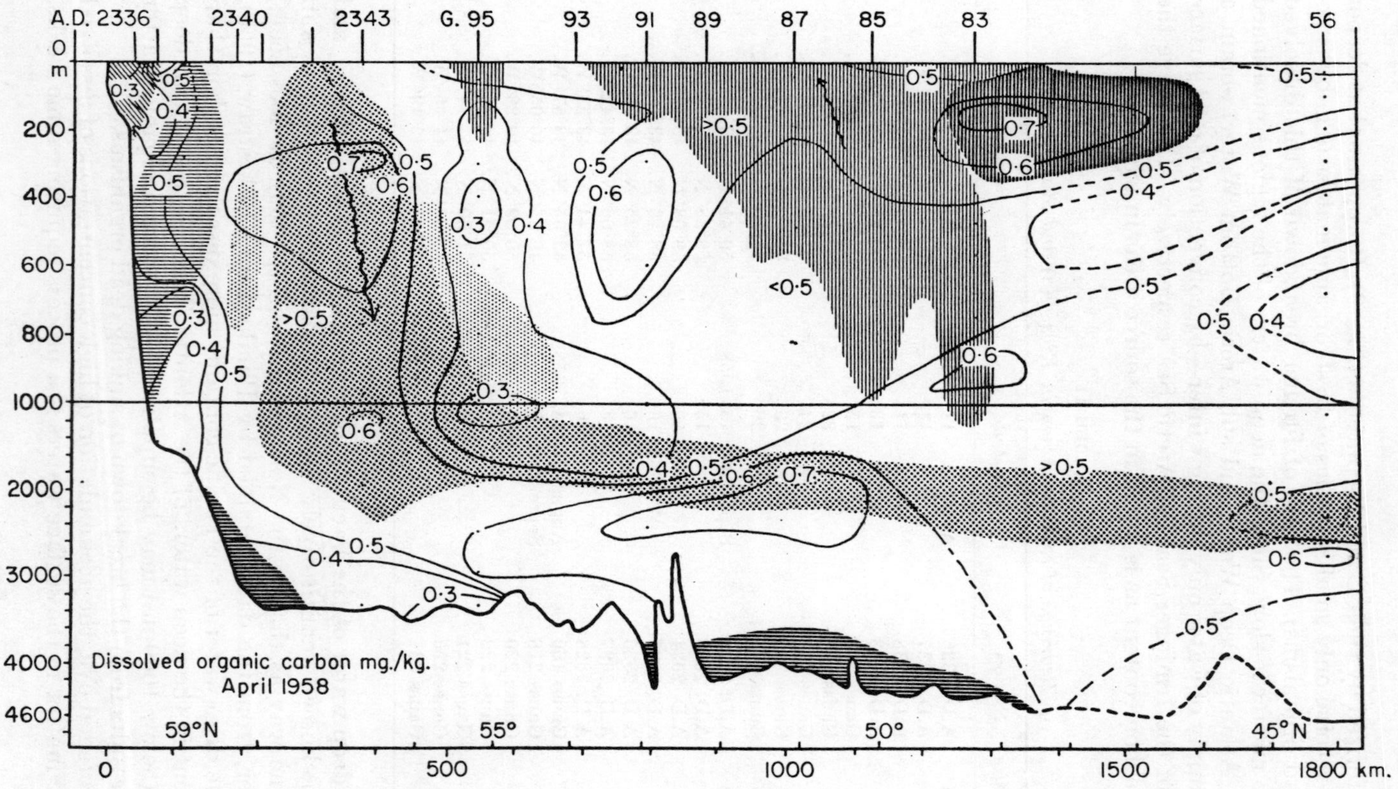

FIG. 9(a). Distribution of dissolved organic carbon in a North Atlantic section (heavy black line in Fig. 5) south of Cape Farewell on Greenland (Duursma, 1961). Stations worked by research vessels *Anton Dohrn* (A.D.) and *Gauss* (G.). For positions of stations see Table II. Reproduced with permission from the Netherland Journal of Sea Research.

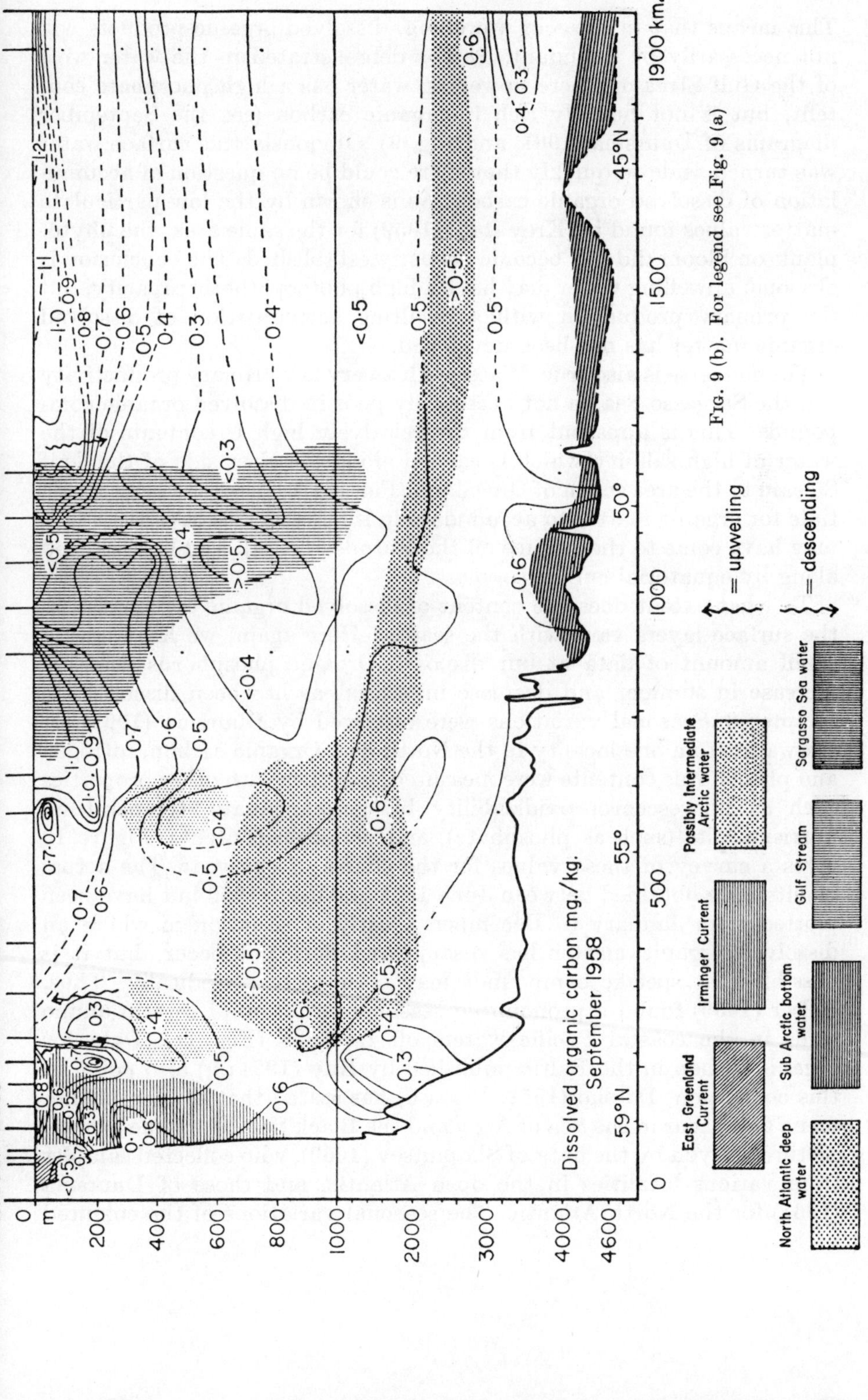

FIG. 9 (b). For legend see Fig. 9 (a)

This means that in areas of upwelling, dissolved organic products will not necessarily be abundant. This is demonstrated in the water west of the Gulf Stream, where upwelling water has a high phosphate content, but is not notably rich in organic carbon (see the September diagrams of Duursma, 1961, and Fig. 9). Obviously this surface water was turned aside so quickly that there could be no question of accumulation of dissolved organic carbon. As is shown by the low particulate matter values found by Krey *et al.* (1959) for the same area, the phytoplankton bloom did not become properly established. The conclusion is obvious: upwelling water may have a high photosynthetic capacity, but the primary production with a resulting accumulation of dissolved organic matter has not been developed.

The converse is also true. Water with a very low primary productivity e.g. the Sargasso Sea, is not necessarily poor in dissolved organic compounds. This is apparent from the relatively high C contents of the water of high salinity which is carried along the East side of the Gulf Stream to the area south of Greenland (Fig. 9(a)). There has been ample time for organic matter to accumulate in this water. Perhaps the water may have come to the surface off the European coast, and been carried along by equatorial currents.

To what extent does the content of dissolved organic compounds in the surface layers vary with the season? Here again, we have only a small amount of data at our disposal. Organic phosphorus contents increase in summer and decrease in winter, as has been discussed in Chapter 8. Seasonal variations were observed by Duursma (1961) for sea water from one locality in the North Sea. Organic carbon, nitrogen and phosphorus contents were measured, together with other properties such as fluorescence, oxidizability by permanganate, temperature, nutrient salts (such as phosphate), and plankton contents. Figure 10 gives a survey of these values for the course of one year. The actual results were obtained between June 1958 and June 1959 but have been plotted from January to December. Clearly, a situation in which all dissolved organic matter has disappeared does not occur, but it is possible that specific compounds may occur only periodically. Thus, Belser (1959) found a pronounced seasonal variation of certain amino acids in the coastal Pacific waters off La Jolla (Fig. 11). Data for organic carbon in the Baltic published by Kay (1954a,b) also point to this conclusion. Datsko (1951, 1955a,b) has stated that similar phenomena also occur in the Sea of Azov and the Black Sea. The same general picture is given by the data of Skopintsev (1960), who collected samples from various localities in the open Atlantic, and those of Duursma (1961) for the North Atlantic. The seasonal variations of the coloured

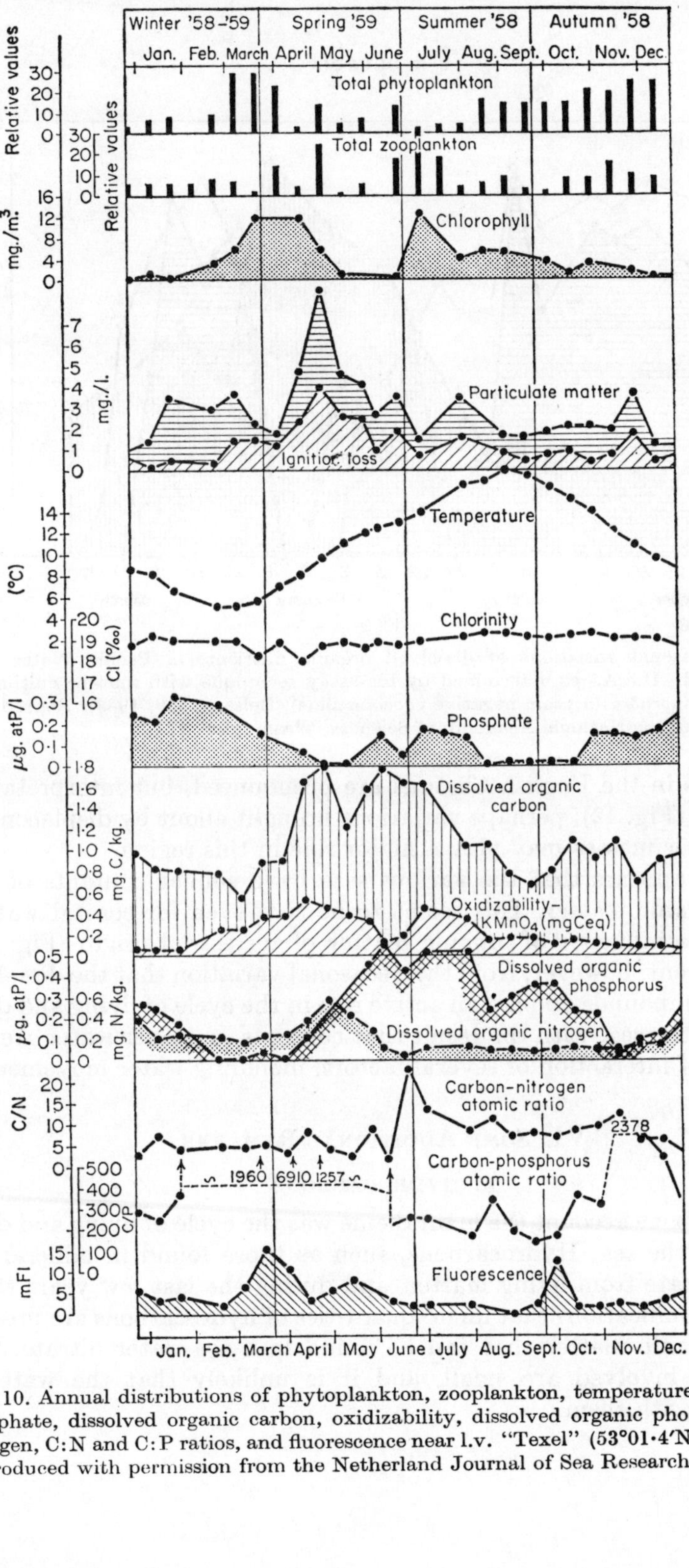

FIG. 10. Annual distributions of phytoplankton, zooplankton, temperature, chlorinity, phosphate, dissolved organic carbon, oxidizability, dissolved organic phosphorus and nitrogen, C:N and C:P ratios, and fluorescence near l.v. "Texel" (53°01·4′N. 4°21·8′E.). Reproduced with permission from the Netherland Journal of Sea Research.

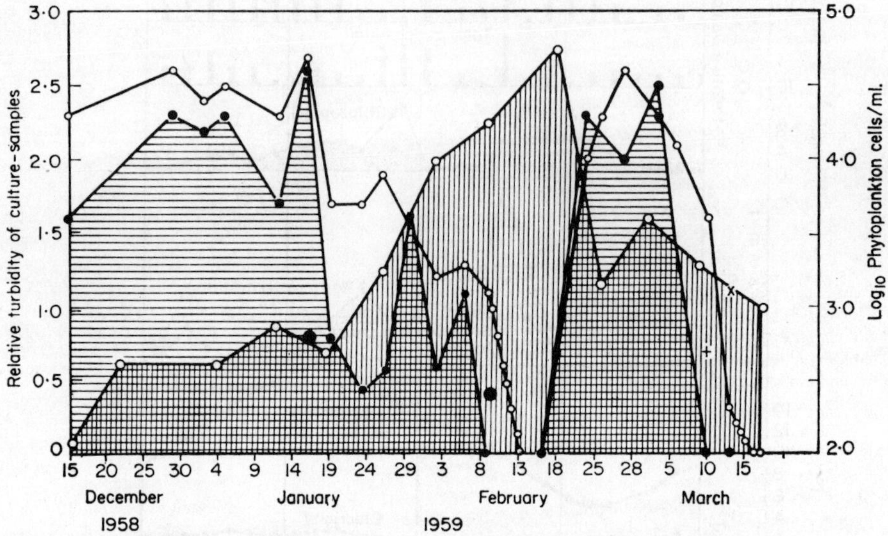

FIG. 11. Seasonal variations of dissolved organic nutrients in Pacific Water near La Jolla, Cal., U.S.A., as determined by bio-assay technique with mutant cultures of *Serratia marinorubra* (a gram-negative coccobacillus), Belser (1959, 1963). Reproduced by courtesy of the National Academy of Sciences, Washington, D.C.

substances in the English Channel are pronounced, but interpretation is difficult (Fig. 12); perhaps variations brought about by displacement of the water masses may play a major role in this region.

It thus appears that the various seas have similar contents of dissolved organic matter. This seems to be true even for coastal waters, despite the influence of organic matter of terrestrial origin (Fig. 13). It is apparent, however, from their seasonal variation that the dissolved organic compounds do play an active role in the cycle of living and dead matter in the sea, with the result that contents in deeper water are the outcome of interaction of several factors, including water movements.

VI. SOME ADDITIONAL REMARKS

A. HYDROCARBONS

In the above account the main theme was the cycle of living and dead matter in the sea. Hydrocarbons, such as those found in mineral oil, may originate from living matter, and during the last few years there have been indications that micro quantities of hydrocarbons are present in the organic matter that can be found in a sea-water filtrate. The quantities involved are small, and it is unlikely that the water is saturated with them.

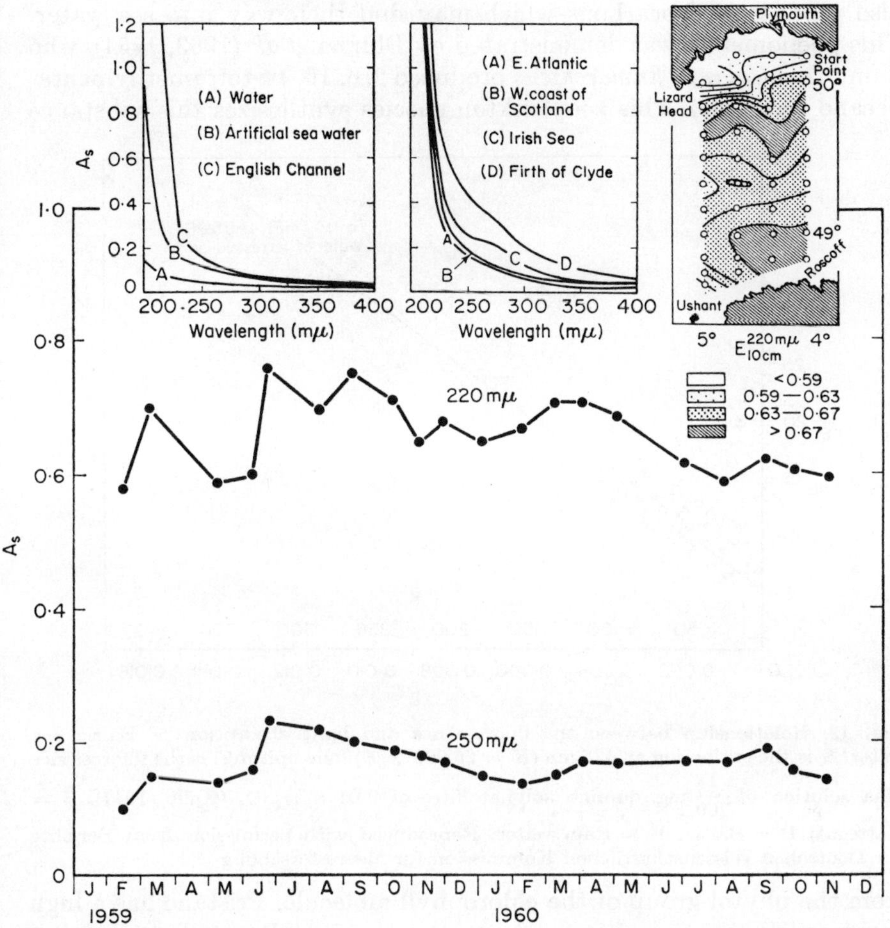

FIG. 12. Seasonal change in light absorbancy of sea water at the International Hydrographic Station E I (50°02′N., 4°22′W.). The greater part of the absorbancy is due to dissolved organic material in the water. Insets: absorbancy as a function of wavelength of waters of various origins, and absorbancy of English Channel waters at 220 mμ. $A_s = \log_{10} \frac{I_0}{I}$ in a 10 cm. cuvette (Armstrong and Boalch, 1961). Reproduced by courtesy of H.M. Stationery Office, London, and the Council of the Marine Biological Association.

The mode of formation of these substances is not fully understood, but it seems likely that the breakdown of the organic matter contained in some marine sediments can lead to their production, and such processes acting on buried sediments for long periods appear to give rise to petroleum hydrocarbons. Certain phytoplankton and zooplankton can also produce hydrocarbons which may find their way into sea water. This phenomenon was demonstrated by Blumer *et al.* (1963, 1964), who found that *Calanus finmarchicus* produced 2, 6, 10, 14-tetramethylpentadecane (pristane). This zooplankton species synthesizes this substance

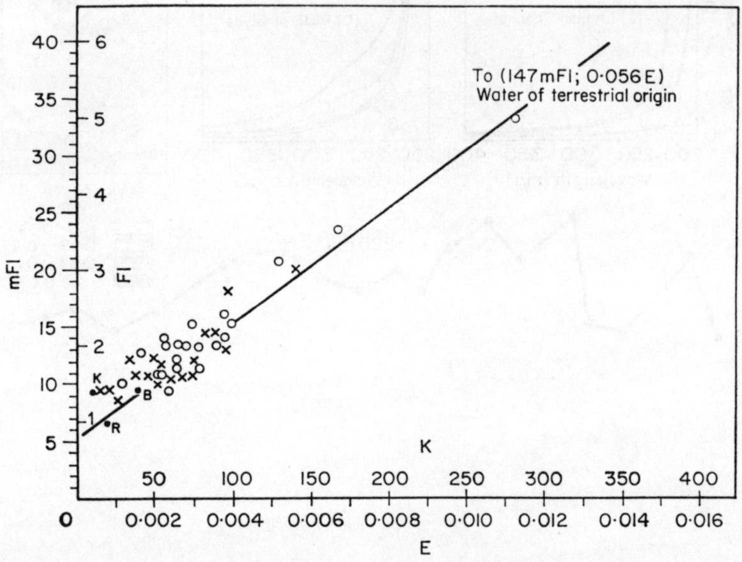

FIG. 13. Relationship between the fluorescence and light absorption of Baltic Sea water. E is the extinction at 425 mμ ($K = 25{,}000 \times E$); one unit mFl is the fluorescence of a solution of $\frac{1}{700}$ mg. quinine sulphate/litre of 0·01 N H_2SO_4 (Kalle, 1947). K = Kattegat; B = Baltic; R = Rain water. Reproduced with permission from Berichte der Deutschen Wissenschaftlichen Kommission fur Meeresforschung.

from the phytol-group of the chlorophyll molecule. Pristane has a high resistance to decomposition and can, therefore, be detected at all stages in the marine food cycle.

B. SOLUBILITY OF ORGANIC COMPOUNDS

In the foregoing discussion we have not considered whether the organic compounds determined are in true solution, or are colloidal. Datsko (1955a,b) has attempted to differentiate between these two possibilities by adsorbing colloidal compounds on freshly precipitated aluminium hydroxide. However, it is probable that many organic compounds which occur in true solution in sea water are also adsorbed on iron and aluminium hydroxides and, in fact, Tatsumoto *et al.* (1961) have been able to obtain high recoveries of dissolved organic matter (polypeptides and free amino acids) through their co-precipitation with ferric hydroxide. It is, therefore, preferable to characterize as colloidal, those organic compounds which are retained by fine membrane filters (see Chapter 21, Volume 2).

It is known (Skopintsev, 1947; Postma and Kalle, 1955) that the high concentrations of colloidal organic matter discharged into the sea by rivers may be coagulated within a short distance of the coast. In the open sea, however, the extreme dilution of organic matter makes the coagulation process less likely. The question of whether the organic material in the sea is colloidal or dissolved is probably not important, but this does not discharge the investigator from defining the term used. Tentatively, dissolved matter can be defined as possessing the following properties:

(1) Not precipitable at pH = 1, within 24 hours in a half-litre bottle, (cf. Duursma, 1961) (only for low particulate matter concentrations);

(2) not retained by a membrane filter;

(3) not readily separated by centrifugation;

(4) not co-precipitate with $Al(OH)_3$, cf. Datsko (1955a,b).

As far as can be judged all the data used in this chapter fall within the first three categories.

REFERENCES

Armstrong, F. A. J., and Boalch, G. T. (1961a). *Nature, Lond.* **192**, 858.

Armstrong, F. A. J., and Boalch, G. T. (1961b). *J. Mar. biol. Ass. U.K.* **41**, 591.

Bader, R. G. (1955). *Geochim. et cosmoch. Acta*, **7**, 205.

Bader, R. G. (1956). *Deep Sea Res.* **4**, 15.

Belser, W. L. (1959). *Proc. nat. Acad. Sci., Wash.* **45**, 1533.

Belser, W. L. (1963). *In* "The Sea" (M. N. Hill, ed.), Vol. **II**, pp. 220–231. Interscience Publishers, New York.

Bentley, Joyce A. (1960). *J. Mar. biol. Ass. U.K.* **39**, 433.

Blumer, M., Mullin, M. M., and Thomas, D. W. (1963). *Science*, **140**, 974.

Blumer, M., Mullin, M. M., and Thomas, D. W. (1964). Helgoländ Wiss. Meeresunters. In the press.

Collier, A. (1956). *Spec. sci. Rep. U.S. Fish. Wildl.* **178**, 7.

Collier, A., Ray, S. M., and Magnitzky, A. W. (1950). *Science*, **111**, 151.

Collier, A., Ray, S. M., Magnitzky, A. W., and Bell, J. O. (1953). *Spec. sci. Rep. U.S. Fish Wildl. Fish. Bull.* **54**, 167.

Cowey, C. B. (1956). *J. Mar. biol. Ass. U.K.* **35**, 609.

Creac'h, P. (1955). *C.R. Acad. Sci., Paris*, **240**, 2551.

Daisley, K. W., and Fisher, L. R. (1958). *J. Mar. biol. Ass. U.K.* **37**, 683.

Datsko, V. G. (1939). *C.R. Acad. Sci. U.R.S.S.* **24**, 294.

Datsko, V. G. (1951). *C.R. Acad. Sci. U.R.S.S.* **77**, 1059.

Datsko, V. G. (1955a). *Trud. Azov–Black Sea sci. Inst. f. Seafishery and Oceanography. VNIRO*, **16**, 25.

Datsko, V. G. (1955b). *Trud. Azov–Black Sea sci. Inst. f. Seafishery and Oceanography, VNIRO*, **16**, 121.
Degens, E. T., Reuter, J. H., and Shaw, K. N. F. (1964). *Geochim. et cosmoch. Acta*, **28**, 45.
Dietrich, G. (1957). *Dtsch. Hydrogr. Z.* **10**, 39.
Duursma, E. K. (1961). *Netherl. J. Sea Res.* **1**, 1.
Duursma, E. K. (1963). *Netherl. J. Sea Res.* **2**, 85.
Enders, C., and Theis, K. (1938). *Brenns. Chemie*, **19**, 360, 404, 439.
Enders, C., Tschapek, M., and Glawe, R. (1948). *Kolloidzschr.* **110**, 240.
Fieser, L. S., and Fieser, Mary (1956). "Organic Chemistry", 1112 pp. Reinhold Publishing Co., New York.
Fogg, G. E. (1958). *Rapp. Cons. Explor. Mer*, **144**, 56.
Fogg, G. E. (1959). *New Biol.* **29**, 31.
Fogg, G. E., and Nalewajko, C. (1963). *Proc. roy. Soc.* **B 157**, 381.
Freudenberg, K. (1956). *Angew. Chem.* **68**, 84, 508.
Gillbricht, M. (1959). *Dtsch. Hydrogr. Z. Ergänz. Reihe B, No. 3*, 90.
Guillard, R. R. L., and Wangersky, P. J. (1958). *Limnol. Oceanogr.* **3**, 449.
Hagmeier, E. (1962). *Kieler Meeresforsch.* **18**, 189.
Harvey, H. W. (1925). *J. Mar. biol. Ass. U.K.* **13**, 953.
Harvey, H. W. (1948). *J. Mar. biol. Ass. U.K.* **27**, 337.
Harvey, H. W. (1955). "The chemistry and fertility of sea waters", 224 pp. Cambridge University Press, Cambridge.
Hood, D. W. (1963). *In* "Oceanography and Marine Biology; an annual review (H. Barnes, ed.), Vol. **I**, p. 478. George Allen & Unwin Ltd., London.
Jeffrey, L. M., and Hood, D. W. (1958). *J. Mar. Res.* **17**, 247.
Jerlov, N. G. (1955). *Göteborgs Vetensk Samh. Handl. F.*6. **B. 6**. (14).
Johnston, R. (1955). *J. Mar. biol. Ass. U.K.* **34**, 185.
Johnston, R. (1963). *J. Mar. biol. Ass. U.K.* **43**, 427.
Kalle, K. (1933). *Ber. Dtsch. Komm. Meeresforsch.* **6**, 273.
Kalle, K. (1949). *Dtsch. Hydrogr. Z.* **2**, 117.
Kalle, K. (1957). *Ber. Dtsch. Komm. Meeresforsch.* **14**, 313.
Kalle, K. (1962). *Kieler Meeresforsch.* **18**, 128.
Karrer, P. (1959). "Lehrbuch der organischen Chemie," 1058 pp. G. Thieme, Stuttgart.
Kashiwada, K., Kakimoto, D., Morita, T., Kanazawa, A., and Kawagoe, K. (1957). *Bull. Jap. Soc. sci. Fish.* **22**, 637.
Kay, H. (1954a). *Kieler Meeresforsch.* **10**, 26.
Kay, H. (1954b). *Kieler Meeresforsch.* **10**, 202.
Koyama, T., and Thompson, T. G. (1959). *Preprints Int. Oceanogr. Congr.* 1959, 925.
Kreps, E. (1934). *In* "James Johnstone Memorial Volume" (A. J. Daniel, ed.), pp. 193–202. Liverpool University Press, Liverpool.
Krey, J., Hantschmann, D., and Wellershaus, St. (1959). *Dtsch. Hydrogr. Z. Ergänz. Reihe B, No.* **3**, 73.
Krogh, A. (1930). *Z. allg. Physiol.* **12**, 668.
Krogh, A. (1931). *Biol. Rev.* **6**, 412.
Krogh, A. (1934a,b). *Ecol. Monogr.* **4**, 421, 430.
Krogh, A., and Keys, A. (1934). *Biol. Bull., Woods Hole*, **67**, 132.
Krogh, A., and Lange, E. (1930). *Biochem. J.* **24**, 1666.
Lewin, R. A. (1956). *Canad. J. Microbiol.* **2**, 665.

Lewis, G. J., and Rakestraw, N. W. (1955). *J. Mar. Res.* **14**, 253.
Mallette, M. F., Althouse, P. M., and Clagett, C. O. (1960). "Biochemistry of plants and animals", 552 pp. John Wiley & Sons, London.
Menzel, D. W. (1964). *Deep Sea Res.* In the press.
Menzel, D. W., and Spaeth, J. P. (1962). *Limnol. Oceanogr.* **7**, 151.
Moore, B., Edie, E. S., Whitley, E., and Dakin, W. J. (1912). *Biochem. J.* **6**, 255.
Natterer, K. (1892–1894). *Denkschr. Akad. Wiss. Wien*, **59**, 1e Reihe, 83; **60**, 2e Reihe, 49; **61**, 3e Reihe, 23.
Palmork, K. H. (1963a). *Acta Chem. scand.* **17**, 1456.
Palmork, K. H. (1963b). *Rept. Norweg. Fish. Invest.* **13**, 120.
Park, K., Williams, W. T., Prescott, J. M., and Hood, D. W. (1962). *Science*, **138**, 531.
Plunkett, M. A., and Rakestraw, N. W. (1955). *Deep Sea Res.* **3** (Suppl.), 12.
Postma, H., and Kalle, K. (1955). *Dtsch. Hydrogr. Z.* **8**, 137.
Provasoli, L. (1963). *In* "The Sea" (M. N. Hill, ed.), Vol. **II**, pp. 165–219. Interscience Publishers, New York.
Pütter, A. (1909). "Die Ernährung der Wassertiere und der Stoffhaushalt der Gewasser,". J. Fischer, Jena.
Raben, E. (1910). *Wiss. Meeresuntersuch.* **11**, 111.
Rippel-Baldes, A. (1955). "Grundriss der Mikrobiologie", 418 pp, Springer, Verlag, Berlin.
Scheffer, F., and Ulrich, B. (1960). "Lehrbuch der Agrikulturchemie und Bodenkunde", Teil III, 266 pp. F. Enke, Stuttgart.
Schmalfuss, K. (1940). *Bodenk. u. Pfl. Ernähr. N.F.* **21/22**, 442.
Seiwell, H. R., and Seiwell, F. E. (1934). *Ann. Hydrogr., Berl.* **62**, 7.
Shapiro, J. (1957). *Limnol. Oceanogr.* **2**, 161.
Skopintsev, B. A. (1947). *C.R. Acad. Sci. U.R.S.S., Investiga*, **11**, 22.
Skopintsev, B. A. (1959). *Preprints Int. Oceanogr. Congr.* 1959, 953.
Skopintsev, B. A. (1960). *Trans. Inst. Marine Hydrophysics, Acad. Sci. U.R.S.S.* **29**, 3.
Slowey, J. F., Jeffrey, L. M., and Hood, D. W. (1962). *Geochim. et cosmoch. Acta*, **26**, 607.
Tatsumoto, M., Williams, W. T., Prescott, J. M., and Hood, D. W. (1961). *J. Mar. Res.* **19**, 89.
Thiele, H., and Kettner, H. (1953). *Kolloid zschr*, **130**, 131.
Tracey, M. V. (1948). "Proteins and life", 154 pp. The Pilot Press Ltd., London.
Vallentyne, J. R. (1957). *J. Fish. Res. Bd Can.* **14**, 33.
Vishniac, H. S., and Riley, G. A. (1961). *Limnol. Oceanogr.* **6**, 36.
Waksman, S. A. (1952). "Soil Microbiology", 356 pp. Wiley, London and New York.
Wangersky, P. J. (1952). *Science*, **115**, 685.
Wanklyn, J. A., Chapman, E. T., and Smith, M. H. (1867). *J. Chem. Soc.* **20**, 445.
Williams, P. M. (1961). *Nature, Lond.* **189**, 219.
Wilson, D. P., and Armstrong, F. A. J. (1952). *J. Mar. biol. Ass. U.K.* **31**, 335.
ZoBell, C. E. (1954). *Int. Un. Biol. Sci.*, Series B. No. 16, 20.

Chapter 12

Production of Organic Matter in the Primary Stages of the Marine Food Chain

J. D. H. STRICKLAND

Institute of Marine Resources
University of California, La Jolla, California, U.S.A.

I. Introduction 478
II. Definitions 479
A. Plankton Terminology 479
B. Productivity Terminology 481
III. Phytoplankton Species 483
A. Cyanophyta 484
B. Chlorophyta 484
C. Pyrrophyta 485
D. Chrysophyta 485
E. Cryptophyceae 486
IV. Chemical Composition of Phytoplankton and Methods for Measuring Plant Bio-mass 487
A. Pigments 487
B. Elementary Composition 502
C. Major Metabolites 506
D. Practical Considerations, Composition Ratios 514
V. Effect of Light, Temperature and Nutrition on Phytoplankton Growth and Composition 518
A. Methods of Measuring Photosynthesis and their Interpretation 519
B. The Effect of Light and Temperature on Photosynthesis 524
C. The Effects of Major and Minor Nutrients on Photosynthesis 536
D. Organic Sources of Nutrition, Heterotrophy and Nitrogen Fixation 542
E. Cell Division Rates and Cellular Composition 545
VI. Field Measurements of Primary Production 551
A. Sampling Techniques 551
B. Time of Sampling and Duration of Experiments 552
C. *In situ* Production and its Prediction from Experiments using Light-incubators 554
D. *In situ* Production and its Prediction from Changes in the Chemical Composition of Natural Waters 560
VII. The Growth and Distribution of Phytoplankton in Nature 561
A. Vertical Distribution 561
B. Geographical Distribution 568
C. Temporal Distribution 576
VIII. Phytoplankton Utilization 585
A. Direct Food Supply 585
B. Plant and Animal Detritus 589
C. Phytoplankton as a Source of Dissolved Organic Matter for Heterotrophs 590
References 595

I. Introduction

The chemistry of marine food production is as far ranging as the science of biochemistry itself. Even if we restrict our consideration to the initial stage in the marine food chain, the production of sea plants, a comprehensive treatise would be too lengthy to have a place in the present work. It is necessary, therefore, to be selective, and the selection made here is largely a personal one based on the author's main interests and experience.

The production of the large attached (benthic) algae is of considerable importance in the ecology of certain shallow areas, but by far the greatest product of aquatic photosynthesis (and probably of photosynthesis on our planet as a whole) is the group of microscopic organisms known collectively as the *phytoplankton*. The following account is restricted almost exclusively to the marine representatives of the phytoplankton, although some reference to freshwater species will be necessary since the bulk of the work on unicellular plant physiology has been restricted to such organisms.

The presence of microscopic living material in sea water, at the concentrations normally encountered, has little effect on its physical properties, other than the attenuation of visible and near-visible radiant energy. Considerable changes may be brought about, however, in the distribution of the minor chemical constituents of the oceans. These changes are significant in studies of physical oceanography, geochemistry and atomic waste disposal. Such processes and the chemistry of the remineralization of marine organisms are described elsewhere (cf. Chapters 8 and 9, and Chapter 22, Volume 2).

Plant production by photosynthesis in the sea is of supreme importance because it initiates the whole marine food chain, which terminates in the larger fishes and sea mammals. The present and future importance of understanding the processes involved does not need to be stressed, and therefore brief mention is made in this chapter of the utilization of organic matter in the early stages of the marine food web.

The chemistry of plant growth and nutrient utilization cannot be discussed without a knowledge of the physiology of phytoplankton and of the effect of physical factors such as light and temperature on their growth.

Some knowledge of algal taxonomy is also essential when describing the chemistry of sea plants. Although this subject of algal taxonomy is clearly beyond the scope of this chapter, an outline of the main classifications of marine phytoplankters has been included for the benefit of the reader who may be unfamiliar with this field.

Finally, the terminology used by those studying marine photosynthesis may be unfamiliar to some readers. It is, at best, ponderous, and precise definitions are often lacking. A brief and self-consistent terminology and a set of definitions are therefore included which, although in no way authoritative, should prove useful in the context of the work under discussion. Many reviews of marine productivity have appeared. The little classic by Harvey (1957) still provides an excellent elementary introduction. Recently, Nielsen (1964) has reviewed primary production, and a comprehensive treatise on the marine plankton, with particular reference to research carried out in Europe and the United Kingdom, is to be found in the text by Raymont (1963). The present chapter attempts to give a comprehensive introduction to the more important literature up to early 1964. The emphasis, however, is on the chemistry and physiology underlying primary production rather than on descriptive work which is covered very adequately by the above mentioned texts, especially that by Raymont.

II. Definitions

A. Plankton Terminology

(1) *The plankton* (*individually, plankters*). These are organisms suspended in water, without mobility or with very limited mobility, which cannot or do not maintain their distribution against the effects of local water movements. The terms *phytoplankton* and *zooplankton* are used for the plant and animal subdivisions of these organisms.

(2) *The nekton.* These are organisms capable of sufficient mobility to maintain their distribution against local water movements. The definitions of plankton and nekton are not sufficiently precise for the unambiguous classification of all organisms under certain conditions, but in general the terminology is adequate.

(3) *Detritus.* This is suspended matter of organic origin, permanently incapable of reproduction. The prefix phytoplankton, zooplankton or nekton can be used to indicate the exact origin of the detritus. Suspended mineral matter is best described as such, and not termed detritus, although organic detritus may often "collect" a considerable amount of inorganic material.

(4) *Terminology used to describe the location of plankton.* The *euphotic zone* is the surface layer of the ocean in which, under the conditions stated, the net rate of photosynthesis of phytoplankton is positive. The depth of this zone is governed mainly by the mean amount of solar radiation and the optical properties of the water. The terms *thermocline* and *halocline* are sometimes relevant to the location of plankton and

refer respectively to the planes in a water column where the temperature and salt content of the water have a maximum rate of change with depth. Temperature generally decreases with depth, and salt content increases, both leading to an increase of density (to give one or more *pycnoclines*) and presenting a barrier to vertical mixing processes. The fairly turbulent water above a pycnocline is known as the *epilimnion*, and the lower, undisturbed waters as the *hypolimnion* when describing lakes.

The term *oceanic* is used for plankton located outside coastal waters and *neritic* for the plankton of coastal areas. It is not clear whether the generally accepted definition of the oceanic province (viz. that area where the water depth is greater than about 200 m.) applies too well when discussing phytoplankton. Metabolites characteristic of inshore waters may be transported well into the "oceanic" area if there is little or no continental shelf, and if offshore currents are strong. From the standpoint of primary productivity, a truly oceanic area is generally at least 100 miles from the nearest part of the continental shelf.

The terms *meroplankton* and *holoplankton* are used to indicate respectively phytoplankters which do and which do not form spores or resting stages in their life cycles. Many of the oceanic plankters are holoplankton, and many neritic plankters are meroplankton, but there are notable exceptions and the terms are not synonymous.

The terms *allochthonous* and *autochthonous* are sometimes used to denote respectively plankton which have developed, or originated outside or within the area under study, i.e. exotic and endemic populations.

(5) *Terminology used to describe the abundance of plankton.* Several rather ill-defined terms, used mainly in limnology, will classify waters according to their content of living matter and biotic potential. Only the terms *eutrophic* and *oligotrophic* are used in this chapter to describe marine conditions, and these indicate respectively the extremes of plankton abundance from very fertile areas (generally coastal waters with pronounced upwelling) to sparsely populated waters (characterized by the tropical oceans).

(6) *Terminology used in the size classification of phytoplankton.* For phytoplankters the terms *microplankton*, *nannoplankton* and *ultraplankton* are all that need be used, and these indicate a size distribution of organisms approximately in the ranges 50–500 μ, 10–50 μ and 0·5–10 μ, respectively, for the largest dimension.

(7) *Terminology used to describe the nutritional requirements of phytoplankters.* The term *autotrophic* is reserved for those phytoplankters that require no organic material whatever for their normal growth and reproduction. *Auxotrophic* organisms can obtain most of their growth

requirements from inorganic sources, but have definite requirements for certain organic molecules, usually in trace amounts, and without these they cannot grow or reproduce in a normal manner. *Heterotrophic* plants may obtain their constituent carbon, and perhaps also the nitrogen, sulphur and phosphorus they need, exclusively from organic sources. Many algae may be heterotrophs in the dark, but behave mainly as autotrophs in the light.

B. PRODUCTIVITY TERMINOLOGY

(1) *Photosynthetic rate.* The rate of photosynthesis can be defined in at least three ways, according to the method of measurement used.

(a). The rate of increase of dry organic matter in plants *that results from photosynthetic activity.* This definition is rarely employed in work with algae, as it is not often practicable to measure such a weight increase, or to decide the fractions of weight increase arising from photosynthesis and from heterotrophic growth.

(b). The rate of increase of the amount of oxygen in the extra-cellular environment of plants *that results from photosynthetic activity.* This rate of increase is not necessarily the same rate as that measured in (a) above, even on a molecular basis, as the plant material may not liberate all photosynthetically formed oxygen immediately, and some uptake of carbon dioxide may occur without simultaneous oxygen liberation. However, for growth periods exceeding a few minutes, the two measurements will be proportional to each other and related by a suitable empirical factor.

(c). The rate of decrease of the amount of carbon dioxide in the extra-cellular environment of plants *that results from photosynthetic activity.* This rate of decrease is not necessarily the same rate as in (a) or (b), as the carbon dioxide taken in by a photosynthesizing system may not all be used at once, or may produce organic compounds of transient stability. However, for periods exceeding a few minutes, provided that a "dark carbon dioxide uptake" blank is deducted, the measurements will be related to one another by a proportionality constant.

The terms *gross* and *net* are used in conjunction with photosynthesis to distinguish between the true synthesis of organic matter resulting from exposure to light (gross), and the net formation of organic matter that is found after allowance has been made for the respiration and other losses that occur simultaneously with the photosynthetic processes.

(2) *Photosynthetic quotient* (*PQ*) *and respiratory quotient* (*RQ*). These quotients are dimensionless numbers, and indicate the relative amounts of oxygen and carbon dioxide involved in the processes of photosynthesis and respiration.

Photosynthetic quotient

$$= \text{PQ} = \frac{\text{Molecules of oxygen liberated during photosynthesis}}{\text{Molecules of carbon dioxide assimilated}}$$

$$= \frac{+\Delta O_2}{-\Delta CO_2}$$

Respiratory quotient

$$= \text{RQ} = \frac{\text{Molecules of carbon dioxide liberated during respiration}}{\text{Molecules of oxygen assimilated}}$$

$$= \frac{+\Delta CO_2}{-\Delta O_2}$$

(3) *Unicellular algal growth.* At *any instant*, the increase in plant material in a system containing phytoplankton will be given by:

$$\frac{dp}{dt} = kp$$

where p is the amount of material; k may be negative.

If k is finite, positive and constant for a significant period, the cell system is said to be in a state of *exponential* growth and k may be obtained experimentally from the integrated form of this equation. It is important to note that whereas k always has the dimension of $(\text{time})^{-1}$, and is best expressed in $(\text{hours})^{-1}$ for most phytoplankton work, its value will depend on the property of the system which is being measured (dry weight, carbon, protein, etc.), and only if the *composition of the growing cells remains unchanged* will k have a common value for all properties.

Most phytoplankters reproduce by simple cell division or by the simultaneous release of a few daughter cells from a parent cell. It is sometimes possible to induce all cells to reproduce in a culture within a relatively short time to give a *synchronous culture*, but generally cell division takes place more or less randomly. A mean value for k in the expression:

$$\frac{dn}{dt} = kn$$

where n is the number of cells per unit volume, can be obtained by integration over any (say) 24-hour period, and the mean division time or doubling time (t_d) for cell over this period is given by:

$$t_d = \frac{0{\cdot}69}{k}$$

or

$$t_d = \frac{0{\cdot}30}{k^1}$$

if k and k^1 are derived from natural and base 10 logarithms respectively.

(4). *Primary product or primary standing crop* (*dimensions:* $M.L^{-3}$ *or* $M.L^{-2}$). This is the *instantaneous value of the amount of living plant material present in water.* It will be expressed in this chapter as organic carbon present in unit volume or beneath a unit area:

$$\text{mg.C/m.}^3 \text{ and mg.C/m.}^2$$

The term standing *stock* is perhaps preferable to standing crop, which implies harvesting, but the latter is now in common usage.

(5). *Net primary productivity* (*dimensions:* $M.L^{-3}.T^{-1}$ *or* $M.L^{-2}.T^{-1}$) is the *net rate of synthesis of the organic constituents of plant material in water.* This is again generally best measured on the basis of organic carbon per unit volume or beneath unit area of sea surface, with the time scale chosen to suit a particular context. The definition is limited to the primary productivity of plant material and, by virtue of the methods used in its determination, generally refers to productivity from inorganic sources of carbon. The measurement does *not* include the rate of loss of primary material by death or predation. The units that will be used in this chapter are:

$$(\text{mg.C/m.}^3)/\text{hour; and } (\text{g.C/m.}^2)/\text{day.}$$

The term *primary production* is perhaps preferable, but both *production* and *productivity* are in common usage to describe this concept. Production can be used for the actual rate in nature and productivity for the maximum possible rate in the same water mass under ideal conditions of illumination and hydrography (Davis, 1963).

(6) *Gross primary productivity*. The definition and units are the same as those used above for net productivity, except that the rate under consideration is the *gross* rate of the photosynthetic increase in plant crop before any corrections for respiration or excretion are made.

III. Phytoplankton Species

There is still no universally accepted classification of the algae, and strong differences of opinion exist amongst taxonomists as to the position of certain species, genera or even classes. The following classification is derived mainly from the treatise by Fritsch (1935, 1945) and the Manual of Phycology (Smith, 1951). Another different arrangement will be found in the new text on algal physiology edited by Lewin (1962).

The marine phytoplankters are found mainly in four out of the seven divisions listed by Smith. The division Phaeophyta contains the large brown seaweeds, only the reproductive cells of which may occasionally be found in the plankton. The division Rhodophyta contains the sessile red algae, but also a very few unicellular forms may appear in the plankton, such as a marine *Porphyridium* which has many similarities to the blue-green algae. Members of the division Euglenophyta are most abundant in fresh water with a high content of organic matter, but brackish and salt-water forms are known. Euglenoids do not appear to contribute much to the bio-mass of true marine phytoplankton, although their occurrence is sometimes noted. The remaining four divisions will be considered in more detail. They are listed in what is probably increasing order of importance as far as their contribution to marine phytoplankton bio-mass is concerned. Some freshwater genera are given, as they are often encountered in works on phytoplankton physiology.

A. CYANOPHYTA

Class I. Myxophyceae (blue-green algae).

Order I. Chroococcales.

Some genera: *Anacystis*, *Chroococcus*, *Synechococcus*.

Order II. Nostocales.

Some genera: *Anabaena*, *Calothrix*, *Nostoc*, *Oscillatoria*, *Trichodesmium*.

The myxophyceae are not very common in the marine phytoplankton, but some species may occur in great abundance at certain times of the year in some nearshore and brackish areas, coastal inlets, etc. Large numbers are sometimes found in warm tropical seas where they may fix nitrogen directly. A similar role is suspected for the large blooms of blue-green algae which occur in many lakes in late summer after a well-established epilimnion is stripped of nutrients. The myxophyceae all reproduce by simple cell division. There is no easily detected cell structure, and the photosynthetic pigments are distributed throughout the cell. These organisms have many morphological and biochemical similarities to certain bacteria.

B. CHLOROPHYTA

Class I. Chlorophyceae (green algae).

Order I. Volvocales.

Some genera: *Carteria*, *Chlamydomonas*, *Haematococcus*, *Dunaliella*, *Tetraselmis*, *Nannochloris*, *Platymonas*.

Order II. Chlorococcales.

Some genera: *Ankistrodesmus*, *Chlorella*, *Scenedesmus*.

The chlorophyceae are predominantly freshwater plants. The planktonic forms are mainly flagellates in the nanno- and ultra-plankton range, occurring chiefly in temperate coastal regions, especially in late summer and early winter. Methods of reproduction are complex, but simple cell division occurs in most of the flagellate forms. *Palmelloid* stages are known in which the same cell exists in a non-motile form (with a mucilaginous sheath), or in an actively swimming form. There is some doubt as to whether chlorophyceae occur to any significant extent in the open ocean beyond the continental shelf.

C. PYRROPHYTA

Class I. Dinophyceae.
Subclass Desmokontae.
Order I. Prorocentrales.
Some genera: *Exuviella*, *Prorocentrum*.
Subclass Dinokontae.
Order II. Gymnodiniales.
Some genera: *Amphidinium*, *Gymnodinium*, *Gyrodinium*.
Order III. Peridiniales.
Some genera: *Ceratium*, *Glenodinium*, *Gonyaulax*, *Peridinium*.

Species in the dinophyceae (dinoflagellates) make up a considerable fraction of the marine phytoplankton at certain times of the year, and "bloom" to form many of the spectacular "red tides". Their morphology and physiology are very diverse, as some species have lost all ability for photosynthesis, and live phagotrophically or symbiotically; all species have two flagella. There are two subclasses which differ in the positioning and orientation of the flagella. The species in the order gymnodiniales have "naked" protoplasts, but most other species are "armoured" and have characteristic patterned envelopes or "theca". Reproduction is by cell division, longitudinally in the desmokontae, but more or less obliquely in the other orders.

D. CHRYSOPHYTA

Class I. Chrysophyceae.
Order I. Chrysomonadales.
Some genera: *Chromulina*, *Coccolithus*, *Isochrysis*, *Monochrysis*, *Ochromonas*, *Syrascosphaera*, *Phaeocystis*, *Silicoflagellates*.

The small flagellate chrysophyceae, and such related organisms as the silicoflagellates, make up a large fraction of the marine nanno- and ultra-plankton. The coccolithophores, in particular, may comprise the bulk of the plankton bio-mass in some open ocean areas. A large number of the unidentified naked and delicate μ-flagellates (ultra-plankton), which are known to be a significant fraction of the plant material in tropical and sub-tropical oceans, may belong to this class.

Class II. Bacillariophyceae (diatoms).

Order I. Pennales.

Some genera: *Asterionella, Bacillaria, Fragilaria, Navicula, Nitzschia, Pinnularia, Pleurosigma.*

Order II. Centrales.

Some genera: *Biddulphia, Chaetoceras, Corethron, Coscinodiscus, Cyclotella, Ditylum, Melosira, Planktoniella, Rhizosolenia, Skeletonema, Stephanopyxis, Thalassiosira.*

The small chrysophycean, *Phaeodactylum tricornutum*, given the name *Nitzschia closterium forma minutissima*, or simply *Nitzschia closterium* in early work, has been one of the most extensively studied "diatoms" in marine phytoplankton physiology. It is, however, not a true diatom but probably the sole known member of a sub-order in the bacillariophyceae, see Lewin *et al.* (1958).

The diatoms, together with the rest of the Chrysophyta, generally furnish the largest fraction of the marine phytoplankton bio-mass. The diatoms are particularly important in temperate coastal waters during the spring and autumn, and in the arctic and antarctic seas during the summer. Numerous species, usually rather small, occur in the open oceans. They reproduce mainly by cell division, but sexual reproduction is known in both pennales and centrales.

There is now a suggestion that the bacillariophyceae should be made a separate Division, the Bacillariophyta, with centrics and pennates as the two classes.

Class III. Xanthophyceae.

Several classes are known in fresh water. If the most commonly occurring marine form, *Halosphaera*, has to be removed from the xanthophyceae (Parke and Adams, 1961), there will be some doubt as to the importance of this class in the marine phytoplankton. *Halosphaera* is often recorded in ocean samples, and may sometimes be abundant.

E. CRYPTOPHYCEAE

This class of organisms is something of a taxonomic enigma. It has been placed in the division Pyrrophyta, but recent work has shown that the cells contain phycobilin-like pigments (see later), which are characteristic of the red and blue-green algae, but are not found in dinoflagellates. A new Division, the Cryptophyta, may therefore be necessary. Not much is known of the occurrence of these organisms in sea water, and they are rarely a major constituent of the phytoplankton. These algae are small red-brown or bluish flagellates, although colourless forms exist. They certainly occur in large numbers at times in inshore waters, and may be a constituent of the ultra-plankton populations in coastal and oceanic waters; very little is known about them.

It is interesting to note how chemical evidence, such as pigment composition, can play a role in taxonomy and phylogeny (see for example Haxo and Fork, 1959; Dougherty and Allen, 1960).

IV. Chemical Composition of Phytoplankton and Methods for Measuring Plant Bio-mass

A considerable body of knowledge has now accumulated about the chemical composition of the Phaeophyta and Rhodophyta, because of the easy harvesting and commercial importance of the benthic algae. Much less is known of the composition of the phytoplankton, especially when we restrict ourselves to marine species and remove such genera as *Chlorella* and *Scenedesmus* from the list. However, enough information has been obtained for a start to be made in relating marine phytoplankton to the chemistry of sea water and to oceanic food production.

It must be stressed from the outset that the composition of phytoplankters not only varies from species to species, but has no absolute value for any one species, unless growth conditions, in particular the concentration of all nutrients in the medium, are exactly defined. Some aspects of this problem will be mentioned later. In this section, we will describe in a semi-quantitative way the composition of unicellular algae, and indicate the use which has been made of such information for the measurement of the standing crop of phytoplankton in nature.

A. Pigments

(1) *Distribution.* Plant pigments play a key role not only in the biochemistry of photosynthesis, but also in the taxonomic identification of phytoplankters with respect to Division, and sometimes Class and Order. Inasmuch as plant pigments are the only compounds which enable us to distinguish plants from animals, their measurement in nature is of great practical importance.

There are four main classes of pigments to be found in the marine phytoplankton: the chlorophylls, the carotenes, the xanthophylls and the biliproteins. Most of the known occurrences are summarized in Table I. This table is constructed after the format given by Strain which has been used by several other authors. It is repeated here for the convenience of *in situ* reference. Data are taken from Strain *et al.* (1944), Strain (1951, 1958), Goodwin (1952, 1955, 1957, 1960), Dales (1960), Allen *et al.* (1960b), OhEocha (1960), Parsons (1961), Jeffrey (1961) and Chapman and Haxo (1963).

TABLE I

Pigments of the principal classes of marine phytoplankters

Pigments	Class					
	Myxophyceae	*Chlorophyceae*	*Cryptophyceae*	*Dinophyceae*	*Chrysophyceae*	*Bacillariophyceae*
Chlorophylls						
Chlorophyll *a*	***	***	***	***	***	***
Chlorophyll *c*	—	—	**	**	**	**
Chlorophyll *b*	—	**	—	—	—	—
Carotenes						
β-carotene	***	***	—	***	***	***
α-carotene	—	*	***	—	—	*
γ-carotene	—	*	—	—	—	—
ε-carotene	*	*?	*	—	—	*
Xanthophylls						
Oscillaxanthin	*	—	—	—	—	—
Myxoxanthophyll	**	—	—	—	—	—
Echinenone (Myxoxanthin)	**	—	—	—	—	—
Antheraxanthin (5,6-epoxy-zeaxanthin)	*	—	—	—	—	—
Lutein (xanthophyll)	—	***	—	—	—	—
Neoxanthin (monoepoxy-zeaxanthin)	—	**	—	—	—	—
Zeaxanthin	*	**	*?	—	—	—
Violoxanthin (5,6, 5′,6′-di-epoxy-zeaxanthin)	—	**	—	—	—	—

TABLE I—*continued*

Pigments	*Myxophyceae*	*Chlorophyceae*	*Cryptophyceae*	*Dinophyceae*	*Chrysophyceae*	*Bacillariophyceae*
			Class			
Xanthophylls – cont.						
Flavoxanthin (5,8-epoxy-lutein)	—	*	—	—	—	—
Astaxanthin	—	*	—	—	—	—
Dinoxanthin	—	—	—	**	*	—
Peridinin	—	—	—	***	—	—
Neoperidinin	—	—	—	**	—	—
Diadinoxanthin	—	—	—	**	**	**
Diatoxanthin	—	—	—	—	*	**
Fucoxanthin	—	—	—	—	***	***
Neofucoxanthin	—	—	—	—	**	**
Unidentified	*	**	**	—	—	—
Biliproteins						
Phycoerythrin	**	*?	**	—	—	—
Phycocyanin	***		**	—	—	—

*** Major pigment.
** Secondary pigment in all or most species.
* Reported in one or two species; position not clear.

The occurrence of a pigment in one species does not guarantee its occurrence in another species of the same order, or even genera. Growth conditions have a marked effect on the pigment composition of a cell, and the concentration of a "minor" pigment may rise above or fall below the detection limit. Such effects, and perhaps a tendency to over-generalize, have caused some confusion in the literature.

Pteridines have been reported from the Cyanophyta, Chlorophyta, Euglenophyta and from the chrysophyceae, sometimes in significant quantities (see Wolf, 1960). Little is known of the function of these pigments, but a role in photosynthesis is suspected.

The amount of pigment present in the algae is very variable, and it is impossible to generalize. On a dry weight basis, chlorophyll *a* may reach about 2% in the chlorophyceae, but is about 0·2–1% in most of the other classes studied. With poor growth conditions, amounts will drop to one-third or less of their maximum values. The amounts of chlorophylls *b* and *c* are less than that of chlorophyll *a*, and vary considerably between species. The total weight of carotenoid pigments in a cell rarely exceeds two-thirds of the weight of the chlorophylls. The xanthophylls predominate, and are normally present at several times the concentration of the carotenes. Some of the few quantitative data for marine phytoplankters are given by Parsons (1961) and Jeffrey (1961). The biliprotein content of the myxophyceae is probably less than the chlorophyll content, and may amount to a few tenths of 1%.

(2) *Chemistry.* Details of the structure and properties of plant pigments will be found in the references cited above, and in the comprehensive monograph by Rabinowitch (1945, 1951, 1956). Some aspects merit a brief summary.

(a) *Chlorophylls.* It is now well established that chlorophylls *a* and *b* have "Fischer" structures which are built from a dihydroporphyrin ring which contains a central non-ionizable magnesium atom. In addition to the four pyrrole nuclei surrounding the magnesium atom, there is a cyclopentanone ring with a methyl ester group attached. A propionic acid side-chain, esterified with phytol (a mono-unsaturated long-chain primary alcohol) is attached to one of the pyrrole nuclei. The structure of chlorophyll *c* has not yet been determined. It is ironical that this pigment, which is one of the most abundant in nature, should only recently have been obtained in a pure state. It is undoubtedly a porphyrin in which the magnesium is more strongly bound than in the other two chlorophylls. It was once thought that phytol was absent, because chlorophyll *c* is relatively strongly *hypophasic* (i.e. preferentially enters the aqueous phase when partitioned between aqueous alcohols or ketones, and hydrocarbons), but the molecular weight appears to be at least as great as that of chlorophyll *a* (see Granick, 1949; Jeffrey, 1962, 1963), which indicates that the phytol fragment, if absent, must have been replaced by one equally large, or by numerous smaller groups.

The chlorophylls are characterized by two main absorption bands, one in the blue and one in the red region of the visible spectrum. With chlorophyll *c*, absorption in the red region is relatively weak, and its solutions are a pale yellow-green. All three pigments have a strong red fluorescence when excited by ultra-violet light.

Both the position and intensity of the blue and red absorption maxima depend upon the solvent in which the chlorophyll is dissolved. In general, shift in the blue maximum is irregular, but the red maximum tends to move to longer wavelengths as the refractive index of the solvent increases. Of considerable practical importance are the very marked changes in intensity and wavelength that occur when even traces of polar substances are present as impurities in non-polar solvents. The use of aqueous polar solvents, such as wet alcohols or ketones, greatly reduces spectral shifts.

The chlorophyll in the chloroplasts of an algal cell may be present in "concentrations" of about 0·1M (10%), and is bonded to protein and lipid. This bonding produces a shift of up to 300 Å towards longer wavelengths for the chlorophyll red absorption peak. New techniques of *in vivo* spectrophotometry have shown that several chlorophyll *a* compounds exist in the living cell, and that each has a peak at a slightly different wavelength. These compounds presumably arise from variations in the bonding of the chlorophyll *a* molecule with protein or lipid, since only one pigment can be extracted by solvents (for example refer to Allen *et al.*, 1960a; Brown and French, 1961; Allen *et al.*, 1962).

Chlorophyll, in solution in alcohols or ketones, can be oxidized by air, when it is said to be "allomerized". The blue maximum is reduced, but there is little change in the red maximum except for a slight displacement to shorter wavelengths. This behaviour is not found with chlorophyll solutions in diethyl ether or in non-polar solvents. In all solvents, however, a slow bleaching of chlorophyll solutions takes place. especially in the light (photo-oxidation). The effect is less evident with crude plant extracts than with pure chlorophyll solutions, and the former are generally stable for several days if kept in the dark at temperatures below 0°C.

Three decomposition products of chlorophyll may be encountered in the sea, only the chlorophyll *a* decomposition products being of practical importance. *Chlorophyllide* is produced from chlorophyll by loss of the phytol group by hydrolysis. This hydrolysis is catalyzed in living or recently dead cells by a lipase enzyme *chlorophyllase* which is extraordinarily active in some species (e.g. *Skeletonema costatum* and *Phaeodactylum tricornutum*). Chlorphyllide *a* has an absorption spectrum identical with that of chlorophyll *a* in most solvents, but it is more hypophasic and moves more slowly on most chromatographic columns. *Pheophytin a* is formed whenever *in vivo* or *in vitro* chlorophyll becomes acidic, when the magnesium atom is irreversibly replaced by hydrogen. The spectrum differs markedly from that of chlorophyll, since it shows much less intense absorption in the red region, and the absorption maximum is shifted to slightly longer wavelengths. Pheophytin, like chlorophyll, is *epiphasic* (i.e. on extraction from polar solvents it preferentially enters the hydrocarbon layer), but travels much more rapidly in normal chromatographic systems. *Pheophorbide* is obtained if phytol is removed from pheophytin, and bears the same relationship to pheophytin as chlorophyllide does to chlorophyll. It has the same visible absorption spectrum as pheophytin, and although hypophasic, its behaviour on a chromatograph column is intermediate between that of chlorophyllide and that of pheophytin.

"Detrital" chlorophyll *a* assumes some importance in marine work, since normally the analysis for chlorophyll is undertaken spectrophotometrically using an absorption measurement near to 6600 Å. The result is assumed to be "living" chlorophyll. However, the chlorophyll in dead cells, chlorophyllide, pheophytin and pheophorbide will also give solutions absorbing in this region of the spectrum. Correction is possible for chlorophyllide which will give the same extinctions as an equivalent weight of chlorophyll, but it is not possible to correct for absorption arising from dead cells. The pheo-pigments give lower extinctions per unit weight than does the parent pigment, but only a chemical separation of each sample can resolve the problem with certainty; however very little work has been done with marine samples. A few experiments by Patterson and Parsons (1963) have indicated that the problem may not be too serious. When suitable precautions were taken to suppress chlorophyllase activity and to prevent acidification of cells and extracts, the chlorophyll decomposition products in coastal sea-water samples, even after heavy grazing, did not exceed 10–20% of the total. Their concentrations were very high, however, in mud samples, and in the digestive tracts of marine animals (see also Currie, 1962), and may be relatively high in the open ocean well below the euphotic depth (Yentsch and Menzel, 1963).

The absorption peak wavelengths and, more particularly, the specific extinction coefficients, $E^{1\%}_{1\,\text{cm.}}$ of the chlorophylls are important analytically, but accurate values are difficult to obtain. Values quoted in the literature vary according to the solvent used and its purity, the instrument employed, and the method of purifying the pigment; different values are obtained according to whether the pigment is dried or not before being dissolved in the solvent. The purification of plant pigments from colourless contaminants can be particularly difficult (see Perkins and Roberts, 1962; Jeffrey, 1963), and recently, Anderson and Calvin (1962) have reported yet another set of specific extinction coefficients for highly purified crystalline chlorophyll *a*. Ninety per cent aqueous acetone has become the most widely used solvent for marine pigment analysis (see later). When directly analyzing plant extracts, extinction values are probably best determined using "undried" chlorophyll. Specific extinction coefficients at various wavelengths for chlorophylls in 90% acetone and in pure ethyl ether are collected in Table II, together with some data on carotenoids, which will be discussed later. These data are thought to be the most reliable available at the time of writing, but will almost certainly be subject to later changes. Values are taken from Smith and Benitez (1955), Vernon (1960), Richards (1952), Jeffrey (1963), Parsons and Strickland (1963) and from other data from the author's laboratory. These values refer to the wavelengths of either the absorption maximum or the analytically useful region.

(b). *The carotenes.* These compounds are yellow-orange, fat-soluble unsaturated hydrocarbons containing a large number of conjugated double bonds. They are associated *in vivo* with proteins and lipids which give rise to a "red shift" of 100–200 Å in their absorption spectra. Solutions of carotene are quickly bleached by light and undergo cis-trans changes, oxidation and other transformations. β-Carotene predominates in all phytoplankton except perhaps the cryptophyceae; γ- and ε-carotenes are relatively rare.

The absorption spectra of the carotenes show two peaks which degenerate to one broad peak in polar solvents. Peak wavelength values in various solvents have been collected by Goodwin (1955), and the data for β-carotene (in 90% aqueous acetone) are given in Table II.

(c) *The xanthophylls.* These compounds have the most complex distribution in the algae and thus are of the greatest value for characterizing their various classes. The xanthophylls are oxygenated products of the carotenes, and there is now evidence for their interconversion and production from carotenes within the living cell (see Claes, 1959; Yamamoto *et al.*, 1962).

For an account of the general chemistry and properties of the xanthophyll pigments the reader is referred to the monographs by Goodwin (1952) and Strain (1958), and to the excellent early paper by Strain *et al.* (1944). Superficially, the absorption spectra of these

TABLE II

Some spectral characteristics of phytoplankton pigments: values for $E^{1\%}_{1\,cm.}$

Wavelength (Å)	*Chlorophyll* a*		*Chlorophyll* b		*Chlorophyll* c	
	Ether	*Acetone*†	*Ether*	*Acetone*†	*Ether*	*Acetone*†
4085						
4090						
4300	1320					
4330		1015				
4460						1400
4500					1690	
4550			1750			
4600				1480		
4800		12		280		47
5050						
5100		21		31		180
5360						
5400						
5800		104		93	157	115
5950				118		
6100						
6150		190				
6280					158	
6300		139		164		195
6440			620			
6450		218		540		43
6480				522		
6620	1010					
6640		910				
6650		890		63		7
6665						
6670						

* Chlorophyllide *a* and pheophorbide *a* have spectra almost identical with those of chlorophyll *a* and pheophytin *a* respectively.

The underlined values are for absorption maxima.

† 90% aqueous acetone. Values in 80 or 90% acetone are generally almost identical. Chlorophyll *c* is most susceptible to the water content of the solvent. Values in 80 to 100% methanol are generally about 25% less than acetone values, with peaks shifting slightly to longer wavelengths.

pigments resemble those of the carotenes by having two peaks in about the same spectral regions. The two maxima found in the spectra of the extracted pigments in non-polar solvents degenerate to one broad peak

TABLE II—*continued*

Wavelength (Å)	*Pheophytin* a*		*β-carotene*	*Fucoxanthin*	*Peridinin*	*Astacin and astaxanthin (approximate values)*
	Ether	*Acetone*†	*Acetone*†	*Acetone*†	*Acetone*†	*Acetone*†
4085	1320					
4090		1310				
4300			1780	731	690	
4330						
4460						
4500				880	770	2200
4550			2500			
4600						
4800			2235	725	840	2500
5050		150				
5100			455	200	490	1700
5360		130				
5400			37			510
5800				Nil	10	20
5950						
6100		119				
6150						
6280						
6300						
6440						
6450						
6480						
6620						
6640						
6650						
6665		566				
6670	637					

extending over several hundred Ångstrom units when xanthophylls are dissolved in polar solvents. *In vivo* spectra show a large "red shift". Spectral data will be found in Goodwin (1955). The spectra of many xanthophylls are almost identical, and positive identification depends therefore on chromatographic and chemical evidence. Lutein, the predominate xanthophyll in the chlorophyceae, has a spectrum and specific extinction coefficient similar to that of the β-carotene shown in Table II. The predominant carotenoids in the Pyrrophyta and Chrysophyta are peridinin and fucoxanthin which are exceptional in having quite low specific extinction coefficients (see Table II).

The treatment of phytoplankton in a manner designed to denature the protein and break the pigment-protein and pigment-lipid bonds (e.g. heat, decay, formaldehyde preservation) makes the xanthophyll absorption maxima move into the blue region, and many plants lose their brown colour, and become green owing to the appearance of unmasked chlorophyll.

The xanthophylls resemble chlorophylls and carotenes in their general instability, especially in dead plant tissue, but may be somewhat more resistant to atmospheric attack in the dark. Solutions rapidly deteriorate if exposed to strong illumination, even at room temperature.

(d) *Biliproteins*. These compounds are relatively unimportant in marine productivity studies as they occur only in the myxophyceae and cryptophyceae, and have not been used for identification purposes. The biliproteins are composed of a chromophore (phycobilin) and a protein fragment. Phycobilins are tetrapyrroles structurally similar to the bile pigments. The two main classes of pigments, are phycoerythrins (red) and phycocyanins (blue). The proportions of these pigments in a cell vary with the algae and with the growth conditions, and cells may range in colour from red to deep blue. The biliproteins are water soluble, and their aqueous solutions have a beautiful fluorescence. Phycoerythrins have absorption peaks at about 5000 Å, 5400 Å, and 5650 Å, whereas phycocyanins have only one main peak near 6100 Å. For a detailed account see OhEocha (1960).

The biliproteins are not extracted from plant cells by acetone or alcohol, and are not used in routine plant analysis. Occasionally, an acetone-soluble blue pigment will be found which interferes with chlorophyll analysis (Strickland and Parsons, 1960). Such a coloration is presumably caused by a phycobilin-like molecule, although this is not necessarily the chromophore of the biliprotein present in the cell (see Fujita and Hattori, 1962).

(3) *Analysis*. The most commonly employed method for the measurement of the standing crop of phytoplankton in a water sample is the determination of the pigment content. The results can be used, as will be described later, to give an estimate of plant bio-mass, as distinct from other particulate organic matter, and can also be used to measure the "photosynthetic potential" of a body of water; for this purpose chlorophyll *a* is measured. A determination of chlorophylls *b* and *c* is used mainly for the assessment of the contribution of the chloro-

phyceae or of the chlorophyll *c*-rich diatoms and dinoflagellates to a crop. A knowledge of the carotenoid content of a sample may indicate the physiological condition of the species present, and can also be used for rough taxonomic sorting as shown by Table I. For the latter purpose it is necessary to separate and identify the major carotenoids.

Three methods may be used to collect phytoplankton from sea water for analysis for pigments, namely: settling, centrifugation and filtration. The use of nets, which collect only the microplankton, has long been abandoned for quantitative marine work. Settling and centrifugation, although satisfactory for certain purposes, are now rarely used in conjunction with chemical analysis. The sampling and separation of marine phytoplankton has been reviewed by Strickland (1960), and will not be discussed further.

Filtration of water through the finer analytical grade papers retains all micro- and nannoplankters. Collodion membrane filters (see Chapter 21, Volume 2, Section IV) are used extensively in marine work, they collect ultraplankters and all micro- and nannoplankters; they often retain appreciably more plant material than does filter paper (Harvey, 1950). The new fine porosity glass filter papers appear to be satisfactory for most ultraplankters, but must be used with only very gentle suction.

Creitz and Richards (1955) recommended an aerosol membrane (Millipore Corp. AA) with a pore size of about 1 μ. The ultraplankter *Nannochloris*, with a cell of 1–2·5 μ diameter, was quantitatively retained by this filter. Most membrane filters dissolve completely in acetone making direct extraction of the plant cells possible without further treatment. Margalef (1962) has described equipment which permits the semi-continuous extraction of a large number of samples without the need to dismantle the filtration apparatus. Double thickness filter paper was used, and therefore losses may have been appreciable with some samples.

The complete extraction of pigments from phytoplankton cells presents a problem. Only ethyl alcohol, methanol and acetone have been employed for this purpose in marine work. Addition of a few per cent of water to the solvents assists the disruption of pigment-protein bonds. A number of species with resistant cell membranes will defy this treatment even with boiling solvent. It is doubtful if any solvent or combination of solvents will avail with the most resistant cells, and although methanol may be somewhat superior to acetone, the latter is in common usage and has the advantage that a more complete set of associated spectral data is available.

Three methods have been used for the disruption of the tougher cell walls of marine phytoplankton: repeated lysis with water (freezing and thawing if necessary), the use of supersonic disintegrators (Nelson, 1960; Lassøe and Hansen, 1961), and the use of a tissue grinder (Yentsch and Menzel, 1963). The last method is probably the best, but none of these approaches is very convenient for routine analysis, although probe-type sonic disintegrators can be used without much difficulty. Fortunately, although incomplete extraction is sometimes encountered with marine samples, it is rarely very serious, especially if filters are not dried, and if at least 24 hours is allowed for extraction. Most of the species giving poor extractions are in the classes chlorophyceae and myxophyceae, and present the greatest problem to limnologists or to those working with tidal creeks, estuaries or salt marshes.

The extraction technique most commonly employed for pigment analysis is a treatment in the cold and dark for 24 hours with a 90% acetone-water mixture (Richards with Thompson, 1952). Filters should be thoroughly drained of sea water to prevent the subsequent development of turbidities when the membranes are dissolved in the solvent. An initial treatment of the algae with boiling solvent or steam has been recommended to remove chlorophyllase and other enzyme activities. It is desirable to add a small amount of inorganic or organic base (commonly magnesium carbonate powder or dimethylaniline) to minimize the formation of pheo-pigments.

The credit for first using the pigment extraction method for the estimation of marine phytoplankton is generally given to Kreps and Verbinskaya (1930), who applied chlorophyll analysis to the measurement of the phytoplankton in net-hauls. Standardization may be made against solutions of pure chlorophyll. Even the commercial pigment can be used, as the purity of the chlorophyll is often a minor consideration compared with other possible sources of error. The main difficulties in the direct use of a chlorophyll standard lie in its preparation and in its instability.

An attempt to overcome these difficulties was made by Harvey (1934a), who used visual matching of pigment extracts against a standard of nickel sulphate and potassium chromate in water (see also Atkins and Parke, 1951). This method led to the recording of mixed chlorophyll-carotenoid pigments in terms of an HPPU (Harvey plant pigment unit). This was the amount of phytoplankton pigment, which, when dissolved in a given volume of 80% aqueous acetone, gave a colour which matched that of a solution of 430 μg.$NiSO_4 \cdot 6H_2O$ and 25 μg.K_2CrO_4 in the same volume of water. The method was only approximate, but its use is of historical importance as the HPPU was used by many workers over a period of nearly 20 years. The HPPU was *very approximately* equal to 0·3 μg. of chlorophyll. The advent of modern spectrophotometers at a cost within the budget of most institutions has made the comparison of chlorophyll extracts, either visually or with colorimeters, rarely necessary, and more sophisticated methods can now be employed.

(a) *Multichromatic spectrophotometry.* Krey (1939) measured the extinction of extracts at two different wavelengths in order to evaluate both chlorophyll *a* and carotenoids separately, and this method was subsequently employed by Gillbricht (1952). It remained, however, for Richards with Thompson (1952), using the spectra data determined by Richards (1952), to popularize a method for the differential analysis of some of the pigments extracted from marine phytoplankton by 90% aqueous acetone.

In theory, it is only necessary to know extinction values at as many wavelengths as there are separate pigments present for the composition of the solution to be determined. However, in practice, the method is limited to pigments which have absorption maxima reasonably separated (by at least 150–200 Å for the normal type of band). For this reason the equations of Richards with Thompson (1952) enable separate estimates to be made of chlorophylls *a*, *b* and *c* only and give some arbitrary measure of total plant carotenoids and of the astacin-type carotenoids found in Crustacea. No satisfactory resolution of the carotenoids into carotenes and xanthophylls is possible, and the precision of the chlorophyll *c* measurement is generally rather poor.

The Richards-Thompson method, especially as modified by Creitz and Richards (1955) for direct use with a membrane filter, has been understandably popular. It suffers from the disadvantage that chlorophyll *c* and total carotenoids are expressed as arbitrary specified pigment units (SPU) and the data used for chlorophyll *a* and *b* are now considered to be in appreciable error. A revised version utilizing the coefficients given in Table II for chlorophyll *a*, chlorophyll *b* and chlorophyll *c* has been given by Parsons and Strickland (1963). These authors have suggested a new SPU for carotenoids, reasoning that as the major carotenoids in most marine phytoplankton are fucoxanthin or peridinin, an appreciably lower extinction coefficient than that assumed by Richards should be used. Further, the correction made from extinction measurements at 5100 Å by Richards is of doubtful value since peridinin, in particular, absorbs quite strongly at this wavelength, and causes an over-correction for plant carotenoids when the Richards-Thompson formula is used. Full analytical details of a modified Creitz and Richards technique have been given by Strickland and Parsons (1960).

Parsons and Strickland have recommended that the following

equations should be used *pro tempore* to replace those given by Richards, on the understanding that further improvements should be attempted when more reliable specific extinction values become available.

$$C_a = 11{\cdot}6\,E_{6650} - 0{\cdot}14\,E_{6300} - 1{\cdot}31\,E_{6450}$$
$$C_b = 20{\cdot}7\,E_{6450} - 4{\cdot}34\,E_{6650} - 4{\cdot}42\,E_{6300}$$
$$C_c = 55{\cdot}0\,E_{6300} - 16{\cdot}3\,E_{6450} - 4{\cdot}64\,E_{6650}$$

Carotenoids for most marine crops (MSPU/l.) $= 10\,E_{4800}$

Carotenoids for chlorophyceae and myxophyceae (MSPU/l.) $= 4{\cdot}0\,E_{4800}$

where E values are extinctions in a 1 cm. cell at the wavelengths shown; C_a, C_b and C_c are the concentrations of chlorophylls in mg./l of 90% aqueous acetone; SPU is the new specified pigment unit which should approximate to 1 g. of dry carotenoids. The MSPU approximates to the milligram.

(b) *Fluorometry*. The characteristic fluorescence of the chlorophylls has been used to recognize intact plant cells by microscopy (Wood, 1962), and has long been advocated by Kalle (1951) as a means of measuring the concentration of chlorophylls in solution. The method has the disadvantage that special equipment is necessary, and that only one pigment or group of pigments is measured. Standardization is difficult, but the fluorescence technique greatly surpasses the spectrophotometric method in sensitivity. For this reason it is to be recommended when working with tropical ocean samples in which the chlorophyll *a* concentration may be less than 0·2 mg./m.^{3} (Yentsch and Menzel, 1963).

(c) *Methods for individual pigments*. These all involve physico-chemical separations. Parsons (1963) has developed a method for the determination of chlorophyll *c* based on its separation from all pigments other than chlorophyllide *a* which illustrates a simple partition technique.

Ten millilitres of pigment extract in 90% aqueous acetone are mixed with 3·5 ml. of 0·05% aqueous sodium chloride solution, and extracted with 13·5 ml. of hexane. A large and fairly constant fraction (about 90%) of the chlorophyll *c* remains in the hypophasic layer, and can either be measured directly or be determined from the change of extinction at 4500 Å which results when the chlorophyll *c* changes to pheophytin *c* on the addition of a little acid. This method is greatly superior to the trichromatic method when dealing with low pigment concentrations, since the trichromatic method is very insensitive to the presence of chlorophyll *c* and is subject to large spectrophotometric errors when only small extinctions are measured.

If the hexane layer from the above separation is shaken with 10 ml. of 90% aqueous methanol, most of the fucoxanthin, peridinin and pheophorbide *a* is extracted, and can be estimated by a trichromatic technique involving measurements at 6650 Å, 5100 Å, and 4800 Å

(Table II). The remaining carotenoids, together with chlorophyll *a* and pheophytin *a*, are left in the hexane and can be separated chromatographically after evaporation of the solvent. Details for the partition of carotenoids are given by Petracek and Zechmeister (1956), and separations by partition chromatography or adsorption chromatography are discussed by Purcell (1958), and Strain (1958). Goodwin (1955) has collected an excellent bibliography. Techniques for the chromatographic separation of chlorophylls and carotenoids can be adapted to the use of micro-columns suitable for work at sea (T. R. Parsons, unpublished work).

A direct separation of carotenoids, chlorophylls and their breakdown products by two-dimensional chromatography was described by Bauer (1952) and by Holden (1962). An excellent method using one- or two-dimensional chromatography has been described by Jeffrey (1961). The separation of nearly all pigments of importance from marine samples is possible in less than two hours.

(d) *Measurements* in vivo. The position and intensity of the pigment absorption bands as measured directly on suspensions of algal cells depends greatly on the size of the cells (or probably on the chloroplasts within the cells), the refractive index of cells and medium, and the angle of observation (see Latimer, 1959). In general, absorption curves are greatly distorted and sensitivity is lost, but valuable information can be obtained about the nature of the pigments *in vivo* (see e.g., Allen *et al.*, 1960a, 1962). If strictly parallel light is used, as in the Cary 14-R recording spectrophotometer, very good spectra can be obtained (Murchio and Allen, 1962). One method of avoiding the distortion of spectra resulting from scattered light is to introduce complete scattering artificially into the system. This is achieved by having opal glass next to the detector in both the sample and blank compartments of a spectrophotometer (Shibata, 1958).

None of the above methods are directly applicable to sea-water samples as pigment concentrations are generally much too low. Yentsch (1957) has described an ingenious method for the measurement of chlorophyll in the phytoplankton filtered onto a Millipore membrane filter. The filter is rendered transparent with cedar oil, and pressed between glass slides in a special holder fitted into the spectrophotometer. The chlorophyll absorption, *in situ* in the plant cells, is measured at 6700 Å and corrected for the scattering caused by the filter by making a further measurement at 7500 Å. The method is calibrated by the trichromatic extraction procedure, but is less sensitive than the latter.

Tyler (1961) has shown that "space light" seen horizontally in sea water can be used to detect phytoplankton if the *in situ* chlorophyll concentrations are greater than about 1–2 mg./m.[3]. Strickland (1961) has suggested that an analogous technique might be possible by making measurements from aircraft to compare the back-scattered light from the sea with the incoming radiation, using narrow wave-band filters and making corrections for the direct surface reflection of skylight. Neither of these procedures would be expected to be practicable outside coastal areas since they require large standing crops of phytoplankton.

B. ELEMENTARY COMPOSITION

Tables showing the mean elementary composition of "plankton" have been given by Fleming (1939a) and others, but in view of the wide range of species present in the phytoplankton, and of possible variations in the physiological state of the organisms, such tables have little value, except for the crudest calculations.

The composition of unicellular algae can be related to several standard states of the sample; *wet weight* (excess water being removed under specified conditions), *packed volume* (cells packed by centrifugation under specified conditions), the true *algal weight* (cell volume multiplied by *in situ* specific gravity), or the *dry weight*. The last two are preferable, and the latter is the most convenient to measure. Drying at 60°C to constant weight in nitrogen is best, although air drying at 105°C probably introduces little error in most determinations.

The use of algal weight (or algal volume) has the advantage that compositions can be related to the natural state of the algae. Difficulties are mainly practical, but have been diminished by the introduction of electronic counters such as the Zeiss and the Fisher blood-cell counters. The Coulter counter has the additional advantage that it makes possible a direct estimate of algal volume and of size distribution (see Maloney *et al.*, 1962). Chain-forming species are counted as units of chains rather than cells. In any sample of sea water or culture, the size distribution of a given species is quite large. The distribution roughly follows a normal Gaussian curve, but 25% of the cells may well have a volume less than half or exceeding twice the mean volume.

The relation of dry weight to algal weight is very variable, and is between 20 and 50%, depending on the species (see Strickland, 1960). The ash content of dry phytoplankton also depends greatly on the species, and is very difficult to estimate experimentally on marine organisms because of the external salts. A minimum value of 5–10% would be expected, and much higher values (30–60%) are found for siliceous diatoms and calcareous coccolithophores. In view of the

*Some analytical data for carbon, nitrogen and phosphorus in phytoplankton**

Phytoplankter	*Carbon %*	*Nitrogen %*	*Phosphorus %*	*Reference*
(Myxophyceae)				
Agmenellum quadruplicatum	45	5·8	1·4	Parsons *et al.* (1961)
Chlorogloea fritschii	—	4·3	—	Fay and Fogg (1962)
Typical analysis	28	4·5	0·8	Vinogradov (1953)
(Chlorophyceae)				
Scenedesmus sp.	46	7·1	2·7	Ketchum and Redfield (1949)
Chlorella	48	6·6	2·35	Ketchum and Redfield (1949)
Chlorella (N and P deficient)	55	2·4	0·9	Ketchum and Redfield (1949)
Tetraselmis maculata	35	8·3	3·3	Parsons *et al.* (1961)
Dunaliella salina	45	9·1	3·3	Parsons *et al.* (1961)
(Dinophyceae)				
"Ceratia"	—	5·0	—	Krey (1958)
Gonyaulax	37	4·0	0·6	Harris and Riley (1956)
Amphidinium carteri	47	4·5	1·1	Parsons *et al.* (1961)
Exuviella sp.	47	5·0	1·3	Parsons *et al.* (1961)
"Peridineae"	40·5	3·25	—	Vinogradov (1953)
(Chrysophyceae)				
Monochrysis lutheri	44·5	7·8	3·0	Parsons *et al.* (1961)
Syracosphaera carterae	35	9·0	1·2	Parsons *et al.* (1961)
(Bacillariophyceae)				
Coscinodiscus sp.	20	4·53	0·66	Harris and Riley (1956)
Phaeodactylum tricornutum	49·5	5·3	2·0	Parsons *et al.* (1961)
Nitzschia closterium	44	5·35	0·83	Ketchum and Redfield (1949)
"Diatoms"	—	5·6	—	Krey (1958)
Skeletonema costatum	31	5·9	1·7	Parsons *et al.* (1961)
Coscinodiscus wailsii	19	2·7	0·4	Parsons *et al.* (1961)

* Percentage of dry weight.

difficulty of obtaining pure dry samples of marine phytoplankton, it is often more convenient to express elementary and other analytical results as ratios. This aspect will be discussed more fully later.

There is still, surprisingly, little information available about the elementary composition of marine phytoplankters. Some of the pioneering work was carried out by Brandt and Raben (1920), and early results have been collected by Vinogradov (1953). The analytical techniques were of uneven quality. Recently Parsons *et al.* (1961) made a systematic study of 11 marine phytoplankters representative of the chlorophyceae, myxophyceae, dinophyceae, chrysophyceae and bacillariophyceae. Some results are collected in Table III, and serve to illustrate the variability to be expected, even when organisms are grown under relatively "ideal" conditions.

(1) *Carbon.* On a dry *ash-free* basis, the carbon content of marine organisms is remarkably constant between 45 and 55% (for example see Ryther, 1956a). The variation of carbon content between 20 and 50% shown in Table III is largely a reflection of changes in ash values.

The carbon content is, in many respects, the best single property with which to measure the standing crop of phytoplankton in nature or in algal cultures. The main analytical problems are connected with the separation of suitable amounts of phytoplankton from sea water and the removal of chloride from the separated samples. In most natural waters, the total particulate organic carbon content (plant and detritus) rarely exceeds 1 mg./l., and may often be one-tenth of this amount. Even when micro or semi-micro methods are used, several litres of water may have to be filtered.

Strickland and Parsons (1960) have shown that the particulate matter in sea water retained by a Millipore filter can be quantitatively washed off this filter, with no contamination, if it is first coated with a little fine magnesium carbonate powder. In recent years, fine porosity glass filter papers (e.g. Whatman GF/C) which will retain most phytoplankton cells from sea water have become available. They have a high filtration capacity and speed, and give rise to negligible "blanks". The plankton on glass filters can be burnt at a low temperature (600°C) and the carbon determined in standard micro-combustion equipment. If wet combustion is used, it is necessary to remove chloride. When carbon dioxide is measured in the evolved gases, chlorine can be removed by potassium iodide solution (see Kay, 1954).

The wet combustion method using dichromate, can be made very rapid if the carbon is measured in terms of the decrease of oxidant (for example see El Wakeel and Riley, 1957). Strickland and Parsons (1960) measured the decrease in absorption of dichromate at 4400 Å with a spectrophotometer, after first removing chloride by a preliminary heating of the separated phytoplankton with syrupy phosphoric acid. The "oxidizable matter" measured in this way is not exactly equal to the true carbon content of the sample, but is a sufficiently good approximation for most purposes, especially in routine field work. El Wakeel and Riley (1957) suggested a correction factor of 1·05 to convert oxidizable carbon to true carbon; Parsons *et al.* (1961) found 1·08 ± 0·12. The variability appears to be a function of the species of algae rather than of the relative amounts of lipid, protein and carbohydrate present in the organic matter.

(2) *Nitrogen.* The nitrogen content of algae varies widely according to the state of nutrition of the phytoplankton. Some results are given in Table III. In early marine work by Von Brand (1935), a combustion method was used, but the Kjeldahl method is now used almost exclusively for the analysis, and a routine technique suitable for sea water has been suggested by Strickland and Parsons (1960). The phytoplankton, separated on a magnesium carbonate coated Millipore filter, is digested with sulphuric acid and the ammonia determined without distillation using Nessler's reagent. It is not certain that all forms of nitrogen are measured by the Kjeldahl technique, and discrepancies are sometimes found which lead one to suppose that some nitrogen may be missed, or that substantial amounts of non-protein nitrogen are present in certain samples (Yentsch and Vaccaro, 1958; McAllister *et al.*, 1961; Parsons *et al.*, 1961).

(3) *Phosphorus.* The determination of this element in phytoplankton presents few analytical problems, but its variability is so great that figures such as those quoted in Table III have very little significance, except in the context of a specific experiment on phosphate metabolism. "Storage" polyphosphate occurs in some, and may be present in all "glutted" cells grown in a medium with excess phosphate present (Sommer and Booth, 1938; Nihei, 1957; Arnon, 1956; Kuhl, 1960). As much as half of the phosphate in the cell may be present as polyphosphate, much of it as RNA-polyphosphate (Correll and Tolbert, 1962).

(4) *Other constituents.* Silicon is a major constituent of the diatoms, although its percentage varies considerably from species to species. For example, the bacillariophycean *Phaeodactylum tricornutum* has only a few tenths of a per cent of silicon, less than appears to be present in some of the chlorophyceae (Parsons *et al.*, 1961). Heavily silicified diatoms may contain up to 25% of silica. The degree of silicification depends *inter alia* on the amount of silica available to the diatoms from the surrounding water, and may show as much as a five-fold variation (Jørgensen, 1955; Lewin, 1957). In natural marine populations, the

Si:C ratio varies from about 0·4–1 (Vinogradov, 1953; Antia *et al.*, 1963), but is probably near to 0·8 in areas where silica is not a limiting nutrient.

The calcium carbonate content of coccolithophores appears to be high, although few analyses have been made. Parsons *et al.* (1961) found 24% of calcium carbonate in *Syracosphaera carterae*. Paasche (1962) has observed that the ratio of coccolith carbon to cell carbon in *Coccolithus huxleyi* has the high value of 0·4. The ratio for coccoliths varies greatly even for a given species, and depends on age and conditions of growth. The morphology and the beautiful microstructure of coccoliths have been described by Halldal and Markali (1954).

Other elements present in the marine algae have little significance in the context of the present chapter (but cf. Chapter 5). The minor elements such as potassium, sodium, magnesium, calcium, and sulphur, which are of importance in fresh-water phytoplankton nutrition, are never in short supply in the sea. Analysis for these elements would be difficult. The amounts to be expected in marine algae are probably similar to the quantities found in fresh-water plants (Krauss, 1956; Scott, 1945). Small concentrations of elements such as manganese, cobalt, zinc and molybdenum are essential as enzyme co-factors, but are generally present in excess of these amounts, owing to adsorptive processes.

Iron has been the subject of some study, as it has been thought to be a limiting nutrient in certain sea areas. Diatoms contain quite a large percentage of this element, but the amount is very variable. The ash of marine chlorophyceae is pronouncedly brown due to the presence of relatively large amounts of iron oxide.

C. Major Metabolites

The chemistry of the benthic algae, in particular that of their carbohydrates, proteins and sterols, has received considerable attention, but it is doubtful if any information concerning the Pheophyta or Rhodophyta can be applied to the phytoplankton; not all the constituents of the benthic Chlorophyta are necessarily present in the unicellular plankton from this Division. Of the phytoplankton, only the fresh-water chlorophyceae (in particular the various species of *Chlorella*) and certain fresh-water myxophyceae have been analysed in much detail. Information about truly marine phytoplankton is very meagre. Some data are given in Table IV, and a similar breakdown of composition into carbohydrate, lipid and protein (and including ash figures) is given for Black Sea diatoms and dinoflagellates by Lanskaya and Pshenina (1961). A useful compilation of existing knowledge of the composition and metabolism of algae is given in the treatise edited by Lewin (1962).

*Some analytical data for major metabolites in phytoplankton**

Phytoplankter	*Carbohydrate*† *Cell*	*Crude fibre*	*Lipid*‡	*Protein* §	*Reference*
(Myxophyceae)					
Anabaena	40	0·6	1	51·5	Birge and Juday (1922)
Agmenellum quadruplicatum	31·5	5·5	13	36	Parsons *et al.* (1961)
(Chlorophyceae)					
Ankistrodesmus	30	9·7	8·8	52	Birge and Juday (1922)
Scenedesmus obliquus	23	—	26·5	50	Ketchum and Redfield (1949)
Chlorella pyrenoidosa	32	—	18	50	Spoehr and Milner (1949)
Tetraselmis maculata	15	1·9	2·9	52	Parsons *et al.* (1961)
Dunaliella salina	31·5	3·1	6·4	57	Parsons *et al.* (1961)
Chlorella ellipsoida	27	—	15	53	Iwamoto and Suzuki (1958)
(Dinophyceae)					
Peridineae (mixed)	22	—	4	50	Brandt and Raben (1920)
Amphidinium carteri	30·5	0·6	18	28	Parsons *et al.* (1961)
Exuviella sp.	37	13·5	15	31	Parsons *et al.* (1961)
(Chrysophyceae)					
Monochrysis lutheri	31·5	1·1	11·5	49	Parsons *et al.* (1961)
Syracosphaera carterae	18	0·3	4·6	56	Parsons *et al.* (1961)
(Bacillariophyceae)					
Fresh-water diatoms	23	1·4	14	23	Birge and Juday (1922)
Marine diatoms	13	—	15·5	15·5	Serenkov and Barashkov (1954)
Skeletonema costatum	21	2·0	4·7	37	Parsons *et al.* (1961)
Skeletonema costatum	—	—	12·1	36·2	Ogino (1963)
Coscinodiscus sp.	4·1	1·15	1·8	17	Parsons *et al.* (1961)
Phaeodactylum tricornutum	24	0·6	6·6	33	Parsons *et al.* (1961)
Chaetoceros sp.	6·6	1·5	6·9	35	Parsons *et al.* (1961)
Chaetoceros simplex	—	—	6·0	12·1	Ogino (1963)

* Percentage of dry weight.
† Carbohydrate results from Parsons *et al.* expressed as equivalent weight of glucose determined by the anthrone method.
‡ Lipid results from Parsons *et al.* expressed as equivalent weight of stearic acid.
§ Most protein results are Nitrogen percentage × 6·25.

(1) *Carbohydrates.* A detailed analysis of phytoplankton carbohydrate may be carried out using paper chromatography after sulphuric acid hydrolysis of the whole cells, but for routine purposes, only total carbohydrate is generally measured, on the assumption that the ether-insoluble non-protein portion of the plankton (corrected for ash) is "carbohydrate". The fraction of this material which is not dissolved by some standard attack with dilute acid and alkali is called "crude fibre", after the practice used in agricultural chemistry. A more direct, sensitive and rapid approach is the determination of the total sugars by a colorimetric procedure, such as the anthrone reaction. A convenient routine procedure for material concentrated on a Millipore filter using anthrone is given by Strickland and Parsons (1960), the carbohydrate being expressed in terms of a glucose standard, as this appears to be the principal sugar unit in all marine plankton so far investigated.

The phytoplankton carbohydrates exist in three main forms: as constituents of the cell wall, as "free" or low molecular weight sugars (mono- and oligo-saccharides) in the cell sap, and as sugar polymers, probably used for energy storage. The cell wall material, which is largely measured by the "crude fibre" method, differs greatly in composition and amount from species to species. The concentrations of free sugars are relatively small. Some data for the total carbohydrate contents of phytoplankton are collected in Table IV. A detailed breakdown of the carbohydrates into their component sugars is shown in Table V. It will be seen that glucose predominates, and that galactose is also present in substantial amounts. In well nourished cells, the amount of carbohydrate determined by direct analysis rarely exceeds that of the protein.

(a) *Myxophyceae.* Kylin (1943a) found the cell walls to be "pectin-like" with sulphate-ester linkages. Glucose, galactose and a pentose were produced by hydrolysis. Sucrose, glucose, mannitol and other free sugars appeared to be absent, but a disaccharide, trehalose, has been reported (Payen, 1938). The glycogen-like material acting as storage product in the myxophyceae has been called "Cyanophycean starch" by Kylin (1943b). The polysaccharide in *Anabaena cylindrica* on hydrolysis gave glucose, xylose, galactose, rhamnose, arabinose, and glucuronic acid in the ratios 5:4:1:1:1:4 (Bishop *et al.*, 1954). Similar results for *Nostoc* and *Oscillatoria* are reported by Hough *et al.* (1952). Considerable work has been done by Frederick on the structure and biosynthesis of the starch in *Oscillatoria* (Frederick, 1954). The myxophycean analysed by Parsons *et al.* (1961) contained fructose and hexosamine.

(b) *Chlorophyceae.* By far the greatest amount of work has been done on members of this class, in particular the genus *Chlorella.* Northcote

TABLE V

Carbohydrate composition of phytoplankton cells. Data from Parsons et al. (*1961*)

Species	*Crude fibre* (*percentage of total carbohydrate*)	*Principal sugars* (*percentage dry weight of cells*)										
		Glucose	Galactose	Mannose	Ribose	Xylose	Arabinose	Rhamnose	Fucose	Fructose	Hexosamine	Hexuronic acids
Chlorophyceae												
Tetraselmis maculata	12·6	11·9	2·3	—	0·95	—	—	—	—	—	—	+
Dunaliella salina	9·8	17·2	11·8	—	1·7	—	—	—	—	—	—	+
Chrysophyceae												
Monochrysis lutheri	3·6	22·1	4·4	—	1·3	3·5	—	—	—	—	—	+
Syracosphaera carterae	1·7	9·2	7·1	—	1·5	0·8	1·9	—	—	—	—	+
Bacillariophyceae												
Chaetoceros sp.	22·8	3·3	1·5	0·79	0·71	0·4	—	2·8	+	—	—	+
Skeletonema costatum	9·6	16·4	1·8	0·87	1·2	—	—	1·0	0·9	—	—	+
Coscinodiscus sp.	29·0	2·1	0·4	0·41	+	—	—	0·7	0·5	—	—	+
Phaeodactylum tricornutum	2·5	10·7	2·7	3·7	0·72	0·7	—	1·5	—	—	—	+
Dinophyceae												
Amphidinium carteri	2·0	19·0	8·4	—	0·9	—	—	+	—	—	—	—
Exuviella sp.	37·0	26·8	8·3	—	+	+	+	+	—	—	—	—
Myxophyceae												
Agmenellum quadruplicatum	17·4	17·4	3·2	—	1·5	—	—	—	—	3·5	0·3	+

\+ Sugars detected but not estimated.
— Sugars not detected.

et al. (1958) have made a masterly study of the composition and structure of the cell wall of *Chlorella pyrenoidosa*, which was found to consist of fibrils of alpha-cellulose in a matrix of glycoprotein and hemicelluloses. Galactose, arabinose and mannose were present in significant amounts. By contrast, Lewin (1958) found that the cell walls of *Platymonas subcordiformis* were non-fibrous, and were composed mainly of galactose and uronic acid, with no glucose present. Low concentrations of free glucose and other simple sugars have been reported in the chlorophyceae, especially in nitrogen-deficient cells (Iwamoto and Sugimoto, 1958). Sucrose is present in *Dunaliella salina* (Serenkov and Pakhomova, 1961); Schükerk (1960) found small amounts of ten reducing and five non-reducing sugars. Six genera of fresh-water chlorophyceae were studied, the various genera showing quantitative and qualitative differences. Olaitan and Northcote (1962) reported a very thorough study of the polysaccharides of *Chlorella pyrenoidosa*, which consisted of two hemicelluloses and a starch; one hemicellulose was high in galactose. The starch-type polysaccharide in a species of *Dunaliella* has been characterized by Eddy *et al.* (1958).

(c) *Dinophyceae*. The dinophyceae are reputed to possess cellulose walls and to store reserve carbohydrates as a starch. As might be expected, the Gymnodinialen *Amphidinium carteri* was found to have a very low crude fibre content by Parsons *et al.* (1961). It contains glucose and mannitol in the free form, with glucose polymers as the "insoluble" carbohydrate (Bidwell, 1957).

(d) *Chrysophyceae*. Little is known about the chemical composition of members of this class, but it is probably very varied. The carbohydrate storage product, which is called "leucosin", was studied in *Ochromonas malhamensis* by Archibald *et al.* (1958). It is a glucose polymer with a high proportion of β-glycosidic linkages. Galactose and mannose polymers are also present. Analyses of the sugars in the rock-pool flagellate *Monochrysis lutheri* and the coccolithophore *Syracosphaera carterae* are given in Table V.

(e) *Bacillariophyceae*. The composition of the frustules of diatoms is under study by several workers at the time of writing, but little has been published up to now. As might be expected, the percentage of crude fibre is relatively high. The pectin-like material on which the sculptured silica (opal) is formed, contains both protein and a complex carbohydrate mixture in which mannose is present (probably giving rise to the mannose recorded in Table V). The polyglucuronate excreted by *Navicula pelliculosa* (Lewin, 1955c) may not be a normal constituent of the cell wall in healthy cells. Barashkov (1956) found that *Thalassiosira gravida* and *T. nordenskiöldii* contained no sucrose but that oligosaccharides of two or three glucose units were present. Small amounts of sucrose and fructose together with glucose (*c.* 0·1% in all) were reported in *Nitzschia closterium* by Rho (1959).

The less soluble fraction of the carbohydrates inside diatoms is mainly a "leucosine". Barashkov (1956) found glucose, galactose, arabinose, xylose, ribose and rhamnose. In addition, Parsons *et al.* (1961) detected fucose, which was absent from all other classes of phytoplankton studied. The leucosine of diatoms resembles the laminarin of the Phaeophyta. To avoid confusion, the name *chrysolaminarin* is best given to this storage material in the Chrysophyta. Beattie *et al.* (1961) have shown that chrysolaminarin contains over 99% glucose, unlike laminarin which had a mannitol group terminating the chain. The chrysolaminarin molecule contains about 20 glucose units, mainly joined by 1:3-linkages.

(2) *Lipids.* The lipids in phytoplankton appear to enter the food chain largely unchanged. Zooplankters and fish have a spectrum of fatty acids very similar to that of the marine plants (Lovern, 1936; Kelly *et al.*, 1958).

Analysis for bulk lipid has generally been made by weighing the ether-extractable material from plant cells, but many substances other than simple fatty acid esters are then recorded as "lipid". Strickland and Parsons (1960) suggested a more specific and sensitive micro-method for use with the phytoplankton collected on a Millipore filter. The particulate matter washed from the filter was hydrolyzed with alcoholic potash, and the liberated long-chain acids were determined spectrophotometrically after bromobenzene extraction of their pinacyanol salts. The method gives a measure of fatty acids with more than about 12 carbon atoms, expressed as an equivalent weight of stearic acid. Some values by this method and the classical technique (which gives higher results) are collected in Table IV. On a dry weight basis, the two dinophyceae and the myxophycean appear to have the highest lipid content, but the relative composition of phytoplankters with respect to lipid depends a lot on their state of nutrition.

As might be expected, the chlorophyceae have been the subject of considerable study, but the spectrum of fatty acids has been determined in relatively few species. The diatoms have been investigated because oily material is noticeable in the cells microscopically, and would appear to be a reserve product. However, diatoms grown with excess nitrogen in the medium do not seem to have a particularly high lipid content (Table IV). The classical work by Clark and Mazur (1941) showed a predominance of palmitic and stearic acids in diatoms (much of which was in the free acid form), together with the presence of the hydrocarbon hentriacontane which also exists in the brown algae. Saturated and unsaturated C_{16}, C_{18}, and C_{20} acids have been reported in chlorophyceae and diatoms by Lovern (1936) and Williams and McMillan (1961). Kayama *et al.* (1963) have reported the complete fatty acid spectrum of *Chaetoceros simplex*, in which C_{14} and C_{16} chains predominate. The quantity of lipid found in chlorophyceae, chrysophyceae and bacillariophyceae is given by Rho (1959), Low (1955), Lovern (1936), Aaronson and Baker (1961), Parsons *et al.* (1961) and others, but the lipid content of the cells can be so variable that this type of data has little absolute significance.

Some studies have been made of the sterols in the phytoplankton, and these show marked changes between classes, or even between orders and genera in the same class. Sterols are said to be absent from the myxophyceae. Ergosterol is present in organisms differing as widely as the chlorophyceae, chrysophyceae and euglenineae. Chondrillasterol occurs in the chlorophyceae and bacillariophyceae (Carter *et al.*, 1940; Low, 1955; Aaronson and Baker, 1961). Clark and Mazur (1941) reported sterols similar to the sitosterols of higher plants in diatoms.

(3) *Proteins and amino acids.* The proteins of phytoplankton contain a surprisingly wide range of "essential" acids; they are rich in aspartic and glutamic acids, with large amounts of glycine, alanine, arginine, and lysine. Histidine is in short supply, as also are tyrosine and proline. More information is needed about the tryptophane, cystine and methionine contents of marine algae.

The distributions and abundances of amino acids do not differ greatly between species. Phytoplankton proteins are remarkably similar to those of zooplankton and fish (Cowey and Corner, 1961). Detailed spectra are given for all classes by Parsons *et al.* (1961), for the diatom *Skeletonema costatum* by Cowey and Corner (1963), for *Skeletonema* and *Chaetoceros simplex* by Ogino (1963), and for the myxophyceae by Fowden (quoted in Fogg, 1953) and Linko *et al.* (1957). α,ϵ-Diaminopimelic acid and citrulline are reported from this class. Details for chlorophyceae are given by Hundley *et al.* (1956), and Eny (1949). Serenkov and Pakhomova (1961) found α-aminobutyric acid and citrulline in *Dunaliella*. Low (1955) has analysed for 19 amino acids in fresh-water and marine diatoms.

Most plant cells contain several per cent of free amino acid or polypeptide nitrogen (see Serenkov and Barashkov, 1954; Krauss, 1956; Rho, 1959). Schükerk (1960) found up to 18 different free amino acids in six species of chlorophyceae. The amount and nature of the free amino acids depend on the organism and on its source of nitrogen (ammonia, urea, nitrate, etc.). Their amount decreases greatly in nitrogen-starved cells (Riesner *et al.*, 1960; Thang, 1961).

Protein is normally determined from total cell nitrogen by multiplication by a conversion factor (usually 6·25). The method is open to criticism, but is standard practice, and is probably satisfactory for most field work. Attempts to make a more direct routine estimate of protein have been undertaken by Krey (1951) and by Strickland and Parsons (1960). The Krey method uses the biuret reaction, after a gentle alkaline hydrolysis, and gives a measure of the more "reactive" proteins. This may give a crude differentiation between living matter and less reactive detritus. Strickland and Parsons employed a prolonged hydrolysis with 6N HCl, followed by a treatment with Ehrlich's reagent and acetonyl acetone (Keeler, 1959). This gives a better measure of the total amino acids and amide nitrogen than do procedures using the biuret reaction or the Folin phenol reagent. A comparison of the Keeler method, using a casein standard, and protein expressed as N $\times$ 6·25 has been given by Parsons *et al.* (1961). The mean ratio of results by the Keeler method to those by the Kjeldahl method was 0·77, for 11 species of algae in 6 Orders.

(4) *Other compounds*. The nucleic acids and related compounds in marine phytoplankton have received relatively little attention. By analogy with other plants, it would be expected that some 20% or more of the cell phosphorus would be combined as phospho-lipo-protein, and that sizeable concentrations of RNA, DNA and nucleotides would also be present. About 6% of RNA + DNA was found in vigorously growing chlorophyceae and myxophyceae (Jeener, 1952; Serenkov and Pakhomova, 1961; Serenkov and Vladychenskaya, 1962; Strange *et al.*, 1962). Barashkov (1962) reported about 5% of nucleic acid in marine diatoms. Much of the DNA is synthesized immediately prior to cell division (Iwamura, 1955).

There are two distinct sources of DNA in *Chlorella*, one in the chloroplast and the other in the "nucleus" (Iwamura, 1960). Ratios of the RNA bases in three marine algae are given by Low (1958), and of ATP to ADP in *Chlorella* by Syrett (1958). Markham (1955) has written an excellent treatise on the determination of nucleic acids in plants.

Ascorbic acid may be present in marine algae to the extent of several per cent (Black, 1953). Droop (1957) and Guillard and Cassie (1963) report a reasonably constant amount of between 3 and 20 molecules of vitamin B_{12} per μ^3 of living plant material in cultures where growth has stopped due to vitamin deficiency. B_{12} contents are given by Darken (1953), Hashimoto (1954), Brown *et al.* (1956) and Cowey (1956). The synthesis of this vitamin is variable, but the amount present in plant cells is of the same order as that found in bacteria, namely, about 0·02 p.p.m. of the dry cell weight. Coombs (1952) has given the concentration range for most of the B-vitamins in dry *Chlorella*, and some data for the benthic algae have been collected by Black (1953).

The above list of metabolites includes by no means all the chemical substances that have been found in phytoplankton, and a large number of possible compounds have not yet been investigated. Free acids of small molecular weight are certain to be present. Sieburth (1961) has reported acrylic acid in *Phaeocystis pouchetii* and Rho (1959) lactic acid in *Nitzschia closterium*.

D. PRACTICAL CONSIDERATIONS, COMPOSITION RATIOS

If water from lakes or seas is passed in volumes of a few litres through clean netting of 150–250 μ mesh size, most of the zooplankton is removed, and phytoplankton and detritus remain in the water. The lower size limit of "particulate" material in water is generally defined by the method of filtration. If adsorptive filters are used (Fox *et al.*, 1953), the amount of organic carbon retained may be as great or greater than the amount left in the filtered water. However, it is now common practice to use membrane filters of 0·5–1 μ pore size. This gives a fairly realistic division of the micro-particulate matter into material containing living organisms, which is retained by the membrane, and non-cellular (colloidal as well as truly soluble) organic matter which passes through the filter. Membrane filters, even if coated with magnesium carbonate (which may adsorb some of the larger protein colloids), retain much less organic carbon than they allow to pass. In the ocean, this "soluble" carbon exceeds the "particulate" carbon by a factor of at least 10, and the ratio is greater in coastal waters and many lakes.

There are several practical disadvantages in weighing the particulate matter retained from sea water on a filter. This method has been used (Krey, 1950; Lisitzin, 1956; Riley, 1959), but the organic matter in the total solids is found to be less than half of the weight present.

The best method for estimating organic particulate matter consists of filtration, followed by the use of specific or semi-specific microanalytical procedures such as have been outlined in the previous sections. Unfortunately, these techniques give no information about the amount of a given substance which is associated with the phytoplankters, as distinct from that associated with the smaller zooplankters or the detritus in a water sample. In the open ocean, the concentration of micro-detritus far exceeds the concentration of phytoplankton (fivefold or more, see

McAllister *et al.*, 1960), and even in coastal waters, where heavy "blooms" of plant material may occur, the fraction of detritus is rarely negligible, and significant amounts of "nanno" zooplankton will be collected along with the phytoplankton.

There is, at present, no known *physical* method for separating plant material from zooplankton and detritus, although the flotation techniques used with bacteria (Gaudin *et al.*, 1960) may prove of value. Direct microscopic observation (Krey, 1961b) is the most satisfactory method for estimating the relative proportions of the various constituents of particulate matter, but is slow and tedious. The percentage of total carbohydrate which is "crude fibre" gives a qualitative indication of the detritus to plant ratio, as the concentration of crude fibre is high in detritus (about 70%), and low (about 15%) in living phytoplankton (Parsons and Strickland, 1962a). The percentage of the total protein which is measured by the biuret reaction, after gentle hydrolysis, should also give an indication of the living, as distinct from detrital matter (Krey *et al.*, 1957). Both the above techniques, although somewhat arbitrary, may be useful and complementary approaches for field investigations, but further experimentation is needed.

Probably the most reliable and rapid method in use is the estimation of plant constituents via the determination of chlorophyll *a*. If the concentration of this pigment is measured, the amount of plant carbon, protein, carbohydrate, phosphorus, etc. in a water sample can be evaluated, if the ratio of these substances to chlorophyll *a* is known. Unfortunately, these ratios are very variable, and depend not only on the species of plant present in the sample, but also on the level of nutrients in the water. The subject has been reviewed by Strickland (1960) who has assessed a few of these factors. Since this review was written, further information has been obtained on the ratios of elements and metabolites to chlorophyll *a* in marine phytoplankters under various conditions of nutrition. This information is summarized in Table VI, which includes unpublished data obtained by the author and results by McAllister *et al.* (1960, 1961), Parsons *et al.* (1961), and Antia *et al.* (1963). Further experiments, especially on *in situ* cultures of oceanic populations in nutrient-depleted water are required.

It is sometimes convenient to express the composition of phytoplankters, in pure culture, or in naturally-occurring mixed populations, in terms of the ratios of various metabolites to plant carbon. This practice enables comparisons to be made of the chemical composition on an effectively ash-free basis, and has the practical advantage that the absolute weight of a sample does not have to be determined. Some examples taken from data given by McAllister *et al.* (1960), Parsons *et al.*

TABLE VI

Some ratios involving chlorophyll a

Species and conditions	Ratios: Carbon / Chlorophyll a	Protein / Chlorophyll a	Carbohydrate / Chlorophyll a	Lipid / Chlorophyll a	N / Chlorophyll a	P / Chlorophyll a
Chlorophyceae in culture Large excess of N and P in medium	25–40	30–55	c. 20	c. 4	5–9	2–4
Chrysophyceae in culture Large excess of N and P in medium	75–110	c. 100	40–80	Very variable 7–35	c. 15	2·5–6·5
Bacillariophyceae in culture Large excess of N and P in medium	25–90	45–90	10–25	7–13	7–15	c. 2·5
Dinophyceae in culture Large excess of N and P in medium	65–80	65–80	45–65	c. 25	c. 9	c. 2
Open ocean crop (coccolithophores) Excess of N and P in water	c. 65	c. 95	c. 60	—	c. 15	c. 2
Coastal crop (mainly diatoms) Excess of N and P in water	c. 25	c. 60	c. 6	3–7	c. 10	c. 1·5
Coastal crop (mainly diatoms) Nutrient-depleted water	c. 60	c. 100	c. 60	10–25	c. 16	c. 2

Note: Carbon as dichromate-oxidizable matter.
Protein as N × 6·25.
Carbohydrate as glucose by anthrone reaction.
Lipid as stearic acid by pinacyanol.
Chlorophyll *a* using revised formula (Parsons and Strickland, 1963).

(1961), and Antia *et al.* (1963) are shown in Table VII. It will be seen that the composition of an open ocean crop containing coccolithophores and other chrysophyceae is very similar to that of the coccolithophore *Syracosphaera carterae*. Ratios involving protein show the least variation between the different species, and are least affected by a change of nutrient conditions.

The ratio of chlorophyll *a* to total carotenoids is remarkably constant in the various species. Both pigments change in a similar manner when the cell concentration of either is reduced because of a low nutrient environment, and the pigment ratio remains nearly constant (Parsons *et al.*, 1961; McAllister *et al.*, 1961; Antia *et al.*, 1963). However, when

TABLE VII

Some ratios involving carbon, nitrogen and phosphorus

Species and conditions	*Protein*/*C*	*Carbohydrate*/*C*	*Lipid*/*C*	(*Atoms*) *C*/*N*	*C*/*P*	*N*/*P*
Dunaliella salina (nutrient-rich medium)	1·43	0·80	0·15	4·3	11	6·1
Monochrysis lutheri (nutrient-rich medium)	0·94	0·59	0·22	6·8	19	5·8
Syracosphaera carterae (nutrient-rich medium)	1·41	0·45	0·12	4·4	33	17·4
Skeletonema costatum (nutrient-rich medium)	1·38	0·79	0·17	4·4	15	7·7
Coscinodiscus wailsii (nutrient-rich medium)	1·08	0·27	0·11	5·9	40	15·0
Amphidinium carteri (nutrient-rich medium)	0·69	0·75	0·44	9·0	37	13·2
Exuviella sp. (nutrient-rich medium)	0·70	0·84	0·34	8·9	34	12·3
Agmenellum quadruplicatum (nutrient-rich medium)	0·86	0·75	0·31	7·2	30	10·6
Open ocean crop (coccolithophores) (nutrient-rich water)	1·45	0·9	—	4·3	33	15·5
Coastal crop (diatoms) (nutrient-rich water)	2·0	0·2	0·15	3	20	14
Coastal crop (diatoms) (nutrient-depleted water)	1·3	1·1	0·35	4·7	33	16·5

Note: Carbon as dichromate-oxidizable matter.
Protein as N × 6·25.
Carbohydrate as glucose.
Lipid as stearic acid.

algae are grown with gross nutrient shortages, the ratio of chlorophyll *a* to total carotenoids will decrease. Chlorophyll concentrations are more sensitive to cell deficiencies than are carotenoid concentrations.

Results obtained with cultures (Ketchum *et al.*, 1958; Yentsch and Vaccaro, 1958) have indicated that pigment ratios can be used to measure both the nitrogen content of phytoplankters, and the decrease of photosynthetic efficiency as reflected by the ratio of net to gross photosynthesis. The observations undoubtedly apply to cultures, but the value of this approach to field work still remains to be proved as it has not been fully substantiated under more natural conditions, with mixed populations, when nitrogen and phosphorus become depleted simultaneously (McAllister *et al.*, 1961; Antia *et al.*, 1963). Very severe treatment of phytoplankters, such as decay or ingestion, causes a preferential destruction of chlorophyll, and abnormally low ratios of chlorophyll to total carotenoids probably indicate a heavy concentration of plant detritus. As the detritus ages, xanthophylls are destroyed in preference to carotenes (Fox, 1950).

V. Effect of Light, Temperature and Nutrition on Phytoplankton Growth and Composition

The unicellular marine phytoplankton, in common with other algae, respond to three main variables in nature. These are light (intensity and duration of illumination), temperature and the chemical composition of the environment. In the open sea, the salinity of the water has generally little effect except to form a barrier for certain species which have narrow salt tolerances. Carbon, sulphur, sodium, potassium, calcium and boron are never in short supply. Nutrient shortages can arise only with nitrogen and phosphorus, certain of the B-group vitamins and perhaps, on occasions, some transitional elements such as iron, manganese, copper, cobalt, zinc and molybdenum. Some idea of the effect of these variables on the growth of marine phytoplankton is essential if the production of plant food in the sea is to be understood. The subject is one of great complexity as the effect of each variable depends on the others, and is different for each plant species.

An account of the biochemistry of the photosynthetic process in algae would be out of place in this chapter, but before discussing the growth of phytoplankters and factors which affect their chemical composition, an outline will be given of the methods that have been used for measuring photosynthesis and the interpretations which have been put upon these measurements.

The net change with time of cell numbers, algal weight, algal dry weight, total algal carbon or of the specific metabolities in a water sample require no further discussion. Such measurements of algal growth are perhaps the least ambiguous, but are rarely used, mainly because of the difficulties and length of the analyses involved, and the comparatively large volumes of sample required. Changes of concentration are generally too small for precise measurements to be made unless experiments are extended for an unreasonable period of time.

A. Methods of Measuring Photosynthesis and their Interpretation

(1) *Oxygen changes, PQ and RQ.* Historically, the most important technique used for measuring marine photosynthesis was the measurement of changes in dissolved oxygen. The Winkler method was used to find the concentration of oxygen in sea water before, and after, a period of exposure of the water plus indigenous flora to light. This procedure was popularized by Gran and others in the period around 1930. B.O.D. bottles of 300 ml. capacity are convenient, and with great care, oxygen changes of the order of 0·02 ml./l. can be detected. The increase of oxygen upon illumination of algal suspensions measures the net photosynthesis. Corrections can be made for respiration using a light-tight bottle, and this enables gross photosynthesis (net photosynthesis plus respiration) to be evaluated. Full experimental details are given by Strickland and Parsons (1960), and the subject is discussed in detail in the review by Strickland (1960), in which many of the practical difficulties are described. The Warburg manometer, with suitable modifications to allow for illumination, can be used with phytoplankton cultures, but the plant concentrations in natural waters are generally too low for this equipment to be used. Several membrane oxygen electrodes which do not affect the cell physiology and which possess the required sensitivity have been developed recently (see Chapter 21, Volume 2, Section VIII, A(4) and Kanwisher, 1959). They are very convenient for certain studies, but do not have as high a precision as does the Winkler method.

In order to estimate from oxygen changes the amount of photosynthesis or respiration in terms of carbon, the photosynthetic quotient (PQ) or respiratory quotient (RQ) must be known (see Section II).

Over periods of several hours, the PQ must, theoretically, equal unity with a cell producing hexoses, or about 1·4 with a cell making only lipid. If a protein is formed, with ammonia as the source of nitrogen, the PQ should equal about 1·05, but it will be as high as 1·6 if nitrate is the source. The form of nitrogen in the medium and the fat content of the final cells are thus the main factors which determine the PQ (Syrett, 1954, 1956), and experimental results generally agree well with those predicted from chemical analysis of the plant cells. In the sea, values of

1·25 or even 1·33 are attained (Ryther, 1956a; Nielsen, 1952a). If cells are nitrogen starved and exposed to high light intensities, the PQ will be around 1·1 while the cells make carbohydrate, but it may rise if fat metabolism predominates. If light intensities are reduced and external nitrogen is supplied, a very high PQ can result because carbohydrate-glutted cells resynthesize protein, partially at the expense of the existing hexose (Myers and Cramer, 1948).

Carbohydrate rather than lipid material appears to form the substrate from most respiratory processes; as witnessed by the fact that the RQ stays near to unity even for starved cells (Myers, 1951; Fogg, 1953). Only when nitrate is taken up by the nitrogen-deficient cells during a period of darkness will the RQ increase substantially; these conditions are probably rare in the natural environment.

(2) *Carbon dioxide changes, the uptake of radioactive carbon.* The change of carbonate concentration in sea water caused by the photosynthesis or respiration of the phytoplankton can be measured by direct gas analysis in a manometer, or by using the Van Slyke or other gas analyzers (cf. Hamm and Thompson, 1941; Saruhashi, 1953). The analytical problems are mainly associated with the difficulty of obtaining a sufficient precision for *small differences* of total carbonate (typically 2–3%) to be measured. A more rapid method is to determine the alkalinity of the sample, and to calculate carbonate changes from changes of pH as described by Harvey (1957). Again, sensitivity is such that only the highest levels of productivity in the marine environment can be determined with worthwhile precision.

Lomis (1958) and Kanwisher (1960) have developed methods in which dissolved carbon dioxide is equilibrated with air and its concentration in the gas phase is measured (see p. 252). Since the partial pressure of carbon dioxide in the gas phase, which is in equilibrium with the liquid phase, is very sensitive to small changes of total carbon dioxide, this approach should have increasing applications to field work.

The low productivity of oligotrophic waters makes it impossible to measure photosynthesis through the detection of changes in concentrations of either dissolved carbon dioxide or oxygen using present techniques. In such circumstances, measurements using radio-carbon give the necessary sensitivity. The credit for the first application of the ^{14}C-isotope to marine productivity studies goes to Nielsen who first established a general shipboard technique (1952a) that has been changed very little by subsequent workers.

A sea-water sample is inoculated with a small volume of radioactive carbonate solution, and the indigenous phytoplankters are allowed to photosynthesize for *N* hours. The plankton is then filtered onto a membrane filter and the radioactivity in the cells is measured with a suitable Geiger counter. If c is the number of counts per minute obtained from the filter, C is the total number of counts per minute which would have been found if all the added radioactivity were present as a weightless source, and W is the concentration of carbonate-carbon in the water sample (as mg./m.^{3}, typically 20,000 to 27,000), then:

$$\text{Photosynthesis (mg.C/m.}^3\text{/hr.)} = \frac{c}{CN} \times W \times f$$

where f is a factor to allow for fractionation of the ^{14}C-isotope during photosynthesis; this is thought to be about 1·06 (Nielsen, 1955; Sorokin, 1960a). To allow for isotopic exchange and other processes of carbon-14 intake not associated with photosynthesis, a dark "blank" must be carried out on a sample of the same water, incubated in a light-tight bottle.

The ^{14}C method is very sensitive, and will measure photosynthetic rates as low as 0·1 mg.C/m.^{3}/hour. Experimental details are given by Strickland and Parsons (1960) (see also McAllister, 1961a), and the subject is discussed more fully by Strickland (1960). The main difficulties are associated with the determination of the total added activity in a "weightless source" (C). The method employed by Jitts and Scott (1961) using a scintillation counter is to be recommended.

(3) *External sources of carbon dioxide.* Carbon dioxide is present in sea water mainly as bicarbonate, but there are appreciable quantities of carbonate if the pH exceeds 8. The free carbon dioxide concentration is around 1% or less of the total carbon dioxide (see Fig. 4, p. 255). The concentration of carbon dioxide is never a limiting factor in marine photosynthesis, but there is some question as to whether the bicarbonate ion can be used directly, or whether only dissolved free carbon dioxide penetrates the plant cell wall. It is most improbable that the carbonate ion could be used as such. The reader is referred to the review by Rabinowitch (1956) for details. Information has been obtained mainly by studying the rate of photosynthesis in solutions containing widely differing concentrations of bicarbonate ion and dissolved carbon dioxide, but the main difficulties arise in the interpretation of the data so obtained. From the work of Osterland, Wittingham, Gaffron, Steemann-Nielsen and others, a clearer picture has recently emerged from which it seems that both carbon dioxide and bicarbonate ions can participate. As usual with these problems of algal physiology, much of the work has been undertaken using fresh-water chlorophyceae, and one is left in considerable doubt as to what generalizations are possible to cover marine conditions.

Dissolved carbon dioxide is undoubtedly the favoured compound for carbon transfer across the cell membrane, but some aquatic plants appear to have acquired the ability to utilize bicarbonate, especially those whose natural habitat is an alkaline environment. Nielsen has suggested that the deciding factor may be the dimension of the liquid diffusion path into the cell. Where this is only 1 μ or less, carbon dioxide is probably transported at an adequate rate, whatever its extra-cellular concentration. With longer path lengths, a plant cell may be able to change from one transport mechanism to the other according to the relative abundance of carbon dioxide and bicarbonate ion. It might be expected that most marine plants would, by now, have adapted themselves to a bicarbonate mechanism and there is evidence that such is the case, although there may be variability among species (Felföldy, 1960a,b; Hood and Park, 1962). Nielsen (1963) still considers direct utilization of carbon dioxide is likely for most marine phytoplankton; this view is supported by Watt and Paasche (1963) who have criticized the evidence and reasoning used by Hood and Park in support of bicarbonate utilization.

The possibility that some of the carbon dioxide in sea water may be present in a readily available carbamino-complex (Neuberg *et al.*, 1957) cannot be overlooked. Such complexes carry some of the carbonate in mammalian blood (Stadie and O'Brien, 1936), and a glutamic acid complex appears to be important in photosynthesis by *Chlorella* (Warburg, 1958). There is recent evidence of carbonate, in the carb-amino form, being used by the marine phytoplankters of inshore waters rich in organic matter (Smith *et al.*, 1960). A serious error could arise in the radio-carbon technique if a substantial fraction of the carbonate taken from the water by plants was in the form of a carbamino-complex, since added $^{14}CO_2$ equilibrates very slowly with such complexes. An inhibition of $^{14}CO_2$ uptake has been reported for cultures in the presence of the alanine carbaminate, but there is still no evidence of any such inhibition under natural conditions (Nielsen, 1963).

(4) *Internal sources of carbon dioxide.* The respiration of marine phyto-plankton has received comparatively little study, and from recent work by Ryther and Guillard (1962) on diatoms, it would appear that few, if any, generalizations are valid. Maximum Q_{10} values* of about 2 are found, but the past history of the cells is important, and species differ greatly and have different temperature ranges and conditioning times. Generally, the respiration rates do not exceed 5–10% of the maximum rate of photosynthesis of the same cell under conditions of healthy growth in adequate light (Nielsen and Hansen, 1959a).

* Rate at + 10°C / rate at t °C.

The respiration of algae and other plants *during* photosynthesis is difficult to estimate, and has only been measured unequivocally since experiments have been conducted using isotopic oxygen and carbon. It now appears that respiration is, in general, the same in the light as in the dark (Van Norman and Brown, 1952; Brown, 1953), although in strong illumination the respiration rate may increase (Brown and Weis, 1959). However, later work is conflicting, and the subject is by no means settled (see Weis and Brown, 1959).

Working with $^{14}CO_2$, Weigl *et al.* (1951) found that respiration in the light was sometimes variable, and depended upon the external concentration of carbon dioxide, suggesting that intra-cellular carbon dioxide was being re-used for photosynthesis before it left the organism. This effect is important, because it is involved in the interpretation of productivity measurements, and merits a fuller discussion. Franck (1953) suggested that photosynthesis could occur from respiratory intermediates, and Ryther (1956c), working with cultures of *Dunaliella euchlora* heavily labelled with ^{14}C, found evidence for the complete re-assimilation of respiratory carbon dioxide before it left the cell. If this were a general phenomenon, the radio-carbon method would measure net photosynthesis, since all respiratory carbon (largely unlabelled in experiments of short duration) would be utilized to the exclusion of a corresponding quantity of labelled extra-cellular carbon dioxide. Nielsen (see Nielsen, 1953, 1955; Nielsen and Al Kholy, 1956) maintains that only some 50–70%, rather than all of the respiratory carbon dioxide, is reassimilated, and this would result in rate measurements by radio-carbon lying somewhere between those for gross and net photosynthesis. A considerable controversy has arisen as to the exact interpretation of the $^{14}CO_2$ uptake data, but there seems to be no reason why the fraction of respired carbon reassimilated should be constant for all species and all growth conditions. At best, the radio-carbon method might be assumed to measure values somewhere between net and gross photosynthesis, but the matter is of no practical importance except in areas of poor light or low nutrient concentrations.

(5) *Excretion of labelled carbon compounds during the measurement of photosynthesis by radio-carbon.* In view of the above uncertainties, it is remarkable how few direct comparisons have been made between the radio-carbon and oxygen methods of assessing photosynthesis on the same population, especially using bacteria-free cultures where plant respiration in the dark can be readily measured by the oxygen technique. Although some radio-carbon photosynthesis rates are in fact comparable with net photosynthesis, many results have been obtained giving ^{14}C rates that are much smaller, especially in eutrophic coastal

waters (Ryther and Yentsch, 1958; McAllister *et al.*, 1961; Antia *et al.*, 1963). Some explanation other than reassimilation, or the preferential use of carbamino-complexes seems necessary.

The most probable reason is that recently synthesized organic matter, labelled with ^{14}C, is excreted by the plant cells almost immediately it is formed. Evidence that this does occur, and that the radio-carbon method gives an approximate measure of the formation of *particulate* organic matter, has been obtained by Antia *et al.* (1963), and is suggested by the work of Duursma (1963) on dissolved organic carbon in the sea. Unpublished work by Strickland and co-workers using bacteria-free cultures has confirmed the general picture, but the excretion of labelled compounds depends on species, nutrient conditions, light intensity and the duration of photosynthesis. Generalizations are dangerous at this stage, and the position is by no means clear. Although the radio-carbon technique of photosynthesis measurement is valuable, and is indeed the only practical method for use in oligotrophic waters, results must still be interpreted with caution.

B. THE EFFECT OF LIGHT AND TEMPERATURE ON PHOTOSYNTHESIS

(1) *Light in the sea and its measurement.* The subject of solar radiation and its attenuation in sea water, with special reference to marine photosynthesis, has been reviewed by Holmes (1957) and Strickland (1958).

(a) *Illumination.* The term illumination is best applied only to light from a standard candle, or a source of closely related spectral distribution. Illumination is flux per unit area, and is commonly expressed in energy terms such as watts per square centimetre, or langlies per minute (ly./min. = g.cal./cm.^{2}/min.). The use of the term "lux" is recommended for illumination and "langlies per minute" for radiation in general.

The use of units based on the standard candle is to be discouraged now that instruments for measuring radiant energy are available. Unambiguous information can be given by stating "illumination" in terms of ly./min., watts/cm.^{2}, etc., *together with* a statement of the wavelength of the radiation used, or the band widths and band "centre of gravity" if a polychromatic light source is employed.

The standard candle unit is useful mainly for visual comparisons of illuminations which have the same spectral characteristics as the standard candle. The illumination efficiency at a given energy rating varies widely with the light source. Factors for converting candle units to energy units are given by Strickland (1958). These are, of necessity, very approximate, and can vary by as much as 3- or 4-fold according to the light source (e.g., various types of fluorescent lamps, incandescent lamps, daylight, etc.).

(b) *Attenuation.* If λ is the wavelength of a monochromatic source, I_0 the initial illumination (or energy per unit time and area), I the attenuated illumination after the radiation has passed through a length dl of attenuating medium, then

$$\mathrm{d}I_\lambda = I_\lambda \cdot \chi \cdot \mathrm{d}l$$

and if the absorption coefficient, χ, is unaltered through a distance l, Lambert's Law states that:

$$\ln I_{0\lambda} - \ln I_\lambda = \chi \cdot l$$

where l is generally expressed in metres for oceanographic measurements. Lambert's Law is not valid when a multichromatic source of diffused light is used, and it is now usual to define a term "vertical absorption coefficient" (k) for submarine measurement. This constant is used in an expression of the same form as Lambert's Law which describes the attenuation of radiation with depth, irrespective of the angle of incidence of the radiation I_0, and the fact that light is attenuated by scattering as well as true absorption. The values of k measured in this way will not be constant near the surface of the sea, but at greater depths, the variation of k is generally quite small within a uniform water mass.

(c) *Solar energy.* The solar energy reaching the sea surface depends on the sun's angle, and on such factors as cloud cover and the dust and the water vapour content of the atmosphere. At midsummer in clear weather at temperate latitudes, the maximum values are about 1·4 ly./min. rising to 1·6 ly./min. in the tropics, but local transient effects may give rise to values exceeding 2 ly./min. In December at latitude 50°, the maximum to be expected is about 0·5 ly./min. Even a moderately overcast sky will reduce values to a third or less of those recorded in clear weather.

About one-half of the total radiant energy from the sun is in the infra-red region of the spectrum, and there is little or no radiation with a wavelength of less than 3000 Å or exceeding 30,000 Å at the sea surface. The fraction in the "photosynthetic range" (3800–7200 Å, see later) will vary with atmospheric conditions and sun altitude. A value of 0·5 for this factor is sufficiently precise for field work of several days duration. If the sun is shining in a clear sky, the fraction will be a few per cent greater, and will be as low as 0·35 when the sun is obscured by cloud in an otherwise clear sky. On overcast days, the factor may rise to 0·8 due to preferential absorption of far infra-red energy by the clouds.

(d) *Distribution of light in the sea.* The vertical attenuation of light in sea water has a complex pattern, and is affected by absorption and scattering by water molecules, dissolved organic matter and particulate organic and inorganic material. The reader is referred to the reviews mentioned above, and to the treatise by Hutchinson (1957) for a more detailed account. The most important features, as far as marine photosynthesis is concerned, are the relatively rapid attenuation of red light near the sea surface, and the weaker attenuation of blue-green light which determines the depth down to which photosynthesis takes place.

For all practical purposes, radiant energy of wavelength exceeding 10,000 Å is completely removed in the first 10 cm. of water. At a depth of 2–3 m., all wavelengths greater than about 7500 Å have been absorbed; this leaves a total illumination of 0·4–0·6 ly./min. of photosynthetically active light in clear water during much of a bright summer day.

The penetration of ultra-violet light may be of considerable ecological importance. Its absorption by pure water is very small even at 3600 Å, and short wavelength components are removed in the sea only by scattering or by absorption by dissolved organic matter. In moderately turbid coastal waters the incident light with a wavelength of 3800 Å or less is almost completely dissipated at a depth of 1–2 m., but in the clearest parts of the ocean or in some oligotrophic lakes, it requires 10–20 m. of water to remove 90% of this radiation.

In the open ocean, the wavelength of maximum light transmission is ill-defined, but is around 4800 Å. For turbid, inshore waters a more pronounced maximum occurs at 5000–5500 Å. With increasing depth, the residual light in the sea has a progressively narrower spectral range with an intensity maximum between about 5000 and 5500 Å, according to the water mass and meteorological conditions. The depth at which 1% of the surface visible light energy remains (which is often assumed to be roughly the depth of the euphotic zone) varies from about 15 m. in moderately turbid coastal waters, to as much as 100 m. in the clearest parts of the ocean.

(e) *Methods of measurement*. A full review of the various techniques for assessing the distribution of radiation in the sea has been given by Strickland (1958). More sophisticated instrumentation has been described since then, but is still not standardized. For work in the euphotic zone, the multi-unit barrier-layer cell equipment described by Holmes and Snodgrass (1961) may be taken as typical. The vertical light at a known depth is compared with the surface illumination at a series of wavelengths obtained by covering the cells with coloured filters. As a plant in the sea is illuminated from all directions, and not just vertically, a spherical collector would perhaps be preferable, and an irradiance meter for measuring the total illumination in this way has been designed by Currie (1960). None of these devices enables the *total photosynthetically active radiant energy* to be assessed at any depth without either making calculations, or making assumptions about the relationship between the attenuation of total energy and the attenuation of light of a given wavelength (see Jerlov, 1947).

The direct measurement of radiation by a thermopile radiometer was employed by Birge and Juday (1929), but does not seem to have found much favour. The equipment has limited sensitivity, but is adequate for work in the euphotic zone. Stephens and Strickland (1962) have recently described a radiometer for the measurement of the absolute value of photosynthetically active light at any depth. The response of thermopile radiometers is slow, and there are other practical disadvantages to their use, but these are offset to some extent by the fact that the photosynthetic radiation is measured directly. A radiometer is of greatest use in the surface layers of the sea where spectral changes are most complex. At depths below which the radiation is reduced to a third or less of its surface value, submarine light is effectively monochromatic, and results similar to those obtained by a radiometer can be more readily obtained using a barrier-layer cell covered with a blue-green filter.

(2) *Photosynthetically active light, action spectra.* Gross photosynthesis is, by definition, zero at zero light intensity. The intensity at which gross photosynthesis equals the respiration rate (i.e. the net photosynthesis is just zero) is known as the *instantaneous compensation intensity.* The value of the compensation light intensity depends on a number of factors such as plant species, temperature and nutrition; very few reliable measurements have been made, especially with bacteria-free cultures of marine organisms. The compensation intensity is almost certainly less than 0·01 ly./min., and is probably somewhat less than 0·005 ly./min. for most natural populations near the bottom of the euphotic zone.

If light intensities become too great, the rate of photosynthesis will decrease. The effect depends upon the duration of illumination, and is accompanied by a bleaching of pigments or an arresting of pigment production, chlorophyll *a* being more readily affected than the carotenoids. If the intensity much exceeds 1 ly./min. its effect is irreversible, and enzyme destruction probably results from photo-oxidation processes (see e.g., Myers and Burr, 1940; Nielsen, 1952b; Sironval and Kandler, 1958; Kandler and Sironval, 1959).

The light energy used in plant photosynthesis was originally thought to be solely that absorbed by chlorophyll. It was soon apparent, however, that light absorbed by some of the carotenoids and other auxiliary pigments must also be effective, presumably via an energy transfer within the chloroplasts. The subject has been investigated with algae by determining *action spectra.* Cell suspensions are illuminated in a special apparatus by monochromatic light of variable wavelength and the relative photosynthesis at each wavelength is measured polarographically by the rate of oxygen evolution. The efficiency per quantum of *absorbed* light is nearly the same throughout the visible part of the spectrum (Tanada, 1951), so that if the relative photosynthetic rate at each wavelength is normalized to the rate per *incident* quantum, the resulting curve follows almost exactly the *in vivo* pigment absorption curve for the same suspension of algal cells. The subject has been reviewed recently by Haxo (1960). If the rates are normalized to unit incident *energy*, which is the more readily measured parameter, the action spectra cannot be made to coincide with the absorption spectra. There are fewer quanta per unit of energy the shorter the wavelength used, and the effect is to *flatten* the action spectra at the blue end of the spectrum. Action spectra are normally measured at very low light intensities, but recently McLeod (1961) has shown that there is a similar behaviour at relatively high light intensities.

It is clear from the above, that the more auxiliary pigments there are in an alga that absorb at wavelengths between the chlorophyll peaks at about 4400 Å and 6700 Å, the more effectively the alga can make use of visible radiation. Protein-bonded fucoxanthin and peridinin enable most marine algae to utilize light from 5000 Å–5500 Å, and the biliproteins very effectively close the "window" in the absorption spectra of carotenoids and chlorophylls.

Although action spectra roughly follow absorption spectra, there are some exceptions; in particular, many algae with biliproteins appear to have much less than the expected quantum efficiency at the chlorophyll and carotenoid peaks. The explanation for this now appears to be associated with the method used for the measurement of action spectra in which the cells are illuminated with *one wavelength at a time*. Action spectra taken whilst the plant is continuously illuminated by light of a second wavelength show fewer anomalies. In particular, illumination by light absorbed by auxiliary pigments such as the biliproteins, chlorophyll *b* or fucoxanthin (but not necessarily all the carotenoids), enhances the action spectra (the Emerson effect) in a manner which suggests high quantum efficiency at all wavelengths. A detailed study of such behaviour is helping to resolve the problem of the mechanism of photosynthesis, and auxiliary pigments are now thought to have a much more complex role than that of simply assisting energy transfers.

Light of wavelength exceeding about 7200 Å is ineffective in photosynthesis, although Rabinowitch and Thomas (1960) have reported a "negative Emerson effect", with inhibition occurring in the photosynthesis of some algae when they are bathed in light of about 7500 Å.

It is not known what, if any, practical significance this observation may have in marine photosynthesis. McLeod and Kanwisher (1962) have extended action spectra to wavelengths shorter than 3000 Å, and have shown that the efficiency does not drop until about 3500 Å. Below 3000 Å, irreversible damage was done to the cells. Although this did not occur at longer wavelengths in the comparatively short time used in their experiments, there is evidence that ultra-violet light can depress or inhibit photosynthesis (Talling, 1961). The effect depends on the age and preconditioning of cells (E. S.-Nielsen and R. A. Vollenweider, personal communications). For practical purposes it is general to assume utilization of light between about 3800 and 7200 Å.

The effectiveness of this utilization clearly depends upon the class of plant, since this determines the pigment systems, and hence the action spectra. A somewhat idealized action spectrum *for unit energy*, which would apply roughly to most chrysophyceae and dinophyceae, is shown in Fig. 1, normalized to 100 at the chlorophyll *a* peak (6750 Å). In the

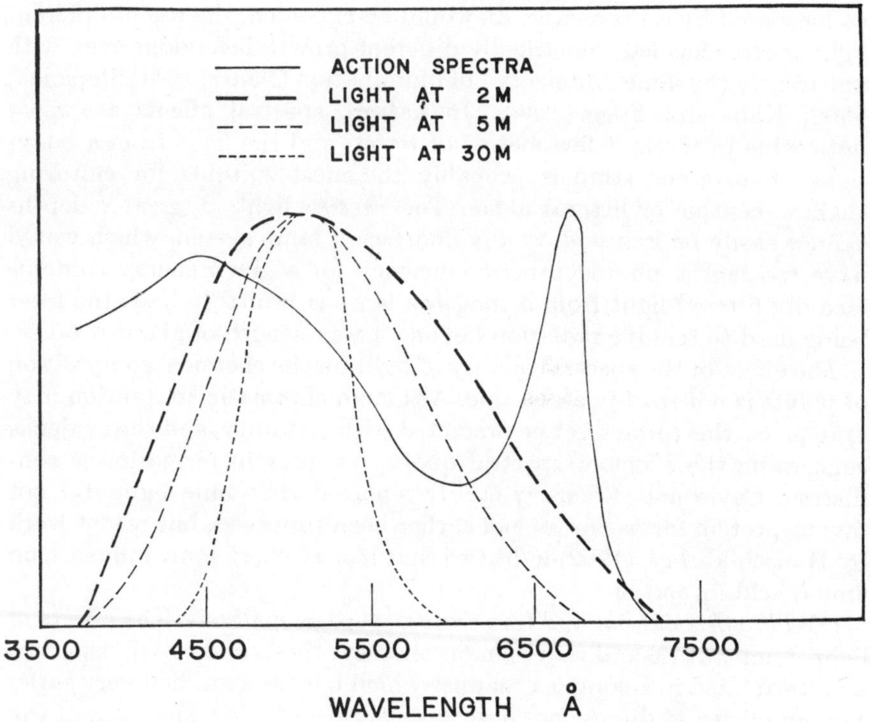

Fig. 1. Spectra for sunlight in sea water and idealized action spectrum for marine phytoplankton.

same figure, the spectral energy curve has been drawn for daylight (sun shining in a cloudless sky) at three depths in clear coastal sea water. The total radiant energy at 2 m. would reach a maximum of about 0·5 ly./min. on a bright summer day at mid-latitudes. The energy at 30 m. on the same day would not exceed about 0·02 ly./min. It will be seen that the chlorophyll *a* has no value as a light-trap, except within a few metres of the sea surface. For most of the euphotic zone, the carotenoids are responsible for light utilization, and as the action spectrum is relatively uniform in the region of peak absorption (this would be nearer 5500 Å in most turbid waters), the efficiency of photosynthesis *per unit radiant energy* should not differ greatly throughout the euphotic zone. Thus, the total energy in the range 4500–6500 Å should be a good index of the effectiveness of a light source in marine photosynthesis, and it may be that chromatic adaptation (Stanbury, 1931) further improves matters.

It is clear from Fig. 1, however, that all light sources could not be expected to be equally effective for the growth of algal cultures, even on the basis of equal energy, let alone of equal "apparent" illumination, as measured by a lux-meter. As would be expected, the use of different light sources has led to markedly different growth behaviour even with apparently the same "intensity" of illumination (Baatz, 1941; Stepanek, 1960; Kain and Fogg, 1960). In nature, spectral effects are most noticeable in the first few metres of water, and the light from a "daylight" fluorescent lamp is probably the most suitable for culturing shallow benthic or littoral algae. The *in situ* light at greater depths cannot easily be imitated by any fluorescent lamp system which would have too high a photosynthetic efficiency for a given energy content. Heavily filtered light from a tungsten lamp is probably best, the filter being used to remove radiation beyond a wavelength of about 6000 Å.

The effect of the spectral quality of light on the chemical composition of plants is not readily assessable. Although chromatic adaptation may take place, this cannot yet be predicted with certainty, and the evidence concerning the effect of spectral quality on protein formation is conflicting. Cayle and Emerson (1957) reported that blue light did not favour protein formation as had earlier been supposed, but recent work by Hauschild *et al.* (1962) indicates that it may exert some influence on amino acid formation.

(3) *The effect of light and temperature on photosynthesis.* The effects of light, temperature and nutrition on photosynthesis and cell division are all interrelated in a complex manner which has been studied very little. For simplicity of discussion, it is probably best to consider each of the main variables separately, with the concurrent effect of other variables being mentioned only where they are of particular significance.

The rate of photosynthesis by algal cells is a function of light intensity, and increases with intensity until a "light saturation" value is reached. Above the saturation value, the rate stays constant until the illumination becomes too great (supra-optimal), after which light inhibits the photosynthetic processes. The shape of the curves relating photosynthetic rate (P) to illumination (I), which we will term *P vs I* curves, depends on the species of phytoplankton, on the temperature, on the method used, and on the time taken to measure photosynthesis. The absolute value of the maximum rate of photosynthesis at light saturation will differ according to the way in which the crop of photosynthetic cells is measured, and it is, therefore, useful for many practical purposes to express the photosynthetic rate P and light intensity I in a *relative* unit f, where f is defined by

$$f = \frac{P}{P_{max.}} \cdot$$

$P_{max.}$ occurs at the saturation light intensity $I_{sat.}$ or I_{max}.

It can be shown (Rabinowitch, 1951) that all theoretical models of the role of light in photosynthesis, i.e. a rate control by enzyme reactions and diffusion processes that cannot eventually keep pace with the primary photosynthetic reaction, lead to the formulation of hyperbolic curves for the relationship between photosynthetic rate and light intensity. In some circumstances, these hyperbolae may be of a simple rectangular form:

$$\frac{f}{1 - f} = KI$$

where K is a constant, but experimental data rarely fit such curves with much precision. The empirical relationship used by Smith (1937) which is of the form:

$$\frac{f}{(1 - f^2)^{\frac{1}{2}}} = KI$$

has been found by other workers to hold quite well. Winokur (1949) used this equation for *Chlorella vulgaris* cultures, and Talling (1957a) claimed great success with it for both cultures and natural populations of a fresh-water *Asterionella*. Steele (1961) has suggested that an equation of the form:

$$f = \alpha I \exp (1 - \alpha . I)$$

where α is a constant, accounts for the inhibiting effects of high light intensities, but it is doubtful whether enough is known about the factors affecting *P vs I* curves for any empirical mathematical representation to have much value, and *in situ* experimental curves should be obtained whenever possible (Talling, 1960).

Some of the earliest *P vs I* curves for marine phytoplankters were obtained for diatoms and dinoflagellates by Barker (1935a,b), and since then many such curves and $I_{max.}$ values for algal cultures and natural populations have been reported (see Jenkins, 1937; Lanskaya and Sivkov, 1950; Nielsen and Jensen, 1957; Nordli, 1957a; Talling, 1957a; Berg, 1958; Kain and Fogg, 1958a; Steele and Baird, 1961). Ryther (1956b) divided the phytoplankton into three groups, the chlorophyceae, with a low $I_{max.}$ and rapid depression at supra-optimal light intensity, the dinophyceae, with a high $I_{max.}$ and relatively little supra-optimal depression and the bacillariophyceae, with intermediate characteristics. It is unlikely, however, that any generalization can be made concerning the depressing effect of high light intensities. Such behaviour is a function of temperature, and depends upon the time of exposure of the algae before the *P vs I* curve is measured (McAllister, 1961b). It also appears that the spectral distribution of the light source must be important (see earlier). McAllister *et al.* (1964) could find no indication of supra-optimal effects with chlorophyceae exposed to over 0·3 ly./min. of radiation when the radiation was free from ultra-violet and red light, in contrast to the effects reported by Ryther, and others, who used sunlight and found supra-optimal light intensities at 0·1 ly./min. or less.

The fact that authors use different light sources and measure illumination intensities by different methods makes the intercomparison of *P vs I* curves and $I_{max.}$ values almost impossible.

The effect of temperature on $P_{max.}$ is generally small over a range of 3°C or more on each side of the "optimum" temperature for each species, especially if the plant cells are allowed to adapt themselves for a day or two at the temperature concerned. At higher temperatures there is an abrupt and generally irreversible decrease in photosynthesis, which may be brought about in part by a greatly increased demand for certain nutrients (Hutner *et al.*, 1957). A large excess of nitrogen and phosphorus extends the temperature tolerance of certain phytoplankters appreciably. For reasons still not understood, the optimum growth temperatures for phytoplankton cultures nearly always seem to be many degrees higher than those which the species would normally be expected to encounter in their natural environments.

The decrease in $P_{max.}$ with decreasing temperature is less dramatic than the effect of high temperatures, and is largely dependent on adaptation time. A moderate degree of cold shock is generally reversible, in contrast to the effect of overheating a culture. Temperature effects are greatest at high light intensities, in particular, the depression of photosynthetic rate by supra-optimal light intensity seems to be assisted by elevated temperatures. The effect of temperature on photosynthesis by marine phytoplankters is referred to in many papers (Barker, 1935a,b; Ichimura and Aruga, 1958; Nielsen and Hansen, 1959b; Curl and McLeod, 1961; Felföldy, 1961; Ichimura *et al.*, 1962).

The use of P *vs* I curves gives no quantitative information about the organisms concerned with photosynthesis, and f *vs* I curves show only *relative* trends. To compare two populations in cultures or in nature, the rate of photosynthesis should be expressed on the basis of carbon present, i.e. (see Section II, B (3))

$$\frac{\text{mg.C/m.}^3\text{/hr.}}{\text{mg.C/m.}^3} = \text{k(hr.}^{-1}\text{)},$$

but this is rarely done since cell carbon is difficult to measure in nature. For field work, the most rapid and convenient way to compare the photosynthesis of two populations is to use the rate of photosynthesis per unit of total photosynthetic pigment present in the same sample or, more generally, per unit of chlorophyll a:

$$\frac{\text{mg.C/m.}^3\text{/hr.}}{\text{mg.Chl.}a\text{/m.}^3} = P(c).$$

The use of $P(c)$ values is dictated mainly by practical considerations, although it happens that $P_{\text{max.}}(c)$ is, in fact, a reasonably good index for comparative plant physiology. The actual value for $P_{\text{max.}}(c)$ will depend on how photosynthesis is measured (gross, net, ^{14}C uptake, etc.) and how near the temperature is to the optimum temperature for the algae concerned. Values from the literature have been collected by Strickland (1960), and vary between 1 and 10, and lie mostly in the range 2 to 6. $P(c)$ values give a good index of population changes, and of the effect of nutrient deficiencies. Nitrate depletion, for example, reduces the $P(c)$ to one-third or less of its value in the presence of excess nutrients (Fleischer, 1935; Ichimura and Aruga, 1958; Curl and McLeod, 1961; McAllister *et al.*, 1961; Patten, 1961; Aruga and Monsi, 1962; Antia *et al.*, 1963). Recently, McAllister *et al.* (1964) have shown that at *low light intensities* $P(c)$ is less constant than is the rate of photosynthesis per unit of *total* plant pigments (cf. Currie, 1958), and that this applies under *all* growth conditions.

For a mixed population of phytoplankters in nature, we would not expect the mean rate of photosynthesis per cell to have much significance. The rate per unit of cell volume will also vary greatly as the carbon and chlorophyll content of cells can differ from species to species. Wright (1959) has shown that, as would be expected, the rate per unit cell volume increases as the amount of chlorophyll per unit cell volume increases. It remained for Paasche (1960a) to demonstrate experimentally that the rate per unit *area* of cell surface is the best parameter in a mixed natural population. This is to be expected, since the amount of plant pigment per unit area is probably more nearly constant than is the amount of pigment per unit volume, especially in the diatoms with large vacuoles.

As mentioned previously, the shape of *P vs I* or *P*(*c*) *vs I* curves depends greatly on the preconditioning of the algae, in particular, on whether plant cells have been reared for some time (one or more days) at low or at high light intensities. This fact is of considerable ecological importance, because the presence of even a transient pycnocline within the euphotic zone can effectively divide a phytoplankton population into lower (shade) forms, and upper (sun) forms which, being above the pycnocline, are able to reach the sea surface and levels of strong illumination. (The terms "sun" and "shade" are borrowed from the description given to certain exposed and sheltered leaves in terrestrial plants.) In winter, even the surface populations in temperate latitudes may develop the characteristics of shade plankton. The characteristic *P*(*c*) *vs I* curves for sun and shade forms have been determined by many workers (Sargent, 1940; Myers, 1946; Sorokin, 1958) and their significance in the marine environment was investigated by Ryther and Menzel (1959), Nielsen and Hansen (1959b), Ichimura *et al.* (1962) and others. Nielsen and co-workers (Nielsen, 1961a, 1962; Nielsen *et al.*, 1962), in particular, have made quite extensive studies, mainly using *Chlorella*. These researches have been of value in discovering general principles, but are open to criticism because the organism used was unrepresentative of marine flora, and the light source (incandescent bulbs) was spectrally very unlike the sunlight in the ocean.

The rather complex behaviour found by these workers is summarized by the (idealized) sun and shade *P vs I* curves in Fig. 2. It will be seen that $I_{max.}$ depends on the light intensity at which a population is first reared. $I_{max.}$ values for sun and shade plankton might be 0·2 and 0·05 ly./min. respectively, but we have no certain knowledge of these values for marine forms in "marine sunlight".

Shade forms are characterized by a low $I_{max.}$ and $P_{max.}(c)$ compared to the cells reared at a high light intensity, but all *P*(*c*) *vs I* curves coincide at low *I* values. However, if we plot *P vs I* curves on the basis of photosynthetic rate per cell, or rate per unit of carbon (these curves will probably be not too dissimilar in shape), we find that the sun and shade cells have practically the *same* $P_{max.}$ values, and that the curves do not coincide at low light intensities. The photosynthetic rate of the shade cells may be twice, or more, that of the sun cells at low light intensities. This is mainly a consequence of the low light forms having a higher pigment content per cell than do the sun forms. At low light intensities, both sun and shade forms have the same photosynthetic efficiency per unit amount of chlorophyll, hence their *P*(*c*) *vs I* curves coincide, and the rate is limited by the efficiency of what Nielsen terms "photochemical" processes. As the intensity of illumination increases, the *P*(*c*) value cannot increase with the shade cells because they are unable to make use of further photosynthetic "potential". The chloroplasts are already photosynthesizing compounds as fast as the cell can make use of them (see the dashed curve in Fig. 2). This restriction by "enzyme" processes has been termed *rate limitation* by Nielsen.

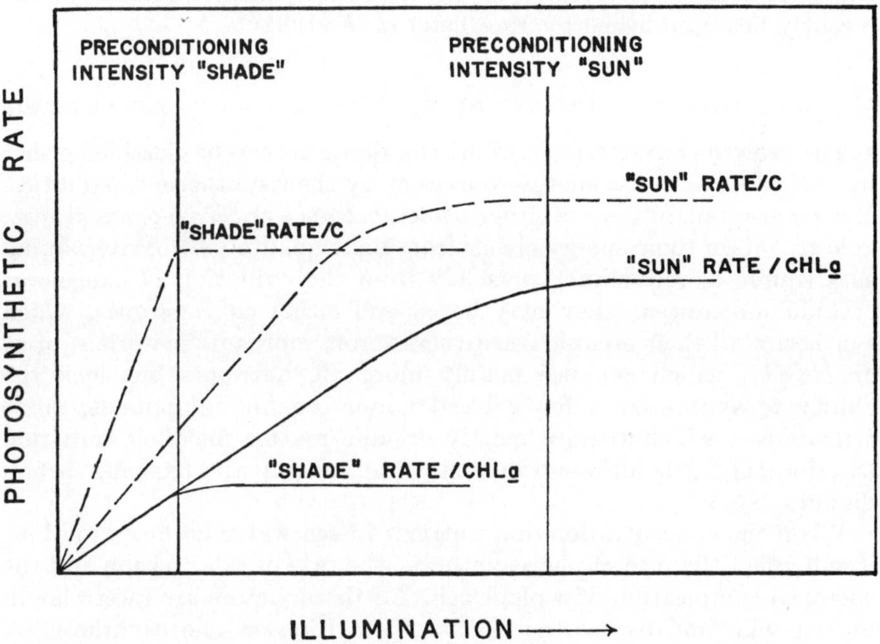

FIG. 2. Effect of light and preconditioning on rate of photosynthesis.

However, this critical rate per cell is not reached with sun cells until much higher light intensities have been attained (owing to these cells containing less pigment), and the *P*(*c*) *vs I* curve continues to rise to a much higher $P(c)_{\max.}$ and $I_{\max}$.

The curves shown in Fig. 2 apply only to short exposures after the period of preconditioning. If sun cells are put in the dark, they retain their characteristics for some time, but when exposed to low light intensities they change, over a period of 24–48 hours, into shade forms. When shade forms are exposed to high light intensities, the chlorophyll content of the cells is reduced, and within a day or two they change to sun forms. The rate of photosynthesis per cell is reduced in shade cells exposed to very bright light even before a significant decrease in the amount of pigment per cell is noted. The reduction of the amount of chlorophyll in cultures of shade forms exposed to bright light is probably caused not by "bleaching", unless intensities are extremely high, but is brought about either because chlorophyll production in new cells is arrested, or because the production rate of chlorophyll decreases relative to the production rate of new cellular material. A detailed study of this phenomenon and of other features of the *P vs I* curves of marine phytoplankters under precisely defined conditions has recently been published by McAllister *et al.* (1964).

C. THE EFFECTS OF MAJOR AND MINOR NUTRIENTS ON PHOTOSYNTHESIS

The growth characteristics of micro-organisms can be classified either by their ability to use energy sources or by their synthetic capabilities. Marine phytoplankton are either *photolithotrophs* or *chemo-organotrophs*, able to obtain their energy either from light quanta (with water acting as a source of reductant), or solely from the oxidation of exogenous organic substances. They may be classed either as *autotrophs*, which synthesize all their organic constituents from inorganic materials, or as *auxotrophs*, which can use mainly inorganic nutrients but lack the ability to synthesize a few critical minor organic compounds, or as *heterotrophs*, which require mostly organic matter for their nutrition (Section II, A (7)) although they may still take up and fix some carbon dioxide.

When the concentration of a nutrient in sea water becomes too low, it will affect the rate of photosynthesis, the rate of cell division and the chemical composition of a plant cell. All these factors are interrelated, but we will first discuss the effects of nutrition on photosynthesis by autotrophs and auxotrophs.

(1) *Major nutrients.* (a) *Nitrogen.* Schreiber (1927) found that ammonia could act as a source of nitrogen for marine phytoplankton. Ammonia is preferentially transported across the plant cell wall in an un-ionized form, and its absorption lowers the pH of soils and cultures, making ammonia a less useful source of nitrogen than nitrate for most plants. In the buffered environment of sea water, such complications do not occur, and the ammonium ion appears to be the preferred nitrogen source for nearly all phytoplankters with, perhaps, some exceptions in the chlorophyceae (Braarud and Føyn, 1930). Nitrate is satisfactory for

diatoms and most other algae, although it is often utilized less rapidly than ammonia, and cannot be utilized at all by some flagellates (see Harvey, 1940; Proctor, 1957a; Harris, 1959; Guillard, 1961; and the reviews by Syrett, 1954 and Provasoli, 1958). Nitrite is used readily by growing phytoplankton, but only in the light (Harvey, 1953a; Harris, 1959; Guillard, 1961; McAllister *et al.*, 1961). Nitrogen-deficient cells can absorb both ammonia and nitrate in the dark, the former more readily than the latter. Hydroxylamine may be available as a nitrogen source for certain organisms (Fogg, 1953), but little is yet known about the occurrence of this compound in sea water.

When the concentration of available nitrogen in sea water decreases below a certain value, the rate of photosynthesis by plant cells will be affected. In nitrogen-starved cells, "enzyme" processes are slowed before "photochemical" processes, and effects are thus most marked at high light intensities. However, much depends on the prehistory of the cell, since, if it initially contains sufficient combined nitrogen, photosynthesis can continue for hours or days without any detectable concentration of nitrate or ammonia being present in the water. For any species, there is doubtless a concentration of available nitrogen which, if continuously maintained, would only just be sufficient to support a maximum rate of photosynthesis at I_{max}. In nature such a constant concentration of nutrients is rarely maintained, except perhaps in the tropics, and we have only a rough indication of the limiting concentration of nitrate. It is probably safe to assume that this is less than 50, and probably less than 10 μg. NO_3^-–N/l.

(b) *Phosphorus.* Inorganic phosphate is readily used by all phytoplankters, and phosphorus exchanges between plant cells and the external medium very rapidly. In lakes, the concentration of phosphate may limit the rate of photosynthesis (Goldman, 1961), but little evidence has been produced that this is the case in the marine environment, where a shortage of nitrogen is usually responsible for stopping growth. Early work by Ketchum (1939) suggested that division rates might be affected if the external phosphate concentration decreased below 15 μg.P/l., but it is doubtful if this figure has any general applicability, and the "limiting" phosphorus concentration in some cultures has been found by the present author to be less than one-third of this amount. The problem is complicated by the fact that auxiliary compounds, the natures of which are still obscure, may affect the availability of phosphate to a plant cell. The classic case concerns the freshwater diatom *Asterionella formosa* which grows well in natural lake waters in the presence of a phosphate concentration which is only a few per cent of the minimum concentration needed to maintain good growth in cultures (Rodhe, 1948; Mackereth, 1953). The shortage of phosphorus in a cell has to be extreme before the "photochemical" or the "enzyme" processes in photosynthesis are affected.

(c) *Silicon*. The silicate concentration of water would be expected to affect only the growth of diatoms for which it is an essential requirement. Using the ^{14}C technique, Ryther and Guillard (1959) and Thomas (1958) have suggested that silicate availability limits phytoplankton photosynthesis in the open sub-tropical oceans, although this work remains to be confirmed.

Studies of silicon metabolism in diatoms have been made by Jørgensen (1953) and Lewin (1955a,b). Only the unpolymerized forms of silicate that react with molybdate are useful for diatom growth, a feature which enables the silicomolybdate method to be used for assaying the physiologically reactive silicate in a water sample. Silicon metabolism is associated in some way with the presence of divalent sulphur, generally in organic combination (Harvey, 1939; Lewin, 1954), but a sufficient amount of such material may always be present in the sea (Curl, 1962). The rate limiting concentration of silicon appears to vary with the species of diatom, but is less than about 50–100 μg.Si/l.

(2) *Trace inorganic elements*. There are two concentrations of a nutrient which may be said to define its effect on the growth of a given plant—the minimum concentration needed to give maximum photosynthesis, and the minimum concentration that must be present before a stimulation of photosynthesis is noted. Neither of these limits has been measured for marine phytoplankton, but an indication of whether or not an element is limiting photosynthesis in a given water mass may be obtained by an elegant procedure using radio-carbon. The water sample and its indigenous plant population is exposed to light with, and without, the addition of the element in question, and any stimulation of growth is shown by an increased uptake of ^{14}C (Ryther and Guillard, 1959; Goldman, 1960). Although such results need to be interpreted with caution, it does seem that trace metals may be responsible for limiting photosynthesis in some of the less fertile sub-tropical ocean areas of the world. The concentrations of trace elements in coastal waters, which are much greater than that in the open ocean, may inhibit the growth of certain oceanic phytoplankton near the coast.

It is most unlikely that the transition elements pass through a plant cell membrane in the free ionic form, or in the form of a complex with a large organic molecule. In experiments with cultures it is common to add a chelating agent, such as citrate or ethylenediamine tetraacetate (EDTA), to enable a large "reserve" of metals to be present at pH values at which precipitation of hydroxides would occur, or at which free ionic forms would be present at inhibitory concentrations. The metal ion, rather than the complex, is probably the form of the element taken into the cell, although membrane transport may involve uncharged or singly charged hydroxy- or chloro-complexes. Chelation sites on the cell wall may compete directly with the external chelator for the free metal ions (Walker, 1954). Such conditions apply to sea water, although ferric iron and tetravalent manganese may not be in true solution. It is unlikely that the *total* concentration of a trace metal in sea water ever limits plant growth, but the form of combination of the metal, about which very little is known may be the governing factor. The subject has been well reviewed by Spencer (1957). Dissolved organic matter may be important to plant growth because of its influence on the availability of iron and other elements (Schelske, 1962). Johnston (1963b) found that the addition of chelators alone to sea water could be as beneficial as the addition of chelated metals.

(a) *Iron.* Iron is needed for the essential haematin compounds of photosynthesis and oxidase systems, and a shortage of it gives rise to chlorotic cells with lowered photosynthetic potential. The most effective "reserve" form for marine phytoplankton appears to be particulate ferric hydroxide, with an optimum particle size distribution for each total iron concentration (Harvey, 1937; Goldberg, 1952). The EDTA-complex at pH 8 appears to promote growth, but we cannot be certain that there is not some photochemical decomposition to hydroxide first. A shortage of iron has apparently been shown to exist in the marine environment, and it may be particularly important in the open ocean (Harvey, 1947; Ryther and Guillard, 1959; Menzel and Ryther, 1961; Tranter and Newell, 1963). Again, caution is needed in the interpretation of experiments. Where insoluble compounds are formed, they may simply stimulate growth by providing an active surface on which other elements are adsorbed.

(b) *Manganese.* A requirement of marine phytoplankters for manganese in sea water has been demonstrated by Harvey (1947), but we have no knowledge of the exact ecological significance of this element, although its essential role in the metabolism of fresh-water algae and terrestrial plants has been amply proved. Manganese acts as a co-factor for nitrate-reducing enzymes (Fogg, 1953) and for enzymes in the fermentation cycle, where it may be replaced by magnesium (Hewitt, 1957). The main function of the element, however, is as a co-factor in the oxygen evolution step, and (possibly) in the reduction step of photosynthesis by autotrophs. Larger concentrations are needed for this purpose than when algae grow heterotrophically, although even then the element is necessary (Reisner and Thompson, 1956; Kessler,

1957; Eyster *et al.*, 1958; Richter, 1961). Walker (1954) found *Chlorella* cells required a minimum of 2·5 p.p.m. of manganese on a dry weight basis. If this is typical, it would appear that the *total* amount of manganese in sea water could never limit plant growth, but its "availability" may be important.

(c) *Other elements*. Elements such as molybdenum, zinc, cobalt, copper and vanadium are known to be co-factors for enzymes essential for plant growth; in particular, molybdenum for reactions involving the reduction of nitrate, zinc in carboxylation and copper in oxidation-reduction cycles. The reader is referred to the reviews by Provasoli (1958) and Hewitt (1959). As far as the author is aware, it has never been shown that a shortage of the above elements limits marine photosynthesis, although they may have an important role in lake waters (e.g. Goldman, 1960 for molybdenum).

The pH of sea water affects plant growth, but probably most phytoplankters can tolerate the extreme range to be expected in the sea (7·5–8·5) (Bachrach and Lucciardi, 1932; Harvey, 1957).

(3) *Trace organics* (*vitamins*). After the significance of vitamin B_{12} and its analogues in the nutrition of protozoa became appreciated, it was not long before an absolute requirement for this vitamin and others was demonstrated for marine phytoplankters, and research has been accelerated by attempts to grow unicellular algae in completely defined artificial media. This work has been pioneered largely by Droop in Millport, Scotland, Provasoli and his co-workers in the Haskins Laboratories, New York, and Guillard at the Woods Hole Oceanographic Institution, although numerous other workers have contributed (see Droop, 1955, 1958; Provasoli, 1958; Droop *et al.*, 1959; Lewin and Lewin, 1960; Guillard and Ryther, 1962; Van Baalen, 1962; Provasoli, 1964). About 70% of the marine phytoplankters thus far examined have been found to be auxotrophic and require cobalamine, thiamine and biotin, either singly or in combination. However, the population studied is undoubtedly a biased one, since much of the research has been undertaken with the kind of organisms that might be expected to show auxotrophy or heterotrophy. The true percentage of auxotrophs in the phytoplankton may well be less than it now appears; the Cyanophyta are almost all autotrophic. Data from the above workers which show the vitamin requirements of some of the marine and fresh-water phytoplankters mentioned in plant physiology literature are collected in Table VIII. Most, but not all, of these species are photolithotrophs.

We have still practically no knowledge of the quantitative effect of vitamin concentrations on the kinetics of photosynthesis. Cobalamine appears to limit rates at concentrations anywhere between 0·1 and

TABLE VIII

Vitamin requirements for some phytoplankters

Class	*Vitamin required*	*Species*
Myxophyceae	None	*Agmenellum quadruplicatum*(?), *Anacystis marina*, *Lyngbya aestuarii*, *Oscillatoria subtilissima*(?)
	B_{12} only	*Phormidium persicinum*, *Synechocystis* sp.
Chlorophyceae	None	*Chlamydomonas moewusii*, *C. reinhardii*, *Chlorella vulgaris*, *Dunaliella salina*, *D. euchlora*, *Haematococcus pluvialis*, *Nannochloris atomus*, *Scenedesmus obliquus*
	B_{12} only	*Chlamydomonas pulsatilla*, *Platymonas tetrathele*
Dinophyceae	B_{12} only	*Exuviella cassubica*, *Glenodinium foliaceum*, *Gonyaulax polyhedra*, *Gymnodinium splendeus*, *Gyrodinium californicum*, *Peridinium trochoideum*
	B_{12} + thiamine + biotin	*Amphidinium carteri*, *Gymnodinium breve*
Cryptophyceae	B_{12} + thiamine	*Chilomonas paramoecium*, *Hemiselmis virescens*, *Rhodomonas lens*
Chrysophyceae	Thiamine only	*Coccolithus huxleyi*, *Dicrateria inornata*, *Syracosphaera* sp., *Pontosphaera roscoffensis*
	B_{12} + thiamine	*Monochrysis lutheri*, *Isochrysis galbana*, *Prymnesium parvum*, *Ochromonas minuta*
	Thiamine + biotin	*Ochromonas danica*
	B_{12} + thiamine + biotin	*Ochromonas malhamensis*
Bacillariophyceae	None	*Melosira* sp., *Asterionella formosa*, *Fragilaria capucina*, *Navicula pelliculosa*, *Phaeodactylum tricornutum*, *Rizosolenia* sp., *Thalassiosira nordenskiöldii*(?)
	B_{12} only	*Chaetoceros simplex*, *Cyclotella nana*, *Skeletonema costatum*, *Stephanopyxis turris*, *Thalassiosira fluviatilis*
	Thiamine only	*Nitzschia closterium*

100 $\mu\mu$g./ml. (Ford, 1958; Droop, 1961a). Research in progress at several laboratories indicates that the data obtained depend very much on growth conditions, and that the results of experiments with laboratory cultures may be difficult to apply to the natural environment.

By measuring photosynthesis in natural waters, with and without added growth factors, some idea of the ecological significance of a substance such as vitamin B_{12} can be obtained. Results to date have not been too spectacular (Ryther and Guillard, 1959; Menzel and Ryther, 1961), but Curl (1962) has shown that there are many similarities between the behaviour of *Skeletonema costatum* in cultures and in the natural environment. In the Sargasso Sea area, small diatoms with a proven B_{12} requirement, "bloom" relative to coccolithophores (needing only thiamine) at the time of year when the B_{12} concentration in the water becomes appreciable (see later).

D. ORGANIC SOURCES OF NUTRITION, HETEROTROPHY AND NITROGEN FIXATION

(1) *Organic nutrition of auxotrophs.* As well as a requirement for certain vitamins, some auxotrophic phytoplankters (and possibly some autotrophs) can make use of organic sources of nitrogen and phosphorus. It is unlikely that any marine phytoplankters would use organic sources of these elements in preference to ammonia, nitrite, nitrate and inorganic phosphate if all these were available, but the *ability* to do so may have some ecological importance. It would enable certain species to continue to grow in sea areas where the inorganic nutrients were entirely depleted, but some dissolved organic matter remained.

As might be expected, the bulk of the experimental work on this subject has been undertaken with fresh-water chlorophyceae. Some results, although probably valid, are open to question because they were obtained in cultures containing bacteria. It is always possible that the bacteria (possibly epigenous and symbiotic) are responsible for mineralizing organic compounds, and that the plants rapidly assimilate the products of bacterial action. Nevertheless, such bacterial associations in nature might still favour one plant species over another.

Finally, it should be realized that a nutrient may satisfy an absolute growth requirement, or it may simply be stimulatory. For example, Arnow *et al.* (1953) showed that *Chlorella vulgaris* grown in the light could use citrulline, arginine or urea as complete substitutes for inorganic nitrogen, but that other amino acids, purines, pyrimidines and nucleic acids only stimulated growth and chlorophyll was not synthesized. In some instances, vitamins have been shown to be stimulatory rather than an absolute requirement for growth.

Phosphorus in the form of glycerophosphate, inositol phosphate, nucleotides, and other similar sources, can be used to varying extents by phytoplankters (Chu, 1946; Harvey, 1953b, 1957; Provasoli, 1958; Pintner and Provasoli, 1961; Provasoli and McLaughlin, 1961). It is, therefore, probable that organic sources are used *per se*, but it is difficult to assess the significance of work in cultures until we know the nature of the organic phosphorus compounds present in natural sea water.

The several reports of amino acids being used directly by marine phytoplankton may be significant since the sea is known to contain appreciable amounts of "dissolved" organic nitrogen in an amino form (see p. 453). Nutrition by amino acids was studied as early as 1927 by Schreiber. In no case has the growth reported with amino acids been better than that with ammonia. Urea is a good source of nitrogen, and this compound and a few amino acids (arginine, citrulline, glycine and glutamic acid) may be as effective as nitrate for the nutrition of some plants. Uric acid and other compounds are assimilated only by neritic and benthic organisms, which would be expected to make use of organic matter. In general, there is little evidence that organic nitrogen has much ecological significance in the open ocean (Harvey, 1957; Guillard, 1961; Pintner and Provasoli, 1961; Provasoli and McLaughlin, 1961).

(2) *Heterotrophy*. An organism growing heterotrophically as a chemo-organotroph in the dark may behave as a photolithotroph in the light. A classical example is the soil *Scenedesmus*, shown by Bristol-Roach (1928) to grow exclusively on glucose in the dark, and phototrophically on carbon dioxide in bright light, with the two mechanisms additive at intermediate light intensities (see also Myers *et al.*, 1947; Samejima and Myers, 1958). It is doubtful if there is any species capable of both heterotrophic and phototrophic growth that will use an organic substrate in the presence of adequate carbon dioxide and light. The photo-assimilation of an organic substrate instead of carbon dioxide, such as that reported by Pringsheim and Wiessner (1960) and Eppley *et al.* (1963), however great the interest physiologically, is not likely to occur to a significant extent in the sea or in lakes, although cases are known (cf. the rock-pool flagellates described by Droop, 1961b).

Heterotrophy is found in all classes of algae, and may or may not occur in different members of the same order. At present, it seems impossible to predict the various substrates that will be utilized by a given species, although glucose, galactose, acetate and lactate seem to be preferred (Lewin, 1953; Kinne-Diettrich, 1955; Lewin and Lewin, 1960; Parker *et al.*, 1961).

Most of the investigations have been restricted to benthic, mud and sewage algae, which would be expected to benefit from heterotrophism.

The possible significance of this mode of growth in planktonic forms is difficult to assess, but cannot be overlooked (Wood, 1956; Parsons and Strickland, 1962b; Kimball *et al.*, 1963). There are reported instances of increases in pigment and bio-mass in a natural population, unaccompanied by photosynthesis (as measured by radio-carbon), which can be reasonably explained on the basis of heterotrophism (Wood, 1959; McAllister *et al.*, 1961; Rodhe, 1962). Despite this, few attempts have been made to demonstrate heterotrophism in marine planktonic algae by direct experimentation. Such work would have to be thorough if it were to be at all conclusive. Many substrates would have to be tried, and it is possible that organic matter may be utilized by some species only when there is illumination, even if this is very low (Lewin, 1953; Fogg and Belcher, 1961). The light may be necessary to ensure cell division (see later).

Finally, it should be remembered that some species in the dinophyceae and chrysophyceae are "plant animals" and can sustain themselves by phagotrophy (Hutner, 1961; Provasoli, 1956). This is a form of heterotrophism which may be of survival value for a cell beneath the euphotic zone, and may enable dissolved organic matter to be utilized after its prior conversion into bacterial bio-mass.

(3) *The fixation of atmospheric nitrogen.* There remains one further mechanism by which phytoplankton can obtain nitrogen, the critical major nutrient of the marine environment, and that is by the direct fixation of this element from the atmosphere. Although nitrogen fixation by algae has been shown to occur only with the myxophyceae, the subsequent decay of the plant cells responsible for this fixation will introduce combined nitrogen into the euphotic layer for the use of other organisms.

Nitrogen fixation appears to be associated with photosynthesis, although it may have a more critical temperature range (Fogg and Than Tun, 1960), and may only be called into play by a cell when all other sources of nitrogen have been depleted (Dugdale and Neess, 1961).

Allen (1959) and Stewart (1962) have reported photosynthetic nitrogen fixation by *Calothrix* and *Nostoc* in the marine environment, and have suggested that there may be a more widespread low-level fixation than had been previously realized. In the great desert areas of the sea, such as the Sargasso, any fixation of nitrogen will be important since the upper euphotic layer is almost devoid of nitrate. The introduction of nitrogen as ammonia in rainfall is significant (Menzel and Spaeth, 1961), and the occurrence of huge surface blooms of *Trichodesmium* in the Sargasso (shown by Dugdale *et al.*, 1961, to be fixing nitrogen, and of *Oscillatoria* in the southern Pacific and Atlantic (Wood, 1961;

Sorokin and Kliashtorin, 1961) may also have some importance in the fertilization of the euphotic zone. It is interesting to note that Stewart (1962) finds that *Nostoc* and *Calothrix* excrete amino acids into the water while they are vigorously fixing nitrogen, and thus efficiently transform the element into substances potentially available for the nutrition of other phytoplankton. At present, there is insufficient data for the rate of nitrogen fixation by these plants to be calculated with certainty.

E. CELL DIVISION RATES AND CELLULAR COMPOSITION

(1) *Cell division kinetics, the effect of temperature, light and nutrition.* (a) *General.* The factors affecting the rate of photosynthesis by a cell will, by and large, affect cell division rates in a like manner, and some of the effects of nutrients, reported in Section V, B above, were actually measured by their effect on cell division and not on photosynthesis *per se.* However, sometimes the ability of a cell to divide and to photosynthesize may be sufficiently different to be important. The dependence of each process on external factors may be different over a short period of time. Clearly, if such a difference persists, either the mean composition of a cell, or its mean size, or both, must be affected. It has been suggested that the ratio of cellular surface area to volume is important in determining division times (Vives and Fraga, 1961).

Sorokin (1960) interpreted the observed cell growth and division of *Chlorella* using a "master reaction theory" which assumed that cell division and photosynthesis are two distinct processes, and have different temperature coefficients. This may well be the case with other unicellular algae. Cell division is slowed by decreasing temperature more rapidly than is photosynthesis; also, a cell does not photosynthesize at a uniform rate during its life history, but slows down before and during cell division. This is only observable in synchronous cultures such as are obtained with certain species of *Chlorella* (Sorokin and Krauss, 1962).

It also appears that a critical light level is necessary for cell division. With *Nostoc muscorum*, Lazaroff and Schiff (1962) found that brief exposure to light was necessary for cell division even when the alga was growing as a chemolithotroph in the dark. Light of wavelength 6500 Å was most effective. Cell division does not necessarily take place during illumination, however, and may be deferred until a subsequent dark period. The effect of light on cell division can be the most critical factor in the growth of algae at low temperatures, since the minimum light intensity needed for cell division is not necessarily the same as the instantaneous compensation intensity (Sorokin and Krauss, 1959, and unpublished work in the author's laboratory).

The time of day at which cell division takes place in nature is uncertain. It is least probable with diatoms during the early hours of the morning according to Lanskaya (1961), but Sweeney and Hastings (1958) found the division of the dinoflagellate *Gonyaulax polyhedra* to be most frequent at the end of a dark period. No regularity can be expected unless phytoplankters have cell division rates which are a whole fraction or multiple of 24 hours. Many appear to have (approximately) such a rate (8 hours, 12 hours, 24 hours, 48 hours) when growing under optimum conditions, but with other combinations of light and temperature intermediate values must be expected. The shortest recorded division time for any alga is about 2 hours for a high temperature *Anacystis* (Kratz and Myers, 1955). For marine species Lanskaya claims a 3-hour doubling time is possible for *Skeletonema*. Surprisingly little work has been done, however, with marine phytoplankters. Values for diatoms are given by Braarud (1937, 1944) and Lanskaya (1961) and some other results for division rates have been collected by Strickland (1960). Baatz (1941) reported that the spectral composition of a light source had a marked effect on the division rates and morphology of marine diatoms and green algae.

Results from work with cultures must be used with caution as serious anomalies have been found. For example, *Skeletonema costatum* grows well in nature at a temperature which is several degrees lower than that which will sustain growth in the laboratory. There is no known explanation for this, but one may surmise that substances are present in natural waters which promote cell division, or perhaps sexual reproduction in nature gives rise to cells having a lower temperature tolerance than any developed in present cultures.

(b) *Combined effect of temperature and light on cell division.* A few studies have been made of the combined effect of temperature and light on cell division kinetics (Spencer, 1954; Mjaaland, 1956; Nordli, 1957a; Kain and Fogg, 1958a,b; Sorokin and Krauss, 1959; McCombie, 1960; Jitts *et al.*, 1964; Maddux and Jones, 1964). The general picture which emerges is one of wide temperature and light tolerance especially if sufficient time (several days) is allowed to offset the effect of "shock" on cells transferred from one culture regime to another. This is most important when cells are changed from a higher to a lower temperature, or from a higher to a lower light intensity. If sufficient time is allowed for adaptation, there may be a temperature range of 10°C or more over which the division time is substantially constant, and not affected by light intensities from 0·1–0·4 ly./min. A more or less sharp temperature minimum is found below which no growth is possible, and there is a critical upper temperature for each species which may depend on the nutritional environment, but little on preconditioning. The observed growth patterns can be interpreted on the assumption that extremes of temperature, light and nutrition give rise to (mutually additive) stresses. For example, the adverse effects of low temperature are exaggerated at low light intensities, and the damaging effects of high light intensities are worse at high temperatures.

In general, cell division appears to be a much more constant phenomenon which is less influenced by light and temperature than is short-term photosynthesis. In a large "culture" of mixed diatoms in the sea, the cell division rate for most species over a period of 10–15 days was found to be little influenced by wide variations in daily illumination and moderate changes in temperature (McAllister *et al.*, 1961; Antia *et al.*, 1963).

A variation in cell size is to be expected if the effect of light and temperature on division and photosynthesis is different. In general, conditions of rapid cell division tend to produce small cells, but it is the relative rates of division and photosynthesis that matter, and if the former is large compared with the latter, small cells will result whatever the division rate. *Skeletonema* cells are small when grown under conditions of low light and high temperature, and large when grown at low temperature with high illumination, despite division rates being similar in both instances. Eventually, photosynthesis must adapt itself to keep pace with cell division, or enormous size variations will be encountered. Alternatively, a cell might excrete some of the material it photosynthesizes in order to maintain a constant size.

(c) *Effect of nutrient concentration on cell division.* Certain elements such as manganese (Brown *et al.*, 1958; Richter, 1961), iron (Goldberg, 1952) and calcium (Rice, 1956) have been found to be necessary for cell division *per se*. Although, eventually, shortages of phosphorus and nitrogen will prevent cell division, there is a very wide range of concentration over which little effect is noted. An upper limit to the concentration of nutrient that can be tolerated exists, but has little practical significance with nitrogen and phosphorus in the marine environment. Phytoplankters seem able to tolerate at least 10–20 times the maximum concentration of phosphate and nitrate found in nature without ill effects.

When nutrient concentrations are very low, cells divide progressively more slowly, but they will continue to divide when the phosphorus concentration of sea water is analytically undetectable. Using radioactive phosphorus, Kuenzler and Ketchum (1962) have recently shown that a *Pheodactylum tricornutum* culture will strip phosphate from the medium down to a level of less than 0·0025 μg. PO_4^{3-}–P/l. The rate of phosphorus intake decreased by only one order of magnitude when the external nutrient concentration decreased by four orders of magnitude.

Similarly, cells first reared with adequate nitrate present can make one or two cell divisions without much effect on growth kinetics after all nitrate has been stripped from solution. This behaviour is well known in cultures, and has been demonstrated in lakes and in the sea under natural conditions (Rodhe, 1948, 1958; McAllister *et al.*, 1961; Antia *et al.*, 1963). However, plants can tolerate a nitrate shortage less readily than a phosphate shortage, and after about two divisions without replenishment of nitrogen, growth comes to an abrupt halt and there is a high cell mortality.

As was mentioned earlier, a shortage of vitamin B_{12} arrests cell division. Diatoms tend to increase in size (Guillard and Cassie, 1963), and a shortage with *Euglena* produces gigantism (Epstein *et al.*, 1962). Some, but not all, algae lose their pigmentation.

(2) *Changes in the chemical composition of phytoplankters brought about by changes of growth conditions.* If a cell continues to divide with no nitrogen and phosphorus present in the external medium, it is obvious that the concentration of these elements per cell will decrease. With ammonia or nitrate absent, each daughter cell that is formed can only contain half the protein of the parent cell, and continued photosynthesis can only increase the relative proportions of carbohydrate and lipid to the remaining protein. Eventually, insufficient nitrogen will be present to furnish the minimal concentration of critical enzymes, and the cell will die or pass into a spore or resting state. Similarly, phosphate concentrations cannot fall below the critical amount needed to give the minimal concentrations of RNA, DNA and ATP for the cell to function. A shortage of any element such as molybdenum or iron, which is involved in one way or another in the production of the proteins, gives rise to cells showing some of the symptoms of nitrogen deficiency, even though nitrate may be present in the medium.

The restoration of nitrogen and phosphorus in starved cells is very rapid when they are transferred to a nutrient-rich environment, and it is nearly independent of the concentration of nutrients in the new solution. It takes place in the light or in the dark, and cell division generally does not recommence until the cell has made good some of its deficiencies. Ammonia is taken in very rapidly, and produces a high soluble-nitrogen fraction in the cell (Scott, 1945; Rodhe, 1948; Harvey, 1953b; Thomas and Krauss, 1955; Al Kholy, 1956).

Very little is known of the effect of light and temperature changes on the chemical composition of marine phytoplankton cells. Chromatic adaptation may take place, and the pigment content of cells will depend upon the illumination under which they are grown. As mentioned previously, very intense illumination will "bleach" the chlorophyll of a cell. The amount of pigment per cell also decreases when a culture is grown at high light intensities because chlorophyll *a* production is arrested in the fresh cellular material that is formed until a new "equilibrium" ratio of cell pigment to cell carbon is reached. This behaviour has been well documented for photosynthetic bacteria by Sistrom (1962). Carotenoids are less affected, and the ratio of carotenoids to chlorophyll therefore increases. Halldal (1958a) has studied the combined effect of varying light and temperature on the pigment formation in myxophyceae, using a cross-gradient growth chamber.

Little is known of the variability possible in the degree of silicification of the cell walls of diatoms. Lewin (1957) found a three-fold change in *Navicula peliculosa*. One would expect the range to vary with species.

Algal cells store a large "excess" of phosphorus when they are grown in a medium containing much phosphate. With phosphate concentrations in the range to be expected in the sea (say 3–100 μg. PO_4^{3-}–P/l.), the "excess" phosphorus in a cell, over and above the *minimum* amount found in a viable cell of that species, is roughly proportional to the external phosphate concentration. As this diminishes (provided excess nitrate is still present), the phosphorus concentration in the cells decreases to 10–20% of its maximum value. Values as low as 3% have been recorded (Ketchum, 1939; Gest and Kamen, 1948; Goldberg *et al.*, 1951; Kuenzler and Ketchum, 1962). The absolute phosphorus content of a phytoplankter has, therefore, little significance, and can probably vary by a factor of 5–10 in the healthy cells of most algae. Phosphorus deficient cells have a relatively high carbohydrate and pigment content unless the deficiency is very severe.

The effect of nitrogen shortage on cell composition has been widely studied, and is more relevant to the marine environment because nitrate is stripped from most sea waters before phosphate. As mentioned above, carbohydrate and lipid increase relative to protein, except with severe nitrogen shortage, when the whole enzyme balance of the cell becomes abnormal and large amounts of lipid are photosynthesized. Some respiratory transfer of carbohydrate to lipid may take place (Iwamoto and Suzuki, 1958). The classical works on *Chlorella* are by Pearsall and Loose (1937) and Spoehr and Milner (1949), and the subject is well reviewed by Fogg (1953, 1957, 1959).

Chlorophyll *a* metabolism is closely associated with nitrogen metabolism. As soon as all the nitrate or ammonia is lost from solution, the production of chlorophyll ceases, and the chlorophyll content per cell begins to decrease as fresh cellular material is formed. If relatively high phosphate concentrations are maintained in solution, the carotenoid synthesis is not immediately affected by an external nitrogen shortage, and the ratio of carotenoid to chlorophyll increases and acts as a convenient indicator of cellular nitrogen deficiency (Yentsch and Vaccaro, 1958). However, when both phosphate and nitrate are depleted (as occurs in the sea), the rate of carotenoid synthesis is also reduced, and the ratio of carotenoid to chlorophyll does not increase to such a marked extent.

An indication of the changes of cell composition which may take place in a marine phytoplankton crop (mainly diatoms) as the external nutrients are depleted, is given by Fig. 3. These curves are taken from data obtained by Antia *et al.* (1963). They show the (approximate) changes on a *relative* scale, with 100 representing the nitrogen, phosphorus, chlorophyll and carbohydrate in a well-nourished crop, and 200 representing the cell carbon in the same crop. It will be seen that cell growth continues, and algal volume increases some three-fold, after nitrate has been stripped from solution. The mean composition of the cells, however, begins to be affected some time before all the nitrate has disappeared. In this experiment, the external nitrate concentration was about 70–100 μg. NO_3^-–N/l. when these changes commenced. There was little variation in the silicon and lipid content of the cells over the period concerned. The increase of carbohydrate was insufficient

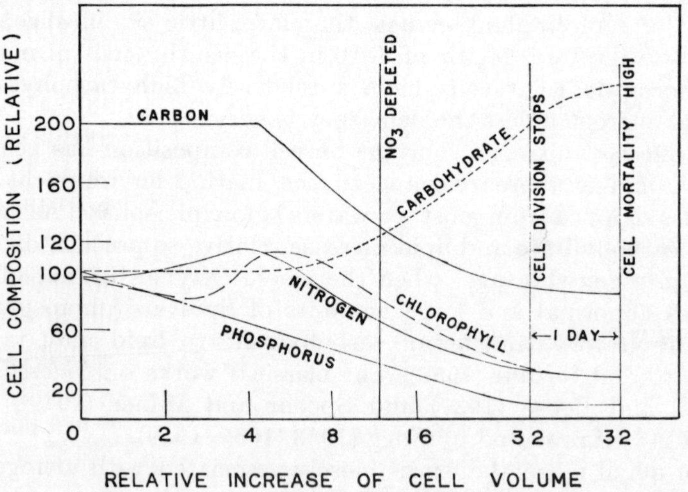

FIG. 3. Relative changes of cell composition of phytoplankton owing to nitrate and phosphorus depletion in sea water.

to offset the loss of protein, and therefore the total carbon per cell decreased to about one-half of its value in the well nourished phytoplankton.

VI. Field Measurements of Primary Production

The main methods for estimating the standing crop of phytoplankters were considered in Section IV. The techniques for measuring photosynthesis by suspended algal cells were outlined in Section V. However, considerable difficulties arise when applying these methods of photosynthesis measurement to the determination of the *in situ* productivity of a natural water body. The subject has been treated at length in recent reviews (Ryther, 1956a,b; Lund and Talling, 1957; Strickland, 1960; Vinberg, 1960; Nielsen, 1964), but it is of such importance that it merits a brief discussion in the present chapter.

A. Sampling Techniques

The foremost problem in field work is to obtain representative samples from the water mass concerned, and to take samples in such a manner that the plant cells are undamaged. The problem of heterogeneity is serious even in small lakes, and becomes almost impossible to solve in most coastal marine areas. Even in the open ocean, there is ample evidence that the amount of plant material can vary greatly over a distance of only a few miles, and it becomes debatable if any productivity measurements are meaningful except to give, *in stricto sensu*, the on-the-spot production at a given instant. Even this limited objective is difficult to attain. There is some value in attempting to make such measurements for certain ecological studies, but it should be realized that, in most instances, the results are little more than crude estimates of the mean productivity of any sea area, often reliable to little better than an order of magnitude.

The collection of large samples from the euphotic zone by pumping, which might be expected to give the most satisfactory answer to the sampling problem, is out of the question as cells are damaged physiologically. The effect varies with the population, and is most serious with the soft-bodied ultra-plankters of tropical waters. Some workers have gone to the length of inoculating samples for radio-carbon work *in situ* in order to ensure the minimum of physical disturbance (Jitts, 1958; Berg, 1958, 1961).

Only non-metallic sampling bottles (glass or plastic) should be used, as the photosynthetic response of phytoplankters is markedly affected by the presence of metallic contamination (Doty and Oguri, 1958, and also Chapter 21, Volume 2).

Samples should be passed through a 150–300 μ net before incubation to ensure the removal of organisms capable of grazing on the algal cells. There is no evidence that this procedure has much effect on the measured rate of photosynthesis. If the rates are to be measured by oxygen evolution, sufficient sample for filling all the bottles needed for each determination should be collected in one container, and the oxygen concentration of this water brought to a value near to saturation by passing air through the water, if necessary. It is generally undesirable to agitate samples very vigorously.

The depth at which samples should be taken presents a problem which has no simple solution. At least four samples should be taken, distributed vertically throughout the euphotic zone. (For productivity purposes the euphotic depth is generally assumed to be the depth at which green light is reduced to about 1% of its surface intensity. This approximates to 3–4 times the Secchi depth.) It is desirable to make a preliminary measurement of the thermal structure of the water. If a marked thermocline (or halocline) exists within the euphotic zone, one extra sample should be taken near to the top of the resulting pycnocline, where extra-high productivity may be expected. If turbidity measurements are made, a sample should be taken in any region of increased turbidity. Surface samples should be collected in a bottle fired slightly below the water surface (1–5 m. according to the euphotic depth), so as to avoid spurious effects which are often noted at the air–sea interface.

B. TIME OF SAMPLING AND DURATION OF EXPERIMENTS

It is now well established that planktonic plants show an intrinsic variation in their rate of photosynthesis throughout the day. Unless measurements of photosynthesis are made for a full 24 hours in the natural environment, the time of day during which a measurement is made will effect the final result.

Doty and Oguri (1957) reported that the $^{14}CO_2$ uptake rate of a natural population taken from tropical waters (at the surface, or even at a depth of 20 m.), varied when the bottles containing the phytoplankters were illuminated for short periods under artificial light of constant intensity. Plants taken from the sea at 08.00 hours local time took up labelled carbon dioxide nearly six times as fast as did the same population when removed from the sea at 19.00 hours local time. A similar behaviour was reported by Shimada (1958), who found that the maximum rate at 06.00–08.00 hours was about four times the minimum rate at 16.00 hours, and by Holmes and Haxo (1958), Ryther *et al.* (1961) and others. The variations reported by Yentsch and Ryther

(1957), Verduin (1957), McAllister (1963) and others are less but show a similar pattern.

Angot (1961) has suggested that the daily periodicity can be represented approximately by the formula:

$$\text{Log (rate)} = \text{constant} + \text{constant. cos } T$$

Where:

$$T = \frac{360}{24} \times \text{(local time of collection} + 3 \text{ hours).}$$

The maximum variation has been assumed to be a function of latitude, and decreases from a 5- to 10-fold change at the equator, to a constant daily rate at latitudes higher than 60–70° (Doty, 1959). Recent experiments by Lorenzen (1963) at one location, at about 40°N., suggest that the ratio is a function of the relative lengths of day and night, and is a maximum in clear weather when there is an equal period of light and dark, and decreases to about unity when either the light or the dark period exceeds some 16 hours. No doubt the exact behaviour depends on the population; the effect is less marked in overcast weather.

The explanation for the observed periodicity is not absolutely clear. The pigment content of the water shows a similar fluctuation to the photosynthetic rate (Yentsch and Ryther, 1957; Shimada, 1958; Yentsch and Scagel, 1958; Lorenzen, 1963), but the *rate of photosynthesis per unit of chlorophyll* also has some periodicity (Hansen, 1962; Lorenzen, 1963). There is undoubtedly an intrinsic daily rhythm in the photosynthetic potential of some phytoplankters as exemplified by the dinoflagellate studied by Hastings *et al.* (1961). Except near the surface of the sea in very bright sunlight, the afternoon decrease of pigment concentration reported by many workers would not be expected to result from bleaching of the chlorophyll in plant cells. Nielsen and Jørgensen (1962) have suggested that plant pigment production is simply arrested in the afternoon, and zooplankton grazing then causes a temporary reduction in cell numbers (see also McAllister, 1963).

In addition to complications from diurnal variations of photosynthetic potential, there is the problem of a suitable duration for experiments designed to measure photosynthesis, especially when the productivity is to be measured using *in situ* illumination. In bright daylight, supra-optimal effects slowly reduce the rate of photosynthesis, and the sum of, say, six 2-hour periods of exposure may give two or three times more photosynthesis than is found by one continuous 12-hour experiment near the sea surface. This effect has been noted by Rodhe (1958) and Verduin (1957) in lakes, but was found to be of little significance with *in situ* illumination of low intensity (McAllister *et al.*, 1961).

It is not clear what the true *in situ* photosynthesis will be near the surface in bright sunlight. Phytoplankton cells are in continuous vertical movement in the sea, and a given cell may not be retained for the whole time at one depth as it is in a glass bottle. The true productivity is likely to be greater in nature than would be deduced from a bottle experiment lasting all day, but may not be as great as the sum of several exposures of a few hours each.

Short exposures have the further disadvantage that extrapolation from one or two hours exposure to the full 24-hour period of production becomes less precise, the briefer the experimental period becomes. Unpublished work by C. D. McAllister carried out on open ocean populations at 50°N. in the Pacific, has indicated that the best approximation to the daily photosynthesis (as measured by the sum of consecutive 2-hourly incubations made throughout the day), is given by twice the photosynthesis measured in one experiment lasting from dawn to solar noon. Twice the value from midday to dusk is markedly lower, and the worst results are obtained from a full day of incubation, or from extrapolations of experiments of short duration made in the morning or afternoon. However, it does not follow that these observations would apply to other sea areas.

Algae removed from their environmental light intensity, and placed in a considerably different one, are unlikely to adjust to the new conditions for many hours, and for this reason short-term experiments are to be avoided. When measuring photosynthesis by oxygen changes, periods of 12 hours incubation should be used. For work using carbon-14 in incubators designed to obtain relative rate values (see below), 2–3 hour exposures are probably best. For *in situ* work with carbon-14, the periods should be greater, although not exceeding half a day unless light intensities are well below 0·02 ly./min.

C. *IN SITU* PRODUCTION AND ITS PREDICTION FROM EXPERIMENTS USING LIGHT-INCUBATORS

(1) *Direct measurement of* in situ *production*. The most obvious method for determining the *in situ* primary productivity in a column of water is to enclose suitable volumes of water in clear glass or plastic containers, and to suspend these containers at the depths from which the samples originated. The photosynthesis is then measured by following oxygen evolution, or carbon dioxide assimilation over a suitable period of time. This technique (using only surface water) was tried out by Gaarder and Gran (1927), and thereafter by numerous workers. In some instances, the suspension of bottles in the euphotic zone has been made directly from the side of a research vessel, or from the end of a pier, but it is

more normal to use a buoy of metal or glass with the string of bottles, or certainly the bottles near the surface, held away from the buoy by some form of boom to avoid direct shading. Free floating buoys may be recognized by a marker flag or by use of radar. In the latter case, a ship may be released for a day, or half a day, to work in another location, steaming back to the buoy to pick up samples at the end of a suitable incubation period.

It is general to enclose samples in 100- to 300-ml. capacity glass bottles clipped directly to a weighted wire or rope or, if many samples are used at each depth, held in some form of tray (Jenkins, 1937). Jitts (1958) sampled the water in adjacent light and dark lucite sampling bottles which were returned to the euphotic zone after the addition of radioactive carbonate. Further details and references are to be found in the review by Strickland (1960).

(2) *Relationship of* in situ *production to production measured in a light-incubator*. Despite the attractiveness of *in situ* productivity measurements, which are probably the most reliable yet devised for estimating the total rate of photosynthesis in the euphotic zone, the method is very limited in its applicability. Except at inshore stations situated conveniently near a marine laboratory, or at a weathership or lightship, the technique is costly and inconvenient, and is useless when the productivity of a wide sea area is to be measured on cruises, where little ship delay is permissible. It is therefore necessary to devise some form of shipboard technique which can be applied to samples withdrawn from the euphotic zone when the ship is on station for hydrographic or similar measurements. To this end, various methods of incubating samples in tanks under artificial or natural illumination have been devised.

(a) *Design of incubators*. Certain aspects of the design and use of light-incubators merit brief discussion.

It is essential that the bottles containing the sea-water samples be cooled in running water to the temperature of the euphotic zone. If a thermocline exists in the euphotic zone, samples taken below it should, ideally, be cooled to the correct temperature, although the elaboration of thermostatically controlled refrigeration may not be worth the trouble in most field investigations.

Two sources of light can be used, natural daylight, or some form of artificial illumination. The spectral distribution of daylight is so unrepresentative of light in the marine environment (except very near the water surface), that some form of filtration should be attempted in order to remove light of wavelength below about 4,000 Å and above 6,000 Å. Failure to do this will almost certainly give misleading results, except for surface samples, or unless the sky is very overcast. For work with daylight, an open tank with bottles resting in blackened compartments is best, and light attenuation can be achieved using neutral filters of fogged photographic emulsion or, more conveniently, fine nylon or metal screens (Berg, 1958; McAllister and Strickland, 1961). H. R. Jitts (personal communication) has suggested the use of a variable number of sheets of coloured window glass, both to simulate the spectral distribution of marine daylight, and to attenuate the light in a more meaningful manner than is possible using neutral filters. Such a system would be ideal if a cheap glass of the correct transmission properties could be obtained, but when a number of glass sheets are used for attenuation the spectral properties of each sheet become quite critical.

Several designs have been used for incubators intended to illuminate samples with light of constant intensity. In general, samples are enclosed in a glass or clear plastic container through which water is circulated, and which is illuminated by strip fluorescent lamps (King *et al.*, 1957; Doty and Oguri, 1958; Jitts, 1958; Saunders *et al.*, 1962). To raise the level of illumination, the lamps may be backed by a reflecting surface and the tank walls may be constructed of polished metal, or painted white. By this means, the amount of light received by a bottle can be increased by up to 50%. To ensure even illumination, the carriage holding the sample bottles may be rotated (Nielsen, 1952a; Corlett, 1958), or may be surrounded by circular fluorescent lighting.

The spectral characteristics of the lamps used in incubators have received too little attention. "Daylight" or "cool-white" fluorescent lamps give illuminations that are markedly different from each other, and from "marine light", which is better imitated by use of mixed light from blue and green fluorescent lamps or an incandescent lamp with a suitable blue filter. The demanded spectral distribution depends on the work and the precision required, but certainly requires a thorough investigation.

(b) *Simulation of* in situ *light*. Probably the most satisfactory method of estimating *in situ* productivity by a shipboard method is to use an incubator to reproduce, as nearly as possible, the light and temperature conditions of the algae at the depths from which they are sampled. Natural illumination is preferable to artificial light in this work. Problems associated with sampling, incubator design and duration of experiments have already been discussed.

(c) *The use of constant light and* P vs I *curves*. The simplest and most reliable incubators are those using constant illumination obtained from artificial light, and much effort has gone into the design of field methods using such equipment. The intensity of light used in these incubators should be on the plateau of the P *vs* I curves so that small fluctuations in light intensity are not important. This is possible, even for tropical phytoplankton, by the use of fluorescent lighting close to the sample bottles, and by paying close attention to the reflectivity of the incubator surfaces. The problem is to translate the rates obtained from bottles illuminated in these incubators into terms of the rates of *in situ* photosynthesis.

If a sea-water sample containing phytoplankton is exposed for a known time to the optimal light intensity ($I_{max.}$) of the population concerned, then the maximum possible primary productivity ($P_{max.}$) may be obtained in terms of mg.C/m.³/hr. Suppose that the *relative* P *vs* I curve (the f *vs* I curve, see Section V, B(3)) is known for this population over a wide range of photosynthetically active light intensities, and that the amount of this photosynthetically effective radiation can be measured at the sea surface throughout the day. Also, let us assume that the light attenuation properties of the sea in the area under study are known sufficiently well for the fraction of the surface radiant energy present at any depth to be estimated. From this information, we may calculate the absolute value of I at any depth (I_z), and from the f *vs* I curve we may thus find the fraction of $P_{max.}$ that applies to this depth. In this way, the absolute rate at any depth can be calculated, and the total rate for the euphotic zone obtained by integration. As f will be unity at a depth where $I_z = I_{max.}$, and may be less in stronger light, the maximum photosynthesis often occurs some distance beneath the sea surface, and the point of maximum photosynthesis will move upwards as the surface light decreases, reaching the surface only when the illumination is quite poor (0·1–0·3 ly./min. of total radiant energy as compared with a midday summer value of 1·4 ly./min. or more).

In most instances it is impracticable to determine f *vs* I curves and I_z values at all stations. Ryther (1956b) has presented a simplified treatment which assumes that an idealized f *vs* I curve can be used with a population of phytoplankton (different curves would be used for temperate, coastal, tropical oceanic populations, etc.) and that a single vertical absorption coefficient, k, applies to the photosynthetically active light in a water column. A factor R may be calculated which is used to relate the maximum rate of production $P_{max.}$ (mg.C/m.³/day) measured in an artificial light incubator at $I_{max.}$ to the total production in the euphotic zone (mg.C/m.²/day).

$$P_{\text{euphotic}} = \frac{R}{k} \cdot P_{\text{max}}.$$

R may be calculated for each day from the assumed or average weekly radiation and the shape of the f *vs* I curve. This method is rapid, but makes some very broad assumptions, especially that the population is uniformly distributed in the euphotic zone, and that its *f vs I* characteristics can be idealized by one curve which applies over a whole season.

A refinement of the above type of approach has been applied with considerable success by the Russian aquatic microbiologist Yu. I. Sorokin. The method was originally designed for use in lakes or reservoirs, but it is appropriate for work in the open ocean in areas where the water turbidity is expected to remain roughly constant. The Sorokin procedure takes into account the uneven distribution of phytoplankters with depth by finding the *relative* productivity of samples taken at intervals throughout the euphotic zone. These samples, including a surface sample, are exposed for a suitable interval in a light-incubator at a light intensity that may have any value which is within the semi-proportional region below $I_{max.}$ At the same time, a duplicate surface sample is exposed at the sea surface, either by suspension from a buoy or by trailing behind the ship. This sample serves to fix the absolute rate of productivity as affected by the incoming solar radiation. A third sample is a composite of samples taken at various depths in the euphotic zone, and is used to find what is, in effect, the *f vs I* response of the population under study. This latter determination is made by exposing aliquots of the composite sample at various depths where the light intensity has known values. The *f vs I* relationship need be redetermined only when the population or light attenuation characteristics of the water are thought to have undergone a marked change.

The ratio of the rate of photosynthesis at each depth to the rate at the surface as measured in the incubator is found. This ratio, K_p, which allows for "population densities", is multiplied by another factor, K_t, obtained from the *f vs I* curve, which relates the rate of photosynthesis at surface light intensities to the rate at the appropriate sub-surface illumination for the sample concerned. This combined factor is then used with the observed surface measurement to estimate the production at each depth. Better K_t factors may be obtained if *in situ* rates are divided by K_p values rather than by using a composite sample. The method fails if the euphotic zone is shared by sun and shade phytoplankton. The method is also of little value if populations and water characteristics change too frequently. It would be better, but less convenient, to relate rates to a sub-surface sample rather than to one exposed at the actual surface where the rate is often depressed to a marked, and possible variable, extent by supra-optimal illumination. The method is described in detail by Sorokin (1956) and Sorokin and Kliashtorin (1961), and was found to be the most satisfactory of several techniques compared recently by Saunders *et al.* (1962).

Talling (1957b) has presented a purely mathematical treatment of the photosynthesis-depth integral; f *vs* I curves were not taken to supra-optimal light intensities in his work. The assumptions made in the Talling treatment are rather extreme, but the agreement between experiment and theory proved to be quite reasonable for the particular case of an isothermal lake in spring. However, attempts to generalize such methods are of doubtful practical value in view of the mathematical complexity of any final expression which attempts to include all natural variables.

(d) *Empirical formulae.* Several empirical formulae have been suggested for converting rates obtained in an incubator using artificial light near to $I_{max.}$, to the rates of photosynthesis in the euphotic zone. Nielsen (1958a) presented an empirical formula which gives a satisfactory prediction of natural productivity if constants are obtained for each location and each incubator used. The equation is of the form:

$$P_{\text{euphotic}} = \frac{1}{5}(2P_{100} + 2P_{10} + P_1) \cdot \frac{D}{2} NK$$

where the P values are the incubator productivity values measured with a sample taken from the surface (P_{100}), from the depths where the photosynthetically reactive light intensity is 10% of the surface value (P_{10}), and where the light is 1% of the surface value (P_1). D is the depth (in metres) where the light is 1% of the surface value, and N is the day length (duration of daylight) in hours. K is a constant which depends on the incubator, sea location and the precise incubation light intensity (which need not be as great as $I_{max.}$). In the tropics, K had a value of unity when samples were illuminated with fluorescent light at 18,000 lux. In some tropical waters, there is an accumulation of phytoplankters at a depth of about 50–75 m. which may not be taken into account if the above equation is used. It might therefore be best to modify the formula to take into account any high productivity layers which may be present in the euphotic zone.

Rodhe *et al.* (1958) have given an empirical formula similar to the Ryther equation for estimating integral photosynthesis in a lake, viz.,

$$P_{\text{euphotic}} = \frac{2 \cdot 5 P_{max.}}{k}$$

where k is the extinction coefficient, and $P_{max.}$ in this equation is the rate of photosynthesis *at the depth of optimum photosynthesis*.

Berg (1961) has suggested a simple relationship between the photosynthesis rate, the illumination and the turbidity, which was measured with a specially designed meter. C. D. McAllister (unpublished work) has shown that even the simple formula:

$$P_{\text{euphotic (day)}} = A.(\text{surface photosynthesis, dawn to midday}) \times (\text{daily total langlies}) + B$$

where A and B are constants, can be used for a rough prediction of production in the north-east Pacific Ocean.

These empirical relationships may have much to commend them on the grounds of simplicity, and may be all that is necessary for most field work, in the present state of our knowledge (Verduin, 1957). Several attempts have been made to avoid any direct measurement of photosynthetic rates by assuming a constant value for $P_{\text{max.}}(c)$ (see Section V, B(3)) i.e.:

$$P_{\text{max.}} = \text{constant} \times \text{mg. chlorophyll } a/\text{m.}^3$$

Ryther and Yentsch (1958) have given a full illustration of this approach, using a value of 3·7 for the constant. Currie (1958) has produced evidence that a measure of total chlorophylls and carotenoids may be better than that of chlorophyll *a* alone (see also McAllister *et al.*, 1960). This would be anticipated from a consideration of action spectra, and has been confirmed by McAllister *et al.* (1964) for low light intensities.

D. *IN SITU* PRODUCTION AND ITS PREDICTION FROM CHANGES IN THE CHEMICAL COMPOSITION OF NATURAL WATERS

In theory, the net productivity of a column of water may be deduced from the change, with time, of the concentration of biotically active elements, notably carbon, oxygen, phosphorus and nitrogen. Ideally, such a method is perhaps the most satisfactory approach to the problem of marine productivity since it can give results integrated over comparatively long periods of time, and over wide sea areas. This approach removes most of the uncertainties attached to "instantaneous" measurements when applied to such a heterogeneous system as marine phytoplankton.

This approach also furnishes the only independent check on measurements made in bottles, and results obtained by Verduin (1960) are disturbing in that they indicate that agreement between the two methods may be poor.

In early investigations, estimates were made of the production during the spring and early summer from changes of pH, oxygen or phosphorus in shallow columns of sea water. The "base line" from which such changes were measured was generally the mid-winter values for the water column or, for oxygen, the saturation concentrations at atmospheric pressure. As a refinement, Buch (1929) suggested that the initial oxygen concentration could be estimated from the concentration of dissolved nitrogen, on the assumption that both elements were originally in equilibrium with the air. In another approach, the observed variations in nutrient concentrations were corrected for the variations to be expected from purely physical transport processes, so as to give the changes which were brought about by plant growth.

Examples of these methods are to be found in the papers by Atkins (1922, 1923), Buch (1929), Kreps and Verbinskaya (1930), Riley (1951), Steele (1956), and Park *et al.* (1958). In very shallow areas, the sea-atmosphere interchange of oxygen must be allowed for when computing respiration or photosynthesis.

The subject is complex, and has been discussed at some length by Strickland (1960). For certain sea areas, the nutrient depletion technique of productivity measurement may be feasible, and may furnish a useful check on other methods, but at present, our mastery of the theory and measurement of the hydrodynamics of the sea is not sufficiently adequate to do full justice to this approach.

VII. The Growth and Distribution of Phytoplankton in Nature

A. Vertical Distribution

(1) *The effect of light: the euphotic zone.* The basic factor determining the distribution of plants in the sea is light, as plants will grow only in the lighted, upper layers of the ocean. The exact depth of the euphotic zone is unknown. The instantaneous compensation illumination will depend on the algal species and their metabolic condition. If the latter is poor, then more light will be necessary before gross photosynthesis equals respiration. Instead of the instantaneous value, we may consider a mean 24-hour figure for the compensation illumination which will be greater because of the respiration taking place during the hours of darkness.

The instantaneous compensation intensity is probably about 0·005 ly./min. for most healthy phytoplankton at their normal growing temperatures. This figure must be nearly doubled for a 24-hour period. Since the photosynthetically active illumination at the sea surface for much of the day during summer is about 0·5 ly./min., the assumptions of "a depth where there is 1% surface light" or "3–4 times the Secchi depth", which are often taken as rule-of-thumb definitions of the euphotic depth, are seen to be reasonable.

Phytoplankton near the base of the euphotic zone, especially if they are "shade" cells trapped in lower cold waters by a pycnocline, may adapt themselves to very low light levels (0·002 ly./min. or less).

The subject is complicated by the fact that algal cells in the sea are generally in continual motion across the level of compensation illumination, and for this reason, viable cells may be expected many metres below the maximum depth of *in situ* growth.

The only satisfactory way to find the extent of the euphotic zone in any location at any specific time is to measure it experimentally, *in situ*, using the water trapped in bottles attached at frequent intervals to a wire suspended from a float. The values reported in the literature vary from somewhat greater than 100 m. in the clearest ocean water, to 5–10 m. in the most productive coastal areas. In exceptionally productive or turbid waters, values as low as 1–3 m. have been recorded. Owing to inhibition by supra-optimal light intensities, the rate of production in the euphotic zone is often at its greatest some distance below the sea surface, where the maximum illumination is about 0·2–0·5 ly./min., but it is impossible to generalize.

When predictions are made for the whole euphotic zone from idealized *P vs I* relationships (for example, the approach used by Ryther, 1956b), it can be shown that the production beneath a unit area does not increase linearly with surface light intensity because of the inhibition of photosynthesis which may occur near the surface of the sea. Changes of total daily radiation are most important when the amount is small; the amount of radiation is, on the average, a function of latitude and time of year. A highly idealized model tells us nothing about actual production, as this depends on nutrients and other factors, but it does enable certain generalizations to be made about the effect of light, *per se*, on productivity. Thus, it can be predicted that little or no mid-winter production is possible at latitudes greater than 55–60°. The growing season in the arctic is restricted to the period from late March to early September, but its mid-summer production is somewhat greater than that at the equator with water of the same transparency. There should be little or no seasonal effects at latitudes less than 30–35°, but marked effects at latitudes 45–55°.

Consider a stable, nutrient-rich column of water with no phytoplankton and with very little other suspended material. Initially, plants would grow in such water down to a depth of 100 m. or greater, the light being absorbed mainly by the water itself, but as the crop grows, the phytoplankton cells increase the light attenuation of the water, and the depth of the euphotic zone decreases. If photosynthesis was proportional to light intensity at all intensities, and if the light was monochromatic and attenuated *solely* by the phytoplankton, then the euphotic depth would be inversely proportional to the concentration of plant cells, and the production beneath unit area would be independent of the standing crop per unit volume. In practice, because of the nature of the *P vs I* relationship, and because red and blue light are absorbed more readily than green, the utilization of sunlight by large concentrations of plant cells is more efficient than in crops where the cell density is low. At high crop densities, the euphotic depth tends to be inversely proportional to the production rate per unit area, instead of being independent of it (Nielsen and Jensen, 1957). However, we often find that the rate of photosynthesis *per unit area* is not very much greater in eutrophic water than in oligotrophic areas. Steele and Menzel (1962) have theorized about the conditions necessary for maximum crop production in a mixed layer of a given depth. In many places, no correlations can be expected because inorganic matter, detritus and dissolved coloured material greatly affect the light attenuation.

(2) *Nutrient supplies and eutrophic and oligotrophic waters.* As we have seen from earlier discussions, nitrogen, as nitrate, is the most important nutrient in the marine environment since this element is generally the first one to be stripped from the water by growing plants. In a few locations phosphate may be exhausted before nitrate, but this is more common in lakes and rivers.

The concentration of dissolved nutrients in the euphotic zone may be said to determine the maximum standing crop that can be produced. This is only strictly true if all the nutrients are converted to plant bio-mass in a stabilized system. In reality, the maximum standing crop in the sea is less than the quantity predicted from the concentrations of nutrients found in the water during winter, or immediately after a period of vigorous upwelling, but it often exceeds half of this theoretical value.

The *rate* of plant growth is also governed by the amount of nitrate, ammonia, phosphate, etc., in the euphotic zone, but in a very complex manner, and it seems unlikely that cell division or rates of photosynthesis are much affected until nutrient levels become very low. As mentioned previously, the first effect of nutrient depletion is to change the chemical composition of plant cells. Eventually, the species composition of the phytoplankton changes to adapt itself to the low levels of nitrate and phosphate, and to what is generally a concomitant increase in the relative amounts of ammonia and soluble organic nitrogen and phosphorus compounds in the sea water.

The rate of photosynthesis $\left(\frac{dp}{dt}\right)$ depends on two factors, k and p in the equation (Section II, B, 3):

$$\frac{dp}{dt} = k \, . \, p.$$

Light and nutrients affect k, and the previous history of the water mass governs p, the standing crop. Thus, it is entirely possible to have a low rate of production in water with a large standing crop either if the nutrients have been stripped from the water, or if much of the light is absorbed by the turbidity arising from the phytoplankters themselves. Such conditions often occur at the end of phytoplankton "blooms". Conversely, if p is kept small by grazing or turbulence, the productivity will be small even though conditions may favour a large growth constant, k. The terms eutrophic and oligotrophic are best thought of in terms of long-term production, but even then they defy any clear definition.

If thermal or salinity stratification stabilizes the water in the euphotic zone so that the rate of nutrient replenishment from deeper regions is very slow, the oligotrophic conditions typical of sub-tropical and other "old waters" are found. At the other extreme, rapid mixing may continually fertilize the euphotic zone, but individual plant cells will be swept from the lighted layer before cell division can take place. In most eutrophic waters, conditions are encountered where the rate of upwelling is optimal (see the theoretical treatment by Riley *et al.*, 1949, and Steele, 1958). Alternatively, we may have periods of stabilization when the plants "bloom" in the euphotic zone, and the waters are then completely remixed by storm or tidal activity, and this replenishes the nutrients and allows the next "bloom" to occur. These last conditions are well illustrated in the sheltered coastal seaways of British Columbia, where the total productivity in a year is affected markedly by the exact pattern of meteorological conditions.

If we consider the two conditions necessary for plant growth, namely, light and nutrients present in a stable surface layer, it is apparent that conditions for optimum growth in the sea will not be common. In the tropics, where the light is always adequate, thermal stabilization will generally prevent adequate fertilization of the water. In the polar regions, nutrients will be present, but either with insufficient light or, where light and nutrients are adequate, the surface waters may lack sufficient stability for a "bloom" to become established. In temperate seas, ideal conditions will be attained only in the spring, when sufficient light and stability both occur in a water mass rich in nutrients. The absence of stability in a water column does not, of course, mean that mixing will necessarily take place, and small "blooms" may start even in December, during spells of calm, bright weather at latitudes between about 40° and 50°.

Where "old water" persists, or in those shallow coastal areas where little mixing occurs during the spring and summer, the production rate will ultimately depend on the rate at which nutrients are remineralized from plants by the agency of bacteria or grazers. A significant turnover of organic matter can occur with only a very small concentration of free nutrients appearing in the water, but we know very little of the quantitative aspects of the nitrogen cycle in these circumstances. Ammonia is probably directly and rapidly exchanged between zooplankton and phytoplankton, and the vertical migration of the former produces a net upward transference of this nutrient which is beneficial to plant growth in the upper parts of the euphotic zone (Harris, 1959).

(3) *Sinking and phototaxis*. Some genera in all divisions of the marine phytoplankton (except the Cyanophyta) have flagellae, and are capable of independent movement. This gives such cells two advantages over immobile species.

Firstly, the plants can overcome gravitational forces and thus remain in the lighted zone. To do this, they must be attracted to the light (positive phototaxis) and, ideally, they should be repelled from the very bright light at the sea surface (negative phototaxis). Both abilities have been demonstrated in flagellates from the classes chlorophyceae, dinophyceae and chrysophyceae, and this ability may exist in nearly all flagellates under the right conditions. The photo-reactive substances in the plants appear to be carotenoid-protein complexes (with definite action spectra) located on the surface of the cells, but not necessarily all in the "eye spot". The behaviour of each species depends on the intensity and the spectral distribution of the light, and on the physiological condition of the plant. It is possibly significant that in temperate seas, the winter phytoplankton is often dominated by flagellates. The reader is referred to the papers by Hasle (1950, 1954a), Halldal (1958b, 1961a,b), Sachs and Mayer (1961) and Pomeroy *et al.* (1956). Many other references are scattered throughout the literature.

Certain benthic diatoms are capable of relatively rapid movement provided that a surface is present, but there is no evidence that self-propulsion occurs to a significant extent in the planktonic forms. Nultsch (1956) has claimed that he has detected that certain plankters show a chemotaxis towards oxygen-rich zones, and certain flagellates have been shown to exhibit chemotaxis towards iron and manganese (H. W. Harvey, personal communication), but it is not known what advantages this behaviour may have in nature. It is possible that cells may be attracted to "marine snow" aggregates of organic and inorganic material (see later). Certain myxophyceae become buoyant because of the formation of gas in the cell tissues, but this is not found in other marine phytoplankters, and although the formation of excess lipid could cause positive buoyancy, the reported instances are rare, and may represent crops in the last stages of metabolic disorder.

The second advantage of independent motion is that a plant can absorb nutrients more efficiently if it is in motion relative to the surrounding water. Such motion decreases the thickness of the diffusion layer around a cell, and thereby enhances the rate of transport of material to and from its surface. This would suggest that flagellates can survive in waters of lower nutrient content than can, say, diatoms, but the evidence is only indirect.

Organisms without flagellae must eventually sink from the euphotic zone unless they can maintain neutral buoyancy. Nevertheless, turbulence will keep the population in the lighted area for some time even though the cells are slightly denser than the surrounding medium. The shape and size of many phytoplankters is such as to delay sinking, and to ensure rotational movements during sinking and thereby facilitate nutrient transfer. The subject is discussed in a semi-quantitative manner in a stimulating paper by Munk and Riley (1952). Certain of the centric diatoms have been shown to maintain their buoyancy in sea water by changing the composition of their vacuole fluid until it has a lower concentration of divalent salts than is present in the surrounding medium (Gross and Zeuthen, 1948; Beklemishev *et al.*, 1961), and it is perhaps significant that large centric diatoms do not occur in the phytoplankton of deep lakes where such a mechanism could not be used.

Sinking rates depend very much on the species of phytoplankter and their state of nutrition. In general, nutrient-starved cells become relatively dense and sink quite rapidly. If the plants, in so doing, enter a layer of water richer in nutrients which is within, or not too far below the euphotic zone, they may recover much of their buoyancy (Steele and Yentsch, 1960).

Healthy diatom crops probably sink only 1–2 m. a day, but rates of 10–20 m. may be expected with nutrient-starved organisms. Few reliable direct observations have been made in nature, and results from the laboratory must be interpreted with caution. A useful summary of the sinking behaviour of phytoplankton has recently been published by Margalef (1961). The sinking rate of the ultraplankters has not been determined, but is, presumably, very small as many of the species in this size range are flagellates.

(4) *Heterotrophic growth of phytoplankton.* This has already been discussed in Section V, D(2). In general, it may be said that the heterotrophic survival of phytoplankton below the euphotic zone is possible, but no quantitative data are available for planktonic forms of marine algae. The subject is discussed by Wood (1956, 1959) and by Parsons and Strickland (1962b). Bernard (1948, 1958a, 1961) has reported the presence of planktonic myxophyceae, dinophyceae and chrysophyceae (coccolithophores), especially the latter, at depths down to several

thousand metres. Their amounts seem to be very variable, and may be associated with sub-surface currents, but occasionally cell concentrations are nearly as high at great depths as at the surface. Heterotrophism is implied by this, rather than proved, but the inference is very strong, and is supported by the marine observations of Ohwada (1960) and Smayda (1958), and by the reports by Rodhe (1955, 1962) of the growth of populations of flagellates in ice-covered arctic lakes during the winter.

A limited ability for heterotrophic maintenance (not necessarily growth) would have a high survival value for marine phytoplankters during temporary excursions beneath the euphotic zone. *Skeletonema costatum* has been reported as surviving for two months in the dark (Matsue, 1954), but we do not know by what mechanism.

(5) *Distribution patterns in the euphotic zone.* Vertical distribution patterns of phytoplankton are very variable, and only a few generalizations can be made. As might be expected with a suspension of small particles of near neutral buoyancy, the distribution is uniform in waters of zero or negative stability, and this is particularly true in some tropical seas with a deep thermocline such as the Sargasso Sea (Riley, 1957). There will, however, be a tendency for sinking plant cells to slow temporarily near the top of any pycnocline where they enter water of increasing density. This is generally a sufficient explanation for the sub-surface maxima in cell concentrations or chlorophyll profiles which have been recorded by field workers (Sorokin, 1960b; Steele and Yentsch, 1960). Many instances could be quoted, but there seems little point in doing so, as a different pattern is found each day in some locations.

Another cause of irregular distribution of phytoplankton with depth is the horizontal advection of plant cells from areas with, initially, higher or lower phytoplankton concentrations. This is very marked in fertile coastal areas with appreciable wind-driven or tidal currents, where the vertical profiles may change radically within a few hours.

Finally, an irregular profile may arise from the successive slow sinking of plants advected to, or blooming at, the sea surface at different times. Quite high concentrations of healthy cells may occur well below the commonly accepted euphotic depth. This might indicate heterotrophism, but could also be the result of a dynamic settling equilibrium (Talling, 1957c). Ryther and Hulbert (1960) have recently reported a heterogeneity of phytoplankton *species* in what appeared, from the physical data, to be a well-mixed euphotic zone. A small centric diatom occurred at only one depth in the Sargasso Sea (Hulburt, 1962). The explanation of these observations is not clear, but presumably such a situation can be expected only after a period of relative calm, and the distribution must arise from the local "grazing" or "blooming" of various species at various depths with perhaps, some positive migration of flagellate forms (Hasle, 1954b). Cassie (1959) has suggested that micro-physical or biological factors are responsible for the uneven distribution of plant cells over distances of less than 1 m. However, part of the observed small-scale patchiness could arise because a fraction of the phytoplankton (over a third of the total in some cases) is aggregated. It is now known that in both coastal and open ocean waters there are aggregates of detritus, bacteria and plant cells known as "marine snow" (Suzuki and Kato, 1953). These aggregates vary greatly in size and shape, but can be several millimetres in length (Nishizawa *et al.*, 1954; Inoue *et al.*, 1955). They are held together by some form of mucilaginous matter of plant or bacterial origin, but are very fragile and break up during the removal of water samples from the sea by ordinary methods. These units, which may have considerable ecological significance, will be discussed later. Riley (1963) has described in some detail the "marine snow" ecology of a near-shore area, and unpublished work has indicated that similar material exists in open ocean waters both at the surface and at depth.

B. GEOGRAPHICAL DISTRIBUTION

(1) *Land-mass effect.* As a broad generalization, it may be said that the primary productivity is greater near to the coasts of continents and islands, or over submerged ridges and banks than it is in the open ocean. This "land-mass" or "island-mass" effect has been noted by many workers (e.g. Gran, 1932; Doty and Oguri, 1956; Ryther and Yentsch, 1958; Hansen, 1961; Paasche and Rom, 1962). Doty and Oguri speculated that near islands in the barren tropical oceans (where there is no marked upwelling or land drainage) the increased production rate could result from a local increase in nutrient concentrations arising from decaying or dislodged benthic flora. However, the phenomenon is observed at all latitudes, and in many areas where there is no noticeable increase of nutrients at the coast. The rate of increased productivity, dp/dt arises mainly from an increase of the standing crop, p, and not of the growth constant k. Thus, McAllister *et al.* (1960) found that the $P(c)$ values of samples taken at Weathership P (50°N., 145°W.) about 500 miles out from the edge of the continental shelf, were similar to those obtained from samples taken in western Canadian coastal waters. In both areas, the sea contained similar high concentrations of nitrate and phosphate. Coblenz-Mishke (1957) suggested that the crops found

in fertile boreal oceans are small because of the rapid loss of cells from a euphotic zone situated within a deep mixed layer. Although this may be the explanation in some regions, it certainly is not true in the northwest Pacific Ocean where the depth of the mixed layer is rarely, if ever, twice the euphotic depth. At Weathership P, in both winter and summer, the phytoplankton crop per unit volume, as measured by chlorophyll, varies less than three-fold (Parsons, 1960), and is less than a tenth of that found near the coast at the same latitude for much of the summer.

The "land-mass" effect is thus of the utmost importance, as the productivity of parts of the world's temperate oceans appears to be an order of magnitude less than it might be, owing to the low standing stock of plant cells. Although the production beneath unit area of sea surface would not be affected as greatly as the production per unit volume because of light attenuation changes (see Section VII, A(1) above), the reduction would still be considerable, and the potential productivity of the sea as a whole is lessened.

(2) *Effect of grazing*. The relatively small size of the standing crop in fertile open ocean areas cannot always be attributed to turbulent mixing, since natural mortality, sinking and grazing are also contributory factors. Mortality and sinking probably reduce the crop only slightly, and it seems that, for some reason, grazing in the open ocean is more continuous and effective than in coastal areas. The relatively poor productivity of temperate oceans can be likened to the poor productivity of a fertile, but over-grazed, pasture; in contrast, the poor productivity of tropical oceans is analogous to the behaviour of an under-fertilized and spent plot of land.

The transition between "coastal" and "oceanic" crop levels appears to coincide, very roughly, with the edge of the continental shelf, and may be associated with the different grazing patterns found in shallow and deep waters. Burkholder and Sieburth (1961) found an inverse relationship between chlorophyll concentrations and water depth in the Antarctic.

Near the coast, the behaviour is usually more "classical", with large planktonic blooms which are followed by periods of grazing and decay. These observations are broad generalizations, and in some parts of the North Atlantic, for example, a distinct spring bloom may occur, although this is usually not so marked and is much later in the year than the bloom found near the coast.

This subject has been discussed at length by Nielsen (1961b) using simple assumptions and models. For blooms to occur, not only must hydrographic and light conditions be suitable, but there must be an initial absence or shortage of grazers, a state of affairs which appears to apply to most types of coastal regions by early spring. In the oceans, on the other hand, the winter zooplankton stock may never be sufficiently depleted, and the animals are thus able to prevent the rapid spring growth of phytoplankton, especially in areas such as the north-east Pacific where plant production can proceed for most of the winter in (salinity) stratified water, and thus sustain an appreciable population of grazers.

For much of the year, in some temperate oceans at least, there must be a steady state between phytoplankton production in the euphotic zone and zooplankton grazing in the whole water column. This suggests that the standing crop of zooplankton carbon *beneath unit area* is several times greater than the standing crop of phytoplankton carbon. In oligotrophic tropical oceans, the growth rates of zooplankters will be slow compared with those of animals inhabiting cooler fertile waters. In these areas also, grazing will be a critical factor in the regeneration of nutrients (see Nielsen, 1958b; Cushing, 1959).

Grazing is responsible for some of the patchiness found in the chlorophyll distribution of surface waters. The surface chlorophyll values can change by 50–100% in distances of only a few miles in water having essentially the same physical and nutrient properties, which suggests that local zooplankton populations imprint grazing patterns upon the already over-grazed phytoplankton (McAllister *et al.*, 1960; Antia *et al.*, 1962). Bainbridge (1957) has given an excellent review of the literature on the shape, size and cell density of phytoplankton patches in the world's oceans.

The theory (Hardy, 1936) that zooplankton avoid large concentrations of plants and graze around the edge of blooms is losing favour. It appears that the distribution of grazers causes the observed phytoplankton concentrations more often than the converse (Nielsen, 1937; Beklemishev, 1957).

Classical studies of plankton production and grazing were made by Lohmann (1908), and much later by Harvey *et al.* (1935). These investigations suggested that the spring outburst of phytoplankton is eventually controlled by grazers, which often destroy plant cells far in excess of their needs.

Attempts have been made to construct mathematical models to describe the interaction of phytoplankton and zooplankton in areas of high productivity by assuming phytoplankton growth to be of the form

$$\frac{dp}{dt} = k \cdot p,$$

and that plant losses occur by physical means (sinking and turbulent dispersion), and by grazing, with rate constants k_L and k_G respectively. A differential equation of the form:

$$\frac{dp}{dt} = p \, . \, (k - k_L - k_G)$$

needs to be solved.

Fleming (1939b) produced a simple model by assuming that k_G was proportional to the zooplankton population at zero time, plus a time-variable factor to allow for an increase in the animal population. A very thorough mathematical treatment was given by Riley *et al.* (1949) in which an attempt was made to evaluate losses by sinking, turbulence and grazing. Steele (1958) has derived a mathematical expression for the production cycle in the northern North Sea which stresses the effects of mixing rates and turbulence. Cushing (1959), using a model of the spring outburst of phytoplankton, laid particular emphasis on the reproduction of grazers, and their methods of destroying phytoplankton. Other workers assumed that herbivorous zooplankton ingested plant cells as a result of the filtration of water, but Cushing suggested that grazing may be largely by encounter, the animal eating the cells it encounters while swimming through the water. This mechanism would enable very large apparent volumes to be filtered in barren water and lesser amounts as the plant population increased. No doubt filtering, grazing on encounter and active hunting for cells all play some part (Beklemishev, 1962).

At present, none of the proposed mathematical models of phytoplankton-zooplankton interrelationships enable precise predictions to be made of the observed productivity in nature. This is partly because the models are, of necessity, greatly oversimplified, and also because completely unknown factors may play a part in the natural environment. Also, we do not, at present, have sufficiently precise data for the parameters used in these expressions, or sufficient observations with which predictions can be tested. However, the attempts by Riley, Cushing, Steele and others to explain factors, other than photosynthesis, which influence the productivity of the sea have been of value in underlining our lack of relevant information about the basic mechanisms of the food chain, and in giving us a greater insight into the nature of the various processes which may govern observed distributions. The subject has been well reviewed by Steele (1959).

(3) *Productivity of the world's oceans.* (a) *Primary standing crop.* Land plants with root systems deep in fertile soil can maintain a standing crop of considerable bio-mass. Phytoplankton, however, have to obtain their nutrients from water in which the concentrations are several orders of magnitude less than those found in most soil solutions, and there is

no mechanism whereby the products of photosynthesis can be retained alive for very long in one location. As a result, the greatest standing crop of phytoplankton found beneath unit area of sea surface probably never exceeds a tenth of a per cent of the plant carbon per unit area of good pasture land.

The concentration of standing crop per unit volume of sea water varies greatly from place to place, and according to the season. Some data have been tabulated by Strickland (1960), but the variability is so great that there seems little point in making detailed lists or charts of such values. Results given in the literature are mostly expressed in terms of chlorophyll *a* concentrations per cubic metre. Figures range from about 0·05 mg./m.^{3} in the most oligotrophic waters, to 10–40 mg./m.^{3} in fertile coastal areas during bloom periods. Average values for temperate ocean areas might be 0·5 mg./m.^{3}, with about 5–10 times this amount near the coast during much of the growing season.

The standing crop of plant carbon beneath *unit area* of the euphotic zone is much less variable because of the reduction of the euphotic depth by high concentrations of cells. A range of 1–2 g.C/m.^{2} is reasonable for most ocean areas (tropical or temperate), and 2–10 g.C/m.^{2} for coastal or other fertile seas for most of the year, except during winter, when not more than 0·5–1 g.C/m.^{2} would be expected.

(b) *Net primary productivity*. If photosynthesis were proportional to illumination at all light intensities and wavelengths, the net production in the euphotic zone would be strictly proportional to the amount of incoming radiation. However, not all wavelengths are equally efficient, since they are absorbed differentially by the water. After a certain level of illumination is reached, extra radiant energy is used less efficiently, and may eventually inhibit photosynthesis near the surface of the sea. The result is that the production of organic matter in the photosynthetic zone increases in proportion to the incoming energy only up to a certain energy; above this value no further increase occurs. This subject is discussed by Ryther (1959) and Strickland (1960). If a reasonable value for the quantum efficiency at each wavelength is assumed, the *theoretical maximum net photosynthesis* in a water column may be estimated as a function of the total amount of incoming solar radiation, and is shown in Fig. 4. taken from Ryther (1959). The maximum rate is around 25 g.C/m.^{2}/day for 600 ly./day of total radiation and decreases at first slowly and then linearly to become zero at the compensation point of *c*. 100 ly./day. In midsummer at latitude 45°, one might expect an average radiation of *c*. 400 ly./day, which corresponds to a maximum net production of about 18 g.C/m.^{2}/day.

These theoretical values are never reached in the sea, but have been approached with cultures of unicellular algae under "ideal" conditions. The highest values for net primary productivity found in very fertile sea areas may reach 5 g.C/m.^{2}/day for short periods, but a more representative average would be nearer to 0·5–1 g.C/m.^{2}/day, i.e. about 5% of the maximum efficiency. In the less fertile oceans, rates rarely exceed 0·2 g.C/m.^{2}/day, and sunlight is used with only about 1% of the theoretical efficiency.

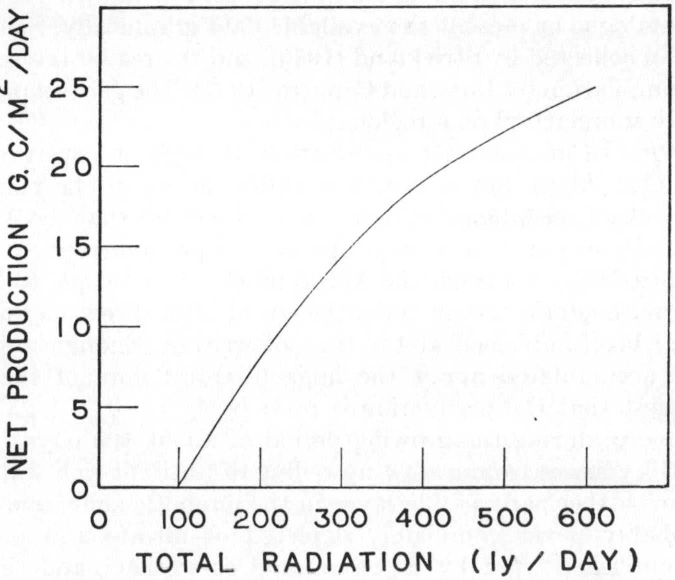

FIG. 4. Theoretical maximum production rate as a function of daily total radiant energy (after Ryther, 1959).

If we compare these estimates with the rates for the more rapidly growing land plants (say, 4 g.C/m.^{2}/day for rice, 1–2 g.C/m.^{2}/day for other cereals at peak growth, 2–3 g.C/m.^{2}/day for a young pine forest) we see that, despite the great difference in standing crop, the marine planktonic vegetation compares favourably with terrestrial plants as a primary producer of organic matter. Indeed, if we remember that the sea surface area exceeds that of moderately productive land by a factor of 10 or more, it is not surprising that the total annual productivity of the sea is thought to equal that of the land (*c.* 10^{10} tons C/year), or to be several times greater (Riley, 1944; Nielsen, 1952c; Ryther, 1959). Recently, Westlake (1963) has given an excellent summary of the average annual net productivity of many aquatic and terrestrial plant communities.

Although benthic algae are not the subject of this chapter, it is worth mentioning that the maximum productivity of a kelp bed can exceed 20 g.C/m.^{2}/day, and even tropical coral reefs produce several grams per day. It is only the comparatively small area of such regions which renders their global contribution insignificant.

For intercomparison purposes, rates are best expressed in terms of g.C/m.^{2}/day. They depend very much on the location and the time of year and, as with the standing crop, there seems little point in attempting to catalogue or present the available data graphically. Some values have been collected by Strickland (1960), and the reader is also referred to the compilation by Doty and Capurro (1961). The general picture will be briefly summarized on a regional basis.

Arctic. In midsummer, production is high in open water (*c.* 1 g.C/m.^{2}/day), but is very low under the ice of the polar basins (*c.* 0·03 g.C/m.^{2}/day maximum), with the result that the Arctic as a whole does not have a very high annual productivity.

Antarctic. By contrast, the Antarctic Ocean is *thought* to be one of the most productive areas in the world. Few direct measurements have been published at the time of writing. Enough information has accumulated about the huge (diatom) flora of the area to suggest that the production is most likely to be *c.* 1 g.C/m.^{2}/day or more, during the growing period of about 100 days. The high fertility arises from a slow upwelling of nutrient-rich water which forms a thermally stable layer in the euphotic zone, and which is probably never completely depleted of nitrate and phosphate. Unpublished work by El Sayed and co-workers, and the results obtained by Burkholder and Sieburth (1961) suggest that the productivity is very similar to that of the north-eastern Pacific Ocean with a very marked land-mass effect as one approaches the Antarctic continental shelf. Bunt (1963) has shown that the enormous crops of diatoms beneath sea ice may contribute considerably to the production of this region.

Equator. The combined action of wind-drift from the Trade Winds and of the Coriolis force causes moderate upwelling over a few degrees on either side of the equator, and gives rise to some of the more fertile tropical ocean areas in the Pacific and Atlantic, i.e. around 0·3–0·4 g.C/m.^{2}/day for most of the year, with a relatively high annual production.

Barren oceans. By contrast, most of the tropical oceans between latitudes 10° and 40°, and stagnant tropical gyres like the Sargasso Sea are very unproductive. The great euphotic depth compensates in part for their low production per unit volume, and at their more temperate extremes a slight seasonal effect may be noted, but for much of the year production does not exceed about 0·1 g.C/m.^{2}/day. However, as this rate persists for most of the year, an annual production of *c.* 50 g.C/m.^{2} can be expected.

Temperate oceans. The productivity of these regions varies considerably according to season and water stability. Rates for the north-east Pacific at 50°N., 145°W., where there is a permanent halocline at about 100 m., have been monitored regularly by the Nanaimo group of the Fisheries Research Board of Canada. These regions are probably typical, and their productivities vary between 0·05 g.C/m.^{2}/day in midwinter, and 0·4 g.C/m.^{2}/day in summer when the sky is semi-permanently overcast. Summer values in parts of the northern Atlantic are somewhat higher. Annual productions may be expected to be *c.* 100–150 g.C/m.^{2}.

Current fronts. Certain ocean areas where strong currents meet or diverge may produce pronounced mixing. A well-known example is the Kuroshio-Oyoshio Current area off Japan. The processes which fertilize the euphotic zone are not fully understood, but can be expected to double or treble the productivity of the areas in which they occur.

Continental shelf upwelling. The effect of prevailing winds and of the Coriolis force on currents off the west coasts of continents may give rise to upwelling, and thus, to great fertility. The best-known areas are off Peru, South-West Africa and California, and although the patterns change locally, the general picture is relatively constant for whole seasons. The position is less clear off the east side of land masses and along the coasts of southern Asia, but the Trade Winds and Monsoons can cause alternating periods of upwelling and of low fertility. At its greatest, production in these areas will exceed 1·5–2 g.C/m.^{2}/day for protracted periods. The effect of upwelling caused by submarine ridges in areas of relatively strong currents can be quite remarkable. Hansen (1961) has found that productivity over the Faroe-Iceland ridge may reach 2·5 g.C/m.^{2}/day, compared with 0·5 g.C/m.^{2}/day or less in the surrounding Atlantic.

Temperate coastal regions. The productivity of these locations is too complex to be generalized. The land-mass effect may reduce the standing crop in a seaward direction at a given latitude, and the growing season will start later in the ocean than over the continental shelf, or in sheltered coastal regions. However, the yearly net primary productivity at the coast and in the open ocean may differ by only about a factor of two (Currie, 1958; Ryther and

Yentsch, 1958). The productivity of the northern North Sea, Norwegian Sea and the coastal areas of Greenland have been particularly well studied (Steele, 1958; Berg, 1958; Nielsen and Hansen, 1961).

It should be remembered that although the total production *beneath unit area* of a barren sea may not be much less than that of a region where there is a higher standing crop per unit volume, the presence of plant cells in a greater concentration may exercise a considerable influence on the efficiency with which they are utilized in the food chain (see later).

C. TEMPORAL DISTRIBUTION

(1) *Spring and other "blooms"*. At mid and high latitudes in winter there is always some phytoplankton growth, but it is generally very slow, owing to the fact that the layer in which a plant cell is randomly transported by mixing processes (the mixed layer D_M) can be many times deeper than the euphotic depth (D_E). With the advent of longer brighter days, the euphotic depth increases and the mixed layer depth decreases owing to the onset of thermal stability. The phytoplankton cells then spend more of their time in the lighted zone, until at some point the growth rate of the population as a whole exceeds rate of loss by respiration, and a very rapid increase in cell numbers is observed during the course of one or two weeks. This is called the "spring bloom".

Spring blooms are of general occurrence in temperate coastal waters, and, as has been mentioned previously, occur to a less marked extent in most temperate oceans. Careful observations have revealed that small blooms take place even in the northern Sargasso Sea (see Hulburt *et al.*, 1960). The timing of the onset of these growth periods is of ecological importance as it may affect the whole cycle of sea life during the early part of the year. Attempts have been made therefore to predict from physical data when a bloom will occur. Gran and Braarud (1935) suggested that, as a rule-of-thumb criterion, a bloom might start when D_M was less than five times D_E. The subject was treated more quantitatively by Riley (1942) and Sverdrup (1953). The formula devised by Sverdrup assumed that plant losses resulted from respiration, and that plant photosynthesis was proportional to light intensity, the attenuation of which could be expressed by a single coefficient (k) applicable to green light. Blooming should occur with:

$$D_M \ll \left[1 - \exp(-k \, . \, D_M) \right] \cdot \frac{I_{\text{surface}}}{I_{\text{comp}} \, . \, k}$$

where I_{surface} and $I_{\text{comp.}}$ are the surface and compensation light intensities respectively. The former may be assumed to be about 20% of the total surface radiation, and $I_{comp.}$ may be put at 0·005 ly./min. Despite the approximations made in deriving this expression, it has proved to be a useful guide for predicting the blooms in certain sea areas. A more refined version has been attempted, in which allowance is made for supra-optimal light effects and for the attenuation of total solar energy rather than of green light (Cushing, 1962).

There is a limit, however, to the value of a too detailed theory at this stage, as the value for $I_{comp.}$ is not known with any certainty, and may well be different for net photosynthesis and for cell division. Also, there is undoubtedly some cell mortality from grazing and sinking, and there may be a time lag for the cell growth, as is commonly found in cultures, which prevents the onset of a bloom even if other conditions are favourable (Riley, 1943; Fogg, 1957). However, it may be worth stressing at this point that the *in situ* or "incubator" measurement of positive photosynthesis in the euphotic zone reveals nothing about the net production of plant material in that sea area, unless the depth of the mixed layer is also measured and found to be comparable to that of the euphotic layer.

The size of the "inoculum" is important. Not all species will find conditions favourable for growth, even when the Sverdrup or similar criteria are satisfied, and the water will require "seeding" with the correct phytoplankton cells or resting spores. Initial numbers may vary greatly, and sometimes be small enough to hinder the onset of a bloom, even if there is immediate activation and vigorous exponential growth of the seed stock. Seeding by meroplankton spores swept up from the sea floor is the most obvious mechanism in shallow water (Wood, 1959), but for deeper water, it seems necessary to assume that haloplankton or meroplankton resting stages remain near the surface throughout the winter (Braarud *et al.*, 1953). However, seeding from spores upwelling from great depths is not too improbable if we accept the evidence given by Kimball *et al.* (1963) for the presence and viability of algal cells in water collected from great depths.

After the initial bloom, others may occur at irregular intervals throughout the year. The classical idea of a spring and autumn bloom with a midsummer minimum (cf., Harvey, 1934b) is an oversimplification. Although spring blooms are general, they may occur in the northern hemisphere between mid-February and late April according to the area. Autumn blooms, which are not inevitable and are very unpredictable, arise when a period of autumn winds and cooling surface waters allows nutrients to mix into the euphotic zone, and is followed by sufficient stability for rapid plant growth to recommence.

In many areas where the correct sequence of meteorological conditions occurs, several blooms may be expected throughout the summer but, generally, the blooms found in the tropics or in nutrient-depleted coastal waters during summer arise either from freak local mixing and upwelling, or from the addition of critical nutrients to the water by land drainage. The organisms in these blooms are often dinoflagellates, and they give rise to the well-known *red tides* in some areas. The right combination of temperature and (perhaps) vitamin-like growth factors becomes ideal for the proliferation of one or more species. There is no evidence that high concentrations of nitrate or phosphate are necessary for red tides, but phototaxis ensures that cells are maintained at the depth of optimal light, where further concentration of cells may be affected by winds and tides. Kierstead and Slobodkin (1953) and Ragotzkie and Pomeroy (1957) have attempted a mathematical treatment of the blooming of dinoflagellates under the adverse influence of turbulence which tends to transport cells from a favourable to a less favourable environment. An excellent bibliography, with suggestions of the possible mechanisms underlying the formation of red tides, is given by Ryther (1955).

(2) *Succession of species*. The remaining topic which merits discussion here is the temporal succession of phytoplankton species. In a given water body free from large-scale advection, the species which predominate in the phytoplankton are continually changing and the species composition rarely repeats itself exactly from year to year in the same locality. The matter has more than a purely botanical importance as the cause of an observed succession and its relation to the physical and chemical environment poses interesting problems, and the succession pattern may be of critical importance for the ecology of the local marine food chain.

Field botanists often examine the plankton at weekly, two-weekly or monthly intervals, but rarely more frequently. The difference in the growth constants of two species which will make one species succeed the other during such a period, say a week, depends on the magnitude of the growth constants. Suppose two species A and B have initially about equal cell numbers and grow exponentially, one with a division time of 20 hours, and the other of 25 hours (which represents fairly rapid growth for most of the diatoms), then, after a week there will be over three times more cells of species A than of species B, and a definite "succession" will be recorded. If the doubling times were 12 and 15 hours (which might be observed for some flagellates), A would outnumber B by over 15 to 1. Thus, for the fast-growing phytoplankton, differences in growth constants of only 20–30% can lead to large changes of population composition, and even for the slower-growing plants (say the 3- to 4-day division rates of many dinoflagellates), one growth constant need be only twice the other for a noticeable succession to occur. Conditions are not, of course, so simple in the sea, and such factors as differential sinking, grazing and lateral advection play a part, but the above examples illustrate how comparatively slight changes of nutrients, light or temperature could be expected to affect growth constants sufficiently for the species composition of the phytoplankton to have the variability often observed.

It is unnecessary to describe or to give references here to the huge amount of descriptive taxonomy and species enumeration which has been undertaken in marine botany. The reader may obtain some idea of the type of work that is involved from the papers by Braarud *et al.* (1953), Conover (1956), Holmes (1956), Smayda (1958, 1959), Hasle (1959), Hasle and Smayda (1960), Hulburt *et al.* (1960), and Paasche (1960b).

A few very broad generalizations can be made. Temperate coastal blooms are generally characterized by the initial development of large diatom communities with a fairly recognizable pattern of genus, but not species, from year to year. *Thalassiosira* societies will be the first to appear, along with, or followed by the ubiquitous diatom *Skeletonema costatum*, followed by *Chaetoceros* societies. As the nutrient level of the water decreases, dinoflagellates and chrysophycean flagellates tend to predominate in the relatively sparse plankton crops of summer, with the diatoms reappearing in large numbers during any autumn bloom brought about by a major nutrient replenishment. Definite associations of species are recognized. Throughout the year there is a roughly constant population of ultraplankters which may constitute a large fraction of the winter phytoplankton bio-mass.

In the less fertile tropical oceans, the picture is different as conditions are more constant, and the changes of species composition are slower. Of the larger organisms, dinoflagellates may prevail, but they generally contribute little to the total bio-mass. A considerable fraction of the phytoplankton consists of unidentified ultraplankton (Hulburt *et al.*, 1960). The nannoplankton contains very small diatoms, but is composed for most of the year of coccolithophores and unknown flagellates which together probably comprise the largest fraction of the total plant mass. This is particularly true of the Mediterranean where Bernard (1958b) has shown that calcareous flagellates prevail at all depths, and that the palmellae of *Coccolithus fragilis* comprise 75–95% by volume of the phytoplankton.

A similar situation is found in more temperate oceans, where such organisms as *Exuviella* sp. and *Coccolithus huxleyi* predominate, and where small diatom outbursts may be expected during the early summer.

In many places, especially tropical coastal areas of only moderate fertility, species succession is not very marked, and is confused by small-scale advection effects which give rise to spatial heterogeneity. In these instances, it is difficult to decide whether or not true succession is taking place, especially if the amount of sampling and identification is to be kept within practicable limits. Some ecologists have suggested that the problem may be made tractable by the use of the information theory approach, and an order-disorder "entropy" concept borrowed from thermodynamics. Margalef (1957) and Patten (1959, 1962) have discussed the matter at length, and have given illustrations.

Most of the factors likely to be responsible for one plant species growing in preference to another have been discussed fully in Section V, and will only be mentioned again briefly in connection with the phenomena observed in nature. The interactions of these factors are not fully understood, and the succession problem presents one of the greatest challenges in marine productivity research. Apart from differential grazing, sinking and lateral advection, the following aspects need consideration.

(a) *Seeding*. The introduction of allochthonous species can be important in starting plankton growth at all times of the year. This is stressed in papers by workers of the Norwegian School, who have demonstrated the influence of surface waters carrying seed stock into various areas of the North Sea and Norwegian Sea. We still have relatively little knowledge of the conditions which lead to auxospore formation in diatoms, and to cysts and resting forms in the main classes of phytoplankton. Even more important, we know little of the factors which stimulate the germination of these spores to give vegetative cells. The classical experiments by Gross (1937) on the behaviour of *Ditylum* and other species in cultures give valuable insight into factors likely to be important. In a recent review, Lewin and Guillard (1963) have given a brief account of most of the literature of diatom sexuality, much of it pertaining to fresh-water forms; the reader is referred to this review for greater details.

(b) *Major nutrient levels*. For no very obvious reason, the presence of large amounts of nitrate and phosphate, such as occur in the Antarctic Ocean or in coastal waters during spring, seems to favour the production of large diatoms, even where other nutrients that are known to permit the rapid growth of flagellates are present. A partial explanation may be that the generally low temperatures associated with nutrient-rich waters favour the growth of bacillariophyceae, and that grazing from nanno-zooplankton simultaneously prevents the outburst of the small flagellates. However, in many areas of the fertile ocean, large crops of large diatoms never appear, and here a differential grazing and sinking pattern must work in favour of the smaller forms. The position is still not clear, but some very significant observations have been made by Menzel *et al.* (1961) on the effect of enriching Sargasso water with nitrogen, phosphorus, silicon and iron, which indicate that the relative proportions of these nutrients have a strongly selective effect on the growth of the endogenous phytoplankters.

It is easy to understand why flagellates should have a greater ability than diatoms to extract nutrients from infertile warm water, especially those species such as the dinoflagellates that do not require silicon, and which are probably able to obtain their nitrogen and phosphorus from organic sources. Some field observations and enrichment experiments have indicated that the growth of dinoflagellates, unlike that of diatoms, may be inhibited by nitrate (Harris, 1959), but this is not always borne out by culture work.

Although not a major nutrient in the normally accepted sense, calcium carbonate is required by coccolithophores, and it is perhaps significant that these organisms constitute a much greater fraction of the phytoplankton in warmer waters where the degree of oversaturation with calcium carbonate is greatest (Braarud *et al.*, 1952).

(c) *The effect of salinity, temperature and light.* The effects of salinity on true marine phytoplankton are likely to be slight, and nearly all marine phytoplankters grow well at a salinity of 25‰. Certain species, especially diatoms, do not grow well at salinities greater than 35‰, which may account for their preference for a coastal habitat (Hulburt and Rodman, 1963). However, the low salinity tolerance of a species will be critical in deciding how far it can penetrate into brackish areas, the low salinity regions around the entrances to large rivers, or seas such as the Baltic. In general, the field evidence is in accord with the meagre amount of experimental work that has been done in cultures (Braarud, 1962).

Sufficient has been said in Section V about the combined effect of light and temperature on cell division to suggest that the exact pattern of the changes of water temperature and light levels in nature will help to determine the succession of species.

Evidence for the inter-dependent effects of light and temperature is obtained from the relatively clear-cut picture of winter, spring, summer and autumn phytoplankton in lakes; these effects are associated with the combinations, low temperature-low light, low temperature-high light, high temperature-high light and high temperature-low light, respectively (see e.g. Vollenweider, 1950; Findenegg, 1943). An organism may be able to stand an extreme of temperature with low light, but not with high light or vice versa. Thus, the different abilities of each plant species to withstand the combined effects of extreme conditions of temperature and light acting together and in conjunction with the nutrient environment, undoubtedly affects the relative growth rates of species in temperate waters during spring and autumn. Water temperature, illumination intensity, nutrient deficiencies and the relative lengths of day and night are all important factors as they control the formation of diatom resting spores and the germination of auxospores. Rodhe (1948) has discussed in detail how the growth patterns in nature of a freshwater *Melosira* species can be explained by assuming the interactions of light and temperature.

The direct evidence for temperature-light control in the sea is limited, but the inference is very strong. Conover (1956) showed that the succession of *Thalassiosira nordenskiöldii* by *Skeletonema costatum* in Long Island Sound could be explained by the preference of the latter species for higher light intensities, and for water with a temperature exceeding 4°C. Recent work with cultures (Jitts *et al.*, 1964) has suggested that high light intensities may be inhibitory to *T. nordenskiöldii* at temperatures below 4°C but not at 15°C when the diatom requires relatively high light intensities. In general, the behaviours of diatoms and dinoflagellates in nature are in accordance with their observed behaviours in cultures (see e.g. Braarud, 1961). The laboratory approach to phytoplankton ecology is proving very fruitful; however, caution is necessary. Not only *Skeletonema costatum*, mentioned earlier, but also *Thalassiosira nordenskiöldii* and *Asterionella japonica* have temperature optima in cultures 10°C or more higher than the temperatures of the seas in which they proliferate, and this seems to be more or less true of most species which have been studied. Furthermore, cultured stocks of *Skeletonema* and *Asterionella* will not grow at the temperatures found in their natural environment. The temperature-light response of clones of the same species differs significantly if the clones are isolated from warm or cold water. Undoubtedly the conditions of culturing are very important, and there is still much to be learned about the effect of preconditioning the cells, the chemical composition of the culture media and the significance of the bacteria and other plants which normally occur in nature along with the species in question.

(d) *Minor nutrients.* We can do little more than speculate on the effect of transition elements (iron, manganese, zinc, etc.) on phytoplankton growth in nature. The concentrations of elements such as copper and manganese are often an order of magnitude greater in coastal waters than in the open ocean, and this might be expected to exercise a selective effect. The work of Ryther and Kramer (1961) has provided some evidence that this might be the case with iron. These authors were unable to show an iron requirement by the oceanic species *Coccolithus huxleyi*, and only a relatively low iron requirement by typical Sargasso Sea diatoms such as *Cyclotella nana*. Coastal species such as *Skeletonema costatum*, on the other hand, would not grow well unless iron was present at ten times the concentration found satisfactory for the oceanic species.

(e) *Humic material from land drainage*. The comparatively large amounts of humic materials found in coastal waters as the result of land drainage almost certainly exercise some effect on the phytoplankton crop. Recent unpublished work from several laboratories indicates that this material improves the growth of diatoms by a mechanism which is not yet understood. The most obvious function of humic acids would be to increase the availability, or reduce the toxicity, of transition elements by chelation processes as first suggested by Gran (1931), but other and more subtle processes may be at work, such as cell membrane activation and the detoxification by humic acids of growth-inhibiting substances excreted by the plants themselves. There is marginal evidence that the decay and excretion products of benthic algae may stimulate the growth of dinoflagellates (Nordli, 1957b).

(f) *Vitamin growth factors*. The vitamin growth factors for phytoplankton development have been discussed in Section V, C. Biotin, thiamine, and B_{12} are apparently the most significant, and all have been detected in sea water (see Chapter 11). What is open to question is whether the species composition of the phytoplankton is materially affected by the concentrations of these substances.

Only vitamin B_{12} has been studied in any detail. Amounts in the sea vary from about 0·1–10 $\mu\mu$g./ml. From the B_{12} concentration of phytoplankton cells, we would not expect these amounts to limit the final bio-mass of any species, except perhaps during heavy blooms (Droop, 1957; Antia *et al.*, 1963), but it is not clear how much the concentration of the vitamin affects growth kinetics. The species composition of a crop may be susceptible to the influence of B_{12} concentrations inasmuch as they determine growth rates even if the total sustainable bio-mass is not limited. There is strong presumptive evidence that the "red tide" blooms of certain dinoflagellates in coastal regions may be stimulated by the presence of vitamin B_{12} which enters the sea from land drainage. It may also be significant that the B_{12} concentration of the Sargasso Sea shows a pronounced maximum (Spaeth and Menzel, 1961) at the time of the short "bloom" in April of diatoms known to require B_{12} (Hulburt *et al.*, 1960). *Coccolithus huxleyi*, which has only a thiamine requirement (see Table VIII), is present for most of the year and shows no such stimulation in spring.

(g) *Plant hormones*. Finally, there is the effect on species succession of the excretion or the decay products of one bloom of cells on the succeeding communities. Much evidence has been obtained from culture work that the cells of one species condition the medium in such a way as to inhibit the growth of another species. This subject has been studied extensively with chlorophyceae (Lefevrè *et al.*, 1949; Prescott, 1960). Proctor (1957b) obtained evidence that the substances involved were long-chain fatty acids. Diatoms are also found to liberate substances which are not only inimicable to the growth of other diatoms, but which also adversely affect their own growth (Levring, 1945; Jørgensen, 1956; Curl and McLeod, 1961; Lanskaya, 1961). *Skeletonema costatum* is well known in this respect. The toxins are all heat labile, and may be slowly destroyed by photo-oxidation. However, it is dangerous to generalize, as Talling (1957d) could find no evidence of mutual inhibition between *Asterionella formosa* and *Fragilaria crotonensis*, which commonly succeeds it in lake flora. Interesting work in progress by J. McLaughlin at the Haskins Laboratories, New York, is showing that certain combinations of dinoflagellates cannot co-exist in cultures except under very specific light intensities.

It is difficult to extrapolate from conditions found with the comparatively dense populations of laboratory cultures to the conditions found in natural water bodies. Although phytoplankton production may be inhibited by dense benthic flora (Grøntved and Nielsen, 1957), interactions at the normal population densities of marine phytoplankton are less certain. However, evidence is accumulating that very dense blooms in fresh water do affect the course of subsequent succession, and it is possible that at the coast the sea is more suitably conditioned for the growth of certain diatoms at one time of year than at another. Johnston (1963b) has found that filtered and briefly sterilized sea water, enriched with a full spectrum of inorganic and organic nutrients, varies in its suitability as a medium for the growth of *Skeletonema costatum*. By studying the growth responses in water taken in different areas at different times of the year, charts can be made showing the distribution of "diatom water", and there are indications that whole sea areas are chemically conditioned for or against one type of flora by the earlier growth of bacteria, plants or zooplankton. Eventually we may be able to recognize "good" or "bad" diatom water on the basis of its fluorescence, ultra-violet absorption spectrum, or by the presence of characteristic bacteria and zooplankton.

Plant hormones are found in, and will react with, diatoms but there has been no conclusive demonstration of their ecological importance (Imamura, 1952; Bentley, 1959). Of the five gibberellins tested by Johnston (1963a) with mixed cultures of phytoplankton, only two had any marked stimulatory effect.

VIII. Phytoplankton Utilization

A. Direct Food Supply

The marine food chain differs from that on land in that there are generally several steps between the plants and the larger invertebrates. The phytoplankton generally enter the food chain as food for small animals, mainly crustacea, known collectively as zooplankton. There is still a serious shortage of quantitative information about the nutritional requirements of marine zooplankters, their grazing habits and digestive processes. Barker-Jørgensen (1962) has recently given an excellent review of the subject.

All species of phytoplankton are not equally suitable for use as food. The most obvious limiting factor is the size and shape of a plant cell in relation to the oesophagus of its predator. The very small marine larvae probably use only those cells which are smaller than about 10 μ (Bruce *et al.*, 1940), and young stages of *Calanus* find diatoms such as *Ditylum brightwelli* and a large *Coscinodiscus* too large for ingestion (Marshall and Orr, 1958). However, there is a definite lower limit to the cell size in the feed of an adult *Calanus*, although Gauld (1951) reported that the animal could utilize a *Chlamydomonas* sp. with cells 6–8 μ in diameter. Cells with a diameter less than about 10 μ are filtered less readily than larger ones, and are used very inefficiently if less than 5 μ in size (Marshall and Orr, 1955a, 1958; Mullin, 1963). The smallest ultraplankters and bacteria are said not to be ingested at all unless first incorporated into larger phagotrophic flagellates.

By contrast, lamellibranchs, in particular oysters, seem to be capable of filtering out particles 1–2 μ or less in diameter by a mechanism which is still not fully understood (MacGinitie, 1945; Barker-Jørgensen, 1960).

All the diatoms tested by Marshall and Orr (1952, 1955a,b) were found to be satisfactory food for *Calanus finmarchicus*, although some were better than others, and two dinoflagellates and a coccolithophore were also utilized. On the other hand, a marine cladoceran studied by Bainbridge (1958) appeared to eschew diatoms in favour of certain tintinnids and dinoflagellates which it captured by hunting, or encountering, rather than by "filtering". Conover (1960) tested the ability of herbivorous zooplankters to feed on diatoms in pure cultures.

Skeletonema costatum is generally agreed to be an excellent diet for both copepods and lamellibranchs, but the latter digest only the smaller diatoms, and prefer ultraplankters (a long list of which is given by Thorson, 1946). Davis and Guillard (1958) have investigated the dietary value of flagellates for oyster and clam larvae using 12 species of algae from 10 genera. Again, some species were better than others. *Isochrysis galbana* and *Monochrysis lutheri* were high on the list, and equally beneficial for both larvae, whereas a *Chlorococcum* was preferred by the clams. A more recent and very thorough study by Walne (1963) on the larvae of *Ostrea edulis* has given a similar picture. *Chlorella* was a poor food, and *Hemiselmis virescens* was indifferent. Good results were obtained, however, with *Isochrysis galbana* and *Monochrysis lutheri*, and with *Phaeodactylum tricornutum* and *Dunaliella tertiolecta* at low cell densities. Somewhat better growth was experienced when a mixture of several algae was fed to the larvae to give a "balanced diet". Similar conclusions were arrived at by Walne (1956) in an earlier study of the feeding of *Ostrea* larvae on pure algal diets, and it seems likely that marine creatures, like those on land, thrive better if given some variety in their food.

Corner (1961) measured the intake of carbohydrates, proteins and lipids from natural sea water, and found that organic matter was ingested in preference to inorganic material in the same sample. Although this and other evidence indicates that some selectivity in feeding may be possible with crustacea (Beklemishev, 1954), there is no doubt that certain lamellibranchs control their diet by selective feeding. Nelson (1947) reported an instance where a *Chaetoceros* was rejected by oysters feeding on *Skeletonema costatum*, and the extraordinary degree of selectivity found by Loosanoff (1949) could only be explained reasonably by assuming the presence of chemo-receptors on the labial palps.

Only by the use of bacteria-free cultures of both plant and animal can the true food value of a phytoplankter for a given zooplankter be correctly evaluated. There are regrettably few such studies. The classical work of this type is by Gibor (1956), who showed that there were very real differences in food value between various species of phytoplankton. Although a single species (*Monochrysis lutheri*) provides a balanced diet for *Tigriopus* (27 generations), it is generally necessary for several different species to be eaten together if the full growth and development of the animal is to be attained (Shiriaishi and Provasoli, 1959). Provasoli *et al.* (1959) have given a list of algal diets for *Artemia salina* and *Tigriopus japonicus*, only a few of which are wholly satisfactory.

Although some species of *Calanus* may be able to break open diatoms with their mandibles (Beklemishev, 1954), plant cells generally enter the animal gut whole, and therefore a prime requirement for the digestion of a plant is that the gut of the animal should contain enzymes capable of disrupting the cell membrane. Remarkable as it may seem, there is evidence to suggest that although amylase enzymes are present, cellulases are in short supply. Unless chlorophyceae and dinophyceae are initially damaged, they are expelled unchanged by oysters (Bruce *et al.*, 1940; Cerruti, 1941; Coe, 1945, 1948). A cellulase shortage has been reported in crustacea by von Dehn (1930), Bond (1934), Hasler (1935), and Huang and Giese (1958), and may be inferred by the absence of attack on many plant cells (see e.g. Gibor, 1956). Whole cells often pass through a copepod's gut unharmed. One cannot generalize, however, as Conover (1960) noted that after *Thalassiosira decipiens* had passed unchanged through part of the digestive tract of *Acartia tonsa*, it disintegrated quite suddenly. The absence of cellulase activity has been questioned by Morton (1958) and Nelson (1947). A plant cell can be digested whatever its membrane if it is small enough for phagocytic scavenging by the blood cells in lamellibranchs (Yonge, 1935; Coe, 1945), or if the intestinal bacterial flora of animals is sufficiently active; very little is known about the latter (see Huang and Giese, 1958).

It is unfortunate that the enzyme systems of neither marine nor fresh-water zooplankters have been much studied. Lipase activity has been demonstrated (Bond, 1934; Hasler, 1935). Proteinase activity at a pH as low as 3·5 was reported by Bond (1934) in *Calanus*, but in *Daphnia magna* and *D. pulex* only trypsin-like enzymes were present (Hasler, 1935). It is not known to what extent the observations by Hasler (1937) of dipeptidase, aminopolypeptidase and carboxypolypeptidase reactions in fresh-water crustacea apply to salt-water species.

The gross efficiency of transfer of bio-mass from marine phytoplankton to zooplankton remains, for all practical purposes, unknown as no one transfer coefficient can be expected. It is necessary to distinguish between the gross efficiency (the ratio of consumed matter to the production of animal protoplasm) and the net efficiency (the ratio of *assimilated* material to protoplasm production). Much of the variation in gross efficiency in nature is caused by changes in the fraction of the ingested food which is excreted or voided as faecal matter. There is a further source of variability, since some of the assimilated material is consumed in providing energy for the animal. The amount used in this way depends on water temperature, the metabolic condition of the zooplankter, and the stage of its life history. Food chain efficiency can be defined in many ways, and it is usual to consider the transfer of energy stored in various forms, rather than the weight of organic matter. There is much to be said, however, for considering the transfer of protein, since protein is generally the limiting metabolite in the marine environment.

No figure for gross efficiency given in the literature can be taken as generally applicable, except perhaps for the crudest of ecological models. Richman (1958) found from a thorough study of the transformation of energy in the fresh-water zooplankter *Daphnia pulex* a net efficiency of *c.* 55% in the pre-adult stages; the efficiency fell to only a few per cent during ovogenesis. Studies with the same organsim by Monakov and Sorokin (1961), using a ^{14}C-labelled nannoplankter, gave a maximum of about 40% gross efficiency when the phytoplankton concentration was optimal, the efficiency decreasing with too high or too low a concentration of plant cells. This picture is general, and Lasker (1960) reported efficiencies as high as 70% for the marine zooplankter *Euphausia pacifica* feeding on carbon-14 labelled *Dunaliella* which decreased to as low as 10% when the animal blocked its gut with cells owing to overfeeding. Gibor (1957), using the brine shrimp *Artemia* fed on *Dunaliella*, found the same effect with about 55% as the maximum gross efficiency. Marshall and Orr (1955b) reported maximum efficiencies of between 50 and 80% for *Calanus finmarchicus* fed on a ^{14}C-labelled diatom and coccolithophore. Efficiencies found by Conover (1962) using *Thalassiosira* sp. were *c.* 15–20%.

The convention of assuming a gross efficiency of 10% at each step in the food chain is thus not justified. Efficiencies are probably much higher under optimum conditions of transfer, but the mean daily efficiency for the entire life history of any organism, or for one of the stages in its development, could well be less. The concentration of plant cells is of great practical significance in determining the transfer coefficients, and the mean standing crop per unit volume of the euphotic zone may, therefore, be as important as the production rate per unit area (see Section VII, B).

There is a good correlation between the abundance of suitable fresh plant food and the egg-laying timing and ability of copepods (see e.g. Marshall and Orr, 1952, 1962; Conover, 1962), which may be of critical importance in marine ecology. However, fresh plant food may play only a minor role in the nutrition of many lamellibranchs, other benthic filter feeders and some of the deep-water crustacea. Detritus, small protozoa, and bacteria appear to constitute the main food supply of these animals, and may supplement the nutritionally inadequate phytoplankton diets of other organisms (see Shiriaishi and Provasoli, 1959). It is interesting to note that as early as 1925 Pütter had suggested that particulate matter might provide the vitamin content of zooplankton diet, although he had assumed, on doubtful evidence, that soluble organic matter was the principal food. The manner by which phytoplankton production indirectly furnishes other organic materials for use in the food-chain is worth considering.

B. PLANT AND ANIMAL DETRITUS

Zooplankters, when presented with an over-abundance of phytoplankton, as is usual for a month or two each year in most fertile coastal seas, consume food far in excess of their needs, excreting half or more as faecal pellets. This semi-digested plant detritus is often a significant fraction of the suspended organic matter in coastal waters, and acts as a valuable source of food for mid-water animals and bottom fauna (Beklemishev, 1962).

Plant detritus would not be expected in such large quantities in the less productive waters of the open ocean, and it is indeed difficult to detect in such localities, but nevertheless, the oceans are not free from detrital organic matter. Several workers have commented on the amount of organic debris shown to be present by microscopic examination. The quantity of detritus may be estimated by the direct determination of particulate organic matter, making allowance for the presence of living phytoplankton and zooplankton (see e.g. Bernard and Fage, 1936; Allen, 1939; Lisitzin, 1959, 1961; Hagmeier, 1960; Hulburt *et al.*, 1960; McAllister *et al.*, 1960; Ryther and Menzel, 1960; Krey, 1961a,b,c; Sushchenya, 1961).

Near the coast, concentrations may reach 1000 mg./m.^{3} dry weight of organic material, but even in the open ocean, quantities of 200 to 500 mg./m.^{3} have been recorded in surface waters. What is more significant is the finding by Parsons and Strickland (1962a) that, although concentrations decrease at depths below a few hundred metres, detritus is present in significant amounts (50–150 mg./m.^{3}) down to the bottom of the ocean. Recent results by other workers (Menzel and Ryther, 1964; Dal Pont and Newell, 1963) confirm these observations. The chemical composition of this material is such that it could serve as a potential source of food for pelagic marine organisms. Only the most limited data are at present available, but there is reason to suppose that in all deep oceans, tropical and temperate, several hundred grams of particulate organic matter exist beneath each square metre of surface compared with a standing crop of phytoplankton which does not exceed some two or three grams. The nature of this material is obscure, but it is almost certainly animal in origin, and is perhaps largely derived from the moults of crustacea.

In coastal regions, detritus, including semi-colloidal material, is known to be used as food by benthic filter feeders and may be used by zooplankton if the plant crop is sparse (see e.g. Lohmann, 1908; Coe, 1948; Harvey, 1950; Marshall and Orr, 1958; Riley, 1959; Fisher and Goldie, 1959; Cowey and Corner, 1963; and the reviews by Fox, 1950, and Krey, 1961b), but its role in the open ocean, especially the ocean depths, is less clear. The food of deep-sea plankton has been discussed recently by Vinogradov (1962), who concluded that detritus played only a secondary role. Tchindonova (1959) found detritus in the gut of deep-sea herbivores and carnivores, although only a small percentage of the species studied were predominantly detritus feeders. Gillbricht (1961) has shown a suggestive inverse correlation between the amounts of detritus and of zooplankton across the North Atlantic.

Although satisfactory quantitative data are lacking, the preponderance of detritus over living matter in a water column appears to be so great that it is difficult to believe that the detritus has no food-chain significance. It is perhaps best to consider that its entry into the food chain affects the overall efficiency of conversion of primary plant material to animal matter, and helps in the distribution of foodstuff both vertically and laterally. There can be little doubt that within the euphotic zone, detritus is of secondary importance for much of the year, but it may be important for the survival of wintering zooplankton stock and as a food for some of the animals which permanently inhabit the deeper regions of the world's oceans.

C. PHYTOPLANKTON AS A SOURCE OF DISSOLVED ORGANIC MATTER FOR HETEROTROPHS

1. *Bacteria and bacteria-detritus aggregates as food.* The most obvious role of bacteria in the sea is in the decomposition of detrital material with the consequent reconversion of elements such as phosphorus and nitrogen to a mineralized form. In addition, bacteria produce most of the vitamins (B_{12}, biotin and thiamine) required for the growth of phytoplankton and zooplankton. However, the bio-mass of bacteria and other heterotrophs may be appreciable, and it is possible that these organisms may act as a direct source of particulate food for secondary producers.

The evidence is still conflicting. Zhukova (1961) suggested that bacteria may be digested and form a significant fraction of the diet of benthic and planktonic animals. Support for this suggestion had already been given by Voroschilova and Dianova (1937), who examined marine zooplankters, and there is little doubt that bottom filter feeders make use of bacteria as food (see e.g. Luck *et al.*, 1931; MacGinitie, 1932; ZoBell and Feltham, 1938; Rodina, 1948). However, other workers have suggested that such a diet alone is inadequate (Fuller and Clarke, 1936; Loosanoff *et al.*, 1951). Davis (1953) could find no utilization of 13 species of bacteria by oysters.

The bacteria are largely associated with detrital particles, and may comprise a significant fraction of the detrital bio-mass (Wood, 1953; Jannasch, 1954; Rodina, 1962, 1963; Jannasch and Jones, 1959). It is probable that heterotrophs most readily enter the food chain when associated with detritus, especially as "marine snow" in which bacteria, detritus and phytoplankters are collected together into one particle (Riley, 1963). There is no direct proof that zooplankters graze on "marine snow", but the possibility cannot be overlooked. Nelson (1947) suggested that an important role for "refractory" detritus, such as cellulose fragments, is to act as a substrate, or at least a base, for the attachment of bacteria, and it is these bacteria which form the food for the lamellibranchs which ingest cellulose fragments. Rodina (1963) states that *Daphnia magna* can easily be reared through all stages exclusively on *fresh* detritus-bacteria aggregates, and Baylor and Sutcliffe (1963) have claimed that *Artemia salina* can be sustained on "synthetic" marine snow made by bubbling air through sea water. If the detritus used by Rodina was heat sterilized, the animals would not breed owing, presumably, to a shortage of vitamins. The bacterial population of the ocean may be "overgrazed" by protozoa in the same manner as the phytoplankton appear to be over-grazed in certain areas by larger zooplankters (Keys *et al.*, 1935).

(2) *The quantity of heterotrophs in the sea.* The sea is known to contain mould fungi and yeasts as well as bacteria. The yeasts are found mainly in coastal waters, where cell counts of 10^6 cells/l. have been recorded—comparable to the counts of phytoplankters in a small bloom—although the cell numbers are normally between 10^3 and 10^4 cells/l. (Vishniac, 1955; Fell and Van Uden, 1961). The lower marine fungi resemble the bacteria in that they are associated with surfaces. Cell counts approaching 5×10^5 cells/l. have been reported, but there is generally less than one-hundredth of this amount in areas well away from the land (Vishniac, 1956).

Jannasch and Jones (1959) have shown that the number of bacterial cells in the sea is greater than was once supposed. In some fresh-water systems, the bacterial bio-mass can approach that of the phytoplankton (see Kuznezow, 1959; Stepanek and Pokorny, 1960; Sorokin, 1961) and this may also apply at times in warm coastal seas. The question of immediate importance to marine ecologists, however, is the relative contribution of photolithotrophic and chemo-organotrophic processes in the open ocean.

Although more data are needed, we can make *very rough* calculations of the likely bio-mass of micro-heterotrophs in the warmer oceans. A figure of *c.* 0·1 mg.C/m.^{3} is obtained, but this may have to be reduced if the most recent Russian work on the number of deep sea heterotrophs (Sorokin, 1962) is confirmed. If one assumes a bacterial growth rate comparable with that of the phytoplankton, the heterotrophic production of organic carbon would be a few tenths of a percent of the average photosynthetic production per unit volume. When it is remembered that the depth of the total water column in an ocean may be 50 to 100 times that of the euphotic zone, it will be seen that the heterotrophic production beneath unit area could be nearly as great as the photosynthetic production. In shallow coastal regions, one would not expect the production of heterotrophs to be significant compared with that of phytoplankton, except during early winter, or immediately after a phytoplankton bloom.

(3) *The production of dissolved organic matter by phytoplankton and its role as a substrate for marine heterotrophs.* The quantity of "dissolved" organic matter in sea water depends on the definition of "dissolved" (see Chapter 11). If this is taken to be the organic matter which passes through a membrane filter of pore size 0·5 μ, then this quantity nearly always exceeds the amount of "particulate" organic material by a factor of 10 or more and 50 or more if we consider only living cells. If only that fraction which eludes an adsorbent filter, such as kieselguhr, is termed soluble, then the amount will be greatly reduced, but will still furnish sufficient substrate for the observed growth of marine heterotrophs. The subject has been dealt with fully in Chapter 11, and it is sufficient to remind the reader that a wide variety of compounds known to act as substrates for the growth of bacteria and algae have been detected in sea water (e.g. carbohydrates, amino acids, glycollic acid and several acids of the tricarboxylic-acid-cycle).

Algae must furnish the bulk of dissolved organic matter. It was assumed by Krogh *et al.* (1930) that the amount of organic matter excreted by phytoplankton was not very significant, but more recent work is reversing this view. Not only may cells excrete at certain stages of their life cycles, often under physiological stress (see e.g. Braarud and Føyn, 1930; Aleyev, 1934; Bishop *et al.*, 1954; Lewin, 1956; Guillard and Wangersky, 1958), but there is also evidence that healthy cells during exponential "growth" may excrete a substantial fraction of the material which they have recently photosynthesized (Tolbert and Zill, 1956; Allen, 1956; Antia *et al.*, 1963). The excretion of glycollate is reversible (at least in *Chlorella*) depending on the carbon dioxide tension and pH (Miller *et al.*, 1963), but we do not know if this applies to marine algae. Recent (unpublished) work in the author's laboratories has indicated that this excretion may be particularly marked with marine diatoms, and that perhaps as much organic material is produced when cells are alive, as by their decay once they are dead.

There is no quantitative information about the equilibrium concentration of substrates in the sea, or about the rate at which they are produced by plants and animals, or arise from the larger reservoir of more inert organic matter. It is not known how rapidly or efficiently these substrates are used by heterotrophs. Parsons and Strickland (1962b) used a ^{14}C-labelled glucose and acetate method for the measurement of the "relative heterotrophic potential" of a water column. This method might, in some circumstances, give absolute rates. This information is desirable because the chemo-organotrophic growth of bacteria, eumycetes, protozoa and algae provide a mode of entry into the food chain for organic matter which might otherwise go to waste. Utilization of heterotrophs and any re-utilization of detritus affect the overall efficiency with which the photolithotrophs of the world's oceans are converted into food for mankind.

The classical theory of Pütter (1909, 1925) (see also Chapter 11) that dissolved organic matter in sea water acts as a direct foodstuff for animals has been a subject of controversy, and although at the present time the main weight of opinion is against Pütter's view (see Krogh, 1931; Bond, 1933; Gellis and Clarke, 1935), it is premature to regard it as definitely disproved, especially since the larger colloidal particles (Gellis and Clarke, 1935; Fox, 1950; Morris, 1955; Stephens and Schinske, 1961) and soluble organic matter may prove to be important as a source of essential growth factors to supplement solid diets. Das (1960) has demonstrated how yeast extract and vitamin B_{12} influence the early survival of fish larvae. Most important, the formation of particulate "marine snow" from surface active dissolved and colloidal matter (as a result of condensation on the surface of bubbles, see Riley, 1963) reveals a mechanism whereby dissolved material can be made available to animals without postulating direct absorption through surface or gut.

Shiriaishi and Provasoli (1959) showed by an elegant series of experiments that growth factors (glutathione and vitamin mixtures) could supplement inadequate algal diets in the feeding of *Tigriopus japonicus*. The significance of this work cannot be overstressed, as it seems probable that certain blooms of phytoplankton may be useless as a source of food for zooplankters, unless an adequate mixture of vitamin-like compounds is already present in the water.

Collier (1953) has described sea water in the following terms: "neritic waters can be viewed much as the mammalian physiologist would view blood; as a transport system charged with living cells and agglomerates of proteinaceous and carbohydrate complexes, as well as dissolved gases. There is good reason for the marine biologist to approach the problem from the point of view of blood chemistry." This may sound extreme, but it emphasizes the exciting possibilities which are opened up to the biochemist when he studies the marine food chain in its earlier stages.

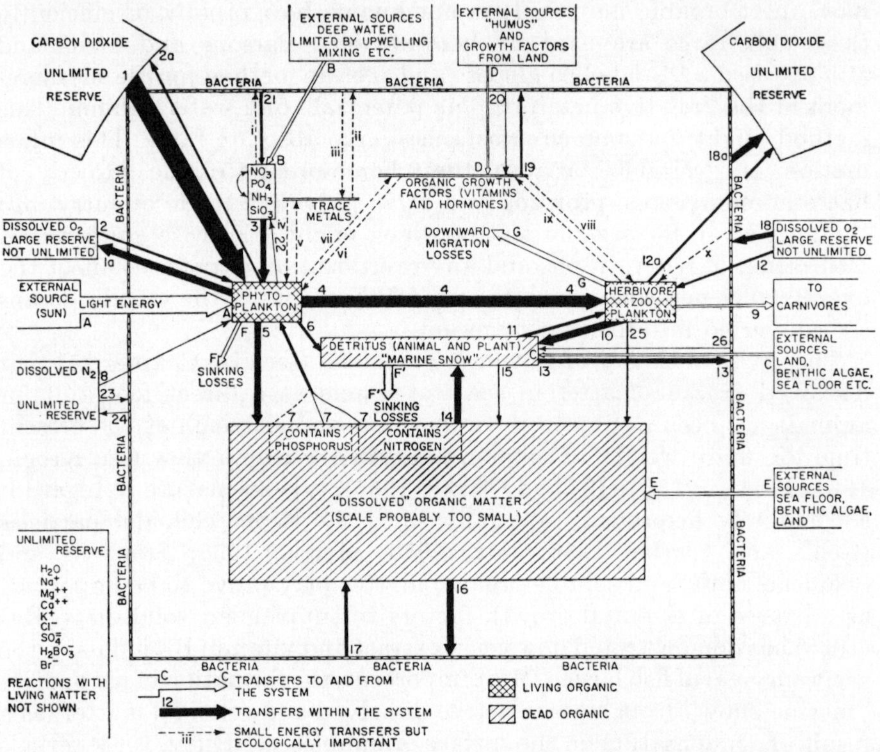

FIG. 5. Chart of idealized food-web in the early stages.

The complexity of the food web is illustrated by Fig. 5 which represents the state of affairs in a hypothetical (large) volume of water in the euphotic zone of a fertile sea area. The area of various blocks are proportional to the standing stock of the components and the pathways for energy and mass transfers (numbered for easy reference) are shown by arrows the widths of which are proportional to flow rates. Although all these pathways (shown here for a condition of dynamic equilibrium) are known, their relative importance and the relative values for standing stocks are based on subjective guesses and are only intended as an illustration.

REFERENCES

Aaronson, S., and Baker, H. (1961). *J. Protozool.* **8**, 274.

Aleyev, B. S. (1934). *Mikrobiologiya,* **3**, 506.

Al Kholy, A. A. (1956). *Physiol. Plant.* **9**, 137.

Allen, M. B. (1956). *Arch. Mikrobiol.* **24**, 163.

Allen, M. B. (1959). Photosynthetic nitrogen fixation in the marine environment. Preprint 194. Int. Oceanogr. Congress, New York.

Allen, M. B., Bendix, S. A., and Murchio, J. C. (1962). *Arch. Mikrobiol.* **42**, 36.

Allen, M. B., French, C. S., and Brown, J. S. (1960a). Native and extractable forms of chlorophyll in various algae groups. *In* "Comparative Biochemistry of Photoreactive Systems" (M. B. Allen, ed.), p. 33. Academic Press, New York.

Allen, M. B., Goodwin, T. W., and Phagpolngarm, S. (1960b). *J. gen. Microbiol.* **23**, 93.

Allen, W. E. (1939). *Science,* **89**, 532.

Anderson, A. F. H., and Calvin, M. (1962). *Nature, Lond.* **194**, 285.

Angot, M. (1961). *Bull. Inst. océanogr. Monaco,* **1200**, 3.

Antia, N. J., McAllister, C. D., Parsons, T. R., Stephens, K., and Strickland, J. D. H. (1963). *Limnol. Oceanogr.* **8**, 166.

Antia, N. J., Stephens, K., Tripp, R. B., Parsons, T. R., and Strickland, J. D. H. (1962). *Fish. Res. Bd Can. M.S. Report (Series Limnol. and Oceanogr.) No. 135.* 37 pp.

Archibald, A. R., Manners, O. J., and Ryley, J. F. (1958). *Chemy. Ind.* **46**, 1516.

Arnon, D. I. (1956). *Annu. Rev. Pl. Physiol.* **7**, 325.

Arnow, P., Oleson, J. J., and Williams, J. H. (1953). *Amer. J. Bot.* **40**, 100.

Aruga, Y., and Monsi, M. (1962). *J. Oceanogr., Kobe,* **18**, 37.

Atkins, W. R. G. (1922). *J. Mar. biol. Ass. U.K.* **12**, 717.

Atkins, W. R. G. (1923). *J. Mar. biol. Ass. U.K.* **13**, 119.

Atkins, W. R. G., and Parke, M. (1951). *J. Mar. biol. Ass. U.K.* **29**, 609.

Baatz, I. (1941). *Planta,* **31**, 726.

Bachrach, E., and Lucciardi, N. (1932). *Rev. algol.* **6**, 251.

Bainbridge, R. (1957). *Biol. Rev.* **32**, 91.

Bainbridge, V. (1958). *J. Mar. biol. Ass. U.K.* **37**, 349.

Barashkov, G. K. (1956). *C.R. Acad. Sci. U.R.S.S.* **111**, 148.

Barashkov, G. K. (1962). *Trud. Murmanskogo Morsk. Biol. Inst. Akad. Nauk S.S.S.R.* **4**, 27. (*C.A.58*, 12860).

Barker, H. A. (1935a). *Arch. Mikrobiol.* **6**, 141.

Barker, H. A. (1935b). *Arch. Mikrobiol.* **6**, 157.

Barker-Jørgensen, C. (1960). *J. Cons. int. Explor. Mer,* **26**, 94.

Barker-Jørgensen, C. (1962). *Cons. Explor. Mer,* **153**, 99.

Bauer, L. (1952). *Naturwissenschaften,* **39**, 88.

Baylor, E. R., and Sutcliffe, W. H. (1963). *Limnol. Oceanogr.* **8**, 369.

Beattie, A., Hirst, E. L., and Percival, E. (1961). *Biochem. J.* **79**, 531.

Beklemishev, K. V. (1954). *Zool. Zh. Akad. Nauk S.S.S.R.* **33**, 1210.

Beklemishev, K. V. (1957). *Trud. Inst. Okeanol. Akad. Nauk S.S.S.R.* **20**, 254.

Beklemishev, K. V. (1962). *Rapp. Cons. Explor. Mer,* **153**, 107.

Beklemishev, K. V., Petrikova, M. N., and Semina, G. I. (1961). *Trud. Inst. Okeanol. Akad. Nauk S.S.S.R.* **51**, 33.

Bentley, J. A. (1959). *J. Mar. biol. Ass. U.K.* **39**, 433.

Berg, G. (1958). *Rapp. Cons. Explor. Mer,* **144**, 85.

Berg, G. (1961). *Rapp. Cons. Explor. Mer,* **149**, 148.

Bernard, F. (1948). *J. Cons. int. Explor. Mer*, **15**, 177.
Bernard, F. (1958a). *C.R. Acad. Sci., Paris*, **247**, 2045.
Bernard, F. (1958b). *Rapp. Cons. Explor. Mer*, **144**, 103.
Bernard, F. (1961). Density of flagellates and myxophyceae in the heterotrophic layers related to environment. *In* "Marine Microbiology" (C. H. Oppenheimer, ed.), p. 215. Charles C. Thomas, Springfield, Illinois, U.S.A.
Bernard, F., and Fage, L. (1936). *Bull. Inst. océanogr. Monaco*, **701**, 1.
Bidwell, R. G. S. (1957). *Canad. J. Res. Bot. Ed.* **35**, 945.
Birge, E. A., and Juday, C. (1922). *Bull. Wis. geol. nat. Hist. Surv.* **64**, (Sci. Ser. No. 13), 222 pp.
Birge, E. A., and Juday, C. (1929). *Bull. Nat. Res. Coun. Wash.* **68**, 61.
Bishop, C. T., Adams, G. A., and Hughes, E. O. (1954). *Canad. J. Res. Chem. Ed.* **32**, 999.
Black, W. A. P. (1953). *Annu. Rep. Progr. Chem.* **50**, 322.
Bond, R. M. (1933). *Bull. Bingham oceanogr. Coll.* **4**, 1.
Bond, R. M. (1934). *Biol. Bull., Woods Hole*, **67**, 461.
Braarud, T. (1937). *J. Cons. int. Explor. Mer*, **12**, 321.
Braarud, T. (1944). *Avh. norske Vidensk Akad.* **10**, 3.
Braarud, T. (1961). Cultivation of marine organisms as a means of understanding environmental influences on populations. *In* "Oceanography", p. 271–298. American Association for the Advancement of Science.
Braarud, T. (1962). *J. Oceanogr., Kobe* (*20th Anniversary Vol.*), 628.
Braarud, T., and Føyn, E. (1930). *Avh. norske Vidensk Akad.* **14**, 1.
Braarud, T., Gaarder, K. R., and Grøntved, J. (1953). *Rapp. Cons. Explor. Mer*, **133**, 7.
Braarud, T., Ringdal, K., Markali, J., and Nordli, E. (1952). *Nyt. Mag. Botan.* **1**, 129.
Brandt, K., and Raben, E. (1920). *Wiss. Meeresuntersuch. abt. Kiel, N.F.* **19**, 177.
Bristol-Roach, B. M. (1928). *Ann. Bot.* **42**, 317.
Brown, A. H. (1953). *Amer. J. Bot.* **40**, 719.
Brown, A. H., and Weis, D. (1959). *Plant. Physiol.* **34**, 224.
Brown, F., Cuthbertson, W. F. J., and Fogg, G. E. (1956). *Nature, Lond.* **177**, 188.
Brown, J. S., and French, C. S. (1961). *Biophys. J.* **1**, 539.
Brown, T. E., Eyster, H. C., and Tanner, H. A. (1958). Physiological effects of manganese deficiency. *In* "Trace Elements" (C. A. Lamb, O. G. Bentley, and J. M. Beattie, eds.), p. 135. Academic Press, New York.
Bruce, J. R., Knight, M., and Parke, M. (1940). *J. Mar. biol. Ass. U.K.* **24**, 337.
Buch, K. (1929). *J. Cons. int. Explor. Mer*, **4**, 162.
Bunt, J. S. (1963). *Nature, Lond.* **200**, 1255.
Burkholder, P. R., and Sieburth, J. M. (1961). *Limnol. Oceanogr.* **6**, 45.
Carter, P. W., Heilbron, I. M., and Lythgoe, B. (1940). *Proc. roy. Soc.* **B. 128**, 82.
Cassie, R. M. (1959). *New Zealand J. Sci.* **2**, 398.
Cayle, T., and Emerson, R. (1957). *Nature, Lond.* **179**, 89.
Cerruti, A. (1941). *Arch. Oceanogr. Limnol., Roma*, **1**, 165.
Chapman, D. J., and Haxo, F. T. (1963). *Plant and Cell Physiol.* **4**, 57.
Chu, S. P. (1946). *J. Mar. biol. Ass. U.K.* **26**, 285.
Claes, H. (1959). *Z. Naturf.* **14b**, 4. (C. A. **53**, 20320(1959)).
Clark, H. T., and Mazur, H. (1941). *J. biol. Chem.* **141**, 283.

Coblentz-Mishke, O. I. (1957). *C.R. Acad. Sci., U.R.S.S.* **116**, 1029.
Coe, W. R. (1945). *J. Exp. Zool.* **99**, 1.
Coe, W. R. (1948). *J. Mar. Res.* **7**, 586.
Collier, A. (1953). *Trans. 18th N. Amer. Wildl. Conf.* **463**.
Conover, R. J. (1960). *Biol. Bull., Woods Hole*, **119**, 399.
Conover, R. J. (1962). *Rapp. Cons. Explor. Mer*, **153**, 190.
Conover, S. A. (1956). *Bull. Bingham oceanogr. Coll.* **15**, 62.
Coombs, C. F. (1952). *Science*, **116**, 453.
Corlett, J. (1958). *Rapp. Cons. Explor. Mer*, **144**, 96.
Corner, E. D. S. (1961). *J. Mar. biol. Ass. U.K.* **41**, 5.
Correll, D. L., and Tolbert, N. E. (1962). *Plant Physiol.* **37**, 627.
Cowey, C. B. (1956). *J. Mar. biol. Ass. U.K.* **35**, 609.
Cowey, C. B., and Corner, E. D. S. (1961). *Rapp. Cons. Explor. Mer*, **153**, 124.
Cowey, C. B., and Corner, E. D. S. (1963). *J. Mar. biol. Ass. U.K.* **43**, 495.
Creitz, G. I., and Richards, F. A. (1955). *J. Mar. Res.* **14**, 211.
Curl, H. (1962). *Limnol. Oceanogr.* **7**, 422.
Curl, H., and Mcleod, G. C. (1961). *J. Mar. Res.* **19**, 70.
Currie, R. I. (1958). *Rapp. Cons. Explor. Mer*, **144**, 96.
Currie, R. I. (1960). Scalar irradiance as a parameter in phytoplankton photosynthesis and a proposed method for its measurement. *In* "Symposium on Radiant Energy in the Sea." (N. G. Jerlov, ed.), p. 107. L'Institut Géographique National, Paris.
Currie, R. I. (1962). *Nature, Lond.* **193**, 956.
Cushing, D. H. (1959). *Fish. Invest., Lond., Ser. II.* **22**: (6) 40 pp.
Cushing, D. H. (1962). *J. Cons. int. Explor. Mer*, **27**, 131.
Dales, R. P. (1960). *J. Mar. biol. Ass. U.K.* **39**, 693.
Dal Pont, G., and Newell, B. (1963). *Aust. J. Mar. Freshw. Res.* **14**, 155.
Darken, M. A. (1953). *Bot. Rev.* **19**, 99.
Das, B. C. (1960). *Canad. J. Res. Biochem. Ed.* **38**, 453.
Davis, H. C. (1953). *Biol. Bull., Woods Hole*, **104**, 334.
Davis, H. C. (1963). *Arch. Hydrobiol.* (*Plankt.*) **59**, 145.
Davis, H. C., and Guillard, R. R. (1958). *Circ. U.S. Fish Wildl. Serv., Fish. Bull. No. 136*, 293.
Dehn, M., von (1930). *Z. vergl. Physiol.* **13**, 334.
Doty, M. S. (1959). *J. Mar. biol. Ass. India*, **1**, 66.
Doty, M. S., and Capurro, L. R. A. (1961). Productivity measurements in the world oceans. I.G.Y. Oceanogr. Rept. No. **4**. Parts I and II. I.G.Y. World Data Centre A. Oceanography. Agr. Mech. College. Texas, U.S.A.
Doty, M. S., and Oguri, M. (1956). *J. Cons. int. Explor. Mer*, **22**, 33.
Doty, M. S., and Oguri, M. (1957). *Limnol. Oceanogr.* **2**, 37.
Doty, M. S., and Oguri, M. (1958). *Rapp. Cons. Explor. Mer*, **144**, 47.
Dougherty, E. C., and Allen, M. B. (1960). Is pigmentation a clue to protistan phylogeny? *In* "Comparative Biochemistry of Photoreactive systems" (M. B. Allen, ed.), p. 129. Academic Press, New York.
Droop, M. R. (1955). *J. Mar. biol. Ass. U.K.* **34**, 229.
Droop, M. R. (1957). *Nature, Lond.* **180**, 1041.
Droop, M. R. (1958). *J. Mar. biol. Ass. U.K.* **37**, 323.
Droop, M. R. (1961a). *J. Mar. biol. Ass. U.K.* **41**, 69.
Droop, M. R. (1961b). *Rev. algol.* **4**, 247.

Droop, M. R., McLaughlin, J. J. A., Pintner, I. J., and Provasoli, L. (1959). Specificity of some protophytes towards Vitamin B_{12}-like compounds. Preprint 916. Int. Oceanogr. Congress, New York.
Dugdale, R. C., Menzel, D. W., and Ryther, J. H. (1961). *Deep Sea Res.* **7**, 297.
Dugdale, R. C., and Neess, J. C. (1961). *R. A. Taft Sanitary Engng Centre Tech. Rept. W. 61–3*, 103.
Duursma, E. K. (1963). *Neth. J. Sea Res.* **2**, 85.
Eddy, B. P., Fleming, I. D., and Manners, D. J. (1958). *J. Chem. Soc.* 2827.
El Wakeel, S. K., and Riley, J. P. (1957). *J. Cons. int. Explor. Mer*, **22**, 180.
Eny, D. (1949). *J. Bacteriol.* **58**, 269.
Eppley, R. W., Gee, R., and Saltman, P. (1963). *Physiol. Plant.* **16**, 777.
Epstein, S. S., Weiss, J. B., Causeley, D., and Bush, P. (1962). *J. Protozool.* **9**, 336.
Eyster, H. C., Brown, T. E., and Tanner, H. A. (1958). Mineral requirements for *Chlorella pyrenoidosa* under autotrophic and heterotrophic conditions. *In* "Trace Elements" (C. A. Lamb, O. G. Bentley, and J. M. Beattie, eds.), p. 157. Academic Press, New York.
Fay, P., and Fogg, G. E. (1962). *Arch. Mikrobiol.* **42**, 310.
Felföldy, L. J. M. (1960a). *Physiol. Plant.* **13**, 487.
Felföldy, L. J. M. (1960b). *Acta biol. hung.* **21**, 176.
Felföldy, L. J. M. (1961). *Mag. Tud. akad. Tihanyi Biol.* **28**, 95.
Fell, J. W., and Uden, N. van (1961). Yeasts in marine environments. *In* "Marine Microbiology" (C. H. Oppenheimer, ed.), p. 329. Charles C. Thomas, Springfield, Illinois.
Findenegg, I. (1943). *Int. Rev. Hydrobiol.* **43**, 368.
Fisher, L. R., and Goldie, E. H. (1959). *J. Mar. biol. Ass. U.K.* **38**, 291.
Fleischer, W. E. (1935). *J. gen. Physiol.* **18**, 573.
Fleming, R. H. (1939a). *Proc. 6th Pacif. Sci. Congr.* **3**, 535.
Fleming, R. H. (1939b). *J. Cons. int. Explor. Mer*, **14**, 210.
Fogg, G. E. (1953). "The Metabolism of Algae," 149 pp. John Wiley and Sons Inc., New York.
Fogg, G. E. (1957). *J. Gen. Microbiol.* **16**, 294.
Fogg, G. E. (1959). Nitrogen nutrition and metabolic patterns in algae. *In* "Utilization of Nitrogen and its Compounds by Plants" (H. K. Porter, ed.), p. 106. (Symposium of the Society for Experimental Biology.) Cambridge University Press, London.
Fogg, G. E., and Belcher, J. H. (1961). *Verh. int. Ver. Limnol.* **24**, 893.
Fogg, G. E., and Than Tun, (1960). *Proc. roy. Soc. Ser. B.* **153**, 111.
Ford, J. E. (1958). *J. Gen. Microbiol.* **19**, 161.
Fox, D. L. (1950). *Ecology*, **31**, 100.
Fox, D. L., Oppenheimer, C. H., and Kittredge, J. S. (1953). *J. Mar. Res.* **12**, 233.
Franck, J. (1953). *Arch. Biochem. Biophys.* **45**, 190.
Fredrick, J. F. (1954). *Physiol. Plant.* **7**, 182.
Fritsch, F. E. (1935) (1945). "The Structure and Reproduction of the Algae." Vol. I, 1935; Vol. II, 1945. Cambridge University Press, London.
Fuller, J. L., and Clarke, G. L. (1936). *Biol. Bull., Woods Hole*, **70**, 308.
Fujita, Y., and Hattori, A. (1962). *J. Biochem., Tokyo*, **51**, 89.
Gaarder, T., and Gran, H. H. (1927). *Rapp. Cons. Explor. Mer*, **42**, 3.
Gaudin, A. M., Mulav, A. L., and O'Connor, R. F. (1960). *Appl. Microbiol.* **8**, 84.
Gauld, D. T. (1951). *J. Mar. biol. Ass. U.K.* **29**, 695.

Gellis, S. S., and Clarke, G. L. (1935). *Physiol. Zoöl.* **8**, 127.
Gest, H., and Kamen, M. D. (1948). *J. biol. Chem.* **176**, 299.
Gibor, A. (1956). *Biol. Bull., Woods Hole*, **111**, 230.
Gibor, A. (1957). *Nature, Lond.* **179**, 1304.
Gillbricht, M. (1952). *Kieler Meeresforsch.* **8**, 173.
Gillbricht, M. (1961). *Rapp. Cons. Explor. Mer*, **149**, 189.
Goldberg, E. D. (1952). *Biol. Bull., Woods Hole*, **102**, 243.
Goldberg, E. D., Walker, T. J., and Whisenand, A. (1951). *Biol. Bull., Woods Hole*, **101**, 274.
Goldman, C. R. (1960). *Science*, **132**, 1016.
Goldman, C. R. (1961). *Verh. int. Ver. Limnol.* **24**, 120.
Goodwin, T. W. (1952). "The Comparative Biochemistry of the Carotenoids," 356 pp. Chapman and Hall Ltd., London.
Goodwin, T. W. (1955). Carotenoids. *In* "Modern Methods of Plant Analysis" (K. Paech and M. V. Tracey, eds.), p. 272, Vol. III. Springer-Verlag, Berlin.
Goodwin, T. W. (1957). *J. Gen. Microbiol.* **17**, 467.
Goodwin, T. W. (1960). Algal carotenoids. *In* "Comparative Biochemistry of Photoreactive Systems" (M. B. Allen, ed.), p. 1. Academic Press, New York.
Gran, H. H. (1931). *Rapp. Cons. Explor. Mer*, **75**, 37.
Gran, H. H. (1932). *J. Cons. int. Explor. Mer*, **7**, 343.
Gran, H. H., and Braarud, T. (1935). *J. biol. Bd Can.* **1**, 279.
Granick, S. (1949). *J. biol. Chem.* **179**, 505.
Gross, F. (1937). *Phil. Trans. B.* **228**, 1.
Gross, F., and Zeuthen, E. (1948). *Proc. roy. Soc. Ser. B*, **135**, 382.
Grøntved, J., and Nielsen, E. Steemann (1957). *Medd. Komm. Havundersög, Kbh., Ser. Plankton*, **5** (**6**), 1.
Guillard, R. R. L. (1961). *Sci. Rept. Bermuda Biol. St.* **1**, *Appendix I*, 11 pp.
Guillard, R. R. L., and Cassie, V. (1963). *Limnol. Oceanog.* **8**, 161.
Guillard, R. R. L., and Ryther, J. H. (1962). *Can. J. Microbiol.* **8**, 229.
Guillard, R. R. L., and Wangersky, P. J. (1958). *Limnol. Oceanog.* **3**, **449**.
Hagmeier, E. (1960). Untersuchungen über die Menge und die Zusammensetzung von Seston und Plankton in Wasserproben von Reisen in die Nordsee und nach Island. Dissertation. University of Kiel. 168 pp.
Halldal, P. (1958a). *Physiol. Plant.* **11**, 401.
Halldal, P. (1958b). *Physiol. Plant.* **11**, 118.
Halldal, P. (1961a). *Physiol. Plant.* **14**, 133.
Halldal, P. (1961b). *Physiol. Plant.* **14**, 558.
Halldal, P., and Markali, J. (1954). *Nyt. Mag. Botan.* **2**, 117.
Hamm, R. E., and Thompson, T. G. (1941). *J. Mar. Res.* **4**, 11.
Hansen, V. Kr. (1961). *Rapp. Cons. Explor. Mer*, **149**, 160.
Hansen, V. Kr. (1962). *Conseil permanent intern. exploration mer. Plankton Committee Draft Report No. 129.*
Hardy, A. C. (1936). *Proc. Linn. Soc. Lond.* **148** (**2**), 64.
Harris, E. (1959). *Bull. Bingham oceanogr. Coll.* **17**, (**1**) 31.
Harris, E., and Riley, G. A. (1956). *Bull. Bingham. oceanogr. Coll* **15**, **315**.
Harvey, H. W. (1934a). *J. Mar. biol. Ass. U.K.* **19**, 761.
Harvey, H. W. (1934b). *J. Mar. biol. Ass. U.K.* **19**, 775.
Harvey, H. W. (1937). *J. Mar. biol. Ass. U.K.* **22**, 205.
Harvey, H. W. (1939). *J. Mar. biol. Ass. U.K.* **23**, 499.
Harvey, H. W. (1940). *J. Mar. biol. Ass. U.K.* **24**, 115.

Harvey, H. W. (1947). *J. Mar. biol. Ass. U.K.* **26**, 562.
Harvey, H. W. (1950). *J. Mar. biol. Ass. U.K.* **29**, 97.
Harvey, H. W. (1953a). *J. Mar. biol. Ass. U.K.* **31**, 477.
Harvey, H. W. (1953b). *J. Mar. biol. Ass. U.K.* **31**, 475.
Harvey, H. W. (1957). "The Chemistry and Fertility of Sea Waters," 234 pp. (2nd Edition). Cambridge University Press, London.
Harvey, H. W., Cooper, L. H. N., Lebour, M. V., and Russell, F. S. (1935). *J. Mar. biol. Ass. U.K.* **20**, 407.
Hashimoto, J. (1954). *J. Vitaminol.* (*Japan*), **1**, 44. (*C. A.* **49**, 7651 (1955).)
Hasle, G. R. (1950). *Oikos*, **2**, 162.
Hasle, G. R. (1954a). *Nyt. Mag. Botan.* **2**, 139.
Hasle, G. R. (1954b). *Nyt. Mag. Botan.* **2**, 121.
Hasle, G. R. (1959). *Deep Sea Res.* **6**, 38.
Hasle, G. R., and Smayda, T. J. (1960). *Nyt. Mag. Botan.* **8**, 53.
Hasler, A. D. (1935). *Biol. Bull., Woods Hole*, **68**, 207.
Hasler, A. D. (1937). *Biol. Bull., Woods Hole*, **72**, 290.
Hastings, J. W., Astrachan, L., and Sweeney, B. M. (1961). *J. gen. Physiol.* **45**, 69.
Hauschild, A. H. W., Nelson, C. D., and Krotkov, G. (1962). *Can. J. Bot.* **40**, 179.
Haxo, F. T. (1960). The wavelength dependence of photosynthesis and the role of accessory pigments. *In* "Comparative Biochemistry of Photoreactive Systems" (M. B. Allen, ed.), p. 339. Academic Press, New York.
Haxo, F. T., and Fork, D. C. (1959). *Nature, Lond.* **184**, 1051.
Hewitt, E. J. (1957). *Nature, Lond.* **180**, 1020.
Hewitt, E. J. (1959). *Biol. Rev.* **34**, 333.
Holden, M. (1962). *Biochim. Biophys. Acta*, **56**, 378.
Holmes, R. W. (1956). *Bull. Bingham oceanogr. Coll.* **16**, (Art. 4) 74 pp.
Holmes, R. W. (1957). Solar radiation, submarine daylight and photosynthesis. *In* "Treatise on Marine Ecology and Paleoecology" (J. W. Hedgpeth, ed.), Geological Society of America, Vol. I, pp. 109–128. Memoir 67, Waverly Press, Maryland, U.S.A.
Holmes, R. W., and Haxo, F. T. (1958). Diurnal variation in the photosynthesis of natural phytoplankton populations in artificial light. *In* "Physical, Chemical and Biological Investigations in the Eastern Tropical Pacific" (R. W. Holmes, ed.), pp. 73–76. *Special Scientific Report No. 279. Scripps Institution of Oceanography to U.S. Fish and Wildlife Service.*
Holmes, R. W., and Snodgrass, J. M. (1961). *J. Mar. Res.* **19**, 40.
Hood, D. W., and Park, K. (1962). *Physiol. Plant.* **15**, 273.
Hough, L., Jones, J. K. N., and Wadman, W. H. (1952). *J. Chem. Soc.* **3393**.
Huang, H., and Giese, A. C. (1958). *Science*, **127**, 475.
Hulburt, E. M. (1962). *Deep Sea Res.* **9**, 72.
Hulburt, E. M., and Rodman, J. (1963). *Limnol. Oceanogr.* **8**, 263.
Hulburt, E. M., Ryther, J. H., and Guillard, R. R. L. (1960). *J. cons. int. Explor. Mer*, **25**, 115.
Hundley, J. M., Ing, R. B., and Krauss, R. W. (1956). *Science*, **124**, 536.
Hutchinson, G. E. (1957). "A Treatise on Limnology," Vol. I. 1015 pp. John Wiley, New York.
Hutner, S. H. (1961). *Bull. Torrey bot. Cl.* **88**, 339.
Hutner, S. H., Baker, H., Aaronson, S., Nathan, H. A., Rodriguez, E., Lockwood, S., Sanders, M., and Petersen, R. A. (1957). *J. Protozool.* **4**, 259.
Ichimura, S., and Aruga, Y. (1958). *Bot. Mag. Tokyo*, **71**, 261.

Ichimura, S., Saijo, Y., and Argua, Y. (1962). *Bot. Mag., Tokyo*, **75**, 212.
Imamura, H. (1952). *Bull. Jap. Soc. sci. Fish.* **17**, 289.
Inoue, N., Nishizawa, S., and Fukudu, M. (1955). *Proceedings UNESCO Symposium on the Physics of Oceanography, Tokyo.* 53–58.
Iwamoto, H., and Sugimoto, H. (1958). *Bull. agric. chem. Soc. Japan*, **22**, 410.
Iwamoto, H., and Suzuki, A. (1958). *Bull. agric. chem. Soc. Japan*, **22**, 420.
Iwamura, T. (1955). *J. Biochem., Tokyo*, **42**, 574.
Iwamura, T. (1960). *Biochim. biophys. Acta*, **42**, 161.
Jannasch, H. W. (1954). *Naturwissenschaften*, **41**, 42.
Jannasch, H. W., and Jones, G. E. (1959). *Limnol. Oceanogr.* **4**, 128.
Jeener, R. (1952). *Biochim. biophys. Acta*, **8**, 125.
Jeffrey, S. W. (1961). *Biochem. J.* **80**, 336.
Jeffrey, S. W. (1962). *Nature, Lond.* **194**, 600.
Jeffrey, S. W. (1963). *Biochem. J.* **86**, 313.
Jenkins, P. G. (1937). *J. Mar. Biol. Ass. U.K.* **22**, 301.
Jerlov, N. G. (1947). Optical studies of ocean waters. *In* "Reports of the Swedish Deep-Sea Expedition" (H. Petterson, ed.), Vol. III, 59 pp. Elanders Boktryckeri Aktiebolag, Göteborg.
Jitts, H. R. (1958). Reports Nos. 7 and 8, Marine Biological Laboratory, Cronulla, Australia.
Jitts, H. R., and Scott, B. D. (1961). *Limnol. Oceanogr.* **6**, 116.
Jitts, H. R., McAllister, C. D., Stephens, K., and Strickland, J. D. H. (1964). *J. Fish. Res. Bd Can.* **21**, 139.
Johnston, R. (1963a). *Limnol. Oceanogr.* **8**, 270.
Johnston, R. (1963b). *J. Mar. Biol. Ass. U.K.* **43**, 427.
Jørgensen, E. G. (1953). *Physiol. Plant.* **6**, 301.
Jørgensen, E. G. (1955). *Physiol. Plant.* **8**, 840.
Jørgensen, E. G. (1956). *Physiol. Plant.* **9**, 712.
Kain, J. M., and Fogg, G. E. (1958a). *J. Mar. biol. Ass. U.K.* **37**, 397.
Kain, J. M., and Fogg, G. E. (1958b). *J. Mar. biol. Ass. U.K.* **37**, 781.
Kain, J. M., and Fogg, G. E. (1960). *J. Mar. biol. Ass. U.K.* **39**, 33.
Kalle, K. (1951). *Dtsch. hydrogr. Z.* **4**, 92.
Kandler, O., and Sironval, C. (1959). *Biochim. biophys. Acta*, **33**, 207.
Kanwisher, J. (1959). *Limnol. Oceanogr.* **4**, 210.
Kanwisher, J. (1960). *Tellus*, **12**, 209.
Kay, H. (1954). *Kieler Meeresforsch.* **10**, 26.
Kayama, M., Tsuchiya, Y., and Mead, J. F. (1963). *Bull. Jap. Soc. sci. Fish.* **29**, 452.
Keeler, R. F. (1959). *Science*, **129**, 1617.
Kelly, P. B., Reiser, R., and Hood, D. W. (1958). *J. Amer. Oil. Chem. Soc.* **35**, 189.
Kessler, E. (1957). *Physiol. Plant.* **49**, 435.
Ketchum, B. H. (1939). *Amer. J. Bot.* **26**, 399.
Ketchum, B. H., and Redfield, A. C. (1949). *J. cell. comp. Physiol.* **33**, 281.
Ketchum, B. H., Ryther, J. H., Yentsch, C. S., and Corwin, N. (1958). *Rapp. Cons. Explor. Mer*, **144**, 132.
Keys, A., Christensen, E., and Krogh, A. (1935). *J. Mar. biol. Ass. U.K.* **20**, 181.
Kierstead, H., and Slobodkin, L. B. (1953). *J. Mar. Res.* **12**, 141.
Kimball, Jr., J. F., Corcoran, E. F., and Wood, E. J. F. (1963). *Bull. Mar. Sci. Gulf Caribb.* **13**, 574.

King, J. E., Austin, T. S., and Doty, M. S. (1957). *Circ. U.S. Fish. Wildl. Serv., Fish. Bull. No. 210.* 155 pp.

Kinne-Diettrich, E. M. (1955). *Kieler Meeresforsch.* **11**, 34.

Kratz, W. A., and Myers, J. (1955). *Amer. J. Bot.* **42**, 282.

Krauss, R. W. (1956). *Industr. Engng. Chem.* **48**, 1449.

Kreps, E., and Verbinskaya, N. (1930). *J. Cons. int. Explor. Mer,* **5**, 329.

Krey, J. (1939). *J. Cons. int. Explor. Mer,* **14**, 201.

Krey, J. (1950). *Kieler Meersforsch.* **7**, 58.

Krey, J. (1951). *Kieler Meeresforsch.* **8**, 16.

Krey, J. (1958). *J. Mar. Res.* **17**, 312.

Krey, J. (1961a). *Kieler Meeresforsch.* **17**, 163.

Krey, J. (1961b). *J. Cons. int. Explor. Mer,* **26**, 263.

Krey, J. (1961c). *Rapp. Cons. Explor. Mer,* **149**, 194.

Krey, J., Banse, K., and Hagmeier, E. (1957). *Kieler Meeresforsch.* **13**, 35.

Krogh, A. (1931). *Biol. Rev.* **6**, 412.

Krogh, A., Lange, E., and Smith, W. (1930). *Biochem. J.* **24**, 1666.

Kuenzler, E. J., and Ketchum, B. H. (1962). *Biol. Bull., Woods Hole,* **123**, 134.

Kuhl, A. (1960). Die Biologie der kondensierten anorganischen Phosphate. *In* "Ergebnisse der Biologie" (H. Autrum, ed.), **23**, 144. Springer, Berlin.

Kuznezow, S. L. (1959). Die Rolle der Mikro-organismen im Stoffkreislauf der Seen. 301 pp. Deutscher Verlag der Wissenschaften, Berlin.

Kylin, H. (1943a). *K. fysiogr. Sällsk. Lund. Förh.* **13**, (**13**) 1.

Kylin, H. (1943b). *K. fysiogr. Sällsk. Lund Förh.* **13** (**17**), 1.

Lanskaya, L. A. (1961). The division rates of plankton algae of the Black Sea in Cultures. *In* "Marine Microbiology" (C. H. Offenheimer, ed.), p. 127. Charles C. Thomas, Springfield, Illinois.

Lanskaya, L. A., and Pshenina, T. I. (1961). *Trav. Sta. biol. Sevastopol,* **14**, 292.

Lanskaya, L. A., and Sivkov, S. J. (1950). *C.R. Acad. Sci., U.R.S.S.* **73**, 581.

Lasker, R. (1960). *Science,* **131**, 1098.

Lassøe, A., and Hansen, V. K. (1961). *J. Cons. int. Explor. Mer. Plankton Committee Draft Report No. 143.*

Latimer, P. (1959). *Plant Physiol.* **34**, 193.

Lazaroff, N., and Schiff, J. (1962). *Science,* **137**, 603.

Lefèvre, M., Nisbet, M., and Jacob, H. (1949). *Verh. int. Ver. Limnol.* **10**, 259.

Levring, T. (1945). *Göteborgs VetenskSamh. Handl. Ser. B,* **3**, 12.

Lewin, J. C. (1953). *J. Gen. Microbiol.* **9**, 305.

Lewin, J. C. (1954). *J. Gen. Physiol.* **37**, 589.

Lewin, J. C. (1955a). *J. Gen. Physiol.* **39**, 1.

Lewin, J. C. (1955b). *Plant Physiol.* **30**, 129.

Lewin, J. C. (1955c). *J. Gen. Microbiol.* **13**, 162.

Lewin, J. C. (1957). *Can. J. Microbiol.* **3**, 427.

Lewin, J. C., Lewin, R. A., and Philpott, D. E. (1958). *J. Gen. Mocribiol.* **18**, 427.

Lewin, J. C., and Guillard, R. R. L. (1963). *Annu. Rev. Microbiol.* **17**, 373.

Lewin, J. C., and Lewin, R. A. (1960). *Can. J. Microbiol.* **6**, 127.

Lewin, R. A. (1956). *Can. J. Microbiol.* **2**, 665.

Lewin, R. A. (1958). *J. Gen. Microbiol.* **19**, 87.

Lewin, R. A. (1962). "Physiology and Biochemistry of Algae" (R. A. Lewin, ed.) 938 pp. Academic Press, New York.

Linko, P., Holm-Hansen, O., Bassham, J. A., and Calvin, M. (1957). *J. exp. Bot.* **8**, 147.

Lisitsin, A. P. (1956). *Trud. Inst. Okeanol. Akad. Nauk S.S.S.R.* **19**, 204.
Lisitsin, A. P. (1959). Preprint Int. Oceanogr. Congress, New York. **470.**
Lisitsin, A. P. (1961). *Oceanological Researches* (Results of Researches on the Programme of the International Geophysical year). *Articles* **5**, 130.
Lohmann, H. (1908). *Wiss. Meeresuntersuch, Abt. Kiel N.F.* **10**, 131.
Lomis, W. F. (1958). *Analyt. Chem.* **30**, 1865.
Loosanoff, V. L. (1949). *Science*, **110**, 122.
Loosanoff, V. L., Miller, W. S., and Smith, P. B. (1951). *J. Mar. Res.* **10**, 59.
Lorenzen, C. J. (1963). *Limnol. Oceanogr.* **8**, 56.
Lovern, J. A. (1936). *Biochem. J.* **30**, 387.
Low, E. M. (1955). *J. Mar. Res.* **14**, 199.
Low, E. M. (1958). *Nature, Lond.* **182**, 1096.
Luck, J. M., Sheets, G., and Thomas, J. O. (1931). *Quart. Rev. Biol.* **6**, 46.
Lund, J. W. G., and Talling, J. F. (1957). *Bot. Rev.* **23**, 489.
MacGinitie, G. E. (1932). *Science*, **76**, 490.
MacGinitie, G. E. (1945). *Biol. Bull., Woods Hole*, **88**, 107.
Mackereth, F. J. (1953). *J. exp. Bot.* **4**, 296.
Maddux, W. S., and Jones, R. F. (1964). *Limnol. Oceanogr.* **9**, 79.
Maloney, T. E., Donovan, E. J., and Robinson, E. L. (1962). *Phycologia.* **2**, 1.
Margalef, R. (1957). *Mem. Real Acad. Cien. Artes. Barcelona*, **23**, 373.
Margalef, R. (1961). *Invest. Pesq.* **18**, 1.
Margalef, R. (1962). *Rapp. Comm. int. Mer Médit.* **16**, 223.
Markham, R. (1955). Nucleic acids, their components and related compounds. *In* "Modern Methods of Plant Analysis" (K. Paech and M. V. Tracey, eds.), Vol. IV, p. 246. Springer-Verlag, Berlin.
Marshall, S. M., and Orr, A. P. (1952). *J. Mar. biol. Ass. U.K.* **30**, 527.
Marshall, S. M., and Orr, A. P. (1955a). *J. Mar. biol. Ass. U.K.* **34**, 495.
Marshall, S. M., and Orr, A. P. (1955b). *Deep Sea Res.* (*Suppl. to Vol. 3*), 110.
Marshall, S. M., and Orr, A. P. (1958). *UNESCO Conference on radio-isotopes in scientific research. UNESCO/NS/RIC/137.* 6 pp.
Marshall, S. M., and Orr, A. P. (1962). *Rapp. Cons. Explor. Mer*, **153**, 92.
Matsue, Y. (1954). On the culture of the marine planktonic diatom *Skeletonema costatum* (Grev.) Cleve. *Review of Fish Science in Japan.* (Japanese Society for the Advancement of Science).
McAllister, C. D. (1961a). *Limnol. Oceanogr.* **6**, 447.
McAllister, C. D. (1961b). *Limnol. Oceanogr.* **6**, 483.
McAllister, C. D. (1963). *Limnol. Oceanogr.* **8**, 289.
McAllister, C. D., and Strickland, J. D. H. (1961). *Limnol. Oceanogr.* **6**, 226.
McAllister, C. D., Parsons, T. R., and Strickland, J. D. H. (1960). *J. Cons. int. Explor. Mer*, **25**, 240.
McAllister, C. D., Parsons, T. R., Stephens, K., and Strickland, J. D. H. (1961). *Limnol. and Oceanog.* **6**, 237.
McAllister, C. D., Shah, N., and Strickland, J. D. H. (1964). *J. Fish. Res. Bd Can.* **21**, 159.
McCombie, A. M. (1960). *J. Fish. Res. Bd Can.* **17**, 871.
McLeod, G. C. (1961). *Plant Physiol.* **36**, 114.
McLeod, G. C., and Kanwisher, J. (1962). *Physiol. Plant.* **15**, 581.
Menzel, D. W., and Ryther, J. H. (1961). *Deep Sea Res.* **7**, 279.
Menzel, D. W., and Ryther, J. H. (1964). *Limnol. Oceanogr.* **9**, 179.

Menzel, D. W., and Spaeth, J. P. (1961). *Sci. Rept. Bermuda Biol. Sta.* **2**, Appendix *I. No. 11*.

Menzel, D. W., Hulburt, E. M., and Ryther, J. H. (1961). *Sci. Rep. Bermuda Biol. Sta.* **2**, *Appendix I. No. 9*. See also *Deep Sea Res.* **10**, 209 (1963).

Miller, R. M., Meyer, C. M., and Tanner, H. A. (1963). *Plant Physiol.* **38**, 184.

Mjaaland, G. (1956). *Oikos*, **7**, 251.

Monakov, A. V., and Sorokin, Yu. I. (1961). *Trud. Inst. Biologii Vodoahranilish*, **4**, 251.

Morris, W. R. (1955). *J. Cons. int. Explor. Mer*, **20**, 255.

Morton, J. E. (1958). "Molluscs", 232 pp. Hutchinson, London.

Mullin, M. M. (1963). *Limnol. Oceanogr.* **8**, 239.

Munk, W. H., and Riley, G. A. (1952). *J. Mar. Res.* **11**, 215.

Murchio, J. C., and Allen, M. B. (1962). *Photochem. Photobiol.* **1**, 259.

Myers, J. (1946). *J. Gen. Physiol.* **29**, 429.

Myers, J. (1951). *Annu. Rev. Microbiol.* **5**, 157.

Myers, J., and Burr, G. O. (1940). *J. Gen. Physiol.* **24**, 45.

Myers, J., and Cramer, M. (1948). *J. Gen. Physiol.* **32**, 103.

Myers, J., Cramer, M., and Johnston, J. (1947). *J. Gen. Physiol.* **30**, 217.

Nelson, D. J. (1960). *Science*, **132**, 351.

Nelson, T. C. (1947). *Ecol. Monogr.* **17**, 337.

Neuberg, C., Grauer, A., Kreidl, M., and Lowry, H. (1957). *Arch. Biochem. Biophys.* **70**, 70.

Nielsen, E. S. (1937). *J. Cons. int. Explor. Mer*, **12**, 147.

Nielsen, E. S. (1952a). *J. Cons. int. Explor. Mer*, **18**, 117.

Nielsen, E. S. (1952b). *Physiol. Plant*, **5**, 334.

Nielsen, E. S. (1952c). *Nature, Lond.* **169**, 956.

Nielsen, E. S. (1953). *Physiol. Plant*, **6**, 316.

Nielsen, E. S. (1955). *Physiol. Plant*, **8**, 945.

Nielsen, E. S. (1958a). *Rapp. Cons. Explor. Mer*, **144**, 38.

Nielsen, E. S. (1958b). *J. Cons. int. Explor. Mer*, **23**, 178.

Nielsen, E. S. (1961a). *Physiol. Plant*, **14**, 868.

Nielsen, E. S. (1961b). *Rapp. Cons. Explor. Mer*, **153**, 178.

Nielsen, E. S. (1962). *Physiol. Plant*, **15**, 161.

Nielsen, E. S. (1963). *Physiol. Plant*, **16**, 466.

Nielsen, E. S. (1964). Productivity, definition and measurement. *In* "The Sea" (M. N. Hill, ed.), Vol. 2, p. 129. Interscience, New York.

Nielsen, E. S., and Al Kholy, A. A. (1956). *Physiol. Plant*, **9**, 144.

Nielsen, E. S., and Hansen, V. Kr. (1959a). *Deep Sea Res.* **5**, 222.

Nielsen, E. S., and Hansen, V. Kr. (1959b). *Physiol. Plant*, **12**, 353.

Nielsen, E. S., and Hansen, V. Kr. (1961). *Rapp. Cons. Explor. Mer*, **149**, 158.

Nielsen, E. S., Hansen, V. Kr., and Jørgensen, E. G. (1962). *Physiol. Plant*, **15**, 505.

Nielsen, E. S., and Jensen, A. (1957). Primary oceanic production. The autotrophic production of organic matter in the ocean. *Galathea Rept.* **1**, 49.

Nielsen, E. S., and Jørgensen, E. G. (1962). *Arch. Hydrobiol.* **58**, 349.

Nihei, T. A. (1957). *J. Biochem., Tokyo*, **44**, 389.

Nishizawa, S., Fukudu, M., and Inoue, N. (1954). *Bull. Fac. Fish. Hokkaido Univ.* **5**, 36.

Nordli, E. (1957a). *Oikos*, **8**, 201.

Nordli, E. (1957b). *Nyt. Mag. Botan.* **5**,13.

Northcote, D. H., Goulding, K. J., and Horne, R. L. (1958). *Biochem. J.* **70**, 391.
Nultsch, W. (1956). *Arch. Protistenk.* **101**, 1. (*C. A.* **50**, 10861).
Ogino, C. (1963). *Bull. Jap. Soc. Sci. Fish.* **29**, 459.
OhEocha, C. (1960). Chemical studies of phycoerythrins and phycocyanins. *In* "Comparative Biochemistry of Photoreactive Systems" (M. B. Allen, ed.), p. 181. Academic Press, New York.
Ohwada, M. (1960). *Mem. Kobe. mar. Observ.* **14**, 69.
Olaitan, S. A., and Northcote, D. A. (1962). *Biochem. J.* **82**, 509.
Paasche, E. (1960a). *J. Cons. int. Explor. Mer*, **26**, 33.
Paasche, E. (1960b). *Fiskeridir. Skr. Havundersøk.* **12**, 11.
Paasche, E. (1962). *Nature, Lond.* **193**, 1094.
Paasche, E., and Rom, A. M. (1962). *Nyt. Mag. Botan.* **9**, 33.
Park, K., Hood, D. W., and Odum, H. T. (1958). *Publ. Univ. Texas Inst. Mar. Sci.* **5**, 47.
Parke, M., and Adams, I. (1961). *Bull. Res. Counc. Israel*, **100**, 94.
Parker, B. C., Bold, H. C., and Deason, T. R. (1961). *Science*, **133**, 761.
Parsons, T. R. (1960). *J. Fish. Res. Bd Can. MS Rept.* (*Oceanog. and Limnol. Ser.*) *No. 81.*
Parsons, T. R. (1961). *J. Fish. Res. Bd Can.* **18**, 1017.
Parsons, T. R. (1963). *J. Mar. Res.* **21**, 164.
Parsons, T. R., and Strickland, J. D. H. (1962a). *Science*, **136**, 313.
Parsons, T. R., and Strickland, J. D. H. (1962b). *Deep Sea Res.* **8**, 211.
Parsons, T. R., and Strickland, J. D. H. (1963). *J. Mar. Res.* **21**, 155.
Parsons, T. R., Stephens, K., and Strickland, J. D. H. (1961). *J. Fish. Res. Bd Can.* **18**, 1001.
Patten, B. C. (1959). *Ecology*, **40**, 221.
Patten, B. C. (1961). *Limnol. Oceanogr.* **6**, 369.
Patten, B. C. (1962). *J. Mar. Res.* **20**, 57.
Patterson, J., and Parsons, T. R. (1963). *Limnol. Oceanogr.* **8**, 355.
Payen, J. (1938). *Rev. algol.* **11**, 1.
Pearsall, W. H., and Loose, L. (1937). *Proc. roy. Soc. Ser. B*, **121**, 451.
Perkins, H. J., and Roberts, D. W. A. (1962). *Biochim. Biophys. Acta*, **58**, 486.
Petracek, F. J., and Zechmeister, L. (1956). *Analyt. Chem.* **28**, 1484.
Pintner, I. J., and Provasoli, L. (1961). Nutritional characteristics of some chrysomonads. *In* "Marine Microbiology" (C. H. Oppenheimer, ed.), p. 114. Charles C. Thomas, Springfield, Illinois.
Pomeroy, L. R., Haskin, H. H., and Ragotzkie, R. A. (1956). *Limnol. Oceanogr.* **1**, 54.
Prescott, G. W. (1960). Biological disturbances resulting from algal populations in standing waters. *In* "The Ecology of Algae", pp. 22–37. *University of Pittsburgh Special Publication No. 2.*
Pringsheim, E. G., and Wiessner, W. (1960). *Nature, Lond.* **188**, 919.
Proctor, V. W. (1957a). *Amer. J. Bot.* **44**, 141.
Proctor, V. W. (1957b). *Limnol. Oceanogr.* **2**, 125.
Provasoli, L. (1956). *Boll. Zool. agr. Bachic.* **22**, 143.
Provasoli, L. (1958). *Annu. Rev. Microbiol.* **12**, 279.
Provasoli, L., and McLaughlin, J. J. A. (1961). Limited heterotrophy of some photosynthetic dinoflagellates. *In* "Marine Microbiology" (C. H. Oppenheimer, ed.), p. 105. Charles C. Thomas, Springfield, Illinois.
Provasoli, L., Shiriaishi, K., and Lance, J. R. (1959). *Ann. N.Y. Acad. Sci.* **77**, 250.

Provasoli, L. (1964). Organic regulation of phytoplankton fertility. *In* "The Sea" (M. N. Hill, ed.), Vol. 2, p. 165. Interscience, New York.
Purcell, A. E. (1958). *Analyt. Chem.* **30**, 1049.
Pütter, A. (1909). Die Ernährung der Wassertiere und der Stoffhaushalt der Gewässer. *Thesis, Jena.*
Pütter, A. (1925). *Arch. Hydrobiol.* **15**, 70.
Rabinowitch, E. I. (1945), (1951), (1956). "Photosynthesis and Related Processes." Vol. I, 1945; Vol. II, Part I, 1951, Vol. II, Part II, 1956. Interscience, New York.
Rabinowitch, E. I., and Thomas, G. J. B. (1960). *Science,* **132**, 422.
Ragotzkie, R. A., and Pomeroy, L. R. (1957). *Limnol. Oceanogr.* **2**, 62.
Raymont, J. E. G. (1963). "Plankton and Productivity in the Oceans", 660 pp. McMillan Co., New York.
Reisner, G. S., and Thompson, J. F. (1956). *Nature, Lond.* **178**, 1473.
Reisner, G. S., Gering, R. K., and Thompson, J. F. (1960). *Plant Physiol.* **35**, 48.
Rice, T. R. (1956). *Limnol. Oceanogr.* **1**, 123.
Richards, F. A. (1952). *J. Mar. Res.* **11**, 147.
Richards, F. A., and Thompson, T. G. (1952). *J. Mar. Res.* **11**, 156.
Richman, S. (1958). *Ecol. Monogr.* **28**, 273.
Richter, G. (1961). *Planta,* **57**, 202.
Riley, G. A. (1942). *J. Mar. Res.* **5**, 67.
Riley, G. A. (1943). *Bull. Bingham. oceanogr. Coll.* **8**, 14.
Riley, G. A. (1944). *Amer. Scient.* **32**, 129.
Riley, G. A. (1951). *Bull. Bingham. oceanogr. Coll.* **13**, 1.
Riley, G. A. (1957). *Limnol. Oceanogr.* **2**, 252.
Riley, G. A. (1959). *Bull. Bingham. oceanogr. Coll.* **17**, 83.
Riley, G. A. (1963). *Limnol. Oceanogr.* **8**, 372.
Riley, G. A., Stommel, H., and Bumpus, D. F. (1949). *Bull. Bingham. oceanogr. Coll.* **12**, 169.
Rho, Joon Hee (1959). Preprint 199. Int. Oceanogr. Congress, New York.
Rodhe, W. (1948). *Symb. bot. upsaliens.* **10**, 1.
Rodhe, W. (1955). *Verh. int. Ver. Limnol.* **12**, 117.
Rodhe, W. (1958). *Rapp. Cons. Explor. Mer,* **144**, 122.
Rodhe, W. (1962). *Mem. Ist. ital. Idrobiol. de Marchi,* **15**, 21.
Rodhe, W., Vollenweider, R., and Nauwerck, A. (1958). The primary production and standing crop of phytoplankton. *In* "Perspectives in Marine Biology" (A. A. Buzzati-Traverso, ed.), p. 299–322. University of California Press, Berkeley.
Rodina, A. G. (1948). *Mikrobiologiya,* **17**, 232.
Rodina, A. G. (1962). *Abstracts of the 15th International Congress on Limnology (Wisconsin)*, p. 92.
Rodina, A. G. (1963). *Limnol. Oceanogr.* **8**, 388.
Ryther, J. H. (1955). Ecology of autotrophic marine dinoflagellates with reference to red water conditions. *In* "Luminescence of Biological Systems" (F. H. Johnson, ed.). AAAS Publication, Washington, D.C.
Ryther, J. H. (1956a). *Limnol. Oceanogr.* **1**, 72.
Ryther, J. H. (1956b). *Limnol. Oceanogr.* **1**, 61.
Ryther, J. H. (1956c). *Nature, Lond.* **178**, 861.
Ryther, J. H. (1959). *Science,* **130**, 602.
Ryther, J. H., and Guillard, R. R. L. (1959). *Deep Sea Res.* **6**, 65.

Ryther, J. H., and Guillard, R. R. L. (1962). *Can. J. Microbiol.* **8**, 447.
Ryther, J. H., and Hulburt, E. M. (1960). *Limnol. Oceanogr.* **5**, 337.
Ryther, J. H., and Kramer, D. D. (1961). *Sci. Rept. Bermuda Biol. Sta.* Final Report, Part 1.
Ryther, J. H., and Menzel, D. W. (1959). *Limnol. Oceanogr.* **4**, 492.
Ryther, J. H., and Menzel, D. W. (1960). *Sci. Rept. Bermuda Biol. Sta.* Final Report, Part 1.
Ryther, J. H., and Yentsch, C. S. (1958). *Limnol. Oceanogr.* **3**, 327.
Ryther, J. H., Menzel, D. W., and Vaccaro, R. F. (1961). *Limnol. Oceanogr.* **6**, 149.
Sachs, T., and Mayer, A. M. (1961). *Phycologia*, **1**, 149.
Samejima, H., and Myers, J. (1958). *J. Gen. Microbiol.* **18**, 107.
Sargent, M. C. (1940). *Plant Physiol.* **15**, 275.
Saruhashi, K. (1953). *Pap. Met. Geophys., Tokyo*, **3**, 202.
Saunders, G. W., Trama, F. B., and Bachmann, R. W. (1962). Evaluation of a modified ^{14}C technique for shipboard estimation of photosynthesis in large lakes. Publication No. 8. Institute of Science and Technology, University of Michigan, Great Lakes Research Division.
Schelske, C. L. (1962). *Science*, **136**, 45.
Schreiber, E. (1927). *Wiss. Meeresuntersuch, Abt. Helgoland, N.F.* **16**, 10.
Schükerk, W. (1960). *Arch. Mikrobiol.* **35**, 279.
Scott, G. T. (1945). *J. cell. comp. Physiol.* **26**, 35.
Serenkov, G. P., and Barashkov, G. K. (1954). *Vestnik. Moskov. Univ.* **9**: *Ser. Fiz. Mat. No. 8.* 95 (*C. A.* **49**, 7657 (1955).)
Serenkov, G. P., and Pakhomova, M. V. (1961). *Vestnik. Moskov. Univ. Biol. Pochvoved. Ser VI*, **16**, 22. (*C. A.* **56**, 9124 (1962).)
Serenkov, G. P., and Vladychenskaya, N. S. (1962). *Nauchn. Dok. Vysshei, Shkoly. Biol. Nauki*, **2**, 147 (*C. A.* **57**, 8906 (1962).)
Shibata, K. (1958). *J. Biochem., Tokyo*, **45**, 599.
Shimada, B. M. (1958). *Limnol. Oceanogr.* **3**, 336.
Shiriaishi, K., and Provasoli, L. (1959). *Tohoku J. agric. Res.* **10**, 89.
Sieburth, J. M. (1961). Bacterial habitats in the arctic environment. *In* "Marine Microbiology" (C. H. Oppenheimer, ed.), p. 533. Charles C. Thomas, Springfield, Illinois.
Sironval, C., and Kandler, O. (1958). *Biochim. Biophys. Acta*, **29**, 359.
Sistrom, W. R. (1962). *J. Gen. Microbiol.* **28**, 607.
Smayda, T. J. (1958). *Nyt. Mag. Botan.* **6**, 75.
Smayda, T. J. (1959). *Oikos*, **9**, 158.
Smith, E. L. (1937). *J. Gen. Physiol.* **20**, 807.
Smith, G. M. (1951). The classification of algae. *In* "Manual of Phycology" (G. M. Smith, ed.), p. 13. Chronica Botanica Co., Waltham, Massachusetts.
Smith, J. B., Tatsumoto, M., and Hood, D. W. (1960). *Limnol. and Oceanog.* **5**, 425.
Smith, J. H. C., and Benitez, A. (1955). Chlorophylls: Analysis in plant materials. *In* "Modern Methods of Plant Analysis" (K. Paech and M. V. Tracey, eds.), Vol. IV, p. 142. Springer-Verlag, Berlin.
Sommer, A. L., and Booth, T. E. (1938). *Plant Physiol.* **13**, 199.
Sorokin, C. (1958). *Physiol. Plant*, **11**, 275.
Sorokin, C. (1960). *Biochem. Biophys. Acta*, **38**, 197.
Sorokin, C., and Krauss, R. W. (1959). *Proc. nat. Acad. Sci., Wash.* **45**, 1740.

Sorokin, C., and Krauss, R. W. (1962). *Plant Physiol.* **37**, 37.
Sorokin, Y. I. (1956). *Trud. vsesoyuz. gidrobiol. Obshch.* **7**, 271.
Sorokin, Y. I. (1960a). *Mikrobiologiya*, **29**, 204.
Sorokin, Y. I. (1960b). *J. Cons. int. Explor. Mer*, **26**, 49.
Sorokin, Y. I. (1961). *Mikrobiologiya*, **30**, 289.
Sorokin, Y. I. (1962). *C.R. Acad. Sci. U.R.S.S.* **45**, 192.
Sorokin, Y. I., and Kliashtorin, L. B. (1961). *Trud. vsesoyuz. gidrobiol. Obshch.* **11**, 265.
Spaeth, J. P., and Menzel, D. W. (1961). *Sci. Rept. Bermuda Biol. Sta.* **2**. Appendix I. No. 16.
Spencer, C. P. (1954). *J. Mar. biol. Ass. U.K.* **33**, 265.
Spencer, C. P. (1957). *J. Gen. Microbiol.* **16**, 282.
Spoehr, H. A., and Milner, H. W. (1949). *Plant Physiol.* **24**, 120.
Stadie, W. C., and O'Brien, H. (1936). *J. Biol. Chem.* **112**, 723.
Stanbury, F. A. (1931). *J. Mar. biol. Ass. U.K.* **17**, 633.
Steele, J. H. (1956). *J. Mar. biol. Ass. U.K.* **35**, 1.
Steele, J. H. (1958). *Publ. Scot. Home Dept., Mar. Res.* **7**, 1.
Steele, J. H. (1959). *Biol. Rev.* **34**, 129.
Steele, J. H. (1961). *Repts. Bermuda Biol. Sta.* **1**, Appendix I. June 1961.
Steele, J. H., and Baird, I. E. (1961). *Limnol. Oceanogr.* **6**, 68.
Steele, J. H., and Menzel, D. W. (1962). *Deep Sea Res.* **9**, 39.
Steele, J. H., and Yentsch, C. S. (1960). *J. Mar. biol. Ass. U.K.* **39**, 217.
Stepanek, M. (1960). *Sci. Pap. Inst. Chem. Technol. Prague*, **4**, 379.
Stepanek, M., and Pokorny, J. (1960). *Sci. Pap. Inst. Chem. Technol. Prague*, **4**, 333.
Stephens, G. C., and Schinske, R. A. (1961). *Limnol. Oceanogr.* **6**, 175.
Stephens, K., and Strickland, J. D. H. (1962). *Limnol. Oceanogr.* **7**, 485.
Stewart, D. P. (1962). *Ann. Bot., Lond.* **26**, 439.
Strain, H. H. (1951). The pigments of algae. *In* "Manual of Phycology" (G. M. Smith, ed.), p. 243. Chronica Botanica Co., Waltham, Massachusetts.
Strain, H. H. (1958). Chloroplast pigments and chromatographic analysis. Thirty-second Annual Priestly Lectures. The Pennsylvania State University, University Park, Pa., U.S.A.
Strain, H. H., Manning, W. M., and Hardin, G. (1944). *Biol. Bull., Woods Hole*, **86**, 169.
Strange, L., Kirk, M., Bennett, E. L., and Calvin, M. (1962). *Biochim. Biophys. Acta*, **61**, 681.
Strickland, J. D. H. (1958). *J. Fish. Res. Bd Can.* **15**, 453.
Strickland, J. D. H. (1960). *Bull. Fish. Res. Bd Can.* **122**, 172.
Strickland, J. D. H. (1961). *Fish. Res. Bd Can. M.S. Report Series* (*Oceanogr. and Limnol.*) *No. 88*. 21 pp.
Strickland, J. D. H., and Parsons, T. R. (1960). *Bull. Fish. Res. Bd Can.* **125**, 185.
Suschenya, L. M. (1961). *Okeanologia* **1** (**4**), 664.
Suzuki, N., and Kato, K. (1953). *Bull. Fac. Fish. Hokkaido Univ.* **4**, 132.
Sverdrup, H. U. (1953). *J. Cons. int. Explor. Mer*, **18**, 287.
Sweeney, B. M., and Hastings, J. W. (1958). *J. Protozool.* **5**, 217.
Syrett, P. J. (1954). Ammonia and nitrate assimilation by green algae (Chlorophyceae). *In* "Autotrophic Micro-Organisms" (B. A. Fry and J. L. Peel, eds.), 4th Symp. Soc. Gen. Microbiol., p. 126. Cambridge University Press, London.
Syrett, P. J. (1956). *Physiol. Plant.* **9**, 28.

Syrett, P. J. (1958). *Arch. Biochem. Biophys.* **75**, 117.
Talling, J. F. (1957a). *New Phytol.* **56**, 29.
Talling, J. F. (1957b). *New Phytol.* **56**, 133.
Talling, J. F. (1957c). *Limnol. Oceanogr.* **2**, 213.
Talling, J. F. (1957d). *Physiol. Plant,* **10**, 215.
Talling, J. F. (1960). *Limnol. Oceanogr.* **5**, 62.
Talling, J. F. (1961). *Annu. Rev. Pl. Physiol.* **12**, 133.
Tanada, T. (1951). *Amer. J. Bot.* **38**, 276.
Tchindonova, J. G. (1959). *Trud. Inst. Okeanol. Akad. Nauk S.S.S.R.* **30**, 166.
Thang, Minh-Nguy. (1961). *Biochem. Biophys. Acta,* **52**, 495.
Thomas, W. H. (1958). *In* Quarterly Progress Report No. 4. Scripps Tuna Oceanography Research Program.
Thomas, W. H., and Krauss, R. W. (1955). *Plant Physiol.* **30**, 113.
Thorson, G. (1946). *Medd. Komm. Havundersøg., Kbh., Ser. Plankton,* **4** (**1**), 1.
Tolbert, N. E., and Zill, L. P. (1956). *J. biol. Chem.* **222**, 895.
Tranter, D. J., and Newell, B. S. (1963). *Deep Sea Res.* **10**, 1.
Tyler, J. E. (1961). *Proc. nat. Acad. Sci., Wash.* **47**, 1726.
Van Baalen, C. (1962). *Botanica Mar.* **4**, 129
Van Norman, R. W., and Brown, A. H. (1952). *Physiol. Plant.* **27**, 691.
Verduin, J. (1957). *Limnol. Oceanogr.* **2**, 333.
Verduin, J. (1960). *Limnol. Oceanogr.* **5**, 372.
Vernon, L. P. (1960). *Analyt. Chem.* **32**, 1144.
Vinberg, G. G. (1960). "The Primary Production of Bodies of Water" (Russian), 328 pp. Institute of Biology, Academy of Sciences of the Byelorussian S.S.R. Minsk.
Vinogradov, A. P. (1953). "The Elementary Chemical Composition of Marine Organisms". 647 pp. Translation by Sears Foundation for Marine Research, Yale University Press, New Haven, Connecticut.
Vinogradov, M. E. (1962). *Rapp. Cons. Explor. Mer,* **153**, 114.
Vishniac, H. S. (1955). *Trans. N.Y. Acad. Sci.* **17**, 352.
Vishniac, H. S. (1956). *Biol. Bull., Woods Hole,* **111**, 410.
Vives, F., and Fraga, F. (1961). *Invest. Pesq.* **19**, 17.
Vollenweider, R. A. (1950). *Schweiz. Z. Hydrol.* **12**, 193.
Von Brand, T. (1935). *Biol. Bull., Woods Hole,* **69**, 221.
Voroschilova, A., and Dianova, E. (1937). *Mikrobiologiya,* **6**, 741.
Walker, J. B. (1954). *Arch. Biochem. Biophys.* **54**, 1.
Walne, P. R. (1956). *Fish. Invest. Lond., Ser. II, 20:* (*9*). 23 pp.
Walne, P. R. (1963). *J. Mar. biol. Ass. U.K.* **43**, 767.
Warburg, O. (1958). *Science,* **128**, 68.
Watt, W. D., and Paasche, E. (1963). *Physiol. Plant.* **16**, 674.
Weigl, J. W., Warrington, P. M., and Calvin, M. (1951). *J. Amer. chem. Soc.* **73**, 5058.
Weis, D., and Brown, A. H. (1959). *Plant Physiol.* **34**, 235.
Westlake, D. F. (1963). *Biol. Rev.* **38**, 385.
Williams, V. R., and McMillan, R. (1961). *Science,* **133**, 459.
Winokur, M. (1949). *Amer. J. Bot.* **36**, 287.
Wolf, F. T. (1960). Pteridine pigments in micro-organisms and higher plants. *In* Comparative Biochemistry of Photoreactive Systems" (M. B. Allen, ed.), p. 53. Academic Press, New York.
Wood, E. J. F. (1953). *Aust. J. Mar. Freshw. Res.* **4**, 160.

Wood, E. J. F. (1956). *Pacif. Sci.* **10**, 377.
Wood, E. J. F. (1959). *J. Mar. Biol. Ass. India,* **1**, 26.
Wood, E. J. F. (1961). The relative importance of groups of algae and protozoa in marine environments in the south-western Pacific. *In* "Marine Microbiology" (C. H. Oppenheimer, ed.), p. 236. Charles C. Thomas, Springfield, Illinois.
Wood, E. J. F. (1962). *Limnol. Oceanogr.* **7**, 32.
Wright, J. C. (1959). *Limnol. Oceanogr.* **4**, 235.
Yamamoto, H. Y., Nakayama, T. O. M., and Chichester, C. O. (1962). *Arch. Biochem. Biophys.* **97**, 168.
Yentsch, C. S. (1957). *Nature, Lond.* **179**, 1302.
Yentsch, C. S., and Menzel, D. W. (1963). *Deep Sea Res.* **10**, 221.
Yentsch, C. S., and Ryther, J. H. (1957). *Limnol. Oceanogr.* **2**, 140.
Yentsch, C. S., and Scagel, R. F. (1958). *J. Mar. Res.* **17**, 567.
Yentsch, C. S., and Vaccaro, R. F. (1958). *Limnol. Oceanogr.* **3**, 443.
Yonge, C. M. (1935). *J. Mar. biol. Ass. U.K.* **20**, 341.
Zhukova, A. I. (1961). Quantitative value of bacteria in the nutrition of marine invertebrates. *In* "Marine Microbiology" (C. H. Oppenheimer, ed.), p. 699. Charles C. Thomas, Springfield, Illinois.
Zobell, C. E., and Feltham, C. B. (1938). *J. Mar. Res.* **1**, 312.

Chapter 13

Anoxic Basins and Fjords*

FRANCIS A. RICHARDS

Department of Oceanography
University of Washington, Seattle, U.S.A.

I. Introduction 611
II. Classification and Examples of Oxygen-deficient and Anoxic Marine Systems 613
A. General 613
B. Oxygen-deficient Environments 614
C. Intermittently Anoxic Systems 615
D. Permanently Anoxic Systems 617
E. Landlocked Fjords Containing Relict Sea Water 619
III. Some Aspects of the Physical Oceanography of Anoxic Basins and Fjords .. 619
A. General 619
B. The Stability across the Sills of Anoxic Basins and Fjords 620
C. Flushing of Anoxic Basins 620
D. Circulation and Exchanges 622
IV. Chemical and Biochemical Characteristics of Anoxic Basins and Fjords 623
A. General 623
B. The Steps in the Oxidation of Organic Matter, and a Stoichiometric Model 623
V. Some Consequences of Anoxic Conditions 634
A. Consequences of Denitrification 634
B. Lowered Redox Potentials 635
C. Accumulation of Organic Decomposition Products 640
D. Toxicity of Sulphides 642
E. Reduction of Sulphate: Chlorinity Ratios 642
Acknowledgments 643
References 644

I. Introduction

At all depths in the seas greater than the photosynthetic compensation depth, there is a net consumption of oxygen as a consequence of the respiratory and biochemical decomposition of organic matter. The organic matter is formed essentially by the photosynthetic fixation of carbon in the photic zone.

* Contribution No. 337 from the Department of Oceanography, University of Washington.

In most of the world's marine waters, circulatory processes replenish the depths with oxygen-bearing water at rates such that the oxidative consumption does not exceed oxygen renewal, and these waters never lose all their dissolved oxygen. However, in special circumstances, generally where circulatory processes are restricted, and frequently where rates of organic production in the photic zone are high, the rate of oxygen consumption outstrips the rate of circulatory renewal of oxygen, and anoxic conditions arise. The word anoxic is used here to designate conditions or environments containing no dissolved oxygen although the other air gases are present. The term anaerobic has a biological connotation (from *an*, not; *aero*, air; Gr. *bios*, life), and is used to describe organisms living in, or physiological processes occurring in, the absence of air or, as it has come to mean, free oxygen.

In this discussion, any basin or fjord, any part of which at any time contains anoxic water, will be considered. Many marine sediments become anoxic (see Chapter 18, Volume 2), but this chapter is concerned primarily with those systems in which the water itself loses all of its oxygen.

Anoxic waters are well known in many marine environments. The Black Sea is the best known example, and anoxic conditions have long been recognized in many fjords and inlets in Norway (Strøm, 1936), and in basins similar in their physiography and circulation to the Norwegian fjords. Such basins are relatively common on the Pacific Coast of British Columbia, Canada, and presumably they exist in similar inlets along the coast of Chile.

Less well known than the above basins, which are invariably characterized by strong haloclines, are a few tropical basins and inlets, which owe their strong pycnoclines to thermoclines rather than to haloclines. The physical restraints to vertical circulation are, however, essentially the same in both types. Three of these tropical basins have been reported in the literature—the Cariaco Trench and the Gulf of Cariaco in the Caribbean Sea, and Kaoe Bay in Indonesia.

Anoxic basins and fjords represent sea water systems in which organic decomposition has completely removed dissolved oxygen because of the amount of organic matter introduced into the lower layers, and because these layers are only weakly circulating or are stagnant. These systems are nutrient traps, and unusually high concentrations of organic decomposition products accumulate in them. The chemical and biochemical processes and the biological populations in these environments differ markedly from those found in environments where oxygen is perpetually present. Among these differences are the following.

(i) Concurrent (or nearly so) with the disappearance of free dissolved oxygen, denitrification begins. During denitrification, nitrate and nitrite ions supply the reducible substrate for the biochemical oxidation of organic matter. Because nitrate is generally formed in the biochemical cycle only in the presence of free dissolved oxygen, denitrification exhausts the supply of nitrate and nitrite ions and leads to the onset of (ii).

(ii) Sulphate reduction, during which sulphate ions are the reducible substrate. Free nitrogen appears to be the product of the denitrification process; this does not radically alter the chemical nature of sea water, which always contains relatively large amounts of free nitrogen from the solution of air. However, during sulphate reduction, sulphide is produced and occurs in the water mainly as undissociated H_2S and HS^-; at pH values below 11 only minute amounts of S^{2-} are present. The introduction of sulphide into the environment is a near-catastrophic event, since sulphides are highly toxic and thus eliminate all organisms except anaerobic bacteria from the ecosystem. Sulphides form highly insoluble compounds with many metal ions at the pH of sea water, and thus not only tend to strip the waters of ions of these metals, but may enrich the sediments with precipitates of them.

(iii) Probably the most significant consequence of the disappearance of oxygen from the marine environment is the lowering of the redox potential, *Eh*. With this reduction (coincident with the disappearance of dissolved oxygen and the appearance of sulphides), the thermodynamic drive is diminished, and organic materials tend to accumulate in the sediments rather than to be oxidized. Thus, sediments laid down under anoxic conditions are considerably richer (up to 10-fold) in organic carbon than are sediments deposited under oxygen-bearing waters. This must have been important in the geological past in connection with the deposition of organically rich sediments which were subsequently metamorphosed into petroleum sources.

II. Classification and Examples of Oxygen-deficient and Anoxic Marine Systems

A. General

It has been pointed out that whether or not a basin becomes anoxic depends on the balance between the circulatory supply of oxygen-bearing waters and the rate of oxygen consumption in the basin. Furthermore, the rate of oxygen consumption depends largely on the rate at which organic matter is introduced into the deeper waters of the system. Certain marine environments exist in which the balance approaches that leading to anoxic conditions, and in which oxygen deficient conditions exist either permanently or intermittently. Examples of oxygen-deficient waters are found in many threshold fjords and, although the region cannot be considered either a basin or a fjord, in large areas of the eastern tropical Pacific Ocean.

B. OXYGEN-DEFICIENT ENVIRONMENTS

Very low oxygen concentrations in the Gulf of Panama were first observed on the *Dana* expedition on Jan. 11th 1922 (Schmidt, 1925). Some extreme examples of low oxygen concentrations in the eastern tropical North Pacific Ocean are shown in Fig. 1. This region is pertinent to this discussion because it appears that one of the initial steps leading to truly anoxic conditions—denitrification—is proceeding in it. The abnormally high concentrations of nitrite ions observed by Brandhorst (1959) in this region (Fig. 1), when viewed in the light of studies of anoxic regions, evidently represent an intermediate step in denitrification. Sulphides have not been demonstrated to exist in the region, but the possibility that they do occur there has not been eliminated.

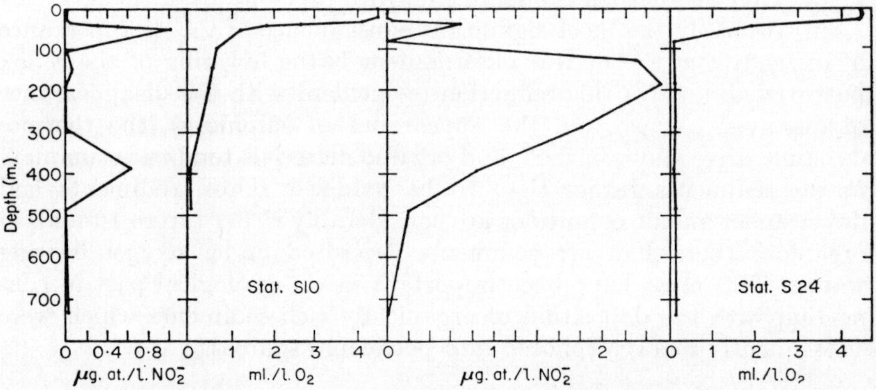

FIG. 1. The vertical distribution of dissolved oxygen and nitrite ions at two stations in the east tropical Pacific Ocean, from Brandhorst (1959). In this part of the ocean, dissolved oxygen disappears almost completely from the water column, and extraordinarily high concentrations of nitrite ions appear. Station S 10 was occupied Nov. 2nd, 1956 at 8°42′N. lat., 86°01·0′W. long. Station S 24 was occupied Dec. 10th, 1956 at 19°30·0′N. lat., 105°52·0′W. long.

Dabob Bay, an arm of Puget Sound in the state of Washington, illustrates a fjord-like basin which is not known to become anoxic, but which has been demonstrated by Barnes and Collias (1958) to retain, in its deeper parts, waters of properties which indicated that it remained *in situ* and without large changes by mixing for up to five months. Barnes and Collias were able to estimate rates of oxygen consumption (Table I) by assuming that these waters remained relatively unmixed.

It can be seen that had these waters not eventually been replaced, their oxygen would have been completely removed in a matter of four to six months. Observations in Dabob Bay have failed to reveal anoxic conditions, but the system is a borderline case, and it appears that even moderate climatic changes would cause it to become permanently or intermittently anoxic.

TABLE I

Rates of oxygen consumption in waters of Puget Sound estimated by Barnes and Collias (1958)

Place	*Period*	*Average rate (ml./l./day)*
Dabob Bay (180 m.)	22 Jan.–22 Sept., 1953	0·012
	25 Mar.–17 June, 1954	0·010
	9 July–30 Sept., 1954	0·005
Port Susan (120 m.)	23 Sept., 1953–12 Jan., 1954	0·017
Lynch Cove (30 m.)	18 Feb.–21 May, 1953	0·020
	26 Mar.–9 July, 1954	0·018

C. INTERMITTENTLY ANOXIC SYSTEMS

Several of the Norwegian fjords are known to become anoxic and to be flushed intermittently. There are recorded cases of the catastrophic effects of the relatively sudden flushing of some of these fjords, with the result that sulphide-bearing water was displaced upwards into the surface layers with consequent mass mortality of marine life.

Dramsfjord, an arm of the Oslofjord in Norway, is an example of an intermittently anoxic system. It is some 18 km. long, from its head at the city of Drammen, where it receives the Drammen River and domestic and industrial pollution, to the sill, some 8 m. deep, at Svelvik. Inside the sill, it deepens to its maximum depth of 117 m., and when observed on 14 November, 1956, it contained anoxic and sulphide-bearing water at depths of between 30 and nearly 55 m. Figure 2 shows the changes observed between that time and the following August. It appears that sometime shortly preceding the first observations, dense, oxygenated water had crossed the sill at Svelvik and displaced the sulphide-bearing water upwards. The figure shows the fate of this stratum through the following months. On 11 December, 1956, sulphides were again encountered in the bottom water. This bottom layer later thickened until it appeared to have joined the upper layer sometime before

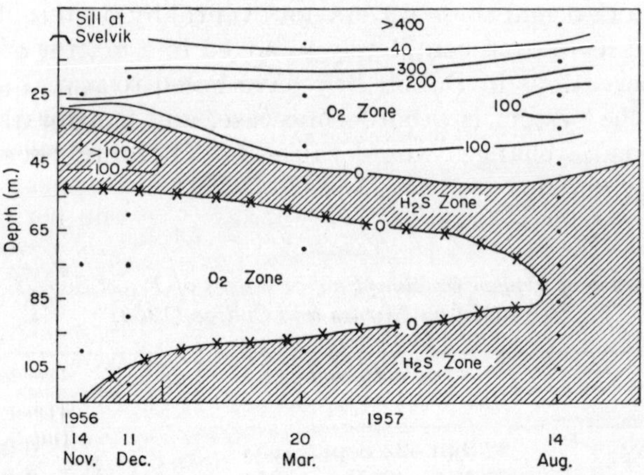

FIG. 2. Regime of dissolved oxygen and H_2S in Dramsfjord, Norway, Nov. 14th, 1956 to Aug. 14th, 1957. Oxygen isopleths are in μg.-at./l. The shaded area represents waters in which H_2S was present, after Richards and Benson (1961). ——, oxygen isopleths; —×—×—, H_2S isopleths.

14 August, 1957. At the time of the August 1957 sampling, the highest concentrations of sulphides were found at 55 m. (0·4 ml. H_2S/l.) and at the deepest sampling depth, 105 m. (0·9 ml./l.), with an intermediate minimum (0·02 ml./l.) at 85 m.

The Gulf of Cariaco (Fig. 3) is in the tropics, and owes its vertical stability to the vertical temperature gradients. Gade (1961) has published a diagram showing the occurrence and disappearance of sulphide-bearing waters in the deepest part of the gulf. Displacement of the

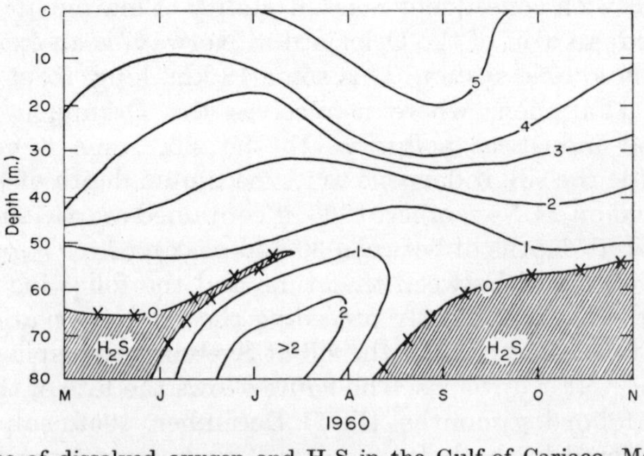

FIG. 3. Regime of dissolved oxygen and H_2S in the Gulf of Cariaco, May–November, 1960, after Gade (1961). Oxygen isopleths are in ml./l., and the shaded area represents waters in which H_2S was present. ——, O_2 isopleths; —×—×—, H_2S isopleths.

deeper water can apparently be accomplished by easterly winds which tend to remove surface waters from the gulf (Richards, 1960); these waters are replaced by a deeper inflow over the sill of water sufficiently dense to lift the bottom water to intermediate depths where it is mixed and advected outwards.

The flushing of Saanich Inlet, which is a typical fjord lying north of Victoria on Vancouver Island, British Columbia, will be referred to later. Gucluer and Gross (1964) have reported on its geomorphology and on some aspects of its sedimentology, and Gross *et al.* (1963) have described cores containing varved sediments from the inlet. Herlinveaux (1962) has reported on its hydrographic regime. The inlet has a sill some 90 m. deep, and deepens to about 240 m. toward its head. It is approximately 22 km. long and has a maximum width of nearly 8 km.

D. PERMANENTLY ANOXIC SYSTEMS

The Black Sea is the world's largest anoxic basin. With its adjunct, the shallow Sea of Azov, it has an area of 413, 500 km.^{2}, is 1149 km. long, and 610 km. wide (Caspers, 1957). It can be thought of as a deep estuary in which precipitation and run-off exceed evaporation. Sverdrup *et al.* (1942) have estimated that its outflow to the Mediterranean Sea, through the Bosphorus, Sea of Marmora, and the Dardanelles, averages 12,600 m.^{3}/second, while an undercurrent of Mediterranean Sea water of 6100 m.^{3}/second flows eastward into the Black Sea. Caspers (1957) has reviewed many of the physical, chemical, and biological features of the system. In many respects, it has been the most intensively studied of all anoxic basins and fjords, having received the scientific attention of a large number of Russian and other workers since the turn of the century.

Oxygen disappears from the waters of the Black Sea at depths varying from 150 m. or less in the central parts, to 250 m. or more along the coasts. At greater depths, sulphides generally increase toward the bottom. Along with the increase in sulphide concentrations with depth, the concentrations of other products of the decomposition of organic matter, inorganic phosphate, silicate, carbon dioxide, and ammonia increase with depth to values much higher than those encountered in freely circulating marine environments. Like other anoxic basins and fjords, it is a nutrient trap, and contains, in its depths, up to 10 times the concentrations of inorganic phosphate and silicate found at comparable depths in the Mediterranean Sea and Atlantic Ocean, in spite of its low salinity (*c.* 20‰). Vertical circulation is severely restricted by the stability of the water column, and although the Mediterranean Sea water which enters the Black Sea must tend to cascade to considerable depths, its volume is too small in relation to the volume of the Black Sea to cause a replacement adequate to ventilate its depths.

The Cariaco Trench (Richards and Vaccaro, 1956), lying north of the Caribbean coast of Venezuela, is a depression up to 1400 m. deep in the continental shelf, and is apparently permanently anoxic in its deeper parts. The maximum sill depth is approximately 150 m. Oxygen disappears and sulphides appear at between 300 and 400 m. The water is essentially isothermal and isohaline from about 600 m. to the bottom. Sulphide concentrations increase to maximum values between 800 and 900 m., suggesting that new water is introduced into the depths, but the data are too sparse to indicate whether this occurs steadily or intermittently.

Lake Nitinat will be considered here as an example of a permanently anoxic fjord, although this point remains to be verified. Some of the characteristics of Lake Nitinat, on the south coast of Vancouver Island, British Columbia, have been reported by Pickard (1963) from observations made by Dr. T. G. Northcote of the University of British Columbia. Although it is called a lake and has a surface layer of brackish to fresh water, its depths contain anoxic sea water and it is a true fjord. Its

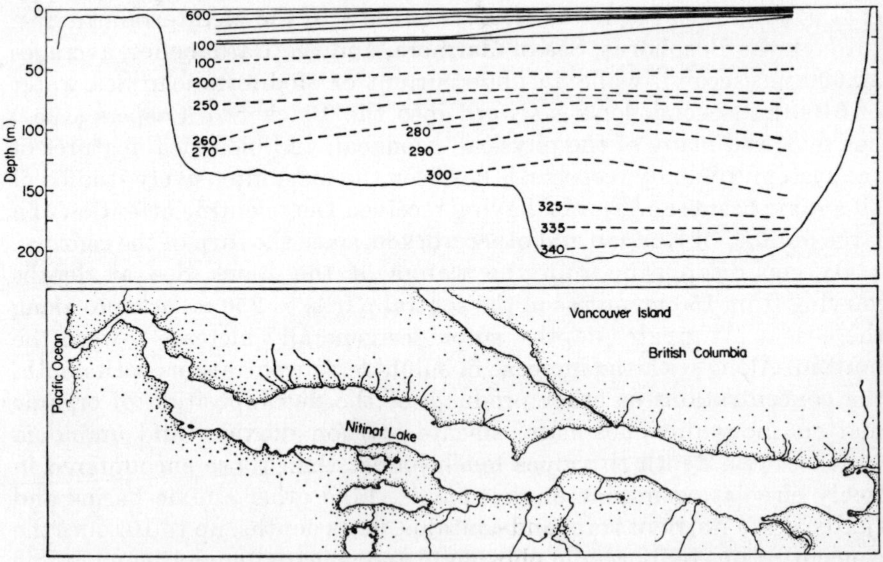

FIG. 4. Map of Lake Nitinat, Vancouver Island, and the distribution of oxygen and sulphides observed there June 10–12th, 1964. Oxygen and total sulphide isopleths are in μg.at./l. The map and profile are reproduced through the courtesy of Dr. T. C. Northcote of the University of British Columbia. ——, O_2 contours, μg.-at./l., – – – –; S_T contours, μg.-at./l.

exchange with the Pacific Ocean is limited by a shallow entrance some 2·7 km. long with several sills which probably prevent the horizontal exchange of water much deeper than 4 m. Inside the entrance (Fig. 4), the depth increases sharply to approximately 140 m., for about one-half of its 22·5 km. length, and then again increases to 210 m. Concentrations of sulphides in the depths of the fjord are about equal to those in the depths of the Black Sea, *c.* 300 μg.-at./l. Chemical observations made by the author and his colleagues in this fjord in June, 1964, will be referred to later.

The data regarding Lake Nitinat are inadequate to describe water exchange in it. Inhabitants of the region report that a fish kill occurred in 1955, but there are no other records to relate this incident to the circulation of the water. Because of its extreme pycnocline and high sulphide concentrations, it is probably permanently anoxic.

E. LANDLOCKED FJORDS CONTAINING RELICT SEA WATER

It is not known definitely whether there are any extant marine basins or fjords in which the anoxic waters are never exchanged. The only systems that are known to contain relict sea water are Lakes Tokke and Botnvatn in Norway (Strøm, 1957, 1961) and Powell Lake, British Columbia (Williams *et al.*, 1961). These lakes are the remains of fjords which have become landlocked by the geological uplift of their sills above sea level. Strøm (1961) suggested that Lake Øvrevatn, a landlocked fjord near Botnvatn, is at the point of being isolated from the sea, and the author's observations of Lake Nitinat indicate that a similar phenomenon is in progress there.

III. Some Aspects of the Physical Oceanography of Anoxic Basins and Fjords

A. GENERAL

The circulation of the deeper water in anoxic basins and fjords is invariably limited horizontally by the geomorphology, and vertically by the stability of the water column.

Most anoxic basins and fjords in the temperate latitudes are estuarine, receive appreciable amounts of fresh-water run-off, and have a precipitation and run-off exceeding evaporation; they are fjord estuaries in the terminology of Cameron and Pritchard (1963), and possess threshold sills. Vertical circulation in them is limited in such a way as to lead to anoxic conditions where the pycnocline lies at a depth near or below the depth of the basin's sill. A pycnocline lying above the sill, or one in which the stability is insufficiently high, will not restrict vertical circulation sufficiently to lead to anoxic conditions.

In anoxic systems in the temperate latitudes, the pycnocline arises largely from a strong halocline, and the downward increase in density may be either augmented or diminished by the vertical temperature gradient. Anoxic basins in the tropics, e.g. the Cariaco Trench and Kaoe Bay (Richards and Vaccaro, 1956) and the Gulf of Cariaco (Richards, 1960; Gade, 1961), may owe their strong pycnoclines almost wholly to the thermocline.

B. THE STABILITY ACROSS THE SILLS OF ANOXIC BASINS AND FJORDS

Richards and Vaccaro (1956) have implied that a certain critical value of the stability of the water column near the depth of the sill must obtain before it imposes a sufficiently large restraint on the vertical circulation to permit anoxic conditions to develop. These values, approximated by values of $\Delta\sigma_t/\Delta z$, the change in sigma-t per unit change of depth, are given in Table II, which is taken in part from Richards and Vaccaro (1956). These examples are inadequate to make the criterion universally useful, and it would also be necessary to develop other criteria, such as the rate of photosynthetic production in the region and a more exact evaluation of the balance between vertical circulation and rates of consumption of dissolved oxygen, to permit the prediction of anoxic conditions.

TABLE II

Approximate stability across the sills of some basins and fjords

Basin	$\Delta\sigma_t/\Delta z \times 10^{-3}m^{-1}$
Oxygen-bearing basins	
Gulf of Mexico	2×10^{-6}
Catalina Basin	$3{\cdot}1 \times 10^{-6}$
Santa Barbara Basin	$4{\cdot}4 \times 10^{-6}$
Santa Monica Basin	$8{\cdot}8 \times 10^{-6}$
Anoxic basins	
Kaoe Bay	$2{\cdot}4 \times 10^{-5}$
Cariaco Trench	$3{\cdot}2 \times 10^{-5}$
Gulf of Cariaco	$3{\cdot}5 \times 10^{-5}$
Saanich Inlet	$1{\cdot}5 \times 10^{-5}$
Black Sea	$3{\cdot}0 \times 10^{-4}$

C. FLUSHING OF ANOXIC BASINS

Some anoxic basins and fjords appear to flush, i.e. to have their water replaced in whole or in part by relatively highly-oxygenated water.

The flushing may be at rather regular intervals—perhaps annually—or it may occur at irregular intervals. Still others, for example, the Black Sea and the Cariaco Trench, are known not to flush completely.

The course of replacement of anoxic by oxygenated water in the Gulf of Cariaco has been described by Gade (1961) (see Fig. 3), in Dramsfjord by Richards and Benson (1961), and in Saanich Inlet by Herlinveaux (1962). Similar observations made by the author for Saanich Inlet in 1961, 1962, and 1963 are shown in Fig. 5.

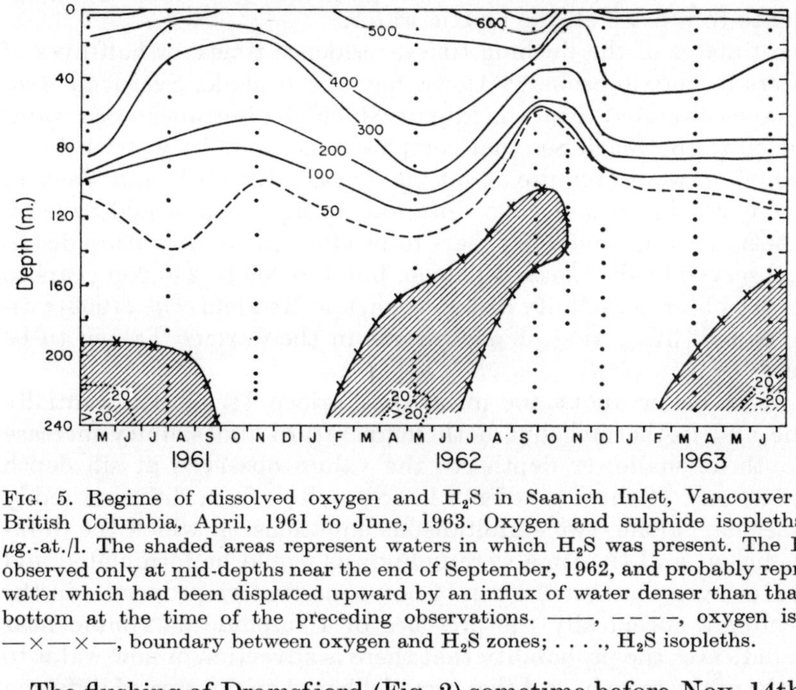

FIG. 5. Regime of dissolved oxygen and H_2S in Saanich Inlet, Vancouver Island, British Columbia, April, 1961 to June, 1963. Oxygen and sulphide isopleths are in μg.-at./l. The shaded areas represent waters in which H_2S was present. The H_2S was observed only at mid-depths near the end of September, 1962, and probably represented water which had been displaced upward by an influx of water denser than that at the bottom at the time of the preceding observations. ——, – – – –, oxygen isopleths; —×—×—, boundary between oxygen and H_2S zones; . . . ., H_2S isopleths.

The flushing of Dramsfjord (Fig. 2) sometime before Nov. 14th, 1956 (Richards and Benson, 1961) may not have cleared the bottom water of sulphides, and probably indicated an influx of water somewhat less dense than the deepest water.

The process of flushing in these basins has not been fully documented, but it is clear that whether or not the anoxic water will be completely replaced depends on the balance between the amount of oxygen-bearing water entering and the rate at which oxygen is consumed by biological and biochemical processes. It is also clear, and chemically important, that what circulation there is within these basins will produce mixtures of sulphide-bearing and oxygen-bearing waters. The chemistry of these mixtures will be returned to later.

D. CIRCULATION AND EXCHANGES

There is probably some vertical water exchange and circulation in all basins—if nothing more than that caused by molecular diffusion—and water below the pycnocline can be replaced by either molecular or eddy diffusion, or by advection, or by all of these (see Chapter 2). It is evident that oxygen-bearing Mediterranean Sea water is introduced into the Black Sea, but the rate is too slow to replace the sulphide-bearing waters. It is probable that the deep waters of the Cariaco Trench receive advective increments of new water from the Caribbean Sea, but they are inadequate to replace all the anoxic water.

Few estimates of the flushing times, residence times, or half-lives of the waters in anoxic basins or fjords have been made. Sverdrup *et al.* (1942) have estimated from the rate at which Mediterranean Sea water is admitted to the Black Sea that complete renewal of its water below a depth of 30 m. would require about 2500 years. Richards and Vaccaro (1956) concluded that molecular diffusion, acting alone, would have required between 1000 and 2000 years to produce the temperature distribution observed in the Cariaco Trench, but 120,000 to 240,000 years to produce the observed salinity distribution, and Redfield *et al.* (1963) estimated the half-life period of phosphorus in the Cariaco Trench to be 142 years.

The water below about 400 m. in the Cariaco Trench is essentially isohaline and isothermal, but both temperature and salinity increase with depth at shallower depths to the values observed at sill depth (about 150 m.). If these gradients have resulted from diffusion (eddy and molecular) alone, two simultaneous equations, in which the diffusion coefficient and the time were unknowns, could be formulated and solved. However, the distribution of non-conservative properties in the water column, specifically the presence of a sulphide maximum near 900 m., indicates the probability that there is advection of new water to the depths of the trench, and that a model based on downward diffusion of properties is not applicable. It is, nevertheless, useful to examine such a model.

Using a model which ignores advective introduction of deep water into the Cariaco Trench, Redfield *et al.* (1963) estimated the half-life period, or the average time a constituent remains below sill depth, for phosphorus. The half-life, τ_N, is $0{\cdot}693x/(\mathrm{d}x/\mathrm{d}t)$, where x is the phosphorus content of the basin and $\mathrm{d}x/\mathrm{d}t$ is the rate at which x is changed with time. The phosphorus content can be expressed as $N \,.\, h$, where N is the average concentration below sill depth and h is the depth of water below the sill. The content is changed at the rate $\mathrm{d}Q_R/\mathrm{d}t$ which is equal to $-\mathrm{D} \,.\, \mathrm{d}N/\mathrm{d}z$, where $\mathrm{d}Q_R/\mathrm{d}t$ is the rate at which the quantity of phosphorus is increased by the decomposition of organic matter, D is the diffusion coefficient, and dN/dz is the vertical gradient of phosphorus concentration in the transition zone. From the above definitions,

$$\tau_N = 0{\cdot}693 \frac{N \,.\, h}{dQ_R/dt} = 0{\cdot}693 \frac{N \,.\, h}{D \,.\, dN/dt}\,.$$

In the Cariaco Trench, $N = 2{\cdot}4 \times 10^{-6}$ mg.-at.P/cm.^{2}, $h = 1{\cdot}2 \times 10^5$ cm. and $dQ_R/dt = 0{\cdot}44 \times 10^{-10}$ mg.-at. P/cm.^{2}/second. The half-life is thus estimated to be 142 years, and is based on the assumption that 1/10 of the organic matter produced in the photic zone sinks below sill depth and decomposes there.

IV. Chemical and Biochemical Characteristics of Anoxic Basins and Fjords

A. General

The decomposition of organic matter in anoxic basins and fjords proceeds to a lesser extent than in other marine environments. This results, after the disappearance of dissolved oxygen, in denitrification and in sulphate reduction with the production of sulphides. Consider an initial state in which an isolated sea-water system saturated with dissolved oxygen begins to receive organic matter of marine origin. The organic matter will, in time, essentially all be remineralized by respiratory and biochemical processes, thus returning its constituents to the water as soluble inorganic materials. As progressively more organic matter is introduced, a reasonably regular and predictable series of events should follow, in which the laws of conservation of mass and stoichiometry will be obeyed. Anoxic basins and fjords approximate to such systems.

B. The Steps in the Oxidation of Organic Matter, and a Stoichiometric Model

(1) *By oxygen*. In oxygen-bearing marine environments, organisms respire and organic matter decomposes at the expense of dissolved oxygen as the hydrogen ion acceptor. The decomposition and respiration are accompanied by the release to the water of inorganic decomposition products. It is convenient to discuss these processes in terms of organic matter which, on average, has a fixed ratio and state of oxidation of carbon, nitrogen, and phosphorus. According to Fleming (1940) and others, this ratio is C : N : P = 106 : 16 : 1, by atoms. It will be assumed that, on average, the oxidation state of carbon is that of carbohydrate (CH_2O), the bound nitrogen is in the amino form, and the

oxidation state of phosphorus is that of orthophosphoric acid. The organic matter can then be formulated as $(CH_2O)_{106}(NH_3)_{16}H_3PO_4$. It will be shown that this model has some empirical justification, but it is apparent that large deviations from it exist in nature. The model ignores all the other elements, but this discussion will be mainly limited to the above three elements and their fate during decomposition, because phosphorus and nitrogen compounds occur in such small concentrations in the ocean that their depletion can limit the growth of organisms. The growth of the organisms in turn alters the mode of combination of much of the total amounts of these elements in the sea.

The complexities of the nitrogen and phosphorus cycles are considered elsewhere in this book (Chapters 8 and 9). Here it will be assumed that the first step of organic decomposition is the hydrolytic release of NH_4^+ and PO_4^{3-} ions, concurrently with the respiratory or biochemical oxidation of CH_2O, represented by the equations

$$(CH_2O)_{106}(NH_3)_{16}H_3PO_4 = 106\ CH_2O + 16\ NH_3 + H_3PO_4$$

$$106\ CH_2O + 106\ O_2 = 106\ CO_2 + 106\ H_2O$$

$$16\ NH_3 + 32\ O_2 = 16\ HNO_3 + 16\ H_2O$$

adding,

$$(CH_2O)_{106}(NH_3)_{16}H_3PO_4 + 138\ O_2$$
$$= 106\ CO_2 + 122\ H_2O + 16\ HNO_3 + H_3PO_4.$$

Thus, the complete oxidation of the organic matter associated with one atom of phosphorus would require 276 atoms of oxygen, and would be accompanied by the release to the water of one phosphate ion, 16 nitrate ions, and 106 molecules of carbon dioxide. The equations apply to complete oxidation, but it is clear that intermediate steps are represented in the ocean by the occurrence of ammonium and nitrite ions and by soluble organic compounds of nitrogen and phosphorus. However, the completely oxidized nitrate ions form by far the most abundant reservoir of nitrogen compounds in the ocean.

Because free oxygen is present in most of the marine environment, the above equations should represent the oxidation of organic matter in most of the ocean, as well as in environments approaching anoxic conditions until all, or nearly all, of the oxygen is consumed.

(2) *Denitrification*. When all, or nearly all, of the free dissolved oxygen has been consumed, nitrate ions are the next most abundant source of free energy for the oxidation of organic matter (McKinney and Conway, 1957). The process in which they serve as the reducible substrate is denitrification, and the evidence indicates that this process produces

nitrite ions as an intermediate, and free molecular nitrogen as a final product. Unusually high concentrations of nitrite ions have been observed in strata just above and just below sulphide-bearing, oxygen-free strata in Saanich Inlet, and Brandhorst (1959) has demonstrated unusual accumulations of nitrite ions in nearly (or perhaps completely) anoxic waters of the east tropical North Pacific Ocean. Richards and Benson (1961) have demonstrated the presence of free nitrogen which they presumed to have originated by way of denitrification in Dramsfjord and the Cariaco Trench, and Broenkow (1963) has shown the same thing in Saanich Inlet. The limiting concentration of oxygen which can be present at the inception of denitrification is uncertain, and probably depends on the nitrate concentration (McKinney and Conway, 1957), but in any event it is relatively low—probably of the order of tenths of a millilitre per litre.

Denitrification can be represented by the equation

$$(CH_2O)_{106}(NH_3)_{16}H_3PO_4 + 84{\cdot}8\ HNO_3$$
$$= 106\ CO_2 + 42{\cdot}4\ N_2 + 148{\cdot}4\ H_2O + 16\ NH_3 + H_3PO_4.$$

This reaction indicates that the ammonia released during denitrification is not oxidized. However, it is possible that it does oxidize according to the equation

$$5NH_3 + 3\ HNO_3 = 4\ N_2 + 9\ H_2O.$$

If this were so, the overall reaction would be

$$(CH_2O)_{106}(NH_3)_{16}H_3PO_4 + 94{\cdot}4\ HNO_3$$
$$= 106\ CO_2 + 55{\cdot}2\ N_2 + 177{\cdot}2\ H_2O + H_3PO_4.$$

In general, chemical observations made on anoxic environments have been too imprecise to elucidate this point, but recently, collected data on the accumulation of ammonia in Lake Nitinat suggest that in this lake about half of the ammonia-nitrogen produced during this step is oxidized to free N_2, and that the other half accumulates as ammonium ions.

(3) *Sulphate reduction.* When all, or nearly all, of the oxygen, nitrate ions and nitrite ions have been consumed, sulphate reduction can begin. Wheatland (1954) concluded from experimental evidence that sulphate reduction is negligible in the presence of more than 0·11 ml./l. (and probably less) of dissolved oxygen.

Organic decomposition by sulphate reduction can be represented by the equation

$$(CH_2O)_{106}(NH_3)_{16}H_3PO_4 + 53\ SO_4^{2-}$$
$$= 106\ CO_2 + 53\ S^{2-} + 16\ NH_3 + 106\ H_2O + H_3PO_4.$$

The evidence is reasonably clear that the ammonia released during this step merely accumulates and is not oxidized in sulphide-bearing environments. This has been suggested by Richards (1964), and is demonstrated by Fig. 6, which shows the accumulation in Lake Nitinat of ammonia and sulphides in the ratio which would be expected from the above equation. The value of the intercept of the least squares regression line suggests the presence of quantities of ammonia at the inception of sulphate reduction which are about one-half of those that would be expected if no ammonia was oxidized during denitrification in this environment.

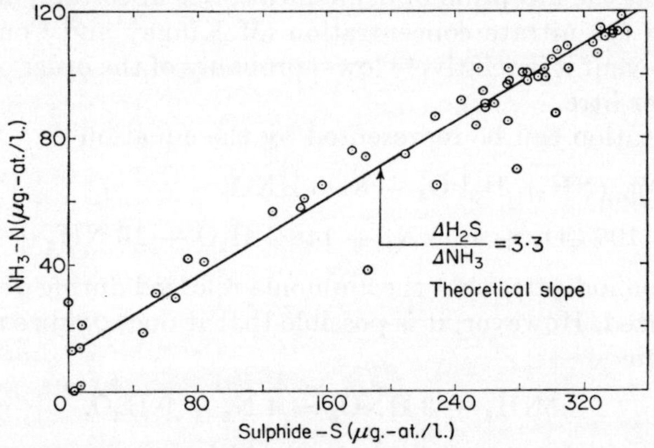

FIG. 6. The correlation between ammonia and sulphides observed in Lake Nitinat, 10–12 June, 1964. The line shows both the linear least squares correlation between the two variables and the slope of the ratio of change in ammonia to the change in sulphide predicted by the stoichiometric model described in the text. The intercept on the NH_3 axis is interpreted as ammonia present at the inception of sulphate reduction.

(4) *Complications*. It can be seen from the above discussion that, under the assumed conditions, there would be stoichiometric relationships between the amounts of organic matter decomposed in a water mass and (a) the amount of dissolved oxygen consumed, which would be directly related to the amount of nitrate ions formed during the oxidation by free oxygen; (b) the amount of free nitrogen formed during denitrification; (c) the amounts of carbon dioxide, phosphate ion, and ammonia released to the water; (d) the amount of sulphide formed; and (e) the amount of ammonia accumulated during sulphate reduction. If these processes all took place in an isolated water mass, the products would occur in definite proportions which could be calculated from the amount of organic matter decomposed and the original oxygen content. These conditions appear to be approximated to in several anoxic basins

and fjords, but there are several complicating factors, in addition to departures from the assumptions on which the model is based concerning the composition of organic matter, which will alter the above relationships. These will be considered before comparing conditions in anoxic basins and fjords with those predicted by the model.

(a) It is assumed that the amount of oxygen consumed can be calculated as the difference between the solubility of oxygen in sea water equilibrated at the *in situ* temperature and salinity with air and the observed *in situ* oxygen concentration, or in other words, that the water was saturated with oxygen and contained no organic breakdown products (specifically phosphate, nitrate, and carbon dioxide) when decomposition began. In nature, large quantities of carbon dioxide will be present (as dissolved CO_2, H_2CO_3, HCO_3^-, and CO_3^{2-}) and appreciable amounts of nitrate and phosphate may be present. These correspond to preformed nutrients as defined by Redfield (1942). The preformed phosphate and carbon dioxide will simply be carried along in addition to that produced during decomposition, as will the nitrate during oxidation by free oxygen, but during denitrification the nitrate will be consumed together with that formed during oxidation by free oxygen.

(b) The estimation of the extent of sulphate reduction is complicated. The amount of sulphate reduced is usually so small in comparison to the total sulphate present in sea water that estimates of decreases in the sulphate : chlorinity ratio are too imprecise to serve this purpose. Redfield *et al.* (1963, Fig. 6) found a linear relationship between sulphide concentrations and decreases in the sulphate : chlorinity ratio in the Black Sea. This may be interpreted as indicating that the sulphide concentrations represented the bulk of the sulphate reduction that had taken place there. However, if sulphide-bearing and oxygenated waters mix, as they do at interfaces, the sulphides will be oxidized to higher oxidation states, although the free nitrogen, phosphate, silicate, and carbon dioxide produced during denitrification and sulphate reduction may persist in the water. Since this oxidation may proceed to sulphate or to intermediate oxidation states, such as free sulphur, sulphite, thiosulphate, or tetrathionate, it would be necessary to have accurate determinations of all compounds of sulphur and of free sulphur to make a precise estimate of the amount of sulphate reduction.

Wheatland (1954) concluded from studies of the oxidation of sulphides in the Thames estuary that the oxidation of sulphides there is purely inorganic. The rate of oxidation is increased by the presence of suspended matter and of iron, and is limited by the rate at which oxygen enters the water. Studies in the author's laboratory by Mr. Dana

Kester have indicated that the reaction between sulphides and dissolved oxygen proceeds according to the equation

$$2\,HS^- + 2\,O_2 = S_2O_3^{2-} + H_2O.$$

He determined that, beginning with concentrations of 30 μg.-at. of total H_2S/l. and oxygen-saturated water, the reaction was first order with respect to sulphur, and had a specific rate constant of 0·23/hour at 25°C and a pH of 8·0. At the same pH at 6·5°C, the constant was 0·063/hour. At the higher temperature, water containing 30 μg.-at./l. of sulphide-sulphur, when mixed with water saturated with oxygen, would require about 30 hours for the complete oxidation of the sulphides. Skopintsev *et al.* (1961) have also carried out experiments on the formation and oxidation of sulphides. They found that the oxidation rates corresponded to first order reactions, with higher rate constants corresponding to higher initial oxygen concentrations. Starting with an initial $O_2 : H_2S$ ratio of 1 : 2, half the added H_2S was oxidized in 60 to 70 hours at room temperature. During the experiment (from 46 to 236 days), the consumption of oxygen per unit of H_2S varied from 0·7 to 1·6. As the initial ratio of $O_2 : H_2S$ was increased, the consumption of oxygen per unit of H_2S also increased. If the ratio of $O_2 : H_2S$ was initially small, thiosulphate and sulphite were formed, although the latter was formed only in small amounts. Skopintsev concluded that the oxidation of sulphide ion took place in two stages: initially it was oxidized to sulphur, and subsequently to sulphate, but the complete oxidation was observed only if excess oxygen was present. Other work indicates that the formation of thiosulphate and sulphite is favoured if the initial concentrations of oxygen are low in relation to initial sulphide concentrations, and that oxidation to sulphate is favoured by relatively high oxygen concentrations.

Skopintsev and Gubin (1955) reported that they had found sulphites and thiosulphates (principally the latter) in 77% of 74 water samples from the Black Sea. These ions exceeded 1·1 mg./l. in only four samples, which contained 1·60 to 2·30 mg./l. The author and his co-workers have recently attempted to determine "total reduced" sulphur by iodometric titration and sulphide-sulphur by colorimetry in water from Lake Nitinat. At depths in the system, the total reduced sulphur appeared to exceed the sulphide-sulphur by about 40 μg.-at./l., thus indicating the presence of appreciable concentrations of sulphite and thiosulphate sulphur. However, it should be realized that other reducing substances in the water could produce the differences between the colorimetric and the titrimetric estimates.

The author has been able to find no observations or determinations of particulate or colloidal elementary sulphur, which might be expected to be produced when sulphide-bearing and oxygen-bearing waters mix. It has been suggested by C. A. Barnes (personal communication) that suspended sulphur may have been responsible for an increase in turbidity which he noted in the oxygenated layer above a pool of sulphide-bearing water in Lake Union, Seattle, Washington.

(c) Hydrogen sulphide may be photosynthetically converted to sulphur by sulphur bacteria

$$CO_2 + 2\,H_2S = CH_2O + 2\,S^0 + H_2O.$$

In the environments discussed in this chapter, the sulphide zones are generally below the photic zone, and therefore this reaction does not complicate matters in them.

(d) It should be pointed out that the estimation of the extent of denitrification from estimates of the quantities of nitrogen present in excess of those expected from the solution of air are not highly satisfactory. Richards and Benson (1961) made the estimate by comparing nitrogen : argon ratios observed in Dramsfjord and in the Cariaco Trench with these ratios in similar sea water equilibrated with air. Broenkow made his estimate (reported by Richards, 1964) by comparison of gasometric analyses of the gases extracted from water samples from Saanich Inlet with the solubility of nitrogen plus argon reported by Rakestraw and Emmel (1938a). These solubility data are in some question, and it is uncertain how valid is the assumption that the water was initially in equilibrium with the atmosphere; even if equilibrium was at one time achieved, the physical conditions under which it was achieved are unknown. Gasometric nitrogen determinations reported in the literature (for example, by Rakestraw and Emmel, 1938b and Hamm and Thompson, 1941) vary by several percent from the equilibrium saturation values of Rakestraw and Emmel (1938a).

(e) In examining the stoichiometric relationships between the various products of organic decomposition, the following relationships have been used to estimate the amount of organic oxidation that has taken place. (i) The number of microgram-atoms of oxygen consumed during oxidation by free oxygen has been estimated as the difference between the actual oxygen content of the water (zero if anoxic) and the solubility of oxygen in sea water of the *in situ* salinity and temperature when equilibrated at the sea surface with one atmosphere of water-saturated air. (ii) The microgram-atoms of oxygen expended from nitrate ions during denitrification have been computed as three times the number of microgram-atoms of biogenic free nitrogen, estimated as described earlier. (iii) The microgram-atoms of oxygen expended from sulphate ions during sulphate reduction has been arrived at by multiplying the sulphide concentrations by four. These estimates of the oxygen equivalents expended have been compared with the phosphate and carbon dioxide added to the water. The amount of carbon dioxide added to the water is probably not simply that arising from the equations presented above, because of the solution of calcium carbonate which takes place when the sea water rich in carbon dioxide comes into contact with $CaCO_3$. The ratio of $\Delta S^{2-}/\Delta CO_2^B$, where CO_2^B is used to indicate carbon dioxide arising from the oxidation of organic carbon compounds, should be 0·5. If, however, each molecule of carbon dioxide arising from this source is balanced by one from the solution of calcium carbonate, the ratio $\Delta S^{2-}/(\Delta CO_2^B + \Delta CO_2^{(CaCO_3)}) = 0{\cdot}25$. Redfield *et al.* (1963) found an intermediate value of 0·36 in their analysis of Skopintsev's data from the Black Sea. The author and his co-workers have found variable values for this ratio in Saanich Inlet and Lake Nitinat which are difficult to interpret; however, exceptionally high values for the alkalinity were observed in Lake Nitinat. When these alkalinity and pH values were used

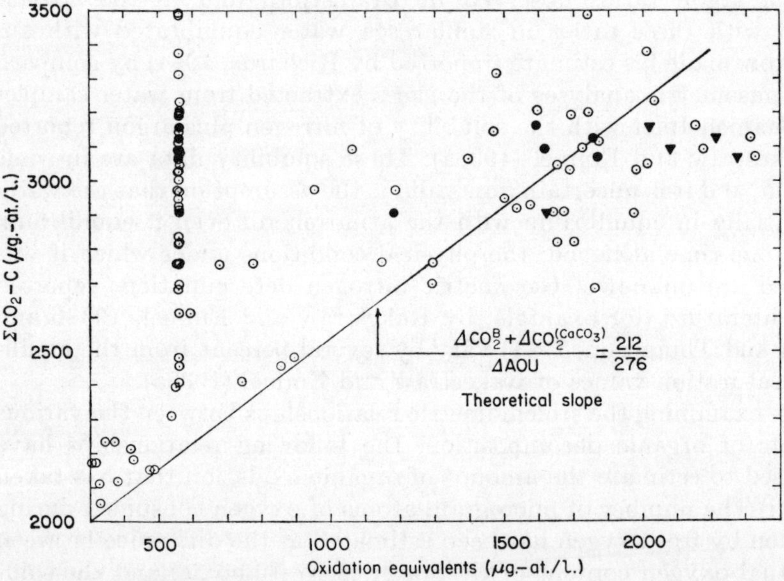

FIG. 7. The relationship between the amount of oxygen consumed (from free oxygen, during denitrification, and sulphate reduction) and the total carbon dioxide observed in the waters of Lake Nitinat. The solid symbols represent samples in which the gas content was determined gasometrically. The oxidation equivalents in excess of *c.* 570 are attributed to denitrification and sulphate reduction. The slope of the line indicates the ratio of the theoretical change in inorganic carbon to the consumption of oxygen (from free O_2, denitrification, and sulphate reduction), assuming that each mole of CO_2 arising from the decomposition of organic matter is augmented by a mole of CO_2 from the solution of calcium carbonate.

to compute total carbon dioxide by Buch's method, as given by Harvey (1955); it appeared that a $\Delta S^{2-}/\Delta CO_2$ ratio of 0·25 fitted the data better than a ratio of 0·50 (Fig. 7; the oxygen equivalents > 1000 μg.-at./l. are $4 \times [S^{2-}]$).

(f) It appears that in both the Black Sea and Lake Nitinat there is less inorganic phosphate present in the deep, sulphide-rich waters than would be predicted from the oxygen equivalents consumed (see Fig. 5, Richards and Vaccaro, 1956, and Fig. 8, this chapter). These negative

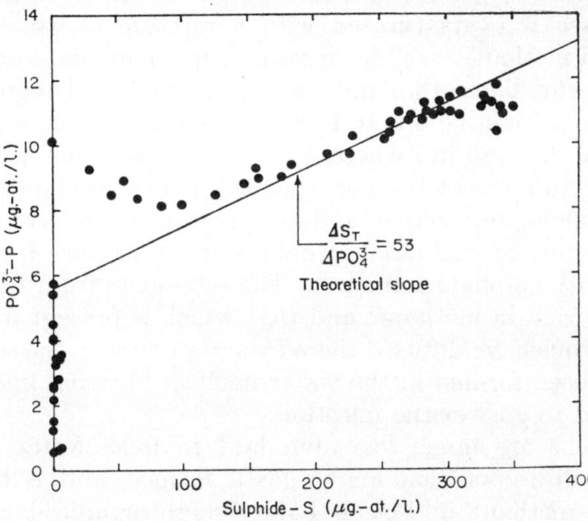

FIG. 8. Correlation between phosphate concentrations and total reduced sulphur compounds, primarily H_2S, observed in Lake Nitinat. Each atom of sulphide-sulphur would be equivalent to four oxygen atoms. The line is the ratio of changes predicted by the stoichiometric model described in the text. The positive PO_4^{3-}-P deviations at the lower sulphide concentrations may represent excess quantities of secondarily oxidized sulphides. Possible explanations of the negative deviations in the higher ranges are discussed in the text, and include the possibility of phosphine formation.

departures are difficult to explain. If they represented phosphate taken up by anaerobic bacteria, one might expect the underlying sediments to be enriched in phosphorus, but this is not generally the case: if anything, sediments under anoxic waters contain a smaller fraction of phosphorus than those laid down under oxygen-bearing waters. Another possible explanation is that the phosphate ion itself may be reduced in these environments. The author is aware of no evidence for the formation of phosphine in anoxic environments, but Datsko (1955) has stated that phosphine is formed at or near the bottom of the Sea of Azov. The formation of phosphine, if it occurs, should proceed according to the equation:

$$2\,CH_2O + H_3PO_4 = 2\,CO_2 + 2\,H_2O + PH_3,$$

and the disappearance of 1 μg.-at./l. of phosphate ion would increase the equivalent oxygen consumption by only 4 μg.-at./l.

(g) Another step in the oxidation of organic matter that must be considered is methane formation, which occurs when carbon dioxide acts as the energy source for the biochemical oxidation of organic matter. Although Kriss (1963, p. 255) has stated that "... water in the depths of the (Black) sea is oversaturated with a mixture of gases: nitrogen, hydrogen and methane, ..." he gives no supporting data or references for the statement. The author and his co-workers have detected approximately 1·5 ml. of methane/l. at 140 m. in Lake Nitinat, and approximately 0·9 ml./l. at 80 m., whereas it was undetectable in the surface water. This is in spite of the fact that sulphate ions should provide a sufficiently rich energy source, and are present in sufficient abundance that carbon dioxide reduction would not be expected to compete favourably with sulphate reduction. The sediments from Lake Nitinat are relatively rich in methane, and that which is present in the water column may thus have diffused there from its source in the sediment, or it may have been formed in the water itself. Additional investigations will be needed to answer the question.

Sulphate ions are much less abundant in fresh water systems in which it is well known that marsh gas is formed, and Williams *et al.* (1961) found methane in Powell Lake, where organic decomposition had reduced all the sulphate ions.

(5) *Tests of the model.* Some tests of the stoichiometric model presented above have been published by Richards (1964), who concluded that there were reasonably good correlations between the amounts of phosphate and total carbon dioxide present in Saanich Inlet and the total number of oxidation equivalents expended during oxidation by free oxygen, during denitrification, and during sulphate reduction. His data indicated that denitrification is relatively more important than sulphate reduction in Saanich Inlet, accounting for the equivalent of nearly 450 microgram atoms of oxygen (150 μg.-at. of N_2) per l., whereas sulphate reduction accounted for the equivalent of 60–80 μg.-at./l. It is probable that the extensive denitrification in Saanich Inlet reflects high values of preformed nitrate in the waters entering the system. In contrast is the situation in Lake Nitinat, where sulphate reduction has proceeded to a greater extent than denitrification. The waters entering this system contain only small amounts of nitrate in comparison with

those of Saanich Inlet. Figure 8 shows the relationship between inorganic phosphate and the oxidation equivalents estimated for each step (oxidation by oxygen, denitrification, and sulphate reduction) in Lake Nitinat, and the line shows the relationships predicted by the model.

Figure 9 illustrates the relationship between oxidation equivalents and phosphate concentrations obtained on four occasions in Saanich Inlet. When both denitrification and sulphate reduction are accounted for, the observed phosphate concentrations are reasonably close to those predicted by the model. The corresponding correlation from Lake Nitinat (Fig. 8) shows a regularly increasing negative departure of phosphate concentrations from those predicted by the model in the increasingly deep water of the system. These departures may reflect the results of the processes discussed in Section IV, B (4f).

Figure 6 shows diagrammatically the relationship between ammonia and sulphide concentrations which the model predicts and the observations from Lake Nitinat. In this figure, the "sulphides" were estimated

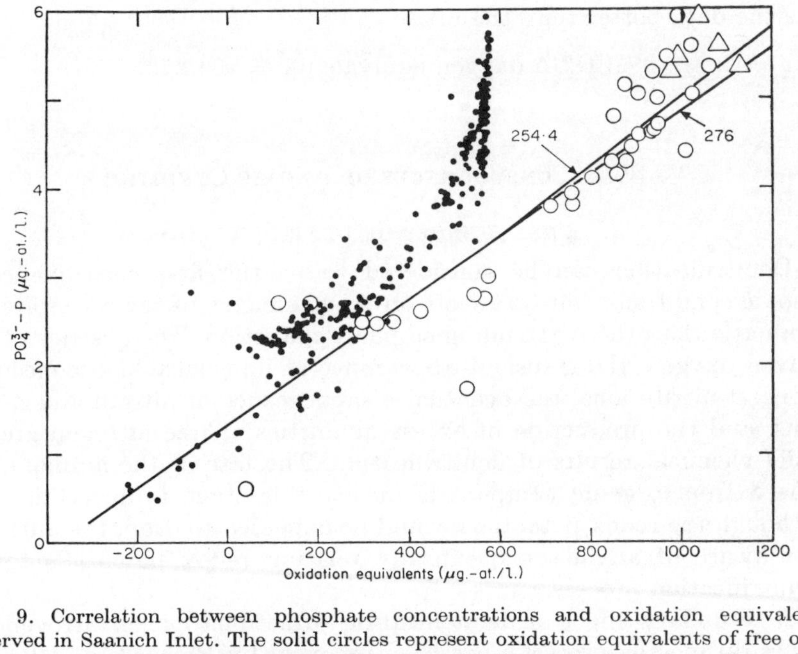

FIG. 9. Correlation between phosphate concentrations and oxidation equivalents observed in Saanich Inlet. The solid circles represent oxidation equivalents of free oxygen; the open circles represent the addition of oxidation equivalents from nitrate ions during denitrification (for samples for which estimates of biogenically produced N_2 were available), and the triangles represent four additional oxidation equivalents for each sulphide-sulphur atom observed. The lines show the theoretical relationships predicted by the stoichiometric model described in the text.

by an iodometric titration, and would, therefore, include such other reduced forms of sulphur as thiosulphate and sulphite. A poorer agreement between the model and nature was observed in Saanich Inlet (Richards, 1964, Fig. 10), where the secondary oxidation of sulphides is probably more important than in Lake Nitinat, and no attempt was made to estimate the other reduced forms of sulphur.

Correlations between oxidation equivalents and total carbon dioxide components are less apparent in the environments that have been studied. The deviations are probably in part the result of imprecision in the measurements of total carbon dioxide, and in part because of the additional contribution of bicarbonate ions resulting from the solution of calcium carbonate. Figure 7 shows this correlation in samples from Lake Nitinat, and it appears that the ratio which is calculated on the assumption that each mole of carbon dioxide added to the water by organic decomposition is augmented by a bicarbonate ion from the solution of calcium carbonate,

$$(\Delta CO_2^B + \Delta CO_2^{(CaCO_3)})/\Delta \text{ oxidation equivalents} = 212/276$$

fits the data better than the ratio

$$\Delta CO_2^B/\Delta \text{ oxygen equivalents} = 106/276.$$

V. Some Consequences of Anoxic Conditions

A. CONSEQUENCES OF DENITRIFICATION

Denitrification can be considered to be the first consequence of anoxic conditions, but its effects on the chemistry of the water are less dramatic than those attending sulphate reduction. The absence of dissolved oxygen, the transient appearance of unusually high concentrations of nitrite ions, the eventual disappearance of nitrate and nitrite ions, and the production of excess quantities of free nitrogen are the chief chemical results of denitrification. The fate of the ammonia released from organic compounds during this stage is uncertain, and although the redox potentials would be expected to drop, the author is not aware of any data specifically relating redox potentials to the denitrification process.

A secondary effect of denitrification—fractionation of nitrogen isotopes during the process—has been reported by Richards and Benson (1961), who observed a higher ratio of $^{15}N^{14}N$ to $^{14}N_2$ in the anoxic waters of the Cariaco Trench and Dramsfjord than in atmospheric nitrogen or in nitrogen contained in "normal" oxygen-bearing sea water.

B. LOWERED REDOX POTENTIALS

Perhaps one of the most fundamental consequences of anoxic conditions is the decreased redox potential of these environments. Skopintsev (1957) has made extensive studies of redox potentials in the Black Sea and their relationship to the regimes of oxygen and hydrogen sulphide. Some of his results are given in Table III. It will be noted that

TABLE III

Redox potentials, oxygen and sulphide concentrations, and $SO_4^{2-}/Cl‰$ ratios in the Black Sea, from Skopintsev (1957) and Skopintsev et al. *(1958)*

Depth (m.)	*Redox potential (mV)*	*Oxygen concentration ml./l.*	*µg.-at./l.*	*Sulphide concentration ml./l.*	*µg.-at./l.*	$SO_4^{2-}/Cl‰$
0	395	5·60	498	—	—	0·1408
25	408	7·06	628	—	—	—
50	404	6·35	565	—	—	0·1418
100	340	1·08	96	—	—	0·1408
150	− 26	0·25	22	0·02	2	0·1402
200	− 88	0·08	7	0·67	60	0·1400
300	−139	0	—	1·74	155	0·1394
500	−170	0	—	3·60	320	0·1387
750	−152	0	—	5·29	471	0·1377
1000	−144	0	—	6·15	547	0·1380
1500	−129	0	—	7·34	653	0·1378
1750	—	—	—	—	—	0·1378
2000	—	—	—	—	—	0·1361

this table shows an overlap of sulphides and dissolved oxygen at 150 and 200 m., and there are no data on nitrate, nitrite, or dissolved nitrogen that might be used to relate redox potential values to a zone of denitrification. The coexistence of sulphides and dissolved oxygen is apparently not a thermodynamically stable state, but may well represent a steady state, reflecting the mixing of oxygenated and sulphide-bearing water.

It is evident from Table III that the redox potential is not highly sensitive to changes in dissolved oxygen or sulphide concentrations except in a transition region in which oxygen concentrations drop to zero and sulphides appear. Subsequent increases in sulphide concentration produce a relatively small reduction in the redox potential.

(1) *Preservation of organic matter in anoxic environments.* A consequence of lowered redox potentials appears to be the tendency for greater accumulations of organic matter to be laid down under anoxic rather than under oxygen-bearing waters. Most pelagic sediments contain

1–2·5% of organic carbon (Sverdrup *et al.*, 1942). However, Richards and Vaccaro (1956) have reported up to 4% organic carbon in sediments from the Cariaco Trench, and Gucluer and Gross (1964) found up to 7% organic carbon in sediments from Saanich Inlet in the parts of the system where anoxic waters occur. In the shallower waters of the inlet, where the sediments are not exposed to anoxic conditions, the sediments contained only 1–2% of organic carbon. Richards (1964) has reported that the apparent chlorophyll *a* content of sediments from Saanich Inlet ranged from 1·5 mg./g. in the shallow sediments that are exposed only to oxygen-bearing waters, up to 84 mg./g. in the deep parts of the inlet which are exposed to anoxic waters for much of the time.

The effect of anoxic conditions on the amount of organic carbon in the water column is less clearly documented than the effect on the organic content of sediments. Skopintsev and Timofeeva (1960) reported that the organic carbon content of Black Sea water slowly decreases with increasing depth from about 3·6 mg./l. at the surface to 2·3 mg./l. at 2000 m. They noted a sudden increase at 200 m.; this effect was attributed to the appearance at that level of H_2S which killed the micro-organisms and increased the concentration of organic carbon there. The same authors (Skopintsev and Timofeeva, 1961) reported that the average concentration of organic carbon in the upper 200 m. of the Black Sea (3·5 mg./l.) is about twice that of Atlantic waters, but this does not appear to be related to the anoxic conditions in the Black Sea. Although these authors reported a gradual decrease with depth, Datsko (1951) found a gradual increase in soluble, and a decrease in colloidal and suspended organic matter with depth. It seems probable that the oxidation of organic matter is slower in anoxic than in oxygen-bearing water, and Skopintsev has reported experiments in which the organic carbon content of sea water decreased from 38·5–11·8 mg./l. (a 26·7 mg./l. decrease) in 220 days in the presence of oxygen, while it decreased to 14·2 mg./l. (a 24·3 mg./l. decrease) under anoxic conditions. This rate difference is not large and may not be significant.

In general, there is no evidence of spectacular increases in the organic carbon content of anoxic water columns. Richards and Vaccaro (1956) found that the total phosphorus in the anoxic waters of the Cariaco Trench was not more than 0·15 μg.-at./l. greater than the inorganic phosphate. According to Ketchum *et al.* (1955), this difference is not significant, but taking it at its face value, it would represent approximately 2 mg. of organic carbon/l., which is of the same order as that which other workers have found in the open ocean.

(2) *Altered oxidation states of metal ions and precipitation of sulphides.*

The lowered redox potentials of anoxic environments appear to result in lowered oxidation states of several elements in sea water. In oxygen-bearing waters, iron is present in the +3 oxidation state, but it is generally reduced to the +2 state in the presence of sulphides. This alters the solubility relationships of iron markedly, and tends to increase its solubility (see Chapter 18, Volume 2). The most insoluble compound of iron in oxygen-bearing water is probably a hydrated ferric oxide, which is considerably less soluble than ferrous sulphide, and the latter may control the concentration of soluble iron in sulphide-bearing waters. The author has observed concentrations of iron of up to about 18 μg./l. in the Cariaco Trench, whereas Lewis and Goldberg (1954) found maximum concentrations of about 5·5 μg./l. in the Pacific Ocean.

Observations from the Cariaco Trench permit the calculation of the ionic product $c_{Fe^{2+}} \times c_{S^{2-}}$. At a station occupied on Nov. 8th and 9th, 1957, the concentrations of iron and of H_2S at 1200 m. were $0{\cdot}32 \times 10^{-6}$ M and 20×10^{-6} M, respectively. The pH was about 7·8, and the salinity 36·2‰. Skopintsev (1957) has quoted the following relationship for the sulphide ion concentration.

$$c_{S^{2-}} = c_{H_2S} \frac{K_2' \cdot K_1'}{c_H^2 + K_1' \cdot c_H}$$

in which $$pK_1' = 7{\cdot}04 - 0{\cdot}4\,(\mu)^{\frac{1}{2}},$$

$$pK_2' = 14{\cdot}92 - 1{\cdot}2\,(\mu)^{\frac{1}{2}},$$

and $$\mu = 0{\cdot}020\ S‰\quad ,$$

where μ is the ionic strength. The calculation yields an ionic product $c_{Fe^{2+}} \cdot c_{S^{2-}}$ of $3{\cdot}54 \times 10^{-18}$. The solubility product for ferrous sulphide in distilled water is given by Lange (1956) as $3{\cdot}7 \times 10^{-19}$. To correct this for the salt effect of sea water with an ionic strength of about 0·7, it should be multiplied by 70 (Krauskopf, 1956), giving $2{\cdot}6 \times 10^{-17}$, or about ten times the ionic product. It thus appears that the anoxic waters of the Cariaco Trench are nearly saturated with ferrous sulphide, and that they may therefore represent a system in which the iron concentration is controlled by the solubility of ferrous sulphide. Krauskopf's (1956) general conclusion was that the concentrations of metal ions in sea water were generally much smaller than those which would be predicted from the solubility products of those compounds which are probably the most insoluble ones of the metal ion involved (see Chapter 5). The nearly equilibrium concentrations of iron found in the Cariaco Trench may be a significant characteristic of sulphide-bearing waters.

The soluble iron concentrations observed in the Cariaco Trench are in fair agreement with the 6–20 μg./l. reported at 1500–2000 m. in the Black Sea by Dobrzhanskaya and Pshenina (1958, 1959). These authors also reported practically no soluble iron to be present in the upper layers of the Black Sea. Skopintsev has recorded similar observations for dissolved iron, and suggested an average of 26 μg. of suspended iron/l. in the surface waters. Dobrzhanskaya and Pshenina (1958, 1959) found the concentration of suspended iron to be variable, but to average about 26 μg./l. near the coasts. They found a regular relationship between the concentration of suspended iron and the redox potential, with a distinct maximum at about 100 m. of some 100 μg./l. There was a rapid decrease to 10 μg./l. at about 150 m. Between 500 and 1500 m., the suspended iron decreased significantly, but it again increased between 1500 and 2000 m.

The ionic product $c_{Fe^{2+}} \cdot c_{S^{2-}}$ in the Black Sea has been calculated using the above range of soluble iron concentrations (6 – 20 μg./l.) and the data reported by Skopintsev (1957) for 750 m. at his station 3a, 1955. The range of ionic products corresponding to the two iron concentrations is $1{\cdot}28$–$4{\cdot}31 \times 10^{-17}$. Since the solubility product $2{\cdot}59 \times 10^{-17}$ as estimated above falls within this range, it appears that the Black Sea is just saturated with ferrous sulphide.

Uranium may be reduced from the hexavalent to the tetravalent state in the H_2S zone of the Black Sea, according to Agamirov (1963), although Kolyadin *et al.* (1960) stated that it occurs in the hexavalent state at all depths in the Black Sea as a carbonate complex, that it cannot be reduced to the tetravalent state and that it cannot be deposited in that state. Agamirov (1963) has used the Nernst equation in the form

$$Eh = E_0 + \frac{RT}{nF} \ln \frac{c^4_{UO_2(CO_3)_3^{4-}}}{c^3_{CO_3^{2-}}}$$

for the reaction

$$[UO_2\,(CO_2)_3]^{4-} + 2e = UO_{2(solid)} + 3\,CO_3^{2-} \quad ,$$

where $E_0 = -0{\cdot}32$V. This equation has been postulated by Naumov, 1959. This system gives values of Eh of $-0{\cdot}16$ to $-0{\cdot}21$V, in agreement with the values of Eh which Skopintsev (1957) has observed for the bottom part of the Black Sea. Agamirov has found that the uranium content of Black Sea sediments increases with the depth of water over the sediments (which would approximately correspond to increasingly large negative values of Eh) as follows: at 1119, 1860, and 2137 m.

depths, the uranium contents of the sediments were 2·6, 4·3, and 6·5 p.p.m., respectively. These values are reported to be 4 to 5 times those in oceanic sediments, and were much greater than those in Black Sea sediments which were laid down under oxygen-bearing waters and contained 0·7–2 p.p.m. Sugimura and Sugimura (1962) found sediments from 50 m. in Matsu Bay to contain 2·6 and 2·1 p.p.m.; 8·7 p.p.m. from 200 m. in Kagoshima Bay; 3·5 p.p.m. from 200 m. in the Japan Sea off Sakata; and 1·3 ± 0·03 p.p.m. in Pacific red clay from 6215 m. Holland and Kulp (1954) stated that shelf sediments, hemipelagic, and pelagic sediments contained 3·0, 2·5, and 1·0–1·8 p.p.m. of uranium, respectively. These observations and analyses of Cariaco Trench sediments lead to the tentative conclusion that sediments laid down in anoxic environments are richer in uranium than those deposited under oxygen-bearing water.

Nikolaev *et al.* (1960) determined the uranium content of 53 samples of water from the Black Sea fluorometrically, and found it to vary between 1·5 and 2·5 μg./l. (an average of 2·0 ± 0·3 μg./l). They were unable to detect any systematic variation of uranium content with depth or location. Lambet and Nikolaev (1962) found 17 and 1·1 μg. of U/l. in membrane-filtered water from the Vityaz and Swash estuaries on the Sea of Azov, respectively. These values should represent the uranium in solution.

Grashchenko *et al.* (1960) found, on average, $1 \pm 0{\cdot}1 \times 10^{-13}$ g.Ra/l. in the Black Sea. This value does not vary widely, and is of the same order as that found in open ocean waters (see Richards, 1957).

Nikolaev *et al.* (1961) found that thorium-232 varied from 4×10^{-9} to 219×10^{-9} g./l. in the Sea of Azov, but no such wide variation was observed in the Black Sea. They considered the variability in the Sea of Azov to be associated with the higher concentrations of suspended matter, particularly since the highest values were found near shores and estuaries.

Skopintsev and Popova (1960) reported finding 30 μg.Cu/l. in Black Sea waters, but it is doubtful if this is a reliable figure. Richards (1964) has reported on the copper content of the waters of the Cariaco Trench, where concentrations between 0·3 and 1·6 μg./l. were found in the upper, oxygen-bearing waters, but no copper could be detected in the H_2S zone by a method (Chow and Thompson, 1952) which should be able to detect 0·25 μg./l. It is probable that copper concentrations are limited by the solubility of CuS in sulphide-bearing waters, but these concentrations would be so low that extensive concentration procedures would be required to effect a determination using conventional analytical methods. However, it can be concluded that the copper content of the H_2S zone of the Cariaco Trench is less than 0·25 μg./l., and probably much less.

M. G. Gross (Geological Society of America, 1963 Annual Meeting, N.Y.) has reported on optical emission spectroscopic analyses of sediments from Saanich Inlet. He found the following concentrations in the varved, diatomaceous, clayey silts laid down under anoxic waters: Sc 14 p.p.m., Y 22 p.p.m., Ti 2700 p.p.m., Zr 92 p.p.m., V 66 p.p.m., Cr 59 p.p.m., Mn 320 p.p.m., Fe 2·2%, Co 7 p.p.m., Ni 27 p.p.m., Cu 43 p.p.m., B 61 p.p.m. Boron was most abundant in the sediments having the highest clay and organic matter contents. Except for boron, these concentrations were similar to those in sediments from the beaches of the inlet and in the nearby Strait of Georgia. Molybdenum, on the other hand, appeared to be enriched in the sediments from the anoxic region which contained up to 70 p.p.m. in contrast to 10 p.p.m. in the sediments from adjacent areas. Gross has suggested that the enrichment may have been caused by precipitation of MoS_2 or as a result of biological activity.

C. ACCUMULATION OF ORGANIC DECOMPOSITION PRODUCTS

(1) *Nutrient concentrations*. In the Cariaco Trench, phosphate ion concentrations reach approximately 2·6 μg.-at./l., an excess of about 0·6 units over maximum concentrations found in the open Caribbean Sea. The corresponding concentrations for "reactive" silicates are about 60 μg.-at./l. in the trench, and 28 μg.-at./l. in the open sea. In the Black Sea, over 8 μg.-at./l. of phosphate is found at 2000 m., whereas in the Mediterranean Sea and Atlantic Ocean at corresponding depths, the concentrations are less than 0·3 and approximately 1·5 μg.-at./l., respectively. Silicates are correspondingly accumulated in the Black Sea, which contains over 150 μg.-at./l. at 2000 m., whereas both the open Atlantic Ocean and Mediterranean Sea have not over 20 μg.-at./l. at 2000 m. Similar high accumulations have been observed in the Gulf of Cariaco, Saanich Inlet, Lake Nitinat, and other anoxic basins and fjords. The carbonate components of sea water are correspondingly high in anoxic waters, but the fractional increase is smaller, because of the large reservoir of carbonate present in all sea waters. In Saanich Inlet, the total carbonate-carbon varies between *c.* 1800 and 2450 μg.-at./l. and most of this range can be attributed to carbon dioxide formed *in situ* rather than to variations in the salinity of the water.

The accumulation of ammonia in anoxic waters has been referred to previously; concentrations encountered are many times higher than those found in the open sea.

(2) *Effect on combined nitrogen.* Almost all the combined inorganic nitrogen in anoxic waters is in the form of ammonium ions; nitrate and nitrite ions have not been observed in them, except near the interface where, like dissolved oxygen, they may be observed transiently. Although Hamm and Thompson (1941) have reported that nitrites and ammonium ions will react according to the equation

$$NO_2^- + NH_4^+ = N_2 + H_2O,$$

and have implied that this may proceed as a purely inorganic process, it appears more likely that nitrite ions are destroyed by bacterial denitrification.

(3) *Effect on alkalinity and pH.* The alkalinity of sea water is a measure of the quantity of anions of weak acids present in it and of the cations balanced against them (Sverdrup *et al.*, 1942, p. 195). In most sea water, the cations of weak bases, specifically ammonium ions, are present in negligible concentrations, and the only anions that need be considered are those of carbonic and boric acids. To consider the alkalinity of anoxic waters, however, the much larger concentrations of phosphate ion, silicate ion, ammonium ion, and the ions HS^- and S^{2-} must also be considered.

The titration alkalinity in most sea water can be formulated (Chapter 7, Equation 7) by the equation

$$\text{Titration alkalinity} = c_{HCO_3^-} + 2\,c_{CO_3^{2-}} + c_{H_2BO_3^-} + (c_{OH^-} - c_{H^+}).$$

For anoxic waters with high concentrations of the anions mentioned above, the equation should be extended:

$$\begin{aligned}\text{Titration alkalinity} = {} & c_{HCO_3^-} + 2\,c_{CO_3^{2-}} + c_{H_2BO_3^-} + c_{HS^-} + c_{S^{2-}} \\ & + c_{H_3SiO_4^-} + c_{H_2SiO_4^{2-}} + c_{H_2PO_4^-} + c_{HPO_4^{2-}} + c_{PO_4^{3-}} - c_{NH_4^+} \\ & + (c_{OH^-} - c_{H^+}).\end{aligned}$$

It appears that ammonium ions will tend to supply some of the ion balance required, but generally, high calcium concentrations would be expected in anoxic waters. However, there is no apparent increase in the $Ca^{2+}/Cl‰$ ratio with depth in the Black Sea, according to Skopintsev *et al.* (1958), who also found a remarkably constant value of the ratio of total anions to the chlorinity. The latter balance appears to be caused by the balancing of the decrease in sulphate/chlorinity ratio by the increase in the bicarbonate/chlorinity ratio, the ionization of H_2S being too small to add much to the total anion concentration.

Silicic and phosphoric acids and ammonia, in view of their low concentrations and dissociation constants, make only small contributions to the alkalinity. The author and his colleagues have observed high apparent values for the alkalinity in Lake Nitinat, but their investigations have so far been inconclusive. In the anoxic zone, the values of the specific alkalinity (titration alkalinity $\times$ 10^3/Cl‰) gradually increased from about 0·155 to 0·200 near the bottom. For comparison, Wattenberg reported a maximum mean value of 0·127 at 5000 m. in the Atlantic Ocean (Harvey, 1955) ,and even though specific alkalinities in the Pacific are generally higher than those in the Atlantic, the values observed in Lake Nitinat are unusual.

D. TOXICITY OF SULPHIDES

The toxicity of sulphides to most living forms results in the elimination of these organisms from sulphide-bearing waters, and restricts the populations to bacteria. Occasionally, sulphides will be mixed into strata containing organisms with resulting mass mortalities. These phenomena have been discussed by Brongersma-Sanders (1957). There are also recorded cases in which sulphide-bearing waters have been introduced into surface layers of arms of Great South Bay, Long Island, New York, to such an extent that hydrogen sulphide was introduced into the air in sufficient concentrations to blacken houses painted with white lead paints.

The elimination of organisms from sulphide-bearing waters has an important effect on the sediments, since benthic organisms are not present to ingest the remains of organisms and destroy them to the degree that they do in most marine environments. Heezen *et al.* (1959) have reported the presence of fish bones in the sediments of the Cariaco Trench, and they have also been observed in the sediments of Lake Nitinat. Twigs, fish scales, and other organic debris have been recovered from sediments from the Cariaco Trench and Saanich Inlet. Such materials are absent from deep-sea sediments.

E. REDUCTION OF SULPHATE: CHLORINITY RATIOS

One result of the reduction of sulphate is to reduce the sulphate: chlorinity ratio correspondingly. In the open oceans, this ratio has a value close to 0·1398 (Chapter 4). According to the data of Skopintsev *et al.* (1958), the ratio is somewhat higher in the upper layers of the Black Sea (0·142 at 100 m.), but decreases rather regularly with depth in the hydrogen sulphide zone (see Table III). Redfield *et al.* (1963, Fig. 6) have shown that this decrease is similar to that to be expected from the reduction of sulphates to sulphides, as estimated from the sulphide concentrations. It is evident from the stoichiometric relationships that the bulk of the sulphides of the Black Sea is formed by the reduction of sulphates.

This is contrary to the conclusions of Kriss (1963), who is of the opinion that they arise chiefly from protoplasmic sulphur. Kriss used sulphate: chlorinity ratios calculated from the data of Danil'chenko and Chigirin (1926) which do not appear to be as reliable as the more recent data of Skopintsev. However, the author does not agree with Kriss' (p. 262) conclusion "that there is no relation between the fluctuations in the relative concentration of sulphates and the hydrogen sulphide content of the Black Sea", because such a relationship is evident even from a plot of the data on which Kriss based his statement. Kriss' argument, which is based on his evaluation of populations of sulphate-reducing bacteria in the Black Sea, also ignores the stoichiometric consequences of attributing the observed accumulations of sulphides to the sulphur contained in organic compounds. As has been pointed out by Redfield *et al.* (1963), the decomposition of these organic compounds would be expected to be accompanied by the release of much larger amounts of phosphorus to the water than have been observed in any of the anoxic systems which have been studied. Danil'chenko and Chigirin (1926, referred to by Kriss, 1963, p. 258) concluded that the hydrogen sulphide in the Black Sea was derived mainly (99·4 to 99·6%) from the reduction of sulphates and only 0·4 to 0·6% was derived from the mineralization of sulphur-containing proteins. This appears to be a more accurate conclusion than that of Kriss.

ACKNOWLEDGMENTS

The author is indebted to many colleagues for stimulating discussions regarding these interesting environments. He is particularly grateful for help on the field work in Saanich Inlet and Lake Nitinat to E. V. Grill, W. W. Broenkow, C. P. Falls, L. Codispoti, J. D. Cline, D. R. Kester, and L. Atkinson, and to R. A. Kletsch, Meredythe Meller, R. W. Riley, and Edwyna von Gal for their splendid technical assistance. Part of the work was supported by the Office of Naval Research Contract Nonr-477(37) Project RR 004-03-01 and National Science Foundation Grant NSF-G 19784.

REFERENCES

Agamirov, S. S. (1963). *Geochemistry,* **1,** 104.
Barnes, C. A., and Collias, E. E. (1958). *J. Mar. Res.* **17,** 68.
Brandhorst, W. (1959). *J. Cons. int. Explor. Mer,* **25,** 3.
Broenkow, W. W. (1963). M.S. Thesis, University of Washington, Seattle.
Brongersma-Sanders, M. (1957). *In* "Treatise on Marine Ecology and Paleoecology" (J. W. Hedgpeth, ed.), Vol. I, p. 941. Memoir 67, Geological Society of America, New York.
Cameron, W. M., and Pritchard, D. W. (1963). *In* "The Sea" (M. N. Hill, ed.), Vol. II, p. 306. Interscience Publishers, New York.
Caspers, H. (1957). *In* "Treatise on Marine Ecology and Paleoecology" (J. W. Hedgpeth, ed.), Vol. I, p. 801. Memoir 67, Geological Society of America, New York.
Chow, T. J., and Thompson, T. G. (1952). *J. Mar. Res.* **11,** 124.
Danil'chenko, P. T., and Chigirin, N. I. (1926). *Tran. sta. biol. Sebastopol, Ser.* **2,** *No.* **10,** 141.
Datsko, V. G. (1951). *C.R. Acad. Sci. U.R.S.S.* **77,** 1059.
Datsko, V. G. (1955). *Gidrokhim. Mater.* **25,** 28.
Dobrzhanskaya, M. A., and Pshenina, T. I. (1958). *C.R. Acad. Sci. U.R.S.S.* **123,** 895.
Dobrzhanskaya, M. A., and Pshenina, T. I. (1959). *Tran. sta. biol. Sebastopol, Akad. Nauk S.S.S.R.* **11,** 316.
Fleming, R. H. (1940). *Proc. Pacif. Sci. Congr.* **3,** 535.
Gade, H. G. (1961). *Bol. Inst. Oceanogr. Univ. Oriente, Cumana, Venez.* **1,** 21.
Grashchenko, S. M., Nikolaev, D. S., Kolyadin, L. B., Kuznetsov, Yu. V., and Lazarev, K. F. (1960). *C.R. Acad. Sci. U.R.S.S.* **132,** 1171.
Gross, M. G., Gucluer, S. M., Creager, J. S., and Dawson, W. A. (1963). *Science,* **141,** 918.
Gucluer, S. M., and Gross, M. G. (1964). *Limnol. Oceanogr.* **9,** 359.
Hamm, R. E., and Thompson, T. G. (1941). *J. Mar. Res.* **4,** 11.
Harvey, H. W. (1955). "The Chemistry and Fertility of Sea Waters", 224 pp. Cambridge University Press, Cambridge.
Heezen, B. C., Menzies, R. J., Broecker, W. S., and Ewing, W. M. (1959). *In* "International Oceanographic Congress Preprints" (Mary Sears, ed.), p. 99. American Association for the Advancement of Science, Washington D.C.
Herlinveaux, R. H. (1962). *J. Fish. Res. Bd. Can.* **19,** 1.
Holland, H. D., and Kulp, J. L. (1954). *Geochim. et cosmoch. Acta,* **5,** 197.
Ketchum, B. H., Corwin, N., and Keen, D. J. (1955). *Deep Sea Res.* **2,** 172.
Kolyadin, L. B., Nikolaev, D. S., Grashchenko, S. M., Kuznetsov, Yu. V., and Lazarev, K. F. (1960). *C.R. Acad. Sci. U.R.S.S.* **132,** 915.
Krauskopf, K. B. (1956). *Geochim. et cosmoch. Acta,* **9,** 1.
Kriss, A. E. (1963). "Marine Microbiology (Deep Sea)" (translated by Shewan, J. M., and Kabata, Z.). Oliver and Boyd, Edinburgh and London.
Lambet, M. S., and Nikolaev, D. S. (1962). *C.R. Acad. Sci. U.R.S.S.* **142,** 681.
Lange, N. A. (ed.) (1956). "Handbook of Chemistry", 9th ed. 1969 pp. Handbook Publishers, Inc., Sandusky, Ohio.
Lewis, G. J., Jr., and Goldberg, E. D. (1954). *J. Mar. Res.* **13,** 183.
McKinney, R. E., and Conway, R. A. (1957). *Sewage industr. Wastes,* **29,** 1097.
Naumov, G. B. (1959). *Geochemistry,* **1,** 5.

Nikolaev, D. S., Korn, O. P., Lazarev, K. F., Kolyadin, L. B., Kuznetsov, Yu. V., and Grashchenko, S. M. (1960). *C.R. Acad. Sci. U.R.S.S.* **132**, 1411.

Nikolaev, D. S., Lazarev, K. F., Grashchenko, S. M. (1961). *C.R. Acad. Sci. U.R.S.S.* **138**, 674.

Pickard, G. L. (1963). *J. Fish. Res. Bd. Can.* **20**, 1109.

Rakestraw, N. W., and Emmel, E. V. M. (1938a). *J. Phys. Chem.* **42**, 1211.

Rakestraw, N. W., and Emmel, E. V. M. (1938b). *J. Mar. Res.* **1**, 207.

Redfield, A. C. (1942). *Pap. phys. Oceanogr.* **9**, 22 pp.

Redfield, A. C., Ketchum, B. H., and Richards, F. A. (1963). *In* "The Sea," (M. N. Hill, ed.), Vol. II, p. 26. Interscience Publishers, New York.

Richards, F. A. (1957). *In* "Progress in Physics and Chemistry of the Earth", Vol. II, p. 77. Pergamon Press, New York.

Richards, F. A. (1960). *Deep Sea Res.* **7**, 163.

Richards, F. A. (1964). *Proc. Second Intern. Conf. Water Pollution Res. 1964* In the press.

Richards, F. A., and Benson, B. B. (1961). *Deep Sea Res.* **7**, 254.

Richards, F. A., and Vaccaro, R. F. (1956). *Deep Sea Res.* **3**, 214.

Schmidt, Johannes. (1925). *Science*, **61**, 592.

Skopintsev, B. A. (1957). *Gidrokhim. Mater.* **27**, 21.

Skopintsev, B. A., and Gubin, F. A. (1955). *Gidrokhim. Mater.* **25**, 16.

Skopintsev, B. A., and Popova, T. P. (1960). *Trudy Morskogo. gidrofiz. Inst., Akad. Nauk S.S.S.R.* **19**, 21.

Skopintsev, B. A., and Timofeeva, S. N. (1960). *C.R. Acad. Sci. U.R.S.S.* **134**, 688.

Skopintsev, B. A., and Timofeeva, S. N. (1961). *Pervichnaya Produktsiya Morei i Vnutr. Vod. 1961*, 60.

Skopintsev, B. A., Karpov, A. V., and Vershinina, O. A. (1961). *Gidrokhim. Mater.* **31**, 127.

Skopintsev, B. A., Gubin, F. A., Vorob'eva, R. V., and Verschinina, O. A. (1958). *C.R. Acad. Sci. U.R.S.S.* **119**, 121.

Strøm, K. M. (1936). *Skr. norske Vidensk. Akad.* **7**, 88 pp.

Strøm, Kaare (1957). *Nature, Lond.* **180**, 982.

Strøm, Kaare (1961). *Nature, Lond.* **189**, 913.

Sugimura, Y., and Sugimura, T. (1962). *Nature, Lond.* **194**, 568.

Sverdrup, H. U., Johnson, M. W., and Fleming, R. H. (1942). "The Oceans", 1087 pp. Prentice Hall, New York.

Wheatland, A. B. (1954). *J. Hyg. Camb.* **52**, 194.

Williams, P. M., Mathews, W. H., and Pickard, G. L. (1961). *Nature, Lond.* **191**, 830.

Appendix

TABLE I

Concentration of the major ions in sea water of various salinities

Salinity (‰)	Concentration in g/kg. of sea water										
	Na^+	Mg^{2+}	Ca^{2+}	K^+	Sr^{2+}	B	Cl^-	SO_4^{2-}	Br^-	F^-	HCO_3^-
10	3·074	0·370	0·118	0·111	0·002	0·001	5·530	0·775	0·019	0·0004	0·041
15	4·611	0·555	0·177	0·167	0·003	0·002	8·294	1·162	0·029	0·0006	0·061
20	6·148	0·739	0·236	0·221	0·004	0·002	11·059	1·550	0·038	0·0008	0·081
25	7·685	0·924	0·295	0·277	0·006	0·003	13·824	1·937	0·048	0·0009	0·101
30	9·222	1·109	0·354	0·332	0·007	0·003	16·589	2·324	0·057	0·0011	0·122
31	9·529	1·146	0·366	0·343	0·007	0·003	17·142	2·402	0·059	0·0012	0·126
32	9·837	1·183	0·377	0·354	0·007	0·004	17·695	2·479	0·061	0·0012	0·130
33	10·144	1·220	0·389	0·365	0·008	0·004	18·248	2·557	0·063	0·0013	0·134
34	10·452	1·257	0·401	0·376	0·008	0·004	18·801	2·634	0·065	0·0013	0·138
35	10·759	1·294	0·413	0·387	0·008	0·004	19·354	2·712	0·067	0·0013	0·142
36	11·066	1·331	0·424	0·398	0·008	0·004	19·907	2·789	0·069	0·0014	0·146
37	11·374	1·368	0·436	0·409	0·008	0·004	20·460	2·867	0·071	0·0014	0·150
38	11·681	1·405	0·448	0·420	0·009	0·004	21·012	2·944	0·073	0·0014	0·154
39	11·989	1·442	0·460	0·431	0·009	0·004	21·565	3·022	0·075	0·0015	0·158
40	12·296	1·479	0·472	0·442	0·009	0·005	22·118	3·099	0·077	0·0015	0·162
41	12·603	1·516	0·483	0·454	0·009	0·005	22·671	3·177	0·078	0·0016	0·166

Table II

Preparation of artificial sea water (Cl = 19·00‰)

Lyman and Fleming (1940)*	(g.)	Kalle (1945)†	(g.)
NaCl	23·477	NaCl	28·014
$MgCl_2$	4·981	$MgCl_2$	3·812
Na_2SO_4	3·917	$MgSO_4$	1·752
$CaCl_2$	1·102	$CaSO_4$	1·283
KCl	0·664	K_2SO_4	0·8163
$NaHCO_3$	0·192	$CaCO_3$	0·1221
KBr	0·096	KBr	0·1013
H_3BO_3	0·026	$SrSO_4$	0·02824
$SrCl_2$	0·024	H_3BO_3	0·02768
NaF	0·003		
Water to	1000	Water to	1000

Note. (i) Allowance must be made for water of crystallization of any of the salts used. (ii) After aeration the pH should lie between 7·9 and 8·3.

* *J. Mar. Res.* **3**, 134.

† *In* "Probleme der kosmischen Physik", 2nd edn., Vol. 23. Leipzig.

Table III

First apparent dissociation constant of boric acid in sea water (expressed as* pK'_B*) (Lyman, 1956) (see p. 251)*

				t°C				
Cl(‰)	*0°*	*5°*	*10°*	*15°*	*20°*	*25°*	*30°*	*35°*
0	9·50	9·44	9·38	9·33	9·28	9·24	9·20	9·16
1	9·40	9·34	9·28	9·23	9·18	9·14	9·10	9·06
4	9·28	9·22	9·16	9·11	9·06	9·02	8·98	8·94
9	9·14	9·08	9·03	8·98	8·93	8·88	8·85	8·82
16	9·00	8·95	8·89	8·84	8·80	8·76	8·72	8·69
17	8·98	8·93	8·88	8·83	8·78	8·74	8·70	8·67
18	8·96	8·91	8·86	8·81	8·76	8·72	8·69	8·66
19	8·95	8·90	8·85	8·80	8·75	8·71	8·67	8·64
20	8·94	8·88	8·83	8·78	8·74	8·69	8·65	8·63
21	8·92	8·87	8·82	8·77	8·72	8·68	8·64	8·61
25	8·85	8·80	8·75	8·70	8·66	8·62	8·59	8·56
36	8·71	8·66	8·61	8·57	8·53	8·49	8·46	8·43
49	8·56	8·52	8·47	8·43	$8·39^5$	8·36	8·33	8·30
64	8·41	8·37	8·33	8·30	$8·26^5$	8·23	8·20	8·17

* Values based on the N.B.S. pH scale.

Table IV

Second apparent dissociation constant of boric acid in sea water (expressed as* pK'_{2B}*) (Lyman, 1956)* *(see p. 251)*

				t°C				
Cl(‰)	*0°*	*5°*	*10°*	*15°*	*20°*	*25°*	*30°*	*35°*
16	11·17	10·91	10·65	10·40	10·16	9·97	9·80	9·64
17	11·14	10·88	10·63	10·38	10·13	9·94	9·78	9·62
18	11·11	10·85	10·60	10·35	10·10	9·92	9·75	9·60
19	11·08	10·82	10·57	10·32	10·08	9·89	9·73	9·58
20	11·05	10·79	10·54	10·30	10·05	9·87	9·71	9·57
21	11·01	10·76	10·52	10·27	10·03	9·85	9·69	9·55
25	10·92	10·68	10·43	10·18	9·96	9·78	9·63	9·49
36	10·71	10·46	10·23	10·01	9·80	9·62	9·48	9·35
49	10·52	10·28	10·04	9·84	9·65	9·48	9·34	9·23
64	10·35	10·10	9·88	9·69	9·52	9·36	9·23	9·11

* Values based on the N.B.S. pH scale.

TABLE V

*Solubility of carbon dioxide in pure water, α_0 and in sea waters, α_s, in moles l./atm** (see p. 256)

Cl‰	$t°(C)$ 0	2	4	6	8	10	12	14	16	18	20	22	24	26	28	30
0	$\alpha_0 = 770$	712	662	619	576	536	502	472	442	417	394	372	351	332	314	299×10^{-4}
15	$\alpha_s = 674$	623	578	538	504	472	442	416	393	371	351	331	314	299	284	270
16	667	617	573	533	499	468	438	413	390	368	348	329	312	297	281	268
17	660	611	567	528	495	464	434	410	387	365	346	327	310	294	279	266
18	653	605	562	524	490	460	431	406	384	362	343	324	307	292	277	264
19	646	599	557	519	486	456	428	403	381	359	340	321	304	289	275	262
20	640	593	551	514	482	452	424	400	377	356	337	319	302	287	273	260
21	633	587	546	509	477	448	421	396	374	354	335	317	300	285	271	258

See Chapter 7 and Lyman (1956).

* From Bohr's data (Buch, 1951).

TABLE VI

First apparent dissociation constant of carbonic acid in sea water (expressed as* pK'_{L1}*) (Lyman, 1956) (see p. 261)*

Cl(‰)	t(°C) 0°	5°	10°	15°	20°	25°	30°	35°
0	6·58	6·52	6·47	6·42	6·38	6·35	6·33	6·31
1	6·47	6·42	6·37	6·33	6·29	6·26	6·24	6·23
4	6·36	6·32	6·28	6·24	6·21	6·18	6·16	6·15
9	6·27	6·23	6·19	6·15	6·13	6·10	6·08	6·07
16	6·18	6·14	6·11	6·07	6·05	6·03	6·01	5·99
17	6·17	6·13	6·10	6·06	6·04	6·02	6·00	5·98
18	6·16	6·12	6·09	6·06	6·03	6·01	5·99	5·97
19	6·15	6·11	6·08	6·05	6·02	6·00	5·98	5·97
20	6·14	6·10	6·07	6·04	6·01	5·99	5·97	5·96
21	6·13	6·09	6·06	6·03	6·00	5·98	5·96	5·95
25	6·09	6·05	6·02	6·00	5·97	5·95	5·93	5·92
36	6·00	5·97	5·94	5·92	5·89	5·87	5·86	5·84
49	5·92	5·88	5·86	5·84	5·82	5·80	5·78	5·77
64	5·84	5·80	5·78	5·76	5·74	5·72	5·71	5·70

* Values based on the N.B.S. pH scale.

TABLE VII

Second apparent dissociation constant of carbonic acid in sea water (expressed as* pK'_2*) (Lyman, 1956) (see p. 261)*

Cl(‰)	t(°C) 0°	5°	10°	15°	20°	25°	30°	35°
0	10·62	10·55	10·49	10·43	10·38	10·33	10·29	10·25
1	10·06	9·99	9·93	9·87	9·81	9·76	9·71	9·66
4	9·78	9·72	9·67	9·61	9·54	9·49	9·43	9·38
9	9·64	9·58	9·52	9·46	9·40	9·34	9·27	9·21
16	9·46	9·40	9·35	9·29	9·23	9·17	9·10	9·02
17	9·44	9·38	9·32	9·27	9·21	9·15	9·08	9·00
18	9·42	9·36	9·30	9·25	9·19	9·12	9·06	8·98
19	9·40	9·34	9·28	9·23	9·17	9·10	9·02	8·95
20	9·38	9·32	9·26	9·21	9·15	9·08	9·01	8·92
21	9·36	9·30	9·25	9·19	9·13	9·06	8·98	8·89
25	9·29	9·23	9·17	9·11	9·05	8·98	8·91	8·82
36	9·12	9·06	8·99	8·93	8·86	8·79	8·72	8·63
49	8·95	8·89	8·82	8·75	8·68	8·61	8·53	8·43
64	8·77	8·71	8·64	8·57	8·50	8·42	8·34	8·23

* Values based on the N.B.S. pH scale.

TABLE VIII

Percentage molar fraction of free carbonic acid (F), bicarbonate ion (B) and carbonate ion (C) at different temperatures and pH values. For fresh water (Saruhashi, 1953) (see p. 269)

pH	t(°C) 0			2			4			6		
	F	B	C	F	B	C	F	B	C	F	B	C
5·0	97·5	2·5	0	97·3	2·7	0	97·2	2·8	0	97·0	3·0	0
5·1	97·0	3·0	0	96·6	3·4	0	96·5	3·5	0	96·3	3·7	0
5·2	96·0	4·0	0	95·8	4·2	0	95·6	4·4	0	95·4	4·6	0
5·3	95·0	5·0	0	94·7	5·3	0	94·5	5·5	0	94·3	5·7	0
5·4	93·9	6·1	0	93·6	6·4	0	93·3	6·7	0	92·9	7·1	0
5·5	92·5	7·5	0	92·0	8·0	0	91·7	8·3	0	91·3	8·7	0
5·6	90·5	9·5	0	90·1	9·9	0	89·5	10·5	0	89·0	11·0	0
5·7	88·5	11·5	0	88·0	12·0	0	87·5	12·5	0	86·5	13·5	0
5·8	85·9	14·1	0	85·5	14·5	0	84·9	15·1	0	84·0	16·0	0
5·9	83·0	17·0	0	82·0	18·0	0	81·5	18·5	0	80·8	19·2	0
6·0	79·4	20·6	0	78·1	21·9	0	77·6	22·4	0	76·3	23·7	0
6·1	75·5	24·5	0	73·8	26·2	0	73·2	26·8	0	72·0	28·0	0
6·2	70·9	29·1	0	69·5	30·5	0	68·6	31·4	0	67·6	32·4	0
6·3	66·0	34·0	0	64·3	35·7	0	63·2	36·8	0	62·0	38·0	0
6·4	60·6	39·4	0	59·0	41·0	0	57·9	42·1	0	56·5	43·5	0
6·5	55·5	44·5	0	53·5	46·5	0	52·0	48·0	0	50·8	49·2	0
6·6	49·0	51·0	0	47·5	52·5	0	46·3	53·7	0	44·9	55·1	0
6·7	43·5	56·5	0	41·5	58·5	0	40·4	59·6	0	39·5	60·5	0
6·8	37·5	62·5	0	36·0	64·0	0	34·9	65·1	0	33·8	66·2	0
6·9	32·5	67·5	0	30·8	69·2	0	29·8	70·2	0	29·0	71·0	0
7·0	27·8	72·2	0	26·2	73·8	0	25·2	74·8	0	24·4	75·6	0
7·1	23·0	77·0	0	22·0	78·0	0	21·2	78·8	0	20·4	79·6	0
7·2	19·5	80·5	0	18·5	81·5	0	17·9	82·1	0	17·2	82·8	0
7·3	16·5	83·5	0	15·5	84·5	0	14·8	85·2	0	14·0	86·0	0
7·4	13·5	86·5	0	12·8	87·2	0	12·1	87·9	0	11·5	88·5	0

7·5	10·5	89·5	0	10·0	90·0	0	9·5	90·5	0	9·0	91·0	0
7·6	8·4	91·5	0·1	7·9	92·0	0·1	7·7	92·2	0·1	7·5	92·4	0·1
7·7	6·9	93·0	0·1	6·1	93·8	0·1	6·0	93·9	0·1	5·9	94·1	0·1
7·8	5·4	94·5	0·1	5·1	94·8	0·1	4·9	95·0	0·1	4·8	95·0	0·2
7·9	4·3	95·5	0·2	4·3	95·5	0·2	4·2	95·6	0·2	4·0	95·8	0·2
8·0	3·8	96·0	0·2	3·7	96·1	0·2	3·6	96·1	0·3	3·4	96·3	0·3
8·1	3·3	96·5	0·2	3·0	96·7	0·3	2·8	96·9	0·3	2·6	97·0	0·4
8·2	2·7	97·0	0·3	2·3	97·3	0·4	2·1	97·5	0·4	2·0	97·5	0·5
8·3	2·1	97·5	0·4	1·8	97·7	0·5	1·7	97·7	0·6	1·6	97·8	0·6
8·4	1·5	98·0	0·5	1·4	98·0	0·6	1·3	98·0	0·7	1·3	98·0	0·7
8·5	1·2	98·1	0·7	1·1	98·1	0·8	1·1	98·0	0·9	1·0	98·0	1·0
8·6	0·9	98·2	0·9	0·9	98·1	1·0	0·9	98·1	1·0	0·8	98·1	1·1
8·7	0·6	98·2	1·2	0·6	98·1	1·3	0·6	98·0	1·4	0·6	98·0	1·4
8·8	0·4	98·2	1·4	0·4	98·0	1·6	0·4	97·9	1·7	0·4	97·8	1·8
8·9	0·3	98·0	1·7	0·4	97·7	1·9	0·4	97·5	2·1	0·4	97·3	2·3
9·0	0·3	97·5	2·2	0·4	97·3	2·3	0·3	97·1	2·6	0·3	96·9	2·8
9·1	0·2	97·0	2·8	0·3	96·7	3·0	0·3	96·4	3·3	0·3	96·2	3·5
9·2	0·2	96·3	3·5	0·2	96·1	3·7	0·2	95·6	4·1	0·2	95·4	4·4
9·3	0·2	95·5	4·3	0·2	95·1	4·7	0·2	94·6	5·2	0·2	94·5	5·3
9·4	0·1	94·5	5·4	0·1	94·1	5·8	0·1	93·7	6·2	0·1	93·3	6·6
9·5	0·1	93·4	6·5	0·1	92·6	7·3	0·1	92·1	7·8	0·1	91·5	8·4
9·6	0·1	91·5	8·4	0·1	91·0	8·9	0·1	90·2	9·7	0·1	89·6	10·3
9·7	0	89·5	10·5	0	88·6	11·4	0	88·0	12·0	0	87·0	13·0
9·8	0	87·2	12·8	0	86·3	13·7	0	85·5	14·5	0	84·7	15·3
9·9	0	84·5	15·5	0	83·3	16·7	0	82·8	17·2	0	81·0	19·0
10·0	0	81·3	18·7	0	80·0	20·0	0	78·8	21·2	0	77·5	22·5
10·1	0	78·0	22·0	0	75·5	24·5	0	74·8	25·2	0	72·9	27·1
10·2	0	73·5	26·5	0	71·5	28·5	0	70·1	29·9	0	68·2	31·8
10·3	0	68·8	31·2	0	66·0	34·0	0	65·5	34·5	0	63·0	37·0
10·4	0	63·2	36·8	0	61·4	38·6	0	59·7	40·3	0	57·7	42·3

TABLE VIII—*continued*

pH	t(°C) 8 F	B	C	10 F	B	C	12 F	B	C	14 F	B	C
5·0	96·9	3·1	0	96·7	3·3	0	96·6	3·4	0	96·5	3·5	0
5·1	96·2	3·8	0	96·0	4·0	0	95·8	4·2	0	95·7	4·3	0
5·2	95·2	4·8	0	94·9	5·1	0	94·7	5·3	0	94·6	5·4	0
5·3	94·0	6·0	0	93·8	6·2	0	93·5	6·5	0	93·4	6·6	0
5·4	92·5	7·5	0	92·1	7·9	0	91·8	8·2	0	91·5	8·5	0
5·5	91·0	9·0	0	90·5	9·5	0	90·0	10·0	0	89·7	10·3	0
5·6	88·5	11·5	0	88·1	11·9	0	87·8	12·2	0	87·4	12·6	0
5·7	86·0	14·0	0	85·5	14·5	0	85·0	15·0	0	84·5	15·5	0
5·8	83·4	16·6	0	82·7	17·3	0	82·0	18·0	0	81·4	18·6	0
5·9	80·0	20·0	0	78·5	21·5	0	78·5	21·5	0	77·5	22·5	0
6·0	75·8	24·2	0	74·6	25·4	0	73·6	26·4	0	73·0	27·0	0
6·1	71·5	28·5	0	69·5	30·5	0	69·1	30·9	0	68·7	31·3	0
6·2	66·2	33·8	0	64·9	35·1	0	64·2	35·8	0	63·3	36·7	0
6·3	61·5	38·5	0	59·5	40·5	0	58·5	41·5	0	57·5	42·5	0
6·4	55·3	44·7	0	53·8	46·2	0	52·9	47·1	0	51·8	48·2	0
6·5	49·7	50·3	0	48·0	52·0	0	47·0	53·0	0	46·0	54·0	0
6·6	43·9	56·1	0	42·6	57·4	0	41·7	58·3	0	40·5	59·5	0
6·7	38·5	61·5	0	36·8	63·2	0	35·8	64·2	0	35·0	65·0	0
6·8	32·8	67·2	0	31·8	68·2	0	31·1	68·9	0	30·0	70·0	0
6·9	28·0	72·0	0	27·0	73·0	0	26·5	73·5	0	25·5	74·5	0
7·0	23·6	76·4	0	22·7	77·3	0	22·0	78·0	0	21·3	78·7	0
7·1	19·8	80·2	0	19·0	81·0	0	18·2	81·8	0	18·0	82·0	0
7·2	16·4	83·6	0	15·7	84·3	0	15·2	84·8	0	14·7	85·3	0
7·3	13·3	86·7	0	12·8	87·2	0	12·4	87·6	0	12·0	88·0	0
7·4	11·0	89·0	0	10·4	89·6	0	10·1	89·9	0	9·6	90·4	0

7·5	8·8	91·2	0	8·6	91·4	0	8·3	91·6	0·1	8·0	91·9	0·1
7·6	7·3	92·6	0·1	6·9	93·0	0·1	6·7	93·2	0·1	6·4	93·5	0·1
7·7	5·5	94·3	0·2	5·4	94·4	0·2	5·3	94·5	0·2	5·1	94·8	0·1
7·8	4·7	95·1	0·2	4·5	95·3	0·2	4·3	95·5	0·2	4·1	95·7	0·2
7·9	3·7	96·0	0·3	3·6	96·1	0·3	3·5	96·2	0·3	3·4	96·4	0·2
8·0	3·0	96·7	0·3	2·9	96·8	0·3	2·8	96·9	0·3	2·7	97·0	0·3
8·1	2·5	97·1	0·4	2·4	97·2	0·4	2·3	97·3	0·4	2·2	97·4	0·4
8·2	1·9	97·6	0·5	1·9	97·6	0·5	1·8	97·7	0·5	1·7	97·8	0·5
8·3	1·5	97·9	0·6	1·5	97·9	0·7	1·4	97·9	0·7	1·4	97·9	0·7
8·4	1·2	98·1	0·7	1·1	98·1	0·8	1·1	98·1	0·8	1·1	98·0	0·9
8·5	1·0	98·1	0·9	0·9	98·1	1·0	0·9	98·0	1·1	0·8	98·0	1·2
8·6	0·8	98·1	1·1	0·7	98·0	1·3	0·7	98·0	1·3	0·6	98·0	1·4
8·7	0·6	97·9	1·5	0·6	97·8	1·7	0·5	97·8	1·7	0·5	97·7	1·8
8·8	0·4	97·7	1·9	0·4	97·6	2·0	0·4	97·5	2·1	0·4	97·4	2·2
8·9	0·4	97·2	2·4	0·4	97·0	2·6	0·4	97·0	2·6	0·4	96·8	2·8
9·0	0·3	96·7	3·0	0·3	96·6	3·1	0·3	96·5	3·2	0·3	96·3	3·4
9·1	0·3	96·0	3·7	0·2	95·9	3·9	0·2	95·7	4·1	0·3	95·3	4·4
9·2	0·2	95·3	4·5	0·1	95·1	4·8	0·1	94·8	5·1	0·2	94·4	5·4
9·3	0·2	94·2	5·6	0·1	93·8	6·1	0·1	93·5	6·4	0·2	92·9	6·9
9·4	0·1	93·0	6·9	0·1	92·5	7·4	0·1	92·1	7·8	0·1	91·5	8·4
9·5	0·1	90·9	9·0	0	90·6	9·4	0	90·2	9·8	0	89·5	10·5
9·6	0	89·4	10·5	0	88·6	11·4	0	88·2	11·8	0	87·5	12·5
9·7	0	86·7	13·3	0	85·9	14·1	0	85·2	14·8	0	84·5	15·5
9·8	0	84·0	16·0	0	83·1	16·9	0	82·3	17·7	0	81·5	18·5
9·9	0	80·5	19·5	0	79·4	20·6	0	78·6	21·4	0	77·5	22·5
10·0	0	76·9	23·1	0	75·7	24·3	0	74·9	25·1	0	73·5	26·5
10·1	0	72·5	27·5	0	71·1	28·9	0	70·2	29·8	0	68·5	31·5
10·2	0	68·0	32·0	0	66·5	33·5	0	65·5	34·5	0	63·5	36·5
10·3	0	62·0	38·0	0	61·0	39·0	0	60·0	40·0	0	57·8	42·2
10·4	0	57·3	42·7	0	55·5	44·5	0	54·2	45·8	0	52·1	47·9

TABLE VIII—*continued*

pH	t(°C) 16			18			20			22		
	F	*B*	*C*	*F*	*B*	*C*	*F*	*B*	*C*	*F*	*B*	*C*
5·0	96·4	3·6	0	96·3	3·7	0	96·2	3·8	0	96·0	4·0	0
5·1	95·6	4·4	0	95·5	4·5	0	95·3	4·7	0	95·0	5·0	0
5·2	94·4	5·6	0	94·2	5·8	0	94·1	5·9	0	93·9	6·1	0
5·3	93·2	6·8	0	93·0	7·0	0	92·7	7·3	0	92·4	7·6	0
5·4	91·3	8·7	0	91·1	8·9	0	90·7	9·3	0	90·5	9·5	0
5·5	89·5	10·5	0	89·3	10·7	0	88·9	11·1	0	88·5	11·5	0
5·6	87·0	13·0	0	86·6	13·4	0	86·2	13·8	0	85·9	14·1	0
5·7	84·0	16·0	0	83·5	16·5	0	83·0	17·0	0	82·9	17·1	0
5·8	80·7	19·3	0	80·3	19·7	0	79·8	20·2	0	79·4	20·6	0
5·9	77·2	22·8	0	76·6	23·4	0	76·0	24·0	0	75·0	25·0	0
6·0	72·5	27·5	0	72·0	28·0	0	71·5	28·5	0	70·4	29·6	0
6·1	68·0	32·0	0	67·3	32·7	0	66·6	33·4	0	65·8	34·2	0
6·2	62·5	37·5	0	61·9	38·1	0	61·4	38·6	0	60·6	39·4	0
6·3	56·9	43·1	0	56·3	43·7	0	55·7	44·3	0	54·8	45·2	0
6·4	51·0	49·0	0	50·0	50·0	0	49·3	50·7	0	48·6	51·4	0
6·5	45·0	55·0	0	44·5	55·5	0	43·8	56·2	0	42·8	57·2	0
6·6	39·8	60·2	0	39·2	60·8	0	38·6	61·4	0	37·6	62·4	0
6·7	34·5	65·5	0	33·8	66·2	0	33·3	66·7	0	32·7	67·3	0
6·8	29·4	70·6	0	28·8	71·2	0	28·3	71·7	0	27·8	72·2	0
6·9	25·0	75·0	0	24·5	75·5	0	24·0	76·0	0	23·5	76·5	0
7·0	20·8	79·2	0	20·4	79·6	0	20·0	80·0	0	19·4	80·6	0
7·1	17·5	82·5	0	17·0	83·0	0	16·7	83·3	0	16·3	83·7	0
7·2	14·3	85·7	0	14·0	86·0	0	13·7	86·3	0	13·3	86·7	0
7·3	11·8	88·2	0	11·5	88·5	0	11·2	88·8	0	10·9	89·1	0
7·4	9·4	90·5	0·1	9·1	90·8	0·1	8·8	91·1	0·1	8·6	91·3	0·1

7·5	7·8	92·7	0·1	7·5	92·4	0·1	7·2	92·6	0·2	7·0	92·8	0·2
7·6	6·2	93·7	0·1	6·0	93·8	0·2	5·9	93·9	0·2	5·6	94·2	0·2
7·7	5·0	94·8	0·2	4·8	94·9	0·3	4·7	95·0	0·3	4·6	95·1	0·3
7·8	4·0	95·8	0·2	3·8	95·9	0·3	3·7	96·0	0·3	3·6	96·1	0·3
7·9	3·3	96·4	0·3	3·1	96·5	0·4	3·0	96·6	0·4	3·0	96·6	0·4
8·0	2·6	97·1	0·3	2·5	97·1	0·4	2·4	97·2	0·4	2·4	97·2	0·4
8·1	2·1	97·4	0·5	2·0	97·5	0·5	2·0	97·5	0·5	1·9	97·6	0·5
8·2	1·7	97·7	0·6	1·6	97·8	0·6	1·6	97·8	0·6	1·5	97·8	0·7
8·3	1·3	97·9	0·8	1·2	98·0	0·8	1·2	98·0	0·8	1·2	97·9	0·9
8·4	1·0	98·1	0·9	1·0	98·0	1·0	1·0	98·0	1·0	0·9	98·0	1·1
8·5	0·8	98·0	1·2	0·7	98·0	1·3	0·7	97·9	1·4	0·7	97·8	1·5
8·6	0·6	97·9	1·5	0·6	97·8	1·6	0·6	97·7	1·7	0·5	97·7	1·8
8·7	0·5	97·6	1·9	0·5	97·4	2·1	0·5	97·3	2·2	0·5	97·2	2·3
8·8	0·4	97·3	2·3	0·4	97·1	2·5	0·4	97·0	2·6	0·4	96·9	2·7
8·9	0·4	96·6	3·0	0·4	96·4	3·2	0·4	96·3	3·3	0·4	96·2	3·5
9·0	0·3	96·1	3·6	0·3	95·9	3·8	0·3	95·7	4·0	0·3	95·4	4·3
9·1	0·3	95·1	4·6	0·3	94·9	4·8	0·3	94·7	5·0	0·3	94·3	5·4
9·2	0·2	94·2	5·6	0·2	94·0	5·8	0·2	93·8	6·0	0·2	93·4	6·4
9·3	0·2	92·7	7·1	0·2	92·3	7·5	0·2	92·2	7·6	0·2	91·8	8·0
9·4	0·1	91·3	8·6	0·1	90·7	9·2	0·1	90·6	9·3	0·1	90·0	9·9
9·5	0	89·2	10·8	0	88·5	11·5	0	88·2	11·8	0	87·8	12·2
9·6	0	87·0	13·0	0	86·3	13·7	0	85·6	14·4	0	85·0	15·0
9·7	0	83·8	16·2	0	83·0	17·0	0	82·4	17·6	0	81·8	18·2
9·8	0	80·6	19·4	0	79·8	20·2	0	79·1	20·9	0	78·0	22·0
9·9	0	76·5	23·5	0	75·6	24·4	0	74·5	25·5	0	73·5	26·5
10·0	0	72·5	27·5	0	71·4	28·6	0	70·6	29·4	0	69·4	30·6
10·1	0	67·5	32·5	0	66·6	33·4	0	65·6	34·4	0	64·6	35·4
10·2	0	62·5	37·5	0	61·9	38·1	0	61·2	38·8	0	58·7	41·6
10·3	0	56·9	43·1	0	55·8	44·2	0	54·6	45·4	0	53·5	46·5
10·4	0	51·3	48·7	0	50·2	49·8	0	49·3	50·7	0	47·5	52·5

TABLE VIII—*continued*

pH	t(°C) 24			26			28			30		
	F	*B*	*C*	*F*	*B*	*C*	*F*	*B*	*C*	*F*	*B*	*C*
5·0	95·9	4·1	0	95·8	4·2	0	95·8	4·2	0	95·7	4·3	0
5·1	94·9	5·1	0	94·8	5·2	0	94·8	5·2	0	94·7	5·3	0
5·2	93·7	6·3	0	93·6	6·4	0	93·5	6·5	0	93·4	6·6	0
5·3	92·1	7·9	0	92·0	8·0	0	91·9	8·1	0	91·8	8·2	0
5·4	90·3	9·7	0	90·1	9·9	0	90·0	10·0	0	89·9	10·1	0
5·5	88·2	11·8	0	87·9	12·1	0	87·7	12·3	0	87·5	12·5	0
5·6	85·5	14·5	0	85·1	14·9	0	85·0	15·0	0	84·8	15·2	0
5·7	82·5	17·5	0	82·2	17·8	0	82·0	18·0	0	81·5	18·5	0
5·8	79·1	20·9	0	78·6	21·4	0	78·3	21·7	0	77·5	22·5	0
5·9	74·7	25·3	0	74·3	25·7	0	74·0	26·0	0	73·5	26·5	0
6·0	70·0	30·0	0	69·5	30·5	0	69·2	30·8	0	69·0	31·0	0
6·1	65·0	35·0	0	64·2	35·8	0	64·0	36·0	0	63·8	36·2	0
6·2	59·9	40·1	0	59·2	40·8	0	58·9	41·1	0	58·5	41·5	0
6·3	54·1	45·9	0	53·2	46·8	0	53·0	47·0	0	52·7	47·3	0
6·4	48·1	51·9	0	47·6	52·4	0	47·2	52·8	0	47·0	53·0	0
6·5	42·4	57·6	0	42·0	58·0	0	41·5	58·5	0	41·2	58·8	0
6·6	36·9	63·1	0	36·4	63·4	0	36·1	63·9	0	35·9	64·1	0
6·7	31·8	68·2	0	31·3	68·7	0	31·0	69·0	0	30·5	69·5	0
6·8	27·4	72·6	0	26·9	73·1	0	26·5	73·5	0	26·0	74·0	0
6·9	23·0	77·0	0	22·3	77·7	0	22·2	77·8	0	21·8	78·2	0
7·0	18·9	81·1	0	18·5	81·5	0	18·4	81·6	0	18·2	81·8	0
7·1	15·8	84·2	0	15·3	84·7	0	15·0	85·0	0	14·8	85·2	0
7·2	13·0	87·0	0	12·7	87·3	0	12·5	87·5	0	12·3	87·7	0
7·3	10·6	89·4	0	10·3	89·7	0	10·1	89·9	0	10·0	90·0	0
7·4	8·4	91·5	0·1	8·3	91·6	0·1	8·2	91·7	0·1	8·1	91·8	0·1

7·5	6·9	92·9	0·2	6·8	93·0	0·2	6·6	93·2	0·2	6·6	93·2	0·2
7·6	5·4	94·4	0·2	5·4	94·4	0·2	5·3	94·5	0·2	5·3	94·5	0·2
7·7	4·5	95·2	0·3	4·4	95·3	0·3	4·4	95·3	0·3	4·4	95·3	0·3
7·8	3·6	96·1	0·3	3·5	96·2	0 3	3·5	96·2	0·3	3·5	96·2	0·3
7·9	2·9	96·7	0·4	2·9	96·7	0·4	2·9	96·7	0·4	2·8	96·8	0·4
8·0	2·3	97·3	0·4	2·2	97·3	0·5	2·2	97·3	0·5	2·1	97·4	0·5
8·1	1·9	97·6	0·5	1·8	97·6	0·6	1·8	97·6	0·6	1·7	97·7	0·6
8·2	1·5	97·8	0·7	1·4	97·9	0·7	1·4	97·9	0·7	1·4	97·8	0·8
8·3	1·2	97·9	0·9	1·1	98·0	0·9	1·1	98·0	0·9	1·1	97·9	1·0
8·4	0·9	98·0	1·1	0·8	98·0	1·2	0·8	98·0	1·2	0·8	97·9	1·3
8·5	0·6	97·9	1·5	0·6	97·9	1·5	0·7	97·8	1·5	0·7	97·7	1·6
8·6	0·5	97·6	1·9	0·5	97·6	1·9	0·6	97·5	1·9	0·5	97·5	2·0
8·7	0·5	97·2	2·3	0·5	97·1	2·4	0·5	97·1	2·4	0·5	97·1	2·4
8·8	0·4	96·7	2·9	0·4	96·6	3·0	0·4	96·5	3·1	0·4	96·4	3·2
8·9	0·4	96·0	3·6	0·4	95·9	3·7	0·4	95·8	3·8	0·4	95·6	4·0
9·0	0·3	95·3	4·4	0·3	95·2	4·5	0·3	95·1	4·6	0·3	94·8	4·9
9·1	0·3	94·1	5·6	0·3	94·0	5·7	0·3	93·8	5·9	0·3	93·6	6·1
9·2	0·2	93·0	6·8	0·2	92·8	7·0	0·2	92·6	7·2	0·2	92·4	7·4
9·3	0·2	91·6	8·2	0·2	91·5	8·3	0·2	91·3	8·5	0·2	91·0	8·8
9·4	0·1	89·6	10·3	0·1	89·1	10·8	0·1	88·9	11·0	0·1	88·6	11·3
9·5	0	87·2	12·8	0	86·9	13·1	0	86·4	13·6	0	86·0	13·5
9·6	0	84·5	15·5	0	83·9	16·1	0	83·5	16·5	0	83·0	17·0
9·7	0	81·2	18·8	0	80·6	19·4	0	80·1	19·9	0	79·7	20·3
9·8	0	77·0	23·0	0	76·3	23·7	0	75·1	24·1	0	75·5	24·5
9·9	0	72·7	27·3	0	72·0	28·0	0	71·5	28·5	0	71·0	29·0
10·0	0	68·5	31·5	0	67·6	32·4	0	66·9	33·1	0	66·0	33·8
10·1	0	63·7	36·3	0	62·4	37·6	0	61·8	38·2	0	61·2	38·8
10·2	0	57·5	42·5	0	56·8	43·2	0	56·1	43·9	0	55·3	44·7
10·3	0	52·5	47·5	0	51·5	48·5	0	50·5	49·5	0	49·8	50·2
10·4	0	46·3	53·7	0	45·1	54·9	0	44·6	55·4	0	43·8	56·2

* Values based on Buch's (1951) constants and Sørensen's pH scale.

Table IX

Percentage molar fraction of free carbonic acid (F), bicarbonate ion (B) and carbonate ion (C) at different temperatures, pH values and chlorinities in sea water (Saruhashi, 1953) (see p. 269)

(a) *Cl* 15‰

pH	t(°C) 0			2			4			6			8		
	F	*B*	*C*	*F*	*B*	*C*	*F*	*B*	*C*	*F*	*B*	*C*	*F*	*B*	*C*
7·4	6·4	92·7	0·9	6·0	93·0	1·0	5·7	93·2	1·1	5·5	93·3	1·2	5·2	93·5	1·3
7·5	5·2	93·6	1·2	4·9	93·9	1·3	4·6	94·0	1·4	4·4	94·1	1·5	4·2	94·2	1·6
7·6	4·1	94·3	1·6	3·8	94·4	1·8	3·7	94·5	1·8	3·5	94·5	2·0	3·4	94·5	2·1
7·7	3·3	94·7	2·0	3·1	94·7	2·2	3·0	94·7	2·3	2·8	94·7	2·5	2·7	94·7	2·6
7·8	2·7	94·8	2·5	2·5	94·8	2·7	2·3	94·8	2·9	2·3	94·6	3·1	2·1	94·6	3·3
7·9	2·1	94·7	3·2	2·0	94·5	3·5	1·9	94·4	3·7	1·8	94·3	3·9	1·7	94·2	4·1
8·0	1·7	94·3	4·0	1·5	94·2	4·3	1·4	94·0	4·6	1·4	93·7	4·9	1·3	93·5	5·2
8·1	1·2	93·7	5·1	1·3	93·3	5·4	1·2	93·1	5·7	1·1	92·9	6·0	1·0	92·6	6·4
8·2	1·0	92·7	6·3	0·9	92·4	6·7	0·8	92·0	7·2	0·8	91·5	7·7	0·8	91·1	8·1
8·3	0·8	91·4	7·8	0·8	90·9	8·3	0·7	90·5	8·8	0·7	90·0	9·3	0·6	89·5	9·9
8·4	0·6	89·7	9·7	0·5	89·2	10·3	0·5	88·6	10·9	0·5	87·8	11·7	0·5	87·2	12·3

pH	t(°C) 10			12			14			16			18		
	F	*B*	*C*	*F*	*B*	*C*	*F*	*B*	*C*	*F*	*B*	*C*	*F*	*B*	*C*
7·4	5·0	93·6	1·4	4·9	93·7	1·4	4·7	93·8	1·5	4·5	93·9	1·6	4·4	93·9	1·7
7·5	4·0	94·3	1·7	3·9	94·3	1·8	3·7	94·4	1·9	3·6	94·4	2·0	3·5	94·4	2·1
7·6	3·2	94·6	2·2	3·1	94·5	2·4	3·0	94·6	2·3	2·9	94·5	2·6	2·8	94·5	2·7
7·7	2·6	94·6	2·8	2·5	94·6	2·9	2·4	94·5	3·1	2·3	94·4	3·3	2·2	94·4	3·4
7·8	2·0	94·4	3·6	1·9	94·3	3·8	1·9	94·2	3·9	1·8	94·1	4·1	1·7	94·0	4·3
7·9	1·6	94·0	4·4	1·5	93·8	4·7	1·5	93·7	4·8	1·5	93·5	5·0	1·4	93·2	5·4
8·0	1·2	93·2	5·6	1·2	92·9	5·9	1·2	92·7	6·1	1·1	92·5	6·4	1·1	92·2	6·7
8·1	1·0	92·2	6·8	1·0	91·8	7·2	0·9	91·4	7·7	0·8	91·2	8·0	0·8	90·9	8·3
8·2	0·8	90·6	8·6	0·7	90·3	9·0	0·7	89·9	9·4	0·6	89·7	9·7	0·7	89·1	10·2
8·3	0·6	89·0	10·4	0·5	88·3	11·2	0·5	88·0	11·5	0·5	87·6	11·9	0·5	87·2	12·3
8·4	0·4	86·5	13·1	0·4	86·0	13·6	0·4	85·5	14·1	0·4	84·9	14·7	0·4	84·2	15·4

pH	20 F	20 B	20 C	22 F	22 B	22 C	24 F	24 B	24 C	26 F	26 B	26 C	28 F	28 B	28 C	30 F	30 B	30 C
7·4	4·3	94·0	1·7	4·1	94·1	1·8	4·0	94·1	1·9	3·9	94·1	2·0	3·8	94·1	2·1	3·7	94·1	2·2
7·5	3·4	94·4	2·2	3·3	94·4	2·3	3·2	94·3	2·5	3·1	94·3	2·6	3·1	94·2	2·7	3·0	94·3	2·7
7·6	2·6	94·5	2·9	2·5	94·5	3·0	2·5	94·4	3·1	2·5	94·3	3·2	2·4	94·2	3·4	2·3	94·2	3·5
7·7	2·2	94·3	3·5	2·0	94·4	3·6	2·0	94·2	3·8	1·8	94·1	4·1	1·8	94·0	4·2	1·8	93·9	4·3
7·8	1·7	93·8	4·5	1·6	93·8	4·6	1·5	93·6	4·9	1·4	93·5	5·1	1·5	93·2	5·3	1·4	93·1	5·5
7·9	1·3	93·1	5·6	1·3	92·9	5·8	1·3	92·7	6·0	1·2	92·6	6·2	1·2	92·4	6·4	1·1	92·3	6·6
8·0	1·0	92·0	7·0	1·0	91·8	7·2	1·0	91·5	7·5	0·9	91·3	7·8	0·9	90·9	8·2	0·9	90·7	8·4
8·1	0·8	90·5	8·7	0·8	90·3	8·9	0·7	90·1	9·2	0·7	89·6	9·7	0·7	89·3	10·0	0·7	88·9	10·4
8·2	0·7	88·7	10·6	0·6	88·4	11·0	0·6	87·9	11·5	0·6	87·4	12·0	0·6	87·1	12·3	0·6	86·7	12·7
8·3	0·5	86·8	12·7	0·5	86·0	13·5	0·5	85·8	13·7	0·5	85·5	14·0	0·5	85·3	14·2	0·5	85·2	14·3
8·4	0·4	83·7	15·9	0·3	83·1	16·6	0·3	82·6	17·1	0·3	82·0	17·7	0·3	81·4	18·3	0·4	80·8	18·8

(Column groups 20–30 are t(°C).)

(b) *Cl 16‰*

pH	0 F	0 B	0 C	2 F	2 B	2 C	4 F	4 B	4 C	6 F	6 B	6 C	8 F	8 B	8 C
7·4	6·3	92·7	1·0	5·9	93·0	1·1	5·6	93·2	1·2	5·4	93·3	1·3	5·2	93·4	1·4
7·5	5·1	93·6	1·3	4·8	93·7	1·5	4·6	93·9	1·5	4·3	94·0	1·7	4·1	94·1	1·8
7·6	4·0	94·2	1·8	3·8	94·4	1·8	3·6	94·4	2·0	3·5	94·5	2·0	3·3	94·5	2·2
7·7	3·3	94·5	2·2	3·1	94·5	2·4	2·9	94·6	2·5	2·7	94·6	2·7	2·7	94·5	2·8
7·8	2·6	94·7	2·7	2·4	94·7	2·9	2·3	94·6	3·1	2·2	94·5	3·3	2·1	94·4	3·5
7·9	2·0	94·5	3·5	1·9	94·4	3·7	1·7	94·3	4·0	1·7	94·2	4·1	1·6	93·9	4·5
8·0	1·6	94·1	4·3	1·5	93·9	4·6	1·4	93·6	5·0	1·4	93·4	5·2	1·3	93·2	5·5
8·1	1·3	93·2	5·5	1·2	93·0	5·8	1·1	92·7	6·2	1·1	92·4	6·5	1·0	92·1	6·9
8·2	1·0	92·3	6·7	0·9	92·0	7·1	0·8	91·5	7·7	0·8	91·2	8·0	0·8	90·6	8·6
8·3	0·8	90·8	8·4	0·7	90·3	9·0	0·7	89·8	9·5	0·6	89·4	10·0	0·6	88·8	10·6
8·4	0·6	89·1	10·3	0·5	88·6	10·9	0·5	87·8	11·7	0·5	87·2	12·3	0·5	86·4	13·1

(Column groups 0–8 are t(°C).)

Table IX—*continued*

pH	*t*(°C) 10 *F*	10 *B*	10 *C*	12 *F*	12 *B*	12 *C*	14 *F*	14 *B*	14 *C*	16 *F*	16 *B*	16 *C*	18 *F*	18 *B*	18 *C*
7·4	5·0	93·5	1·5	4·8	93·7	1·5	4·6	93·7	1·7	4·4	93·8	1·8	4·3	93·9	1·8
7·5	4·0	94·1	1·9	3·9	94·2	1·9	3·7	94·2	2·1	3·6	94·2	2·2	3·5	94·2	2·3
7·6	3·2	94·5	2·3	3·0	94·5	2·5	2·9	94·4	2·7	2·8	94·4	2·8	2·7	94·3	3·0
7·7	2·6	94·5	2·9	2·4	94·5	3·1	2·3	94·4	3·3	2·5	94·3	3·5	2·2	94·2	3·6
7·8	2·0	94·3	3·7	1·9	94·1	4·0	1·8	94·0	4·2	1·7	93·9	4·4	1·7	93·7	4·6
7·9	1·6	93·7	4·7	1·5	93·5	5·0	1·4	93·2	5·4	1·4	93·1	5·5	1·3	92·9	5·8
8·0	1·3	92·9	5·8	1·2	92·6	6·2	1·2	92·3	6·5	1·1	92·1	6·8	1·0	91·8	7·2
8·1	1·0	91·7	7·3	1·0	91·4	7·6	0·9	91·0	8·1	0·8	90·7	8·5	0·9	90·2	8·9
8·2	0·7	90·3	9·0	0·7	89·8	9·5	0·7	89·3	10·0	0·7	88·9	10·4	0·6	88·5	10·9
8·3	0·5	88·3	11·2	0·5	87·6	11·9	0·5	87·1	12·4	0·5	86·6	12·9	0·5	86·0	13·5
8·4	0·4	85·9	13·7	0·4	85·3	14·3	0·4	84·5	15·1	0·4	84·0	15·6	0·4	83·3	16·3

pH	*t*(°C) 20 *F*	20 *B*	20 *C*	22 *F*	22 *B*	22 *C*	24 *F*	24 *B*	24 *C*	26 *F*	26 *B*	26 *C*	28 *F*	28 *B*	28 *C*	30 *F*	30 *B*	30 *C*
7·4	4·1	94·0	1·9	4·1	94·0	1·9	3·9	94·0	2·1	3·8	94·1	2·1	3·7	94·1	2·2	3·6	94·1	2·3
7·5	3·4	94·2	2·4	3·2	94·2	2·6	3·1	94·2	2·7	3·0	94·2	2·8	3·0	94·1	2·9	2·8	94·1	3·1
7·6	2·6	94·3	3·1	2·6	94·2	3·2	2·4	94·3	3·3	2·3	94·2	3·5	2·3	94·1	3·6	2·2	94·1	3·7
7·7	2·1	94·0	3·9	2·1	93·9	4·0	2·0	93·9	4·1	1·9	93·8	4·3	1·8	93·7	4·5	1·7	93·7	4·6
7·8	1·6	93·6	4·8	1·6	93·4	5·0	1·5	93·3	5·2	1·5	93·1	5·4	1·4	93·0	5·6	1·4	92·8	5·8
7·9	1·3	92·6	6·1	1·2	92·6	6·2	1·2	92·3	6·5	1·2	92·0	6·8	1·2	91·9	6·9	1·1	91·7	7·2
8·0	1·0	91·6	7·4	0·9	91·3	7·8	0·9	91·1	8·0	0·9	90·8	8·3	0·8	90·5	8·7	0·8	90·3	8·9
8·1	0·8	89·8	9·4	0·8	89·4	9·8	0·7	89·2	10·1	0·7	89·0	10·3	0·7	88·7	10·6	0·7	88·5	10·8
8·2	0·6	88·1	11·3	0·6	87·6	11·8	0·6	87·3	12·1	0·5	86·9	12·6	0·5	86·4	13·1	0·5	86·0	13·5
8·3	0·5	85·5	14·0	0·5	85·1	14·4	0·4	84·7	14·9	0·4	84·2	15·4	0·4	83·7	15·9	0·4	83·2	16·4
8·4	0·4	82·7	16·9	0·4	82·2	17·4	0·3	81·6	18·1	0·3	81·0	18·7	0·3	80·4	19·3	0·3	79·9	19·8

(c) *Cl 17‰*

pH	t(°C) 0			2			4			6			8		
	F	*B*	*C*	*F*	*B*	*C*	*F*	*B*	*C*	*F*	*B*	*C*	*F*	*B*	*C*
7·4	6·1	92·8	1·1	5·8	93·0	1·2	5·6	93·1	1·3	5·4	93·2	1·4	5·1	93·4	1·5
7·5	5·0	93·5	1·5	4·6	93·9	1·5	4·5	94·2	1·5	4·3	94·0	1·7	4·1	94·3	1·6
7·6	3·9	94·3	1·8	3·7	94·4	1·9	3·5	94·4	2·1	3·4	94·4	2·2	3·3	94·4	2·3
7·7	3·2	94·5	2·3	3·0	94·7	2·3	2·8	94·6	2·6	2·7	94·6	2·7	2·6	94·5	2·9
7·8	2·5	94·6	2·9	2·4	94·6	3·0	2·2	94·4	3·4	2·1	94·4	3·5	2·0	94·3	3·7
7·9	2·0	94·5	3·5	1·9	94·3	3·8	1·8	94·1	4·1	1·7	94·1	4·2	1·6	93·9	4·5
8·0	1·6	93·9	4·5	1·5	93·7	4·8	1·4	93·5	5·1	1·3	93·2	5·5	1·3	92·9	5·8
8·1	1·2	93·2	5·6	1·1	93·0	5·9	1·1	92·7	6·2	1·0	92·4	6·6	1·0	92·0	7·0
8·2	1·0	92·1	6·9	0·9	91·7	7·4	0·9	91·2	7·9	0·8	90·7	8·5	0·8	90·3	8·9
8·3	0·8	90·9	8·3	0·8	90·2	9·0	0·7	89·5	9·8	0·7	89·0	10·3	0·7	88·4	10·9
8·4	0·5	88·7	10·8	0·5	88·1	11·4	0·5	87·4	12·1	0·5	86·7	12·8	0·5	85·9	13·6

pH	t(°C) 10			12			14			16			18		
	F	*B*	*C*	*F*	*B*	*C*	*F*	*B*	*C*	*F*	*B*	*C*	*F*	*B*	*C*
7·4	4·9	93·5	1·6	4·7	93·6	1·7	4·5	93·7	1·8	4·3	93·9	1·8	4·2	93·9	1·9
7·5	3·9	94·3	1·8	3·7	94·3	2·0	3·6	94·2	2·2	3·5	94·1	2·4	3·4	94·2	2·4
7·6	3·1	94·4	2·5	3·0	94·3	2·7	2·9	94·3	2·8	2·7	94·2	3·1	2·7	94·1	3·2
7·7	2·5	94·5	3·0	2·4	94·4	3·1	2·3	94·5	3·2	2·2	94·3	3·5	2·1	94·1	3·8
7·8	2·0	94·1	3·9	1·9	93·9	4·2	1·8	93·8	4·4	1·7	93·6	4·7	1·7	93·5	4·8
7·9	1·5	93·7	4·8	1·5	93·3	5·2	1·4	93·2	5·4	1·4	93·1	5·5	1·3	92·8	5·9
8·0	1·2	92·6	6·2	1·2	92·3	6·5	1·1	92·0	6·9	1·1	91·8	7·1	1·0	91·5	7·5
8·1	1·0	91·6	7·4	0·9	91·3	7·8	0·9	91·1	8·0	0·8	90·7	8·5	0·8	90·3	8·8
8·2	0·7	89·8	9·5	0·7	89·3	10·0	0·7	88·9	10·4	0·6	88·5	10·9	0·6	88·0	11·4
8·3	0·6	87·7	11·3	0·6	87·1	12·3	0·6	86·6	12·8	0·5	86·3	13·2	0·5	86·0	13·5
8·4	0·5	85·2	14·3	0·5	84·5	15·0	0·4	84·0	15·6	0·4	83·3	16·3	0·4	82·6	17·0

TABLE IX—*continued*

pH	t(°C) 20			22			24			26			28			30		
	F	*B*	*C*	*F*	*B*	*C*	*F*	*B*	*C*	*F*	*B*	*C*	*F*	*B*	*C*	*F*	*B*	*C*
7·4	4·1	93·9	2·0	4·0	94·0	2·1	3·9	94·0	2·1	3·8	94·0	2·2	3·7	94·0	2·3	3·6	94·0	2·4
7·5	3·3	94·2	2·5	3·2	94·3	2·5	3·0	94·2	2·8	3·0	94·0	3·0	2·9	94·1	3·0	2·9	94·0	3·1
7·6	2·6	94·1	3·3	2·5	94·0	3·5	2·4	94·1	3·5	2·4	94·1	3·5	2·4	94·0	3·6	2·2	93·7	4·1
7·7	2·0	94·1	3·9	2·0	93·9	4·1	1·9	93·9	4·2	1·9	93·6	4·5	1·9	93·6	4·6	1·8	93·5	4·7
7·8	1·6	93·3	5·1	1·6	93·1	5·3	1·5	93·1	5·4	1·5	92·8	5·7	1·4	92·7	5·9	1·5	92·4	6·1
7·9	1·3	92·5	6·2	1·2	92·4	6·4	1·2	92·2	6·6	1·1	92·0	6·9	1·1	91·7	7·2	1·1	91·5	7·4
8·0	1·0	91·2	7·8	1·0	90·8	8·2	0·9	90·7	8·5	0·8	90·4	8·8	0·8	90·1	9·1	0·8	89·8	9·4
8·1	0·8	89·8	9·4	0·7	89·4	9·9	0·7	89·1	10·2	0·7	88·7	10·6	0·7	88·4	10·9	0·7	88·0	11·3
8·2	0·6	87·5	11·9	0·6	87·1	12·3	0·5	86·7	12·8	0·5	86·7	13·3	0·6	85·7	13·7	0·5	85·3	14·2
8·3	0·5	85·5	14·0	0·4	85·2	14·4	0·4	84·1	15·5	0·4	83·7	15·9	0·4	83·3	16·3	0·4	82·5	17·1
8·4	0·4	81·9	17·7	0·3	81·3	18·4	0·3	80·8	18·9	0·3	80·2	19·5	0·3	79·5	20·2	0·3	78·9	20·8

(d) *Cl 18‰*

pH	t(°C) 0			2			4			6			8		
	F	*B*	*C*	*F*	*B*	*C*	*F*	*B*	*C*	*F*	*B*	*C*	*F*	*B*	*C*
7·4	6·0	92·7	1·3	5·7	93·0	1·3	5·5	93·2	1·3	5·3	93·2	1·5	5·0	93·5	1·5
7·5	4·9	93·7	1·4	4·6	93·6	1·8	4·4	93·9	1·7	4·2	94·1	1·7	4·1	94·0	1·9
7·6	3·9	94·2	1·9	3·7	94·3	2·0	3·5	94·4	2·1	3·4	94·3	2·3	3·2	94·3	2·5
7·7	3·1	94·3	2·6	2·9	94·5	2·6	2·8	94·5	2·3	2·7	94·4	2·9	2·6	94·4	3·0
7·8	2·5	94·6	2·9	2·3	94·4	3·3	2·2	94·3	3·5	2·1	94·1	3·8	2·0	93·9	4·1
7·9	2·0	94·2	3·8	1·8	94·1	4·1	1·8	93·9	4·3	1·7	93·8	4·5	1·6	93·6	4·8
8·0	1·5	93·7	4·8	1·4	93·4	5·2	1·3	93·2	5·5	1·4	92·8	5·8	1·3	92·5	6·2
8·1	1·2	92·8	6·0	1·1	92·6	6·3	1·1	92·2	6·7	1·0	91·7	7·3	0·9	91·4	7·7
8·2	0·9	91·7	7·4	0·9	91·2	7·9	0·8	90·7	8·5	0·8	90·2	9·0	0·8	89·7	9·5
8·3	0·8	90·4	8·8	0·7	89·6	9·7	0·7	88·8	10·5	0·6	88·1	11·3	0·6	87·6	11·8
8·4	0·6	88·1	11·3	0·5	87·4	12·1	0·5	86·7	12·8	0·4	85·9	13·7	0·5	85·2	14·3

pH	t(°C) 10			12			14			16			18		
	F	*B*	*C*	*F*	*B*	*C*	*F*	*B*	*C*	*F*	*B*	*C*	*F*	*B*	*C*
7·4	4·9	93·7	1·5	4·6	93·7	1·7	4·5	93·8	1·7	4·3	93·8	1·9	4·1	93·9	2·0
7·5	3·9	94·0	2·1	3·8	94·0	2·2	3·7	94·0	2·3	3·6	94·0	2·4	3·4	94·0	2·6
7·6	3·1	94·2	2·7	2·9	94·2	2·9	2·8	94·1	3·1	2·7	94·0	3·3	2·6	94·0	3·4
7·7	2·4	94·4	3·2	2·4	94·2	3·4	2·3	94·0	3·7	2·2	93·8	4·0	2·1	93·8	4·1
7·8	1·9	93·8	4·3	1·8	93·8	4·4	1·8	93·6	4·6	1·7	93·4	4·9	1·6	93·3	5·1
7·9	1·6	93·3	5·1	1·4	93·1	5·5	1·4	93·0	5·6	1·4	92·7	5·9	1·3	92·4	6·3
8·0	1·2	92·2	6·6	1·1	92·0	6·9	1·1	91·7	7·2	1·0	91·4	7·6	1·0	91·1	7·9
8·1	0·9	91·0	8·1	0·9	90·6	8·5	0·8	90·3	8·9	0·8	90·0	9·2	0·8	89·6	9·6
8·2	0·7	89·3	10·0	0·7	88·8	10·5	0·7	88·3	11·0	0·6	87·9	11·5	0·6	87·3	12·1
8·3	0·6	87·4	12·0	0·6	86·8	12·6	0·6	85·9	13·5	0·5	85·4	14·1	0·5	85·0	14·5
8·4	0·4	84·5	15·1	0·4	83·8	15·8	0·4	83·1	16·5	0·4	82·4	17·2	0·3	81·7	18·0

pH	t(°C) 20			22			24			26			28			30		
	F	*B*	*C*	*F*	*B*	*C*	*F*	*B*	*C*	*F*	*B*	*C*	*F*	*B*	*C*	*F*	*B*	*C*
7·4	4·0	93·9	2·1	3·9	93·9	2·2	3·8	93·9	2·3	3·7	93·9	2·4	3·6	93·9	2·5	3·5	94·0	2·5
7·5	3·2	94·1	2·7	3·1	94·1	2·8	3·0	94·1	2·9	3·0	94·1	2·9	2·9	94·0	3·1	2·8	94·0	3·2
7·6	2·5	94·0	3·5	2·5	93·9	3·6	2·4	93·8	3·8	2·3	93·8	3·9	2·3	93·7	4·0	2·2	93·6	4·2
7·7	2·0	93·7	4·3	2·0	93·5	4·5	1·9	93·5	4·6	1·9	93·5	4·4	1·9	93·2	4·9	1·8	93·1	5·1
7·8	1·5	93·2	5·3	1·5	92·9	5·6	1·5	92·7	5·8	1·4	92·5	6·1	1·4	92·4	6·2	1·4	92·1	6·5
7·9	1·2	92·1	6·7	1·2	91·9	6·9	1·2	91·7	7·1	1·1	91·5	7·4	1·1	91·3	7·6	1·1	91·3	7·6
8·0	1·0	90·7	8·3	0·9	90·5	8·6	0·9	90·2	8·9	0·8	89·9	9·3	0·8	89·6	9·6	0·8	89·2	10·0
8·1	0·7	89·2	10·1	0·7	88·9	10·4	0·7	88·7	10·6	0·6	88·5	10·9	0·6	88·0	11·4	0·6	87·6	11·8
8·2	0·6	86·9	12·5	0·6	86·4	13·0	0·5	86·0	13·5	0·5	85·6	13·9	0·5	85·1	14·4	0·5	84·6	14·9
8·3	0·5	84·5	15·0	0·5	84·4	15·1	0·4	84·1	15·5	0·4	83·0	16·6	0·4	82·7	16·9	0·4	82·1	17·5
8·4	0·3	81·1	18·6	0·3	80·5	19·2	0·3	79·8	19·9	0·3	79·2	20·5	0·3	78·6	21·1	0·3	78·0	21·7

TABLE IX—*continued*

(e) *Cl* 19‰

pH	t(°C) 0 *F*	0 *B*	0 *C*	2 *F*	2 *B*	2 *C*	4 *F*	4 *B*	4 *C*	6 *F*	6 *B*	6 *C*	8 *F*	8 *B*	8 *C*
7·4	6·0	92·9	1·1	5·6	93·1	1·3	5·4	93·1	1·4	5·2	93·3	1·5	4·9	93·5	1·6
7·5	4·7	93·9	1·4	4·6	93·9	1·5	4·3	94·0	1·7	4·1	94·0	1·9	3·9	94·0	2·1
7·6	3·8	94·3	1·9	3·6	94·3	2·1	3·4	94·2	2·4	3·3	94·1	2·6	3·1	94·1	2·8
7·7	3·1	94·5	2·4	2·9	94·5	2·6	2·8	94·3	2·9	2·6	94·3	3·1	2·5	94·2	3·3
7·8	2·4	94·4	3·2	2·3	94·3	3·4	2·2	94·1	3·7	2·1	94·0	3·9	2·0	93·9	4·1
7·9	1·8	94·1	4·1	1·8	93·9	4·3	1·7	93·6	4·7	1·6	93·5	4·9	1·6	93·2	5·2
8·0	1·6	93·4	5·0	1·4	93·2	5·4	1·3	92·9	5·8	1·3	92·6	6·1	1·2	92·2	6·6
8·1	1·1	92·6	6·3	1·1	92·3	6·6	1·1	91·9	7·0	1·0	91·5	7·5	1·0	91·0	8·0
8·2	0·9	91·3	7·8	0·8	90·8	8·4	0·8	90·3	8·9	0·7	89·9	9·4	0·7	89·3	10·0
8·3	0·7	89·7	9·6	0·6	89·1	10·3	0·6	88·6	10·8	0·6	87·9	11·5	0·6	87·2	12·2
8·4	0·6	87·5	11·9	0·5	86·8	12·7	0·5	86·1	13·4	0·5	85·4	14·1	0·5	84·4	15·1

pH	t(°C) 10 *F*	10 *B*	10 *C*	12 *F*	12 *B*	12 *C*	14 *F*	14 *B*	14 *C*	16 *F*	16 *B*	16 *C*	18 *F*	18 *B*	18 *C*
7·4	4·7	93·6	1·7	4·6	93·6	1·8	4·4	93·7	1·9	4·2	93·7	2·1	4·1	93·7	2·2
7·5	3·7	94·0	2·3	3·6	93·9	2·5	3·5	93·9	2·6	3·4	94·0	2·6	3·3	94·0	2·7
7·6	3·0	94·2	2·8	2·9	94·2	2·9	2·7	94·2	3·1	2·7	94·1	3·2	2·6	94·1	3·4
7·7	2·3	94·2	3·5	2·3	94·1	3·6	2·1	94·0	3·9	2·2	93·8	4·0	2·1	93·8	4·1
7·8	1·9	93·8	4·3	1·8	93·7	4·5	1·7	93·4	4·9	1·8	93·1	5·1	1·6	93·0	5·4
7·9	1·5	93·0	5·5	1·4	92·9	5·7	1·4	92·6	6·0	1·2	92·4	6·4	1·3	92·1	6·6
8·0	1·1	92·0	6·9	1·2	91·6	7·2	1·1	91·3	7·6	1·0	91·0	8·0	1·0	90·7	8·3
8·1	0·9	90·7	8·4	0·8	90·3	8·9	0·8	90·0	9·2	0·8	89·4	9·8	0·8	89·0	10·2
8·2	0·7	88·7	10·6	0·7	88·3	11·1	0·7	87·8	11·5	0·6	87·3	12·1	0·5	86·8	12·7
8·3	0·5	86·5	13·0	0·5	86·1	13·4	0·5	85·4	14·1	0·5	84·7	14·8	0·4	84·0	15·6
8·4	0·4	83·8	15·8	0·4	83·1	16·5	0·4	82·4	17·2	0·4	81·6	18·0	0·3	81·0	18·7

pH	t(°C) 20			22			24			26			28			30		
	F	*B*	*C*	*F*	*B*	*C*	*F*	*B*	*C*	*F*	*B*	*C*	*F*	*B*	*C*	*F*	*B*	*C*
7·4	3·9	93·7	2·4	3·8	93·7	2·5	3·7	93·7	2·6	3·6	93·8	2·6	3·5	93·8	2·7	3·5	93·8	2·7
7·5	3·2	94·0	2·8	3·1	94·0	2·9	2·9	93·9	3·2	2·7	93·9	3·3	2·7	93·9	3·4	2·7	93·8	3·5
7·6	2·5	93·8	3·7	2·4	93·8	3·8	2·3	93·7	4·0	2·2	93·7	4·1	2·1	93·6	4·3	2·0	93·6	4·4
7·7	2·0	93·6	4·4	2·0	93·3	4·7	2·0	93·2	4·8	1·9	93·1	5·0	1·8	93·0	5·2	1·6	93·0	5·4
7·8	1·5	92·9	5·6	1·4	92·7	5·9	1·4	92·5	6·1	1·4	92·2	6·4	1·4	92·0	6·6	1·3	91·9	6·8
7·9	1·2	91·9	6·9	1·2	91·6	7·2	1·1	91·4	7·5	1·1	91·1	7·8	1·0	91·0	8·0	1·0	90·7	8·3
8·0	0·9	90·3	8·7	0·9	90·0	9·1	0·9	89·7	8·4	0·8	89·4	9·8	0·8	89·0	10·2	0·8	88·8	10·4
8·1	0·7	88·6	10·7	0·7	88·3	11·0	0·7	87·9	11·4	0·7	87·5	11·8	0·7	87·1	12·2	0·6	86·6	12·8
8·2	0·5	86·4	13·1	0·5	85·8	13·7	0·5	85·3	14·2	0·5	84·8	14·7	0·5	84·3	15·2	0·5	83·8	15·7
8·3	0·4	83·7	15·9	0·4	83·0	16·6	0·4	82·4	17·2	0·4	81·8	17·8	0·4	81·3	18·3	0·4	80·7	18·9
8·4	0·3	80·3	19·4	0·3	79·5	20·2	0·3	78·9	20·8	0·3	78·2	21·5	0·3	77·5	22·2	0·3	76·9	22·8

(f) *Cl* 20‰

pH	t(°C) 0			2			4			6			8		
	F	*B*	*C*	*F*	*B*	*C*	*F*	*B*	*C*	*F*	*B*	*C*	*F*	*B*	*C*
7·4	5·9	93·0	1·1	5·6	93·1	1·3	5·3	93·1	1·6	5·1	93·1	1·8	4·9	93·2	1·9
7·5	4·7	93·5	1·8	4·5	93·6	1·9	4·3	93·7	2·0	4·1	93·8	2·1	3·9	93·9	2·2
7·6	3·8	94·0	2·2	3·6	94·1	2·3	3·4	94·2	2·4	3·2	94·2	2·5	3·1	94·2	2·7
7·7	3·1	94·2	2·7	3·0	94·2	2·8	2·8	94·1	3·1	2·6	94·1	3·3	2·4	94·1	3·5
7·8	2·3	94·3	3·4	2·2	94·2	3·6	2·1	94·1	3·8	2·1	93·7	4·2	2·0	93·6	4·4
7·9	1·9	93·9	4·2	1·7	93·7	4·6	1·6	93·5	4·9	1·7	93·1	5·2	1·5	93·0	5·5
8·0	1·5	93·2	5·3	1·5	92·8	5·7	1·3	92·6	6·1	1·3	92·2	6·5	1·2	91·9	6·9
8·1	1·1	92·3	6·6	1·2	91·7	7·1	1·0	91·5	7·5	1·0	91·1	7·9	0·9	90·7	8·4
8·2	0·9	90·9	8·2	0·8	90·3	8·9	0·8	89·9	9·3	0·8	89·2	10·0	0·8	88·7	10·5
8·3	0·7	89·0	10·3	0·7	88·3	11·0	0·6	88·0	11·4	0·6	87·2	12·2	0·6	86·4	13·0
8·4	0·6	87·0	12·4	0·5	86·1	13·4	0·5	85·3	14·2	0·5	84·4	15·1	0·5	83·7	15·8

TABLE IX—*continued*

pH	t(°C) 10			12			14			16			18		
	F	*B*	*C*	*F*	*B*	*C*	*F*	*B*	*C*	*F*	*B*	*C*	*F*	*B*	*C*
7·4	4·7	93·3	2·0	4·5	93·4	2·1	4·3	93·5	2·2	4·1	93·6	2·3	4·0	93·6	2·4
7·5	3·8	93·9	2·3	3·7	93·9	2·4	3·5	94·0	2·5	3·4	94·0	2·6	3·2	93·9	2·9
7·6	2·9	94·1	3·0	2·8	94·0	3·2	2·7	94·0	3·3	2·5	94·0	3·5	2·5	93·8	3·7
7·7	2·2	94·0	3·8	2·2	93·9	3·9	2·2	93·7	4·1	2·2	93·5	4·3	2·0	93·4	4·6
7·8	2·0	93·4	4·6	1·9	93·2	4·9	1·8	93·0	5·2	1·7	92·9	5·4	1·6	92·7	5·7
7·9	1·4	92·8	5·8	1·4	92·5	6·1	1·3	92·3	6·4	1·3	92·0	6·7	1·2	91·7	7·1
8·0	1·1	91·6	7·3	1·1	91·2	7·7	1·0	90·9	8·1	1·0	90·6	8·4	1·0	90·2	8·8
8·1	0·9	90·1	9·0	0·8	89·8	9·4	0·8	89·4	9·8	0·8	89·0	10·2	0·7	88·6	10·7
8·2	0·7	88·1	11·2	0·7	87·7	11·6	0·7	87·1	12·2	0·7	86·6	12·7	0·6	86·1	13·3
8·3	0·5	85·8	13·7	0·5	85·4	14·1	0·5	84·7	14·8	0·5	84·0	15·5	0·4	83·3	16·3
8·4	0·4	82·9	16·7	0·4	82·2	17·4	0·4	81·4	18·2	0·4	80·8	18·8	0·3	80·0	19·7

pH	t(°C) 20			22			24			26			28			30		
	F	*B*	*C*	*F*	*B*	*C*	*F*	*B*	*C*	*F*	*B*	*C*	*F*	*B*	*C*	*F*	*B*	*C*
7·4	3·9	93·7	2·4	3·8	93·7	2·5	3·7	93·7	2·6	3·6	93·7	2·7	3·5	93·6	2·9	3·4	93·5	3·1
7·5	3·1	93·9	3·0	3·0	93·8	3·2	2·9	93·7	3·4	2·8	93·7	3·5	2·8	93·6	3·6	2·8	93·5	3·7
7·6	2·4	93·7	3·9	2·3	93·7	4·0	2·2	93·6	4·2	2·2	93·5	4·3	2·2	93·3	4·5	2·1	93·2	4·7
7·7	1·9	93·3	4·8	1·8	93·2	5·0	1·8	93·0	5·2	1·7	92·9	5·4	1·7	92·7	5·6	1·7	92·6	5·7
7·8	1·5	92·6	5·9	1·4	92·4	6·2	1·4	92·2	6·4	1·4	91·9	6·7	1·3	91·7	7·0	1·3	91·4	7·3
7·9	1·2	91·5	7·3	1·1	91·3	7·6	1·1	90·9	8·0	1·1	90·6	8·3	1·0	90·3	8·7	1·0	90·0	9·0
8·0	0·9	89·9	9·2	0·8	89·6	9·6	0·8	89·2	10·0	0·8	88·9	10·3	0·7	88·5	10·8	0·7	88·1	11·2
8·1	0·7	88·2	11·1	0·7	87·5	11·8	0·6	87·2	12·2	0·6	86·8	12·6	0·6	86·4	13·0	0·6	86·1	13·3
8·2	0·6	85·6	13·8	0·5	85·0	14·5	0·5	84·5	15·0	0·5	84·1	15·4	0·5	83·5	16·0	0·5	82·9	16·6
8·3	0·4	82·9	16·7	0·4	82·3	17·3	0·4	81·4	18·2	0·4	80·7	18·9	0·4	80·1	19·5	0·4	79·7	19·9
8·4	0·3	79·4	20·3	0·3	78·5	21·2	0·3	77·8	21·9	0·3	77·2	22·5	0·3	76·4	23·3	0·3	75·7	24·0

pH	t(°C) 0			2			4			6			8		
	F	B	C	F	B	C	F	B	C	F	B	C	F	B	C
7·4	5·8	92·8	1·4	5·5	93·0	1·5	5·3	93·1	1·6	5·1	93·2	1·7	4·8	93·3	1·9
7·5	4·7	93·5	1·8	4·4	93·7	1·9	4·2	93·8	2·0	4·0	93·9	2·1	3·8	93·9	2·3
7·6	3·7	94·1	2·2	3·5	94·1	2·4	3·3	94·0	2·7	3·2	94·0	2·8	3·1	94·0	2·9
7·7	3·0	94·2	2·8	2·8	94·2	3·0	2·7	94·1	3·2	2·6	94·0	3·4	2·4	94·0	3·6
7·8	2·3	94·1	3·6	2·3	94·0	3·7	2·1	93·8	4·1	2·0	93·7	4·3	1·9	93·6	4·5
7·9	1·9	93·7	4·4	1·7	93·5	4·8	1·7	93·3	5·0	1·6	93·1	5·3	1·5	92·9	5·6
8·0	1·4	93·0	5·6	1·4	92·6	6·0	1·3	92·3	6·4	1·2	92·0	6·8	1·2	91·6	7·2
8·1	1·1	91·9	7·0	1·1	91·5	7·4	1·0	91·2	7·8	1·0	90·7	8·3	0·9	90·3	8·8
8·2	0·9	90·5	8·6	0·8	90·0	9·2	0·8	89·5	9·7	0·7	88·9	10·4	0·7	88·3	11·0
8·3	0·7	88·7	10·6	0·7	87·9	11·4	0·7	87·3	12·0	0·6	86·6	12·8	0·6	86·1	13·3
8·4	0·5	86·4	13·1	0·5	85·6	13·9	0·5	84·8	14·7	0·5	83·9	15·6	0·5	83·1	16·4

pH	t(°C) 10			12			14			16			18		
	F	B	C	F	B	C	F	B	C	F	B	C	F	B	C
7·4	4·6	93·4	2·0	4·4	93·5	2·1	4·2	93·6	2·2	4·1	93·6	2·3	3·9	93·7	2·4
7·5	3·7	93·9	2·4	3·6	93·9	2·5	3·4	93·9	2·7	3·2	93·9	2·9	3·1	93·9	3·0
7·6	3·0	93·9	3·1	2·8	93·9	3·3	2·7	93·8	3·5	2·6	93·7	3·7	2·5	93·7	3·8
7·7	2·3	93·9	3·8	2·2	93·7	4·1	2·1	93·5	4·4	2·0	93·3	4·7	1·9	93·3	4·8
7·8	1·9	93·3	4·8	1·8	93·0	5·2	1·7	92·9	5·4	1·6	92·7	5·7	1·6	92·4	6·0
7·9	1·4	92·5	6·1	1·3	92·2	6·5	1·3	91·9	6·8	1·2	91·6	7·2	1·2	91·3	7·5
8·0	1·2	91·2	7·6	1·1	90·8	8·1	1·1	90·5	8·4	1·0	90·2	8·8	1·0	89·8	9·2
8·1	0·9	89·7	9·4	0·8	89·2	10·0	0·7	88·8	10·5	0·7	88·2	11·1	0·7	87·9	11·4
8·2	0·7	87·7	11·6	0·7	87·0	12·3	0·6	86·7	12·7	0·6	86·1	13·3	0·6	85·5	13·9
8·3	0·6	85·4	14·0	0·5	84·3	15·2	0·5	83·6	15·9	0·5	83·2	16·3	0·5	82·6	16·9
8·4	0·5	82·3	17·2	0·4	81·4	18·2	0·4	80·7	18·9	0·3	80·0	19·7	0·3	79·2	20·5

TABLE IX—*continued*

pH	20 *F*	20 *B*	20 *C*	22 *F*	22 *B*	22 *C*	24 *F*	24 *B*	24 *C*	26 *F*	26 *B*	26 *C*	28 *F*	28 *B*	28 *C*	30 *F*	30 *B*	30 *C*
7·4	3·8	93·7	2·5	3·7	93·7	2·6	3·6	93·7	2·7	3·5	93·7	2·8	3·4	93·6	3·0	3·4	93·5	3·1
7·5	3·0	93·8	3·2	2·9	93·8	3·3	2·7	93·8	3·5	2·7	93·7	3·6	2·6	93·6	3·8	2·6	93·5	3·9
7·6	2·4	93·6	4·0	2·3	93·5	4·2	2·2	93·4	4·4	2·1	93·3	4·6	2·1	93·1	4·8	2·1	93·0	4·9
7·7	1·9	93·1	5·0	1·9	93·0	5·2	1·7	92·9	5·4	1·7	92·7	5·6	1·7	92·6	5·7	1·6	92·4	6·0
7·8	1·5	92·2	6·3	1·4	92·1	6·5	1·4	91·9	6·7	1·3	91·6	7·1	1·3	91·4	7·3	1·2	91·1	7·7
7·9	1·2	91·1	7·7	1·1	90·9	8·0	1·1	90·6	8·3	1·1	90·3	8·6	1·0	90·0	9·0	1·0	89·6	9·4
8·0	0·9	89·4	9·7	0·9	89·1	10·0	0·8	88·7	10·5	0·8	88·3	10·9	0·8	87·9	11·3	0·8	87·6	11·6
8·1	0·7	87·5	11·8	0·7	87·1	12·2	0·7	86·8	12·5	0·6	86·4	13·0	0·6	85·7	13·7	0·6	85·4	14·0
8·2	0·6	84·9	14·5	0·5	84·4	15·1	0·5	83·8	15·7	0·5	83·3	16·2	0·5	82·8	16·7	0·4	82·3	17·3
8·3	0·4	82·0	17·6	0·4	81·3	18·3	0·4	80·7	18·9	0·4	80·0	19·6	0·4	79·4	20·2	0·3	78·8	20·9
8·4	0·3	78·4	21·3	0·3	77·7	22·0	0·3	76·9	22·8	0·3	76·1	23·6	0·3	75·5	24·2	0·3	74·7	25·0

Column heads: $t(°C)$ = 20, 22, 24, 26, 28, 30; each subdivided into *F*, *B*, *C*.

Values based on Buch's (1951) constants and Sørensen's pH scale.

TABLE X

Collected conversion factors

Conversion	Factor	Reciprocal
μg. NO_3^- ⟶ μg. N	0·2259	4·427
μg. NO_2^- ⟶ μg. N	0·3045	3·286
μg. NH_3 ⟶ μg. N	0·8225	1·216
μg. NH_4^+ ⟶ μg. N	0·7764	1·287
μg. PO_4^{3-} ⟶ μg. P	0·3261	3·066
μg. P_2O_5 ⟶ μg. P	0·4364	2·291
μg. SiO_2 ⟶ μg. Si	0·4675	2·139
μg. SiO_4^{4-} ⟶ μg. Si	0·3050	3·278
μg. N ⟶ μg.-at. N	0·07138	14·008
μg. P ⟶ μg.-at. P	0·03228	30·975
μg. Si ⟶ μg.-at. Si	0·03560	28·09

TABLE XI

Table for conversion of weights of nitrogen, phosphorus and silicon expressed in terms of μg./l. into μg.-at./l.

μg. N, P, or Si/l.	*μg.-at. N/l.*	*μg.-at. P/l.*	*μg.-at. Si/l.*
1	0·071	0·036	0·032
2	0·143	0·071	0·065
3	0·214	0·107	0·097
4	0·286	0·142	0·129
5	0·357	0·178	0·161
6	0·428	0·214	0·194
7	0·500	0·249	0·226
8	0·571	0·284	0·258
9	0·643	0·320	0·291
10	0·714	0·356	0·323
20	1·428	0·712	0·646
30	2·142	1·068	0·968
40	2·856	1·424	1·291
50	3·569	1·780	1·614
60	4·283	2·136	1·937
70	4·997	2·492	2·260
80	5·711	2·848	2·582
90	6·425	3·204	2·905
100	7·139	3·560	3·228

TABLE XII

Solubility of oxygen in sea water (after C. J. J. Fox, 1909 and Montgomery, Thom and Cockburn, 1964†)*

Expressed in ml. (at standard temperature and pressure) per l. of water from a free dry atmosphere at 760 mm. pressure

	After Fox (1909)									Montgomery et al. (1964)
	Chlorinity ‰									
t(°C)	0	13	14	15	16	17	18	19	20	19·35
0	10·29	8·78	8·66	8·55	8·43	8·32	8·20	8·08	7·97	7·98
1	10·02	8·56	8·44	8·33	8·22	8·11	8·00	7·88	7·77	—
2	9·75	8·34	8·23	8·11	8·02	7·91	7·80	7·69	7·58	—
3	9·50	8·14	8·03	7·93	7·82	7·72	7·61	7·51	7·40	—
4	9·26	7·94	7·84	7·74	7·64	7·53	7·43	7·33	7·23	7·22
5	9·03	7·75	7·65	7·56	7·46	7·36	7·26	7·16	7·07	—
6	8·81	7·57	7·48	7·38	7·28	7·20	7·10	7·01	6·91	—
7	8·60	7·40	7·31	7·22	7·13	7·04	6·95	6·85	6·76	—
8	8·40	7·24	7·15	7·06	6·97	6·89	6·80	6·71	6·62	6·58
9	8·21	7·09	7·00	6·91	6·83	6·74	6·66	6·57	6·48	—
10	8·02	6·94	6·85	6·77	6·69	6·60	6·52	6·44	6·35	—
11	7·84	6·80	6·71	6·63	6·55	6·47	6·39	6·31	6·23	—
12	7·68	6·66	6·58	6·50	6·43	6·35	6·27	6·19	6·11	6·03
13	7·52	6·55	6·46	6·38	6·31	6·23	6·15	6·08	6·00	—
14	7·37	6·41	6·34	6·26	6·19	6·11	6·04	5·97	5·89	—
15	7·22	6·29	6·22	6·14	6·07	6·00	5·93	5·86	5·79	—
16	7·08	6·17	6·10	6·03	5·96	5·89	5·82	5·76	5·69	5·57
17	6·94	6·06	5·99	5·93	5·86	5·79	5·72	5·66	5·59	—
18	6·81	5·96	5·89	5·83	5·76	5·69	5·63	5·56	5·49	—
19	6·69	5·86	5·79	5·73	5·66	5·60	5·53	5·47	5·40	—
20	6·57	5·76	5·69	5·63	5·56	5·50	5·44	5·38	5·31	5·18
21	6·46	5·66	5·59	5·53	5·47	5·41	5·35	5·29	5·22	—
22	6·35	5·56	5·50	5·44	5·38	5·32	5·26	5·20	5·13	—
23	6·24	5·47	5·41	5·35	5·29	5·23	5·17	5·11	5·04	—
24	6·14	5·38	5·32	5·26	5·20	5·14	5·09	5·03	4·95	4·84
25	6·04	5·29	5·23	5·17	5·12	5·06	5·00	4·95	4·86	—
26	5·94	5·20	5·14	5·09	5·03	4·97	4·92	4·86	4·78	—
27	5·84	5·11	5·06	5·00	4·94	4·89	4·83	4·78	4·70	—
28	5·75	5·02	4·97	4·91	4·86	4·80	4·76	4·69	4·62	4·55
29	5·66	4·93	4·88	4·83	4·77	4·71	4·66	4·60	4·54	—
30	5·57	4·85	4·79	4·74	4·68	4·63	4·58	4·52	4·46	—

* *Trans. Faraday Soc.* **5**, 68.

† *J. Appl. Chem.* **14**, 280.

Table XIII

Solubility of nitrogen in sea water (Rakestraw and Emmel, 1938)*
Expressed in ml. (at standard temperature and pressure) per l. of water, from a normal atmosphere of 760 mm.

t(°C)	Chlorinity (‰)						
	15	16	17	18	19	20	21
0	15·22	15·02	14·82	14·61	14·40	14·21	14·01
1	14·81	14·61	14·42	14·23	14·04	13·85	13·66
2	14·43	14·24	14·06	13·88	13·69	13·51	13·33
3	14·07	13·89	13·72	13·54	13·36	13·19	13·01
4	13·73	13·56	13·40	13·24	13·06	12·89	12·72
5	13·43	13·26	13·10	12·94	12·78	12·62	12·45
6	13·14	12·98	12·82	12·67	12·51	12·35	12·19
7	12·88	12·72	12·56	12·41	12·26	12·10	11·95
8	12·62	12·47	12·32	12·17	12·02	11·87	11·71
9	12·38	12·23	12·08	11·94	11·79	11·64	11·49
10	12·15	12·00	11·86	11·71	11·56	11·42	11·27
11	11·92	11·78	11·63	11·49	11·34	11·20	11·06
12	11·69	11·56	11·42	11·28	11·13	10·99	10·85
13	11·47	11·34	11·21	11·07	10·93	10·79	10·65
14	11·25	11·13	10·99	10·86	10·73	10·59	10·45
15	11·04	10·92	10·79	10·66	10·53	10·39	10·26
16	10·83	10·72	10·59	10·47	10·34	10·21	10·08
17	10·63	10·52	10·40	10·28	10·15	10·03	9·91
18	10·44	10·33	10·21	10·10	9·98	9·86	9·74
19	10·26	10·15	10·03	9·92	9·81	9·69	9·58
20	10·08	9·98	9·87	9·76	9·65	9·54	9·43
21	9·91	9·82	9·71	9·61	9·50	9·39	9·28
22	9·75	9·66	9·56	9·46	9·35	9·24	9·14
23	9·60	9·51	9·41	9·31	9·20	9·10	9·00
24	9·45	9·36	9·26	9·16	9·06	8·96	8·86
25	9·30	9·21	9·11	9·02	8·92	8·82	8·73
26	9·16	9·07	8·97	8·88	8·79	8·60	8·60
27	9·02	8·94	8·84	8·75	8·66	8·56	8·47
28	8·89	8·84	8·72	8·62	8·53	8·44	8·35

* *J. Phys. Chem.* **42**, 1211.

Author Index

(*Numbers in italics refer to pages in the References at the end of each Chapter*)

A

Aaronson, S., 512, 532, *595*, *600*
Adam, N. K., 117, *117*
Adams, G. A., 508, 592, *596*
Adams, I., 486, *605*
Adams, J. A. S., 193, *194*
Adams, R., 245, 248, 272, *319*
Adams, R. W., 253, *320*
Adel, A., 373, *404*
Agamirov, A., 193, *194*
Agamirov, S. S., 638, *644*
Agulhon, H., 149, *158*
Alexander, G. B., 410, 411, *430*
Aleyev, B. S., 592, *595*
Al Kholy, A. A., 523, 548, *595*, *604*
Allen, E. C., 22, *41*, 241, 247, 259, 281, *319*, *320*
Allen, M. B., 371, 372, *404*, 487, 491, 501, 544, 592, *595*, *597*, *604*
Allen, W. E., 589, *595*
Althouse, P. M., 438, 442, *475*
Altschuler, Z. S., 189, *194*
Amagat, E. H., 111, *117*
Anderegg, G., 130, *158*
Anderson, A. F. H., 493, *595*
Anderson, D. H., 247, 248, *317*, *321*
Anderson, E. C., 70, 71, 229, 296, 297, 308, *317*, *319*
Anderson, T. G., 144, 145, 146, 147, *158*
Angot, M., 553, *595*
Anselm, C. D., 139, *158*
Antia, N. J., 34, *38*, 506, 515, 517, 518, 524, 533, 547, 548, 550, 570, 583, 592, *595*
Archibald, A. R., 510, *595*
Armstrong, F. A. J., 325, 326, 330, 361, *362*, 412, 418, 419, 423, *430*, 455, 462, 471, *473*, *475*
Armstrong, R. L., 186, *196*
Arnold, J. R., 70, *71*, 229, 295, 297, 308, *317*, *319*
Arnon, D. I., 505, *595*
Arnow, P., 542, *595*
Arons, A. B., 262, *317*
Arrhenius, G. O. S., 26, *39*, 173, 180, 187, 190, 191, *194*, *195*, 414, *430*
Aruga, Y., 533, 534, *595*, *600*
Astrachan, L., 553. *600*
Atkins, W. R. G., 28, *38*, 335, 340, *362*, 389, *404*, 412, 416, 417, 423, *430*, 498, 561, *595*
Ault, W. U., 158, *158*
Austin, K. H., 324, 329, 356, *364*
Austin, T. S., 556, *601*

B

Baas Becking, L. G. M., 372, *404*
Baatz, I., 530, 546, *595*
Bachmann, R. W., 556, 558, *607*
Bachrach, E., 540, *595*
Bader, R. G., 439, 459, *473*
Bainbridge, R., 570, *595*
Bainbridge, V., 585, *595*
Baird, I. E., 532, *608*
Baker, H., 512, 532, *595*, *600*
Baker, M., 412, *431*
Balard, A. J., 13, *38*
Bandurkin, G. A., 194, *196*
Banewicz, J. J., 127, *158*
Bangham, D. H., 411, *431*
Banse, K., 515, *602*
Barashkov, G. K., 507, 510, 511, 512, 513, *595*, *607*
Barker, H. A., 532, 533, *595*
Barker-Jørgensen, C., 585, *595*
Barnes, C. A., 216, *223*, 426, *431*, 614, *644*
Barnes, H., 251, 259, *317*, 378, *404*
Barnes, W. J., 137, 150, *158*
Barth, T. W., 172, 173, *194*
Bassham, J. A., 512, *602*
Bates, R. G., 233, 235, 236, 238, *317*
Bather, J. M., 135, 146, 147, *158*
Bauer, L., 501, *595*
Bavendamm, W., 372, *404*

Bayer, E., 183, *194*
Baylor, E. R., 591, *595*
Beattie, A., 511, *595*
Becker, H. G., 315, *317*
Bein, W., 99, 107, *117*
Beklemishev, K. V., 566, 570, 571, 586, 587, 589, *595*
Belcher, J. H., 544, *598*
Belcher, R., 130, 134, 139, *158*
Bell, J. O., 439, 450, *473*
Belser, W. L., 453, 468, 470, *473*
Bendix, S. A., 491, 501, *595*
Benecke, W., 370, *404*
Benitez, A., 493, *607*
Bennett, E. L., 513, *608*
Benson, B. B., 202, 204, 206, 208, 217, 222, *223*, *224*, 368, 373, *404*, *407*, 616, 621, 625, 629, 634, *645*
Bentley, J. A., 448, 449, 455, *473*, 585, *595*
Berg, D., 253, 257, *317*
Berg, G., 532, 551, 556, 559, 576, *595*, *596*
Berglund, E., 138, 151, 152, *158*
Bergman, T. O., 9, 21, *38*, 123, *159*
Berkshire, D., 245, 248, 272, *319*
Bernard, F., 566, 579, 589, *596*
Berthelot, D., 373, *404*
Bertrand, G., 128, 149, *158*, *159*
Beyers, R. J., 245, *317*
Bhavanarayana, P. V., 421, *431*
Bidwell, R. G. S., 510, *596*
Bien, G. S., 20, *40*, 217, *224*, 301, 302, 303, 304, 310, *317*, 412, *430*
Bigelow, H. B., 329, *362*, 384, *404*, 423, *430*
Billett, F., 256, *320*
Bird, J., 286, 288, 304, *317*
Birge, E. A., 507, 527, *596*
Bischof, W., 290, *317*
Bishop, C. T., 508, 592, *596*
Bjerknes, V., 102, *117*
Bjerrum, J., 171, *195*
Black, W. A. P., 185, *195*, 513, 514, *596*
Blumer, M., 455, 472, *473*
Boalch, G. T., 455, 462, 471, *473*
Bohr, C., 256, 309, 312, *317*
Bold, H. C., 543, *605*
Bolin, B., 282, 283, 290, 291, 292, 293, 294, 296, 308, 309, 313, 316, *317*
Bond, R. M., 587, 593, *596*
Bongers, L. H. J., 376, *404*
Booth, T. E., 505, *607*
Booth, U. H., 253, *321*
Bouillon-Lagrange, E. J., 9, *38*
Boulton, R., 5, *38*
Bowden, K. F., 64, *71*
Bowen, H. J. M., 136, 148, *159*
Bowen, V. T., 183, 185, *195*, *196*
Boyle, R., 7, *38*
Braarud, T., 389, *404*, 536, 546, 576, 577, 579, 581, 582, 592, *596*, *599*
Bradley, W. F., 414, *431*
Bradshaw, A., 106, *119*
Bradshaw, B. C., 104, *118*
Bramlette, M., 187, *194*
Brander, E., 266, *317*
Brandhorst, W., 373, 391, *405*, 614, 625, *644*
Brandt, K., 26, 27, 28, *39*, 329, *362*, 374, *405*, 417, *430*, 504, 507, *596*
Brannon, H. R., 295, 297, 308, *317*
Bray, D., 294, *317*
Bray, V. B., 279, *321*
Brennecke, W., 20, *39*
Brice, B. A., 99, *117*
Brinkley, S. R., 266, *320*
Brisou, J., 372, *405*
Bristol-Roach, B. M., 543, *596*
Britton, H. T. S., 236, *317*
Brodia, H. P., 368, *405*
Brodie, J. W., 301, *317*
Broecker, W. S., 71, *71*, 286, 287, 288, 298, 299, 300, 301, 304, 305, 306, 308, 309, 310, 311, *317*, *318*, 642, *644*
Broenkow, W. W., 625, *644*
Brongersma-Sanders, M., 642, *644*
Brown, A. H., 523, *596*, *609*
Brown, F., 513, *596*
Brown, J. S., 491, 501, *595*, *596*
Brown, N. L., 90, 104, 106, *117*
Brown, T. E., 540, 547, *596*, *598*
Bruce, H. E., 334, *362*
Bruce, J. R., 585, 587, *596*
Bruneau, L., 248, *318*
Brust, A. F., 334, *363*
Bruun, A. F., 413, *430*
Buch, K., 22, *39*, 150, *159*, 222, *223*, 237, 239, 241, 244, 250, 251, 252, 256,

259, 260, 261, 264, 266, 267, 268, 270, 281, 290, 291, *318*, *319*, 329, *362*, 560, 561, *596*, 650, 659, 670
Buchanan, J. Y., 15, *39*
Buljan, M., 26, *39*
Bumpus, D. F., 390, *407*, 564, 571, *606*
Bunt, J. S., 574, *596*
Burkholder, P. R., 569, 574, *596*
Burling, R. W., 301, 310, *317*, *318*
Burns, F., 179, *195*
Burr, G. O., 527, *604*
Buser, W., 192, *195*
Bush, P., 548, *598*
Butler, E. I., 330, 361, *362*, 418, 419, *430*

C

Calamai, 124, *159*
Caldwell, W. E., 37, *39*
Cali, L. J., 253, *320*
Callendar, G. S., 294, *318*
Calvin, M., 493, 512, 513, 523, *595*, *602*, *608*, *609*
Cameron, A. T., 152, 153, *159*
Cameron, W. M., 619, *644*
Capurro, L. R. A., 574, *597*
Carey, C. L., 29, *41*, 370, 376, 378, *405*, *408*
Carles, P., 153, *159*
Carlisle, D. B., 183, *195*
Carnot, A., 139, 153, *159*
Carpenter, J. H., 80, *118*, 125, 127, 130, 131, 136, 140, 141, 142, 143, *159*
Carpenter, W. B., 20, *39*
Carrier, G. F., 47, *72*
Carritt, D. E., 80, *118*, 125, *159*, 200, 210, 222, *223*, 333, 335, *362*, *363*, 413, *430*
Carroll, D., 369, *405*
Carruthers, J. N., 38, *39*
Carter, D. B., 400, *405*
Carter, P. W., 512, *596*
Caspers, H., 617, *644*
Cassie, R. M., 568, *596*
Cassie, V., 513, 548, *599*
Causeley, D., 548, *598*
Cayle, T., 530, *596*
Cerruti, A., 587, *596*
Chacko, P. I., 332, *362*
Chapman, D. J., 487, *596*
Chapman, E. T., 435, *475*
Chapman, H. W., 290, *318*
Chapman, M. W., 368, *405*
Chatterjee, B., 411, *430*
Cheek, C. H., 210, *224*
Chelle, L., 138, 151, 152, *159*
Chevallier, A., 116, *119*
Chichester, C. O., 493, *610*
Chigirin, N. I., 643, *644*
Chow, D. T. W., 411, *430*
Chow, T. J., 127, 132, 136, 140, 141, 148, 149, *159*, 179, 180, *195*, 639, *644*
Chown, H. B., 324, *362*
Christensen, E., 591, *601*
Chu, S. P., 543, *596*
Claes, H., 493, *596*
Clagett, C. O., 438, 442, *475*
Clark, H. T., 511, 512, *597*
Clark, M. W., 236, *318*
Clarke, F. W., 172, *195*, 403, *405*, 412, *430*
Clarke, G. L., 112, 113, *118*, 591, 593, *598*
Clarke, R. S., 189, *194*
Clausman, P., 139, 153, *159*
Close, R. A., 130, *158*
Cloud, P. E., 228, 259, *318*
Clowes, A. J., 352, 353, 354, 355, 358, *362*, 420, 421, 423, 427, 428, *430*
Coblentz-Mishke, O. I., 568, *597*
Cockburn, A., 205, *224*, 672
Coe, W. R., 587, 590, *597*
Cohn, E. J., 22, *40*
Colefax, A. N., 331, *362*
Collias, E. E., 216, *223*, 614, *644*
Collier, A., 439, 450, *473*, 593, *597*
Collins, C. B., 158, *159*
Conger, P. S., 416, *430*
Conover, R. J., 585, 587, 588, *597*
Conover, S. A. M., 329, 361, *363*, 384, *407*, 579, 582, *597*
Contois, D. E., 412, *430*
Conway, E. J., 173, *195*
Conway, R. A., 401, *406*, 624, 625, *644*
Coombs, C. F., 514, *597*
Cooper, L. H. N., 28, 29, *39*, 242, 270, 287, *318*, 324, 325, 361, *362*, 372, 377, 378, 383, 384, 387, 389, *405*, 416, 418, 423, *430*, 570, *600*
Corcoran, E. F., 544, 577, *601*
Corlett, J., 556, *597*
Corner, E. D. S., 512, 586, 590, *597*

Cornell, D. L., 505, *597*
Correns, C., 414, *430*
Corwin, N., 203, 205, 209, *224*, 325, 328, 361, *363*, 384, 385, 386, *406*, *474*, 518, *601*, 636, *644*
Cowey, C. B., 455, *473*, 512, 513, 590, *597*
Cox, J. D., 256, *318*
Cox, R. A., 74, 79, 80, 82, 85, 88, 90, 93, 98, 106, 111, 116, *118*, 125, *159*
Craig, H., 70, *71*, 155, *159*, 208, *224*, 229, 287, 288, 298, 301, 306, 307, 309, *318*
Cramer, M., 520, 543, *604*
Creac'h, P., 454, *473*
Creager, J. S., 617, *644*
Crease, J., 111, 112, *118*
Creitz, G. I., 497, 499, *597*
Cromwell, T., 356, 357, *364*, 427, *431*
Crowe, C., 286, *318*
Culkin, F., 74, 79, 98, *118*, 125, *159*
Cummings, N. W., 97, *118*
Curl, H., 183, 185, *196*, 533, 538, 542, 584, *597*
Currie, R. I., 492, 526, 533, 560, 575, *597*
Cushing, D. H., 570, 571, 577, *597*
Cuthbertson, W. F. J., 513, *596*

D

Daisley, K. W., 455, *473*
Dakin, W. J., 331, *362*, 436, *475*
Dales, R. P., 487, *597*
Dal Pont, G., 589, *597*
Danckwerts, P. V., 253, *318*
Danil'chenko, P. T., 643, *644*
Darken, M. A., 513, *597*
Das, B. C., 593, *597*
Datsko, V. E., 468, *474*
Datsko, V. G., 462, 468, 472, *473*, *474*, 631, 636, *644*
Daubrée, A., 24, *39*
Daughtry, A. C., 295, 297, 308, *317*
Davidson. V., 416, *431*
Davies, C. W., 278, *318*
Davis, H. C., 483, 586, 591, *597*
Davis, R., 256, 258, *319*
Davy, J., 12, *39*
Dawson, W. A., 617, *644*
Deacon, G. E. R., 31, *39*, 352, 355, *362*
Deason, T. R., 543, *605*
Defant, A., 44, 49, 53, 54, 57, 64, 66, *71*, 397, *405*
Degens, E. T., 447, 450, 451, 452, 453, 455, 460, *474*
Delsman, H. C., 331, *362*
Denigès, H., 138, *159*
Department of Scientific and Industrial Research, 205, *223*
Desbois, C., 37, *41*
Desgrez, M. A., 148, *159*
de Sousa, A., 130, *161*
de Vries, H., 286, 287, 298, 300, *318*, *320*
Dianova, E., 590, *609*
Diéhert, F., 417, *431*
Dietrich, G., 230, *318*, 458, 459, *474*
Dietz, R. S., 324, *362*, 414, *431*
Dieulafait, L., 24, *39*
Diller, V. M., 375, 389, *407*
Dingle, H. N., 308, *318*
Disteche, A., 238, *318*
Dittmar, W., 14, 15, 19, 20, 21, *39*, 75, 79, *118*, 124, 126, 132, 133, 138, 140, 141, 142, 143, 144, 145, 146, 151, 152, 153, *159*
Dobrzhanskaya, M. A., 638, *644*
Dole, M., 156, *159*, 221, *224*, 236, *319*, 368, 369, *405*
Donovan, E. J., 502, *603*
Dorrestein, R., 94, *118*
Doty, M. S., 551, 552, 553, 556, 568, 574, *597*, *601*
Dougherty, E. C., 487, *597*
Douglas, E., 208, *223*
Downing, A. L., 203, 204, 205, 208, 209, *225*
Drew, G. H., 372, *405*
Droop, M. R., 513, 540, 542, 543, 583, *597*, *598*
Dubuisson, M., 238, *318*
Duce, R. A., 179, *195*
Dugdale, R. C., 211, 217, *223*, 371, *405*, *406*, 544, *598*
Dugdale, V. A., 217, *223*, 371, *405*, *406*
Durocher, 24, *40*
Durum, W. H., 173, *195*
Duursma, E. K., 371, *405*, 437, 459, 460, 462, 466, 468, 473, *474*, 524, *598*

E

Eckart, C., 111, *118*
Edie, E. S., 436, *475*
Eddy, B. P., 510, *598*
Edsall, J. T., 253, *319*
Edwards, G. A., 314, *322*
Ehhalt, D., 157, *159*
Ekman, V. W., 111, 116, *118*
Ellis, A. J., 266, *319*
Ellis, B. A., 128, 142, 143, *159*
Elmore, H. L., 204, *223*
Emerson, R., 530, *596*
Emery, K. O., 189, 195, 324, *362*, 373, 403, 404, *405*, *407*, 412, *431*
Emmel, E. V. M., 200, 203, 205, 208, 210, *224*, 368, *406*, 629, *645*, 673
Enders, C., 445, *474*
Engel, K., 255, *320*
Engelhardt, W., 414, *430*
Eny, D., 512, *598*
Epperson, A. W., 128, *159*
Eppley, R. W., 543, *598*
Epstein, S., 155, 156, *159*
Epstein, S. S., 548, *598*
Eriksson, E., 278, 282, 283, 296, 308, *317*, *319*
Esterson, G. L., 94, *118*
Ethe, H., 5, *39*
Evans, B. S., 138, *159*
Eve, A. S., 34, *39*
Ewing, W. M., 286, 287, 288, 298, 301, 305, 306, 308, 310, *317*, *318*, 642, *644*
Eyster, H. C., 540, 547, *596*, *598*

F

Fairbairn, H. W., 413, *431*
Fairbridge, R., 267, *321*
Fage, L., 589, *596*
Farren, G. P., 370, *405*
Faurholt, C., 253, *319*
Faust, R. C., 100, *118*
Fay, P., 503, *598*
Feitel, R., 27, *39*
Feldman, I., 236, *319*
Felföldy, L. J. M., 522, 533, *598*
Fell, J. W., 591, *598*
Feltham, C. B., 590, *610*
Ferguson, J., 253, *320*
Fergusson, G. J., 229, 296, 297, 299, 301, 308, 310, *319*, *321*
Fieser, L. S., 438, *474*
Fieser, Mary, 438, *474*
Findenegg, I., 581, *598*
Fisher, L. R., 455, *473*, 590, *598*
Fleischer, W. E., 533, *598*
Fleming, I. D., 510, *598*
Fleming, J., 12, *39*
Fleming, R. H., 14, *40*, 49, *72*, 102, 117, *118*, *119*, *159*, 201, 215, 216, *223*, *224*, 279, *321*, 337, 351, 361, *362*, *364*, 374, 378, 395, 403, *405*, *407*, 502, 571, *598*, 617, 622, 623, 636, 641, *644*
Fleming, W. H., 140, 144, 146, 158, *159*
Fogg, G. E., 371, *405*, 439, 440, 447, 459, *474*, 503, 512, 513, 520, 530, 532, 537, 539, 544, 546, 549, 577, *596*, *598*, *601*
Folkard, A. R., 106, 117, *118*, *119*
Fonselius, S., 252, 304, 310, *319*
Forch, C., 18, *39*, 75, 97, 101, 102, *118*
Forchhammer, G., 13, 14, 23, 24, 26, 34, *39*, 75, *118*, 124, 139, 153, *159*
Ford, H. T., 369, *405*
Ford, J. E., 542, *598*
Fork, D. C., 487, *600*
Forsberg, K., 144, 145, *159*
Fox, C. J. J., 21, *39*, 203, 204, 205, 208, 209, *223*, 368, *405*, 672
Fox, D. L., 325, *362*, 412, *431*, 514, 518, 590, 593, *598*
Føyn, E., 212, *223*, 536, 592, *596*
Fraga, F., 545, *609*
Franck, J., 523, *598*
Frank, E. R., 253, *320*
Frederick, J. F., 508, 598
Frederikson, A. F., 137, 150, *159*
Frémy, 19, *39*
French, C. S., 491, 501, *595*, *596*
French, D. M., 253, *322*
Frenklikh, M. S., 193, *196*
Freudenberg, K., 439, *474*
Friedman, I., 155, 156, 157, *159*
Fritsch, F. E., 483, *598*
Frysinger, G. R., 106, *118*
Fujita, Y., 496, *598*
Fukai, R., 134, 140, 142, 144, 145, 146, 147, *159*
Fukase, S., 394, *405*

Fukudu, M., 568, *601*, *604*
Fuller, J. L., 591, *598*

G

Gaarder, K. R., 577, 579, *596*
Gaarder, T., 554, *598*
Gade, H. G., 616, 620, 621, *644*
Ganapati, P. N., 421, *431*
Garner, D. M., 301, 310, *318*, *319*
Garrels, R. M., 121, *159*, 167, 178, 189, *195*, 255, 278, 279, 280, 281, 282, *319*
Gast, J. A., 137, 150, 151, *159*, 166, 179, *195*, 250, *319*
Gaudechon, H., 373, *404*
Gaudin, A. M., 515, *598*
Gauld, D. T., 585, *598*
Gautier, A., 139, 153, *159*
Gay-Lussac, J. L., 12, *39*
Gazert, H., 27, *39*
Gee, R., 543, *598*
Gellis, S. S., 593, *598*
George, P. C., 332, *362*
Gerard, R., 286, 287, 288, 298, 301, 305, 306, 308, 310, *317*, *318*
Gering, R. K., 513, *606*
German, C., 95, *118*
Gest, H., 370, *406*, 549, *599*
Gibor, A., 586, 587, 588, *599*
Gibson, R. O., 152, 153, *159*
Giese, A. C., 587, *600*
Gillbricht, M., 462, 464, *474*, 499, 590, *599*
Gilson, H. C., 373, 391, *405*
Girin, Yu. P., 194, *196*
Glawe, R., 445, *474*
Gleason, L. S., 290, *318*
Gleuckauf, E., 198, *223*
Göbel, F., 13, *39*
Goering, J. J., 217, *223*, 371, *406*
Gold, V., 233, 236, *319*
Goldberg, E. D., 8, 24, 26, 35, *39*, *41*, 169, 173, 174, 179, 180, 183, 184, 185, 186, 187, 190, 191, 192, 193, 194, *195*, *196*, 412, 413, 414, 415, *431*, 539, 547, 549, *599*, 637, *644*
Goldie, E. H., 590, *598*
Goldman, C. R., 537, 538, 540, *599*
Goldschmidt, V. M., 25, *39*, 149, *159*
Gololobov, Y. K., 400, 401, *405*
Goodgal, S., 333, *362*, 413, *430*
Goodwin, T. W., 487, 493, 495, 501, *595*, *599*
Gorgy, S., 325, *362*
Goto, K., 410, *431*
Goulding, K. J., 508, *604*
Graf, D. L., 186, *195*
Gräf, O., 27, *40*
Graham, H. W., 355, *362*, 426, *431*
Graham, J., 192, *195*
Graham, M., 335, 336, *362*
Gran, H. H., 554, 568, 576, 583, *598*, *599*
Granick, S., 491, *599*
Graschenko, S. M., 182, 193, *195*, *196*, 638, 639, *644*, *645*
Grauer, A., 522, *604*
Greenberg, D. M., 22, *41*, 241, 247, 281, *320*
Greenberg, G., 259, *319*
Greenhalgh, R., 74, 79, 98, *118*, 125, 138, 139, 154, *159*, *160*
Greenwald, I., 278, *319*
Griffin, J. J., 193, *195*, 413, 414, *431*
Grim, R. W., 174, *195*, 414, *431*
Gripenberg, S., 22, *39*, 126, 127, 140, 141, 142, 143, *160*, 241, 244, 247, 249, 250, 251, 256, 259, 260, 261, 266, 267, 270, 281, *318*, *319*
Grøntved, J., 577, 579, 584, *596*, *599*
Gross, F., 566, 580, *599*
Gross, M. G., 617, 636, *644*
Grütter, A., 192, *195*
Gubin, F. A., 628, 635, 641, 642, *645*
Gucluer, S. M., 617, 636, *644*
Guenther, G. R., 356, *362*
Guillard, R. R. L., 376, *405*, 439, 447, *474*, 513, 522, 537, 538, 539, 540, 542, 543, 548, 576, 579, 580, 583, 586, 589, 592, *597*, *599*, *600*, *602*, *606*
Gunnerson, C. G., 412, *431*
Guntz, A. A., 76, *118*
Guyer, A., 315, *319*
Gysling, H., 129, *160*

H

Haber, F., 37, *40*
Haffty, J., 173, *195*
Hagmeier, E., 437, *474*, 515, 589, *599*, *602*
Halldal, P., 506, 549, 565, *599*
Halwer, H., 99, *117*

Hamada, A., 256, *320*
Hamberg, A., 21, *40*
Hamilton, R. D., 383, *405*
Hamm, R. E., 210, *224*, 520, *599*, 629, 641, *644*
Hamon, B. V., 88, 89, 90, 93, 95, 104, 106, *117*, *118*
Hansen, A. L., 325, *362*
Hansen, V. Kr., 498, 522, 533, 534, 553, 568, 575, 576, *599*, *602*, *604*
Hantschmann, D., 468, *474*
Hanya, T., 269, 314, *319*
Hardin, G., 487, 493, *608*
Harding, M. W., 150, *160*
Hardman, Y., 29, *41*
Hardy, A. C., 570, *599*
Harned, H. S., 178, *195*, 232, 233, 258, *319*
Harris, E., 377, 384, 387, *405*, 503, 537, 565, 581, *599*
Harris, J., 157, *159*
Harrison, J. L., 414, *431*
Hart, S. M., 413, *431*
Harvey, H. W., 22, 28, 29, *38*, *39*, *40*, 239, 241, 245, 251, 256, 259, 260, 261, 264, 270, 276, 281, *318*, *319*, 325, 326, 328, 360, *362*, 376, 377, 379, 384, 387, 389, *405*, 423, *430*, 455, *474*, 479, 497, 498, 520, 537, 538, 539, 540, 543, 548, 570, 577, 590, *599*, *600*, 631, 642, *644*
Hashimoto, J., 513, *600*
Haskin, H. H., 565, *605*
Haslam, J., 138, 152, 153, *159*, *160*
Hasle, G. R., 565, 568, 579, *600*
Hasler, A. D., 587, *600*
Hastings, J. W., 546, 553, *600*, *608*
Hattori, A., 496, *598*
Hauschild, A. H. W., 530, *600*
Haxo, F. T., 487, 528, 552, *596*, *600*
Hayes, F. N., 297, *319*
Hayes, F. R., 324, *364*
Hayes, T. W., 204, *223*
Hays, E. E., 101, *119*
Head, A. J., 256, *318*
Headridge, J. B., 129, *160*
Heezen, B. C., 286, 287, 288, 298, 301, 305, 306, 308, 310, *317*, *318*, 642, *644*·
Heilbron, I. M., 512, *596*
Henderson, L. J., 22, *40*
Hendey, N. I., 417, *431*
Hendricks, S. B., 414, *431*
Henry, B. S., 376, *406*
Hercher, R., 7, *40*
Herlinveaux, R. H., 617, 621, *644*
Hernegger, F., 35, *40*
Heston, W. M., 410, 411, *430*
Hewitt, E. J., 539, 540, *600*
Hidaka, K., 426, *431*
Hildebrand, G. P., 130, *160*
Hindman, J. C., 259, *319*
Hinkelmann, H., 96, *118*
Hin Lew, 93, *118*
Hirsekorn, H., 99, 107, *117*
Hirst, E. L., 511, *595*
Hoering, T., 368, 369, *405*
Holden, M., 501, *600*
Hollaender, A., 383, *406*
Holland, H. D., 639, *644*
Holme-Hansen, O., 512, *602*
Holmes, R. W., 524, 526, 552, 579, *600*
Holser, W. T., 111, *119*
Holt, L. E., 324, *362*
Hood, D. W., 137, 150, 151, *160*, 235, 238, 241, 242, 244, 245, 246, 248, 250, 252, 255, 269, 270, 272, 273, 276, 314, 315, *319*, *320*, *321*, 334, *362*, 447, 450, 451, 453, 472, *473*, *474*, *475*, 511, 522, 561, *600*, *601*, *605*, *607*
Hooke, R., 5, *40*
Horibe, Y., 155, 156, *160*
Horne, R. A., 106, *118*
Horne, R. L., 508, *604*
Hosohara, K., 180, *195*
Hotchkiss, M., 370, 378, *408*
Hough, L., 508, *600*
Huang, H., 587, *600*
Huber, B., 290, *319*
Hughes, E. O., 508, 592, *596*
Hummell, R. W., 136, 148, *160*
Hulburt, E. M., 567, 576, 579, 581, 583, 589, *600*, *604*, *606*
Hundley, J. M., 512, *600*
Hurley, P. M., 413, *431*
Hutchinson, G. E., 242, 296, 297, *319*, 367, 369, 373, 389, 403, *405*, *406*, 526, *600*
Hutner, S. H., 532, 544, *600*

I

Ibert, E. R., 246, 252, 269, 270, 273, 276, 314, 315, *319*, *321*
Ichimura, S., 533, 534, *600*
Igelsrud, I., 150, *160*
Iler, R. K., 410, 411, *430*, *431*
Imamura, H., 585, *601*
Ing, R. B., 512, *600*
Ingerson, E., 228, *319*
Ingri, N., 411, *431*
Inoue, N., 568, *601*, *604*
Irving, L., 314, *322*
Iselin, C, O'D., 392, *406*
Ishibashi, M., 26, *40*, 325, *362*
Ishibashi, Y., 325, *362*
Ishii, J., 412, *431*
Ishikawa, T., 412, *431*
Ishiwatari, R., 269, 314, *319*
Issatchenko, B. L., 370, *406*
Iwamoto, H., 507, 510, 549, *601*
Iwamura, T., 513, *601*

J

Jacob, H., 584, *602*
Jacobsen, A. W., 94, *118*
Jacobsen, J. P., 18, 20, *39*, *40*, 77, *118*
Jacobsen, O., 19, *40*
James, H. R., 112, 113, *118*
Jannasch, H. W., 591, *601*
Jayaraman, R., 332, 333, *363*, *364*, 420, 421, *431*
Jeener, R., 513, *601*
Jeffrey, L. M., 241, *321*, 450, 453, *474*, *475*
Jeffrey, S. W., 487, 490, 491, 493, 501, *601*
Jenkins, P. G., 412, *430*, 532, 555, *601*
Jennings, P. P., 253, *319*
Jensen, A., 532, 563, *604*
Jentoft, R. E., 133, 145, 146, *160*
Jerlov, N. G., 248, *318*, 455, 462, *474*, 527, *601*
Jitts, H. R., 334, *363*, 521, 546, 551, 555, 556, 582, *601*
Johnson, M. W., 49, *72*, 102, *119*, 201, *224*, 279, *321*, 337, 351, *364*, 378, 395, 403, *407*, 418, *431*, 617, 622, 636, 641, *645*
Johnson, W. R., 146, 147, *160*
Johnston, J., 279, *321*, 543, *604*
Johnston, R., 447, 448, 455, *474*, 539, 584, 585, *601*
Joly, J., 25, 34, *40*
Jones, G., 104, *118*
Jones, G. E., 591, *601*
Jones, J. K. N., 508, *600*
Jones, R. F., 546, *603*
Jørgensen, E. G., 415, 416, *431*, 505, 534, 538, 553, 584, *601*, *604*
Joseph, J., 69, *71*, 93, *120*
Juday, C., 507, 527, *596*
Junge, C. E., 292, *320*, 369, *406*

K

Kain, J. M., 530, 532, 546, *601*
Kakimoto, D., 455, *474*
Kalle, K., 98, *119*, 205, *224*, 325, 326, 340, *363*, 436, 449, 455, 456, 462, 463, 473, *474*, *475*, 500, *601*, 648
Kamen, M. D., 370, *406*, 549, *599*
Kampitsch, E., 128, *160*
Kanazawa, A., 455, *474*
Kandler, O., 527, *601*, *607*
Kanwisher, J., 200, 211, 212, 213, *224*, 252, 278, 284, 285, 309, 314, 315, *320*, 519, 520, 529, *601*, *603*
Kaplan, L. D., 228, *320*
Karlick, B., 35, *40*
Karpov, A. V., 628, *645*
Karrer, P., 438, *474*
Karsten, G., 27, 28, *40*
Kashiwada, K., 455, *474*
Kato, K., 568, *608*
Kawagoe, K., 455, *474*
Kawamoto, T., 326, 356, *363*
Kawasaki, N., 135, 140, 141, 148, 149, *160*
Kay, H., 462, 468, *474*, 504, *601*
Kayama, M., 512, *601*
Keeler, R. F., 513, *601*
Keeling, C. D., 272, 276, 288, 289, 290 291, 292, 293, 294, 316, *317*, *320*
Keen, D. J., 328, *363*, *474*, 636, *644*
Kelly, P. B., 511, *601*
Kennedy, G. C., 111, *119*
Kenner, C. T., 127, *158*
Kentner, J., 370, *404*
Kern, D. M., 254, *320*
Kersten, H. J., 375, 389, *407*

Kessler, E., 376, 389, *406*, 539, *601*
Ketchum, B. H., 215, *224*, 323, 325, 326, 327, 328, 361, *363*, 374, 375, 384, 385, 386, 393, 400, *406*, *407*, 503, 507, 518, 537, 547, 549, *601*, *602*, 622, 627, 630, 636, 642, 643, *644*, *645*
Kettner, H., 445, *475*
Keys, A. B., 377, 387, *407*, 436, *474*, 591, *601*
Kientzler, C. F., 262, *317*
Kierstead, H., 578, *601*
Kimball, J. F., Jr., 544, 577, *601*
King, E. J., 416, *431*
King, H. H., 253, *320*
King, J. E., 556, *601*
Kinne-Diettrich, E. M., 543, *602*
Kink, M., 513, *608*
Kirk, P. L., 126, 127, 140, 141, *160*
Kirkpatrick, S. D., 37, *40*
Kitano, Y., 131, *160*
Kittredge, J. S., 514, *598*
Klem, A., 389, *404*
Kliashtorin, L. B., 545, 558, *608*
Knapman, F. W., 133, 144, *160*
Knauss, J. A., 58, *71*
Knight, L. W., 111, *119*
Knight, M., 585, 587, *596*
Knocke, W., 34, *40*
Knott, K., 157, *159*
Knudsen, M., 18, *39*, *40*, 75, 76, 77, 81, 97, 101, 102, 115, *118*, *119*
Kobanov, V. V., 130, *160*
Kobayakawa, M., 155, 156, *160*
Kobe, K. A., 256, *320*
Kocher, J., 76, *118*
Koczy, F. F., 68, *71*, 248, *318*, *320*
Koefoed, J., 255, *320*
Koide, M., 24, *39*, 169, 174, 180, 192, 193, 194, *195*
Kolyadin, L. B., 182, *195*, *196*, 638, 639, *644*, *645*
Kolotoff, S., 145, *160*
Kolthoff, I. M., 138, 144, *160*, 236, *320*
König, H., 20, *40*, 206, 207, 208, 217, *224*
Korn, O. P., 182, *196*, 639, *645*
Koroleff, F., 252, *319*
Korpi, E., 138, 152, 153, *160*
Koyama, T., 454, *474*
Kramer, D. D., 583, *607*
Kratz, W. A., 546, *602*
Krauskopf, K. B., 175, 176, 178, *195*, 410, *431*, 637, *644*
Krauss, R. W., 375, *407*, 506, 512, 545, 546, 548, *600*, *602*, *607*, *609*
Kreidl, M., 522, *604*
Kreps, E., 325, 326, 337, 339, *363*, 383, *406*, 436, *474*, 498, 561, *602*
Krey, J., 468, *474*, 499, 503, 513, 514, 515, 589, 590, *602*
Kriss, A. E., 373, 401, *406*, 632, 643, *644*
Krogh, A., 21, *40*, 252, *320*, 377, 403, *406*, 436, *474*, 591, 592, 593, *601*, *602*
Krotkov, G., 530, *600*
Krumbein, W. C., 189, *195*
Krümmel, O., 117, *119*
Kuenzler, E. J., 547, 549, *602*
Kuhl, A., 505, *602*
Kulp, J. L., 158, *158*, 286, 287, 298, *318*, 639, *644*
Kusnetsov, J., 182, *196*
Kuznetsov, Y. V., 193, *196*, 638, 639, *644*, *645*
Kuznezow, S. L., 591, *602*
Kuztetzou, Y. U., 182, *195*
Kylin, H., 508, *602*

L

Lacombe, H., 60, *72*
La Fond, E. C., 421, *431*
Laitinen, H. A., 236, *320*
Lambet, M. S., 639, *644*
LaMer, V. K., 324, *362*
Lamm, O., 253, *320*
Lance, J. R., 586, *605*
Lane, G. A., 368, 369, *405*
Lang, A. G., 334, *363*
Lang, J. W., 146, 147, *158*
Lange, E., 436, *474*, 592, *602*
Lange, N. A., 637, *644*
Langer, E., 413, *430*
Lanskaya, L. A., 506, 532, 546, 584, *602*
Lao, R. G., 361, *363*
Lasarev, K. F., 182, *196*
Lasker, R., 588, *602*
Lassøe, A., 498, *602*
Latimer, P., 501, *602*
Latimer, W. M., 178, *195*
Laurens, G., 23, *40*

Lavoisier, A. L., 7, 8, 20, *40*
Lazarev, K. F., 182, *195*, 638, 639, *644*, *645*
Lazaroff, N., 545, *602*
Lebour, M. V., 377, 387, *405*, 570, *600*
Lefèvre, M., 584, *602*
Legendre, R., 20, *40*
Leonard, M. A., 139, *158*
Leslie, M., 423, *430*
Levine, E. P., 183, *195*
Levring, T., 584, *602*
Lewin, J. C., 411, 415, 416, *431*, 486, 505, 510, 538, 540, 543, 544, 549, 580, *602*
Lewin, R. A., 439, 447, *474*, 483, 486, 506, 540, 543, 592, *602*
Lewis, G. J., 147, *160*, 174, *195*, 450, *474*, 637, *644*
Libby, W. F., 35, *40*, 286, *320*
Lichtenberg, F. D., 21, *40*
Lieth, H., 289, 290, *320*
Lillick, L. C., 329, *362*, 384, *404*
Linko, P., 512, *602*
Linnenbom, V. J., 210, *224*
Lisitsin, A. P., 412, 413, *431*, 514, 589, *602*, *603*
Liversidge, A., 24, *40*
Ljunggren, S., 253, *320*
Lockwood, S., 532, *600*
Lodge, J. P., 253, *320*
Lohmann, H., 570, 590, *603*
Loomis, W. F., 253, 290, *318*, *320*, 520, *603*
Loosanoff, V. L., 586, 591, *603*
Loose, L., 549, *605*
Lorenzen, C. J., 553, *603*
Loveland, J. W., 253, *320*
Lovern, J. A., 511, 512, *603*
Low, E. M., 512, 513, *603*
Lowden, G. F., 203, 204, 205, 208, 209, *225*
Lowenstam, H. A., 186, *195*
Lowman, F. G., 185
Lowndes, A. G., 412, *431*
Lowry, H., 522, *604*
Lucciardi, N., 540, *595*
Luck, J. M., 590, *603*
Ludwig, C. A., 375, 389, *406*
Lumby, J. R., 117, *119*
Lund, J. W. G., 551, *603*
Lyman, J., 14, *40*, 80, *119*, 140, 146, *159*, 250, 251, 253, 256, 259, 260, 261, *320*, *321*, 648, 649, 650, 651
Lythgoe, B., 512, *596*

M

Macadam, S., 23, *40*
Macallum, A. B., 145, *160*
McAllister, C. D., 34, *38*, 505, 506, 515, 517, 518, 521, 524, 532, 533, 536, 537, 544, 546, 547, 548, 550, 553, 556, 560, 568, 570, 582, 583, 589, 592, *595*, *601*, *603*
McClendon, J. F., 22, *40*
McCombie, A. M., 546, *603*
MacDonald, A. J., Jr., *406*
McGill, D. A., 393, 399, *406*
MacGinitie, G. E., 585, 590, *603*
McInnes, D. A., 258, *321*
MacKay, C., 285, *320*
McKelvey, V. E., 189, *195*
Mackereth, F. J., 538, *603*
McKinney, R. E., 401, *406*, 624, 625, *644*
McLaughlin, J. J. A., 540, 543, *598*, *605*
Mcleod, G. C., 528, 529, 533, 584, *597*, *603*
McMillan, R., 512, *609*
Macnamara, J., 158, *159*
Macquer, P. J., 7, *40*
Maddux, W. S., 546, *603*
Maeda, H., 413, *431*
Maeda, M., 146, 147, *160*
Magliocca, A., 335, *363*
Magnitzky, A. W., 439, 450, *473*
Mahncke, H. E., 150, 151, *160*
Makin, C. J. S., 144, 145, 152, *160*
Malaguti, 24, *40*
Mallette, M. F., 438, 442, *475*
Maloney, T. E., 502, *603*
Manheim, F. T., 187, *195*
Manners, D. J., 510, *595*, *598*
Manning, P. D. V., 36, *40*
Manning, W. M., 487, 493, *608*
Marcet, A., 10, 11, 12, 13, 23, *40*, 74, 75, *119*, 123, *160*
Marchand, E., 23, 24, *40*
Margalef, R., 497, 566, 580, *603*
Markali, J., 506, 581, *596*, *599*
Markham, A. E., 256, *320*

Markham, R., 513, *603*
Marrinan, H. J., 100, *118*
Marshall, S. M., *363*, 387, *406*, 585, 588, 590, *603*
Marsilli, L. F., 7, *40*
Matida, Y., 126, 140, 142, 146, 152, 153, *160*
Mathews, W. H., 619, 632, *645*
Matsudaira, Y., 426, *431*
Matsue, Y., 567, *603*
Matsuo, S., 230, 272, *320*
Matthews, D. J., 27, *41*, 109, 111, *119*, 128, 142, 143, *159*, 329, *363*
Maury, M. F., 124, *160*
Mayeda, T., 155, 156, *159*
Mayer, A. M., 565, *607*
Mazor, E., 208, *224*
Mazur, H., 511, 512, *597*
Mead, J. F., 512, *601*
Megia, T. G., 361, *363*
Melkersson, K. H., 253, *318*
Menard, H. W., 175, 190, 191, *196*
Menzel, D. W., 211, 217, *223*, 361, *363*, 370, 371, *405*, *406*, 455, 461, *475*, 492, 498, 500, 534, 539, 542, 544, 552, 563, 581, 583, 589, *598*, *603*, *604*, *607*, *608*, *610*
Menzies, R. J., 642, *644*
Meunier, J., 148, *159*
Meyer, C. M., 592, *604*
Migdisov, A. A., 194, *196*
Miller, R. M., 592, *604*
Miller, W. S., 591, *603*
Millot, G., 414, *431*
Mills, A., 253, 257, *320*
Milner, H. W., 375, *407*, 507, 549, *608*
Minkina, A. L., 340, *363*
Mitchell, R. L., 185, *195*
Miyake, Y., 107, *119*, 135, 138, 140, 142, 144, 145, 146, 148, 150, 151, 152, 153, *160*, 179, *196*, 230, 272, *320*
Mjaaland, G., 546, *604*
Moberg, E. G., 22, 41, 79, *119*, 126, 127, 140, 141, 150, *160*, 241, 247, 259, 281, *319*, *320*, 355, *362*, 423, 426, *431*
Moeller, Th., 169, *196*
Mokyevskaya, V. V., 182, *196*
Molins, L. R., 130, *160*
Möller, F., 228, *320*
Möller, L., 99, 107, *117*
Monakov, A. V., 588, *604*
Monsi, M., 533, *595*
Montgomery, H. A. C., 205, *224*, 672
Moore, B., 436, *475*
Moore, H. B., 333, *363*
Moorey, J., 88, 93, *118*
Morita, T., 455, *474*
Morren, S., 19, *41*
Morris, A. W., 76, *119*, 139
Morris, W. R., 593, *604*
Morrison, T. J., 256, *320*
Mortimer, C. H., 333, *363*
Morton, E. H., 13, *41*, 144, 146, 147, 152, *160*
Morton, J. E., 587, *604*
Moses, G., 138, *160*
Mulav, A. L., 515, *598*
Mullin, J. B., 184, *196*
Mullin, M. M., 455, 472, *473*, 585, *604*
Munk, W. H., 47, *72*, 566, *604*
Munnich, K. O., 286, 299, *320*
Münster, C. A., 24, *41*
Murchio, J. C., 491, 501, *595*, *604*
Murphy, J., 325, *363*
Murray, H. H., 414, *431*
Murray, J., 10, 12, 14, 16, 38, *41*, 413, *431*
Murray, R. G. E., 378, *406*
Myers, J., 520, 527, 534, 543, 546, *602*, *604*, *607*

N

Nagel, J. F., 157, *159*
Nakagawa, Y., 325, *362*
Nakayama, T. O. M., 493, *610*
Nalewajko, C., 447, 459, *474*
Nansen, F., 98, *119*
Nasanen, R., 258, *320*
Nathan, H. A., 532, *600*
Nathanson, A., 27, *41*
Natterer, K., 144, 145, *160*, 434, *475*
Naudé, S. M., 368, *406*
Naumov, G. B., 638, *644*
Nauwerck, A., 559, *606*
Neess, J. C., 217, *223*, 371, *406*, 544, *598*
Neill, J. M., 251, *321*
Nelson, C. D., 530, *606*
Nelson, D. J., 498, *604*
Nelson, T. C., 586, 587, 591, *604*
Neuberg, C., 522, *604*

Newcombe, C. L., 334, *363*
Newell, B. S., 331, 332, *363*, 539, 589, 597, *609*
Nicholas, D. J. D., 375, *406*
Nicholls, G. D., 183, 185, *196*
Nielsen, E. Steemann, 31, *41*, 371, *407*, 479, 520, 521, 522, 523, 527, 532, 533, 534, 551, 553, 556, 559, 563, 569, 570, 573, 576, 584, *599*, *604*
Nielsen, J. N., 398, *406*
Nihei, T. A., 505, *604*
Nikolaev, D. S., 182, *195*, *196*, 638, 639, *644*, *645*
Nisbet, M., 584, *602*
Nisikawa, T., 146, 147, *160*
Nishizawa, S., 568, *601*, *604*
Noakes, J. E., 137, 150, 151, *160*, 250, *320*
Noddack, I., 24, *41*
Noddack, W., 24, *41*
Noll, W., 148, *160*
Nordli, E., 532, 546, 581, 583, *596*, *604*
Norin, E., 414, *431*
Northcote, D. H., 508, 510, *604*, *605*
Nowak, F. A., 253, *320*
Nultsch, W., 565, *604*
Nutman, P. S., 370, 371, *406*
Nynäs, O., 239, *318*

O

O'Brien, H., 522, *608*
Occhipinti, A. G., 335, *363*
O'Connor, R. F., 515, *598*
Odum, H. T., 135, 148, 149, *161*, 242, 244, 245, 246, *317*, *320*, 561, *605*
Ogata, Y., 146, 147, *160*
Ogino, C., 507, 512, *605*
Oguri, M., 551, 552, 556, 568, *597*
OhEocha, C., 487, 496, *605*
Ohwada, M., 567, *605*
Okamoto, G., 410, *431*
Okubo, A., 69, *72*
Okuno, T., 146, 147, *160*
Okura, T., 410, *431*
Olaitan, S. A., 510, *605*
Oleson, J. J., 542, *595*
Olson, E. A., 299, 304, 305, *317*
Oppenheimer, C. H., 514, *598*
Orr, A. P., 244, *320*, 331, 337, *363*, 375, 387, 389, *406*, 585, 588, 590, *603*
Orr, W. L., 373, 403, 404, *405*, *407*
Osadchih, M., 325, 326, *363*
Osborn, E. M., 253, *319*
Osmond, J. K., 193, *194*
Östlund, G., 158, *161*, 286, 304, 310, *319*, *320*
Ostroff, R., 376, *406*
Owen, B. B., 178, *195*, 232, 233, 266, *319*, *320*

P

Paasche, E., 506, 522, 534, 568, 579, *605*, *609*
Pachadzhanov, D. N., 194, *196*
Pakhomova, M. V., 510, 512, 513, *607*
Palitzsch, S., 22, *41*
Palmork, K. H., 452, *475*
Pandow, M., 285, *320*
Pape, C., 99, *119*
Paquette, R. G., 89, *119*
Park, K., 242, 244, 246, *320*, 450, *475*, 522, 561, *600*, *605*
Parke, M., 486, 498, 585, 587, *595*, *596*, *605*
Parker, C. A., 137, 150, *158*
Parker, B. C., 543, *605*
Parker, P. D. M., 202, 204, 206, 208, 217, *223*, 368, *404*
Parker, R. H., 193, *195*
Parr, A. E., 399, *406*
Parsons, T. R., 34, *38*, 251, *321*, 325, *364*, 487, 490, 492, 493, 496, 499, 500, 503, 504, 505, 506, 507, 508, 509, 510, 511, 512, 513, 515, 516, 517, 518, 519, 521, 524, 533, 537, 544, 547, 548, 550, 553, 560, 566, 568, 569, 570, 583, 589, 592, 593, *595*, *603*, *605*, *608*
Pate, B. D., 325, *364*
Pate, J. B., 130, 140, 141, 142, 143, *161*
Patten, B. C., 533, 580, *605*
Patterson, C. C., 181, *196*
Patterson, A., 253, 257, *317*, *322*
Patterson, J., 492, *605*
Pauly, H., 413, *430*
Payen, J., 508, *605*
Pearsall, W. H., 549, *605*
Pearson, L., 253, *320*
Pensar, G., 131, *161*
Percival, E., 511, *595*
Perkins, H. J., 493, *605*

Perry, D., 295, 297, 308, *317*
Peters, Cl., 149, *159*
Petersen, R. A., 532, *600*
Peterson, M. N. A., 193, *195*
Petracek, F. J., 501, *605*
Petrikova, M. N., 566, *595*
Petterson, H., 34, 35, *41*
Pfaff, C., 13, *41*
Phagpolngarm, S., 487, *595*
Phelps, A., 418, 423, *431*
Phifer, L. D., 361, *363*, 387, *404*, 418, *431*
Philpott, D. E., 486, *602*
Picciotto, E., 187, 193, *194*, *195*
Pickard, G. L., 618, 619, 632, *645*
Piggot, C. S., 34, *41*
Pillai, C. M., 332, *362*
Pinsent, B. R. W., 253, *320*
Pinson, W. H., 413, *431*
Pintner, I. J., 540, 543, *598*, *605*
Plass, G. N., 228, 308, *320*
Plunkett, M. A., 371, 377, *404*, 462, 465, *475*
Pobiner, H., 253, *321*
Pokorny, J., 591, *608*
Pollak, M. J., 101, 104, *119*, 399, *406*
Pomeroy, L. R., 565, 578, *605*, *606*
Popova, T. P., 639, *645*
Portmann, J. E., 325, *363*
Portratz, H. A., 35, *41*
Postma, H., 215, *224*, 242, 245, 249, 293, *321*, 358, *363*, 473, *475*
Prasad, R. R., 421, *431*
Pratt, D. M., 334, *363*
Prescott, G. W., 584, *605*
Prescott, J. M., 450, 451, 472, *475*
Price, J. B., 130, *161*
Priddy, R. R., 130, *161*
Pringsheim, E. G., 543, *605*
Pritchard, D. W., 619, *644*
Proctor, V. W., 537, 584, *605*
Proudman, J., 56, 66, *72*
Provasoli, L., 434, 447, 455, *475*, 537, 540, 543, 544, 586, 588, 593, *598*, *605*, *607*
Pshenina, T. I., 506, *602*, 638, *644*
Puff, A., 27, *41*
Purcell, A. E., 501, *606*
Putnam, G. L., 37, *41*
Pütter, A., 435, *475*, 588, 593, *606*
Pytkowicz, R. M., 267, *321*

R

Raben, E., 374, *405*, 417, *431*, 435, *475*, 504, 507, *596*
Rabinowitch, E. I., 490, 521, 528, 531, *606*
Radczewski, O. E., 413, *431*
Rafter, T. A., 301, 308, 310, *321*
Ragotzkie, R. A., 565, 578, *605*, *606*
Rakestraw, N. W., 20, 29, *38*, *40*, *41*, 150, 151, *160*, 200, 203, 205, 208, 210, 217, 221, *224*, 246, 272, 276, 290, 301, 302, 303, 304, 310, 316, *317*, *320*, *321*, 325, 334, 335, *362*, *363*, 368, 371, 377, 378, 380, 383, 384, 389, 390, 399, 400, *406*, *408*, 450, 462, 465, *474*, 629, *645*, 673
Rama, 180, *196*
Ramage, H., 148, *161*
Ramamurthi, S., 332, *363*
Ramsey, W. L., 213, *224*
Rankama, K., 403, *406*
Rao, S. V. S., 326
Ratmanoff, G. E., 138, 151, 152, *161*
Ray, S. M., 439, 450, *473*
Raymont, J. E. G., 378, *406*, 479, *606*
Redfield, A. C., 204, 212, 215, 217, 221, *224*, 323, 325, 326, 327, 334, 350, 360, *363*, 374, 375, 377, 387, 393, 395, 400, *406*, *407*, 503, 507, *601*, 622, 627, 630, 642, 643, *645*
Reilley, C. N., 129, 130, *160*, *161*
Reiser, R., 511, *601*
Reisner, G. A., 513, 539, *606*
Renard, A. F., 413, *431*
Renn, C. E., 378, 380, *408*
Reuszer, H., 29, *41*, 378, *408*
Reuter, J. H., 447, 450, 451, 452, 453, 455, 460, *474*
Revelle, R. R., 22, *41*, 79, 117, *118*, *119*, 201, 213, 218, 223, *224*, 229, 241, 259, 267, 278, 281, 295, 296, 297, 301, 308, 309, *320*, *321*
Rex, R. W., 413, 414, *431*
Reynolds, R. C., 137, 150, *159*
Rho, Joon Hee, 511, 512, 514, *606*
Rial, J. R. B., 130, *160*
Rice, T. R., 547, *606*
Richards, F. A., 31, *41*, 182, 203, 205, 209, 212, 215, 216, 219, 220, 221, 222, *224*, 361, *364*, 373, 393, 400, *407*, 413,

419, 424, 425, 429, *431*, 493, 497, 498, 499, *597*, *606*, 616, 617, 618, 620, 621, 622, 625, 626, 627, 629, 630, 631, 632, 634, 636, 639, 642, 643, *645*
Richardson, W. S., 97, *119*
Richman, S., 588, *606*
Richter, G., 540, 547, *606*
Richter, O., 28, *41*
Riley, G. A., 216, *224*, 329, 350, 361, *363*, 371, 377, 384, 390, 393, 399, *405*, *407*, 455, *475*, 503, 514, 561, 564, 566, 567, 568, 571, 573, 576, 577, 590, 591, 593, *599*, *604*, *606*
Riley, J. P., 74, 76, 79, 98, *118*, *119*, 125, 135, 138, 139, 146, 147, 154, *158*, *159*, 184, *196*, 325, *363*, 504, 505, *609*
Ringbom, A., 131, *161*
Ringdal, K., 581, *596*
Rippel-Baldes, A., 439, 440, *475*
Rittenberg, S. C., 373, 403, 404, *405*, *407*
Roberts, D. W. A., 493, *605*
Robinson, A., 71, *72*
Robinson, C., 37, *41*
Robinson, J. W., 134, *158*
Robinson, E. L., 502, *603*
Robinson, R. A., 166, *196*, 262, 278, *321*
Robinson, R. J., 130, 133, 139, 140, 141, 142, 143, 144, 145, 146, 148, 149, *158*, *160*, *161*, 247, 248, *317*, 325, 329, *362*, *364*, 411, *430*
Rochford, D. J., 98, *119*, 328, 333, 359, *364*
Rodhe, W., 538, 544, 548, 553, 559, 567, 582, *606*
Rodina, A. G., 590, 591, *606*
Rodman, J., 581, *600*
Rodriguez, E., 532, *600*
Rogall, E., 415, 416, 417, *431*
Rogers, J. J. W., 193, *194*
Rom, A. M., 568, *605*
Ross, C. S., 414, *431*
Roughton, F. J. W., 253, 257, *320*, *321*
Rubey, W. W., 367, *407*
Rudd, DeF. P., 221, *224*, 368, 369, *405*
Russell, F. S., 377, 387, *405*, 570, *600*
Ruud, J. T., 352, *364*
Ryley, J. F., 510, *595*
Ryther, J. H., 211, 214, 217, *223*, *224*, 361, *363*, 371, 374, 375, 389, 401, 403, *405*, *407*, 504, 518, 520, 522, 523, 524, 532, 534, 538, 539, 540, 542, 544, 551, 552, 553, 557, 560, 562, 567, 568, 572, 573, 575, 576, 578, 579, 581, 583, 589, *598*, *599*, *600*, *601*, *603*, *604*, *606*, *607*, *610*

S

Sachs, T., 565, *607*
Sackett, W. M., 35, *41*
Sage, B. G., 7, *40*
Sahama, T. G., 403, *406*
Saijo, Y., 533, 534, *600*
Sakurai, S., 150, 151, *160*
Saltman, P., 543, *598*
Samejima, H., 543, *607*
Sanders, M., 532, *600*
Sandström, J. W., 102, *117*
Sargent, M. C., 534, *607*
Saruhashi, K., 252, 255, 259, 268, *321*, 520, *607*, 652, 660
Satterly, J., 34, *41*
Saunders, G. W., 556, 558, *607*
Scagel, R. F., 553, *610*
Scheffer, F., 438, *475*
Schelske, C. L., 539, *607*
Schiff, J., 545, *602*
Schinske, R. A., 593, *608*
Schleicher, K., 106, *119*
Schloesing, T., 144, 145, *161*
Schmalfuss, K., 438, 439, *475*
Schmelck, L., 14, 15, *41*, 128, 144, 145, *161*
Schmid, R. W., 129, 130, *161*
Schmidt, C., 24, *41*, 144, 152, *161*
Schmidt, Johannes, 614, *645*
Schmidt, R. A., 24, *39*, 169, 180, 194, *195*
Schoen, B., 157, *159*
Scholander, P. F., 314, *322*
Scholes, S. R., 258, *319*
Schreiber, E., 536, *607*
Schükerk, W., 510, 512, *607*
Schuler, J. F., 375, 389, *407*
Schwarzenbach, G., 128, 130, *158*, *160*, 166, 171, *195*, *196*
Scott, B. D., 521, *601*
Scott, G. T., 506, 548, *607*
Scripps Institution of Oceanography (1961), 397, *407*

Sears, M., 329, *362*, 384, *404*
Seaton, M. Y., 36, *41*
Seiwell, F. E., 436, *475*
Seiwell, G. E., 350, *364*
Seiwell, H. R., 335, 341, 342, 344, 345, 346, 347, 348, 349, 350, *364*, 377, *407*, 436, *475*
Semina, G. I., 566, *595*
Sendner, H., 69, *71*
Sendroy, J., 251, *321*
Senn, H., 130, *158*
Serenkov, G. P., 507, 510, 512, 513, *607*
Seshappa, G., 333, *364*
Shaffer, J. D., 101, *119*
Shah, N., 532, 533, 536, 560, *603*
Shapiro, J., 449, *475*
Shaw, K. N. F., 447, 450, 451, 452, 453, 455, 460, *473*
Shedlovsky, T., 258, *321*
Sheets, G., 590, *603*
Sheldon, R. F., 189, *195*, *196*
Shepard, F. C., 324, *362*
Shibata, K., 501, *607*
Shigematsu, T., 26, *40*, 325, *362*
Shimada, B. M., 552, 553, *607*
Shiokawa, F., 134, 140, 142, 144, 145, 146, 147, *159*
Shipek, C. J., 175, *196*
Shiriaishi, K., 586, 588, 593, *605*, *607*
Shtukovskaya, L. A., 130, *161*
Sieburth, J. M., 514, 569, 574, *596*, *607*
Siedler, G., 96, *119*
Siever, R., 255, 278, 279, 280, 281, 282, *319*
Sillén, L. G., 168, 171, 178, 194, *195*, *196*, 409, *431*
Sironval, C., 527, *601*, *607*
Sirs, J. A., 253, *321*
Sistrom, W. R., 549, *607*
Sivkov, S. J., 532, *602*
Skopintsev, B. A., 130, *160*, *161*, 462, 468, 473, *475*, 628, 635, 636, 637, 638, 639, 641, 642, *645*
Slobodkin, L. B., 578, *601*
Slocum, G., 294, 295, *321*
Slowey, J. E., 241, *321*, 453, *475*
Smales, A. A., 136, 148, *160*, *161*, 325, *364*
Smayda, T. J., 567, 579, *600*, *607*
Smith, E. H., 78, 89, *120*
Smith, E. L., *607*
Smith, G. M., 483, *607*
Smith, H. P., 325, 326, 327, 334, *363*, 399, 400, *406*
Smith, J. B., 255, *321*, 522, *607*
Smith, J. H. C., 493, *607*
Smith, M. B., *473*
Smith, M. H., 435, *475*
Smith, N. D., 111, 116, *118*
Smith, P. B., 591, *603*
Smith, R. H., 24, *39*, 169, 180, 194, *195*
Smith, V. N., 252, *321*
Smith, W., 592, *602*
Smith, W. H., 235, 238, *321*
Snodgrass, J. M., 526, *600*
Sommer, A. L., 505, *607*
Sonstadt, E., 24, *41*
Soot-Ryen, T., 389, *407*
Sørensen, S. P. L., 18, 22, *39*, *41*, 75, 97, 101, 102, *118*, 233, *321*
Sorokin, C., 534, 545, 546, *607*
Sorolin, Yu. I., 521, 558, 567, 588, 591, 592, *604*, *607*, *608*
Soule, F. M., 78, 89, *120*
Spaeth, J. P., 370, *406*, 455, *475*, 544, 583, *603*, *608*
Spencer, C. P., 539, 546, *608*
Spoehr, H. A., 375, *407*, 507, 549, *608*
Sporek, K. F., 134, 145, *161*
Sprenger, A., 4, *41*
Stadie, W. C., 522, *608*
Stanbury, F. A., 530, *608*
Starik, I. E., 193, *196*
Starkey, R. L., 370, *408*
Steele, J. H., 337, *364*, 531, 532, 561, 563, 564, 566, 567, 576, *608*
Steen, H., 204, *224*
Stefánsson, U., 215, 221, *224*, 361, *364*, 413, 419, *431*
Steiger, G., 144, 145, *161*
Steiner, R., 128, *160*
Stenius, S., 115, *119*
Stepanek, M., 530, 591, *608*
Stephens, G. C., 593, *608*
Stephens, K., 34, 38, 503, 504, 505, 506, 507, 508, 509, 510, 511, 512, 513, 515, 517, 518, 524, 527, 533, 537, 544, 546, 547, 548, 550, 553, 570, 582, 583, 592, *595*, *601*, *603*, *605*, *608*
Stewart, D. F., 544, 545, *608*

Stewart, L. C., 36, *41*
Stich, H., 324, *364*
Stokes, R. H., 166, *196*, 278, *321*
Stommel, H., 47, 58, 64, 71, *72*, 390, *407*, 564, 571, *606*
Stott, V., 98, *119*
Strain, H. H., 487, 493, 501, *608*
Strange, L., 513, *608*
Strecker, E., 13, *41*
Strickland, J. D. H., 34, *38*, 251, *321*, 324, 325, 329, 356, *364*, 493, 496, 497, 499, 502, 503, 504, 505, 506, 507, 508, 509, 510, 511, 512, 513, 515, 516, 517, 518, 519, 521, 524, 526, 527, 532, 533, 536, 537, 544, 546, 547, 548, 550, 551, 553, 555, 556, 560, 561, 566, 568, 570, 572, 574, 582, 583, 589, 592, 593, *595*, *601*, *603*, *605*, *608*
Strøm, Kaare, 619, *645*
Strøm, K. M., 612, *645*
Strutt, R. J., 34, *41*
Suda, K., 426, *431*
Suess, H. E., 20, *40*, 201, 213, 217, 218, 223, *224*, 229, 278, 295, 296, 297, 301, 302, 303, 304, 308, 309, 310, *317*, *321*
Sugawara, K., 135, 140, 141, 148, 149, *160*
Sugimoto, H., 510, *601*
Sugimura, T., 639, *645*
Sugimura, Y., 639, *645*
Sugiura, Y., 314, 315, *321*
Sullivan, J. P., 210, *224*
Supernaw, I., 245, 248, 272, *319*
Suschenya, L. M., 589, *608*
Sutcliffe, W. H., 591, *595*
Sutton, D., 185, *195*
Svenska Geofysika Förenigen (1955, 1956), 369, *407*
Suzuki, A., 507, 549, *601*
Suzuki, N., 568, *608*
Sverdrup, H. U., 45, 49, 50, *72*, 102, *119*, 201, 216, *224*, 279, *321*, 337, 351, *364*, 378, 395, 403, *407*, 576, *608*, 617, 622, 636, 641, *645*
Swallow, J. C., 53, *72*
Swan, E. F., 149, *161*
Swanson, R. W., 189, *195*
Swartout, J. A., 156, *159*
Sweeney, B. M., 546, 553, *600*, *608*
Swinnerton, J. W., 210, *224*
Syrett, P. J., 375, 376, *407*, 513, 519, 537, *608*

T

Takahashi, T., 229, 246, 252, 270, 271, 272, 273, 290, *321*
Talling, J. F., 529, 531, 532, 551, 559, 567, 584, *603*, *608*, *609*
Tanada, T., 528, *609*
Tanner, H. A., 540, 547, 592, *596*, *598*, *604*
Tatsumoto, M., 181, 187, 193, *196*, 255, *321*, 450, 451, 472, *475*, 522, *607*
Tauber, H., 299, *321*
Tausch, E. H., 180, *196*
Taylor, H. J., 139, 153, *161*
Tchindonova, J. G., 590, *609*
Teixera, C., 335, *363*
Tham, A. K., 331, *364*
Thang, Minh-Nguy, 513, *609*
Than-Tun, 371, *405*, 544, *598*
Theis, K., 445, *474*
Thiele, H., 445, *475*
Thode, H. G., 158, *159*
Thom, N. S., 205, *224*, 672
Thomas, B. D., 104, 107, 108, *119*, 148, *161*, 426, *431*
Thomas, D. W., 455, 472, *473*
Thomas, G. J. B., 528, *606*
Thomas, J. O., 590, *603*
Thomas, W. H., 375, *407*, 412, *430*, 538, 548, *609*
Thompson, J. F., 513, 539, *606*
Thompson, M. E., 121, *159*, 167, 178, *195*, 255, 278, 279, 280, 281, 282, *319*
Thompson, T. G., 18, 31, 38, *41*, 102, 104, 107, 108, *119*, 126, 127, 128, 132, 136, 137, 138, 139, 140, 141, 142, 143, 144, 145, 146, 147, 148, 149, 150, 151, 152, 153, *158*, *159*, *160*, *161*, 210, *224*, 247, 250, *319*, *321*, 329, 361, *363*, *364*, 387, *406*, 418, 426, *431*, 454, *474*, 498, 499, 520, *599*, *606*, 629, 630, 641, *644*
Thomsen, H., 335, 355, 359, *364*, 398, *407*
Thorpe, T. E., 13, *41*, 144, 146, 147, 152, *160*
Thorson, G., 586, *609*
Thoulet, J., 116, *119*
Thuras, A. L., 93, *119*

Thurber, D. L., 192, *196*
Timofeeva, S. N., 636, *645*
Tobler, B., 315, *319*
Tolbert, N. E., 505, 592, *597*, *609*
Tørnoe, H., 21, *41*
Tourky, A. R., 411, *431*
Tracy, M. V., 438, *475*
Trama, F. B., 556, 558, *607*
Tranter, D. J., 539, *609*
Trask, P. D., 403, *407*
Tripp, R. B., 570, *595*
Trofimov, A., 158, *161*
Truesdale, G. A., 203, 204, 205, 208, 209, *225*
Tschapek, M., 445, *474*
Tschiegg, C. E., 101, *119*
Tsubota, H., 131, *160*
Tsuchiya, Y., 512, *601*
Tsunogai, S., 179, *196*
Turekian, K. K., 180, 186, *196*
Tyler, J. E., 501, *609*

U

Uden, N. Yan., 591, *598*
Ulrich, B., 438, *475*
Unesco, 80, *119*
Underwood, A. L., 253, *321*
Urey, H. C., 253, 257, *320*
U. S. Navy, 110, *119*
Utterback, C. L., 104, 107, 108, *119*

V

Vaccaro, R. F., 325, 361, *363*, *364*, 373, 375, 382, 383, 384, 385, 386, 388, 389, 400, 401, 403, *406*, *407*, 505, 518, 550, 552, *607*, *610*, 618, 620, 622, 631, 636, *645*
Vallentyne, J. R., 437, *475*
Valsan, A. P., 332, *362*
Van Baalen, C., 540, *609*
Van Norman, R. W., 523, *609*
Van Slyke, D. D., 240, 251, *321*
Vargues, N., 372, *405*
Vasil'ev, V. V., 138, 151, 152, *161*
Vel'Mozhnaya, Yu. A., 99, *119*
Verduin, J., 245, *321*, 553, 560, *609*
Verjbinskaya, N., 337, 339, *363*, 389, *407*, 498, 561, *602*
Vernon, L. P., 493, *609*
Vershinina, O. A., 628, 635, 641, 642, *645*
Vihman, E., *406*
Vinberg, G. G., 551, *609*
Vinogradov, A. P., 148, *161*, 184, *196*, 374, 403, *407*, 506, *609*
Vinogradov, M. E., 503, 504, 590, *609*
Vishniac, H. S., 455, *475*, 591, *609*
Vives, F., 545, *609*
Vladychenskaya, N. S., 513, *607*
Vogel, H. A., 9, *38*
Vogel, J. C., 157, *159*, 299, *320*
Voipio, A., 130, 140, 141. 142, 143, *161*, 413, 422, *431*
Vollenweider, R. A., 559, 581, *606*, *609*
Von Arx, W. S., 49, 52, 63, *71*
Von Brand, T., 29, *38*, 378, 380, *406*, *408*, 505, *609*
Von Bibra, 124, 152, *159*
Von Dehn, M., 587, *597*
Vorob'eva, R. V., 130, *161*, 635, 641, 642, *645*
Voroschilova, A., 590, *609*

W

Wadman, W. H., 508, *600*
Wakeel, S. K., El, 504, 505, *609*
Waksman, S. A., 29, *41*, 370, 376, 378, *408*, 439, 441, *475*
Walker, A. C., 279, *321*
Walker, J. B., 539, 540, *609*
Walker, T. J., 549, *599*
Walne, P. R., 586, *609*
Walton, A., 300, *317*
Wandenbulcke, F., 417, *431*
Wangersky, P. J., 439, 447, 450, *474*, *475*, 592, *599*
Wanke, H., 20, *40*, 217, *224*
Wanklyn, J. A., 435, *475*
Wänninen, E., 131, *161*
Warburg, O., 522, *609*
Wärme, K. E., 252, *319*
Warren, F. J., 412, *430*
Warrington, P. M., 523, *609*
Wasserburg, G. J., 208, *224*
Wasson, J. T., 179, *195*
Waterman, L. S., 270, 272, 273, 275, 276, 277, 290, 316, *320*, *321*
Watson, S. W., 378, *406*, *408*
Watt, W. D., 324, *364*, 522, *609*

Wattenberg, H., 22, *39*, 93, *120*, 208, *224*, *225*, 241, 248, 249, 251, 256, 259, 260, 261, 270, 281, *318*, *322*, 377, *408*
Weaver, C. E., 413, *431*
Webb, D. A., 133, 144, 145, *161*
Wedepohl, K. H., 189, *196*
Weigl, J. W., 523, *609*
Weis, D., 523, *596*, *609*
Weiss, J. B., 548, *598*
Welander, P., 71, *72*
Wellershaus, St., 468, *474*
Wenner, F., 78, 89, *120*
Werby, R. T., 370, *406*
West, T. S., 130, 139, *158*
Westlake, D. F., 573, *609*
Weyl, P., 104, *120*
Wheatland, A. B., 625, 627, *645*
Wheeler, A. S., 144, 145, *161*
Whipple, G. C., 203, 204, *225*
Whipple, M. C., 203, 204, *225*
Whisenaud, A., 549, *599*
Whitaker, W. W., 295, 297, 308, *317*
Whitley, E., 436, *475*
Wiessner, W., 543, *605*
Williams, J. H., 542, *595*
Williams, M., 295, 297, 308, *317*
Williams, P. M., 454, *475*, 619, 632, *645*
Williams, V. R., 512, *609*
Williams, W. T., 450, 451, 472, *475*
Willis, E. H., 299, *322*
Wilson, D. P., 455, *475*
Wilson, W. D., 109, 111, *120*
Winchester, J. W., 179, *195*
Winkler, L. W., 20, 23, *41*, 138, 151, 152, 161, 199, *225*
Winokur, M., 531, *609*
Wirth, H. E., 102, *119*, 146, 147, *160*
Wiseman, J. D. H., 417, *431*
Wissburn, K. F., 253, *322*
Witting, R., 202, *225*
Wolf, F. T., 490, *609*
Wolfgang, R., 285, *320*
Wood, E. J. F., 500, 544, 566, 577, 591, *601*, *609*, *610*
Wooster, W. S., 356, 357, *364*, 427, *431*
Worthington, L. V., 53, *72*, 287, *322*
Wright, C. C., 126, 127, 128, 140, 142, 143, *161*
Wright, J. C., 533, *610*
Wüst, G., 31, *41*, 208, *225*, 398, 399, *408*
Wyman, J., 253, 314, *319*, *322*
Wyrtki, K., 216, 220, *225*

Y

Yamamoto, H. Y., 493, *610*
Yamauchi, N., 126, 140, 142, 152, *160*
Yentsch, C. S., 492, 498, 500, 501, 505, 518, 524, 550, 552, 553, 560, 566, 567, 568, 576, *601*, *607*, *608*, *610*
Yonge, C. M., 587, *610*
Young, E. J., 189, *194*
Yutzy, H., 138, *160*

Z

Zabor, J. W., 378, 380, *408*
Zaukelies, D. A., 368, 369, *405*
Zechmeister, L., 501, *605*
Zen, E. A., 414, *431*
Zeuthen, E., 566, *599*
Zhukova, A. I., 590, *610*
Zill, L. P., 592, *609*
Zobell, C. E., 372, 375, 378, 383, 389, 408, 443, *475*, 590, *610*
Zobell, C. E., 590, *610*
Zwicker, B. M. G., 150, *160*

Subject Index

(*Numbers in bold type indicate the page on which a subject is treated most fully.*)

A

Abundances of elements in sea water, 164–165
Acartia tonsa, 587
Acid-forming substances in sea water, 241
Acrylic acid in *Phaeocystis pouchetii* 514
Action spectra, **527–530**
Activity coefficients of ions in sea water, **177–178**, 234, 236, 278–282
Advective processes, 43
  effect on dissolved gases, 218
Aeolian sediments, 413
Agmenellum quadriplicatum, 503, 507, 509, 517, 541
Algae
  blooms of, 464, 564, 569, 576–578
  chemicals derived from, 36
  elementary composition of, 374
  N/P ratio of, 375
  nitrate assimilation by, 375
  nitrogen fixation by, 371
  trace metal concentration by, 185
Algal taxonomy, 478, **483–487**
"Algal weight", 502
Alkali metals in sea water, 174
  ion-exchange on clay minerals of, 174
Alkaline earth metals in sea water, 174
Alkalinity, 9, 21, **247–248**, 249
  anoxic systems, occurrence in, 641
Allochthonous plankton, 280
Aluminium
  diatoms, uptake by, 417
  effect on solubility of silica, 410
  occurrence in sea water, 23, 164, 172
  residence time in sea water, 164, 173
Aluminosilicates, 412
  anthigenic formation of, 414
Amino acids
  excretion during nitrogen fixation, 545
  formation by breakdown of protein, 441
  nutrient value of, 543
  occurrence in marine organisms, 512
  occurrence in sea water, 447
α-Aminobutyric acid in *Dunaliella*, 512
Ammonia
  anoxic systems, occurrence in, 626
  excretion of, 377, 387
  formation from organic nitrogen, 376
  marine cycle of, 402
  nutrient value of, 26, 375, 387, 536
  occurrence in rain water, 369, 544
  occurrence in sea water, 23
  oxidation of (bacterial), 377–383
  oxidation of (non-bacterial), 383
  seasonal variations of, 387–390
Amphidinium carteri, 503, 507, 509, 510, 517, 541
Amylase, 587
Anabaena sp., 507
Anabaena cylindrica
  carbohydrate content, 508
Anacystis marina, 541
Anchovetta
  trace element concentration by, 185
Anoxic basins and fjords
  ammonia content of, 626
  circulation and exchanges in, 622
  denitrification in, 612, 624
  flushing of, 615, **620–621**
  intermittent anoxia in, 615
  nitrogen content of, 400, 629
  nutrient content of, 400, 640
  redox potentials of, 613
  sediments of, **187–189**, 636, 642
  stability across sills of, 620
  sulphate reduction in, 158, 613, 625, 627
  trace element distribution in, 181–183, 640
Antarctic Intermediate Water, 53, 355, 393

Antarctic (Southern) Ocean
nitrate distribution in, 395–398
phosphorus distribution in, 352–355, 428
primary productivity of, 574
silicon distribution in, 420, 428
surface currents of, 47
vertical circulation in, 352
Antimony
abundance in sea water, 165
residence time in sea water, 165
Apatites (phosphorites), 193
formation of, 188–189
"Apparent oxygen utilization" (A.O.U.), 204, 222, 393, 426
Arabian scientists, early contributions to the chemistry of the sea, 3
Archaeological dating by radio-carbon, 286
Arctic Ocean
phosphorus distribution in, 337–341
primary productivity of, 574
Argon in sea water
abundance, 164
conservative nature of, 222
ratio to nitrogen, 206, 217, 629
solubility of, 205–206, 207, 208
Argon-40 in sea water
formation from ^{40}K, 201, 214, 218
Arsenic
abundance in sea water, 164
occurrence in sea water, 24, 325
Artemia salina, 586, 591
Artificial sea water, preparation of, 648
Ascorbic acid in marine algae, 513
Askintrodesmus sp., 507
Asterionella formosa, 537, 541, 584
Asterionella japonica, 582
Atlantic Ocean
C/N ratio in, 459
carbon-14 activity in, 304
carbon dioxide, partial pressure in, 277
deep-water circulations in, 53–55, 57
denitrification in, 373
N_2/Ar ratio, 217
nitrate distribution in, 395–398
nitrite distribution in, 390–392
organic carbon content of, 465, 468
oxygen distribution in, 344–347
phosphorus distribution in, 328, 342–352, 360
silicon distribution in, 423, 424
specific alkalinity in, 248
surface currents of, 44
trace elements in sediments of, 190
Atlantis R.V., 342, 350, 390
Atmosphere
composition of, 198, 201
isotopic speciation in, 199
nitrous oxide, occurrence in, 198, 373
solubility of gases in sea water, 202
variability of composition of, 199
Atmospheric carbon dioxide, *see* Carbon dioxide, Atmospheric
Attenuation of light in sea water, **525–526**
Authigenic minerals, 174, **188–194**, 414
Autochthonous plankton, 480
Autotrophic plankton, 480
Auxospores, 29, 580
Auxotrophic organisms, 480
organic nutrition of, **542–543**
Azotobacter
nitrogen fixation by, 369, 370

B

Bacillaria paradoxa
silicon uptake by, 415
Bacillariophyceae
carbohydrate content of, 507, 509
carbon content of, 503
lipid content of, 507
nitrogen content of, 503
phosphorus content of, 503
pigment composition of, 488
protein content of, 507
vitamin requirement of, 541
Bacteria, 542, 590
anaerobic, 372, 401, 631, 642
bio-mass (oceanic) of, 591
denitrification by, 27, 372–373, 401
detritus, occurrence on, 591
food source, use as, 590
nitrification by, **377–383**, 401
nitrogen cycle, role in, 401
protein breakdown by, 440
protein synthesis by, 443
sulphate-reducing, 147, 158, 643

Bahama Banks
calcium carbonate precipitation in, 142
Baltic Sea
alkalinity in, 249
calcium to chlorinity ratio in, 143
magnesium to chlorinity ratio in, 143
organic carbon content of, 468
silicon content of, 422
trace element distribution in, 188
Barents Sea
calcium to chlorinity ratio in, 141
magnesium to chlorinity ratio in, 143
phosphate distribution in, 338–340
Barium
abundance in sea water, 165
depth distribution in sea water, 180
occurrence in marine organisms, 186
occurrence in sea water, 23, 179
residence time in sea water, 165, 173
Barytes formation in sea water, 180
"Bathysonde", 96
Bay of Bengal
phosphorus content of, 332
silicate content of, 421
Benguela Current, 46, 56
Benthic organisms, 506, 565
heterotrophism of, 543
productivity of, 574
Beryllium
occurrence in sea water, 164
residence time in sea water, 164, 174
Bicarbonate ion in sea water, 264
activity coefficient of, 279–281
effect of pH and temperature on, 660–671
ion-pair formation by, 167, 278
phytoplankton utilization, 522
Biliproteins in marine phytoplankton, 489, **496**
Biological activity
isotope fractionation by, 221
relation to oxygen consumption, 214–216, 220–222, 519
Biological concentration of trace metals, **183–186**
selectivity of, 183
Biologically-active substances occurrence in sea water, 447, 455
Bio-mass, 496, 544, 579, 583
Biosphere–hydrosphere interactions, **179–187**
Bismuth
abundance in sea water, 165
residence time in sea water, 165
Black Sea
anoxic condition of, 182, 188, 612, 617
denitrification processes in, 373
flushing of, 622
inorganic nitrogen distribution in, 400
nutrients in, 617
organic carbon content of, 468, 636
phosphorus content of, 631
redox potential in, 635
sulphide content of, 617, 643
"Bleaching" of pigments, 527, 536, 549
"Blooms", 464, 564, 569, **576–578**
Boric acid in sea water
buffering action, 21, 22
complex formation by, 250
dissociation constants of, 251, 649
volatility of, 179
Boron in sea water
abundance, 164
complexation with organic matter, 137, 150, 250
determination, **136–138**
fluorometric with benzoin, 137
photometric with curcumin, 137
spectrographic, 137
volumetric with mannitol, 137
historical, 13
principal determinations, 150–151
ratio to chlorinity, 150–151
Boron in sediments, 640
Boundary layer in gas transfer, 211, 313
Brazil Current, 44
Bromine in sea water
abundance, 164
determination, 138
extraction from, 36
historical, 13
principal determinations, 151–153
ratio to chlorinity, 125, 151–153
Bubbles
effect on gas exchange across sea surface, 200, 213
Buch's Tables, 256

Buffer capacity, 240
Buffer effect in sea water, 21, 22, **239–246**

C

Cadmium in sea water
  abundance, 165
  complex formation by, 170
  residence time, 165
Caesium
  abundance in sea water, 165
  residence time in sea water, 165
Calanus sp., food requirement of, 585
Calanus finmarchicus, 585
  hydrocarbon production by, 472
Calcareous shells
  strontium content of, 149
Calcium carbonate
  marine organisms, occurrence in, 142, 506, 581
  precipitation of, 142, 372
  re-solution of, 630
Calcium in sea water
  abundance, 164
  cell division, influence on, 547
  determination, 13, 126–131
    flame photometric, 132
    gravimetric as oxalate, 126–128
    volumetric with E.D.T.A., 129
    volumetric with E.G.T.A., 130
  ion-pair formation by, 167, 278
  principal determinations, 140–142
  ratio to chlorinity, 125, 140–142, 641
  residence time, 164, 173
  river supply of, 142
California Current, 46
Calothrix, nitrogen fixation by, 371, 544
Canaries Current, 44
Carbamino complexes
  occurrence in sea water, 522
Carbohydrate
  breakdown products of, 438–440
  marine organisms, content of, **508–511**
  nitrogen fixation, consumption of, 371
  occurrence in sea water, 450
Carbon-13
  occurrence of, 285
  permille enrichment ($\delta^{13}C$), 287
Carbon-14 activity (atmospheric)
  permille enrichment values of, 288
  production by nuclear explosion, 287, 299
  Suess (dilution) effect on, 287, 297
  variations of, 297–300
Carbon-14 activity in sea water, **300–305**
  variations with depth of, 304
  variations, latitudinal of, 303
Carbon-14 uptake in productivity measurement, 520, 552
Carbon dioxide (atmospheric), 70, 282–284, **288–300**
  carbon-14 activity of, 297–300
  earth's temperature effect on, 228
  interactions with sea water, 282–284
  invasion and evasion rates, 315
  partial pressure of, 252
  production from fossil fuels, 282, 286–287, 295–296
  residence time of, 305–312
  solubility in sea water, 282–284, 650
  variations, latitudinal, 292
  variations, long-term, 294–297
  variations, short-term, 288–294
Carbon dioxide in sea water
  amino complexation by, 522
  anoxic systems, occurrence of, 630, 634
  atmospheric interaction by, 282–284, 312–316
  chemical potential of, 262
  hydration of, 312–314
  invasion and evasion rates of, 315
  isotopic ratio $^{14}C/^{12}C$ of, 70
  partial pressure of, 252, 265, 269–277
  photosynthetic fixation of, 376, 520
  residence time of, 305–312
  solubility of, 256, 650
  solution reactions of, 253, 255–257
  transfer across cell membranes, 521
Carbon dioxide, total, in sea water (ΣCO_2), 251, 255, 269
Carbon dioxide transfer
  boundary layer in, 313
  cell membrane, 521
  gas–liquid interface, **312–316**
Carbon
  natural cycle of, 230

Carbon *contd.*
natural isotopes of, **285–288**
Carbon, natural exchange reservoirs
idealized exchange models of, 291–294, 298, 305–312
isotopic abundances in, 285–288
isotopic fractionation in, 287, 301
radio-carbon content of, 229
residence times in, 164, 305–312
Carbon, organic
C/N ratio in sediments, 459
determination of, 504
occurrence in marine organisms, 374, 503, 504
occurrence in sediments, 613, 616
Carbon, organic, in sea water
C/N ratio, **456–462**
vertical distribution of, 464
See also Organic matter in sea water
Carbonate alkalinity, **250–251**, 264
Carbonate ion in sea water, 265
activity coefficient of, 279–281
effect of pH and temperature on, 660–671
ion-pair formation by, 167, 278
utilization by marine organisms, 522
Carbonate sediments
$^{13}C/^{12}C$ ratio of, 301
Carbonic acid equilibria in fresh water
effect of temperature and pH, 652–659
Carbonic acid equilibria in sea water, 22, 228, 231, 253, **257–262**
anoxic systems, effect on, 640
buffer capacity of, 241
pH, effect on, 255
pressure (depth), effect on, 266–268
Carbonic acid in sea water
dissociation constants of, 21–22, 257, **258, 259–262**
apparent, 259–262, 651
true, 257, 258
effect of pH and temperature on, 660–671
Carbonic anhydrase, 254, 315
Carboxylic acids in sea water
aliphatic, 453
hydroxy, 453
Cariaco Trench, 399
anoxic condition of, 182, 612, 618
circulation model of, 622
denitrification in, **373**
silicon content of, 425
sulphide content of, 618
trace elements in, 182
Caribbean Sea
inorganic nitrogen content of, **399**
phosphate content of, **334**
Carnegie R.V., 355, 426
Carotenes in marine phytoplankton 488, **493**
Carotenoids
occurrence in sea water, 455
Cartilage of vertebrates, 442
Cell division, 482, **544–551**, 570
cell size and rates of, 547
nutritional conditions and, 548–551
rate of, 546
temperature and light, effect on, 546–548
Cellulase, 587
Cerium
occurrence in sediments, 194
Chaetoceros sp., 507, 579, 586
Chaetoceros simplex, 507, 541
amino acids of, 512
fatty acids of, 512
Chain R.V., 398
Challenger H.M.S., 14, 15, 19, 21, 79, 124, 220, 413
Chemo-organotrophic organisms, 536, 593
Chemotaxis, 565
Chesapeake Bay, 334
Chilomonas paramoecium, 541
Chitin of invertebrates, 442
Chlamydomonas moewusii, 541
Chlamydomonas pulsatilla, 541
Chlamydomonas reinhardii, 541
Chlorella sp., 487, 503, 506, 508, 534, 542, 545, 549, 586
carbohydrate content of, 510
magnesium requirement of, 540
nucleic acids of, 513
Chlorella ellipsoida, 507
Chlorella pyrenoidosa, 507
cell-wall structure of, 510
elementary composition of, 375
Chlorella vulgaris, 541, 542
Chloride in sea water, 13, 164

Chloride in sea water *contd*
  complexation by, 169
  See also Chlorinity
Chlorinity
  definition of, 18, 76, **77**
  determination of, 80
  relation to salinity of, 18, 76
Chlorite, 413
Chlorococcum sp., 586
Chlorogloea fritschii, 503
Chlorophyceae
  carbohydrate content of, 507–508
  carbon content of, 503
  elementary composition of, 374
  lipid content of, 507
  nitrogen content of, 503
  phosphorus content of, 503
  pigment composition of, 488
  protein content of, 507
  vitamin requirements of, 541
Chlorophyll *a*, 490
  composition ratios of phytoplankton involving, 516
  determination of, 496
  metabolism in cells, 549
  P(c) values, 533
  plant carbon, measure of, **515–518**, 560
Chlorophylls in marine phytoplankton, 488, **490–493**
  absorption bands of, 491
  detritus, occurrence of, 492
  extinction coefficients of, 493
Chlorophyllase, 492
Chlorophyllide, 491
Chlorophyta, 484
Chlorosity, 78
Chondrillasterol in phytoplankton, 512
Chromium
  abundance in sea water, 164
  residence time in sea water, 164
Chrysolaminarin in Chrysophyta, 511
Chrysophyceae
  carbohydrate content of, 507, 509
  carbon content of, 503
  lipid content of, 507
  nitrogen content of, 503
  phosphorus content of, 503
  pigment composition of, 488
  protein content of, 507
  vitamin requirements of, 541
Chrysophyta, 485, 511
Citrulline in Myxophyceae, 512
Clams
  barium uptake by, 186
  diet selectivity of, 586
  ^{15}N enrichment in, 368
Claustridia, nitrogen fixation by, 370
Clay minerals
  diagenesis of, 414
  ion-exchange properties of, 146, 174
  suspensions in sea water of, 413
Cobalt
  abundance in sea water, 164
  occurrence in sediments, 193
  residence time in sea water, 164, 175
Coccolithophores, 485
  calcium carbonate content of, 506, 581
  composition ratios of, 517
Coccolithus fragilis, 579
Coccolithus huxleyi, 506, 541, 579, 583
Columbia River plume
  nutrients and biological activity of, 222, 419
Common salt
  extraction from sea water of, 36
Compensation depth, 214, 390, 611
Complexation in sea water, **166**–171
  classification of, 167
  ion-pair formation as, 167
Composition of sea water, **9–19**, 122
  constancy of, 9–19, **123–125**
  early investigations of, 7–9, 9–18
Compressibility of sea water, **111–112**
Convection currents, 56
Copper
  abundance in sea water, 164
  anoxic systems, occurrence in, 182
  occurrence in sea water, 639
  residence time in sea water, 164, 173
Coriolis effect, 47
Coscinodiscus sp., 503, 507, 509, 585
Coscinodiscus wailsii, 503, 517
Cosmic spherules, 413
"Crude fibre", **508–509**, 515
Cryptophyceae, 486
  pigment composition of, 488
  vitamin requirements of, 541

Cultures
  illumination of, 530
Currents in the oceans
  density, 59
  surface, 44–49
  tidal, 59
  wind-driven, 59
Cyanophyta, 484
Cyanocobalamin, *see* Vitamin B_{12}
Cyclotella nana, 541, 583

D

Dabob Bay, 614
Dana Cruises, 350, 355, 359, 614
Daphnia magna, 587, 591
Daphnia pulex, 587, 588
Deep-water circulations, 51–59
Denitrification, bacterial, 27, **372–373**, 401
  in anoxic systems, 612, 624, 629, 634
Density currents, 59
Density of sea water, 74, **101–103**, 125
  by hydrometry, 98
  early determinations of, 10, 12
  in situ measurement of, 97
Depth of frictional influence, 62
Desalting of sea water, **37**
Deoxynucleic acid in phytoplankton, 513
Detritus, 146, 479, 515, **589–590**
  chlorophyll *a*, in, 492
  See also "Marine snow"
Deuterium in sea water, 155–156
Deuterium/hydrogen ratio in sea water, 156
Diagenesis of clay minerals, 414
α,ε-Diaminopimelic acid in Myxophyceae, 512
Diatoms, 337, 428, 486, 549, 584. *See* also under individual species
  aluminium in, 417
  amino acid content, 512
  composition ratios of, 517
  elementary composition, 374, 503, 507
  frustules, 415
  lipid content, 511
  ooze, 415
  re-solution of, 415–417
  silicon content, 409, 505
  silicon uptake by, 415, 538
  sinking rate of, 566
Dicrateria inornata, 541
Diffusive processes, 47
  effect on dissolved gases, 218
Dinoflagellate "blooms", 578
Dinophyceae
  carbohydrate content, 507, 509
  carbon content, 503
  lipid content, 507
  nitrogen content, 503
  phosphorus content, 503
  pigment composition, 488
  protein content, 507
  vitamin requirement, 541
Discovery II, 31, 78, 350, 352
Dissolved gases in sea water, historical, **19–23**
  See also under individual gases
Ditylum brightwelli, 580, 585
Diurnal variations
  atmospheric carbon dioxide (p_{CO_2}), 288, 289
  cell division, 546
  dissolved carbon dioxide (P_{CO_2}), 270–272
  dissolved oxygen, 20, 244
  pH, 242, 246
  phosphate, 334
  primary production, 552–554
Downwind Cruise, 291, 316
Dramsfjord, 615, 616, 621
Dunaliella sp., 512, 523
  carbohydrate content, 510
Dunaliella euchlora, 523, 541
Dunaliella salina, 503, 507, 509, 510, 517, 541
Dunaliella tertiolecta, 586

E

Eddies, 64
Eddy diffusion, 64
Ekman's Theory, 61
El Niño, 48
Electrical conductivity methods for salinity determination, **81–97**

Electrical conductivity of sea water, 78, **103–107**
effect of pressure on, 107
effect of temperature on, 106
Electrolytes in sea water
associated, 166
ion-pairs, 166
weak electrolytes, 166
non-associated, 166
Emerson effect, 528
English Channel
ammonia content, 387
organic matter content, 468
phosphorus content, 326, 329, 330
silicate content, 330, 418, 419, 422
Enzymes
amylase, 587
carbonic anhydrase, 254, 315
cellulase, 587
chlorophyllase, 492
humic substances interaction with, 443
lipase, 587
metal co-factors, 506, 540
peptidase, 587
protease, 440
Epilimnion, 480
Equatorial Counter Current, 46
Equatorial undercurrents, 58
Ergosterol in phytoplankton, 512
Estuarine muds
phosphorus content of, 333–334
Ethmodiscus rex, 417
Eudistoma ritteri
titanium enrichment in, 183
Euglenophyta, 484, 548
Euphausia pacifica, 588
Euphotic zone, 329, 479, 526, 552, **561**
nutrient supply to, 564
photosynthesis in, 559
vertical plankton distribution in, 567
Evaporation of sea water
variation of isotope ratios due to, 157
Excretion
by phytoplankton, 390, 439, 447, 524, 592
by zooplankton, 377, 387
Exuviella sp., 503, 507, 509, 517
Exuviella cassubica, 541

F

Faecal matter, 587, 589
Fatty acids, in sea water, 447, 453
in marine organisms, 511
Feldspars
weathering of, 414
Ferromanganese minerals, 37, 175, 188, **190–194**
composition, 191
enrichment of cerium in, 194
rate of growth, 192
Filtration of sea water, 497, 514
Fishing grounds
North Sea, 336
Fjords, 398, 612
landlocked, with relict sea water, 619
sills and water stability of, 620
Flagellates, 581
blooms of, 578
phototaxis of, 565
Flame photometric determinations
calcium, 131
strontium, 136
Fluorapatite, 187
Fluorescence, relation to light absorbancy, 472
Fluoride in sea water
abundance, 164
complexation of, 168–169
detection of, 13
determination of, **139**, 153–154
ratio to chlorinity, 153–154
Flushing of anoxic systems, 615, **620–621**
Food chain in the oceans, 478, 585, 594
Fossil fuel combustion, effect on carbon dioxide balance, 282, 286–287, 295–296
Fossil marine organisms, $^{14}C/^{12}C$ ratios in, 305
Fractionation of isotopes, 221, 287, 301, 367, 369, 634
Fragilaria capucina, 541
Fragilaria crotonensis, 584
Freezing of sea water
effect on isotopic ratios, 157
effect on sulphate to chlorinity ratio, 147
Freezing point of sea water, 115
Frictional influence, depth of, 62
Fungi in sea water, 591

G

Galathea Cruise, 31
Gallium
abundance in sea water, 164
residence time in sea water, 164
Gases in sea water. *See* also under individual gases
atmospheric exchange of, 200
equilibrium at sea surface, 208–211
historical, **19–23**, 177
partial pressures of, 202
"percent saturation", 204
seasonal variations of, 210
solubility, 202–208
solubility coefficients, 203
transfer across sea surface, 200, **211–214**
effect of bubbles on, 200, 213
effect of sea state and wind on, 213
variations *in situ*, 201, **214–218**
"Gelbstoffe", 436, 448, 455
Geochemical balance, 24–25
Germanium
abundance in sea water, 164
residence time in sea water, 164
Glass electrode, 237
Glauconite, 414
Glenodinium foliaceum, 541
Glycollic acid
as excretion product, 447, 592
Gold
abundance in sea water, 165
occurrence in sea water, 24, 37
residence time in sea water, 165
Gonyaulax sp., 503
Gonyaulax polyhedra, 541, 546
Grazing by zooplankton, 34, 377, 387, **569–571**
Greek philosophers, theories and observations relating to oceanography, 2
Gulf of Cariaco, 612, 616
Gulf of Maine
ammonia content of, 387
phosphorus content of, 327
Gulf of Mexico
inorganic nitrogen distribution in, 399
partial pressure of carbon dioxide in, 273
Gulf of Venezuela
phosphorus content of, 334
Gulf Stream, 44, 47, 221, 462
Gymnodinium breve, 541
Gymnodinium splendens, 541
Gyrodinium californicum, 541

H

H.P.P.U. (Harvey plant pigment unit), 498
Haematococcus pluvialis, 541
Halide complexes in sea water, 168–171
Halocline, 479
Halosphaera sp., 486
Helium
abundance in atmosphere, 198
abundance in sea water, 164
occurrence in sea water, 213
solubility in sea water, 207
Hemiselmis virescens, 541, 586
Henry's Law, 202, 256, 367
Heterotrophic organisms, 481, 566, 591
organic nutrition of, 543–544
Holoplankton, 480, 577
Humic substances, **443–446**, 448, 456
effect on phytoplankton growth, 583
See also Gelbstoffe
Hydration of carbon dioxide in sea water, 313–314
Hydrocarbons in sea water, 455, 470
Hydrogen electrode, 236
Hydrogen sulphide in sea water, 23, 242, 613, 643
Hydrometry, 5, 7, **97–98**
Hydroxy-carboxylic acids occurrence in sea water, 453
Hydroxyl ion complexation in sea water, 168, 170
Hydroxylamine in sea water
as nitrogen source, 537
Hypolimnion, 480

I

Illite, 413
Illumination, **524**
distribution of, in sea, 525
effect on photosynthesis, 530–536
measurement of, 526–527
P vs I curves, 531

Incubators for productivity measurement, 556
Indian Ocean
  carbon-14 activity in, 303
  carbon dioxide partial pressure in, 275
  denitrification processes in, 373
  nitrate distribution in, 390
  nitrite distribution in, 395–398
  organic carbon content, 464
  phosphorus distribution in, 328, 358–359, 360
  silicon distribution in, 427–428
Indium
  abundance in sea water, 165
Indole and its derivatives, biological importance, 441
Inductive conductivity cell, 87–88
Inductive salinometers, 94–96
Industrial carbon dioxide production, 282, 286, 287, 295–296, 297
Inert gases, *see* Noble gases
Interferometer
  as salinometer, 99
  "fringe jumping", 100
Iodine
  abundance in sea water, 165
  complexation of, 169
  historical, 23
  volatility in sea water, 179
Ion-exchange
  separation of alkaline earth metals, 131
  by clay minerals, 174, 413
Ion-pair formation, **166–168**, **278–282**
Ionic activities, *see* activity coefficients
Ionic potential, 25
Ionic product of water, 232
  effect of pressure on, 232
Ionic strength of solutions, 236
Ionic theory of Arrhenius, 9, 22
Irminger Sea
  phosphate content of, 340
Iron in sea water
  abundance, 164
  anoxic systems, occurrence in, 182, 637–638
  growth limiting power, 29, 506, 539
  historical, 23
  mode of occurrence, 174, 539
  organic compounds, 455
  residence time, 164
Iron in sediments, *see* ferromanganese minerals
"Irving-Williams order", 171, 184
Isochrysis galbana, 541, 586
Isotope fractionation, 157, 158, 221, 287, 301, 367, 369, 634
Isotopic variations of sea water, **154–158**

K

Kaoe Bay, 612, 620
Kaolinite, 413
Knudsen's Tables, 77, 102
Krypton
  abundance in atmosphere, 198
  abundance in sea water, 164
  solubility in sea water, 207
Kuroshio Current, 46, 575

L

Lactic acid in *Nitzschia closterium*, 514
Lake Nitinat
  as anoxic system, 618, 630, 632
Lambert's Law, 525
Lamellibranchs, food requirement, 586
Land-mass effect, 568
Lanthanides, *see* Rare earths
Lead
  abundance in sea water, 165
  enrichment in ferromanganese minerals, 194
  occurrence in sea water, 181
  originating from lead tetraethyl, 181
  residence time in the sea, 165, 173
Lead-210, 181
Leucosin, 510
Leucosine, 511
Light absorbancy of sea water, 112–114
  relation to fluorescence, 472
  seasonal variation, 471
Light *in situ*, simulation of with incubator, 556
Light in the sea
  absorption, 112–114
  attenuation, 471, 526
  effect on cell division rates, 546
  effect on photosynthesis, 526–530
  effect on phytoplankton distribution, 561

Light in the sea *contd.*
effect on phytoplankton growth, 561
measurement, 526–527
seasonal variation, 576
supra-optimal effect on photosynthesis, 553, 558
wave-length of maximum transmission, 526
Lightning, nitrogen fixation by, 369
Lignin, 440
Lipase, 587
Lipids, 443
determination of in phytoplankton, 511
occurrence in marine organisms, 511
Lithium in sea water
abundance, 164
historical, 24
residence time, 164, 173
Lithosphere-hydrosphere interactions, **187–194**
Lusiad Expedition, 273, 317
Lyngbya aestuarii, 541

M

Magnesium in sea water
abundance, 164
anoxic systems, occurrence in, 182
commercial extraction of, 36
determination of, 13, **126–131**
gravimetric as oxinate, 128
gravimetric as magnesium ammonium phosphate, 126
volumetric with E.D.T.A., 129
fluoride complex of, 169
ion-pair formation by, 167, 278
principal determinations of, 142–143
ratio to chlorinity, 142–143
residence time, 164, 173
Major dissolved constituents of sea water
abundances of, 9–18, 122
constancy of relative proportions, 15, 74, 79, **122–125**
Methods of determination, **126–139**
principal determinations **140–154**,
Manganese in sea water, 23, 182
abundance, 164
anoxic systems occurrence in, 182
growth limiting power of, 29, 539
residence time, 164, 175
Manganese in sediments, *see* ferromanganese minerals
Manganese nodules, *see* ferromanganese minerals
"Marine snow", 565, 568, 591, 593
Mediterranean Sea
inorganic nitrogen content of, 398
Melanin formation, 444
Melanoidines in sea water, 456
Melospira sp., 541, 582
Mercury
abundance in sea water, 165
occurrence in sea water, 180
residence time in sea water, 165
Meroplankton, 480, 577
Metabolites in marine organisms, **506–514**
carbohydrates, 508
lipids, 511
proteins and amino acids, 512
Meteor R.V., 29, 30, 37, 78, 99, 208, 342
Methane
anoxic waters, occurrence in, 632
atmosphere, occurrence in, 198
Microplankton, 480, 497
Mineral resources of the sea, 35–38
Molybdenum
abundance in sea water of, 164
residence time in sea water of, 164, 173
Monochrysis lutheri, 503, 507, 509, 510, 517, 541, 586
Monosaccharides
decomposition products of, 438
Monsoon Cruise, 291, 316
Monsoons
influence on phosphate concentrations, 333
influence on silicate concentrations, 421
Montmorillonite, 412
Myxophyceae
carbohydrate content of, 507, 508
carbon content of, 503
lipid content of, 507
nitrogen content of, 503
phosphorus content of, 503
pigment composition of, 488

Myxophyceae *contd.*
protein content of, 507
vitamin requirements of, 541

N

N.I.O. Salinometer, 90
Nannochloris sp., 497
Nannochloris atomus, 541
Nannoplankton, 480, 484, 497
Navicula pelliculosa, 541, 549
excretion by, 510
frustule re-solution of, 416
silicon uptake by, 415
Nekton, 479
Neon
atmospheric concentration, 198
isotope composition in atmosphere, 199
sea water concentration, 164
solubility in sea water, 207
Nickel, in sea water
abundance, 164
residence time, 164, 173
Niobium
abundance in sea water, 164
residence time in sea water, 164
Nitrate in rain water, 369, 544
Nitrate in sea water
anoxic conditions, effect on, 400–401, 613
distribution in coastal waters, 383–387
distribution in oceans, 391–398, 614
effect on cell division, 548
formation by nitrification, 377
historical, 26–27
marine cycle, 402
maximum layer, 393
nutrient value, 26, 375, 383, 536
seasonal variation, 27, 284, 389
transport, 391
vertical distribution, 388
Nitrification, **377–383**, 401
Nitrite in sea water
anoxic systems, 400, 625, 634
distribution in oceans, 390–391
extracellular production, 375, 389
formation by nitrification, 377
marine cycle, 402
maximum layer, 390–391
nutrient value, 26, 375, 387, 537
seasonal variation, 387–390
Nitrobacter, 378
Nitrocystis oceanus Watson, 378
Nitrogen budget of the earth, 403
Nitrogen, combined inorganic, in oceanic rain, 369, 544
Nitrogen fixation, 201, 211, 217, **370–373**, 369, 544
and carbohydrate consumption, 371
Nitrogen in marine organisms, 374–376, 503, 505, 548, 549
determination, 505
Nitrogen (dissolved) in sea water, 19–20, 200, 210
changes *in situ*, 217
conservative nature, 222
fixation, 217, 371, 544
historical, 19–20
ratio to argon, 206, 217, 629
ratio to oxygen, 206
solubility, **205–206**, 368, 673
Nitrogen, organic, in sea water
as nutrient source, 543
Nitrogen, terrestrial fixation, 217, 373
Nitrogen: phosphorus ratio, **360–361**
Nitrosocystis
nitrification by, 378
Nitrosomonas sp.,
nitrification by, 378
Nitrous oxide, in atmosphere, 198, 373
Nitzschia closterium, 514, 541
Nitzschia linearis
silicon re-solution from, 416
Nitzschia palea
silicon uptake by, 415
Noble gases in sea water
changes *in situ*, 217
historical, 20
solubility, 206–207
Noctiluca, 421
North Atlantic Current, 44
North Equatorial Current
in Atlantic, 44
in Pacific, 46
North Pacific Current, 46
North Sea
fishing grounds, 336
phosphorus distribution, 335–337

Norwegian fjords
as anoxic systems, 612
inorganic nitrogen distribution, 398
Nostoc
carbohydrate content, 508
nitrogen fixation by, 544
Nuclear explosions
carbon-14 production by, 287, 299
Nucleic acids in phytoplankton, 513
Nucleoproteins, 441
Nutrients of biological importance
effect on cell composition, 548, 550
effect on cell division rates, 547
effect on photosynthesis, **536–542**
effect on phytoplankton species, 580
historical, 26–34
in anoxic systems, 640
oceanic comparison, 220
preformed nutrients, 393, 627
regeneration, 29, **376–383**
relation to productivity, 560–561

O

Ob R.V., 242, 249
Ochromonas danica, 541
Ochromonas malhamensis, 510, 541
Ochromonas minuta, 541
Oligotrophic waters, 480, 520, 563
Organic matter and decomposition products, **437–446**
humic substances, 443–446
lipids, 443
nitrogenous (including protein and amino acids), 440–443
Organic matter in sea water
anoxic systems, occurrence in, 636
boric acid, complexation of, 150, 250
C/N ratios of, 459
decomposition of, 437
determination of, 434, 436, 456
elementary composition of, **456–462**
light attenuation by, 471, 526
nutrient value of, 485, **542–544**
oxidation of, stoichiometric model, **623–634**, 636
phytoplankton, production of, 592
seasonal variation of, **462–470**
solubility of, 472
stability of, 437
vertical distribution of, **462–470**
Oscillatoria sp., 544
carbohydrate content, 508
Oscillatoria subtilissima, 541
Osmotic pressure of sea water, 115
Ostrea edulis, 586
Oxygen-deficient environments, **614–615**
Oxygen-18 in sea water, 156–157
Oxygen in sea water, 19–20, 199, 209
biological activity, relation to, 214–216, 221, 519
consumption rate of, 216
diurnal variations of, 20, 244
in situ changes of, 214–216
isotopic variations, 155–157, 221, 369
minimum layer, 218, 220, 350
photosynthetic production of, 214–215, 519
respiratory consumption of, 214–216
solubility of, 204–205, 672
vertical distribution of, 218–220, 342
Oysters, food selectivity, 586

P

P vs I curves, in productivity work, 531–536, 557
Pacific Ocean
carbon-14 activity in, 301
denitrification processes in, 373
nitrate distribution in, **395–398**
nitrite distribution in, 390, 614
organic matter content of, 465, 468
oxygen distribution, 614
partial pressure of carbon dioxide in, 274
phosphorus distribution in, 328, **355–358**, 360
silicon distribution in, 426–427
surface currents of, 46
trace elements in sediments of, 190
Palladium
abundance in sea water, 165
Palmelloid forms, 485
Partial pressure of carbon dioxide in atmosphere,
See carbon dioxide (atmospheric)
Partial pressure of carbon dioxide in sea water, 252, 265, **269–277**
effect of temperature, 276–278
diurnal variations, 270–272
latitudinal variations, 270–272

Partial pressure of gases in sea water, 202
Particulate matter in sea water, 324 412–415, 472, 514, 524
Pelagic sediments, 189, 193
Peptidase, 587
Peptides and Polypeptides in sea water, 450
Peridineans, 503, 507
  elementary composition, 374
"Percent saturation" of gases in sea water, 204
Peridinium trochoideum, 541
Peru Current, 46, 56, 356, 427
pH of sea water
  anoxic systems, 641
  effect on carbonic acid equilibria, 660–671
  effect of photosynthesis, 242, 245
  effect of solar radiation, 242
  historical, 22
  measurement, 22
    colorimetric, 22, 238
    electrometric, 236
  rainfall effect on, 246
  Sørensen scale, 236
  temperature coefficient, 239, 245
  variation (diurnal), 242, 246
  variation with depth, 244, 245, 267
pH standards, 235
Phaeocystis pouchetii, 514
Phaeodactylum tricornutum, 380, 486, 492, 503, 505, 507, 509, 541, 547, 586
Phaeophyta, 483
Phagotrophy, 544
Phenolic compounds in sea water, 455
Pheophorbide, 492
Pheophytin, 492
Phillipsite, 414
Phormidium persicinum, 541
Phosphate in sea water
  determination, 23, 28, 324–325
  distribution in oceans, 359
  distribution in coastal waters, 329–335, 388
  distribution, vertical, 388
  effect on cell division, 547
  historical, 26–27
  in anoxic systems, 640
  maximum layer, 350, 358, 360
  nutrient value, 329, 537
Phosphine in anoxic sea water, 631
Phosphorites
  formation of, 188–189
  Th/U ratios in, 193
Phosphorus in marine organisms, 375, 503, 505, 547, 549
  determination, 505
Phosphorus (organic) in sea water, 324, 329, 338
  nutrient value, 543
  in nucleoproteins, 441
  ratio to nitrogen, 360–362, 375, 462
  ratio to silicon, 426
Phosphorus in sea water
  determination, 23, 28, 324–325
  dissolved species, 324
  distribution, 329–359
  distribution, coastal waters, 329–335
  distribution, ocean waters, 335
  distribution model, idealized, 350–352
  diurnal changes, 334
  exchange with bottom deposits, 333–334
  historical, 23, 26–28
  oxidative ratio, 245, 382
  ratio to nitrogen, 360–361
  seasonal variations, 27, 329–335
  suspended species, 324
  vertical transport by eddy conductivity, 350
  inter-ocean comparison, 359
Photochemical reactions in the nitrogen cycle, 383
Photolithotrophic organisms, 536, 540
Photosynthesis
  effect of light and temperature on, 530–536
  effect on pH, 242, 245
  fixation of carbon dioxide by, 376, 520
  measurement, **519–524**
  nutritional effects, 536–542
  P vs I curves, 531, **556–560**
  production of oxygen by, 201, 214–216, 519
  vitamin requirements during, 540–542

Photosynthetic potential, 496, 535
Photosynthetic quotient (PQ), 481, 519
Photosynthetic rate, 214, 563
  definition, 481
  measurement, **519–524**
Photosynthetically active light, **527–536**
Phototaxis, 565
Phytoplankton, 437, 478, 479
  See also under individual species
  ash content, 502
  bacterial decomposition, 376
  blooms, 464, 569, 576
  cell disruption, 497
  cell division, 546–548
    effect of light and temperature, 546–548
    effect of nutrient shortage, 548–551
  chemical composition, **487–514**, 516
    carbohydrate, 507–510
    elementary composition, 502–506
    lipid, 507, 511–512
    protein and amino acid, 507, 512–513
    sterols, 512
    other compounds, 513–514
  chemotaxis, 565
  classification, 483–486
  composition ratios, 516, 517
  distribution,
    geographical, 568–569
    land-mass effect, 569
    effect of nutrients, 580
    vertical, 561
  effect of humic substances on growth, 583
  effect of light and temperature, 524–536
  effect of minor nutrients, 538–542
  effect of nutrient deficiency, 548–551
  excretion of organic compounds, 390, 439, 447, 524, 592
  filtration, 497, 514
  heterotrophy, 543, 566
  inorganic nitrogen utilization, 365, 370, **373–376**, 384
  nitrogen fixation, 217, 377, 544
  phosphorus utilization, 333, 334, 338, 360
  salinity tolerance, 581
  seasonal variations, 576–583
  seeding by meroplankton, 580
  silicate utilization, 416, 417–421, 429
  sinking and phototaxis, 565–566
  species succession, 578–580
  "sun" and "shade" forms, 534
  terminology, 479
  utilization (direct) 585–589
  utilization (as detritus), 589–590
  vitamin factors, 583
Pigments of phytoplankton, **487–502**
  action spectra, 527–530
  analysis, 496–502
    fluorometry, 500
    individual pigments, 500
    in vivo measurements, 501
    multichromatic spectrophotometry, 499
  "bleaching", 527, 549
  concentrations, 490
  Emerson effect, 528
  species distributions, 487–489
  spectral characteristics, 494
  taxonomic index properties, 497
Pinacyanol method for determination of lipid, 511
Planet Expedition, 27, 200
Plankton, 479
  See also phytoplankton
  holoplankton, 480, 577
  meroplankton, 480, 577
  neritic, 480
  oceanic, 480
  size distribution, 480
  trace metals concentration by, 185
Plant hormones
  as phytoplankton growth factors, 441, 584
Platymonas subcordiformus, 510
Platymonas tetrathele, 541
Polar R.V., 435
Polarization, in conductivity determination, 84
Pomerania R.V., 19
Pontosphaera roscoffensis, 541
Porcupine, R.V., 199, 220
Porphyridium sp., 484
Potassium in sea water
  abundance, 164

Potassium in sea water *contd.*
determination of
gravimetric
as chloroplatinate, 133
as cobaltinitrite, 134
as perchlorate, 134
as tetraphenylboron salt, 134
historical, 13
ion-pair formation, 167, 278
principal determinations, 145–146
ratio to chlorinity, 145–146
residence time, 164, 173
Potassium-40, 201, 214
"Pre-formed nutrients", 393, 627
Primary organic productivity, 354, 468
biologically-active substances, effect of, 560
carbon-14 uptake during, 31, 520
field measurement, duration of, 552
sampling technique, 551
gross values of, 483
in situ measurement, 554–555
incubator measurement, 555–560
net values of, 483, 572
P vs I curves, use of, 557
regional variations of, 574–576
Primary product, 483, *see* also standing crop
Productivity, *see* primary organic productivity
Protease, 440
Protein
bacterial synthesis, 443
breakdown products, 440, 443
and derivatives in sea water, 450–453
in marine organisms, 512
Prymnesium parvum, 541
Pteridines, as phytoplankton pigments, 490
Purines, 442
Pyrrophyta, 485

Q

Quartz, solubility in sea water, 410
Quinhydrone electrode, 237

R

Radio-activity in sea water, **34–35**
vertical mixing of, 67
Radio-carbon, *see* carbon-14
Radium
abundance in sea water, 165
concentration in sediments, 34
Radium-226 in sea water, 34, 639
Radon
abundance in sea water, 165
Rainfall
inorganic nitrogen content of, 369, 403, 544
isotopic ratios, effect on, 157
pH of sea water, effect on, 246
Rare earth elements in sea water
abundances, 165
complexation of, 169
distribution of, 181
historical, 24
residence times, 165
Reactivities of elements in the oceans, 171–172
relation to degree of undersaturation, 175–177
Red clay, 34
"Red tides", 485, 578, 583
Redox potential (Eh) of sea water, 193, 613, **635**
effect on uranium in sediments, 182, 188
Reducing environments, *see* Anoxic basins and fjords
Refractive index of sea water, **107–109**
as salinometer, 99
Refractometer, 99
Residence time of elements, 26, 71, 164–165, **172–175**
Resistance bridge for conductivity determination, 83–85
Respiration, 523
consumption of oxygen by, 215–216
Respiratory quotient (RQ), 481, 519
Rhodomonas lens, 541
Rhodophyta, 483
Ribonucleic acid in phytoplankton, 513
Rivers, supply of dissolved substances to the oceans, 13, 172, 173, 473
calcium content, 142
nitrogen content, 403
organic matter in, 473
silicate content, 412, 419, 422
sulphate content, 147

Rizosolenia sp., 541
Roman philosophers, theories and observations relating to oceanography, 2
Rubidium
  abundance in sea water, 164
  detection in sea salt of, 24
  residence time in sea water, 164, 173

S

S.P.U. (Specified pigment unit), 499
Saanich Inlet
  as anoxic basin, 617, 621, 632
Sagitta hispida
  ammonia and phosphate excretion, 377
Sahara Desert
  as source of marine silica, 423
Salinity
  chlorinity, relation to, 18, 76
  definition of, 75, 77, 80
  determination of by
    conductivity, **81–97**
    density, 97–98
    direct, 76
    refractive index, 98–101
    velocity of sound, 101
  distribution in surface layers, 49–51
  historical, 5, 13
  major components of, 15
  phytoplankton tolerance, 581
Salinometers
  bench type, 87–91
  conductivity instruments, 81–97
  inductive type, 94–96
  in situ type, 91–97
  salinity-temperature-depth instruments, 94–97
Sapropel, 188
Sargasso Sea, 221, 372
  productivity of, 468, 574, 576, 581
Scandium
  abundance in sea water, 164
  residence time in sea water, 164
Scendesmus sp., 487, 503
  heterotrophism of, 543
Scendesmus obliquus, 375, 507, 541
Sea surface
  gas exchange across, **211–214**, 282–284, 312–316
Sea water
  chlorinity of, **76, 77**
  compressibility of, **111–112**
  density of, 74, 97, **101–103**
  electrical conductivity of, 78, **103–107**
  freezing point of, 15
  isotope variations of, **154–158**
  osmotic pressure of, 115
  salinity of, 75, 77, 121
  sound velocity in, 101, **109–110**
  specific heat of, 116
  surface tension of, 117
  transparency of, **112–115**
Seasonal variations
  ammonia, 387–390
  CO_2 (atmospheric), 288–294
  gases, 210
  light, 471, 576
  nitrate, 27, 284, 359
  phosphate, 273, **329–335**
  silicate, 330, **417–421**
  surface currents, 48
Seaweeds
  concentration of trace metals in, 185
Secchi depth, 552, 561
Sediments
  aeolian, 413
  anoxic conditions, occurrence in, **187–189**, 636, 642
  estuarine deposits, 333
  $^{13}C/^{12}C$ ratio in carbonates of, 301
  nitrogen deposition, 403
  organic carbon in, 613, 636
  oxygenated waters, occurrence in, **189–194**
  rare earth content, 169
  Th/U ratios of, 193
  trace elements in, **187–194**
"Seeding" and phytoplankton growth, 580
Selenium
  abundance in sea water, 164
Ship's drift, 49
Silicate in river water, 412, 419, 422
Silica in sea water, 23, 410
  river water input of, 412, 419, 422
  solubility of, 410
  suspended particulate, 412–415
  "unreactive" (colloidal), 411, 412

Silicate in sea water
  annual consumption of, 423
  anoxic systems, occurrence in, 640
  determination of, 411
  distribution, coastal waters, **421–423**
    oceanic waters, 304, **423–430**
    with depth, 424, 425, 429
  inter-ocean comparison of, 429
  nutrient value of, 28, 415, **417–421**, **538**
  regeneration of, 415–417
  river supply of, 412, 419, 422
  seasonal variations of, 330, **417–421**
Silicate minerals in suspension, 412
Silicon in marine organisms, 505, 549, *see* also Diatoms
Silicon of biogenic origin, 415
Silicon, organic, in sea water, 409, 412
  ratio to phosphorus, 426
  ratio to nitrogen, 426
Silver
  abundance in sea water of, 165
  occurrence in sea water of, 24
  residence time in sea water of, 165, 173
Singapore Straits
  phosphate content of, 331
Sinking of phytoplankton, 565–566
Skeletal materials, 442
  trace element composition of, 186–187
Skeletonema costatum, 492, 503, 507, 517, 542, 546, 579, 582, 584, 586
  amino acid content of, 512
  heterotrophism of, 567
  iron requirement of, 583
Sodium in sea water
  abundance, 164
  determination
    uranyl acetate, 133
  historical, 13
  ion-pair formation of, 167, 278
  principal determinations of, 144
  ratio to chlorinity of, 144
  residence time, 164, 173
Solar energy at sea surface, 525
  photosynthetic range of, 525
Solar evaporation, 2
Solar radiation, effect on pH, 242
Solubility of gases in sea water, **202–208**
Somali Current, 48
Sound velocity in sea water, **109–110**
  as salinometer, 101
  effect of salinity on, 110
South Atlantic Current (West Wind Drift), 44
South Equatorial Current
  in Atlantic Ocean, 44
  in Pacific Ocean, 46
South Pacific Current (West Wind Drift), 46
South Polar Expedition, 27
Specific alkalinity, 248–249
  in anoxic systems, 642
Specific conductivity of sea water, 104–105
Specific gravity of sea water, 5, **74**
  anomaly of, 103
Specific heat of sea water, **116**
Sponges
  silicon uptake by, 410
  trace element concentration by, 185
Spring "blooms" of phytoplankton, 576–578
"Standard Mean Ocean Water" (SMOW), 155
Standard Sea-Water Service, 77, 80, 122
Standing crop, 31, 483, 571
Stephanopyxis turris, 541
Sterols in phytoplankton, 512
Storage of sea-water samples, 122
Strontium in marine organisms, 149
Strontium in sea water
  abundance, 164
  determination
    flame photometric, 136
    radio-chemical, 136
  historical, 13
  principal determinations of, 148–149
  ratio to chlorinity, 148–149
  residence time, 164, 173
Suess (dilution) effect, 287, 297
Sugars in marine organisms, 509
Sugars, in sea water, 439, 447
Sulphate in sea water
  determination, gravimetric as barium salt, 135
  historical, 13
  ion-pair formation of, 167, 170

Sulphate in sea water *contd.*
principal determinations of, 146–148
ratio to chlorinity, 23, 125, 146–148, 642
reduction in anoxic systems, 23, 613, 625, 627
Sulphate-reducing bacteria, 147, 158, 643
Sulphide in anoxic systems, 617–619, 628, 642
precipitation by, 636
toxicity of, 642
Sulphites in sea water, 628
Sulphur
isotopic variations of, 158
sea water, occurrence of, 164, 624
Surface currents of the oceans, **44–49**
seasonal variations of, 48
irregular variations of, 48
Surface tension of sea water, **117**
Synechocystis sp., 541
Synchronous culture, 482
Syracosphaera sp., 541
Syracosphaera carterae, 503, 507, 509, 517
calcium carbonate content of, 506
sugars in, 510

T

Temperature
effect on carbonic acid equilibria in sea water, 660–671
effect on cell division kinetics, 545–547
effect on electrical conductivity of sea water, 106
effect on photosynthesis, 532, 581–583
effect on phytoplankton species, 581–583
Temperature–Chlorinity bridge, 88–89, 93
Terrestrial organic life substances, 438
Tetraselmis maculata, 503, 507, 509
Thalassiosira sp., 579, 582, 587
Thalassiosira decipiens, 587
Thalassiosira fluviatilis, 541
Thalassiosira gravida, 510
Thalassiosira nana
silicon re-solution from, 416
Thalassiosira nordenskiöldii, 511, 541, 582
Thermocline, 62, 70, 230, 329, 479
Thermodynamic Processes
in oceanic circulation, 56
Thermopile radiometer for light *in situ*, 527
Thiosulphate in sea water, 628
Thorium
abundance in sea water, 165
occurrence in sea water, 174, 639
residence time in sea water of, 165
Thorium isotopes in sea water and sediments, 35
Thorium in sediments, 193
Thorium/Uranium ratio
as environmental index, 193
Tidal curents, 59
Tigriopus japonicus, 586, 593
Tin
abundance in sea water, 165
residence time in sea water, 165
Titanium
abundance in sea water, 164
residence time in sea water, 164
Trace elements
hydrosphere–biosphere interactions, 179–187
hydrosphere–lithosphere interactions, 187–194
Trace elements in sea water
abundances of, 164–165
anoxic systems, occurrence of, 181–183, 640
biological concentration of, 181, **183–186**
enzyme co-factor function of, 540
reactivities of, 171
residence times of, 164–165, **172–175**
Trace elements in sediments
oxidizing environments, 189–194
reducing environments, 188–189, 640
Trace elements in skeletal material, 186–187
Transformer bridge for conductivity determination, 85–86
Transition metal complex formation in sea water, 171
Transparency of sea water, **112–115**

Trichodesmium thiebautii
nitrogen fixation, 217, 371
Tuna
trace element concentration by, 185
Tungsten
abundance in sea water, 165
residence time in sea water, 165
Turbulence
effect on carbon dioxide transfer, 314–316
Turbulent mixing, **61–69**
horizontal, 63
quantitative treatment of, 64–69
sinking of phytoplanton, relation to, 566
vertical, 61

U

Ultraplankton, 480, 485, 497
Unicellular algal growth, 482, 570
grazing, effect on, 550
nutrient depletion, effect on, 569–571
Undersaturation of elements in the oceans, 175–177
Upwelling, 56
cause of high productivity, 27, 386, 575
Uranium in sea water
abundance, 165
anoxic systems, occurrence in, 182, 188, 638–639
decay series of, 35
historical, 35
Uranium in sediments, 188–189, 193, 638–639
Uranium in skeletal material, 187

V

Vanadium
abundance time in sea water, 164
residence time in sea water, 164
Vapour pressure of sea water, 202
Vegetable organic matter
rate of decomposition, 439
Velocity of sound in sea water, *see* Sound velocity in sea water
Vitamins B in phytoplankton, 514
Vitamin B_1 in sea water, 455
Vitamin B_{12} in sea water, 455, 583
in marine organisms, 513, 548, 583
Vitamin C in marine algae, 513
Vitamins, essential for photosynthesis, **540–542**
bacterial production of, 590
growth factor property of, 583
Volcanic activity
introduction of elements to the oceans, 174, 413
submarine, 154

W

Wagner earth circuit, 84
Water, dissociation equilibria, 231
William Scoresby, R.V., 359
Wind-borne sedimentary particles, 413

X

Xanthophylls in marine phytoplankton, 488, **493–496**
Xenon
abundance in atmosphere, 198
abundance in sea water, 165
solubility in sea water, 207

Y

Yeasts in sea water, 370, 591
"Yellow substances", *see* "Gelbstoffe"
Yttrium
abundance in sea water, 164
residence time in sea water, 164

Z

Zinc
abundance in sea water, 164
concentration by fish, 184
occurrence in sea water, 24
residence time in sea water, 164, 175
Zooplankton, 34, 377, 387, 479, 585
dietary requirements of, 586
grazing by, 569–571
gross feeding efficiency of, 588
Zostera, elementary composition, 374